MONTE CARLO AND MOLECULAR DYNAMICS SIMULATIONS IN POLYMER SCIENCE

Monte Carlo and Molecular Dynamics Simulations in Polymer Science

KURT BINDER

Institut für Physik
Johannes-Gutenberg-Universität Mainz

New York Oxford
OXFORD UNIVERSITY PRESS
1995

Oxford University Press

Oxford New York
Athens Auckland Bangkok Bombay
Calcutta Cape Town Dar es Salaam Delhi
Florence Hong Kong Istanbul Karachi
Kuala Lumpur Madras Madrid Melbourne
Mexico City Nairobi Paris Singapore
Taipei Tokyo Toronto

and associated companies in
Berlin Ibadan

Published by Oxford University Press, Inc.,
198 Madison Avenue, New York, New York 10016

Library of Congress Cataloging-in-Publication Data
Monte Carlo and molecular dynamics simulations in polymer sciences/
[edited by] Kurt Binder.
p. cm. ISBN 0-19-509438-7
1. Polymers—Computer simulation.
2. Molecular dynamics—Computer simulation.
3. Monte Carlo method. I. Binder, K. (Kurt), 1944–.
QD381.9.E4M66 1995
541.2/254/0113—dc20
94-35391

9 8 7 6 5 4 3 2 1

Printed in the United States of America
on acid-free paper

PREFACE

Computer simulation has become an established method of research in science and a useful tool in solving certain engineering problems. With the introduction of powerful workstations on the desk of the scientist, the impact of the application of computer simulation to many problems will increase enormously in the next couple of years.

Polymer science profits from this development in a particular way: the complexity of macromolecular chemical architecture and geometrical structure, the huge variability of physical properties and the widespread range of applications involves many intricate scientific questions, considering the fact that theoretical methods usually imply crude mathematical approximations, the validity of which it is hard to judge in general. Together with the limitation that unknown parameters introduce, the predictive power of such work is often rather limited. In contrast, computer simulation can study a model of a complex many-body system in full detail without invoking such mathematical approximations: in principle, we thus can check the validity of approximate calculations and methods, without unknown parameters obscuring a meaningful comparison. At the same time, comparing with experiment helps to validate and systematically improve the model. In fact, use of computer simulation in this way is an iterative process by which the understanding of complex materials and processes can be significantly improved step by step.

In view of these distinct conceptual and principal advantages, computer simulations in macromolecular materials have aroused considerable interest, and various complementary techniques have been developed. It must be noted, though, that simulations of polymers pose particular challenges, considering the enormous spread of length scales and time scales involved: even the simple case of a single flexible polyumer coil exhibits geometrical structure from the scale of a chemical bond (1Å) to the scale of the gyration radius (100Å), and collective length scales in dense materials often are even much larger. Simultaneously, times scales range from bond vibration times (10^{-13} sec) to macroscopic times (10^3 sec), characterizing interdiffusion or relaxation near the glass transition, etc. Hence the naive use of "molecular modeling" software packages as a "black box" cannot be relied upon as a problem solver — what is needed, of course, is a more basic understanding of what computer simulation is, what methods are implemented and where their strengths and limitations lie. It is one of the aims of the present book to provide the reader with such background knowledge, which will enable him to apply such software with mature judgement in a useful way. Some of the chapters of this book therefore discuss deliberately methodical aspects of Monte Carlo and molecular dynamics simulations and describe the role of "model building". Given the inevitable uncertainties about the force fields for complex polymeric materials, and the difficulty in explicitly relating these forces and other details of the geometrical and chemical structure to phenomenological simplified models, the book pays much attention to the question "which model is adequate for a considered problem?" and clearly distinguishes questions where one can obtain sensible answers from simulations from those where one cannot.

The emphasis of this book is on polymer physics rather then polymer chemistry, and it considers in the main amorphous polymers (in solution, melt, or solid state); crystallized polymers, as well as polymer crystallization processes, etc., are not discussed here, nor are biopolymers treated—although many of the general comments about polymer simulation that can be found in this book are presumably useful for these other fields as well. Such a restriction of scope was necessary in order to keep the length of the text manageable. Taking the input from quantum chemistry (force fields, etc.) as given wherever necessary, the book does contain chapters dealing with fairly realistic models containing much chemical detail (e.g., the chapter by J.H.R. Clarke on simulations of the elastic properties of glassy polymers and that by D.Y. Yoon *et al.* on interfacial properties in thin polymeric layers). Most of the book, though, deals with more "mesoscopic" properties, attempting to bridge the gap between atomistic structure and macroscopic properties: structure and elastic response of polymer networks, forces acting between polymer brushes, phase diagrams of polymer blends, etc. Of course, the structure and dynamics of polymer coils under the various conditions is a central theme that can be found in almost all chapters: Chapter 2 describes what is known about the treatment of excluded volume interaction in dilute solution; Chapter 3 emphasizes electrostatic and hydrodynamic forces; stretching of chains in deformed rubbers is discussed in Chapter 4; stretching of chains due to thermodynamic forces in block copoymer mesophases in Chapter 7; and of tethered chains in Chapter 9. Thus, a wide variety of problems encountered with synthetic polymers is addressed. Studying the book will give a broad overview of almost the complete field of polymer physics and its concepts; and thus certainly will be useful to students and researchers in that field. Besides providing such an introduction, it presents up-to-date reviews from the leading experts on the various applications that are covered here, and it is expected that it will play a stimulating role in research, pushing further the frontier of new developments in academia and industry.

Mainz, September 1994 Kurt Binder

CONTENTS

CONTRIBUTORS

Jörg Baschnagel
Institut für Physik
Johannes-Gutenberg-Universität Mainz
WA 331/339
D-55099 Mainz
Germany

Kurt Binder
Institut für Physik
Johannes-Gutenberg-Universität Mainz
WA 331/339
D-55099 Mainz
Germany

Julian H. R. Clarke
Chemistry Department
University of Manchester Institute of Science and Technology
PO Box 88
Manchester M60 1QD
U.K.

Burkhard Dünweg
Institut für Physik
Johannes-Gutenberg-Universität Mainz
WA 331/339
D-55099 Mainz
Germany

Gary S. Grest
Corporate Research Science Laboratories
Exxon Research and Engineering Co.
Annandale, NJ 08801

Kurt Kremer
Institut für Festkörperforschung
Forschungszentrum Jülich
Postfach 1913
D-52425 Jülich
Germany

Michael Murat
Department of Physics and Applied Mathematics
Soreq Nuclear Research Center
Yavne 70600
Israel

Wolfgang Paul
Institut für Physik
Johannes-Gutenberg-Universität Mainz
WA 331/339
D-55099 Mainz
Germany

Grant. D. Smith
Department of Chemical Engineering
University of Missouri-Columbia
Columbia, MO 65211

Alan D. Sokal
New York University
Department of Physics
4 Washington Place
New York, NY 10003

Mark Stevens
Corporate Research Science Laboratory
Exxon Research and Engineering Co.
Annandale, NJ 08801

M. Vacatello
Department of Chemistry
University of Naples
Via Mezzocannone 4
80134 Napoli
Italy

D.Y. Yoon
IBM Research Division
Almaden Research Center
650 Harry Road
San José, CA 95120-6099

MONTE CARLO AND MOLECULAR DYNAMICS SIMULATIONS IN POLYMER SCIENCE

1

INTRODUCTION: GENERAL ASPECTS OF COMPUTER SIMULATION TECHNIQUES AND THEIR APPLICATIONS IN POLYMER PHYSICS

Kurt Binder

1.1 Why is the computer simulation of polymeric materials a challenge?

In recent years computer simulation has become a major tool in polymer science, complementing both analytical theory and experiment. This interest is due both to the many fundamental scientific questions that polymer systems pose and to the technological importance of polymeric materials. At the same time, computer simulation of polymers meets stringent difficulties, and despite huge progress (as documented in previous reviews[1–7]) many problems are still either completely unsolved or under current study. In the following pages these difficulties are briefly discussed.

1.1.1 Length scales

For standard problems in the physics and chemistry of condensed matter, such as simple fluids containing rare gas atoms or diatomic molecules, etc., computer simulation considers a small region of matter in full atomistic detail.[8–13] For example, for a simple fluid it often is sufficient to simulate a small box containing of the order of 10^3 atoms, which interact with each other with chemically realistic forces. These methods work because simple fluids are homogeneous on a scale of 10 Å already; the oscillations in the pair distribution function then are damped out under most circumstances. Also reliable models for the effective forces are usually available from quantum chemistry methods.

For long flexible polymers we encounter a different situation (Fig. 1.1)[14]: already a single chain exhibits structure from the scale of a single chemical bond ($\approx$ 1 Å) to the persistence length[15] ($\approx$ 10 Å) to the coil radius ($\approx$ 100 Å). Additional length scales occur in a dense polymer solution or melt. Some of these length scales are smaller than the coil radius, such as the screening length ξ_{ev} in semidilute solutions,[16] which describes the range over which excluded volume forces are effective, or the tube diameter in melts,[17] which constrains the motion of a chain (due to entanglements with other chains) in a direction along its own (coarse-grained) contour (see Chapter 4, where estimation of this length from simulations is discussed).

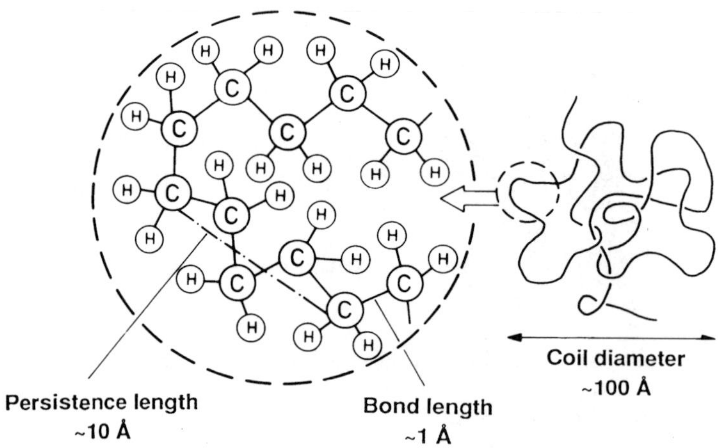

Fig. 1.1 Length scales characterizing the structure of a long polymer coil (polyethylene is used as an example). (From Binder.[14])

Collective phenomena may even lead to much larger lengths: e.g., in a *polymer brush* (i.e., a layer of polymers anchoring with a special end group at an otherwise repulsive wall) the height h of the brush is predicted to scale with degree of polymerization N_p as[18] $h \propto N_p$, while the coil gyration radius R_g scales only as $R_g \propto \sqrt{N_p}$ in a Θ-solution or dense melt[16,19] or $R_g \propto N_p^{\nu}$ with $\nu \approx 0.59$ in a good solvent.[16,20] Thus for N_p of the order 10^3 to 10^4 and a sufficiently high grafting density one expects h to be of order 10^3 Å (see also Chapter 9, where the simulation of polymer brushes is treated further).

Another large length scale occurs in polymer blends near the critical point of unmixing, namely the correlation length ξ of concentration fluctuations.[21] One expects that this length is of the order of R_g, far away from the critical point, but approaching the *spinodal curve* it is enhanced by a factor $|1 - T/T_{sp}(\phi)|^{-1/2}$, where T is the temperature of the blend and $T_{sp}(\phi)$ the spinodal temperature at volume fraction ϕ of one component of the blend. This enhancement factor is a mean-field result,[16,21,22] and close to the critical temperature T_c one expects an even faster growth of critical correlations,[21,23] $\xi \propto \sigma N_p^{(1-\nu_I)}(T/T_c - 1)^{-\nu_I}$, where $\nu_I \approx 0.63$ is the Ising model correlation length exponent[24] and σ the size of a segment of the polymer chain. Thus, near T_c correlation lengths of the order of 10^3 Å are predicted[21–23] and observed[25]. Similar large length scales are predicted[21,22] and observed[21,26,27] in the spinodal decomposition of polymer blends that are quenched into the unstable region of the phase diagram and hence start phase separation (see also Chapter 7, where the simulation of polymer blends is treated further).

This list of characteristic lengths in polymeric materials is far from being exhaustive. Without detailed explanation we here simply mention also the characteristic thickness of lamellae $\lambda \propto \sigma N_p^{2/3}$ in the strongly segregated lamellar mesophase of block copolymers,[28] the Bjerrum length, the electrostatic persistence length, and the Debye–Hückel screening length in polyelectrolyte solutions[29–32] (see Chapter 3 for simulations of such systems), the distance between crosslinks in polymer networks (see Chapter 4), characteristic coil sizes of polymers exposed to shear flow,[33] and so on.

Since a valid computer simulation must choose a system size with linear dimensions L larger than the characteristic lengths of the problem, one finds that for many problems of interest one would need to simulate systems containing of the order of at least 10^6 atoms or even much more. The situation becomes aggravated since the effective potentials for polymers are much more complicated and hence difficult to use in simulations than the pair potential for simple fluids. Let us consider polyethylene (PE), the chemically simplest organic polymer, as an example. A very popular model ignores the hydrogen atoms of the (CH_2) groups and replaces them with united atoms[34–44] (Fig. 1.2; see also Chapter 5 for more details). Neither the bond lengths nor the bond angles in Fig. 1.2 are treated as rigid, and one uses harmonic potentials for bond length and bond angle vibrations, while nonbonded interactions between effective monomers of different chains (or monomers of the same chain if they are separated by more than three bonds along the chain) are represented by a Lennard-Jones form. Thus already for the simplified model of Fig. 1.2 a complicated Hamiltonian with many parameters results:

$$\mathcal{H} = \sum_{j=1}^{N_p-1} \mathcal{H}_\ell(\ell_j) + \sum_{j=1}^{N_p-2} \mathcal{H}_\Theta(\Theta_j) + \sum_{j=2}^{N_p-2} \mathcal{H}_\phi(\phi_j) + \sum_{i \neq j} \mathcal{H}_{LJ}, \tag{1.1}$$

where

$$\mathcal{H}_\ell(\ell_j) = \frac{f_\ell}{2}(\ell_j - \ell_0)^2, \mathcal{H}_\Theta(\Theta_j) = \frac{f_\Theta}{2}(\cos\Theta_j - \cos\Theta_0)^2, \tag{1.2}$$

$$\mathcal{H}_\phi(\phi_j) = f_\phi \sum_{n=0}^{5} a_n \cos^n \phi_j, \mathcal{H}_{LJ} = 4\epsilon[(\sigma/r_{ij})^{12} - (\sigma/r_{ij})^6]. \tag{1.3}$$

The appropriate choice of the parameters f_ℓ, ℓ_0, f_Θ, $\cos\Theta_0$, f_ϕ, $a_1, \ldots, a_5$ ($a_0 \equiv 1$), ϵ and σ is discussed in the literature[34–45]: at this point we note that even for the torsional potential $\mathcal{H}_\phi(\phi_j)$ and the nonbonded interaction $\mathcal{H}_{LJ}$ there is no complete agreement between different authors. Thus for many other polymers where the monomers are chemically more complicated the force fields are even less well known. Furthermore, the nonbonded interaction for computational convenience is often truncated at rather short distances r_{ij} between monomers i, j such as[39] $r_{ij} = 1.5\sigma$; experience

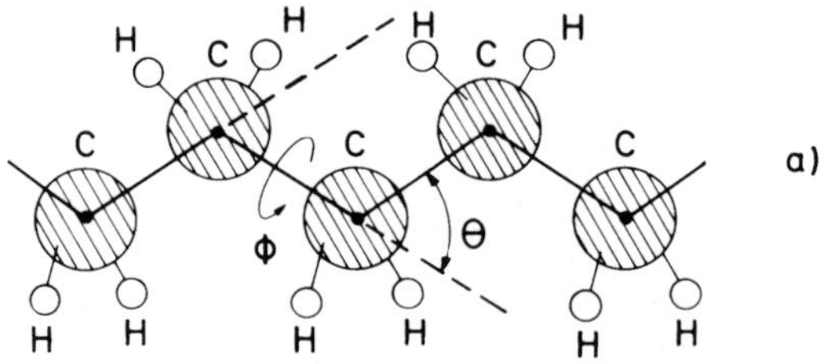

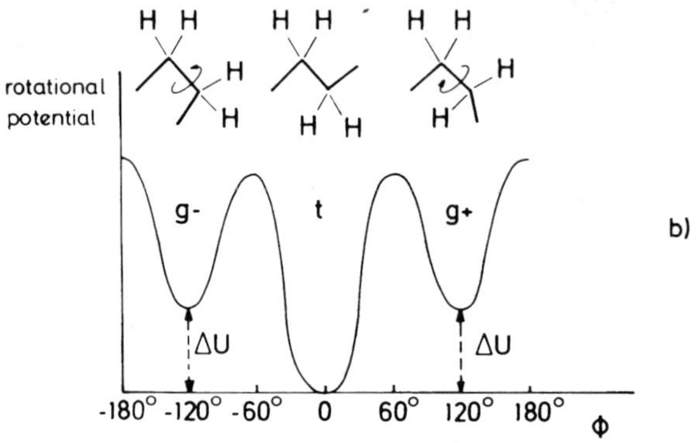

Fig. 1.2 (a) Schematic model of a piece of the polyethylene chain. The hydrogen atoms (H) are not treated explicitly in the "united atom" approximation, but rather one introduces effective spherical segments (shaded) representing a whole CH_2 unit. The segments are connected by harmonic bonds (shown as straight lines), the bond lengths being ℓ_i (segments being labeled consequently by an index i, $i = 0$ to $N_p - 1$). Three successive segments $i-1$, i, $i+1$ define a bond angle θ_i and four successive segments $\{i-2, i-1, i, i+1\}$ a torsional angle ϕ_i, namely the angle between the plane spanned by the bonds formed from the segments $\{i-2, i-1, i\}$ and the plane built from $\{i-1, i, i+1\}$. Note that the ϕ_i values are zero in the all-trans configuration drawn here. (b) Qualitative sketch of the torsional potential for alkane chains, indicating the three energetically prefered states, gauche– (g^-), trans (t), and gauche+ (g^+). The minimum of the trans configuration is deeper by an amount ΔU. (From Kremer and Binder.[2])

with monatomic fluids[46] shows that the equation of state of the fluid is rather sensitive to the value of this cut-off. Note that usually no distinction is made for the CH_3 end-groups of the chain.

In view of this discussion, it is clear that even the detailed model of eqs (1.1)–(1.3) should not be taken as a faithful description of polyethylene (PE), but rather as a prototypical schematic model of linear polymers. In the context of simulations of lipid monolayers,[47] it has been suggested that it is necessary to shift the center of gravity of the united atom off the position of the carbon atom at the chain backbone (anisotropic united atom model). Very recent work[48] (see also Chapter 8) suggests that it is more satisfactory to include the hydrogen atoms explicitly, if one wishes to describe PE properly. For the reasons quoted above, such work is restricted to relatively small degree of polymerization N_p, and thermodynamic states far off from any phase transitions.

1.1.2 Time scales

When one considers the second factor that controls simulation feasibility, namely the time scales involved, one realizes that the situation is even worse. While in simple fluids at states far from phase transitions all fluctuations decay on time scales of a few picoseconds, and hence molecular dynamics (MD) simulations can easily produce well-equilibrated system configurations, this is not so for polymers, since motions occur on many very different time scales. Fast motions such as the vibrations of the length ℓ of a C–C bond (Fig. 1.2) or the bond angle $\ominus$ (Fig. 1.2) may take times of the order of $\tau_{\text{vib}} \approx 10^{-13}$ s or even less. However, already the reorientation jumps in the torsional potential ($t \leftrightarrow g_+, g_-$; Fig. 1.2) are significantly slower; the average time τ_l between two such jumps is typically of the order of $\tau_l \approx 10^{-11}$ s, for polymer melts at the temperatures of interest. Since the time integration step δt of a MD simulation has to be much smaller than the shortest characteristic time, it has become common to either take the bond length as rigid[38] or to soften the spring constant f_ℓ in eq. (1.2) by a factor of seven arbitrarily,[39–42] to alleviate this problem. Using 10 chains with $N_p = 50$, a time-step of $\delta_t = 10^{-14}$ s has been chosen[39] and a CPU time of 0.02 s on a CRAY-XMP supercomputer was reported.[39] Thus, even at high temperatures far above the glass transition one needs 20 s to have a *single* jump in the torsional potential, and near the glass transition—where τ_l increases proportional to $\tau_l \propto \tau_{\text{vib}} \exp(\Delta E/k_B T)$, ΔE being the energy barrier of the torsional potential in Fig. 1.2—even a much longer simulation would hardly observe any $t \rightleftharpoons g_+$, g_- transitions. It should also be noted that the choice of $\delta t = 10^{-14}$ s is rather large, which may also lead to uncontrolled errors (other work[38] makes the more conservative choice of $\delta t = 2.10^{-15}$ s). Choosing MC instead of MD to equilibrate the system configurations is of little help here, however, since for most MC moves the acceptance rates are very small.

Of course, having just a few jumps in the torsional potential suffices to produce equilibrium on a very short length scale, but is not enough to bring the persistence length and the gyration radius R_g to their equilibrium values. Let us assume that we deal with a melt of relatively short, nonentangled chains, i.e., $N_p < N_p^e$ where typically[17] $N_p^e \approx 10^2$, so that the Rouse model[17] holds. Then the relaxation time necessary to relax the global configuration of a coil (Fig. 1.1) varies as

$$\tau_{N_p} = \tau_l N_p^2. \tag{1.4}$$

Thus for $N_p \approx 30$ a relaxation time $\tau_{N_p} \approx 10^{-8}$ s results, 10^6 times more than the time-step of Rigby and Roe.[39] This number shows that even at high temperatures it is at best possible to equilibrate melts of very short chains, and this indeed has recently become feasible.[48] (see also Chapter 4 for a more detailed discussion).

Of course, the situation is much worse for melts of long chains in a mutually entangled state, where reptation theory[16,17] predicts a relaxation time proportional to the third power of N,

$$\tau_{N_\mathrm{p}} \approx \tau_1 (N_\mathrm{p}/N_\mathrm{p}^e) N_\mathrm{p}^2. \tag{1.5}$$

For $N_\mathrm{p} = 500$ one then estimates that $\tau_{N_\mathrm{p}} \approx 10^{-5}$ s, *eight orders of magnitude more* than the vibration time τ_{vib}! Therefore a straightforward application of MD to study the dynamics of entangled melts is clearly impossible—the algorithms lack stability for the prohibitively large number of integration time-steps δ_t, and would also require enormous computer time. In this situation, one must abandon the idea of studying well-equilibrated polymeric materials of a high degree of polymerization N_p in full chemical detail. Several strategies are then conceivable:

- Restrict the study to both fairly high temperatures and rather small N_p. Although this approach is certainly feasible (see Chapter 8 and Refs 38, 48, 49) and yields interesting information on the behavior on small length scales and small time scales, many problems of interest are excluded. Even on small length scales temperatures close to the glass transition temperature T_g are inaccessible; note, however, that most polymeric systems are studied experimentally at temperatures not far above T_g. Simulations are valuable in particular to complement experiments in a temperature range which would be too high for experiments (e.g., one can study the gas–liquid critical point of alkane chains for $N_\mathrm{p} = 20$ to $N_\mathrm{p} = 48$, while the chains would chemically decompose at these high temperatures; but nevertheless these critical-point data are a useful input for the equation of state of alkane fluids).
- Restrict the study to the small length-scale behavior but abandon the quest for good thermal equilibrium. Of course, in such an approach it is no longer sensible to ask how properties depend on N_p, because properties on the length scale of R_g certainly need not be very reliable. This approach is followed in studying the behavior of amorphous polymers in the solid state and their mechanical response (Chapter 5 and Refs 37, 42, 50–58) and certainly is appropriate for studying solid polymers in the crystalline state.[53,58] The latter problem is not our main concern here, as we concentrate on flexible polymers in the random coil state (solution, melt, amorphous solid). Of course, the "preparation" of the simulation "sample" is then very questionable. Cooling from the melt produces rapidly quenched samples which may differ in many physical properties from their slowly cooled experimental counterparts—at least

one must expect this from experience with coarse-grained models (see Chapter 6 and Refs 59, 60). An elegant alternative is the *molecular mechanics* technique of generating a dense configuration of amorphous polymers such that one folds a single (very long) chain back into the simulation box with the help of periodic boundary condition until the desired (experimentally given) density is reached, and minimizing then the potential energy.[50,51,54,55] Although the results of this procedure are sometimes extremely encouraging, one must admit that the systematic errors are uncontrolled: while Monte Carlo (MC) and molecular dynamics methods in principle get better and better the more computing time is invested, e.g., for making cooling slower and slower, it is not clear to what extent this procedure of Theodorou and Suter[50] corresponds to a physical "sample preparation." This method will not be pursued further in this chapter.

- Restrict the study to long wavelength properties of the problem by abandoning chemical detail and introduction of a coarse-grained model. It is this approach which will be emphasized in this chapter and many following ones (Chapters 2–4, 6, 7, 9). It is clear that this coarse-graining approach makes sense only for "universal" properties, and what universal means depends on the problem under study. For example, for dilute polymer solutions one expects that the mean-square gyration radius of a chain varies with N_p as $\langle R_g^2 \rangle^{1/2} = C_p N_p^{\nu}$ for $N_p \to \infty$, where the exponent ν for good solvents is universal, i.e., independent of the chemical nature of both polymer and solvent, while the prefactor C_p clearly is not. In a coarse-grained model of a polymer chain we can integrate n chemical monomers into one effective subunit (cf. section 1.3) and thus consider an equivalent chain of $N = N_p/n$ "segments." While we still expect $\langle R_g^2 \rangle^{1/2} = CN^{\nu}$, the explicit relation between C and C_p is lost; while it is known that the exponent ν found from the coarse-grained model coincides with the exponent ν of real polymers, the prefactor C_p cannot be predicted from the coarse-grained model. Similarly, if we consider unmixing in polymer blends we can consider the exponent β describing the shape of the coexistence curve (binodal) between the unmixed phases (Chapter 7; see also Refs 21, 23, 61–63), the resulting crossover from Ising-like critical behavior to mean-field-like critical behavior when N increases,[23] and consider the question how the critical temperature $T_c(N)$ scales with N in a symmetrical polymer mixture ("symmetrical" means $N_A = N_B = N$ for both types (A, B) of polymer chains). However, we cannot predict the absolute location of $T_c(N)$ nor explain why one pair of polymers (A, B) such as polystyrene–polymethylvinylether[64] has a lower critical point while another pair such as polyisoprene–poly(ethylene–propylene) has an upper critical solution temperature (UCST).[65]

Thus important goals of molecular simulations, namely predicting physical properties of particular materials quantitatively from a knowledge of atomistic potentials, to a large extent is lost; however, as emphasized above, this consequence is inevitable due to the complexity of polymeric systems. Nevertheless, polymer science is rich with questions which are universal, as discussed above, and the large size of polymer coils and the fact that by variation of N_p one has a control parameter at hand which leaves all chemical details invariant (at least to a good approximation) are plausible good reasons that this should be expected. Of course, for problems such as the glass transition[66] presumably both the behavior on small scales (atoms trapped in "cages" formed by their environment[67]) and on large scales (relaxation due to cooperatively rearranging regions[68]) is simultaneously important—neither the coarse-grained modeling (Chapter 6; References 59, 60, 69) nor the atomic modeling (Chapter 5; Refs 37, 39–42) then should be expected to be fully satisfactory! While first attempts to reintroduce atomistic information from quantum-chemical calculations[45] into coarse-grained models[70–72] look encouraging, more work clearly is necessary to show that the errors in this approach really are under control.

1.2 Survey of simplified models

There is no unique way to construct coarse-grained models of polymer systems. In fact, the choice of model very much depends on the physical problems that one may wish to address, and also many details are fixed from the desire to construct computationally efficient simulation algorithms.[2,7,73–77] Thus many variants of models for polymer chains exist, and there may still be the need to invent new ones! This section is also not intended to be exhaustive, but rather restricts attention only to the most common models which are widely used in various contexts, as will become apparent in later chapters of this book.

1.2.1 Off-lattice models

In section 1.1.2 it was already mentioned that one may wish to simplify models for polymers such as polyethylene by replacing CH_2 groups by "united atoms." If we simplify the problem further, replacing n successive CH_2 groups by an effective bond between some effective units, we may end up with a model of the type shown[78] in Fig. 1.3.

In the model of a freely jointed chain (Fig. 1.3[a]) each polymer hence is modeled by a succession of N rigid bonds of length ℓ jointed together at arbitrary angles. The steps of the Monte Carlo procedure then consist in rotations of beads. For example, bead i is rotated by a randomly selected angle φ_i from its old position to its new position. If the chains are treated as completely noninteracting, the static and dynamic properties of such a

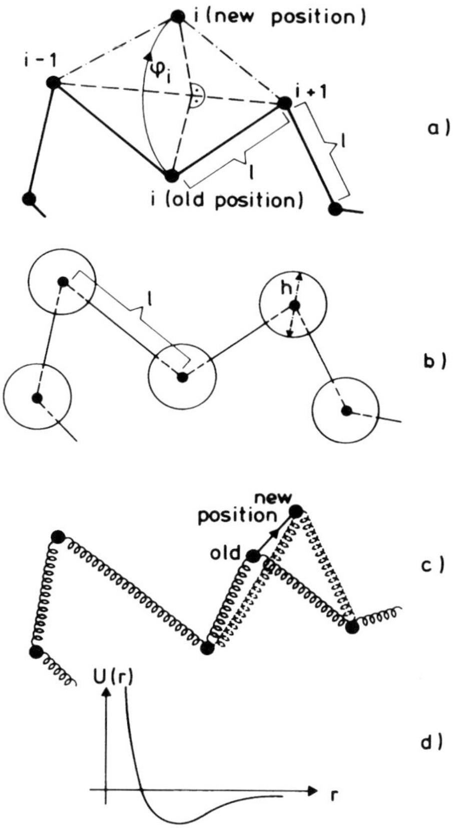

Fig. 1.3 Off-lattice models for polymer chains. In the freely jointed chain (a) rigid links of length ℓ are jointed at beads by dots and may there have arbitrary angles with each other. The stochastic chain conformational changes are modeled by random rotation of bonds around the axis connecting the nearest-neighbor beads along the chain, as indicated. The new position of a bead i may be chosen by assigning an angle φ_i chosen randomly from the interval $[-\Delta\varphi, +\Delta\varphi]$ with $\Delta\varphi \leq \pi$. For the simulation of melts, the freely jointed chain is often supplemented by a Lennard-Jones type potential (d) between any pairs of beads.[80] An alternative model is the pearl necklace model (b), where the beads are at the center of hard spheres of diameter h, which must not intersect each other. By varying the ratio h/ℓ one can control to some extent the persistence length of the chain. This model has also proven useful for the simulation of melts.[82] Still another alternative is the bead-spring model (c), which has also been used both for MC simulations as indicated[76,77] and for MD simulations for solutions[83,84] and melts.[85,86] (From Binder.[78])

model are trivial: the configurations of the chains for N large are Gaussian distributed, and their relaxation is described by the simple Rouse model.[17,79] Of course, one can introduce into this model various interactions, such as a Lennard-Jones (LJ) interaction between all beads[80] (see eq. [1.3]) and/or a potential controlling the angle between subsequent bonds along a chain, etc. These potentials then enter the transition probability $W(\varphi_i) = \text{Min}\{\exp(-\delta\mathscr{H}/k_B T), 1\}$, where $\delta\mathscr{H}$ is the energy change produced by the rotation with φ_i. According to the standard Metropolis[10–12,81] sampling, the trial move is only carried out if $W(\varphi_i)$ exceeds a random number ζ distributed uniformly in the interval from zero to one, since then one generates an ensemble of configurations (when equilibrium is reached) distributed according to the canonical Boltzmann weights.

One can also set the transition probability to zero if the attempted move leads to some intersection of bonds, in an attempt to simulate entanglement restrictions as they operate in the dynamics of polymer melts.[80] Forbidding such intersections thus takes into account that in the course of their motions chains cannot cross each other. However, the resulting algorithm is rather slow and has been applied to very short chains only.[80] While at high temperatures (using an LJ interaction) surprisingly it was found that even in a dense melt the chains relax according to Rouse dynamics,[17] at low temperatures the chains were frozen into a glass-like configuration.[80]

The pearl necklace model (Fig. 1.3[b]) is a somewhat more useful model, although it is strictly athermal; but, by a proper choice of the ration h/ℓ between the radius h of the excluded volume sphere around each bead and the bond length ℓ, one can ensure[82] automatically that chains cannot cross each other if they respect excluded volume restrictions (no spheres are allowed to overlap apart, possibly, from subsequent ones if one chooses $\ell/2 < h < \ell$).

The most popular and efficient off-lattice models are of bead-spring type (Fig. 1.3[c]) and are not only used for MC but also for MD and Brownian dynamics (BD) simulations.[7,76,77,83–86,87–94] It often is advantageous not to use a simple harmonic potential for the bond lengths as in eq. (1.2) but rather allow only a finite extensibility of the chains. In the MD work one works with the so-called FENE potential[95]

$$\mathscr{H}_\ell(\ell_i) = \begin{cases} -0.5\, k\ell_{\max}^2 \ln[1 - (\ell_i/\ell_{\max})^2], & \ell_i < \ell_{\max} \\ \infty & , \ell_i > \ell_{\max} \end{cases} \tag{1.6}$$

where one typically chooses[7] the constants k, $\ell_{\max}$ in relation to the parameters ϵ, σ of the LJ potential between the effective beads (eq. [1.3]) as $\ell_{\max} = 1.5\sigma$, $k = 30\epsilon/\sigma^2$. Note also that one often[76,83–86] does not work with the full LJ potential but with an LJ potential that is cut off and shifted, to avoid a δ-function singularity of the force at the cut-off:

$$\mathscr{H}_{\text{LJ}}^{\text{cut-off}}(r) = \begin{cases} 4\epsilon[(\sigma/r)^{12} - (\sigma/r)^6 - (\sigma/r_c)^{12} + (\sigma/r_c)^6], & r < r_c \\ 0 & , r > r_c \end{cases} \tag{1.7}$$

Choosing[7] $r_c = 2^{1/6}\sigma$ the potential is purely repulsive and also the force is nonsingular at $r = r_c$. Note that eq. (1.7) acts between all pairs of beads, including bonded ones.

With this choice of potentials it is also automatically ensured that an intersection of bonds (in the course of MD or BD simulations) is prevented. For MC simulations, a slightly different choice of potentials was recently proposed,[76] namely

$$\mathscr{H}_\ell(\ell_i) = \begin{cases} \mathrm{k}(\ell_i - \ell_0)^2, & \ell_{\min} < \ell_i < \ell_{\max} \\ \infty, & \ell_i \leq \ell_{\min}, \ell_i \geq \ell_{\max} \end{cases} \tag{1.8}$$

choosing the same non-bonded potential as in eq. (1.7) and $\ell_{\max} = r_c$, $\ell_{\min} = 0.4\,\ell_{\max}$, $\ell_0 = 0.7\,\ell_{\max}$, $k/k_BT = 10$, $k_BT/\epsilon = 10$. With this choice of parameters, chain intersection also is practically forbidden even if one uses large attempted displacements of the effective monomers (in each update, a monomer is attempted to be displaced into a randomly chosen direction and the step width is chosen randomly from a cube with linear dimensions equal to $\ell_{\max}$, the old position being in the center of the cube). This model can be simulated rather efficiently by applying link-cell methods.[8,76,96] Due to the large value of the ratio k/k_BT one probes in practice only the harmonic part of the potential eq. (1.8), the cut-off at $\ell_i = \ell_{\min}, \ell_{\max}$ has very little effect on the properties.

Choosing a model with purely repulsive interactions one expects neither the occurence of a Θ-temperature in solution, nor a reasonable equation of state in the melt (note that in recent MD works[84,97] solvent molecules are included explicitly while in MC simulations one simply reduces the density of monomers in the box[76,77] and reinterprets density as the *volume fraction* of monomers in the solution). Since the LJ potential (eq. [1.3]) for a coarse-grained model lacks any fundamental quantum-chemical foundation, one may choose any other short-range potential if this is computationally convenient. With this reasoning, Milchev *et al.*[77] have proposed to use a Morse potential

$$U_{\mathrm{M}}(r)/\epsilon = \exp[-2\alpha(r - r_{\min})] - 2\exp[-\alpha(r - r_{\min})], \tag{1.9}$$

which has a rather sharp minimum at $r = r_{\min}$ if one takes α large. Choosing $\alpha = 24$, $r_{\min} = 0.8$ (in units of lengths where $\ell_{\max} = 1$), one sees that $U_{\mathrm{M}}(r)/\epsilon$ is essentially zero for $r \geq 1.25\, r_{\min} = 1$, i.e., the same link-cell techniques can be applied as in the purely repulsive case. Indeed in this model the Θ-point has been located and the full regime of concentrations (from dilute solution to dense melts) has been found accessible, both in the regime of the Θ-point and in the good solvent regime.

Finally it should be mentioned that in the MD work it is advantageous not to choose a strictly microcanonical ensemble (where velocity rescalings are needed to fix the desired temperatures[8,9]) but rather to introduce also a coupling to a heat bath.[83] Thus the equations of motions that are integrated are

$$m\frac{\mathrm{d}^2\vec{r}_i}{\mathrm{d}t^2} = -\nabla\mathscr{H}_i - \Gamma\frac{\mathrm{d}\vec{r}_i}{\mathrm{d}t} + \vec{\zeta}_i(t), \tag{1.10}$$

where m is the mass of the effective monomer, $\vec{r}_i$ its position, t the time, $\mathscr{H}_i$ the total potential energy of monomer i (this may include external potentials, of course), Γ is a bead friction that couples the monomers to the heat bath, and $\vec{\zeta}_i(t)$ describes the random force acting on bead i. This random force is related to the friction coefficient via a fluctuation–dissipation relation,

$$\langle\vec{\zeta}_i(t)\cdot\vec{\zeta}_j(t')\rangle = \delta_{ij}\delta(t-t')\,6k_BT\,\Gamma, \tag{1.11}$$

assuming Gaussian white noise. More details about the MD approach can be found in Chapters 3, 4 and 9; here we only mention that this model defined by eqs (1.6), (1.7), (1.10) and (1.11) is not restricted to linear polymers, but is easily extended to both star polymers[98,99] and polymer networks.[100–102] Note that the dynamical properties of the model are affected by the random forces at least at large times; but this problem can be avoided by using eqs (1.10) and (1.11) only to generate an equilibrated ensemble of starting configurations for strictly microcanonical MD runs.[84] This is only of relevance, if one studies effects of hydrodynamics.[84] Finally, one obtains the BD approach considering the overdamped limit (strong friction) by simply omitting the inertial term $\left(m\frac{\mathrm{d}^2\vec{r}_i}{\mathrm{d}t^2}\right)$ in eq. (1.10). The advantage of the BD approach is that it is most closely related to the theoretical descriptions of polymer dynamics, such as Rouse and Zimm models,[17] which are cast in a framework of Langevin equations or the (equivalent) Smoluchowski equation for the probability distribution of the configuration of the effective monomers.

1.2.2 Lattice models

The idea of simplifying the problem even further and representing polymer chains by random walks on a lattice (or, better, by self-avoiding walks [SAW]) is very old and has been used, for instance, for discussing the configurational entropy of polymers in solution and of polymer blends.[19,103,104] For discussing the configurational statistics of single polymer chains (with excluded volume interaction) this has been a classic problem for which a rich literature exists that is reviewed in Chapter 2. Here we shall confine ourselves therefore only to a very brief discussion (see also Ref. 2 for more details on these models).

The simplest model considers a regular lattice where each effective bead of the polymer takes a single lattice site, and a bond connecting two beads is just a nearest-neighbor link on the lattice (Fig. 1.4).[105–113] Since each lattice site can at most be occupied by one bead, the walk cannot intersect itself (or other walks, respectively) and thus an excluded volume interaction is auto-

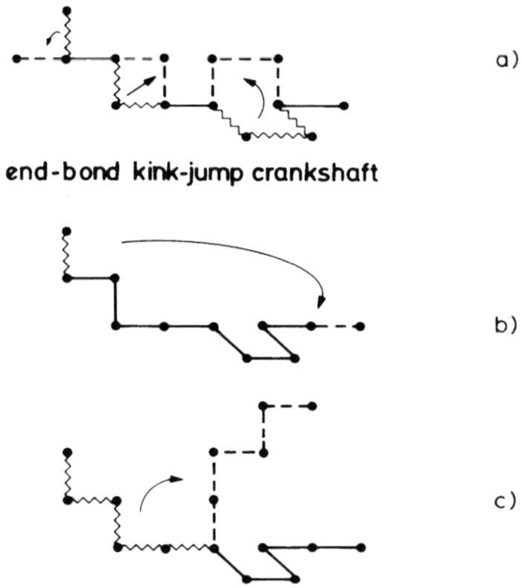

Fig. 1.4 Various examples of dynamic Monte Carlo algorithms for SAWs: sites taken by beads are shown by dots, and bonds connecting the bead are shown by lines. Bonds that are moved are shown as a wavy line (before the move) or broken line (after the move), while bonds that are not moved are shown as full lines. (a) Generalized Verdier–Stockmayer algorithm[105–107] on the simple cubic lattice showing three type of motions: end-bond motion, kink-jump motion, 90° crankshaft rotation; (b) slithering snake algorithm[108,109] (c) pivot algorithm.[110–113] (From Kremer and Binder.[2])

matically included. While the arbitrary angles between successive links of an off-lattice coarse-grained chain are reasonable, if we interpret these links as the *Kuhn effective segments* formed from groups of several monomers along the chemical sequence of the chain, the rigid bond angles (only $\Theta = 0°$ and $\Theta = 90°$ occur at the square and simple cubic lattice) are a further idealization. But, as will be discussed in Chapter 2, for certain universal large-scale properties of polymer chains this does not matter. Of course, other lattices (such as face-centered cubic (fcc) or diamond lattices) are also occasionally used.[2]

We now briefly discuss various dynamic algorithms for this model, and start with the moves shown in Fig. 1.4(a). If one used only the end rotation and the "kink jump" move as done in the original Verdier–Stockmayer algorithm[105] for ideal random walks without excluded volume, a very slow algorithm would result[106]: since kink jumps can be interpreted as an exchange of two neighboring bond vectors, new bond vectors are created only at the ends and must slowly diffuse into the interior of the chain to equilibrate its configuration.[107] So the set of moves in Fig. 1.4(a) (which must be tried at random to avoid biased sampling) are the minimal set that is useful,[62,114] and the relaxation may be facilitated by allowing even for additional moves rotating larger groups,[115] but this makes the algorithm more complicated. It is also clear that the algorithm will have a very small acceptance rate at high volume fraction ϕ of occupied sites, since the vacant sites needed for an acceptable move are then very rare (this holds even more so if one includes additional moves which need even more vacancies[115]).

Another problem with the algorithm of Fig. 1.4(a) is that it can be proven to be nonergodic[113,116,117]: one can easily find locally compact configurations of a chain that cannot relax by the motions shown. Since the algorithm must satisfy detailed balance, of course, the chain cannot get into these configurations from other, nonblocked, configurations, and thus there is a part of the configurational space that simply is not included in the sampling at all! Although this does not seem to be a problem in practice (comparing data from different types of algorithms yields equivalent results within the statistical errors[62]), one must be aware of this drawback if one either tries to study much longer chains or tries to improve significantly the accuracy.

Due to the need for sufficient vacancies this algorithm has only been used[61,62,114] for volume fractions $\phi \leq 0.8$. The "slithering snake' algorithm (Fig. 1.4[b]), on the other hand, needs only one vacant site near a chain end for a successful move and thus can be used for significantly denser systems ($\phi = 0.976$ was successfully used[118,119]). It also suffers from the problem that it is not strictly ergodic,[117] although again these "blocked configurations" do not seem to affect the accuracy to a practically relevant extent.[2] If one is interested in static chain properties, an advantage of the slithering snake algorithm is that it relaxes distinctly faster[2,120] than the generalized Verdier algorithms. The latter (for single chains) produce a Rouse-type[17,79] relaxation and must be used if one is interested in dynamical chain properties, whereas the slithering snake move (in which one tries to remove a randomly chosen end-link of a chain and add it in a randomly chosen direction on the other chain-end) obviously has no counterpart in the dynamics of real chains.

For most static properties of single chains, the algorithm of choice is clearly the pivot algorithm[110–113] (Fig. 1.4[c]), where one randomly chooses a link in the chain and then rotates this link together with the rest of the chain to a randomly chosen new orientation on the lattice. Of course, the new configuration is accepted only if it does not violate the excluded volume constraint. The advantage of such moves is that one rapidly generates new chain configurations, which are not "dynamically correlated" with their predecessors.[2,11,78,113] Of course, this algorithm cannot then be used for a study of dynamical properties of chains, and it is also not useful for dense polymer systems.

An algorithm that incorporates large nonlocal moves of bonds and works for dense polymer systems (even without any vacancies, $\phi = 1$) is the "collective motion" algorithm[121–123] where one transports beads from kinks or chain ends along the chain contour to another position along the chain, for several chains simultaneously, so that in this way this rearrangement exchanges some of the sites taken by their beads. Due to the nonlocal collective rearrangements of several chains, the algorithm is rather complicated and is not straightforwardly suited to vectorization or parallelization. Thus only if the vacancy concentration $\phi_v = 1 - \phi$ is very small, does this algorithm have an advantage (in comparison with the algorithms of Figs 1.4[a,b]) since these other algorithms then have a very small acceptance rate. Also one clearly cannot associate a physical meaning to the time variable in this algorithm, although this sometimes was attempted.[121] Since real polymer melts do have a nonzero compressibility, which in the framework of simple lattice models (SAWs on the sc lattice) is roughly reproduced with a vacancy content[124] of about $\phi_v = 0.1$, one also should not claim that a model with strictly $\phi_v = 0$ is more realistic than the models that contain vacancies. If one defines a *van der Waals density* in terms of the repulsive part of the interatomic potential in real polymers, one obtains a van der Waals density of only about 50%. Thus, in our view, the collective motion algorithm poses more disadvantages than advantages.

The lattice algorithm that is now most widely used for the simulation of many-chain systems is the bond fluctuation model (Fig. 1.5).[73,125–128] It has been used to model the dynamics of both two-dimensional[125] and three-dimensional polymer melts,[127,128] the glass transition (see also Chapter 6),[59,60,69–72] polymer blends (see also Chapter 7),[23,63,129] polymer networks,[130] gel electrophoresis,[131] polymer brushes (see Chapter 9) and so on, and it was attempted to map it on to real materials,[70–72] This model is in a sense intermediate between the simple SAW model of Fig. 1.4 and the off-lattice models (Fig. 1.3), because the vector that connects two monomers can take 36 values (in $d = 2$ dimensions) or 108 values (in $d = 3$ dimensions), rather than four or six (square or simple cubic lattice, respectively). While thus the continuum behavior is almost approximated, one still enjoys the advantages of lattice algorithms (integer arithmetics, excluded volume is checked via the occupancy of lattice sites, etc.) The restriction in the allowed bond lengths assures that the excluded volume interactions simultaneously maintain the constraint that bonds cannot cross each other in the course of random motion of the monomers. The fact that there is a single type of motion that is attempted (random displacement of an effective monomer, i.e., an elementary plaquette of the square lattice [$d = 2$] or an elementary cube of the simple cubic lattice [$d = 3$] in a randomly chosen lattice direction) allows very efficient implementations of this algorithm, including vectorization.[126] This algorithm also suffers less from ergodicity problems than the algorithms of Figs 1.4(a) and (b), although it is not strictly ergodic either. For single chains, from this algorithm a Rouse-model type dynamics results,[17,73,79,127] as desired, while in dense melts (i.e., systems where $\phi \geq 0.5$[127]) a crossover occurs[127,128] from Rouse-like behavior (for $N \lesssim 50$) to reptation-like behavior[16,17,132] for long enough chains that then are mutually entangled.

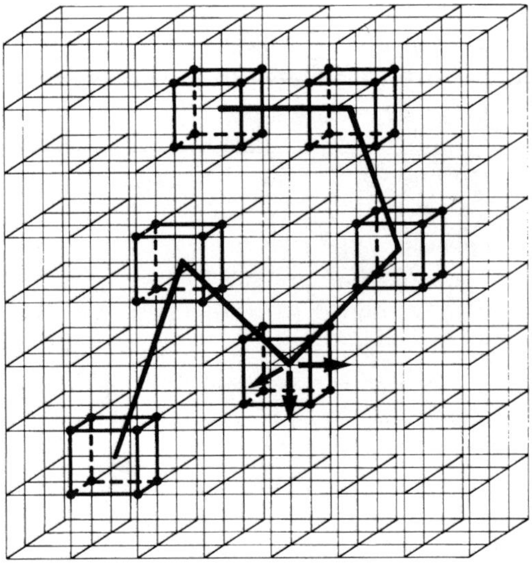

Fig. 1.5 Schematic illustration of the bond fluctuation model in three dimensions. An effective monomer blocks a cube containing eight lattice sites for occupation by other monomers. The length ℓ of the bonds connecting two neighboring cubes along the chain must be taken from the set $\ell = 2, \sqrt{5}, \sqrt{6}, 3, \sqrt{10}$. Chain configurations relax by random diffusive hops of the effective monomers by one lattice spacing in a randomly chosen lattice direction. (From Deutsch and Binder.[129])

While at first sight it may seem pathological that the length of an effective bond may vary over a wide range ($2 \leq \ell \leq \sqrt{13}$ in $d = 2$, $2 \leq \ell \leq \sqrt{10}$ in $d = 3$), this is natural when one recalls that each effective bonds represents a group of n C–C bonds along the backbone of the chemically realistic chain (Fig. 1.1), and, depending on the conformation of this group its end-to-end distance may vary significantly. This idea will be explored in more detail in Section 1.3 by attempting to develop effective potentials for the length ℓ of effective bonds and the angle $\ominus$ between two consecutive such bonds from the underlying microscopic (i.e., chemically realistic) model of the polymer chain. The fact that one can have such potentials for the length of the bond vector easily allows the modeling of the glass transition of polymer melts (see Chapter 6) which is less straightforward on the basis of the model of Fig. 1.4.

Finally we end this survey by emphasizing that there is no unique answer to the question "which is the best model of a polymer chain"; depending on the types of application and the questions that are asked different models and different algorithms may be useful, and there is still room for inventing new algorithms. For example, for a study of ionomers, where chains stick together at certain sites due to electrostatic forces, it was advantageous to use an algorithm which allowed the "stickers" to exchange partners and to translate parts of the chain rigidly.[133] Polymer chains in $d = 2$ dimensions can be simulated by a particularly fast but locally very unrealistic algorithm.[74,75] We also emphasize that we have discussed dynamic algorithms only—there also exist useful algorithms to construct SAW configurations by growing them in a stepwise manner by various techniques. For a review of such "static" methods, see Ref. 2 and Chapter 2. These static methods are particularly useful if one wishes to estimate the number of configurations of the polymers and the associated exponents. Note that such techniques are not only applicable to linear polymers but can be extended to different polymer architectures, such as many-arm star polymers.[134,135]

1.3 Taking the idea of coarse-graining literally

In the previous section, we have seen that a variety of crudely simplified models of polymer chains is available. In this section we discuss the extent to which such models can be connected with more microscopic, chemically realistic descriptions, and how one should proceed when comparing results from different model calculations with each other.

1.3.1 Effective potentials for the bond fluctuation model

The coarse-grained model can be obtained by combining n successive covalent bonds along the backbone of a polymer chain into one effective segment (see Fig. 1.6 where $n = 3$ is chosen). In principle, such a procedure can be carried out for any polymer (e.g., in Refs 45, 70–72 an application to bisphenole-A-polycarbonate (BPA-PC) is discussed). In order to make close contact with reality, one may wish to carry out this mapping such that the large-scale geometrical structure of the polymer coil is left invariant, i.e., properties such as the gyration radius of the coil and the probability distribution of its end-to-end distance should be the same for the coarse-grained model in Fig. 1.6 as for the chemically detailed model.

The idea of Refs 70–72 is that this invariance of long wavelength properties is indeed realized if we introduce suitable potentials in the coarse-grained model which control bond lengths of the effective bonds, angles between effective bonds along the sequence of the coarse-grained chain, etc. In practice, it was proposed[70–72] to use harmonic potentials both for the length ℓ of an effective bond and for the cosine of this effective bond angle, i.e.,

$$U_{\mathrm{eff}}(\ell) = U_0(\ell - \ell_0)^2, \quad V_{\mathrm{eff}}(\Theta) = V_0(\cos\Theta - \cos\Theta_0)^2, \tag{1.12}$$

where U_0, ℓ_0, V_0, $\cos\Theta_0$ are four adjustable parameters (which may depend on the thermodynamic state of the considered polymer melt, of course, i.e., temperature and pressure). It is thought that these potentials "mimick" the

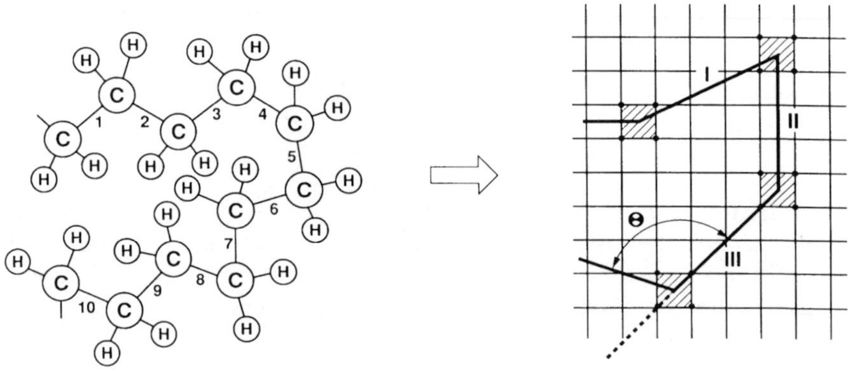

Fig. 1.6 Use of the bond fluctuation model on the lattice as a coarse-grained model of a chemically realistic polymer chain (symbolized again by polyethylene). In the example shown $n = 3$ covalent bonds form one "effective bond" between "effective" monomers: chemical bonds 1,2,3 correspond to the effective bond I, chemical bonds 4,5,6 to the effective bond II, etc. (From Baschnagel *et al.*,[45])

effect of the potentials of the microscopic model, such as eqs (1.1)–(1.3) describing potentials for the lengths of chemical bonds and their bond angles and torsional angles. So the information on these potentials and hence the chemical structure is not completely lost in the coarse-graining, but at least some caricature of it is still found in the simplified model (Figs 1.5, 1.6) via the potentials in eq. (1.12).

We do not only wish to map the behavior on large length scales but also on large time scales. Since a move of an effective monomer as shown in Fig. 1.5 requires transitions from one minimum of the torsional potential to another, on the scale of the chemically realistic model (Fig. 1.2), information on the barrier heights of this torsional potential must also be incorporated into $U_{\text{eff}}(\ell)$, $V_{\text{eff}}(\ominus)$, since these potentials govern the transition probability of the Monte Carlo sampling process.[136] The corresponding explicit construction of the parameters in eq. (1.12) from the available quantum-chemical information on potentials such as written in eqs (1.1)–(1.3) is rather tedious and difficult[136] and will not be discussed further here.

If the potentials $U_{\text{eff}}(\ell)$, $V_{\text{eff}}(\ominus)$ are known, their basic effect will be to generate distributions according to the Boltzmann weight

$$P_n(\ell) \propto \exp[-U_{\text{eff}}(\ell)/k_B T] = \exp[-U_0(\ell - \ell_0)^2/k_B T], \qquad (1.13)$$

and similarly for $P_n(\ominus)$. These distributions can also be obtained directly from the chemically realistic model of an isolated chain (Fig. 1.7)[43,45] and checking that the model distribution (eq. [1.13]) represents the actual distribution faithfully enough hence is an important consistency check of the description. One can infer that for $n = 5$ (which is a reasonable choice, since

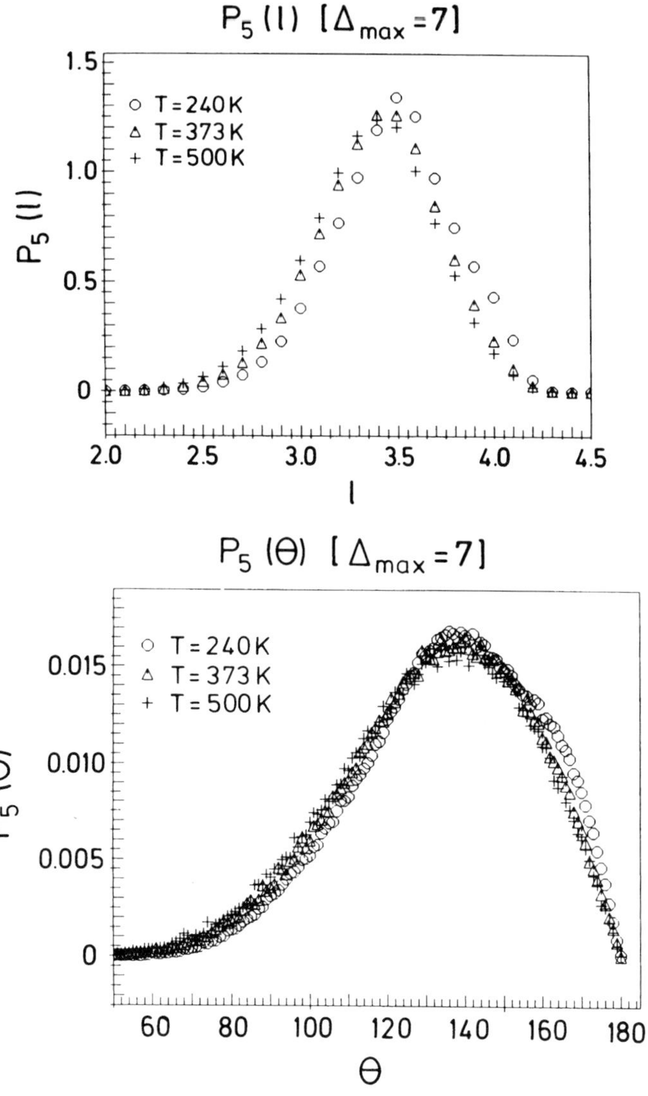

(a)

(b)

Fig. 1.7 (a) Distribution function $P_5(\ell)$ vs. ℓ (measured in units of $\ell_0 = 1.52$Å) for $n = 5$ subsequent CH_2 units integrated into one effective bond. Three temperatures are shown as indicated. The nonbonded LJ interaction was only included up to the seventh neighbor along the chain, to ensure Gaussian chain statistics at large distances, as it occurs in dense melts where the LJ interactions are screened out. (b) same as (a) but $P_5(\theta)$ plotted vs. angle θ between subsequent effective bonds. (From Baschnagel *et al.*[43])

n must be large enough to include several torsional degrees of freedom, but small enough to still contain information on the scale of the persistence length) the distributions for $P_n(\ell)$, $P_n(\cos\ominus)$ are indeed reasonably well approximated as Gaussian: this finding, in fact, was the motivation for the choice of the simple quadratic potentials, eq. (1.12). The weak temperature dependence that one can see in Fig 1.7 simply reflects the gradual tendency of the PE chain to stretch (increase of the persistence length and the characteristic ratio $C_\infty = \lim_{N_p \to \infty} \langle R^2 \rangle / N_p \langle \ell_i^2 \rangle)$ as the temperature is lowered.

A consequence of the mapping in Fig. 1.6 being taken literally is that the (otherwise not explicitly defined) length and time scales of the coarse-grained model get a physical meaning. For example, for BPA-PC one obtains (when one monomer with 12 chemical bonds along the backbone is mapped onto three effective bonds) that a lattice spacing at $T = 570\,\mathrm{K}$ corresponds to 2.3 Å, and one Monte Carlo step per monomer (attempted monomer hop, which has an acceptance rate of about 1% only due to the restrictive potentials) corresponds to 10^{-13} s, a successful hop occurs at about a time of 10^{-11} s. Obviously, this is physically reasonable, and relevant information has been lost on a very small length scale ($\lesssim 3$ Å) only. Choosing[70–72] lattices of a typical linear dimension of $L = 40$ lattice spacings then means that systems of a linear dimension of about 10^2 Å are simulated, which clearly is somewhat larger than the sizes accessible in the study of microscopically realistic models.[36–42, 50–55] Note that the largest lattice used so far[23] (for the study of the critical behavior of polymer blends, see also Chapter 7) used $L = 160$ and did contain 256 000 effective monomers, which with $n = 5$ (Fig. 1.7) would correspond to 1 280 000 CH_2 units (and $N = 512$ would translate into $N_\mathrm{p} = 2560$, a reasonably large degree of polymerization[23]).

1.3.2 How different coarse-grained models can be compared

From the previous discussions it should be clear that there is no unique model description of a polymer chain system; in fact, for different physical questions somewhat different coarse-grained models are optimal. For example, while the bond fluctuation model[73,126–129] is very well suited to the study of polymer melt dynamics,[125–128] polymer blends,[23,63] the glass transition,[59,60,69] polymer brushes in good solvents[137] and so on, for other problems it is less well suited: lattice structure artefacts appear in the study of collapsed polymer brushes in bad solvents[138] and in models for dense lipid monolayers at the air–water interface[139]; in the presence of frozen-in obstacles, locked-in chain configurations must be expected.[76] For such problems, off-lattice models clearly are better suited. How then can results from different models be combined and compared quantitatively?

This question is similar to the question of how to compare simulations and experiment[85] or to compare Monte Carlo work with MD work,[127] of course. But a need for such a conversion arises even when somewhat different Monte Carlo models need to be compared. This question was addressed by Milchev *et al.*[76] who compared data for their off-lattice bead spring model, where results for various chain lengths N and volume fractions ϕ were obtained, to results for the bond fluctuation model[127] where data for a variety of values for chain lengths $\tilde{N}$ and volume fractions $\tilde{\phi}$ are available. While on the lattice the term "volume fraction" is not ambiguous, it is simply the percentage of lattice sites blocked by the effective monomers; there clearly is some arbitrariness for the off-lattice case. Gerroff *et al.*[76,77] define it as the number of effective monomers per cell of size $\ell^3_{\max}$: thus ϕ (unlike $\tilde{\phi}$) can even be larger than unity, and the question of translating the variable ϕ to the variable $\tilde{\phi}$ arises. Similarly, the models may exhibit different degrees of local chain stiffness, and hence there also is no one-to-one correspondence between N and $\tilde{N}$. This point is sometimes overlooked in the comparison of different models.[140]

Gerroff *et al.*[76] argue that it is the universal scaling limit that can be extracted from different models and which must be identical; so, when one considers the scaling limit $N \to \infty$, $\phi \to 0$ (or $\tilde{N} \to \infty$, $\tilde{\phi} \to 0$), one must obtain the same scaling function for $\langle R_g^2(\phi)\rangle / \langle R_g^2(\phi = 0)\rangle$ versus $N/N_{\text{blob}}(\phi) \propto N\phi^{1/(3\nu-1)}$, which describes the crossover from the dilute to the semidilute limit (Fig. 1.8). This scaling function is universal (for good solvents), because on large length scales compared to microscopic length scales (such as bond lengths) the chains are self-similar. The only relevant length scale for the distribution of intramolecular distances can be taken to be the radius of gyration measured in units of the microscopic length scale.

Differences in the nonuniversal prefactors like the persistence length lead to the necessity of mapping different chain lengths onto one another. In the dilute limit, we write for the mean square gyration radius $\langle R_g^2\rangle$

$$\langle R_g^2\rangle / \langle \ell^2\rangle = CN^{2\nu}, \quad N \to \infty, \tag{1.14a}$$

where C is a dimensionless (nonuniversal) constant and[24] $\nu = 0.59$. Similarly, for the bond fluctuation model an analogous relation holds:

$$\langle R_g^2\rangle_{\text{BF}} / \langle \ell^2\rangle_{\text{BF}} = C_{\text{BF}}\tilde{N}^{2\nu}, \tilde{N} \to \infty. \tag{1.14b}$$

Now the two models are mapped onto each other defining $N = \alpha\tilde{N}$ and fixing the conversion factor α such that the amplitude factors in eqs (1.14a) and (1.14b) then are equal, $C\alpha^{2\nu} = C_{\text{BF}}$. From the numerical results $C = 0.24$, $C_{\text{BF}} = 0.164$ it was found that $\alpha \approx 0.274$, i.e., the range from $N = 8$ to $N = 64$ was mapped to a range from $\tilde{N} \approx 11$ to $\tilde{N} \approx 88$ in the bond fluctuation model.

The conversion from ϕ to $\tilde{\phi}$ is achieved similarly, noting that for $\phi \gg \phi^*$ (ϕ^* is the overlap concentration in the solution, defined by $N_{\text{B}}(\phi^*) = N$ where $N_{\text{B}}(\phi)$ is the number of segments per blob, i.e., $N/\langle R_g^2\rangle^{3/2} = \phi^* \propto N^{-(3\nu-1)}$) the behavior of R_g is classical:

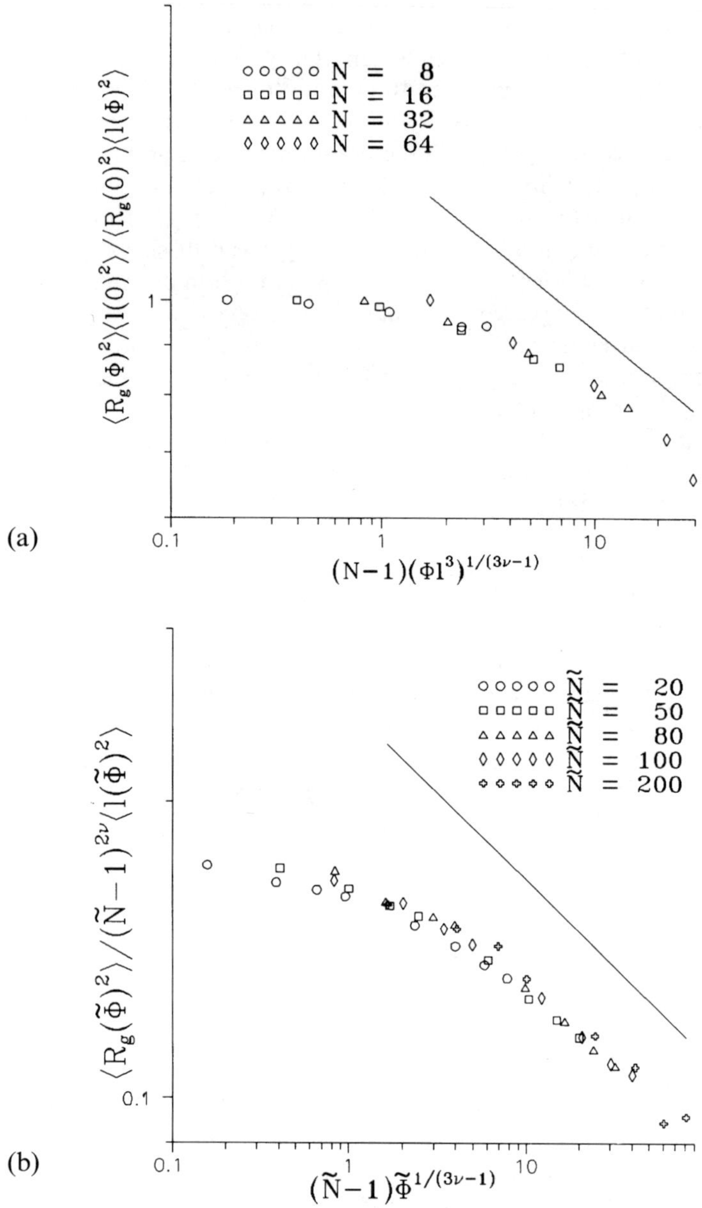

Fig. 1.8 (a) Log–log plot of the reduced mean square gyration radius, $\langle R_g^2(\phi)\rangle / \langle R_g^2(0)\rangle$, vs. rescaled chain length $(N-1)(\phi\ell^3)^{1/(3\nu-1)}$, where ℓ is the root mean-square bond length, and the theoretical value[20,24] for the exponent $\nu(\nu = 0.59)$ is used. All data refer to a bead-spring model with stiff repulsive Lennard-Jones interaction, as described in Section 1.2.1 (Eq. [1.7]). (b) Same as (a) but for the bond fluctuation model. In both (a) and (b) the straight line indicates the asymptotic slope of the crossover scaling function, resulting from eq. (1.15). (From Gerroff *et al.*[76])

$$\langle R_g^2 \rangle / \langle \ell^2 \rangle = C' \phi^{-(2\nu-1)/(3\nu-1)} N, \; N \gg N_B(\phi), \tag{1.15a}$$

or alternatively in the bond fluctuation model

$$\langle R_g^2 \rangle / \langle \ell^2 \rangle = C'_{BF} \, \tilde{\phi}^{-(2\nu-1)/(3\nu-1)} \tilde{N}. \tag{1.15b}$$

Using then $N = \alpha \tilde{N}$ and $\phi = \beta \tilde{\phi}$ and requiring that $\langle R_g^2 \rangle / \langle \ell^2 \rangle$ is the same in both models yields $C'_{BF} = C' \alpha \beta^{-(2\nu-1)/(3\nu-1)}$ as a condition that both scaling functions in Figs 1.8(a) and (b) superimpose. One finds for the model of Gerroff *et al.*[76] that $\beta = 0.3$, i.e., $\phi = 1.5$ in the off-lattice model corresponds to $\tilde{\phi} = 0.5$ in the bond fluctuation model.

It is possible to carry over such considerations to dynamic quantities as well (see also Chapter 4). This is of interest, of course, for comparing the efficiency of different algorithms. It then does not make sense to compare the time it takes to carry out one Monte Carlo step (MCS), since step width and acceptance rate of the moves also matter: what must be compared are physical relaxation times. One such relaxation time for a polymer chain can be defined[128] as the time τ_1 needed for an inner monomer of a chain to travel a distance equal to the gyration radius. Defining mean square dynamic displacements of inner monomers as follows,

$$g_1(t) \equiv \langle [\vec{r}_{N/2}(t) - \vec{r}_{N/2}(0)]^2 \rangle \tag{1.16a}$$

it also is of interest to consider an analogous quantity in the center of mass system of the chain ($\vec{r}_{CM}(t)$ being the position of its center of mass at time t)[127,128]

$$g_2(t) \equiv \langle [\vec{r}_{N/2}(t) - \vec{r}_{CM}(t) - \vec{r}_{N/2}(0) + \vec{r}_{CM}(0)]^2 \rangle, \tag{1.16b}$$

and the mean-square displacement of the center of mass itself,

$$g_3(t) \equiv \langle [\vec{r}_{CM}(t) - \vec{r}_{CM}(0)]^2 \rangle. \tag{1.16c}$$

Finally, mean-square displacements of monomers at the free ends of the chains are defined as

$$g_4 \equiv \langle \vec{r}_{end}(t) - \vec{r}_{end}(0)]^2 \rangle, \tag{1.16d}$$

$$g_5(t) \equiv \langle \vec{r}_{end}(t) - \vec{r}_{CM}(t) - \vec{r}_{end}(0) + \vec{r}_{CM}(0)]^2 \rangle. \tag{1.16e}$$

Now relaxation times τ_1, τ_2, τ_3, τ_4 can be defined from requiring that

$$g_1(\tau_1) = \langle R_g^2 \rangle, g_2(\tau_2) = \frac{2}{3} \langle R_g^2 \rangle, g_3(\tau_3) = g_2(\tau_2), g_5(\tau_4) = \langle R_g^2 \rangle. \tag{1.17}$$

Figure 1.9 illustrates the variation of these mean-square displacements with time and the estimation of the resulting relaxation times τ_1, τ_2, τ_3, τ_4 for three different models: the bond fluctuation model,[127,128] the off-lattice bead-spring model of Gerroff *et al.*,[76] and the molecular dynamics simulation of a realistic model for short polyethylene chains ($N_p = 50$) in the melt

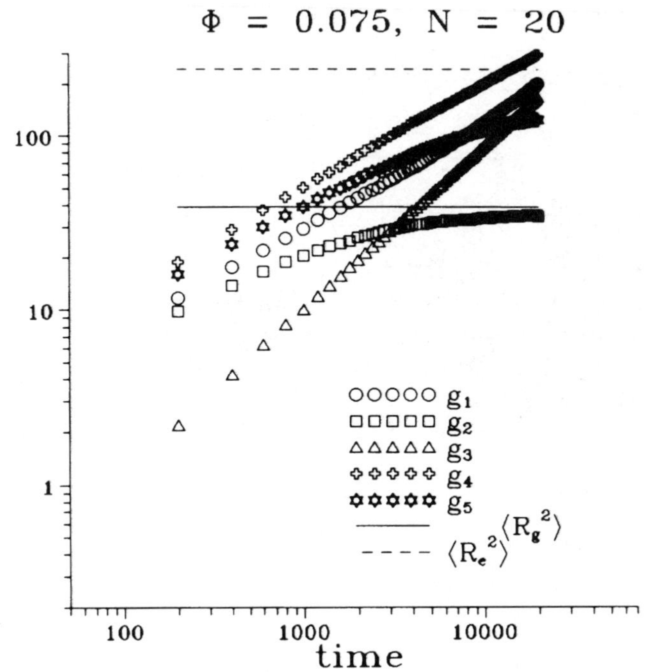
Φ = 0.075, N = 20
100
10
1
100
1000
10000
time
g1
g2
g3
g4
g5
⟨Rg²⟩
⟨Re²⟩

(a)

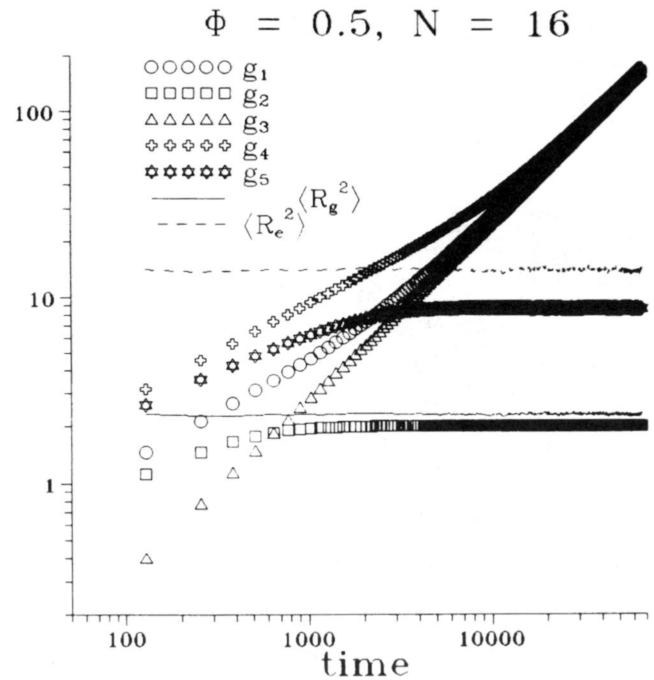
Φ = 0.5, N = 16
100
10
1
100
1000
10000
time
g1
g2
g3
g4
g5
⟨Rg²⟩
⟨Re²⟩

(b)

(c)

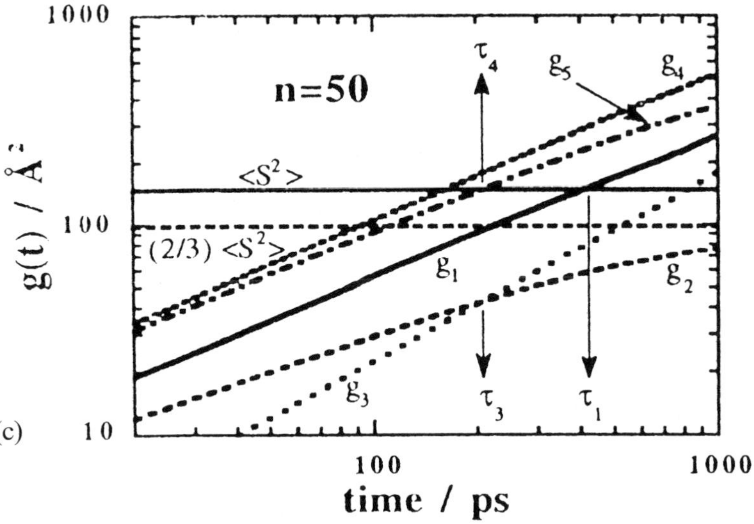

Fig. 1.9 (a) Log–log plot of mean-square displacements versus time (measured in attempted Monte Carlo steps per monomer) for the bond fluctuation model at parameters corresponding to a dilute solution ($\tilde{N} = 20, \tilde{\phi} = 0.075$). (b) Log–log plot of mean-square displacement versus time (measured in attempted Monte Carlo steps per monomer) for the bead-spring model of Gerroff *et al.*[76] at parameters corresponding to a dilute solution ($\phi = 0.5, N = 16$). Broken and straight lines shows the measurement of $\langle R_g^2 \rangle$ and the mean-square end-to-end distance $\langle R_g^2 \rangle$, respectively, and the $g_i(t)$ are defined in eq. (1.16). (From Gerroff *et al.*.[76]) (c) Log–log plot of mean-square displacements versus time (measured in picoseconds) from the MD simulation of 64 systems run in parallel containing 20 PE chains with degree of polymerization $N_p = 50$ and melt densities. The mean-square gyration radius is denoted as $\langle S^2 \rangle$ in this figure. (From Brown *et al.*[38])

(at temperature $T = 500$K).[38] It is seen that the general behavior is rather similar in all three models. Of course, only the chemically realistic model of Brown *et al.*[38] has physical scales (picoseconds for time and Å for lengths, respectively); the Monte Carlo work of the coarse-grained model measures time in attempted moves per monomer and length in lattice spacings. Remembering that one MCS may correspond to about 10^{-13} s in the bond fluctuation model, a time-scale of 10^3 in Fig. 1.9(a) corresponds to about 100 ps, indeed a comparable time to Fig. 1.9(c). Note that $N_p = 50$ (shown in Fig. 1.9[c]) would correspond to about $N = 16$ in Figs 1.9[a],[b]) with a mapping (Fig. 1.6) $N_p = Nn$ with $n = 3$. Remembering that a lattice spacing corresponds to about 2.3 Å (Fig. 1.9[a]) would imply $\langle R_g^2 \rangle \approx 200$ Å, again comparable to the realistic model (Fig. 1.9[c]).

Of course, strictly speaking the different models need be equivalent only in the scaling limit $N \to \infty$, $\phi \to 0$ (for the realistic model of a polymer in solution a different dynamics would then result, described by the Zimm model,[141] due to hydrodynamic forces mediated by the solvent molecules (see also Chapter 3).[7,17,84] Only in the absence of solvent molecules may the Rouse model result also from an MD simulation[83]). Another universal regime results for dense melts: then it is the entanglement molecular weight N_e of the different models that needs to be mapped for a quantitative comparison (see also Chapter 4).[7,85,127] Since Figs 1.9(a), (b) refer to solution while Fig. 1.9(c) refers to a melt, one should not expect more than a similarity of qualitative character here. Since chains renew their configuration only on time scales larger than these characteristic times τ_1, τ_2, τ_3, τ_4, Fig. 1.9(c) provides practical evidence for our estimates of section 1.1.2, that times exceeding a nanosecond are needed to equilibrate melts of non-entangled short chains at high temperatures.

Only for the coarse-grained models can one so far estimate the variation of the relaxation times over a significantly wide range of N, $\tilde{N}$ (Fig. 1.10). One finds the expected power-law behavior for both models. A particularly interesting feature is found when one compares the absolute value of the relaxation times for the same chain length: e.g., for $N = 30$ we have $\tau_1 \approx 1200$ in the off-lattice model but $\tau_1 \approx 3600$ in the bond fluctuation model. Thus the off-lattice model needs a factor of three less MCS to reach the same physical relaxation time. This fact partially offsets the disadvantage that the off-lattice algorithm performs distinctly slower.

Thus the general conclusion of this section is that one must think carefully about the conversion of scales (for length, time, molecular weight) when one compares physical results from different models, or the efficiency of various algorithms. It is hoped that the above examples serve as a useful guideline of how to proceed in practice.

1.4 Selected issues on computational techniques

In this section are briefly reviewed some technical problems of the simulation of dense many-chain systems, such as the sampling of intensive variables such as chemical potential, pressure etc., but also entropy, which are not straightforward to obtain as averages of "simple" quantities. Some of the standard recipes developed for computer simulation of condensed phases in general[8–13] have difficulties here, due to the fact that the primary unit, the polymer chain, is already a large object and not a point particle. But knowledge of quantities such as the chemical potentials are necessary, e.g., for a study of phase equilibria in polymer solutions.[42]

1.4.1 Sampling the chemical potential in NVT simulations

This problem has been brilliantly reviewed by Kumar in a recent book[142] and hence we summarize only the most salient features here. For small molecule systems, sampling of the chemical potential rests on the Widom test particle insertion method[143]

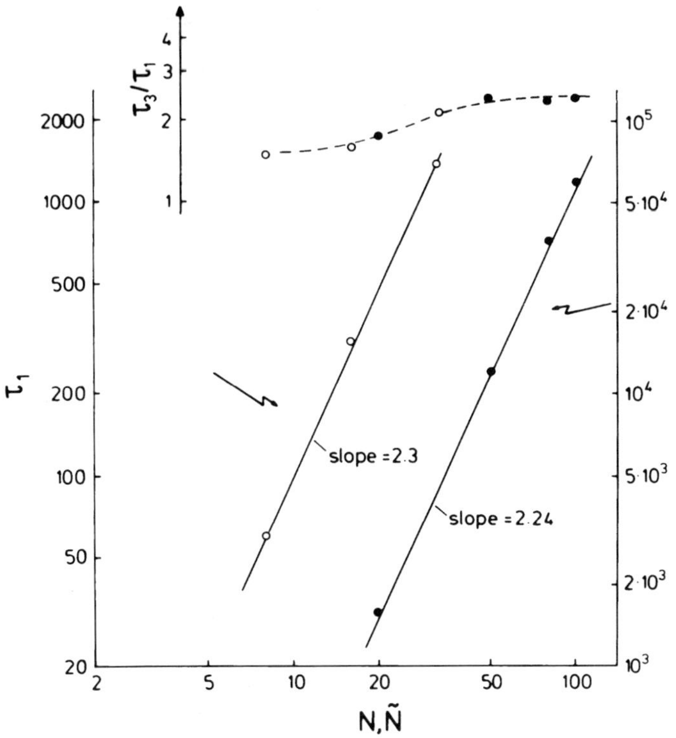

Fig. 1.10 Log–log plot of relaxation time τ_1 vs. chain length N, for the bead-spring model with soft Lennard-Jones repulsion[76] and the bond fluctuation model.[128] Open circles (and left ordinate scale) refer to off-lattice model at $\phi = 0.0625$, full dots (and right scale) to the bond fluctuation model at $\tilde{\phi} = 0.05$ (data taken from Ref. 128). Straight lines indicated the power laws $\tau_1 \propto N^z$, where the exponent $z = 2.3$ or 2.24, respectively, is reasonably compatible with the theoretical prediction[16,17] $z = 2\nu + 1 \approx 2.18$. Insert shows the ratio τ_3/τ_1. It is seen that both models give mutually compatible results for the N-dependence of this dimensionless ratio (which should settle down at some universal constant for $N \to \infty$). In this figure the distinction between N and $\tilde{N}$ (which is only a rather small shift on the logarithmic scale) is disregarded. (From Gerroff *et al.*.[76])

$$-\Delta\mu/k_{\rm B}T = -(F_{\mathcal{N}+1} - F_{\mathcal{N}})/k_{\rm B}T = \ln\left\langle \exp(-U/k_{\rm B}T)\right\rangle_{\mathcal{N}}, \qquad (1.18)$$

where $\Delta\mu$ is the chemical potential difference relative to the chemical potential of an ideal gas at a same temperature and density, $F_{\mathcal{N}}$ is the Helmholtz free energy of a system containing $\mathcal{N}$ particles, and $\langle \ldots \rangle_{\mathcal{N}}$ represents a canonical ensemble average. This test particle insertion method involves the insertion of a ghost particle into a *frozen equilibrium snapshot* of a system containing $\mathcal{N}$ particles, and U denotes the total potential energy experienced by this test particle. Averaging the appropriate Boltzmann factor over many different configurations (frozen snapshots) of the system, the chemical potential is obtained from eq. (1.18), and this method works well in practice for small molecule fluids (for examples see Refs 144, 145). Now, for polymers the insertion of a polymer in a frozen equilibrium snapshot has a very low acceptance probability, and this probability decreases exponentially with increasing chain length. Hence this method has been restricted to $N \leq 20$ for lattice models[146,147] and to $N \leq 15$ for pearl necklace off-lattice chains.[148–151]

Several schemes have been devised by various authors, some of them relying in one form or another on the biased sampling scheme of Rosenbluth and Rosenbluth[152] and others on *thermodynamic integration* methods.[8–13] The Rosenbluth–Rosenbluth method was devised originally as a sampling scheme for generating configurations of SAWs on a lattice (see Chapter 2) that avoids the "attrition problem" (i.e., the loss of chain configurations that have to be abandoned because they are overlapping). In this scheme one grows the SAW step by step and checks at each step which sites are available for the next step without violating the SAW constraint. One of these steps is then selected at random. Since relative to the simple sampling of SAWs this method creates a bias,[2,153] one has to keep track of the probability of each configuration relative to the unbiased simple sampling, and weigh the generated chain configurations with this probability accordingly.

An approximate generalization of this method to multichain systems due to Meirovitch is called "scanning future steps".[154,155] Suppose we wish to put $\mathcal{N}$ chains of N monomers each on a simple cubic lattice of L^3 sites. A starting point for the first polymer is selected out of the L^3 lattice sites with probability L^{-3} and occupied by a monomer. The first chain is then grown by a method where one scans b future steps: once the first k monomers have been placed, one counts for each of the six neighbors of the last site the allowed continuations consisting of b further steps (for the monomers $k+1$ to $k+b$) which start at this last site. The probability for selecting one of the six neighbors is chosen proportional to the number of allowed continuations starting at this site. Then the $(k+1)^{\text{th}}$ monomer is placed on the selected site, and so on. In this way one has to place N monomers for the first polymer. If at any step no continuation is possible, the construction is abandoned and one starts a new polymer from a new starting point.

Once the first polymer is generated on the lattice, a starting point for the second polymer is selected out of the remaining $L^3 - N$ sites with equal probability, and the further $N - 1$ monomers of the second polymer are placed on the lattice according to the same method as described above. The excluded volume interaction is taken into account with respect to the first chain and the already grown parts of the second chain. This procedure is continued until the desired number of chains on the lattice has been reached.

The fraction of successful construction attempts is not an exponentially decreasing function of the number of chains $\mathcal{N}$, but stays approximately constant at unity until a critical value that depends on N and b.[154–156] The larger N and/or $\mathcal{N}$ is, the larger one should use b; however, since the number of future steps that need to be scanned increases with b, in practice one is again limited to rather small N.

Since one knows at each step of the construction of a configuration the probability for selecting a lattice site for the next monomer, one can multiply all these single step probabilities in order to obtain the probability P_ν of constructing the multiple chain configuration ν. Then the partition function Z is estimated from a sampling of the inverse of P_ν

$$Z(N, \mathcal{N}) = \frac{1}{M} \sum_{\nu=1}^{M} (1/P_\nu). \tag{1.19}$$

From the partition function the free energy $F_{\mathcal{N}}$ follows and hence all thermodynamic quantities of interest can be estimated (entropy, chemical potential, osmotic pressure...). Öttinger[156] applied this technique to test the osmotic equation of state for dilute and semidilute polymer solutions for $N \leq 60$. Extension of this technique to off-lattice systems has also been made.[157,158]

A variant of the Rosenbluth–Rosenbluth method tailored to overcome the test chain insertion problem in the Widom method[143] (Eq. [1.8]) has been developed by Frenkel *et al.*[159–163] and is known as *configurational bias Monte Carlo* (CBMC). They rewrite eq. (1.18), using the fact that $U = \sum_{j=1}^{N} U_{j,\mathcal{N}+1}$, the energy of a test chain of length N inserted into a system of $\mathcal{N}$ other chains, can be written as a sum of energies $U_{j,\mathcal{N}+1}$ of the individual beads,

$$\exp(-\Delta\mu/k_{\mathrm{B}}T) = \left\langle \prod_{j=1}^{N} \exp\left(-\frac{1}{k_{\mathrm{B}}T} U_{j,\mathcal{N}+1}\right)\right\rangle. \tag{1.20}$$

Equation (1.20) suggests inserting the test chain bead by bead, and to overcome the sampling problems created by the relatively small probability of randomly inserting a test chain, without overlap, in a frozen snapshot of the system at liquid-like densities. Frenkel *et al.*[159–163] use a biased insertion procedure which favors low energy conformations of the inserted chain. The first bead is inserted at random and the interaction energy of this bead with the rest of the system ($U_{1,\mathcal{N}+1}$) tabulated. Then $k(1 \leq k \leq \infty)$ trial positions are generated for the next bead, obeying any geometric constraints imposed by chain architecture. The energy of each of these trial positions ($U^{\ell}_{2,\mathcal{N}+1}$) is calculated, and one position (ℓ) is randomly chosen according to a weight w_ℓ,

$$w_\ell = \exp(-U^{\ell}_{2,\mathcal{N}+1}/k_{\mathrm{B}}T) / \sum_{\ell=1}^{k} \exp(-U^{\ell}_{2,\mathcal{N}+1}/k_{\mathrm{B}}T) \tag{1.21}$$

Subsequent beads of the test chain are grown similarly until one arrives at the desired chain length.

One now has to correctly weigh the states generated by this biased insertion procedure when one calculates the chemical potential from eq. (1.20): since we generated states of the canonic ensemble modified with a weighting function, w, we have to correct for this weighting function as follows[142,163]

$$\langle f_0 \rangle = \langle f/w \rangle_w / \langle 1/w \rangle_w, \tag{1.22}$$

where $\langle f_0 \rangle$ represents the desired average of an observable f in the canonic ensemble, and $\langle \ldots \rangle_w$ in the weighted ensemble. Applying this to eq. (1.20) yields

$$\exp(-\Delta\mu/k_B T) = \left\langle \exp\left(-\frac{1}{k_B T} U_{1,\mathcal{N}+1}\right) \prod_{j=2}^{N} \frac{1}{w_j} \exp\left(-\frac{1}{k_B T} U_{j,\mathcal{N}+1}\right) \right\rangle_w \Bigg/ \left\langle \prod_{j=2}^{N} \frac{1}{w_j} \right\rangle_w \tag{1.23}$$

noting that no bias needs to be corrected for the first segment. Substituting eq. (1.21) in eq. (1.23) finally yields

$$\exp(-\Delta\mu/k_B T) = \left\langle \exp\left(-\frac{1}{k_B T} U_{1,\mathcal{N}+1}\right) \prod_{j=2}^{N} \left[\exp\left(-\frac{1}{k_B T} U^{\ell}_{j,\mathcal{N}+1}\right) \right] \right\rangle_w / k^{N-1}. \tag{1.24}$$

It is important to emphasize the distinction between the CBMC method [159–163] and the original Rosenbluth scheme.[152] As is well known,[153] the latter generates an unrepresentative sample of all polymer conformations, i.e., the probability that a particular conformation is generated is not proportional to the Boltzmann weight of that conformation, and thus one has to correct for the difference in weights and thus arrives at a biased sampling scheme which has problems for large N.[153] In the CBMC scheme, on the other hand, the Rosenbluth weight is used to bias the acceptance of trial conformations that are generated with the Rosenbluth scheme. Therefore all conformations occur with their correct Boltzmann weight. This is achieved by computing the Rosenbluth weights w_{trial} and w_{old} of the trial conformation and of the old conformation (in the trial conformation one may regrow an entire polymer molecule or only a part thereof). Finally the trial move is accepted only with a probability $\min\{w_{\text{trial}}/w_{\text{old}}, 1\}$. As explained by Frenkel,[163] this method is also readily applied to off-lattice chains.

At this point, we note that the chemical potential defined from a stepwise insertion procedure as described above can also be written as

$$\mu(N) = \sum_{j=1}^{N} [\mu(j) - \mu(j-1)] \equiv \sum_{j=1}^{N} \mu_r(j), \tag{1.25}$$

$\mu_r(j)$ being the incremental chemical potential to add a bead to a chain of length $j-1$. Equation (1.25) is the basis of the chain increment method of Kumar *et al.*[142,164–167] One now can prove[166] an analog of the Widom formula, eq. (1.18), for $\mu_r(j)$,

$$-\mu_r(j)/k_{\mathrm{B}}T = \ln\left\langle \exp\left[-\frac{1}{k_{\mathrm{B}}T}U(\vec{r}_j)\right]\right\rangle_{(\mathcal{N}-1)N,j-1}, \tag{1.26}$$

where the ensemble that is considered comprises $\mathcal{N}-1$ chains of length N and one chain of length $j-1$, with $1 \le j \le N$. One considers a frozen snapshot of this system, inserts a bead onto one of the ends of the chain of length $j-1$ and evaluates $U(\vec{r}_j)$, the test bead energy. The incremental chemical potential $\mu_r(j)$ is then computed by averaging the appropriate Boltzmann factor over many different realizations of the simulated system. In the end one is interested in $\mu_r(N)$, of course, so this procedure has to be carried out for a whole sequence of simulations, where j runs from 1 to N. This restricts the technique in practice to relatively short chain lengths, but the advantage of this scheme in comparison to the CBMC method is that one avoids the sampling problems that biased sampling methods often inevitably have.[153]

An alternative method has recently been proposed by Müller and Paul[168] using the concept of thermodynamic integration. The idea is to circumvent the problem of the vanishingly small acceptance probability for inserting a chain respecting excluded volume constraints in a dense system by allowing overlap of the inserted "ghost chain" with the other chains which are strictly mutually- and self-avoiding. If the ghost chain overlaps with N_0 monomers, the statistical weight of the configuration is λ^{N_0}. One has to carry out this insertion many times in statistically independent configurations of the many-chain systems and carry out a sampling of $\langle N_0(\lambda, \phi)\rangle$ to find the excess chemical potential of the polymers (relative to an ideal noninteracting polymer gas):

$$\begin{aligned}\mu_{\mathrm{ex}} &= \ln Z(\mathcal{N}, N, V, \lambda = 1) - \ln Z(\mathcal{N}, N, V, \lambda = 0)\\ &= \int_0^1 d\lambda \frac{\mathrm{d}}{\mathrm{d}\lambda} \ln Z(\mathcal{N}, N, V, \lambda) = \int_0^1 d\lambda \langle N_0(\lambda, \phi)\rangle/\lambda.\end{aligned} \tag{1.27}$$

It was found useful to carry out the integration in eq. (1.27) by performing simulations at about nine distinct values of λ, which are used as input into a multihistogram analysis which yields a very good estimate of $\langle N_0(\lambda, \phi)\rangle$

over the whole range of the auxiliary parameter λ.[168] It was found that this method works very well even for parameters such as $N = 80$, $\phi = 0.5$, where the insertion probability that one would have to sample with the Widom method[143] would be as small as 10^{-76}. For long chains the applicability of this method is only limited by the requirement that one must have a means of producing a sufficient number of equilibrated and statistically independent configurations in which the ghost chain is immersed to measure the overlap.[168]

1.4.2 Calculation of pressure in dynamic Monte Carlo methods

If a polymer solution is modeled by an assembly of self-avoiding walks on a lattice, a basic physical quantity is the osmotic pressure Π. Carrying out a simulation with a fixed number $\mathcal{N}$ of chains of length N at a lattice of volume V with one of the dynamic algorithms described in Section 1.2.2, the osmotic pressure is not straightforward to sample. If one had methods that yielded the excess chemical potential $\Delta\mu$ and the Helmholtz free energy $F_{\mathcal{N}}$, one would find Π from the thermodynamic relation

$$\Pi V = \Delta\mu \mathcal{N} - F_{\mathcal{N}}. \tag{1.28}$$

Noting that $\Delta\mu = F_{\mathcal{N}+1} - F_{\mathcal{N}}$ (eq. [1.18]) and remembering $F_{\mathcal{N}} = -k_B T \ln Z(\mathcal{N}, N, V)$ where $Z(\mathcal{N}, N, V)$ is the partition function of $\mathcal{N}$ chains of length N in the volume V, it is convenient to relate the insertion probability $p(\mathcal{N}, N, V)$ to a ratio of partition functions,

$$p(\mathcal{N}, N, V) = \frac{(\mathcal{N}+1)Z(\mathcal{N}+1, N, V)}{Z(\mathcal{N}, N, V)Z(1, N, V)}. \tag{1.29}$$

This quantity describes the probability that a randomly chosen N-mer, placed at random into a randomly chosen configuration of $\mathcal{N}$ N-mers on a lattice of volume V, does not overlap any of the $\mathcal{N}$ chains. From eqs (1.18) and (1.29) one derives the relation for the excess chemical potential in terms of this insertion probability,

$$\Delta\mu / k_B T = -\ln p(\mathcal{N}, N, V) - \ln Z(1, N, V) + \ln(\mathcal{N}+1), \tag{1.30}$$

which can be used to derive eq. (1.27). Since

$$\sum_{j=1}^{\mathcal{N}-1} \ln p(\mathcal{N}, N, V) = \sum_{j=1}^{\mathcal{N}-1} \ln j + \ln Z(\mathcal{N}, N, V) - \mathcal{N} \ln Z(1, N, V), \tag{1.31}$$

eq. (1.28) can be rewritten as[147]

$$\begin{aligned}\frac{\Pi V}{k_{\mathrm{B}}T} &= -\mathcal{N}\ln p(\mathcal{N},N,V) + \mathcal{N}\ln Z(1,N,V) + \ln Z(\mathcal{N},N,V) - \mathcal{N}\ln(\mathcal{N}+1)\\ &= \sum_{j=1}^{\mathcal{N}-1}\ln p(j,N,V) - \mathcal{N}[\ln p(\mathcal{N},N,V) - \ln(\mathcal{N}+1)] - \ln(\mathcal{N}!).\end{aligned} \tag{1.32}$$

In the thermodynamic limit, the summation over the number of chains can be replaced by a thermodynamic integration over the volume fraction ϕ of occupied sites ($\phi = N\mathcal{N}/V$ measuring lengths in units of the lattice spacing) to find [$p(\phi, N) = \lim_{N\to\infty, V\to\infty} p(\mathcal{N}, N, V)$ at fixed ϕ]

$$\Pi/k_{\mathrm{B}}T = \frac{\phi}{N}[1 - \ln p(\phi, N)] + \frac{1}{N}\int_0^{\phi} \ln p(\phi', N)\, \mathrm{d}\phi'. \tag{1.33}$$

This result shows that the osmotic pressure can be obtained from a thermodynamic integration if the insertion probability $p(\phi', N)$ is sampled over a range of values from $\phi' = 0$ to $\phi' = \phi$. This method has been applied in conjunction with some of the methods of the previous subsection where the estimation of the chemical potential via the insertion probability was discussed.[147,168]

An interesting alternative method[169,170] relates the pressure of the system to the segment density at a repulsive wall. While usually in simulations one considers a d-dimensional cubic box with all linear dimensions equal to L and periodic boundary conditions, in this method one applies a lattice of length L in $d-1$ dimensions and of length H in the remaining direction, with which one associates the coordinate x. There is an infinite repulsive potential at $x = 0$ and $x = H + 1$, while in the other directions periodic boundary conditions apply. The partition function of $\mathcal{N}N$-mers on the lattice then is $Z(\mathcal{N}, N, L, H) = (\mathcal{N}!)^{-1}\sum \exp(-U/k_{\mathrm{B}}T)$, where the sum runs over all configurations on the lattice, and the potential U incorporates restrictions which define then chain structure, prohibit overlaps, etc. While for a model in continuous space the pressure is

$$\begin{aligned}\Pi/k_{\mathrm{B}}T &= -\frac{1}{k_{\mathrm{B}}T}\frac{\partial}{\partial V}\ln F = \frac{\partial}{\partial V}\ln Z(\mathcal{N}, N, L, H)\\ &= L^{-(d-1)}\frac{\partial}{\partial H}\ln Z(\mathcal{N}, N, L, H),\end{aligned} \tag{1.34}$$

the lattice analog for this expression is

$$\Pi/k_{\mathrm{B}}T = L^{-(d-1)}[\ln Z(\mathcal{N}, n, L, H) - \ln Z(\mathcal{N}, n, L, H-1)]. \tag{1.35}$$

The difference in free energies required here is calculated by introducing a parameter λ. $0 < \lambda < 1$, which enters as a statistical weight for each monomer in the plane $x = H$: it may be viewed as being due to an additional finite repulsive potential next to the wall. Denoting the number of occupied sites in the plane $x = H$ as N_H, the statistical weight factor due to this auxiliary potential is λ^{N_H}, and hence the partition function becomes

$$Z(\mathcal{N}, N, L, H, \lambda) = (\mathcal{N}!)^{-1} \sum \exp[-U/k_B T)]\lambda^{N_H}. \tag{1.36}$$

Note that $Z(\mathcal{N}, N, L, H, 1) = Z(\mathcal{N}, N, L, H)$ and that $Z(\mathcal{N}, N, L, H, 0) = Z(\mathcal{N}, N, L, H-1)$, since for $\lambda = 0$ there are no monomers allowed in the plane $x = H$; effectively the repulsive wall now is at $x = H$ rather than at $x = H + 1$. This yields

$$\Pi/k_B T = L^{-(d-1)} \int_0^1 (\partial \ln Z/\partial\lambda)\, d\lambda = L^{-(d-1)} \int_0^1 \langle N_H \rangle_\lambda \, d\lambda/\lambda \tag{1.37}$$

Thus one must carry out simulations for several values of λ to sample $\langle N_H \rangle_\lambda$, the average number of occupied sites in the plane $x = H$, in order to perform the above integration numerically.[169,170]

We now describe, as an example, a few applications of these methods. Figure 1.11 compares simulation results[170] for the compressibility factor

$$z \equiv N\Pi/(k_B T\phi) \tag{1.38}$$

with predictions of various equations of state, namely of Flory[104]

$$z_F = 1 - N[\ln(1-\phi) + \phi]/\phi, \tag{1.39}$$

of the Flory–Huggins theory[103] (q is the coordination number of the lattice)

$$z_{FH} = z_F + \frac{Nq}{2\phi}\left\{\ln\left[1 - \frac{2\phi}{q}\left(1 - \frac{1}{N}\right)\right] + \frac{2\phi}{q}\left(1 - \frac{1}{N}\right)\right\}, \tag{1.40}$$

and of the Bawendi–Freed theory[171]

$$\begin{aligned} z_{BF} = z_F - \phi\left\{\frac{(N-1)^2}{qN} + \frac{3(N-1)^4 - 8(N-1)^2 - 4(N-1) + 1}{q^2 N^3}\right\} \\ - \phi^2\left\{\frac{20(N-1)^4 + 4(N-1)^3 + 24(N-1)^2}{3q^2 N^3}\right\} - \phi^3\left\{\frac{6(N-1)^4}{q^2 N^3}\right\}. \end{aligned} \tag{1.41}$$

It is seen that the Flory approximation is inaccurate, while both other approximations describe the equation of state well at high volume fractions ϕ. At small volume fractions, however, neither of these approximations is very accurate, as expected, since in the dilute and semidilute concentration regime a scaling description[16,20] of the equation of state is needed.

While Fig. 1.11 refers to the simplest lattice model where polymers are described as SAWs (Fig. 1.4), the above techniques are straightforwardly generalized to more sophisticated lattice models such as the bond fluctuation model (Fig. 1.12).[166,172] It is seen that the repulsive wall method and the

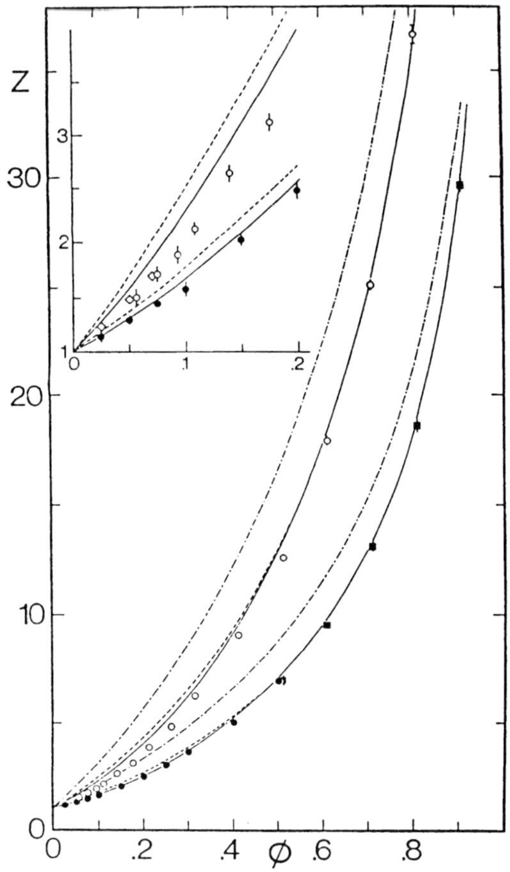

Fig. 1.11 Compressibility factor z plotted vs. volume fraction, for self- and mutually-avoiding walks on the simple cubic lattice, and two chain lengths: $N = 20$ (filled symbols) or $N = 40$ (open symbols), respectively. The Flory theory[104] is shown as a dash–dotted curve, Flory–Huggins theory[103] as broken curve, and the Bawendi–Freed theory[171] as full curve. Circles represent data obtained from the repulsive wall method, while squares or diamonds are obtained from the test-chain insertion method. (From Hertanto and Dickman.[170])

insertion method, where one integrates over the strength of excluded volume interaction with the inserted ghost chain[168] are in reasonable agreement.

In off-lattice simulations in the NVT ensemble the (excess) pressure Δp is usually calculated from the Virial theorem[173–175]

$$\Delta p = \frac{1}{6V} \sum_{i \neq j} r_{ij} \, \partial U(r_{ij})/\partial r_{ij}. \qquad (1.42)$$

Again the kinetic energy term $p_{\text{kin}} = \tilde{\mathcal{N}} k_B T/V$ where $\tilde{\mathcal{N}}$ is the number of atoms per volume V in the system, is omitted throughout, and the

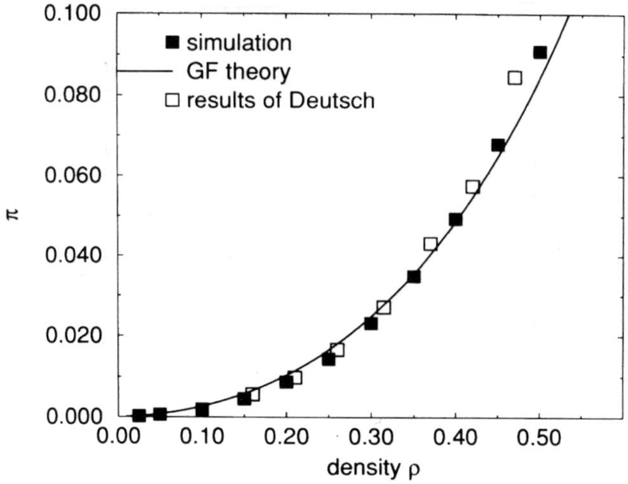

Fig. 1.12 Osmotic pressure $\Pi V/k_B T$ plotted vs. volume fraction ϕ, for the athermal bond fluctuation model on the simple cubic lattice, $N = 20$. Open squares are obtained by Deutsch and Dickman[172] with the repulsive wall method; full squares are based on thermodynamic integration over a variable excluded volume interaction between the inserted "ghost chain" and the other chains.[16] Curve shows the pressure according to the "Generalized Flory" equation of state of Ref. 172, $\Pi(\phi, N)/k_B T = \phi/N + (1/N)[v(N)/v(1)][\Pi(\phi, 1)/k_B T - \phi]$, where $v(N)$ is the exclusion volume of an N-mer. (From Müller and Paul.[168])

summations i, j run over all effective monomers in the system (we use a convention where all pairs are counted twice), U being the total potential energy. One may split eq. (1.42) into three parts: a "covalent" part due to (harmonic) interactions along the chains, an intra-chain part due to non-bonded interactions, and the inter-chain contribution

$$\Delta p = p_{\text{harm}} + p_{\text{intra}}^{\text{nonbonded}} + p_{\text{inter}}, \tag{1.43}$$

$$p_{\text{harm}} = -\frac{1}{6V}\sum_{i\neq j} r_{ij}\partial U_h(r_{ij})/\partial r_{ij}, \tag{1.44}$$

$$p_{\text{intra}}^{\text{nonbonded}} = -\frac{1}{6V}\sum_{i\neq j\in \text{chain}} r_{ij}\;\partial U_{\text{nonbonded}}(r_{ij})/\partial r_{ij} \tag{1.45}$$

and

$$p_{\text{inter}} = -\frac{1}{6V}\sum_{i\neq j\in \text{different chains}} r_{ij}\quad \partial U_{\text{nonbonded}}(r_{ij})/\partial r_{ij}. \tag{1.46}$$

Of course, this separation does not imply that the "springs" of the bead-spring model must be harmonic; it works for anharmonic forces along the chain as well.

Gao and Weiner[174,175] call this pressure contribution Δp due to monomers of the polymers the "atomic pressure" and suggest that it is this quantity that one should consider in the polymer melt. They suggest that at the Θ-temperature the covalent part and the nonbonded intrachain part of Δp should cancel, and then the atomic pressure would reduce simply to the osmotic pressure of a polymer solution. Milchev and Binder[176a] attempted to check this, but it would be interesting to clarify this problem by a comparative study of several other models.

A potentially very useful method to obtain entropy, pressure and chemical potential of many-chain systems is the *scanning method* of Meirovitch.[176b] Lack of space prevents us from discussing it here.

1.5 Final remarks

The field of computer simulation in polymer science is a very active area of research and many developments of simulation methodology are either very recent or even still under study: this will become even more evident when the reader proceeds to the later chapters in this book. But although applications to many problems in polymer physics have been started just a few years ago—such as large-scale simulations of polymer networks, polymer electrolyte solutions, polymer brushes under various solvent conditions, block copolymer mesophase ordering, and so on—even these very first attempts to simulate complex polymeric materials have already been very useful and given a lot of insight. The main direction of research has not been directed towards the prediction of materials parameters for specific polymers—as discussed in Section 1.1 of the present chapter, such a task is difficult and to a large extent not yet feasible with controlled errors—but towards the test of general concepts (such as various "scaling" ideas developed for the various systems of interest) as well as of specific theories. A huge advantage of the simulations is that one can adjust the model that is simulated very closely to the model that the theory considers: e.g., the Flory–Huggins theory of polymer blend thermodynamics uses a very simple lattice model and then the simulations can provide a stringent test by studying exactly that lattice model (see Chapter 7). On the other hand, the *polymer reference interaction site model* (PRISM) theory of polymer melts considers idealized bead-spring type off-lattice models of polymer chains, and thus is tested most stringently by a comparison to corresponding molecular dynamics simulations.[177] As will be described in later chapters, such comparisons have indeed been very illuminating.

At this stage, the comparison between simulation and experiment is somewhat more restricted: either one restricts attention to very short chains of simple enough polymers to allow the treatment of a model including detailed chemistry (Chapters 5, 8) or one has to focus on universal properties. Then a nontrivial comparison between simulation and experiment is still possible, if one compares suitable dimensionless quantities. As an example (more details on this problem will be found in Chapter 4) consider the chain-length-dependence of the self-diffusion coefficient of polymer melts: for short chains one expects that the Rouse model[16, 17,79] holds, i.e., the self-diffusion constant D_N varies inversely with chain length, $D_{\text{Rouse}} \propto 1/N$. For chain lengths exceeding the entanglement chain-length N_e reptation should hold and hence[16,17,132] $D_N \propto 1/N^2$. However, the materials-dependent (or model-dependent, respectively) prefactors in these relations are absorbed if one plots the dimensionless variable D_N/D_{Rouse} where $D_{\text{Rouse}} = \lim_{N\to 0}(ND_N)\}$ versus N/N_e (see Fig. 1.13).[127] It is seen then that both data from MD simulation,[85] MC simulations[127] and experiment[178] superpose on a common curve. The entanglement chain length N_e has been estimated independently[85,127,178] and thus the comparison in Fig. 1.13 does not involve any adjustable parameter whatsoever! The agreement seen in Fig. 1.13 hence is significant and a relevant test of the reptation ideas is indeed provided by these simulations[85,127], as will be discussed in more detail in Chapter 4. On the other hand, the very interesting question of how a parameter such as N_e is related to the detailed chemical structure of polymers escapes the tractability of simulational approaches so far.

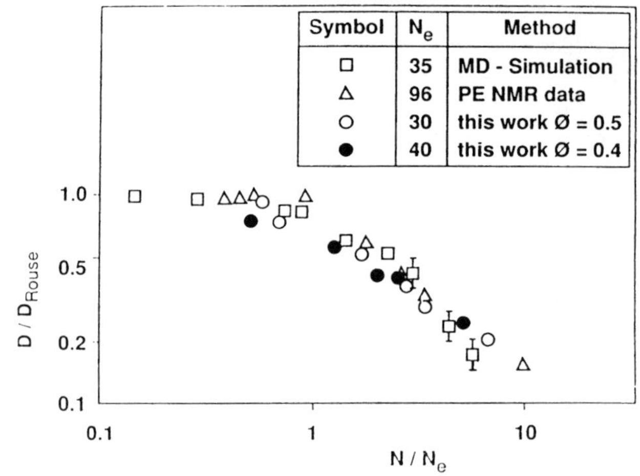

Fig. 1.13 Log–log plot of the self-diffusion constant D of polymer melts vs. chain length. D is normalized by the diffusion constant of the Rouse limit, D_{Rouse}, which is reached for short chain lengths. N is normalized by N_e. Experimental data for polyethylene (PE)[178] and MD results[85] are included. (From Paul *et al.*[127])

Acknowledgments

In this chapter research work performed in collaboration with J. Baschnagel, H.-P. Deutsch, I. Gerroff, D. W. Heermann, K. Kremer, A. Milchev, W. Paul, and K. Qin was used to illustrate some of the main points. It is a pleasure to thank them for a pleasant and fruitful collaboration. The author is also greatly indebted to J. Clarke, R. Dickman, and M. Müller for being allowed to show some of their recent research results (Figs 1.9(c), 1.11, 1.12). It is also a pleasure to thank K. Kremer and R. Dickman for their useful comments on this manuscript.

References

1. A. Baumgärtner, in *Applications of the Monte Carlo Method in Statistical Physics*, edited by K. Binder (Springer, Berlin, 1984), Ch. 5.
2. K. Kremer and K. Binder, *Computer Repts* **7**, 259 (1988).
3. R. J. Roe (ed.) *Computer Simulations of Polymers* (Prentice Hall, Englewood Cliffs, NJ, 1991).
4. J. Bicerano (ed.) *Computational Modelling of Polymers* (M. Dekker, New York, 1992).
5. A. Baumgärtner, in *Monte Carlo Methods in Condensed Matter Physics*, edited by K. Binder (Springer, Berlin, 1992), Ch. 9.
6. E. A. Colbourn (ed.) *Computer Simulation of Polymers* (Longman, Harlow, 1993).
7. K. Kremer, in *Computer Simulation in Chemical Physics*, edited by M. P. Allen and D. J. Tildesley (Kluwer Academic Publishers, Dordrecht, 1993).
8. M. P. Allen and D. J. Tildesley, *Computer Simulation of Liquids* (Clarendon Press, Oxford, 1987).
9. G. Ciccotti and W. G. Hoover (eds) *Molecular Dynamics of Statistical Mechanical Systems* (North–Holland, Amsterdam, 1986).
10. D. W. Heermann, *Introduction to Computer Simulation Methods in Theoretical Physics* (Springer, Berlin, 1986).
11. K. Binder and D. W. Heermann, *Monte Carlo Simulation in Statistical Physics: an Introduction* (Springer, Berlin, 1988).
12. K. Binder (ed.) *Monte Carlo Methods in Condensed Matter Physics* (Springer, Berlin, 1992).
13. M. P. Allen and D. J. Tildesley, *Computer Simulation in Chemical Physics* (Kluwer Academic Publishers, Dordrecht, 1993).
14. K. Binder, *Makromol. Chem., Macromol. Symp.* **50**, 1 (1991).
15. P. J. Flory, *Statistical Mechanics of Chain Molecules* (Interscience, New York, 1969).
16. P. G. de Gennes, *Scaling Concepts in Polymer Physics* (Cornell University Press, Ithaca, NY, 1979).
17. M. Doi and S. F. Edwards, *Theory of Polymer Dynamics* (Clarendon Press, Oxford, 1986).
18. A. Halperin, M. Tirrell, and T. P. Lodge, *Adv. Polym. Sci.* **100**, 31 (1991).
19. P. J. Flory, *Principles of Polymer Chemistry* (Cornell University Press, Ithaca, 1953).
20. J. des Cloizeaux and G. Jannink, *Polymers in Solution: their Modelling and Structure* (Oxford University Press, Oxford, 1990).
21. K. Binder, *Adv. Polym. Sci.* **112**, 181 (1994).
22. K. Binder, *J. Chem. Phys.* **79**, 6387 (1983).

23. H. P. Deutsch and K. Binder, *J. Phys. (France) II* **3**, 1049 (1993).
24. J. C. Le Guillou and J. Zinn–Justin, *Phys. Rev.* **B21**, 3976 (1980).
25. G. Meier, D. Schwahn, K. Mortensen, and S. Janssen, *Europhys.Lett.* **22**, 577 (1993).
26. F. S. Bates and P. Wiltzius, *J. Chem. Phys.* **91**, 3258 (1989).
27. T. Hashimoto, in *Materials Science and Technology, Vol. 12: Structure and Properties of Polymers*, edited by E. L. Thomas (VCH, Weinheim, 1993), p. 251.
28. F. S. Bates and G. H. Fredrickson, *Ann. Rev. Phys. Chem.* **41**, 525 (1990).
29. P. G. de Gennes, P. Pincus, and R. Velasco, *J. Phys.* (Paris) **37**, 1461 (1976).
30. J. Skolnick and M. Fixman, *Macromolecules* **10**, 944 (1977).
31. T. Odijk, *J. Polym. Sci., Polym. Phys. Ed.* **15**, 477 (1977); *Polymer* **19**, 989 (1978).
32. J. Hayter, G. Jannink, F. Brochard–Wyart, and P. G. de Gennes, *J. Phys. (Paris) Lett.* **41**, 451 (1980).
33. P. Y. Lai and K. Binder, *J. Chem. Phys.* **98**, 2366 (1993), and references therein.
34. J. P. Ryckaert and A. Bellemans, *Chem. Phys. Lett.* **30**, 123 (1975).
35. J. P. Ryckaert and A. Bellemans, *Discuss. Faraday. Soc.* **66**, 95 (1978).
36. J. H. R. Clarke and D. Brown, *Molec. Phys.* **58**, 815 (1986).
37. J. H. R. Clarke and D. Brown, *Molec. Simul.* **3**, 27 (1989).
38. D. Brown, J. H. R. Clarke, M. Okuda, and T. Yamazaki, *J. Chem. Phys.* **100**, 1684 (1994).
39. D. J. Rigby and R. J. Roe, *J. Chem. Phys.* **87**, 7285 (1987).
40. D. J. Rigby and R. J. Roe, *J. Chem. Phys.* **88**, 5280 (1988).
41. D. J. Rigby and R. J. Roe, *Macromolecules* **22**, 2259 (1989); **23**, 5312 (1990).
42. H. Takeuchi and R. J. Roe, *J. Chem. Phys.* **94**, 7446, 7458 (1991); R. J. Roe, D. Rigby, H. Furuya, and T. Takeuchi, *Comput. Polym. Sci.* **2**, 32 (1992).
43. J. Baschnagel, K. Qin, W. Paul, and K. Binder, *Macromolecules* **25**, 3117 (1992).
44. A. Sariban, J. Brickmann, J. van Ruiten, and R. J. Meier, *Macromolecules* **25**, 5950 (1992).
45. J. Baschnagel, K. Binder, W. Paul et al., *J. Chem. Phys.* **95**, 6014 (1991).
46. B. Smit and D. Frenkel, *J. Chem. Phys.* **94**, 5663 (1991).
47. S. Karaborni, S. Toxvaerd, and O. H. Olsen, *J. Phys. Chem.* **96**, 4965 (1992).
48. D. Y. Yoon, G. D. Smith, and T. Matsuda, *J. Chem. Phys.* **98**, 10037 (1993); G. D. Smith, R. L. Jaffe, and D. Y. Yoon, *Macromolecules* **26**, 293 (1993).
49. B. Smit, S. Karaborni, and J. Siepmann, *Macromol.Symp.* **81**, 343 (1994) (paper presented at the First International Conference on the Statistical Mechanics of Polymer Systems, Theory and Simulations, Mainz, Germany Oct 4–6, 1993).
50. D. N. Theodorou and U. W. Suter, *Macromolecules* **18**, 1467 (1985); **19**, 139 (1986); ibid **19**, 379 (1986).
51. K. F. Mansfield and D. N. Theodorou, in *Computer Simulations of Polymers* (Prentice Hall, Englewood Cliffs, NJ, 1991), p.122; *Macromolecules* **24**, 6283 (1991).
52. M. F. Sylvester, S. Yip, and A. S. Argon, in *Computer Simulations of Polymers* (Prentice Hall, Englewood Cliffs, NJ, 1991), p. 105.

53. G. C. Rutledge and U. W. Suter, *Polymer* **32**, 2179 (1991); *Macromolecules* **24**, 1921 (1991).
54. M. Hutnik, F. T. Gentile, P. J. Ludovice, U. W. Suter, and A. S. Argon, *Macromolecules* **24**, 5962 (1991); M. Hutnik, A. S. Argon, and U. W. Suter, *Macromolecules* **24**, 5956 (1991).
55. P. J. Ludovice and U. W. Suter, in *Computational Modelling of Polymers* (M. Dekker, New York, 1992), p. 401.
56. D. B. Adolf and M. D. Ediger, in *Computer Simulations of Polymers* (Prentice Hall, Englewood Cliffs, NJ, 1991), p. 154.
57. R. H. Boyd and K. Pant, in *Computer Simulations of Polymers* (Prentice Hall, Englewood Cliffs, NJ, 1991), p. 94.
58. B. G. Sumpter, D. W. Noid, B. Wunderlich, and S. Z. D. Cheng, in *Computer Simulations of Polymers* (Prentice Hall, Englewood Cliffs, NJ, 1991), p. 311.
59. J. Baschnagel, K. Binder, and H. P. Wittmann, *J. Phys. Condens. Matter* **5**, 1597 (1993).
60. J. Baschnagel and K. Binder, *Physica A* **204**, 47 (1994).
61. A. Sariban and K. Binder, *J. Chem. Phys.* **86**, 5859 (1987).
62. A. Sariban and K. Binder, *Macromolecules* **21**, 711 (1988).
63. H. P. Deutsch and K. Binder, *Macromolecules* **25**, 6214 (1992).
64. H. Snyder, S. Reich, and P. Meakin, *Macromolecules* **16**, 757 (1983).
65. A. Cumming, P. Wiltzius, and S. F. Bates, *Phys. Rev. Lett.* **65**, 863 (1990).
66. J. Jäckle, *Reports Progr. Phys.* **49**, 171 (1986).
67. W. Götze, in *Liquids, Freezing and the Glass Transition*, edited by J. P. Hansen, D. Levesque and J. Zinn–Justin (North Holland, Amsterdam, 1990).
68. G. Adam and J. H. Gibbs, *J. Chem. Phys.* **43**, 139 (1965).
69. H. P. Wittmann, K. Kremer, and K. Binder, *J. Chem. Phys.* **96**, 6291 (1992).
70. W. Paul, K. Binder, K. Kremer, and D. W. Heermann, *Macromolecules* **24**, 6332 (1991).
71. W. Paul, *AIP Conf. Proc.* **256**, 145 (1992).
72. W. Paul, K. Binder, J. Batoulis, B. Pittel, and K. H. Sommer, *Makromol. Chem., Macromol. Symp.* **65**, 1 (1993).
73. I. Carmesin and K. Kremer, *Macromolecules* **21**, 2819 (1988).
74. Y. Bar–Yam, Y. Rabin, and M. A. Smith, *Macromolecules* **25**, 2985 (1992).
75. M. A. Smith, Y. Bar–Yam, B. Ostrowsky et al., *Comput. Polym. Sci.* **2**, 165 (1992).
76. I. Gerroff, A. Milchev, K. Binder, and W. Paul, *J. Chem. Phys.* **98**, 6526 (1993).
77. A. Milchev, W. Paul and K. Binder, *J. Chem. Phys.* **99**, 4786 (1993).
78. K. Binder, in *Computational Modelling of Polymers* (M. Dekker, New York, 1992), p. 221.
79. P. E. Rouse, *J. Chem. Phys.* **21**, 127 (1953).
80. A. Baumgärtner and K. Binder, *J. Chem. Phys.* **75**, 2994 (1981).
81. N. Metropolis, A. W. Rosenbluth, M. N. Rosenbluth, A. N. Teller, and E. Teller, *J. Chem. Phys.* **21**, 1087 (1953).
82. A. Baumgärtner, *Ann. Rev. Phys. Chem.* **35**, 419 (1984).
83. G. S. Grest and K. Kremer, *Phys. Rev.* **A33**, 3628 (1986).

84. B. Dünweg and K. Kremer, *Phys. Rev. Lett.* **66**, 2996 (1991); *J. Chem. Phys.* **99**, 6983 (1993); B. Dünweg, *J. Chem. Phys.* **99**, 6977 (1993).
85. K. Kremer and G. S. Grest, *J. Chem. Phys.* **92**, 5057 (1990).
86. K. Kremer and G. S. Grest, *J. Chem. Soc. Faraday Trans.* **88**, 1707 (1992), and in *Computer Simulations of Polymers* (Prentice Hall, Englewood Cliffs, NJ, 1991), p. 167.
87. M. Bishop, D. Ceperley, H. L. Frisch, and M. H. Kalos, *J. Chem. Phys.* **76**, 1557 (1982).
88. T. A. Weber, *J. Chem. Phys.* **69**, 2347 (1978); **70**, 4277 (1979).
89. T. A. Weber and A. Helfand, *J. Chem. Phys.* **71**, 4760 (1979); **87**, 2881 (1983).
90. E. Helfand, Z. Wasserman, and T. Weber, *Macromolecules* **13**, 526 (1980).
91. D. Ceperley, M. H. Kalos, and J. L. Lebowitz, *Phys. Rev. Lett.* **41**, 313 (1978); *Macromolecules* **14**, 1472 (1981).
92. A. A. Darinskii, Yu–Ya Gotlib, A. V. Ljutin, L. I. Khushin, and I. M. Neelov, *Polym Sci.* (USSR) **32**, 2289 (1990).
93. A. A. Darinskii, M. N. Lukjanov, Yu–Ya Gotlib, and I. M. Neelov, *J. Phys. Chem.* (USSR) **57**, 954 (1981).
94. A. A. Darinskii, Yu–Ya Gotlib, A. V. Ljutin, L. I. Khushin, and I. M. Neelov *Polym Sci.* (USSR) **33**, 1211 (1991).
95. R. B. Bird, R. C. Armstrong, and D. Hassager, *Dynamics of Polymeric Liquids* (J. Wiley, New York, 1971).
96. G. S. Grest, B. Dünweg, and K. Kremer, *Comp. Phys. Commun.* **55**, 269 (1989); R. Everaers and K. Kremer, *Comp. Phys. Commun.* **81**, 19 (1994).
97. C. Pierleoni and J. P. Ryckaert, *Phys. Rev. Lett.* **66**, 2992 (1991); *J. Chem. Phys.* **96**, 8539 (1992).
98. G. S. Grest, K. Kremer, and T. A. Witten, *Macromolecules* **20**, 1376 (1987).
99. G. S. Grest, K. Kremer, S. T. Milner, and T. A. Witten, *Macromolecules* **22**, 1904 (1989).
100. E. R. Duering, K. Kremer, and G. S. Grest, *Phys. Rev. Lett.* **67**, 3531 (1991); *Macromolecules* **26**, 3241 (1993); G. S. Grest, K. Kremer, and E. R. Duering, *Physica* **A194**, 330 (1993).
101. G. S. Grest and K. Kremer, *J. Phys.* (France) **51**, 2829 (1990); *Macromolecules* **23**, 4994 (1990).
102. G. S. Grest, K. Kremer, and E. R. Duering, *Europhys. Lett.* **19**, 195 (1992).
103. M. J. Huggins, *J. Chem. Phys.* **9**, 440 (1941).
104. P. J. Flory, *J. Chem. Phys.* **9**, 660 (1941).
105. P. H. Verdier and W. H. Stockmayer, *J. Chem. Phys.* **36**, 227 (1962).
106. P. H. Verdier, *J. Chem. Phys.* **45**, 2122 (1966); **52**, 5512 (1970); **59**, 6119 (1973).
107. H. J. Hilhorst and J. M. Deutch, *J. Chem. Phys.* **63**, 5153 (1975); H. Boots and J. M. Deutch, **67**, 4608 (1977).
108. A. K. Kron, *Polym Sci.* USSR **7**, 1361 (1965); A. K. Kron and O. B. Ptitsyn, *Polym Sci.* USSR **9**, 847 (1967).
109. F. T. Wall and F. Mandel, *J. Chem. Phys.* **63**, 4592 (1975).
110. M. Lal, *Molec. Phys.* **17**, 57 (1969).
111. O. F. Olaj and K. H. Pelinka, *Makromol. Chem.* **177**, 3413 (1976).
112. B. MacDonald, N. Jan, D. L. Hunter, and M. O. Steinitz, *J. Phys.* **A 18**, 2627 (1985).

113. N. Madras and A. D. Sokal, *J. Stat. Phys.* **50**, 109 (1988).
114. M. T. Gurler, C. C. Crabb, D. M. Dahlin, and J. Kovac, *Macromolecules* **16**, 389 (1983).
115. J. Skolnick, R. Yaris, and A. Kolinski, *J. Chem. Phys.* **88**, 1407 (1988).
116. N. Madras, A. Orlitsky, and L. A. Shepp, *J. Stat. Phys.* **58**, 159 (1990).
117. N. Madras and A. D. Sokal, *J. Stat. Phys.* **47**, 573 (1987).
118. A. Baumgärtner, *J.Phys.* **A 17**, L971 (1984).
119. A. Baumgärtner and D. W. Heermann, *Polymer* **27**, 1777 (1986).
120. A. Beretti and A. D. Sokal, *J. Stat. Phys. 40*, 483 (1985).
121. T. Pakula, *Macromolecules* **20**, 679 (1987); T. Pakula and S. Geyler, *Macromolecules* **20**, 2909 (1987).
122. S. Geyler, T. Pakula, and J. Reiter, *J. Chem. Phys.* **92**, 2676 (1990).
123. J. Reiter, T. Edling, and T. Pakula, *J. Chem. Phys.* **93**, 837 (1990).
124. P. Cifra, F. E. Karasz, and W. J. MacKnight, *Macromolecules* **25**, 4895 (1992).
125. I. Carmesin and K. Kremer, *J. Phys. (France)* **51**, 915 (1990).
126. H.-P. Wittmann and K. Kremer, *Comp. Phys. Commun.* **61**, 309 (1990); *Comp. Phys. Commun.* **71**, 343 (1992), erratum.
127. W. Paul, K. Binder, D. W. Heermann, and K. Kremer, *J. Phys. II (France)* **1**, 37 (1991).
128. W. Paul, K. Binder, D. W. Heermann, and K. Kremer, *J. Chem. Phys.* **95**, 7726 (1991).
129. H. P. Deutsch and K. Binder, *J. Chem. Phys.* **94**, 2294 (1991).
130. M. Schulz and J.-U. Sommer, *J. Chem. Phys.* **96**, 7102 (1992); M. Schulz and K. Binder, *J. Chem. Phys.* **98**, 655 (1993).
131. J. Batoulis, N. Pistoor, K. Kremer, and H. L. Frisch, *Electrophoresis* **10**, 442 (1989).
132. P. G. de Gennes, *J. Chem. Phys.* **55**, 572 (1971).
133. M. Murat and T. A. Witten, *Macromolecules* **23**, 520 (1990); A. R. C. Baljon–Haakman and T. A. Witten, *Macromolecules* **25**, 2969 (1992).
134. J. Batoulis and K. Kremer, *Europhys. Lett.* **7**, 683 (1988); *Macromolecules* **22**, 4277 (1989).
135. K. Ohno and K. Binder, *J. Stat. Phys.* **64**, 781 (1991); K. Ohno, X. Hu, and Y. Kawazoe, in *Computer-Aided Innovation of New Materials II*, edited by M. Doyama, J. Kihara, M. Tanaka, and R. Yamamoto (Elsevier, Amsterdam, 1993), p. 315.
136. N. Pistoor and W. Paul, *Macromolecules* **27**, 1249 (1994).
137. P.-Y. Lai and K. Binder, *J. Chem. Phys.* **95**, 9288 (1991); P.-Y. Lai, *J. Chem. Phys.* **98**, 669 (1993).
138. P.-Y. Lai and K. Binder, *J. Chem. Phys.* **97**, 586 (1992).
139. F. Haas, P.-Y. Lai, and K. Binder, *Makromol. Chem., Theory & Simul.* **2**, 889 (1993).
140. A. Yethiraj and R. Dickman, *J. Chem. Phys.* **97**, 4468 (1992).
141. B. Zimm, *J. Chem. Phys.* **24**, 269 (1956).
142. S. Kumar, in *Computer Simulation of Polymers*, edited by E. A. Colbourn (Longman, Harlow, U.K., 1993) Chapter 8.
143. B. Widom, *J. Chem. Phys.* **39**, 2808 (1962).
144. K. S. Shing and K. E. Gubbins, *Mol. Phys.* **43**, 717 (1981); **46**, 1109 (1982).

145. J. G. Powles, W. A. B. Evans, and N. Quirke, *Mol. Phys.* **46**, 1347 (1982).
146. H. Okamoto, *J. Chem. Phys.* **64**, 2868 (1976); **79**, 3976 (1983); **83**, 2587 (1986).
147. H. Okamoto, K. Itoh, and T. Araki, *J. Chem. Phys.* **78**, 985 (1983); R. Dickman and C. K. Hall, *J. Chem. Phys.* **85**, 3023 (1986).
148. R. Dickman and C. K. Hall, *J. Chem. Phys.* **89**, 3168 (1988).
149. K. G. Honnell, R. Dickman, and C. K. Hall, *J. Chem. Phys.* **87**, 664 (1987).
150. K. G. Honnell and C. K. Hall, *J. Chem. Phys.* **90**, 1841 (1987).
151. C. A. Croxton, *Phys. Lett.* **A70**, 441 (1979).
152. M. N. Rosenbluth and A. W. Rosenbluth, *J. Chem. Phys.* **23**, 356 (1955).
153. I. Batoulis and K. Kremer, *J. Phys.* **A21**, 127 (1988).
154. H. Meirovitch, *J. Chem. Phys.* **79**, 502 (1983).
155. H. Meirovitch, *Macromolecules* **16**, 249 (1983); *Macromolecules* **16**, 1628 (1983).
156. H. C. Öttinger, *Macromolecules* **18**, 93 (1985); **18**, 1348 (1985).
157. H. Meirovitch, *Phys. Rev.* **A32**, 3699 (1985).
158. J. Harris and S. A. Rice, *J. Chem. Phys.* **88**, 1292 (1988).
159. G. C. A. Mooij and D. Frenkel, *Mol. Phys.* **74**, 41 (1991); J. I. Siepmann, *Mol. Phys.* **70**, 1145 (1990).
160. D. Frenkel and B. Smit, *Mol. Phys.* **75**, 983 (1992).
161. D. Frenkel, G. C. A. M. Mooij, and B. Smit, *J. Phys. Condens. Matter* **4**, 3053 (1992).
162. J. J. de Pablo, M. Laso, and U. W. Suter, *J. Chem. Phys.* **96**, 6157 (1992).
163. D. Frenkel, in *Computer Simulation in Chemical Physics* (Kluwer Academic Publishers, Dordrecht, 1993), p. 93.
164. S. K. Kumar, I. Szleifer and A. Z. Panagiotopoulos, *Phys. Rev. Lett.* **66**, 2935 (1991).
165. S. K. Kumar, *J. Chem. Phys.* **96**, 1490 (1992).
166. S. K. Kumar, I. Szleifer, and A. Z. Panagiotopoulos, *Phys. Rev. Lett.* **68**, 3658 (1992).
167. I. Szleifer and A. Z. Panagiotopoulos, *J. Chem. Phys.* **97**, 6666 (1992).
168. M. Müller and W. Paul, *J. Chem. Phys.* **100**, 719 (1994).
169. R. Dickman, *J. Chem. Phys.* **86**, 2246 (1987).
170. A. Hertanto and R. Dickman, *J. Chem. Phys.* **89**, 7577 (1988).
171. M. G. Bawendi and K. F. Freed, *J. Chem. Phys.* **88**, 2741 (1988).
172. H. P. Deutsch and R. Dickman, *J. Chem. Phys.* **93**, 8983 (1990).
173. C. G. Cray and K. I. Gubbins, *Theory of Molecular Fluids* (Clarendon Press, Oxford, 1982).
174. J. Gao and J. H. Weiner, *J. Chem. Phys.* **90**, 6749 (1989).
175. J. Gao and J. H. Weiner, *J. Chem. Phys.* **91**, 3168 (1989).
176a. A. Milchev and K. Binder, *Macromol. Theory Simul.* **3**, 915 (1994).
176b. H. Meirovitch, *J. Chem. Phys.* **97**, 5803, 5816 (1992), and references therein.
177. J. G. Curro, K. S. Schweizer, G. S. Grest, and K. Kremer, *J. Chem. Phys.* **91**, 1357 (1989).
178. D. S. Pearson, G. Verstrate, E. von Meerwall, and F. C. Schilling, *Macromolecules* **20**, 1133 (1987).

2

MONTE CARLO METHODS FOR THE SELF-AVOIDING WALK

Alan D. Sokal

2.1 Introduction

2.1.1 Why is the SAW a sensible model?

The self-avoiding walk (SAW) was first proposed nearly half a century ago as a model of a linear polymer molecule in a good solvent.[1,2] At first glance it seems to be a ridiculously crude model, almost a caricature: real polymer molecules* live in continuous space and have tetrahedral (109.47°) bond angles, a non-trivial energy surface for the bond rotation angles, and a rather complicated monomer–monomer interaction potential. By contrast, the self-avoiding walk lives on a discrete lattice and has non-tetrahedral bond angles (e.g., 90° and 180° on the simple cubic lattice), an energy independent of the bond rotation angles, and a repulsive hard-core monomer–monomer potential.

In spite of these rather extreme simplifications, there is now little doubt that the self-avoiding walk is not merely an excellent but in fact a *perfect* model for some (but not all!) aspects of the behavior of linear polymers in a good solvent.† This apparent miracle arises from *universality*, which plays a central role in the modern theory of critical phenomena.[3,4] In brief, critical statistical–mechanical systems are divided into a small number of *universality classes*, which are typically characterized by spatial dimensionality, symmetries and other rather general properties. In the vicinity of a critical point (and only there), the leading asymptotic behavior is *exactly* the same (modulo some system-dependent scale factors) for all systems of a given universality class; the details of chemical structure, interaction energies and so forth are completely irrelevant (except for setting the nonuniversal scale factors). Moreover, this universal behavior is given by simple *scaling laws*, in which the dependent variables are generalized homogeneous functions of the parameters which measure the deviation from criticality.

*More precisely, linear polymers whose backbones consist solely of carbon–carbon single bonds.

†Here "good solvent" means "at any temperature strictly above the theta temperature for the given polymer–solvent pair".

The key question, therefore, is to determine for each physical problem which quantities are universal and which are nonuniversal.

To compute the nonuniversal quantities, one employs the traditional methods of theoretical physics and chemistry: semi-realistic models followed by a process of successive refinement. All predictions from such models must be expected to be only approximate, *even if the mathematical model is solved exactly*, because the mathematical model is itself only a crude approximation to reality.

To compute the universal quantities, by contrast, a very different approach is available: one may choose *any* mathematical model (the simpler the better) belonging to the same universality class as the system under study, and by solving it determine the *exact* values of universal quantities. Of course, it may not be feasible to solve this mathematical model exactly, so further approximations (or numerical simulations) may be required in practice; but these latter approximations are the *only* sources of error in the computation of universal quantities. At a subsequent stage it is prudent to test variants and refinements of the original model, but solely for the purpose of determining the boundaries of the universality class: if the refined model belongs to the same universality class as the original model, then the refinement has *zero* effect on the universal quantities.

The behavior of polymer molecules as the chain length tends to infinity is, it turns out, a critical phenomenon in the above sense.[5] Thus, it is found empirically—although the existing experimental evidence is admittedly far from perfect[6–10]—that the mean-square radius of gyration $\langle R_g^2 \rangle$ of a linear polymer molecule consisting of N monomer units has the leading asymptotic behavior

$$\langle R_g^2 \rangle = AN^{2\nu}[1 + O(N^{-\Delta})] \tag{2.1}$$

as $N \to \infty$, where the *critical exponent* $\nu \approx 0.588$ is universal, i.e. exactly the same for all polymers, solvents and temperatures (provided only that the temperature is above the theta temperature for the given polymer-solvent pair). The *critical amplitude* A is nonuniversal, i.e., it depends on the polymer, solvent, and temperature, and this dependence is *not* expected to be simple.

There is therefore good reason to believe that any (real or mathematical) linear polymer chain which exhibits some flexibility and has short-range,* predominantly repulsive† monomer–monomer interactions lies in the same

---

*Here I mean that the potential is short-range *in physical space*. It is of course—and this is a crucial point—long-range *along the polymer chain*, in the sense that the interaction between two monomers depends only on their positions in physical space and is essentially independent of the locations of those monomers along the chain.

†Here "predominantly repulsive" means "repulsive enough so that the temperature is strictly above the theta temperature for the given polymer–solvent pair".

universality class as the self-avoiding walk. This belief should, of course, be checked carefully by both numerical simulations and laboratory experiments; but at present there is, to my knowledge, no credible numerical or experimental evidence that would call it into question.

2.1.2 Numerical methods for the self-avoiding walk

Over the decades, the SAW has been studied extensively by a variety of methods. *Rigorous methods* have thus far yielded only fairly weak results;[11] the SAW is, to put it mildly, an extremely difficult mathematical problem. *Non-rigorous analytical methods*, such as perturbation theory and self-consistent-field theory, typically break down in precisely the region of interest, namely long chains.[12] The exceptions are methods based on the renormalization group (RG),[13–15] which have yielded reasonably accurate estimates for critical exponents and for some universal amplitude ratios.[16–24] However, the conceptual foundations of the renormalization-group methods have not yet been completely elucidated;[25,26] and high-precision RG calculations are not always feasible. Thus, considerable work has been devoted to developing *numerical methods* for the study of long SAWs. These methods fall essentially into two categories: exact enumeration and Monte Carlo.

In an *exact-enumeration* study, one first generates a complete list of all SAWs up to a certain length (usually $N \approx 15$–35 steps), keeping track of the properties of interest such as the number of such walks or their squared end-to-end distances.[27] One then performs an extrapolation to the limit $N \to \infty$, using techniques such as the ratio method, Padé approximants or differential approximants.[28–30] Inherent in any such extrapolation is an assumption about the behavior of the coefficients beyond those actually computed. Sometimes this assumption is fairly explicit; other times it is hidden in the details of the extrapolation method. In either case, the assumptions made have a profound effect on the numerical results obtained.[25] For this reason, the claimed error bars in exact-enumeration/extrapolation studies should be viewed with a healthy skepticism.

In a *Monte Carlo* study, by contrast, one aims to probe directly the regime of fairly long SAWs (usually $N \approx 10^2$–10^5 steps). Complete enumeration is unfeasible, so one generates instead a random sample. The raw data then contain statistical errors, just as in a laboratory experiment. These errors can, however, be estimated—sometimes even *a priori* (see Section 2.7.3)—and they can in principle be reduced to an arbitrarily low level by the use of sufficient computer time. An extrapolation to the regime of extremely long SAWs is still required, but this extrapolation is much less severe than in the case of exact-enumeration studies, because the point of departure is already much closer to the asymptotic regime.

Monte Carlo studies of the self-avoiding walk go back to the early 1950s,[31,32] and indeed these simulations were among the first applications of a new invention—the "high-speed electronic digital computer"—to pure science.* These studies continued throughout the 1960s and 1970s, and benefited from the increasingly powerful computers that became available. However, progress was slowed by the high computational complexity of the algorithms then being employed, which typically required a CPU time of order at least $N^{2+2\nu} = N^{\approx 3.2}$ to produce one "effectively independent" N-step SAW. This rapid growth with N of the autocorrelation time—called *critical slowing-down*†—made it difficult in practice to do high-precision simulations with N greater than about 30–100.

In the past decade—since 1981 or so—vast progress has been made in the development of new and more efficient algorithms for simulating the self-avoiding walk. These new algorithms reduce the CPU time for generating an "effectively independent" N-step SAW from $\sim N^{\approx 3.2}$ to $\sim N^{\approx 2}$ or even $\sim N$. The latter is quite impressive, and indeed is the best possible order of magnitude, since it takes a time of order N merely to *write down* an N-step walk! As a practical matter, the new algorithms have made possible high-precision simulations at chain lengths N up to nearly 10^5.[39]

The purpose of this chapter is thus to give a comprehensive overview of Monte Carlo methods for the self-avoiding walk, with emphasis on the extraordinarily efficient algorithms developed since 1981. I shall also discuss briefly some of the physical results which have been obtained from this work.

The plan of this chapter is as follows: I begin by presenting background material on the self-avoiding walk (Section 2.2) and on Monte Carlo methods (Section 2.3). In Section 2.4 I discuss *static* Monte Carlo methods for the generation of SAWs: simple sampling and its variants, inversely restricted sampling (Rosenbluth–Rosenbluth algorithm) and its variants, and dimerization. In Section 2.5 I discuss *quasi-static* Monte Carlo methods: enrichment and incomplete enumeration (Redner–Reynolds algorithm). In Section 2.6 I discuss *dynamic* Monte Carlo methods: the methods are classified according to whether they are local or non-local, whether they are N-conserving or N-changing, and whether they are endpoint-conserving or endpoint-changing. In Section 2.7 I discuss some miscellaneous algorithmic and statistical issues. In Section 2.8 I review some preliminary physical results which have been obtained using these new algorithms. I conclude

---

*Here "pure" means "not useful in the sense of Hardy": "a science is said to be useful if its development tends to accentuate the existing inequalities in the distribution of wealth, or more directly promotes the destruction of human life" [Ref. 33, p. 120*n*].

†For a general introduction to critical slowing-down in Monte Carlo simulations, see Refs 34–37. See also Ref. 38 for a pioneering treatment of critical slowing-down in the context of dynamic critical phenomena.

(Section 2.9) with a brief summary of practical recommendations and a listing of open problems.

For previous reviews of Monte Carlo methods for the self-avoiding walk, see Kremer and Binder[40] and Madras and Slade (Ref. 11, Chapter 9).

2.2 The self-avoiding walk (SAW)

2.2.1 Background and notation

In this section we review briefly the basic facts and conjectures about the SAW that will be used in the remainder of this chapter. A comprehensive survey of the SAW, with emphasis on rigorous mathematical results, can be found in the excellent new book by Madras and Slade.[11]

Real polymers live in spatial dimension $d = 3$ (ordinary polymer solutions) or in some cases in $d = 2$ (polymer monolayers confined to an interface[41,42]). Nevertheless, it is of great conceptual value to define and study the mathematical models—in particular, the SAW—in a general dimension d. This permits us to distinguish clearly between the *general* features of polymer behavior (in any dimension) and the *special* features of polymers in dimension $d = 3$.* The use of arbitrary dimensionality also makes available to theorists some useful technical tools (e.g., dimensional regularization) and some valuable approximation schemes (e.g., expansion in $d = 4 - \epsilon$).[15]

So let $\mathscr{L}$ be some regular d-dimensional lattice. Then an *N-step self-avoiding walk*† (SAW) ω on $\mathscr{L}$ is a sequence of *distinct* points $\omega_0, \omega_1, \ldots, \omega_N$ in $\mathscr{L}$ such that each point is a nearest neighbor of its predecessor.‡ We denote by $|\omega| \equiv N$ the number of steps in ω. For simplicity we shall restrict attention to the simple (hyper-)cubic lattice $\mathbb{Z}^d$; similar ideas apply with minor alterations to other regular lattices. We assume all walks to begin at the origin ($\omega_0 = 0$) unless stated otherwise, and we let $\mathscr{S}_N$ (resp. $\mathscr{S}_N(x)$) be the set of all N-step SAWs starting at the origin and ending anywhere (resp. ending at x).

---

*It turns out that $d = 3$ is *very* special, because it is the upper critical dimension for tricritical behavior. This is the deep reason underlying the fact that polymers at the theta point in $d = 3$ are "quasi-ideal" (i.e., have size exponent $\nu = \frac{1}{2}$ and have all dimensionless virial coefficients vanishing in the limit of infinite chain length). In dimension $d < 3$, polymers at the theta point are *not* quasi-ideal.[43–47]

†The term "walk" is a misnomer. The SAW should *not* be thought of as the path of a particle which "walks" (over time). Rather, it should be thought of as the configuration of a polymer chain *at one instant of time*. (In mathematical terms, the SAW is not a stochastic process [not even a non-Markovian one]: the trouble is that the equal-weight distributions on N-step and $(N + 1)$-step SAWs are not consistent.)

‡Note that a SAW is an *oriented* object, i.e., we distinguish between the starting point (ω_0) and the ending point (ω_N). However, all probability distributions and all observables that we shall consider are invariant under reversal of orientation ($\tilde{\omega}_i \equiv \omega_{N-i}$). This is necessary if the SAW is to be a sensible model of a real homopolymer molecule, which is of course (neglecting end-group effects) *unoriented*.

First we define the quantities relating to the *number* (or "entropy") of SAWs: Let c_N (resp. $c_N(x)$) be the number of N-step SAWs on $\mathbb{Z}^d$ starting at the origin and ending anywhere (resp. ending at x). Then c_N and $c_N(x)$ are believed to have the asymptotic behavior

$$c_N \sim \mu^N N^{\gamma-1} \tag{2.2}$$

$$c_N(x) \sim \mu^N N^{\alpha_{sing}-2} \quad (x \text{ fixed} \neq 0) \tag{2.3}$$

as $N \to \infty$; here μ is called the *connective constant* of the lattice, and γ and α_{sing} are *critical exponents*. The connective constant is definitely lattice-dependent, while the critical exponents are believed to be universal among lattices of a given dimension d. (For rigorous results concerning the asymptotic behavior of c_N and $c_N(x)$, see Refs 11, 48–51.)

Next we define several measures of the *size* of an N-step SAW:

- The *squared end-to-end distance*

$$R_e^2 = (\omega_N - \omega_0)^2. \tag{2.4}$$

- The *squared radius of gyration*

$$R_g^2 = \frac{1}{N+1}\sum_{i=0}^{N}\left(\omega_i - \frac{1}{N+1}\sum_{j=0}^{N}\omega_j\right)^2 \tag{2.5a}$$

$$= \frac{1}{N+1}\sum_{i=0}^{N}\omega_i^2 - \left(\frac{1}{N+1}\sum_{i=0}^{N}\omega_i\right)^2 \tag{2.5b}$$

$$= \frac{1}{2(N+1)^2}\sum_{i,j=0}^{N}(\omega_i - \omega_j)^2. \tag{2.5c}$$

- The *mean-square distance of a monomer from the endpoints*

$$R_m^2 = \frac{1}{2(N+1)}\sum_{i=0}^{N}[(\omega_i - \omega_0)^2 + (\omega_i - \omega_N)^2]. \tag{2.6}$$

We then consider the mean values $\langle R_e^2\rangle_N$, $\langle R_g^2\rangle_N$ and $\langle R_m^2\rangle_N$ in the probability distribution which gives equal weight to each N-step SAW. Very little has been proven rigorously about these mean values, but they are believed to have the asymptotic behavior

$$\langle R_e^2\rangle_N, \langle R_g^2\rangle_N, \langle R_m^2\rangle_N \sim N^{2\nu} \tag{2.7}$$

as $N \to \infty$, where ν is another (universal) critical exponent. Moreover, the amplitude ratios

$$A_N = \frac{\langle R_g^2 \rangle_N}{\langle R_e^2 \rangle_N} \tag{2.8}$$

$$B_N = \frac{\langle R_m^2 \rangle_N}{\langle R_e^2 \rangle_N} \tag{2.9}$$

are expected to approach universal values in the limit $N \to \infty$.*,†

Finally, let c_{N_1,N_2} be the number of pairs $(\omega^{(1)}, \omega^{(2)})$ such that $\omega^{(1)}$ is an N_1-step SAW starting at the origin, $\omega^{(2)}$ is an N_2-step SAW starting *anywhere*, and $\omega^{(1)}$ and $\omega^{(2)}$ have at least one point in common (i.e., $\omega^{(1)} \cap \omega^{(2)} \neq \emptyset$). Then it is believed that

$$c_{N_1,N_2} \sim \mu^{N_1+N_2}(N_1 N_2)^{(2\Delta_4+\gamma-2)/2} g(N_1/N_2) \tag{2.10}$$

as $N_1, N_2 \to \infty$, where Δ_4 is yet another (universal) critical exponent and g is a (universal) scaling function.

The quantity c_{N_1,N_2} is closely related to the second virial coefficient. To see this, consider a rather general theory in which "molecules" of various types interact. Let the molecules of type i have a set S_i of "internal states", so that the complete state of such a molecule is given by a pair (x, s) where $x \in \mathbb{Z}^d$ is its position and $s \in S_i$ is its internal state. Let us assign Boltzmann weights (or "fugacities") $W_i(s)$ $[s \in S_i]$ to the internal states, normalized so that $\sum_{s \in S_i} W_i(s) = 1$; and let us assign an interaction energy $\mathscr{V}_{ij}((x,s),(x',s'))$ $[x, x' \in \mathbb{Z}^d, s \in S_i, s' \in S_j]$ to a molecule of type i at (x,s) interacting with one of type j at (x',s'). Then the second virial coefficient between a molecule of type i and one of type j is

$$B_2^{(ij)} = \frac{1}{2} \sum_{\substack{s \in S_i \\ s' \in S_j}} \sum_{x' \in \mathbb{Z}^d} W_i(s) W_j(s') \left[1 - e^{-\mathscr{V}_{ij}\left((0,s),(x',s')\right)}\right]. \tag{2.11}$$

In the SAW case, the types are the different lengths N, the internal states are the conformations $\omega \in \mathscr{S}_N$ starting at the origin, the Boltzmann weights are $W_N(\omega) = 1/c_N$ for each $\omega \in \mathscr{S}_N$, and the interaction energies are hard-core repulsions

$$\mathscr{V}_{NN'}((x,\omega),(x',\omega')) = \begin{cases} +\infty & \text{if } (\omega + x) \cap (\omega' + x') \neq \emptyset \\ 0 & \text{otherwise} \end{cases} \tag{2.12}$$

*For a general discussion of universal amplitude ratios in the theory of critical phenomena, see Ref. 52.

†Very recently, Hara and Slade[48,49] have proven that the SAW in dimension $d \geq 5$ converges weakly to Brownian motion when $N \to \infty$ with lengths rescaled by $CN^{1/2}$ for a suitable (nonuniversal) constant C. It follows from this that eq. (2.7) holds with $\nu = \frac{1}{2}$, and also that eqs (2.8)/(2.9) have the limiting values $A_\infty = \frac{1}{6}$, $B_\infty = \frac{1}{2}$. Earlier, Slade[53–55] had proven these results for sufficiently high dimension d. See also Ref. 11.

It follows immediately that

$$B_2^{(N_1,N_2)} = \frac{c_{N_1,N_2}}{2c_{N_1}c_{N_2}}. \tag{2.13}$$

The second virial coefficient $B_2^{(N_1,N_2)}$ is a measure of the "excluded volume" between a pair of SAWs. It is useful to define a *dimensionless* quantity by normalizing $B_2^{(N_1,N_2)}$ by some measure of the "size" of these SAWs. Theorists prefer $\langle R_e^2 \rangle$ as the measure of size, while experimentalists prefer $\langle R_g^2 \rangle$ since it can be measured by light scattering. We follow the experimentalists and define the *interpenetration ratio*

$$\Psi_N \equiv 2(d/12\pi)^{d/2} \frac{B_2^{(N,N)}}{\langle R_g^2 \rangle_N^{d/2}} \tag{2.14a}$$

$$= (d/12\pi)^{d/2} \frac{c_{N,N}}{c_N^2 \langle R_g^2 \rangle_N^{d/2}} \tag{2.14b}$$

(for simplicity we consider only $N_1 = N_2 = N$). The numerical prefactor is a convention that arose historically for reasons not worth explaining here. Crudely speaking, Ψ measures the degree of "hardness" of a SAW in its interactions with other SAWs.*

Ψ_N is expected to approach a universal value Ψ^* in the limit $N \to \infty$. A deep question is whether Ψ^* is nonzero (this is called *hyperscaling*). It is now known that hyperscaling fails for SAWs in dimension $d > 4$.[11,48,49] It is believed that hyperscaling holds for SAWs in dimension $d < 4$, but the theoretical justification of this fact is a key unsolved problem in the theory of critical phenomena (see e.g., Ref. 39).†

Higher virial coefficients can be defined analogously, but the details will not be needed here.

Remark The critical exponents defined here for the SAW are precise analogues of the critical exponents as conventionally defined for ferromagnetic spin systems.[57,58] Indeed, the generating functions of the SAW are *equal* to the correlation functions of the n-vector spin model analytically

---

*A useful standard of comparison is the hard sphere of constant density:

$$\Psi_{hard-sphere} = \frac{2}{d\Gamma(d/2)} \left(\frac{d+2}{3}\right)^{d/2} = \begin{cases} \approx 1.128\,38 & \text{in } d = 1 \\ 4/3 & \text{in } d = 2 \\ \approx 1.618\,59 & \text{in } d = 3 \\ 2 & \text{in } d = 4 \end{cases}$$

†A very beautiful heuristic argument concerning hyperscaling for SAWs was given by des Cloizeaux.[56] Note first from eq. (2.14b) that Ψ measures, roughly speaking, the probability of intersection of two independent SAWs that start a distance of order $\langle R_g^2 \rangle^{1/2} \sim N^\nu$ apart. Now, by eq. (2.7), we can interpret a long SAW as an object with "fractal dimension" $1/\nu$. Two independent such objects will "generically" intersect if and only if the sum of their fractal dimensions is at least as large as the dimension of the ambient space. So we expect Ψ^* to be nonzero if and only if $1/\nu + 1/\nu \geq d$, i.e., $d\nu \leq 2$. This occurs for $d < 4$. (For $d = 4$ we believe that $d\nu =$ "$2 +$ logs", and thus expect a logarithmic violation of hyperscaling.)

continued to $n = 0$.[11,59–62] This "polymer–magnet correspondence"* is very useful in polymer theory; but we shall not need it in this chapter.

2.2.2 *The ensembles*

Different aspects of the SAW can be probed in four different ensembles†:

- Fixed-length, fixed-endpoint ensemble (fixed N, fixed x)
- Fixed-length, free-endpoint ensemble (fixed N, variable x)
- Variable-length, fixed-endpoint ensemble (variable N, fixed x)
- Variable-length, free-endpoint ensemble (variable N, variable x)

The fixed-length ensembles are best suited for studying the critical exponents ν and $2\Delta_4 - \gamma$, while the variable-length ensembles are best suited for studying the connective constant μ and the critical exponents α_{sing} (fixed-endpoint) or γ (free-endpoint). Physically, the free-endpoint ensembles correspond to linear polymers, while the fixed-endpoint ensembles with $|x| = 1$ correspond to ring polymers.

All these ensembles give equal weight to all walks of a given length; but the variable-length ensembles have considerable freedom in choosing the relative weights of different chain lengths N. The details are as follows:

Fixed-N, fixed-x ensemble. The state space is $\mathscr{S}_N(x)$, and the probability distribution is $\pi(\omega) = 1/c_N(x)$ for each $\omega \in \mathscr{S}_N(x)$.

Fixed-N, variable-x ensemble. The state space is $\mathscr{S}_N$, and the probability distribution is $\pi(\omega) = 1/c_N$ for each $\omega \in \mathscr{S}_N$.

Variable-N, fixed-x ensemble. The state space is $\mathscr{S}(x) \equiv \bigcup_{N=0}^{\infty} \mathscr{S}_N(x)$, and the probability distribution is generally taken to be

$$\pi_{\beta,p}(\omega) = \beta^{|\omega|} |\omega|^p / Z(\beta, p; x) \quad \text{for each } \omega \in \mathscr{S}(x) \tag{2.15}$$

where

$$Z(\beta, p; x) = \sum_{N=0}^{\infty} \beta^N N^p c_N(x). \tag{2.16}$$

*It is sometimes called the "polymer-magnet analogy", but this phrase is misleading: at least for SAWs (athermal linear polymers), the correspondence is an *exact mathematical identity* (Ref. 11, Section 2.3), not merely an "analogy".

†The proper terminology for these ensembles is unclear to me. The fixed-length and variable-length ensembles are sometimes called "canonical" and "grand canonical", respectively (based on considering the *monomers* as particles). On the other hand, it might be better to call these ensembles "microcanonical" and "canonical", respectively (considering the *polymers* as particles and the chain length as an "energy"), reserving the term "grand canonical" for ensembles of *many* SAWs. My current preference is to avoid entirely these ambiguous terms, and simply say what one means: "fixed-length", "variable-length", etc.

Here $p \geq 0$ is a fixed number (usually 0 or 1), and β is a *monomer fugacity* that can be varied between 0 and $\beta_c \equiv 1/\mu$. By tuning β we can control the distribution of walk lengths N. Indeed, from eq. (2.3) we have

$$\langle N \rangle \approx \frac{p + \alpha_{sing} - 1}{1 - \beta\mu} \tag{2.17}$$

as $\beta \uparrow \beta_c$, provided that $p + \alpha_{sing} > 1$.* Therefore, to generate a distribution of predominantly long (but not *too* long) walks, it suffices to choose β slightly less than (but not *too* close to) β_c.

Variable-N, variable-x ensemble. The state space is $\mathcal{S} \equiv \bigcup_{N=0}^{\infty} \mathcal{S}_N$, and the probability distribution is generally taken to be

$$\pi_{\beta,p}(\omega) = \beta^{|\omega|} |\omega|^p / Z(\beta, p) \quad \text{for each } \omega \in \mathcal{S} \tag{2.18}$$

where

$$Z(\beta, p) = \sum_{N=0}^{\infty} \beta^N N^p c_N. \tag{2.19}$$

p and β are as before, and from eq. (2.2) we have

$$\langle N \rangle \approx \frac{p + \gamma}{1 - \beta\mu} \tag{2.20}$$

as $\beta \uparrow \beta_c$. (Here the condition $p + \gamma > 0$ is automatically satisfied, as a result of the rigorous theorem $\gamma \geq 1$.[11])

An unusual two-SAW ensemble is employed in the join-and-cut algorithm, as will be discussed in Section 2.6.6.2.

2.3 Monte Carlo methods: a review

Monte Carlo methods can be classified as static, quasi-static or dynamic. *Static* methods are those that generate a sequence of statistically independent samples from the desired probability distribution π. *Quasi-static* methods are those that generate a sequence of statistically independent batches of samples from the desired probability distribution π; the correlations within a batch are often difficult to describe. *Dynamic* methods are those that generate a sequence of correlated samples from some stochastic process (usually a Markov process) having the desired probability distribution π as its unique equilibrium distribution.

In this section we review briefly the principles of both static and dynamic Monte Carlo methods, with emphasis on the issues that determine the statistical efficiency of an algorithm.

---

*If $0 < p + \alpha_{sing} < 1$, then $\langle N \rangle \sim (1 - \beta\mu)^{-(p+\alpha_{sing})}$ as $\beta \uparrow \beta_c$, with logarithmic corrections when $p + \alpha_{sing} = 0, 1$. If $p + \alpha_{sing} < 0$, then $\langle N \rangle$ remains bounded as $\beta \uparrow \beta_c$.

2.3.1 Static Monte Carlo methods

Consider a system with *state space (configuration space)* S; for notational simplicity, let us assume that S is discrete (i.e., finite or countably infinite). Now let $\pi = \{\pi_x\}_{x \in S}$ be a probability distribution on S, and let $A = \{A(x)\}_{x \in S}$ be a real-valued observable. Our goal is to devise a Monte Carlo algorithm for estimating the expectation value

$$\langle A \rangle_\pi \equiv \sum_{x \in S} \pi_x A(x). \tag{2.21}$$

The most straightforward approach (*standard Monte Carlo*) is to generate independent random samples $X_1, \ldots, X_n$ from the distribution π (if one can!), and use the *sample mean*

$$\bar{A} \equiv \frac{1}{n} \sum_{i=1}^{n} A(X_i) \tag{2.22}$$

as an estimate of $\langle A \rangle_\pi$. This estimate is *unbiased*, i.e.,

$$\langle \bar{A} \rangle = \langle A \rangle_\pi. \tag{2.23}$$

Its *variance* is

$$\begin{aligned} \operatorname{var}\langle \bar{A} \rangle &\equiv \langle \bar{A}^2 \rangle - \langle \bar{A} \rangle^2 \\ &= \frac{1}{n} \operatorname{var}_\pi(A) \\ &\equiv \frac{1}{n} \left[\langle A^2 \rangle_\pi - \langle A \rangle_\pi^2 \right]. \end{aligned} \tag{2.24}$$

However, it is also legitimate to generate samples $X_1, \ldots, X_n$ from *any* probability distribution ν, and then use *weights* $W(x) \equiv \pi_x / \nu_x$. There are two reasons one might want to sample from ν rather than π. Firstly, it might be *unfeasible* to generate (efficiently) random samples from π, so one may be *obliged* to sample instead from some simpler distribution ν. This situation is the typical one in statistical mechanics. Secondly, one might aspire to improve the efficiency (i.e., reduce the variance) by sampling from a cleverly chosen distribution ν.

There are two cases to consider, depending on how well one knows the function $W(x)$:

(a) $W(x)$ is known exactly. (Note that $\sum_{x \in S} \nu_x W(x) = 1$ and $\sum_{x \in S} \pi_x W(x)^{-1} = 1$.)

(b) $W(x)$ is known *except for an unknown multiplicative constant* (normalization factor). This case is common in statistical mechanics: if $\pi_x = Z_\beta^{-1} e^{-\beta H(x)}$ and $\nu_x = Z_{\beta'}^{-1} e^{-\beta' H(x)}$, then $W(x) = (Z_{\beta'}/Z_\beta) e^{-(\beta - \beta')H(x)}$ but we are unlikely to know the ratio of partition functions.

In the first case, we can use as our estimator the *weighted sample mean*

$$\bar{A}^{(W)} \equiv \frac{1}{n}\sum_{i=1}^{n} W(X_i)A(X_i). \tag{2.25}$$

This estimate is unbiased, since

$$\langle \bar{A}^{(W)} \rangle = \langle WA \rangle_\nu = \langle A \rangle_\pi. \tag{2.26}$$

Its variance is

$$\begin{aligned} \operatorname{var}(\bar{A}^{(W)}) &= \frac{1}{n}\left[\langle (WA)^2 \rangle_\nu - \langle WA \rangle_\nu^2\right] \\ &= \frac{1}{n}\left[\langle WA^2 \rangle_\pi - \langle A \rangle_\pi^2\right]. \end{aligned} \tag{2.27}$$

This estimate can be either better or worse than standard Monte Carlo, depending on the choice of ν. The optimal choice is the one that minimizes $\langle WA^2 \rangle_\pi$ subject to the constraint $\langle W^{-1} \rangle_\pi = 1$, namely

$$W(x)^{-1} = \frac{|A(x)|}{\sum_{x \in S} \pi_x |A(x)|}, \tag{2.28}$$

or in other words $\nu_x = \text{const} \times |A(x)|\, \pi_x$. In particular, if $A(x) \geq 0$ the resulting estimate has *zero* variance. But it is impractical: in order to know $W(x)$ we must know the denominator in eq. (2.28), which is the quantity we were trying to estimate in the first place! Nevertheless, this result offers some practical guidance: we should choose $W(x)^{-1}$ to mimic $|A(x)|$ as closely as possible, *subject to the constraint that* $\sum_{x \in S} \pi_x W(x)^{-1}$ *be calculable analytically* (and equal to 1).

In the second case, we have to use a *ratio estimator*

$$\bar{A}^{(W,ratio)} \equiv \frac{\sum_{i=1}^{n} W(X_i)A(X_i)}{\sum_{i=1}^{n} W(X_i)}; \tag{2.29}$$

here the unknown normalization factor in W cancels out. This estimate is slightly biased: using the small-fluctuations approximation

$$\left\langle \frac{Y}{Z} \right\rangle \approx \frac{\langle Y \rangle}{\langle Z \rangle}\left[1 - \frac{\operatorname{cov}(Y,Z)}{\langle Y \rangle \langle Z \rangle} + \frac{\operatorname{var}(Z)}{\langle Z \rangle^2}\right], \tag{2.30}$$

we obtain

$$\langle \bar{A}^{(W,ratio)} \rangle = \langle A \rangle_\pi - \frac{1}{n}\left[\langle WA \rangle_\pi - \langle W \rangle_\pi \langle A \rangle_\pi\right] + O\left(\frac{1}{n^2}\right). \tag{2.31}$$

Since the bias is of order $1/n$, while the standard deviation ($\equiv$ square root of the variance) is of order $1/\sqrt{n}$, the bias is normally negligible compared to the statistical fluctuation.* The variance can also be computed by the small-fluctuations approximation

$$\operatorname{var}\left(\frac{Y}{Z}\right) = \frac{\langle Y\rangle^2}{\langle Z\rangle^2}\operatorname{var}\left(\frac{Y}{\langle Y\rangle} - \frac{Z}{\langle Z\rangle}\right) \tag{2.32a}$$

$$= \frac{\operatorname{var}(Y)}{\langle Z\rangle^2} - \frac{2\langle Y\rangle}{\langle Z\rangle^3}\operatorname{cov}(Y,Z) + \frac{\langle Y\rangle^2}{\langle Z\rangle^4}\operatorname{var}(Z); \tag{2.32b}$$

it is

$$\operatorname{var}(\bar{A}^{(W,ratio)}) = \frac{1}{n}\left\langle W(A - \langle A\rangle_\pi)^2\right\rangle_\pi + O\left(\frac{1}{n^2}\right). \tag{2.33}$$

The optimal choice of ν is the one that minimizes $\left\langle W(A - \langle A\rangle_\pi)^2\right\rangle_\pi$ subject to the constraint $\langle W^{-1}\rangle_\pi = 1$, namely

$$W(x)^{-1} = \frac{|A(x) - \langle A\rangle_\pi|}{\sum\limits_{x\in S} \pi_x\, |A(x) - \langle A\rangle_\pi|}. \tag{2.34}$$

Let us now try to interpret these formulae. First note that

$$\langle W\rangle_\pi - 1 = \langle W^2\rangle_\nu - \langle W\rangle_\nu^2 \equiv \operatorname{var}_\nu(W) \geq 0, \tag{2.35}$$

with equality only if $\nu = \pi$. So $\langle W\rangle_\pi - 1$ measures, in a rough sense, the "mismatch" (or "distance") between ν and π. Now assume for simplicity that A is a bounded observable, i.e., $|A(x)| \leq M$ for all $x \in S$. Then it is immediate from eqs (2.27) and (2.33) that

$$\operatorname{var}(\bar{A}^{(W)}) \leq \frac{M^2}{n}\langle W\rangle_\pi \tag{2.36}$$

$$\operatorname{var}(\bar{A}^{(W,ratio)}) \leq \frac{4M^2}{n}\langle W\rangle_\pi + O\left(\frac{1}{n^2}\right) \tag{2.37}$$

So the variances cannot get large unless $\langle W\rangle_\pi \gg 1$, i.e., ν is very distant from π; and in this case it is easy to see that the variances *can* get large. The

---

* Note that

$$|\langle WA\rangle_\pi - \langle W\rangle_\pi\langle A\rangle_\pi| \leq \langle W(A - \langle A\rangle_\pi)^2\rangle_\pi^{1/2}(\langle W\rangle_\pi - 1)^{1/2}$$

(with equality if and only if $A = c_1 + c_2 W^{-1}$) by the Schwarz inequality with measure ν applied to the functions $W - 1$ and $W(A - \langle A\rangle_\pi)$. Therefore, from eqs (2.31) and (2.33) we have (to leading order in $1/n$)

$$|\operatorname{bias}(\bar{A}^{(W,ratio)})| \leq n^{-1/2}\operatorname{var}(\bar{A}^{(W,ratio)})^{1/2}(\langle W\rangle_\pi - 1)^{1/2}.$$

So the bias is $\ll$ the standard deviation unless $\langle W\rangle_\pi$ is enormous.

moral is this: when the probability distribution actually simulated (ν) differs considerably from the distribution of interest (π), the variance of the Monte Carlo estimates can be vastly higher than one might expect naively for the given sample size n. Heuristically, this is because the states (configurations) that are "typical" for π are "rare" for ν, so the *useful* sample size is much smaller than the *total* sample size.

Here is a concrete example: Let S be the set of all N-step walks (not necessarily self-avoiding) starting at the origin. Let π be uniform measure on *self-avoiding* walks, i.e.

$$\pi_\omega = \begin{cases} 1/c_N & \text{if } \omega \text{ is self-avoiding} \\ 0 & \text{otherwise} \end{cases} \tag{2.38}$$

Unfortunately, it is not easy to generate (efficiently) random samples from π (that is the subject of this chapter!). So let us instead generate *ordinary* random walks, i.e., random samples from

$$\nu_\omega = (2d)^{-N} \quad \text{for all } \omega \in S, \tag{2.39}$$

and then apply the weights $W(\omega) = \pi_\omega / \nu_\omega$. Clearly we have

$$\langle W \rangle_\pi = \frac{(2d)^N}{c_N} \approx \left(\frac{2d}{\mu}\right)^N, \tag{2.40}$$

which grows *exponentially* for large N. Therefore, the efficiency of this algorithm deteriorates exponentially as N grows.

The reader is referred to Chapter 5 of Ref. 63 for some more sophisticated static Monte Carlo techniques. It would be interesting to know whether any of them can be applied usefully to the self-avoiding walk.

2.3.2 Dynamic Monte Carlo methods

In this subsection we review briefly the principles of dynamic Monte Carlo methods, and define some quantities (autocorrelation times) that will play an important role in the remainder of this article.

The idea of *dynamic* Monte Carlo methods is to invent a *stochastic process* with state space S having π as its unique equilibrium distribution. We then simulate this stochastic process, starting from an arbitrary initial configuration; once the system has reached equilibrium, we measure time averages, which converge (as the run time tends to infinity) to π-averages. In physical terms, we are inventing a *stochastic time evolution* for the given system. It must be emphasized, however, that this time evolution *need not correspond to any real "physical" dynamics*: rather, the dynamics is simply a numerical algorithm, and it is to be chosen, like all numerical algorithms, on the basis of its computational efficiency.

In practice, the stochastic process is always taken to be a *Markov process*. We assume that the reader is familiar with the elementary theory of discrete-time Markov chains.*

For simplicity let us assume that the state space S is discrete (i.e. finite or countably infinite); this is the case in nearly all the applications considered in this chapter. Consider a Markov chain with state space S and transition probability matrix $P = \{p(x \to y)\} = \{p_{xy}\}$ satisfying the following two conditions:

(A) For each pair $x, y \in S$, there exists an $n \geq 0$ for which $p_{xy}^{(n)} > 0$. Here $p_{xy}^{(n)} \equiv (P^n)_{xy}$ is the n-step transition probability from x to y. [This condition is called *irreducibility* (or *ergodicity*); it asserts that each state can eventually be reached from each other state.]

(B) For each $y \in S$,

$$\sum_{x \in S} \pi_x p_{xy} = \pi_y. \tag{2.41}$$

(This condition asserts that π is a *stationary distribution* [or *equilibrium distribution*] for the Markov chain $P = \{p_{xy}\}$.)

In this case it can be shown[66] that π is the *unique* stationary distribution for the Markov chain $P = \{p_{xy}\}$, and that the occupation-time distribution over long time intervals converges (with probability 1) to π, irrespective of the initial state of the system. If, in addition, P is *aperiodic* [this means that for each pair $x, y \in S$, $p_{xy}^{(n)} > 0$ for *all* sufficiently large n], then the probability distribution at any single time in the far future also converges to π, irrespective of the initial state—that is, $\lim_{n \to \infty} p_{xy}^{(n)} = \pi_y$ for all x.

Thus, simulation of the Markov chain P provides a legitimate Monte Carlo method for estimating averages with respect to π. However, since the successive states $X_0, X_1, \ldots$ of the Markov chain are in general highly correlated, the variance of estimates produced in this way may be much higher than in independent sampling. To make this precise, let $A = \{A(x)\}_{x \in S}$ be a real-valued function defined on the state space S (i.e., a real-valued observable) that is square-integrable with respect to π. Now consider the *stationary* Markov chain (i.e., start the system in the stationary distribution π, or equivalently, "thermalize" it for a very long time prior to observing the system). Then $\{A_t\} \equiv \{A(X_t)\}$ is a stationary stochastic process with mean

$$\mu_A \equiv \langle A_t \rangle = \sum_{x \in S} \pi_x A(x) \tag{2.42}$$

*The books of Kemeny and Snell[64] and Iosifescu[65] are excellent references on the theory of Markov chains with *finite* state space. At a somewhat higher mathematical level, the books of Chung[66] and Nummelin[67] deal with the cases of *countable* and *general* state space, respectively.

and *unnormalized autocorrelation function**

$$\begin{aligned} C_{AA}(t) &\equiv \langle A_s A_{s+t} \rangle - \mu_A^2 \\ &= \sum_{x,y \in S} A(x)[\pi_x p_{xy}^{(|t|)} - \pi_x \pi_y] A(y). \end{aligned} \tag{2.43}$$

The *normalized autocorrelation function* is then

$$\rho_{AA}(t) \equiv C_{AA}(t)/C_{AA}(0). \tag{2.44}$$

Typically $\rho_{AA}(t)$ decays exponentially ($\sim e^{-|t|/\tau}$) for large t; we define the *exponential autocorrelation time*

$$\tau_{exp,A} = \limsup_{t \to \infty} \frac{t}{-\log|\rho_{AA}(t)|} \tag{2.45}$$

and

$$\tau_{exp} = \sup_A \tau_{exp,A}. \tag{2.46}$$

Thus, τ_{exp} is the relaxation time of the slowest mode in the system. (If the state space is infinite, τ_{exp} might be $+\infty$!)†

On the other hand, for a given observable A we define the *integrated autocorrelation time*

$$\tau_{int,A} = \frac{1}{2} \sum_{t=-\infty}^{\infty} \rho_{AA}(t) \tag{2.47a}$$

$$= \frac{1}{2} + \sum_{t=1}^{\infty} \rho_{AA}(t) \tag{2.47b}$$

---

*In the statistics literature, this is called the *autocovariance function.*

†An equivalent definition, which is useful for rigorous analysis, involves considering the spectrum of the transition probability matrix P considered as an operator on the Hilbert space $l^2(\pi)$. [$l^2(\pi)$ is the space of complex-valued functions on S that are square-integrable with respect to π: $||A|| \equiv (\sum_{x \in S} \pi_x |A(x)|^2)^{1/2} < \infty$. The inner product is given by $(A, B) \equiv \sum_{x \in S} \pi_x A(x)^* B(x)$.] It is not hard to prove the following facts about P:

(a) The operator P is a contraction. (In particular, its spectrum lies in the closed unit disk.)
(b) 1 is a simple eigenvalue of P, as well as of its adjoint P^*, with eigenvector equal to the constant function **1**.
(c) If the Markov chain is aperiodic, then 1 is the only eigenvalue of P (and of P^*) on the unit circle.
(d) Let R be the spectral radius of P acting on the orthogonal complement of the constant functions:

$$R \equiv \inf \{r : \text{spec}(P \restriction \mathbf{1}^{\perp}) \subset \{\lambda : |\lambda| \leq \mathbf{r}\}\}.$$

Then $R = e^{-1/\tau_{exp}}$.

Facts (a)–(c) are a generalized Perron–Frobenius theorem[68]; fact (d) is a consequence of a generalized spectral radius formula.[69] Note that the worst-case rate of convergence to equilibrium from an initial nonequilibrium distribution is controlled by R, and hence by τ_{exp}.

(The factor of $\frac{1}{2}$ is purely a matter of convention; it is inserted so that $\tau_{int,A} \approx \tau_{exp,A}$ if $\rho_{AA}(t) \sim e^{-|t|/\tau}$ with $\tau \gg 1$.) The integrated autocorrelation time controls the statistical error in Monte Carlo estimates of $\langle A \rangle$. More precisely, the sample mean

$$\bar{A} \equiv \frac{1}{n} \sum_{t=1}^{n} A_t \tag{2.48}$$

has variance

$$\mathrm{var}(\bar{A}) = \frac{1}{n^2} \sum_{r,s=1}^{n} C_{AA}(r-s) \tag{2.49}$$

$$= \frac{1}{n} \sum_{t=-(n-1)}^{n-1} \left(1 - \frac{|t|}{n}\right) C_{AA}(t) \tag{2.50}$$

$$\approx \frac{1}{n} (2\tau_{int,A}) C_{AA}(0) \quad \text{for } n \gg \tau \tag{2.51}$$

Thus, the variance of $\bar{A}$ is a factor $2\tau_{int,A}$ larger than it would be if the $\{A_t\}$ were statistically independent. Stated differently, the number of "effectively independent samples" in a run of length n is roughly $n/2\tau_{int,A}$.

In summary, the autocorrelation times τ_{exp} and $\tau_{int,A}$ play different roles in Monte Carlo simulations. τ_{exp} controls the relaxation of the *slowest* mode in the system; in particular, it places an upper bound on the number of iterations n_{disc} which should be discarded at the beginning of the run, before the system has attained equilibrium (e.g., $n_{disc} \approx 20\tau_{exp}$ is usually more than adequate). On the other hand, $\tau_{int,A}$ determines the statistical errors in Monte Carlo estimates of $\langle A \rangle$, once equilibrium has been attained.

Most commonly it is assumed that τ_{exp} and $\tau_{int,A}$ are of the same order of magnitude, at least for "reasonable" observables A. But this is *not* true in general. In fact, one usually expects the autocorrelation function $\rho_{AA}(t)$ to obey a dynamic scaling law[70] of the form

$$\rho_{AA}(t;\beta) \approx |t|^{-a} F\left((\beta - \beta_c)|t|^b\right) \tag{2.52}$$

valid in the limit

$$\beta - \beta_c \to 0, \quad |t| \to \infty, \quad x \equiv (\beta - \beta_c)|t|^b \text{ fixed.} \tag{2.53}$$

Here $a, b > 0$ are dynamic critical exponents and F is a suitable scaling function; β is some "temperature-like" parameter, and β_c is the critical point. Now suppose that F is continuous and strictly positive, with $F(x)$ decaying rapidly (e.g., exponentially) as $|x| \to \infty$. Then it is not hard to see that

$$\tau_{exp,A} \sim |\beta - \beta_c|^{-1/b} \tag{2.54}$$

$$\tau_{int,A} \sim |\beta - \beta_c|^{-(1-a)/b} \quad (\text{if } a < 1) \tag{2.55}$$

$$\rho_{AA}(t; \beta = \beta_c) \sim |t|^{-a} \tag{2.56}$$

so that $\tau_{exp,A}$ and $\tau_{int,A}$ have *different* critical exponents unless $a = 0$.* Actually, this should not be surprising: replacing "time" by "space", we see that $\tau_{exp,A}$ is the analogue of a correlation length, while $\tau_{int,A}$ is the analogue of a susceptibility; and eqs (2.54)–(2.56) are the analogue of the well-known scaling law $\gamma = (2 - \eta)\nu$—clearly $\gamma \neq \nu$ in general! So it is crucial to distinguish between the two types of autocorrelation time.

Returning to the general theory, we note that one convenient way of satisfying the stationarity condition (B) is to satisfy the following *stronger* condition:

$$(\mathrm{B}') \quad \text{For each pair } x, y \in S, \ \pi_x p_{xy} = \pi_y p_{yx}. \tag{2.57}$$

(Summing (B′) over x, we recover (B).) (B′) is called the *detailed-balance condition*; a Markov chain satisfying (B′) is called *reversible*.† (B′) is equivalent to the *self-adjointness* of P as on operator on the space $l^2(\pi)$. In this case, it follows from the spectral theorem that the autocorrelation function $C_{AA}(t)$ has a spectral representation

$$C_{AA}(t) = \int_{-1}^{1} \lambda^{|t|} \, d\sigma_{AA}(\lambda) \tag{2.58}$$

with a *nonnegative* spectral weight $d\sigma_{AA}(\lambda)$ supported on the interval $[-e^{-1/\tau_{exp,A}}, e^{-1/\tau_{exp,A}}]$. It follows that

$$\tau_{int,A} \le \frac{1}{2}\left(\frac{1 + e^{-1/\tau_{exp,A}}}{1 - e^{-1/\tau_{exp,A}}}\right) \le \frac{1}{2}\left(\frac{1 + e^{-1/\tau_{exp}}}{1 - e^{-1/\tau_{exp}}}\right) \approx \tau_{exp}. \tag{2.59}$$

There is no particular *advantage* to algorithms satisfying detailed balance (rather than merely satisfying stationarity), but they are easier to analyze mathematically.

Finally, let us make a remark about transition probabilities P that are "built up out of" other transition probabilities $P_1, P_2, \ldots, P_n$:

(a) If $P_1, P_2, \ldots, P_n$ satisfy the stationarity condition (resp. the detailed-balance condition) for π, then so does any convex combination $P = \sum_{i=1}^{n} \lambda_i P_i$. Here $\lambda_i \ge 0$ and $\sum_{i=1}^{n} \lambda_i = 1$.

---

*Our discussion of this topic in Ref. 71 is incorrect. A correct discussion can be found in Ref. 72.

†For the physical significance of this term, see Kemeny and Snell (Ref. 64, section 5.3) or Iosifescu (Ref. 65, section 4.5).

(b) If $P_1, P_2, \ldots, P_n$ satisfy the stationarity condition for π, then so does the product $P = P_1 P_2 \cdots P_n$. (Note, however, that P does *not* in general satisfy the detailed-balance condition, even if the individual P_i do.*)

Algorithmically, the convex combination amounts to choosing *randomly*, with probabilities $\{\lambda_i\}$, from among the "elementary operations" P_i. (It is crucial here that the λ_i are *constants*, independent of the current configuration of the system; only in this case does P leave π stationary in general.) Similarly, the product corresponds to performing *sequentially* the operations $P_1, P_2, \ldots, P_n$.

2.4 Static Monte Carlo methods for the SAW

2.4.1 *Simple sampling and its variants*

The most obvious static technique for generating a random N-step SAW is *simple sampling*: just generate a random N-step *ordinary* random walk (ORW), and reject it if it is not self-avoiding; keep trying until success. It is easy to see that this algorithm produces each N-step SAW with equal probability. Of course, to save time we should check the self-avoidance as we go along, and reject the walk as soon as a self-intersection is detected. (Methods for testing self-avoidance are discussed in Section 2.7.1.2.) The algorithm is thus:

```
        title Simple sampling.
        function ssamp (N)
        comment This routine returns a random N-step SAW.

        ω₀ ← 0
start:  for i = 1 to N do
            ωᵢ ← a random nearest neighbor of ωᵢ₋₁
            if ωᵢ ∈ {ω₀, ..., ωᵢ₋₁} goto start
        enddo
        return ω
```

(Here and in what follows, we will express algorithms in "pseudocode". Translation to your favorite language—Fortran, C or whatever—is almost always trivial.)

The trouble with this algorithm is, of course, the exponentially rapid sample attrition for long walks. Clearly, the probability of an N-step walk being self-avoiding is $c_N/(2d)^N$, which behaves for large N as

*Recall that if A and B are self-adjoint operators, then AB is self-adjoint *if and only if* A and B commute.

$$\frac{c_N}{(2d)^N} \sim (\mu/2d)^N \, N^{\gamma-1} \tag{2.60a}$$

$$\sim e^{-\lambda N} N^{\gamma-1} \tag{2.60b}$$

where

$$\lambda = \log(2d/\mu) \tag{2.61}$$

is called the *attrition constant*. Therefore, the mean number of attempts required to generate an N-step SAW is $(2d)^N/c_N$, which grows roughly as $e^{\lambda N}$. And the mean CPU time per attempt is of order $\min(1/\lambda, N)$. So this method is extremely inefficient in generating SAWs of length $N \gtrsim 10/\lambda$. For the simple (hyper-)cubic lattices in dimensions 2, 3 and 4, the values of λ are approximately 0.42, 0.25 and 0.17, respectively (see Table 2.1). So it is unfeasible to generate SAWs of length more than $\approx$ 20–60 steps by simple sampling. All alternative SAW Monte Carlo techniques are aimed essentially at alleviating this attrition problem—hopefully without introducing other problems of equal or greater severity!

Some improvement can be obtained by modifying the walk-generation process so as to produce only walks without immediate reversals (such walks are called *non-reversal random walks* (NRRWs) or *memory-2 walks*). The algorithm is thus:

title Non-reversal simple sampling.
function `nrssamp` (N)
comment This routine returns a random N-step SAW.

$\omega_0 \leftarrow 0$
$\omega_1 \leftarrow$ a random nearest neighbor of 0
start: **for** $i = 2$ **to** N **do**
  $\omega_i \leftarrow$ a random nearest neighbor of ω_{i-1}, not equal to ω_{i-2}
  if $\omega_i \in \{\omega_0, \ldots, \omega_{i-1}\}$ **goto** start
enddo
return ω

Table 2.1 *Connective constant μ and attrition constants λ and λ' for simple (hyper-)cubic lattices in dimensions $2 \le d \le 6$ and $d \to \infty$. Estimated errors in the last digit(s) are shown in parentheses.*

d	μ	λ	λ'
2	2.638 158 5 (10)[27,73]	0.416	0.129
3	4.683 907 (22)[74]	0.248	0.065
4	6.772 0 (5)[75]	0.167	0.033
5	8.838 6 (8)[76]	0.123	0.018
6	10.878 8 (9)[76]	0.098	0.011
$d \to \infty$	$2d - 1 - (2d)^{-1} - \ldots$[50,51,77–79]	$(2d)^{-1} + \ldots$	$(2d)^{-2} + \ldots$

Then $(2d)^N$ is replaced by $2d(2d-1)^{N-1}$, and the attrition rate is

$$\lambda' = \log \frac{2d-1}{\mu}. \tag{2.62}$$

For comparison, λ' is approximately 0.13, 0.07 and 0.03, respectively, for $d = 2, 3, 4$. This is much smaller than in the unmodified scheme, but the exponential attrition is still prohibitive for walks of length more than $\approx$ 80–300 steps.

The logical next step is to modify the walk-generation process so that walks with loops of length $\leq r$ are automatically absent. Let us start by building the walk out of *strides* of r steps.[80]* That is, let us enumerate in advance all the r-step SAWs—call them $\omega^{(1)}, \ldots, \omega^{(c_r)}$. (Obviously this takes a memory of order rc_r, and so is feasible only if r is not too large.) We then build up the walk by repeated concatenation of strides. For simplicity let us assume that N is a multiple of r:

title Simple r-step stride sampling.
function `simstride` (r, k)
comment This routine returns a random kr-step SAW.

start: $\omega \leftarrow \{0\}$ (zero-step SAW at the origin)
    for $i = 1$ **to** k **do**
        $\alpha \leftarrow$ a random integer from the set $\{1, \ldots, c_r\}$
        $\omega \leftarrow \omega \circ \omega^{(\alpha)}$ (concatenation)
        if ω is not self-avoiding **goto** start
    enddo
    return ω

The probability of surviving to length $N = kr$ is

$$\frac{c_N}{(c_r)^k} \sim \left(\frac{\mu}{c_r^{1/r}}\right)^N. \tag{2.63}$$

There is still exponential attrition (since $c_r > \mu^r$), but this attrition can in principle be made arbitrarily small by taking r large (since $\lim_{r\to\infty} c_r^{1/r} = \mu$). In practice we can probably handle rc_r of order 10^6 on a modern-day workstation.† The resulting attrition rates $\lambda^{(r)}$ are shown in Table 2.2. They are far from spectacular; the trouble is that $c_r^{1/r}$ converges rather slowly to μ.‡

Of course, we can do better by choosing $\omega^{(\alpha)}$ from among only those r-step walks whose first step is not opposite to the last step of the current

---

*The treatment in the remainder of this section relies heavily on Ref. 11, Section 9.3.1, which is in turn an explication of Ref. 63, p. 129.

† Of course, we can reduce the memory requirements by at least a factor $2d$ by exploiting symmetry, e.g., storing only those r-step SAWs whose first step is in some particular direction. But this only increases the feasible r by about 1.

‡From eq. (2.2) we have $c_r^{1/r} \approx \mu\left[1 + \frac{(\gamma-1)\log r}{r} + O\left(\frac{1}{r}\right)\right]$

Table 2.2 *Attrition constants $\lambda^{(r)}$ and $\lambda'^{(r)}$ for simple and non-reversal r-stride sampling. For each d, we have taken the largest r such that $rc_r \leq 10^6$.*

d	r	c_r	rc_r	$\lambda^{(r)} \equiv \log(c_r^{1/r}/\mu)$	$\lambda'^{(r)} \equiv \log\left(\left(\frac{2d-1}{2d}c_r\right)^{1/r}/\mu\right)$
2	10	44 100	441 000	0.099	0.071
3	7	81 390	569 730	0.071	0.045
4	6	127 160	762 960	0.046	0.024
5	5	64 250	321 250	0.035	0.014
6	5	173 172	865 860	0.026	0.008

ω. In this *non-reversal r-step stride sampling*, the probability of surviving to length $N = kr$ is

$$\frac{c_N}{c_r\left(\dfrac{2d-1}{2d}c_r\right)^{k-1}} \sim \left(\frac{\mu}{\left(\dfrac{2d-1}{2d}c_r\right)^{1/r}}\right)^N . \tag{2.64}$$

The resulting attrition rates $\lambda'^{(r)}$ are shown in the last column of Table 2.2.

Neither of these algorithms in fact eliminates all loops of length $\leq r$, because such loops can be formed by the concatenation of two r-step strides. But we can eliminate such loops if we are willing to pre-compute the list of legal pairs of strides. That is, for each index α $(1 \leq \alpha \leq c_r)$, we make a list L_α containing those indices β such that $\omega^{(\alpha)} \circ \omega^{(\beta)}$ is self-avoiding. (This takes a memory of order c_r^2.) Now, it would *not* be correct to choose at each stage of the algorithm a random SAW from the appropriate list L_α; the trouble is that the lists do not all have the same number of elements, and as a result the walks would not be generated with uniform probability (see also Section 2.4.2). Instead, we must allow the possibility of "rejections". Let $c_r^* \equiv \max_{1\leq\alpha\leq c_r} |L_\alpha|$ be the number of elements in the largest list. We can then perform:

title Super-duper r-step stride sampling.
function `supstride` (r, k)
comment This routine returns a random kr–step SAW.

$\alpha_1 \leftarrow$ a random integer from the set $\{1, \ldots, c_r\}$
start: $\omega \leftarrow \omega^{(\alpha_1)}$
for $i = 2$ **to** k **do**
  $m \leftarrow$ a random integer from the set $\{1, \ldots, c_r^*\}$
  if $m > |L_{\alpha_{i-1}}|$ **goto** start (this is the "rejection")
  $\alpha_i \leftarrow$ the mth element from $L_{\alpha_{i-1}}$
  $\omega \leftarrow \omega \circ \omega^{(\alpha_i)}$
  if ω is not self-avoiding **goto** start
enddo
return ω

Clearly the probability of surviving to length $N = kr$ is

$$\frac{c_N}{c_r(c_r^*)^{k-1}} \sim \left(\frac{\mu}{(c_r^*)^{1/r}}\right)^N. \qquad (2.65)$$

Little is known about c_r^*, but it is certainly $\leq \frac{2d-1}{2d} c_r$, and probably not much less. In practice, the extravagant memory requirements of this method limit r to very small values; for a given amount of memory, non-reversal stride sampling probably works better.*

2.4.2 Inversely restricted sampling (Rosenbluth–Rosenbluth algorithm)

The exponential attrition of simple sampling and its variants arises from the fact that each new step of the walk might lead to a self-intersection. So it is tempting to envisage an algorithm in which one chooses randomly (with equal probability) from among only those next steps which do *not* lead to a self-intersection (assuming such steps exist). Unfortunately, this means that SAWs are not generated with uniform probability; rather, the probability that this algorithm generates a given N-step SAW ω is

$$P(\omega) = \text{const} \times \prod_{i=1}^{N} \frac{1}{k_i}, \qquad (2.66)$$

where $k_i \equiv k_i(\omega_0, \ldots, \omega_{i-1})$ is the number of choices available at step i.† Therefore, each walk must be assigned *a weight* $W(\omega) \sim 1/P(\omega)$, and the mean value of an observable $\mathcal{O}(\omega)$ must be estimated from a sample of walks $\omega^{(1)}, \ldots, \omega^{(n)}$ by a ratio of weighted averages:

$$\langle \mathcal{O}(\omega) \rangle \approx \frac{\sum_{j=1}^{n} W(\omega^{(j)}) \mathcal{O}(\omega^{(j)})}{\sum_{j=1}^{n} W(\omega^{(j)})}. \qquad (2.67)$$

This method is known as *inversely restricted sampling*:[81,82]

title Inversely restricted sampling.
function `irsamp` (N)
comment This routine returns an N-step SAW and its weight factor.

---

*Actually, it suffices to store the list *lengths* $|L_\alpha|$, and not the lists themselves. One can then choose α_i by repeatedly trying to find a stride compatible with $\omega^{(\alpha_{i-1})}$, together with a probability $1 - |L_{\alpha_{i-1}}|/c_r^*$ of giving up ("rejection") at each try.

†Here "const" is in fact the reciprocal of the probability of surviving in this algorithm to N steps without getting "trapped".

```
    ω_0 ← 0
start: weight ← 1/[2d(2d − 1)^(N-1)]   (this is merely a convenient normalization)
    for i = 1 to N do
        S_i ← set of all nearest neighbors of ω_{i-1} not contained in {ω_0, ..., ω_{i-1}}
        if S_i = ∅ goto start   (the walk is "trapped")
        ω_i ← a random element of S_i
        weight ← weight × |S_i|
    enddo
    return (ω, weight)
```

This method has several difficulties: Firstly, there is still exponential sample attrition for long walks (although at a much slower rate than in simple sampling): it is caused now not by mere self-intersection, but by "trapping". This is most serious in $d = 2$.[83–86] Secondly, a ratio estimator (2.67) is slightly biased; however, as discussed in Section 2.3.1, this difficulty is negligible for large sample size n.[87] The most serious difficulty is that "the weights are almost certain to get out of hand, a few of them being very much larger than all the rest. This means that the greater part of the data, corresponding to the negligible weights, gets ignored" (Ref. 63, p. 131). Thus, the variance of the estimates will be *vastly* higher than one might expect naively for the given sample size n. This is in fact a general problem in any Monte Carlo work in which the probability distribution actually simulated differs considerably from the distribution of interest (see Section 2.3.1). In the case at hand, one expects that the discrepancy between the two distributions will grow *exponentially* as the chain length N gets large. This has been verified numerically by Batoulis and Kremer,[86] who conclude that for large N inversely restricted sampling is inferior to non-reversal simple sampling.

Fraser and Winnik[88] have proposed a generalization of inversely restricted sampling, based on strides with cleverly chosen probabilities. Meirovitch[89–92] has introduced a slightly different generalization (which he calls the "scanning method"), in which the algorithm looks ahead more than one step. But neither of these methods appears to avoid the exponential explosion of weights, although they may reduce it.

2.4.3 Dimerization

The dimerization algorithm[93] is an implementation of the computer scientists' principle of "divide and conquer" (Ref. 94, Section 1.3).[95] To generate an N-step self-avoiding walk, we generate two independent $(N/2)$-step SAWs ("dimers") and attempt to concatenate them. If the result is self-avoiding, we are done; otherwise, we discard the two walks and start again from scratch. This procedure can now be repeated recursively: to generate each of the $(N/2)$-step SAWs, we generate a pair of $(N/4)$-step SAWs and attempt to concatenate them, and so on. For $N \leq$ some cutoff

N_0, we stop the recursion and generate the SAWs by some primitive method, such as non-reversal simple sampling. The dimerization algorithm can thus be written recursively as follows:

```
title Dimerization (recursive version).
function dim(N)
comment This routine returns a random N-step SAW.

if N ≤ N_0 then
        ω ← nrssamp(N)
        return ω

else
    N_1 ← ⌊N/2⌋   (integer part)
    N_2 ← N − N_1
    start:
      ω^(1) ← dim(N_1)
      ω^(2) ← dim(N_2)
      ω ← ω^(1) ∘ ω^(2) (concatenation)
      if ω is not self-avoiding goto start
      return ω
endif
```

It is easy to prove inductively that algorithm `dim` produces each N-step SAW with equal probability, using the fact that the subroutine `nrssamp` does so. It is crucial here that after a failure we discard *both* walks and start again *from scratch*.

A non-recursive description of this same algorithm is given by Suzuki.[93]

Let us analyze[96] the efficiency of the dimerization algorithm under the scaling hypothesis

$$c_N \approx A\mu^N N^{\gamma-1} \tag{2.68}$$

(cf. eq. (2.2)). Let T_N be the mean CPU time needed to generate an N-step SAW by algorithm `dim`. Now, the probability that the concatenation of two random $(N/2)$-step SAWs yields an N-step SAW is

$$p_N = \frac{c_N}{(c_{N/2})^2} \approx B^{-1} N^{-(\gamma-1)}, \tag{2.69}$$

where $B = A/4^{\gamma-1}$. We will need to generate, on average, $1/p_N$ pairs of $(N/2)$-step SAWs in order to get a single N-step SAW; hence

$$T_N \approx BN^{\gamma-1} 2T_{N/2}. \tag{2.70}$$

(We have neglected here the time needed for checking the intersections of the two dimers; this time is linear in N, which, as will be seen shortly, is negligible compared to the time $2T_{N/2}$ for generating the two dimers.) Iterating this k times, where $k = \log_2(N/N_0)$ is the number of levels, we obtain

$$T_N \approx \frac{(2BN^{\gamma-1})^k}{2^{(\gamma-1)k(k-1)/2}} T_{N_0} = C_0 N^{C_1 \log_2 N + C_2}, \qquad (2.71)$$

where

$$C_1 = \frac{\gamma - 1}{2} \qquad (2.72a)$$

$$C_2 = \frac{\gamma - 1}{2} + \log_2(2B) = \frac{5 - 3\gamma}{2} + \log_2 A \qquad (2.72b)$$

and C_0 depends on N_0. Thus, the growth of T_N is slower than exponential in N; but if $\gamma > 1$ (which occurs for $d < 4$) it is faster than any polynomial in N. Fortunately, however, the constants C_1 and C_2 are very small: see Table 2.3. For $d = 2$ (resp. $d = 3$) this means that in practice T_N behaves like $N^{\approx 2-3}$ up to N of order several thousand (resp. several million). In $d = 4$ the behavior is presumably $\sim N^{\tilde{C}_1 \log \log N + \tilde{C}_2}$. For $d \geq 5$ we have $C_1 = 0$ and C_2 only very slightly larger than 1, so dimerization is extraordinarily efficient: a random N-step SAW can be produced in a CPU time that grows only slightly faster than linearly in N.

The efficiency can be improved further by choosing the first step of $\omega^{(2)}$ to avoid reversing the last step of $\omega^{(1)}$. This effectively replaces A by $[(2d-1)/2d]$ A in eq. (2.72b), and hence C_2 by $C_2' \equiv C_2 - \log_2[2d/(2d-1)]$. See Table 2.3 for the effect of this change.

It is an open question whether for $d \leq 4$ there exists any static Monte Carlo algorithm for generating a random N-step SAW (with exactly uniform distribution) in a mean CPU time that is bounded by a polynomial in N. (For discussion of some statistical issues related to dimerization, see Ref. 11, Section 9.3.2.)

Table 2.3 *Efficiency of dimerization algorithm as a function of lattice dimension d. Estimates are obtained by extrapolation of the available counts c_N (Ref. 11, Tables C.1 and C.4). The last line is a rigorous bound valid for all $d \geq 5$ (Ref. 11, p. 172). The case $d = 4$ is somewhat anomalous, as it is believed[14] that $c_N \sim \mu^N (\log N)^{1/4}$ in contrast to the usual power-law behavior* (eq. (2.2)).

d	μ (est.)	γ (est.)	A (est.)	C_1	C_2	C_2'
2	2.638 158 5	43/32	1.177 5	11/64 $\approx$ 0.172	0.72	0.31
3	4.683 906 6	1.162	1.184 5	0.081	1.00	0.74
4	6.772 0	1 ($\times \log^{1/4}$)	—	—	—	—
5	8.838 61	1	1.25	0	1.32	1.17
6	10.878 79	1	1.16	0	1.21	1.09
all $d \geq 5$	—	1	$\leq$ 1.493	0	$\leq$ 1.58	$\leq$ 1.58

Remark A slightly different version of the dimerization algorithm was invented independently by Alexandrowicz[97,98] and used subsequently by many others. In this version, one begins by producing a large initial batch of M-step SAWs, where M is some small number ($\approx N_0$); one then joins randomly chosen pairs to form walks of length $2M$, and so forth. The trouble with this method is that the same subchains of length M, $2M$, $4M$, etc. tend to occur repeatedly in the SAWs produced. Thus, the sample of SAWs produced is both slightly biased (SAWs with two or more identical subchains of length M are favored) and somewhat correlated, but it is difficult to assess these effects quantitatively. Both effects can be reduced by using an extremely large initial batch of SAWs, and by making not too many dimerization attempts per batch, but this is likely to be inefficient or unreliable or both. In my opinion Alexandrowicz' version of dimerization should *not* be used; algorithm `dim` is simpler, more efficient, and—above all—is correct.

2.5 Quasi-static Monte Carlo methods for the SAW

2.5.1 Quasi-static simple sampling

In each of the foregoing static methods, a slight improvement in efficiency can be obtained by working with several different values of N at once, and noticing that "a walk that intersects itself for the first time at the Mth step provides [unbiased] instances of N-step self-avoiding walks for all $N < M$" (Ref. 63, p. 129). The resulting method is quasi-static in our classification: each pass through the algorithm produces a batch of SAWs (of various lengths) which are highly correlated among themselves, although successive batches are independent of each other. Unfortunately—and this seems to be characteristic of quasi-static methods—it appears difficult to estimate quantitatively the degree of correlation between the various SAWs in a given batch, and therefore difficult to estimate the actual statistical efficiency of the method.

2.5.2 Enrichment

One method of generating long SAWs with much less attrition than (non-reversal) simple sampling is the *enrichment technique*[99]: if a walk survives to s steps, then several (t) copies are made of it and each copy is used independently as a starting point for further attempts to add steps; and likewise at $2s$, $3s$, etc. (The same idea was used earlier in Monte Carlo work on neutron-transport problems [Ref. 63, Section 8.2].) The free parameters s and t must be chosen judiciously (see below). This method is quasi-static: the SAWs produced in a single pass through the algorithm (i.e., the progeny of a single s-step SAW) are manifestly correlated, since they all have the same initial s steps and many of them have the same initial $2s$, $3s$, ... steps as well. However, as before, it is difficult to assess this correlation quantitatively. (Indeed, the quasi-static simple sampling method can be considered to be the special case of the enrichment method with $s = t = 1$.)

The enrichment algorithm can be analyzed semi-quantitatively as follows[100] (Ref. 11, Section 9.3.3): Let M_{ns} (a random variable) be the number of ns-step walks that are produced in a single pass through the algorithm; by definition of "single pass" we have $M_s = 1$. (The data produced by successive passes through the algorithm are obviously independent.) A SAW of length ns gives rise to t copies, each of which survives to length $(n+1)s$ with (average) probability*

$$\alpha_n \equiv \frac{c_{(n+1)s}}{(2d-1)^s c_{ns}} \approx \left(\frac{\mu}{2d-1}\right)^s. \tag{2.73}$$

So we can regard $M_s, M_{2s}, \ldots$ as a *branching process*[102] in which M_{ns} is the number of "individuals" alive in the nth generation. We assume that each individual reproduces independently, producing a number of "children" which is a binomial random variable with parameters t and $\alpha = [\mu/(2d-1)]^s$.†

It is easy to see that $\langle M_{ns} \rangle = (t\alpha)^n$. Thus, if $t\alpha < 1$ there is exponential sample attrition, just as in simple sampling. If $t\alpha > 1$, some (although not all) starts lead to an exponential explosion of progeny; this is undesirable, as great computational effort will be expended in producing these samples, but the information contained in them is less than proportional to that effort, because they are highly correlated. The most interesting case is $t\alpha = 1$: every start dies out eventually, but the mean lifetime of a start is infinite. More precisely, it can be shown[102] that

$$\text{Prob}(M_{ns} = 0) \approx 1 - \frac{2}{\sigma^2 n} \tag{2.74}$$

$$\text{Prob}(M_{ns} > \beta n) \approx \frac{2}{\sigma^2 n} e^{-2\beta/\sigma^2} \quad (\beta \geq 0) \tag{2.75}$$

where $\sigma^2 \equiv (1-\alpha)t = t - 1$. Thus, M_{ns} is nonzero only with probability of order $1/n$; but when it is nonzero it is typically of order n.‡ On the other

* For large n, one has (from eq. (2.2)) $\alpha_n \approx (\frac{\mu}{2d-1})^s (1 + \frac{1}{n})^{\gamma-1} (\frac{\mu}{2d-1})^s$ as $n \to \infty$, irrespective of the value of s or of γ. For *even s* this convergence is actually a rigorous theorem.[101]

†This assumption is not really correct: the trouble is that some SAWs will have higher or lower "fertility" than others (i.e., be harder or easier to intersect with); and these fertilities are somewhat *correlated* between different walks in the process, as all these walks share some common segments (the degree of correlation obviously depends on the relative location of the two SAWs in the "family tree"). Nevertheless, this assumption seems to be a reasonable approximation.

‡This prediction is strikingly confirmed by Grishman's[103] empirical observations for $t\alpha \approx 1$ and large but fixed n: "if the enrichment parameters are adjusted to make the total number of generated walks approximately equal to the number of starts, one will find that most of the starts 'die out' before generating any walks, while a few starts each lead to a large number of walks."

hand, it is not hard to see that the mean CPU time per start is (when $t\alpha = 1$) of order n times the CPU time needed to produce a single s-step segment.* So even if we make the over-pessimistic assumption that all the walks in the nth generation of a given start are *perfectly* correlated (and thus carry only as much information as *one* walk), it follows that we can obtain one independent ns-step walk in a CPU time of order n^2 (i.e., $\sim n$ starts each taking a CPU time $\sim n$). This $\sim N^2$ behavior is exceptionally good: it is better than all known static algorithms (e.g., dimerization) in dimension $d \leq 4$, and it is comparable to most of the new dynamic algorithms (see Section 2.6). The enrichment algorithm definitely deserves a systematic theoretical and empirical study.

Remarks

1. The integer t can also be a random variable; in this case the role of $t\alpha$ is played by $\langle t \rangle \alpha$. This generalization is useful in permitting fine-tuning of $\langle t \rangle \alpha$.

2. The $\sim N^2$ obtained here may really be $\sim N^{\gamma+1}$ if one takes account of the corrections to the $\approx$ sign in eq. (2.73) (see also Sections 2.5.3 and 2.6.6.1).

2.5.3 Incomplete enumeration (Redner–Reynolds algorithm)

Incomplete enumeration[104,105] is a quasi-static algorithm that generates a batch of SAWs from the variable-N, variable-x ensemble (eq. (2.18)) with $p = 0$. The idea is to take a standard algorithm for systematically enumerating all self-avoiding walks up to some length N_{max}, and modify it so that it pursues each branch of the "SAW tree" only with some probability $\beta < 1$:

```
title Incomplete enumeration (recursive version).
subroutine incenum (ω, β, N_max)
comment This routine performs an incomplete enumeration of the SAW tree
        beginning at ω, with probability parameter β, up to maximum length N_max.

append ω to output list
if |ω| ≥ N_max return
for i = 1 to 2d do
        U ← random number uniformly distributed on [0, 1]
        if U ≤ β then
            ω' ← one-step extension of ω in direction i
            if ω' is self-avoiding incenum (ω', β, N_max)
        endif
enddo
return
```

*Here I assume that one tests self-avoidance using a method that takes a CPU time of order 1 per added step (see Section 2.7.1.2).

To perform an incomplete enumeration of the SAW tree beginning at the "root", one simply invokes incenum($\{0\}, \beta, N_{max}$), where $\{0\}$ is the zero-step walk at the origin. Typically one chooses $\beta < \beta_c \equiv 1/\mu$; in this case it is safe to set $N_{max} = \infty$. (If one sets $N_{max} = \infty$ when $\beta > \beta_c$, then with nonzero probability the algorithm will run forever!)

A non-recursive implementation of incomplete enumeration can be found in Ref. 11, Section 9.3.3.

Incomplete enumeration is very closely related to enrichment: indeed, it is nearly identical to the enrichment algorithm in which $s = 1$ and t is a binomial random variable with parameters $2d$ and β. The only difference is that the enrichment algorithm performs "sampling with replacement" among the various one-step extensions of ω (i.e., the same ω' could occur more than once), while incomplete enumeration performs "sampling without replacement" (each ω' occurs at most once). So one expects incomplete enumeration to be slightly more efficient, in that the resulting batch of SAWs will be less correlated. (Indeed, there may even be some *anti*correlation arising from the sampling without replacement: this certainly occurs if, for example, $\beta = 1$. On the other hand, for $\beta \ll 1$ one still has strong positive correlation for the same reason as in the enrichment algorithm: many walks share the same initial segments.)

The CPU time for one invocation of incomplete enumeration is obviously proportional to the total number of walks encountered during the enumeration.* On the average this is

$$Z(\beta) \equiv \sum_{N=0}^{\infty} \beta^N c_N \sim (1 - \beta\mu)^{-\gamma} \sim \langle N \rangle^{\gamma}. \tag{2.76}$$

On the other hand, it is reasonable to guess that, as in enrichment,

$$\text{Prob(at least one } N\text{-step SAW is produced)} \sim 1/N \quad \text{for } N \sim \langle N \rangle. \tag{2.77}$$

So if we make the over-pessimistic assumption that all the walks of length N in a given batch are *perfectly* correlated (and thus carry only as much information as *one* walk), it follows that we can obtain one independent N-step walk in a CPU time of order $N^{\gamma+1}$ (i.e., choose β so that $\langle N \rangle \sim N$, then make $\sim N$ starts each taking a CPU time $\sim N^{\gamma}$).

This differs from the estimate made previously for enrichment, because γ is in general slightly larger than 1 (namely 43/32 in dimension $d = 2$, ≈ 1.16 in $d = 3$, and 1 in $d \geq 4$). However, I suspect that the correct answer in both cases is $N^{\gamma+1}$, and that this would be obtained also in the branching-process analysis if one were to take account of the n-dependence of α_n [i.e. $\alpha_n \approx (\frac{\mu}{2d-1})^s (1 + \frac{1}{n})^{\gamma+1}$]. The incomplete-enumeration algorithm is also closely related to the slithering-tortoise algorithm discussed in Section 2.6.6.1, and the same $\langle N \rangle^2$ versus $\langle N \rangle^{\gamma+1}$ issue arises there.

*Here I assume that one tests self-avoidance using an $O(1)$ algorithm (see Section 2.7.1.2).

In any case, the incomplete-enumeration algorithm is potentially quite efficient, and deserves a systematic theoretical and empirical study.

2.6 Dynamic Monte Carlo methods for the SAW

In this section I discuss dynamic Monte Carlo methods for generating SAWs in various ensembles. I emphasize that the stochastic dynamics is merely a numerical algorithm, whose goal is to provide statistically independent samples from the desired distribution π in the smallest CPU time possible. *It need not correspond to any real "physical" dynamics!* Indeed, the most efficient algorithms typically make non-local moves which would be impossible for real polymer molecules.

2.6.1 General considerations

This subsection contains some exceedingly pedantic—but I hope useful—general considerations on dynamic Monte Carlo algorithms.

The study of a Monte Carlo algorithm can be divided into three stages:

- *specification* of the algorithm;
- verification of the algorithm's *validity*; and
- study (by mathematical, heuristic and/or numerical means) of the algorithm's *efficiency*.

Let us consider these aspects in turn:

To *specify* a dynamic Monte Carlo algorithm, we must specify three things:

1. The state space S.
2. The desired equilibrium measure π.
3. The transition probability matrix $P = \{p(\omega \to \omega')\}$.

Here S and π specify the *model* that we wish to study, while P specifies the *numerical method* by which we propose to study it.* With regard to P, it is useful to subdivide the issue further, by specifying successively:

3a. The set of allowed *elementary moves*, i.e., the transitions $\omega \to \omega'$ for which $p(\omega \to \omega') > 0$.

3b. The *probabilities* $p(\omega \to \omega')$ for the allowed elementary moves.

*Of course, the choice of P trivially determines S; and if P is ergodic (as it should be if the algorithm is to be valid), then it also determines π. But it is nevertheless useful conceptually to distinguish the three ingredients, which typically correspond to the chronological stages in inventing a new Monte Carlo algorithm.

After specifying the algorithm, we must prove the algorithm's *validity*. As discussed in Section 2.3.2, this has two parts:

(A) We must prove *ergodicity (irreducibility)*, i.e., we must prove that we can get from any state $\omega \in S$ to any other state $\omega' \in S$ by some finite sequence of allowed elementary moves.
(B) We must prove the *stationarity* of π with respect to P.

Ergodicity is a combinatorial problem; it depends only on the set of allowed elementary moves, not on their precise probabilities (this is the motivation for distinguishing sub-questions 3a and 3b). For many SAW algorithms, the proof of ergodicity is highly nontrivial. Indeed, in several embarrassing cases, an algorithm was used for many years before it was realized to be *non*-ergodic. As for stationarity, this is usually (though not always) verified by showing that P satisfies detailed balance for π (eq. (2.57)) or is built up out of constituents $P_1, \ldots, P_n$ which do so (as discussed at the end of Section 2.3.2).

Once the algorithm is known to be valid, we can investigate its *computational efficiency*, as measured by the amount of CPU time it takes to generate one "statistically independent" sample from the distribution π. This study also has two parts:

(i) Investigate the *autocorrelation times* $\tau_{int,A}$ (for observables A of interest) and τ_{exp}.
(ii) Investigate the *computational complexity* of the algorithm, i.e., the mean CPU time per iteration (henceforth denoted T_{CPU}).

The CPU time per "statistically independent" sample is then $2\tau_{int,A}T_{CPU}$, and this provides a criterion for comparing algorithms: for given S and π, the best algorithm is the one with the smallest product $\tau_{int,A}T_{CPU}$. (This criterion may of course depend on the observable A.) Both aspects (i) and (ii) should be studied by all the methods at our disposal: rigorous mathematical analysis, heuristic physical reasoning, and numerical experimentation. Since we are primarily interested in *long* self-avoiding walks, we will emphasize the asymptotic behavior of $\tau_{int,A}$, τ_{exp} and T_{CPU} as $N \to \infty$ (or $\langle N \rangle \to \infty$ for the variable-N algorithms). Usually these quantities behave as some power of N or $\langle N \rangle$, in which case we can identify a *dynamic critical exponent*.

Remarks

1. We are here always considering the relaxation time of the autocorrelation functions *in equilibrium*, as it is this relaxation time which is of primary importance for determining the statistical accuracy of the Monte Carlo method (see Section 2.3.2). Some studies[106–109] have focused instead on the relaxation *to equilibrium* from a highly nonequilibrium initial state; but as Kranbuehl and Verdier[110] have pointed out, this quantity may well

have a strictly smaller dynamic critical exponent. An analogous situation arises in the Glauber dynamics for the Ising model.[111]*

2. In this chapter we always measure time in units of attempted elementary moves. Much of the literature on dynamic SAW models uses a time scale of attempted elementary moves *per bead*; autocorrelation times expressed in this way should be multiplied by N (actually $N + 1$) before comparing them to the present paper.

2.6.2 *Classification of moves*

The elementary moves in a SAW Monte Carlo algorithm can be classified according to whether they are

- N-conserving or N-changing
- endpoint-conserving or endpoint-changing
- local or non-local

Obviously, fixed-N algorithms must use only N-conserving moves, while variable-N algorithms are free to use both N-conserving and/or N-changing moves (and indeed *must* use some of the latter in order to satisfy ergodicity). An analogous statement holds for fixed-x and variable-x algorithms with regard to endpoint-conserving and endpoint-changing moves. The distinction between local and non-local moves will be explained later.

Within these limitations, the elementary moves to be discussed below can be mixed more or less freely to make "hybrid" algorithms. The art is, of course, to find a *useful* combination. (A cocktail can be made by mixing any set of liqueurs in any proportions, but there is no guarantee that the resulting concoction will taste good! In particular, it may or may not taste better than the individual ingredients taken separately.) Thus, a hybrid algorithm is useful when its performance is superior to that of either of its "pure" components, in the sense of having a smaller product $\tau_{int,A} T_{CPU}$. Roughly speaking, this occurs when the "slow modes" under one type of move are speeded up by the other type of move. (An extreme case of this is when the combined algorithm is ergodic while both pure algorithms are nonergodic.)

Note that a hybrid algorithm always contains one or more free parameters—namely the relative probabilities of the different types of moves—which can be tuned to optimize the computational efficiency. (Often this

*In principle, the *asymptotic* behavior as $t \to \infty$ of the relaxation from nonequilibrium is controlled by the same τ_{exp} as the autocorrelation in equilibrium. But to observe this asymptotic relaxation in practice would require enormous statistics (e.g., tens of thousands of repeat runs). Most of these studies have focused instead on the initial relaxation—e.g., the time to relax to within $1/e$,[106,109] the time(s) in a phenomenological fit,[107,108] or the area under the relaxation curve[111–113]—and all of these are likely to exhibit a smaller dynamic critical exponent than either τ_{exp} or $\tau_{int,A}$.

requires a trade-off between autocorrelation time and computational complexity, particularly if one move is local while the other is non-local.) This optimization is a nontrivial problem, and it must be approached systematically if the results are to be of any use. (The behavior of the algorithm at some *arbitrarily chosen* set of parameter values may be completely misleading.) *A priori* there are three possibilities:

(a) One of the pure algorithms is superior to any nontrivial hybrid.
(b) The optimal hybrid is better than either of the pure algorithms, but only by a bounded factor (e.g., 2 or 10 or whatever).
(c) The optimal hybrid has a better dynamic critical exponent than either of the pure algorithms, so that its superiority factor grows without bound as $N \to \infty$.

The most interesting hybrid algorithms are of course those of type (c). But those of type (b) should not be sniffed at, if the gain is large enough (and the human effort required to find an optimal or nearly-optimal mixture is not too large).

Most of the elementary moves to be discussed below have the property that the resulting walk ω' is not *guaranteed* to be self-avoiding. Therefore, it is necessary to *test* whether ω' is self-avoiding (see Section 2.7.1.2 for methods for doing this). If ω' is self-avoiding, then the proposed move is accepted; otherwise, it is rejected, and the old walk is counted once again in the sample. This procedure can be understood as the Metropolis criterion for the energy function

$$E(\omega) = \begin{cases} 0 & \text{if } \omega \text{ is self-avoiding} \\ +\infty & \text{otherwise} \end{cases} \tag{2.78}$$

One final remark: Up to now we have adopted the convention that all walks start at the origin ($\omega_0 = 0$) unless specified otherwise. However, some of the moves to be discussed below may alter the initial site of the walk so that it is no longer at the origin. One could, of course, imagine that the resulting walk is then translated back to the origin. But a more convenient approach is to consider all translates of a given SAW to be *equivalent*; then we need not worry where the initial point is.* This is how the algorithms are most efficiently implemented in practice. (Every once in a while one should translate the walk back to the origin, just as a precaution against integer overflow.)

---

*In mathematical language, we redefine $\mathcal{S}_N$ to be the set of *equivalence classes* of N-step SAWs modulo translation. (And likewise for the other state spaces.) Note also that the observables (2.4)–(2.6) are unaffected by translation of the walk, hence they are well-defined on equivalence classes.

2.6.3 *Examples of moves*

2.6.3.1 *Local N-conserving moves*

A local move is one that alters only a few consecutive sites ("beads") of the SAW, leaving the other sites unchanged. Otherwise put, a local move excises a small piece from the original SAW and splices in a new local conformation in its place. (Of course, it is always necessary to verify that the proposed new walk is indeed self-avoiding.) In this subsection we concentrate on *N-conserving* local moves, i.e., those in which the excised and spliced-in subwalks have the same number of beads.

More precisely, let ω and ω' be N-step SAWs, and let k be an integer ≥ 1. We say that *ω and ω' are interconvertible by a k-bead move* if there exists an index i_{min} $(0 \leq i_{min} \leq N - k + 1)$ such that $\omega_i = \omega_i'$ for all i except possibly $i = i_{min}, i_{min} + 1, \ldots, i_{min} + k - 1$. (We shall also assume that $\omega_{i_{min}} \neq \omega'_{i_{min}}$ and $\omega_{i_{min}+k-1} \neq \omega'_{i_{min}+k-1}$, since otherwise ω and ω' would be interconvertible also by some move of *less than* k beads.) If $i_{min} = 0$ or $N - k + 1$, we call this an *end-group* k-bead move; otherwise we call it an *internal* k-bead move. Clearly, internal moves are endpoint-conserving, while end-group moves are not.

In Fig. 2.1 we show all the possible one-bead moves (on a hypercubic lattice). Move A is a "one-bead flip" (also called "kink-jump"); it is the only one-bead internal move. Moves B and C are end-bond rotations.

In Fig. 2.2 we show all the possible *internal* two-bead moves. Move D is a "180° crankshaft". Move E is a "90° crankshaft"; of course it is possible only in dimension $d \geq 3$. Move F is a "two-bead L-flip". Move G permutes three successive mutually perpendicular steps (which lie along the edges of a cube); again this is possible only in dimension $d \geq 3$.

We leave it to the reader to construct the list of two-bead end-group moves. There are numerous three-bead moves; a few interesting ones are shown in Fig. 2.3.

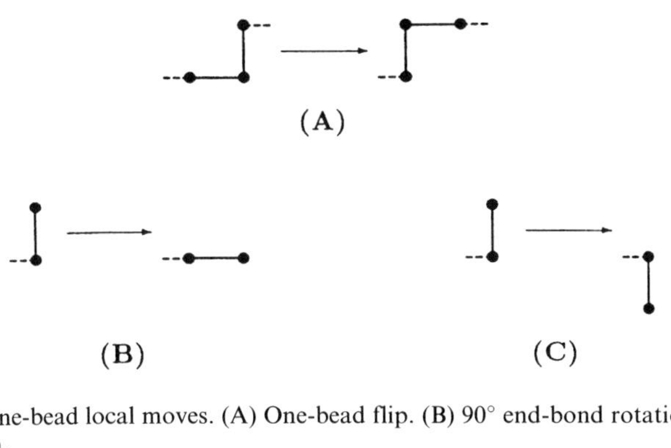

Fig. 2.1 All one-bead local moves. (A) One-bead flip. (B) 90° end-bond rotation. (C) 180° end-bond rotation.

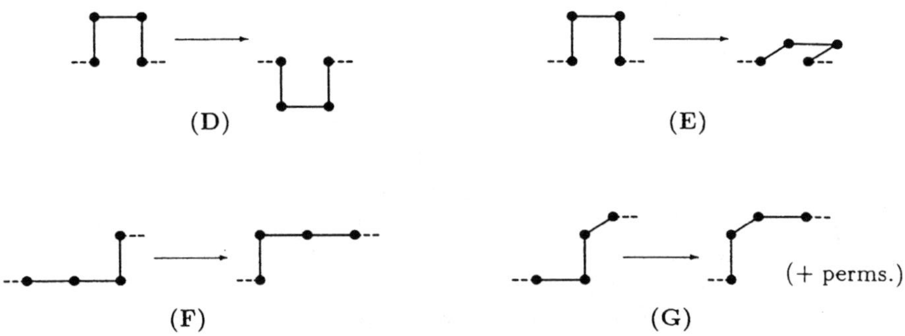

Fig. 2.2 All internal two-bead local moves. (D) 180° crankshaft. (E) 90° crankshaft ($d \geq 3$ only). (F) Two-bead L-flip. (G) Cube permutation ($d \geq 3$ only).

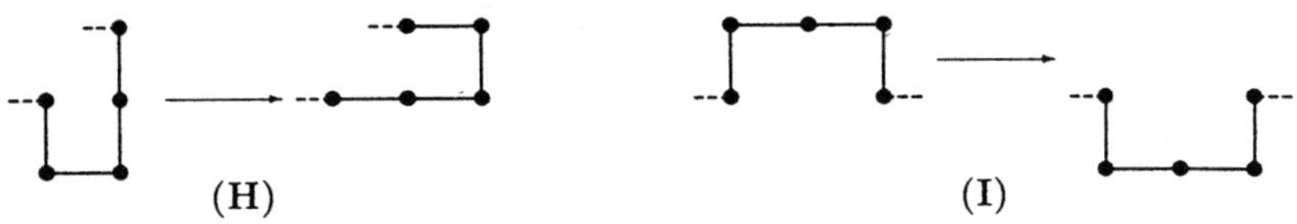

Fig. 2.3 Some selected internal three-bead local moves. (H) Three-bead J-flip. (I) Three-bead 180° crankshaft.

2.6.3.2 Bilocal N-conserving moves

A *bilocal* move is one that alters *two* disjoint small groups of consecutive sites (or steps) of the walk; these two groups may in general be very far from each other. One *trivial* way of making an N-conserving bilocal move is to make two independent (nonoverlapping) N-conserving local moves. Here we are interested in the *nontrivial* N-conserving bilocal moves, i.e., those in which one subwalk loses sites and the other subwalk gains them. Instead of formalizing the concept, let us simply give some examples:

- The *slithering-snake* (or *reptation*) move, which deletes a bond from one end of the walk and appends a new bond (in an arbitrary direction) at the other end (Fig. 2.4).
- The *kink transport* move, which deletes a kink at one location along the walk and inserts a kink (in an arbitrary orientation) at another location (Fig. 2.5).
- The *kink-end reptation* move, which deletes a kink at one location along the walk and appends two new bonds (in arbitrary directions) at one of the ends of the walk (Fig. 2.6→).
- The *end-kink reptation* move, which deletes two bonds from one of the ends of the walk and inserts a kink (in an arbitrary orientation) at some location along the walk (Fig. 2.6←).

Fig. 2.4 The slithering-snake (reptation) move. Head of the walk is indicated by a cross. Dashed lines indicate the proposed new step (resp. the just-abandoned old step).

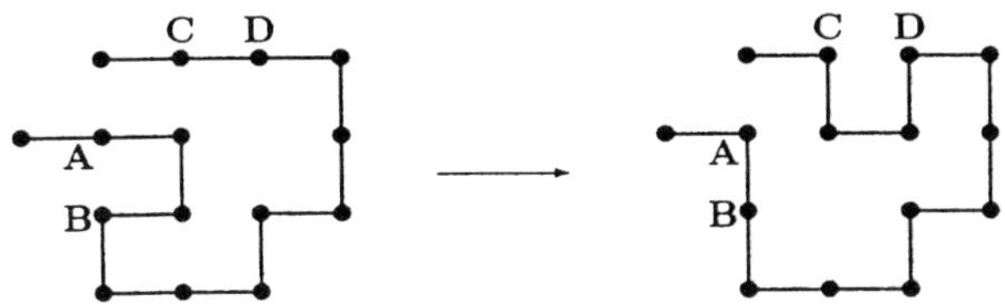

Fig. 2.5 The kink-transport move. A kink has been cleaved from AB and attached at CD. Note that the new kink is permitted to occupy one or both of the sites abandoned by the old kink.

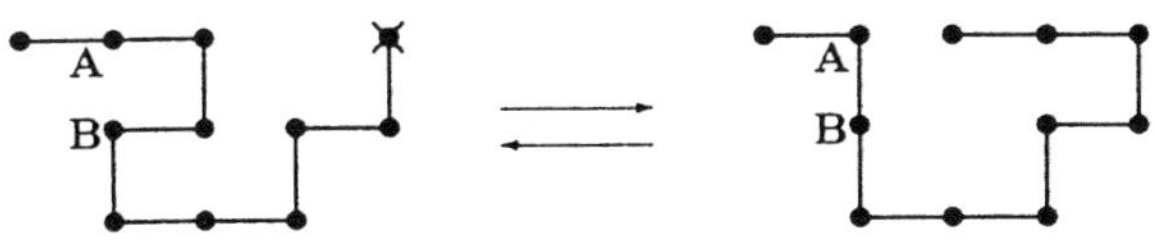

Fig. 2.6 The kink-end reptation (→) and end-kink reptation (←) moves. In →, a kink has been cleaved from AB and two new steps have been attached at the end marked with a cross. Note that the new end steps are permitted to occupy one or both of the sites abandoned by the kink.

2.6.3.3 Non-local N-conserving moves

Here we move definitively out of the realm of systematic classification and into the realm of ingenuity. The possibilities for non-local moves are almost endless, but it is very difficult to find one which is *useful* in a Monte Carlo algorithm. There are two reasons for this: Firstly, since a non-local move is very radical, the proposed new walk usually violates the self-avoidance constraint. (If you move a large number of beads around, it becomes very likely that *somewhere* along the walk two beads will collide.) It is therefore a nontrivial problem to invent a non-local move whose acceptance probability does not go to zero too rapidly as $N \to \infty$. Secondly, a non-local move usually costs a CPU time of order N (or in any case N^p with $p > 0$), in contrast to order 1 for a local or bilocal move. It is nontrivial to find moves whose effects justify this expenditure (by reducing $\tau_{int,A}$ more than they increase T_{CPU}).

The following paragraphs are, therefore, nothing more than a brief listing of those non-local moves which, *as of 1993*, have been demonstrated to be useful for Monte Carlo algorithms in at least one of the SAW ensembles.

Tomorrow someone could invent a new and even better non-local move; I hope that some reader of this chapter will be moved (pardon the pun) to do so. This is a wide-open area of research.

Only two broad types of useful non-local moves are known at present: *pivot* moves, and *cut-and-paste* moves.

In a *pivot* move, we choose some site ω_k along the walk as a pivot point, and apply some symmetry operation of the lattice (e.g., rotation or reflection) to the part of the walk subsequent to the pivot point, using the pivot point as the origin (Fig. 2.7). That is, the proposed new walk is

$$\omega_i' = \begin{cases} \omega_i & \text{for } 0 \le i \le k \\ \omega_k + g(\omega_i - \omega_k) & \text{for } k + 1 \le i \le N \end{cases} \tag{2.79}$$

where g is the chosen symmetry operation.

In a *cut-and-paste* move, we cut the walk into two or more pieces, invert and/or reflect and/or permute the pieces, and finally reassemble the pieces. For example, one may invert the subwalk $\omega^{[k,l]}$ (Fig. 2.8):

$$\omega_i' = \begin{cases} \omega_k + \omega_l - \omega_{k+l-i} & \text{for } k \le i \le l \\ \omega_i & \text{otherwise} \end{cases} \tag{2.80}$$

More generally, one may apply to $\omega^{[k,l]}$ any lattice symmetry operation that leaves ω_k and ω_l invariant; or any lattice symmetry operation that interchanges ω_k and ω_l, followed by the inversion (2.80). Such symmetry operations exist if $\omega_l - \omega_k$ lies on certain axis or diagonal hyperplanes.

Empirically, the pivot and cut-and-paste moves have a reasonable acceptance probability even at large N, of order $\sim N^{-p}$ with p typically of order 0.1–0.4. The heuristic reason is that these moves conserve most of the chain structure (and hence its self-avoidance), except near the pivot point or cut point(s). As a result, the acceptance probability for a move involving a *single* pivot or cut point is roughly similar to that encountered when concatenating

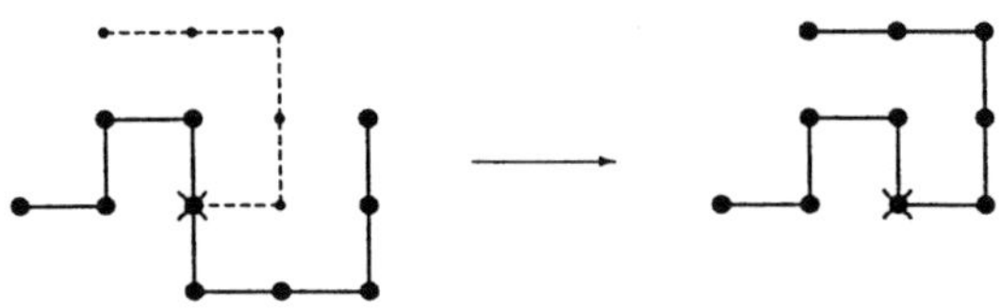

Fig. 2.7 The pivot move (here a +90° rotation). The pivot site is indicated with a cross. Dashed lines indicate the proposed new segment.

Fig. 2.8 Inversion of the subwalk $\omega^{[k,l]}$.

two independent $(N/2)$-step walks, namely $c_N/c_{N/2}^2 \sim N^{-(\gamma-1)}$. By the same reasoning, a move with n cut points might be expected to have an acceptance probability roughly like $\sim N^{-n(\gamma-1)}$.

2.6.3.4 *Local N-changing moves*

Recall that a local move is one that excises a small piece from the original SAW and splices in a new local conformation in its place. An *N-changing* local move has the freedom to splice in a piece with fewer or more sites than the original piece.

The simplest *internal* local N-changing moves are kink insertion and kink deletion (J and K in Fig. 2.9); these have $\Delta N = +2$ and $\Delta N = -2$, respectively. The simplest *end-group* local N-changing moves are end-bond addition and end-bond deletion (L and M in Fig. 2.9); these have $\Delta N = +1$ and $\Delta N = -1$, respectively.

Note that each local N-changing move is simply "one half" of some bilocal move.

2.6.4 *Fixed-N, variable-x algorithms*

2.6.4.1 *Local algorithms*

Historically the earliest dynamic Monte Carlo algorithms for the SAW were local N-conserving algorithms: they date back to the work of Delbrück[114] and Verdier and Stockmayer[106], both published in 1962. During the subsequent two decades, numerous variants on this theme were proposed (see Table 2.4). Most of these algorithms employ some subset of moves A–F from Figs 2.1 and 2.2.

Unfortunately, all local N-conserving algorithms have a fatal flaw: they are *nonergodic*. The nonergodicity of the Verdier–Stockmayer algorithm was noticed as early as 1968,[107,115] but this fact did not seem to deter usage of the algorithm, or to provoke serious discussion about the ergodicity or nonergodicity of related algorithms (one exception is Ref. 108). Finally, in 1987 Madras and Sokal[125] proved that *all* local N-conserving algorithms for SAWs in dimensions $d = 2, 3$ are nonergodic for sufficiently large N.*

Fig. 2.9 Some local N-changing moves. (J) Kink insertion ($\Delta N = +2$). (K) Kink deletion ($\Delta N = -2$). (L) End-bond addition ($\Delta N = +1$). (M) End-bond deletion ($\Delta N = -1$).

*For algorithms based on moves of k or fewer beads, nonergodicity arises in dimension $d = 2$ for all $N \geq 16k + 63$, and for quite a few smaller N as well (Ref. 125, Theorem 1).

Table 2.4 *Some local N-conserving algorithms.*

Scheme	References	Elementary moves (see Figs 2.1 and 2.2)	Autocorrelation time τ (in elementary moves)
Verdier–Stockmayer	106, 115, 116	A, B	$\sim N^{\approx 3+2\nu}$ (?)
Modified Verdier–Stockmayer	109, 116	A, B, C	$\sim N^{\approx 3+2\nu}$ (?)
Heilmann II	107, 108, 117	A, B, E	$\sim N^{\approx 2+2\nu}$ (?)
Birshtein *et al.*/ Heilmann–Rotne 3	108, 118	A, B, D	$\sim N^{\approx 2+2\nu}$ (?)
Taran–Stroganov	119	A, B, D, E	$\sim N^{\approx 2+2\nu}$ (?)
Verdier–Kranbuehl	120, 121	B, D, F	$\sim N^{\approx 2+2\nu}$ (?)
Kranbuehl–Verdier	110, 122, 123	A, B, D, F	$\sim N^{\approx 2+2\nu}$ (?)
Delbrück	114	most one- and two-bead	$\sim N^{\approx 2+2\nu}$ (?)
Meirovitch	124	most three-bead	$\sim N^{\approx 2+2\nu}$ (?)
Lal *et al.*	126, 127	all three-bead	$\sim N^{\approx 2+2\nu}$ (?)
Gény–Monnerie/ Kremer *et al.*	128–131	some two- and three-bead (tetrahedral lattice)	$\sim N^{\approx 2+2\nu}$ (?)

Furthermore, they proved that for large N, each ergodic class contains only an *exponentially small fraction* of the SAW configuration space. (These results are probably true also in dimension $d \geq 4$, but have not yet been proven.)

This nonergodicity is in fact quite easy to see: consider the *double cul-de-sac* configuration shown in Fig. 2.10(a). This SAW is completely "frozen" under elementary moves A, B, D and F: it cannot transform itself into any other state, nor can it be reached from any other state. It follows that the original Verdier–Stockmayer algorithm[106,115] and most of its generalizations[107,108,110,117–120,122,123] are nonergodic (in $d = 2$) already for $N = 11$. If move C is allowed, then the configuration of Fig. 2.10(a) is no longer frozen, but that of Fig. 2.10(b) still is. Thus, any algorithm based on moves A–D and F is nonergodic (in $d = 2$) for $N = 15$. If two-bead end-group moves are allowed, then the configuration of Fig. 2.10(b) is no longer frozen, but that of Fig. 2.10(c) still is. Thus, any algorithm based on one-bead and two-bead moves is nonergodic (in $d = 2$) for $N = 19$.

When three-bead moves are allowed, it is not sufficient simply to make the double cul-de-sac taller. Indeed, *any* double cul-de-sac of the kind shown in Fig. 2.10, no matter how tall, can be unfolded into a straight rod by repreated use of the moves A, B and H. (The reader might find it amusing to work out the required sequence of moves.) But only one additional trick is needed: by folding the double cul-de-sac once more, as in Fig. 2.11, a

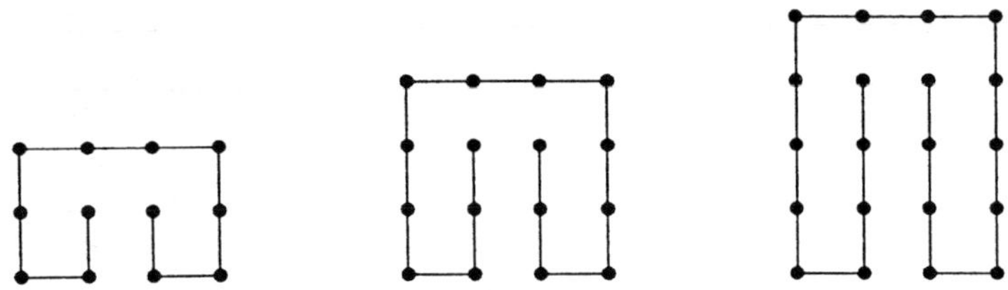

Fig. 2.10 Some "double cul-de-sac" configurations, frozen in the Verdier–Stockmayer algorithm and its generalizations.

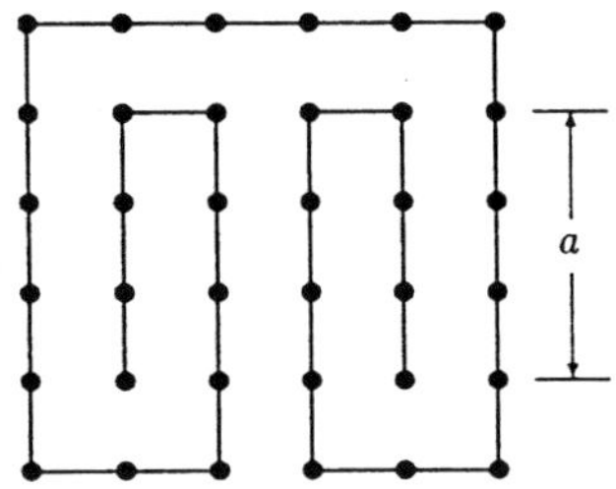

Fig. 2.11 A "folded double cul-de-sac" configuration, drawn here for $a = 3$. Such configurations are frozen under all k-bead local N-conserving moves with $k \leq a$.

frozen configuration can be obtained for the k-bead algorithm for arbitrary k. This is the idea behind the Madras–Sokal proof.

The nonergodicity of the Verdier–Stockmayer algorithm due to double culs-de-sac was noticed already by Verdier[115] in 1969.

An entirely different type of nonergodicity arises in dimension $d = 3$ (and only there) because of the possibility of *knots*, as was first pointed out by Heilmann[107] in 1968. The simplest knotted configuration* is shown in Fig. 2.12: it has $N = 18$, and although it is not completely "frozen", it never-

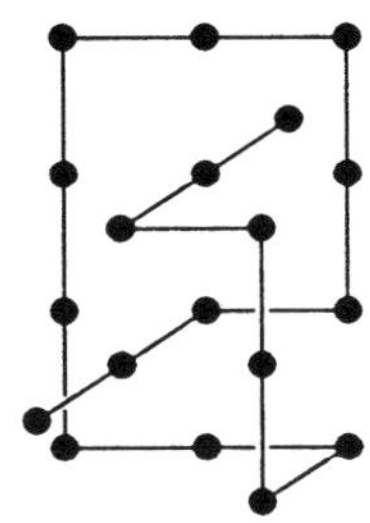

Fig. 2.12 A "knot" that cannot be deformed to a straight rod using moves A–F.

*This is not a knot in the true topological sense, since true knots can occur only for *closed* walks (ring polymers). We are therefore here using the word "knot" in a loose sense to describe the general shape of the chain.

theless cannot be deformed to a straight rod using moves A–F. It is likely that analogous knots can be constructed for the k-bead algorithm for arbitrary k.

For additional historical discussion, along with extensive discussion of the practical implications of nonergodicity, see Ref. 125, Sections 3 and 4.

Since the local N-conserving algorithms are nonergodic, their autocorrelation times τ_{exp} and $\tau_{int,A}$ are by definition $+\infty$: the simulation *never* converges to the correct ensemble average. Nevertheless, we can imagine starting the simulation in some particular configuration (e.g., a straight rod), and ask how long it takes to explore adequately the ergodic class of that particular configuration. Here is a heuristic argument[71] giving a crude estimate of τ for such a restricted simulation:

Let us consider, as a typical quantity, the evolution of the squared radius of gyration of the chain. At each elementary move, a few beads of the chain move a distance of order 1; it follows from eq. (2.5c) that the change in R_g^2 is of order $N^{\nu-1}$. In order to traverse its equilibrium distribution, R_g^2 must change by something of order its standard deviation, which is $\sim N^{2\nu}$. Assuming that R_g^2 executes a random walk, it takes about $(N^{2\nu}/N^{\nu-1})^2 \sim N^{2+2\nu}$ elementary moves for this to occur. So we predict $\tau \sim N^{2+2\nu}$.* This basic estimate ought to be applicable to the dynamics of ordinary random walks (ORWs, free-flight chains) and non-reversal random walks (NRRWs) as well as SAWs.

This reasoning can easily be converted into a more-or-less rigorous proof of the *lower bound* $\tau_{exp} \gtrsim \tau_{int,R_g^2} \geq \text{const} \times N^{2+2\nu}$ (see Ref. 132, p.51 for a slight variant). However, the true τ could be significantly *larger* than this estimate if there are modes which relax essentially more slowly (i.e., with a larger dynamic critical exponent) than the radius of gyration. It also could be wrong if there are, in a particular model, special conservation laws or quasi-conservation laws which inhibit the relaxation of the radius of gyration; this indeed can occur (see the next paragraph). Finally, it is not clear whether the result $\tau \sim N^{2+2\nu}$ is exact—and if so, for *which* τ—or is merely a reasonable first approximation. Usually dynamic critical exponents are *not* expressible in terms of static ones, except for trivial (Gaussian-like) models. It is thus probable that $\tau \sim N^{2+2\nu}$ is not exact for the SAW, although it may be exact for the NRRW; it is known to be exact for the ORW, at least in the Verdier–Stockmayer dynamics.[133–140] The numerical evidence is inconclusive (see references in Ref. 71).

The Verdier–Stockmayer (pure one-bead) and Kranbuehl–Verdier (pure two-bead) algorithms for the SAW or NRRW (but not the ORW) have peculiar conservation laws which inhibit the relaxation of the chain, thereby

*The same order of magnitude is obtained if one considers the evolution of the end-to-end distance vector, the vector from the center-of-mass to an endpoint, or any similar global quantity.

increasing the relaxation time above the usual $N^{2+2\nu}$ by at least an extra factor of N.[116, 121*] These conservation laws can be broken by mixing one-bead and two-bead moves.

2.6.4.2 *Bilocal algorithms*

The *slithering-snake (reptation) algorithm*[†] was invented independently by Kron[144,145] and Wall and Mandel.[146,147‡] The elementary moves of this algorithm are "slithering" motions: one step is appended at one end of the walk, and one step is simultaneously deleted from the other end (Fig. 2.4). Such a move is N-conserving and *bi*local (but not local).

Unfortunately, the reptation algorithm is *nonergodic*, as was observed already by the original inventors.[145,146] In this algorithm, frozen configurations occur when both ends of the walk are trapped in culs-de-sac. The simplest such configuration is shown (for $d = 2$) in Fig. 2.13, and has $N = 14$.[§]

As with the local N-conserving algorithms, we can inquire about the autocorrelation times of a simulation *within* a particular ergodic class. A plausible heuristic argument[147] suggests that $\tau \sim N^2$. This is because the SAW transforms itself by random back-and-forth slithering along the chain; after $\sim N^2$ moves, this slithering will have carried it N steps, and thus all the original bonds of the chain will have disappeared and been

---

*Hilhorst, Deutch and Boots[116,121] have given a very beautiful analysis of these dynamics, using a mapping onto a gas of one-dimensional random walkers interacting via exclusion. It would be interesting to reexamine these questions using mathematical results on this latter model[141–143]; it might be possible to compute the exact asymptotic behavior, at least in the NRRW case.

†The "slithering-snake (reptation) *algorithm*" should not be confused with the "reptation *conjecture*" of de Gennes (Ref. 5, Section VIII.2), who argues that the *most rapid* modes of polymer diffusion in the true *physical* dynamics (in a melt consisting of many entangled polymer chains) are the slithering modes.

‡There are at least three variants of the slithering-snake scheme. In the first version of Wall and Mandel (Ref. 146, pp. 4592–4593), the "head" and "tail" of the chain are interchanged only when an attempted move is rejected; this satisfies the stationarity condition but not the detailed-balance condition. (In fact, one step of the algorithm *followed by a head-tail interchange* satisfies detailed balance.[148] This is analogous to the situation in the Hamilton/Langevin hybrid algorithm for spin models [Ref. 37, p. 237, footnote 18].) In the second version of Wall and Mandel (Ref. 146, top of p. 4594), adopted also by Webman *et al.*,[149] the "head" and "tail" are chosen anew at each cycle, randomly with 50%–50% probability; this satisfies detailed balance. The original algorithm proposed by Kron[144] is much more complicated than either of the preceding versions; it appears to satisfy stationarity but not detailed balance.

§The superficial resemblance between Figs 2.10, 2.11, and 2.13 is, however, very misleading: the nonergodicities in the two types of algorithm are of radically different natures. In the local N-conserving algorithms, nonergodicity is caused by the occurrence of a frozen conformation *anywhere* along the walk. In the reptation algorithm, by contrast, nonergodicity occurs only if both of the *ends* of the walk are trapped in culs-de-sac. It turns out that the ergodic class of the straight rod in the reptation algorithm contains at least a fraction $N^{-(\gamma-1)/2}$ of the SAW configuration space, whereas in the local algorithms it contains only an exponentially small fraction. See Ref. 125, Section 3 for further discussion.

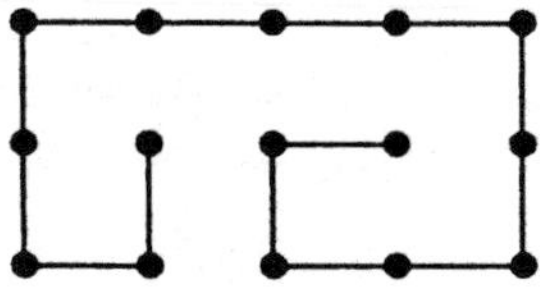

Fig. 2.13 A "double cul-de-sac" configuration which is frozen in the reptation algorithm.

replaced by random new ones. (Alternatively, R_g^2 changes by an amount of order $N^{2\nu-1}$ per elementary move; thus, it takes $\sim N^2$ elementary moves for R_g^2 to traverse a distribution of width $\sim N^{2\nu}$.) It is easy to see that this argument is exact for ORWs and NRRWs; it is not clear whether it is exact or merely approximate for SAWs.

The nonergodicity of the reptation algorithm can be cured by adjoining additional bilocal moves to the algorithm.[150] Indeed, the following bilocal (or mixed local/bilocal) algorithms are known to be ergodic:

- reptation + kink-end and end-kink reptation[150]
- kink-end and end-kink reptation + 90° end-bond rotation[148]
- reptation + kink transport + one-bead flip, at least in dimension $d = 2$[151]
- kink transport + one-bead flip + 90° and 180° end-bond rotations, at least in $d = 2$[151]

All of these algorithms deserve systematic study, in particular of their dynamic critical behavior. One would like to know whether they all lie in the same dynamic universality class, and whether the conjecture $\tau \sim N^2$ is exact, approximate or wrong.

An alternative way of making the reptation algorithm ergodic is to switch to the variable-N ensemble and introduce $\Delta N = \pm 1$ moves (L and M in Fig. 2.9), as proposed by Kron *et al.*[145] But once one does this, there is no great reason to retain the "slithering-snake" moves; one can just as well use *only* the $\Delta N = \pm 1$ moves. This leads to the slithering-tortoise (Berretti–Sokal) algorithm (see Section 2.6.6.1).

2.6.4.3 Pivot algorithm

The *pivot algorithm* has a curious history. It was invented by Lal[152] in 1969, and then promptly forgotten by almost everyone.* It was reinvented in 1985 by MacDonald *et al.*,[161–163] and reinvented a short time later by Madras. This third reinvention led to a comprehensive analytical and numerical study by Madras and Sokal,[96] which showed that the pivot algo-

---

*The only exceptions I know of are Olaj and Pelinka[153] and Clark and Lal.[154] Continuum analogues of the pivot algorithm were used by Stellman and Gans,[155,156] Freire and Horta,[157] Curro,[158,159] Scott[160] and possibly others. For additional history, see Ref. 96, footnote 3.

rithm is *by far the most efficient algorithm yet invented* for simulating the fixed-N, variable-x SAW ensemble. After that the applications took off.*

The elementary move of the pivot algorithm is as follows: Choose at random a pivot point k along the walk ($0 \leq k \leq N-1$); choose at random an element g of the symmetry group of the lattice (rotation or reflection or a combination thereof); then apply g to the part of the walk subsequent to the pivot point (namely $\omega_{k+1}, \ldots, \omega_N$), using ω_k as the temporary "origin" (cf. eq. (2.79)). The resulting walk is accepted if it is self-avoiding; otherwise, it is rejected, and the old walk ω is counted once more in the sample. It is easy to see that this algorithm satisfies detailed balance for the standard equal-weight SAW distribution. Ergodicity is less obvious, but it can be proven.[96,179]

At first thought this seems to be a terrible algorithm: for N large, nearly all the proposed moves will get rejected. In fact, this latter statement is true, but the hasty conclusion drawn from it is radically false! The acceptance fraction f does indeed go to zero as $N \to \infty$, roughly like N^{-p}; empirically, it is found that the exponent p is ≈ 0.19 in $d = 2$[96] and ≈ 0.11 in $d = 3$.[96,167,178] But this means that roughly once every N^p moves one gets an acceptance. And the pivot moves are very radical: one might surmise that after very few accepted moves (say, five or 10) the SAW will have reached an "essentially new" configuration. One conjectures, therefore, that the autocorrelation time τ of the pivot algorithm behaves as $\sim N^p$. Things are in fact somewhat more subtle (see below), but roughly speaking (and modulo a possible logarithm) this conjecture appears to be true. On the other hand, a careful analysis of the computational complexity of the pivot algorithm (see also below) shows that one accepted move can be produced in a computer time of order N. Combining these two facts, we conclude that one "effectively independent" sample (at least as regards *global* observables) can be produced in a computer time of order N (or perhaps $N \log N$). This is vastly better than the $N^{2+2\nu} \sim N^{\approx 3.2}$ of the local N-conserving algorithms and the $N^{\approx 2}$ of the bilocal algorithms. Indeed, this order of efficiency cannot be surpassed by any algorithm which computes each site on successive SAWs, for it takes a time of order N simply to *write down* an N-step walk!

Let us mention briefly some relevant issues; the reader is referred to Ref. 96 for a fuller discussion.

Variants of the pivot algorithm. Different variants of the pivot algorithm are obtained by specifying different distributions when we "choose at random":

(1) The pivot point k can be chosen according to any pre-set family of strictly positive probabilities $p_0, \ldots, p_{N-1}$. The strict positivity ($p_k > 0$ for

---

*Here is a very incomplete list: Caracciolo, Madras, Pelissetto, Sokal and collaborators,[39,132,164–166] Zifferer,[167–172] Bishop and collaborators,[173–176] Chorin,[177] Eizenberg and Klafter.[178]

all k) is necessary to ensure the ergodicity of the algorithm. Most work has used a uniform distribution, but there could be some (probably minor) advantage in using a non-uniform distribution.

(2) Let G be the group of orthogonal transformations (about the origin) which leave invariant the lattice $\mathbb{Z}^d$. Then the symmetry operation $g \in G$ can be chosen according to any pre-set probability distribution $\{p_g\}_{g \in G}$ which satisfies $p_g = p_{g^{-1}}$ for all g, and which has enough nonzero entries to ensure ergodicity (see below). The condition $p_g = p_{g^{-1}}$ is easily seen to be both necessary and sufficient to ensure detailed balance with respect to the equal-weight distribution π.

In dimension $d = 2$, G is the dihedral group D_4, which has eight elements: the identity, two 90° rotations, the 180° rotation, two axis reflections, and two diagonal reflections. Detailed balance holds provided that $p_{+90^\circ rot} = p_{-90^\circ rot}$. Ergodicity holds if the probabilities p_g are nonzero for either

- both diagonal reflections[179]; or
- $\pm 90^\circ$ rotations and the 180° rotation[96]; or
- $\pm 90^\circ$ rotations and both axis reflections.[96]

Most work has used a uniform distribution over the seven non-identity operations, but some gain could probably be achieved by using a non-uniform distribution.

In general dimension d, the symmetry group of $\mathbb{Z}^d$ is the group $O(d, \mathbb{Z})$ of orthogonal matrices with integer entries. In fact every such matrix is of the form

$$g_{ij} = \sigma_i \delta_{i,\pi(j)}, \tag{2.81}$$

where $\sigma_1, \ldots, \sigma_d = \pm 1$ and π is a permutation of $\{1, \ldots, d\}$. Using this description, the pivot algorithm can be programmed very easily in any dimension.

Acceptance fraction and autocorrelation time. Suppose we know that the acceptance fraction f in the pivot algorithm behaves as $f \sim N^{-p}$ as $N \to \infty$. Then, as argued above, after a few successful pivots—i.e., a time of order $1/f \sim N^p$—the *global* conformation of the walk should have reached an "essentially new" state. Thus, we expect that for observables A which measure the *global* properties of the walk (such as the squared end-to-end distance R_g^2 or the squared radius of gyration R_g^2) the autocorrelation time $\tau_{int,A}$ should be a few times $1/f$. This is confirmed numerically (Ref. 96, Section 4.3). On the other hand, it is important to recognize that *local* observables—such as the angle between the 17th and 18th steps of the walk—may evolve a factor of N more slowly than global observables. For example, the observable mentioned in the preceding sentence changes only when ω_{17} serves as a successful pivot point; and this happens, on average, only once every N/f

attempted moves. Thus, for *local* observables A we expect $\tau_{int,A}$ to be of order N/f. By eq. (2.59), τ_{exp} must be of at least this order; and if we have not overlooked any slow modes in the system, then τ_{exp} should be of exactly this order. Finally, even the global observables are unlikely to be precisely orthogonal [in $l^2(\pi)$] to the slowest mode; so it is reasonable to expect that $\tau_{exp,A}$ be of order N/f for these observables too. In other words, for global observables A we expect the autocorrelation function $\rho_{AA}(t)$ to have an extremely-slowly-decaying tail which, however, contributes little to the area under the curve. This behavior is illustrated by the exact solution of the pivot dynamics for the case of ordinary random walk (Ref. 96, Section 3.3), and by numerical calculations for the SAW.

The foregoing heuristic argument is, of course, far from a rigorous proof. It is not in general possible to find upper bounds on the autocorrelation time in terms of the acceptance fraction; the problem is that the state space could contain "bottlenecks" through which passage is unusually difficult. No one has suggested any reason why such bottlenecks should occur in the pivot algorithm, but neither does there exist any proof of their nonexistence.

A heuristic argument (Ref. 96, Section 3.2) suggests that $f \sim N^{-p}$ with $p = \gamma - 1$ ($= 11/32$ in $d = 2$, ≈ 0.16 in $d = 3$, and 0 in $d \geq 4$). Quantitatively this prediction is incorrect; numerical experiments yield $p \approx 0.19$ in $d = 2$[96] and $p \approx 0.11$ in $d = 3$.[96,167,178] However, the argument does correctly predict that p is small and that it gets smaller in higher dimension. It is an open question to find a better theoretical prediction for p, and in particular to express it in terms of static exponents for the SAW.

Computational complexity. A very important issue in any algorithm—but especially in a non-local one—is the CPU time per iteration. By using a hash table (see Section 2.7.1.2), the self-avoidance of a proposed new walk can be checked in a time of order N. But one can do even better: by starting the checking at the pivot point and working outwards, *failures* can be detected in a mean time of order N^{1-p} (Ref. 96, Sections 3.4 and 4.4). The mean CPU time *per successful pivot* is therefore $\sim N^{1-p}$ for each of $\sim N^p$ failures, plus $\sim N$ for one success, or $\sim N$ in all. Combining this with the observations made previously, we conclude that one "effectively independent" sample—as regards *global* observables—can be produced in a computer time of order N.

Initialization. There are two main approaches:

(1) *Equilibrium start*. Generate the initial configuration by dimerization (Section 2.4.3); then the Markov chain is in equilibrium from the beginning, and no data need be discarded. This approach is feasible (and recommended) at least up to N of order a few thousand. There is no harm in spending even days of CPU time on this step, provided that this time is small compared to the rest of the run; after all, the algorithm need only be initialized once.

(2) *"Thermalization"*. Start in an arbitrary initial configuration, and then discard the first $n_{disc} \gg \tau_{exp} \sim N/f$ iterations. This is painful, because τ_{exp} is a factor $\sim N$ larger than $\tau_{int,A}$ for global observables A; thus, for very large $N (\gtrsim 10^5)$, the CPU time of the algorithm could end up being dominated by the thermalization. Nevertheless, one must resist the temptation to cut corners here, as even a small initialization bias can lead to systematically erroneous results, especially if the statistical error is small.[39] Some modest gain can probably be obtained by using closer-to-equilibrium initial configurations (e.g. Ref. 168), but it is still prudent to take n_{disc} at least several times N/f.

Initialization will become a more important issue in the future, as faster computers permit simulations at ever-larger chain lengths.

2.6.5 Fixed-N, fixed-x algorithms

This ensemble has not been considered until very recently; it seems difficult to devise algorithms which are ergodic under the severe constraints of fixed chain length and fixed endpoints.

2.6.5.1 Bilocal algorithms

I do not know whether they are any bilocal (or mixed local/bilocal) algorithms that are ergodic for this ensemble.

2.6.5.2 Cut-and-paste algorithms

These algorithms were devised by Dubins, Madras, Orlitsky, Reeds and Shepp.[179,180] In dimension $d = 2$ the simplest cut-and-paste algorithm uses two moves:

- Inversion of the subwalk $\omega^{[k,l]}$ (Fig. 2.8).
- If $\omega_l - \omega_k$ lies on a $\pm 45°$ diagonal line, then one can reflect the subwalk $\omega^{[k,l]}$ through the perpendicular bisector of the segment $\omega_k\omega_l$, followed by an inversion.

This algorithm is ergodic.[179] In dimension $d \geq 3$ one must adjoin coordinate-interchange moves, and the algorithm is then again ergodic.[179]

The dynamic critical behavior of this algorithm is unclear. One's first guess is a behavior similar to that of the pivot algorithm, i.e., an acceptance fraction $f \sim N^{-q}$ for some small power q, and an autocorrelation time $\tau_{int,A} \sim N^q$ for *global* observables. However, one should remember that the diagonal-reflection moves—which are *necessary* for ergodicity*—are

*Without them, the numbers $N_1, \ldots, N_d$ of steps along each axis would be *separately* conserved.

possibly only for a fraction $\sim N^{-\nu}$ of pairs k, l. So perhaps one should expect $\tau_{int,A} \sim N^{q+\nu}$. (Numerically, this might be seen most easily by looking at an observable which is sensitive to the diagonal-reflection moves, e.g., the numbers $N_1, \ldots, N_d$ of steps along each axis.) On the other hand, failures of diagonal reflection due to disallowed $\omega_l - \omega_k$ can be detected in a CPU time of order 1; so such failures may not affect the scaling of the product $\tau_{int,A} T_{CPU}$.

See Ref. 181 for a first study of the cut-and-paste algorithm (on the face-centered-cubic lattice).

2.6.6 *Variable-N, variable-x algorithms*

2.6.6.1 *Slithering-tortoise (Berretti–Sokal) algorithm*

The slithering-tortoise algorithm*[182] is a Markov chain with state space $\mathcal{S} \equiv \cup_{N=0}^{\infty} \mathcal{S}_N$ and invariant probability distribution (2.18) with $p = 0$, i.e.,

$$\pi_\beta(\omega) = \text{const} \times \beta^{|\omega|}. \tag{2.82}$$

The algorithm's elementary moves are as follows: either one attempts to append a new step to the walk, with equal probability in each of the $2d$ possible directions; or else one deletes the last step from the walk. In the former case, one must check that the proposed new walk is self-avoiding; if it is not, then the attempted move is rejected and the old configuration is counted again in the sample ("null transition"). If an attempt is made to delete a step from an already-empty walk, then a null transition is also made. The relative probabilities of $\Delta N = +1$ and $\Delta N = -1$ attempts are chosen to be

$$P(\Delta N = +1 \text{ attempt}) = \frac{2d\beta}{1 + 2d\beta} \tag{2.83}$$

$$P(\Delta N = -1 \text{ attempt}) = \frac{1}{1 + 2d\beta} \tag{2.84}$$

It is easily verified that this transition matrix satisfies detailed balance for the distribution π_β. It is also easy to see that the algorithm is ergodic: to get from a walk ω to another walk ω', it suffices to use $\Delta N = -1$ moves to transform ω into the empty walk, and then use $\Delta N = +1$ moves to build up the walk ω'.

Regarding the critical slowing-down of the slithering tortoise algorithm, we can argue heuristically that

*Because the walk sticks out and retracts its "head", like a tortoise. This algorithm is also known as the Berretti-Sokal algorithm, or BS for short.

$$\tau \sim \langle N \rangle^2. \tag{2.85}$$

To see this, consider the quantity $N(t) = |\omega|(t)$, the number of steps in the walk at time t. This quantity executes, crudely speaking, a random walk (with drift) on the nonnegative integers; the average time to go from some point N to the point 0 (i.e., the empty walk) is of order N^2. Now, each time the empty walk is reached, all memory of the past is erased; future walks are then independent of past ones. Thus, the autocorrelation time ought to be of order $\langle N^2 \rangle$, or equivalently $\langle N \rangle^2$.

This heuristic argument can be turned into a rigorous proof of a *lower bound* $\tau_{exp} \gtrsim \tau_{int,N} \geq \text{const} \times \langle N \rangle^2$.[69] However, as an argument for an *upper bound* of the same form, it is not entirely convincing, as it assumes without proof that the slowest mode is the one represented by $N(t)$. With considerably more work, it is possible to prove an upper bound on τ that is only slightly weaker than the heuristic prediction: $\tau_{exp} \leq \text{const} \times \langle N \rangle^{1+\gamma}$.[69,183,184] (Note that the critical exponent γ is believed to equal 43/32 in dimension $d = 2$, ≈ 1.16 in $d = 3$, and 1 in $d \geq 4$.) In fact, there is reason to believe[185] that the true behavior is $\tau \sim \langle N \rangle^p$ for some exponent p *strictly between* 2 and $1 + \gamma$. A deeper understanding of the dynamic critical behavior of the slithering-tortoise algorithm would be of definite value.

2.6.6.2 Join-and-cut algorithm

The behavior $\tau \gtrsim \langle N \rangle^2$ of the slithering-tortoise algorithm is in fact characteristic of *any* variable-N algorithm whose elementary moves make *bounded* changes in N: roughly speaking, N must perform a random walk on the nonnegative integers, and the autocorrelation time of such a random walk satisfies

$$\tau \gtrsim \text{var}(N) \equiv \langle N^2 \rangle - \langle N \rangle^2 \sim \langle N \rangle^2 \tag{2.86}$$

(Ref. 132, Theorems A.6 and A.7). Therefore, if one wants to do better than $\tau \gtrsim \langle N \rangle^2$, it is necessary to make *unbounded* changes in N.

An amazingly simple way of doing this was proposed recently by Caracciolo, Pelissetto and Sokal.[165] Their algorithm works in the unorthdox ensemble $\mathscr{T}_{N_{tot}}$ consisting of all *pairs* of SAWs $(\omega^{(1)}, \omega^{(2)})$ (each walk starts at the origin and ends anywhere) such that the *total* number of steps in the two walks is some fixed number N_{tot}:

$$\begin{aligned}\mathscr{T}_{N_{tot}} &\equiv \{(\omega^{(1)}, \omega^{(2)}) : \omega^{(1)}, \omega^{(2)} \text{ are self-avoiding, with } |\omega^{(1)}| + |\omega^{(2)}| = N_{tot}\} \\ &\equiv \bigcup_{N_1=0}^{N_{tot}} (\mathscr{S}_{N_1} \times \mathscr{S}_{N_{tot}-N_1}).\end{aligned}$$

Each pair $(\omega^{(1)}, \omega^{(2)})$ in this ensemble is given equal weight; therefore, the two walks are non-interacting except for the constraint on the sum of their lengths. One sweep of the algorithm consists of two steps:

(a) We update independently each of the two walks, using some N-conserving ergodic algorithm (e.g., the pivot algorithm).
(b) We perform a *join-and-cut* move: we concatenate the two walks $\omega^{(1)}$ and $\omega^{(2)}$, forming a new (not necessarily self-avoiding) walk $\omega^{(1)} \circ \omega^{(2)}$; then we cut $\omega^{(1)} \circ \omega^{(2)}$ at a random position, creating two new walks $\omega'^{(1)}$ and $\omega'^{(2)}$. If $\omega'^{(1)}$ and $\omega'^{(2)}$ are both self-avoiding, we keep them; otherwise the move is rejected and we stay with $\omega^{(1)}$ and $\omega^{(2)}$.

The join-and-cut move is illustrated in Fig. 2.14.

Since the algorithm used in step (a) is ergodic in the ensemble of fixed-length walks, it is easy to see that the full algorithm is ergodic. (If $\omega^{(1)}$ and $\omega^{(2)}$ are perpendicular rods, then the join-and-cut move will always succeed.) It is also easy to see that the algorithm satisfies the detailed-balance condition with respect to the equal-weight measure

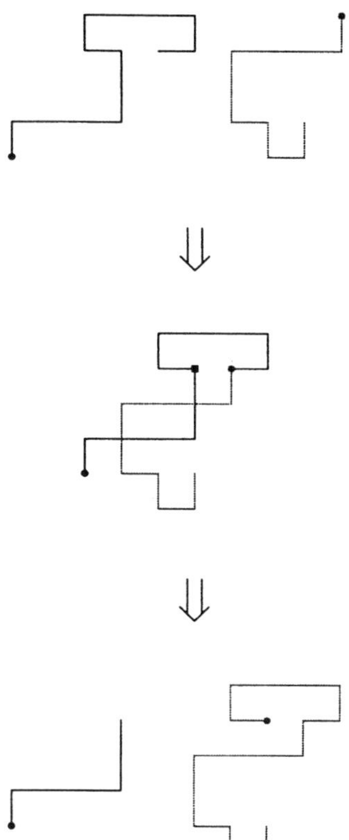

Fig. 2.14 Join-and-cut move: the two SAWs (upper figure) are concatenated (middle figure) and then cut at the point marked with a square (lower figure). Note that the concatenated walk *need not* be self-avoiding.

$$\pi(\omega^{(1)}, \omega^{(2)}) = \frac{1}{Z(N_{tot})} \quad \text{for each}(\omega^{(1)}, \omega^{(2)}) \in \mathcal{T}_{N_{tot}}, \tag{2.88}$$

where

$$Z(N_{tot}) = \sum_{N_1=0}^{N_{tot}} c_{N_1} c_{N_{tot}-N_1}. \tag{2.89}$$

Therefore, from the probability distribution of the random variable $N_1 \equiv |\omega^{(1)}|$ (the length of the first walk) in the measure (2.88), we can obtain estimates of the critical exponent γ by the maximum-likelihood method (see Section 2.7.3). Since the join-and-cut move can make large jumps in N_1 in a single step, this evades the bound (2.86).

The dynamic critical behavior of the pivot + join-and-cut algorithm was studied in Ref. 165 by both analytical and numerical methods. For the relevant observable $X \equiv \log[N_1(N_{tot} - N_1)]$, the autocorrelation time in units of elementary moves is found to grow as $\tau_{int,X} \sim N_{tot}^{\approx 0.70}$ in $d = 2$. On the other hand, the mean CPU time per elementary move behaves as $T_{CPU} \sim N^{\approx 0.81}$, just as in the pivot algorithm. Hence the autocorrelation time in CPU units behaves as $\tau_{int,X} T_{CPU} \sim N^{\approx 1.51}$, which is a significant improvement over the $\tau \sim N^{\approx 2}$ of the slithering-tortoise algorithm. Moreover, the behavior is expected to be even better in higher dimensions.

2.6.7 *Variable-N, fixed-x algorithms*

2.6.7.1 *BFACF algorithm*

The BFACF algorithm was invented by Berg and Foerster[186] and Aragão de Carvalho, Caracciolo and Fröhlich.[61,187]* It is a Markov chain with state space $\mathcal{S}(x) \equiv \bigcup_{N=0}^{\infty} \mathcal{S}_N(x)$ and invariant probability distribution (2.15) with $p = 1$, i.e.,

$$\pi_\beta(\omega) = \text{const} \times |\omega| \beta^{|\omega|}. \tag{2.90}$$

The elementary moves of the BFACF algorithm are the following local deformations:

- The one-bead flip (move A of Fig. 2.1), which has $\Delta N = 0$.
- Kink insertion (move J of Fig. 2.9), which has $\Delta N = +2$.
- Kink deletion (move K of Fig. 2.9), which has $\Delta N = -2$.

Note that all these moves can be generated by displacing the middle bond of a three-bond group by one lattice unit perpendicular to itself in one of the

*The exposition in the original paper[187] suffers from an unfortunate confusion regarding the meaning of $p(\Delta N)$. Here I follow the corrected exposition in Ref. 132, Section 3.1.

$2d - 2$ possible directions. Therefore, one iteration of the BFACF algorithm consists of the following operations:

(1) Choose at random a bond of the current walk ω (with equal probability for each bond).
(2) Enumerate the $2d - 2$ possible deformations of that bond; choose randomly among these deformations, giving each deformation a probability $p(\Delta N)$ depending only on $\Delta N \equiv |\omega'| - |\omega|$. (If the sum of these probabilities is $s < 1$, then make a "null transition" $\omega \to \omega$ with probability $1 - s$.) The probabilities $p(\Delta N)$ will be specified below.
(3) Check whether the proposed new walk ω' is self-avoiding. If it is, keep it; otherwise, make a null transition.

This algorithm satisfies detailed balance for π_β provided that

$$p(+2) = \beta^2 p(-2). \tag{2.91}$$

On the other hand, the sum of the probabilities must in all cases be ≤ 1; this imposes the inequalities[132]

$$[1 + (2d - 3)\beta^2]p(+2) \leq \beta^2 \tag{2.92a}$$

$$2p(0) + (2d - 4)p(+2) \leq 1 \tag{2.92b}$$

A nearly optimal choice is

$$p(-2) = \frac{1}{1 + (2d - 3)\beta^2} \tag{2.93a}$$

$$p(0) = \frac{1 + \beta^2}{2[1 + (2d - 3)\beta^2]} \tag{2.93b}$$

$$p(+2) = \frac{\beta^2}{1 + (2d - 3)\beta^2} \tag{2.93c}$$

(see Ref. 132 for further discussion).

The *ergodicity* of the BFACF algorithm is an extremely subtle problem. The known results are the following:

- In dimension $d = 2$, the BFACF algorithm is *ergodic*, for all choices of x.[11]
- In dimension $d = 3$, the algorithm is *non-ergodic* if $||x||_\infty \equiv \max(|x_1|, |x_2|, |x_3|) = 1$, due to knotted configurations which cannot be untied.[188]* However, it can be made ergodic† by adjoining the three-bead crankshaft move (I of Fig. 2.3).[188]

*For the BFACF algorithm applied to *unrooted polygons* (ring polymers), the ergodic classes are precisely the knot classes.[188] This is probably true also for the BFACF algorithm applied to *walks* with $||x||_\infty = 1$, but it has apparently not yet been proven.

†At least in the case of unrooted polygons. It is probably true also for the BFACF algorithm applied to *walks* with $||x||_\infty = 1$, but it has apparently not yet been proven.

- In dimension $d = 3$, the algorithm is *ergodic* if $||x||_\infty \geq 2$.[189] (It is then possible to disentangle any knot, no matter how large and complicated, by a motion which passes one strand at a time between the endpoints.)
- In dimension $d \geq 4$, the algorithm is presumably ergodic for all x, but a rigorous proof appears to be lacking.

The *dynamical behavior* of the BFACF algorithm is also an extremely subtle problem. The known results are the following:

- $\tau_{exp} = +\infty$ for *all* $\beta > 0$.[190] This surprising result arises from the existence of arbitrarily slowly-relaxing modes associated with sequences of transitions $\omega \to \ldots \to \omega'$ with $\mathscr{A}(\omega, \omega') \gg \max(|\omega|, |\omega'|)$, where $\mathscr{A}(\omega, \omega')$ is the minimum surface area spanned by the union of ω and ω'.
- $\tau_{int,\mathscr{A}} \geq \text{const} \times [\langle \mathscr{A}^2 \rangle - \langle \mathscr{A} \rangle^2]$, where $\mathscr{A} \equiv \mathscr{A}(\omega, \omega^\circ)$ for some fixed walk ω° from 0 to x.[132] Assuming the usual scaling behavior,[191] this implies $\tau_{int,\mathscr{A}} \gtrsim \langle N \rangle^{4\nu}$.
- Numerically, it appears that $\tau_{int,N} \sim \langle N \rangle^p$ with p equal or very nearly equal to 4ν.[191]

However, in dimension $d = 3$ (for $||x||_\infty \geq 2$) the relaxation of long chains may (possibly) be further slowed down by the appearance of "quasi-knots", i.e., configurations whose decay is extremely slow due to the complicated sequence of moves needed to untie them. Since untying a quasi-knot requires passing the end of the chain through the knotted region (which may be near the middle of the chain), this process becomes more time-consuming the larger $\langle N \rangle$ is. This effect might possibly increase the dynamic critical exponent above 4ν, at least for observables that are sensitive to (quasi-)knots.*

2.6.7.2 BFACF + cut-and-paste algorithm

Caracciolo, Pelissetto and Sokal[132] have recently proposed to speed up the BFACF algorithm by adjoining some non-local (cut-and-paste) moves. The hope is that these non-local moves will destabilize the long-lived (metastable) configurations that are responsible for the slow equilibration of the BFACF algorithm. Thus, the algorithm is a hybrid in which the non-local moves hopefully assure the rapid equilibration *within* subspaces of fixed N, while the local (BFACF) moves assure equilibration *between* different N (and in particular make the algorithm ergodic). The algorithm has a free parameter p_{nl}—the percentage of non-local moves—which can be tuned as a function of $\langle N \rangle$ to optimize the computational efficiency.

---

*For example, any one of the standard knot invariants[192,193] applied to $\omega \circ \omega^\circ$, where ω° is some fixed path from x to 0 in $\mathbb{R}^d \setminus \mathbb{Z}^d$.

The best can hope for is an autocorrelation time $\tau \sim \langle N \rangle^2$: for even if the non-local moves were to cause instant equilibration at fixed N, the local moves would still have to carry out a random walk in N. Such a behavior, if achieved, would be a significant improvement over the pure BFACF algorithm. This estimate refers, however, to *physical* time units; since the non-local moves require a mean CPU time per move that grows as a fractional power of $\langle N \rangle$, it is a subtle matter to choose p_{nl} so as to minimize the autocorrelation time as measured in *computer* (CPU) time units.

Numerical experiments in $d = 2$[132] confirm that the autocorrelation time (in units of elementary moves) at *fixed* $p_{nl} > 0$ scales as $\tau_{int,N} \sim \langle N \rangle^{\approx 2}$. Taking into account the CPU time, it is found that the optimal p_{nl} scales as $\sim 1/\langle N \rangle^{\approx 0.8}$, and the autocorrelation time in CPU units then scales as $\tau_{int,N} T_{CPU} \sim \langle N \rangle^{\approx 2.3}$. In practice, at $\langle N \rangle \approx 100$ it is found that the physical (resp. CPU) autocorrelation time of the hybrid algorithm with $p_{nl} = 0.05$ is a factor 6 (resp. 4) smaller than that of the pure BFACF algorithm. The BFACF + cut-and-paste algorithm provides, therefore, a significant (though not exactly earth-shaking) improvement over previous algorithms for variable-N, fixed-endpoint SAWs. But more research is needed to establish conclusively the dynamic critical behavior of this algorithm.

2.7 Miscellaneous issues

2.7.1 Data structures

The new algorithms described in Section 2.6 are potentially very efficient; but in order to realize this potential, it is necessary to choose carefully the data structures and the low-level "bookkeeping" algorithms. For example, a local or bilocal update can be carried out, with *suitable* data structures, in a CPU time of order 1 (as one would expect); but a *naive* choice of data structures might force us into "garbage collection" (mass copying of data from one storage location to another) costing a time of order N. Likewise, checking the global self-avoidance of a walk ω (as arises e.g., in the pivot algorithm) can be carried out, with suitable data structures, in a CPU time of order N; but the naive method (comparing ω_i to ω_j for each pair i,j) takes a time of order N^2—which would nullify the advantages of the pivot algorithm.

We divide the discussion of data structures into two parts: those needed in *updating the walk*, and those needed in *testing self-avoidance*. As will be seen, it is usually necessary to maintain *redundant* data structures, in order that both operations can be carried out efficiently. For the first task, one typically uses *linear (or circular) lists* of various kinds; for the second task, one uses *bit tables* or *hash tables*. A lucid discussion of these structures can be found in Ref. 94, Chapters 11 and 12.

2.7.1.1 Updating the walk

Each of the algorithms described in Section 2.6 requires one or more of the following operations in order to update the walk:

1. Choose a random site (or a random step) of the walk.
2. Update the coordinates of a given site of the walk.
3. Find the successor (or predecessor) of a given site on the walk.
4. Insert one or more new sites
   (a) at the beginning of the walk
   (b) at the end of the walk
   (c) in the interior of the walk
5. Delete one or more sites from (a), (b) or (c).

Depending on the particular operations needed, we will choose to represent the walk as one or another type of list.

The simplest type of list is a *sequentially allocated linear list*: the walk coordinates $\omega_0, \ldots, \omega_N$ are stored in order in successive memory locations (e.g., a Fortran array). Clearly this structure permits the first three operations to be carried out in a time of order 1. In particular, this suffices for local N-conserving moves, pivot moves, and cut-and-paste moves.* Insertion and deletion at the beginning and end of the walk can also be carried out in a time of order 1, *provided that* an upper bound is known in advance on how far the walk can grow in each direction (so that sufficient storage can be allocated). This suffices for the slithering-tortoise algorithm, because one can take, e.g., $N_{max} \approx 70\langle N\rangle$ and be virtually certain that this walk length will never be exceeded (at least not before the sun runs out of hydrogen and engulfs the earth).

For the slithering-snake (reptation) algorithm one should use a *sequentially allocated circular list*: the walk coordinates are stored in sequential memory locations, but in general in a cyclically permuted order, i.e. $\omega_j, \omega_{j+1}, \ldots, \omega_N, \omega_0, \omega_1, \ldots, \omega_{j-1}$. A pointer then indicates which element is ω_0. (A flag may also indicate which direction is "forward": this is useful in implementing the first version of the algorithm (Ref. 146, pp. 4592–93), in which "head" and "tail" are sometimes interchanged.)

None of the foregoing structures is adequate for those algorithms which insert or delete sites in the *interior* of the walk (BFACF, most bilocal algorithms, etc.): insertion or deletion of even one site would necessitate shifting a large part of the list ("garbage collection"), which would take a time of

*In the case of pivot or cut-and-paste moves, one maintains *two* linear lists—one "active" and one "scratch"—together with a flag saying which is currently which. The coordinates of a *proposed* new walk are successively calculated and placed in the "scratch" list, and self-avoidance is simultaneously tested (see Section 2.7.1.2); if the self-avoidance test is passed, the flag is flipped and the "scratch" list now becomes "active".

order N. A more flexible data structure is the *doubly linked list*: here the walk coordinates $\omega_0, \ldots, \omega_N$ may be stored anywhere in memory (neither contiguously nor in order); but together with each coordinate ω_i we maintain link fields $\ell^-(i)$ and $\ell^+(i)$, which indicate the storage locations where the ω_{i-1} and ω_{i+1}, respectively, are to be found. Unused but available memory is also stored in a linked list, the so-called *free list* (here the *two-way* linking is unnecessary). It is easy to insert or delete into the interior of a doubly linked lists, by using the link fields $\ell^-(i)$ and $\ell^+(i)$ to locate the predecessor and successor sites (which take part in the relinking). On the other hand, since the $\{\omega_i\}$ may be scattered throughout memory, it is not easy to choose a random site or step of the walk. (One would have to "thread through" the list sequentially, taking a time of order N.)

To get the virtues of both sequential allocation and linked list, one can use a *contiguously allocated doubly linked linear list* (see e.g., Ref. 132, Section 3.3). That is, the walk coordinates are stored in a contiguous array $\{s(i)\}_{i=0}^N$, but *not in any particular order*; that is, $s(i)$ is *some* site of the walk, but *not* necessarily ω_i. To keep track of the sequence of steps along the walk, we use forward pointers $\{\ell^+(i)\}_{i=0}^N$ and backward pointers $\{\ell^-(i)\}_{i=0}^N$; here $\ell^+(i)$ [resp. $\ell^-(i)$] is the index corresponding to the site following (resp. preceding) the site whose index is i, or -1 if no such site exists. It is often convenient that the initial and final points of the walk be assigned to indices 0 and 1, respectively. Therefore,

$$\begin{aligned} s(0) &= \omega_0 \ (= 0) \\ s(\ell^+(0)) &= \omega_1 \\ s(\ell^+(\ell^+(0))) &= \omega_2 \\ &\vdots \end{aligned}$$

and so on. Likewise,

$$\begin{aligned} s(1) &= \omega_N \ (= x) \\ s(\ell^-(1)) &= \omega_{N-1} \\ s(\ell^-(\ell^-(1))) &= \omega_{N-2} \\ &\vdots \end{aligned}$$

and so on. It is then trivial to choose a random site of the walk, as this is equivalent to choosing a random index i.* It is also trivial to insert new sites: just put them in locations $N+1$, $N+2$, etc. and then relink. Deletion requires "garbage collection" to maintain the contiguous allocation, but

*Note that this works only for choice with *uniform* probability (or uniform excluding one or both endpoints). It would *not* work if one wanted to choose sites with a nonuniform probability depending on their location along the walk.

this also can be performed in a time of order 1: move the entries from locations N, $N-1$, etc. into the just-vacated locations, with appropriate updates to the link fields. So the contiguously allocated doubly linked linear list is the appropriate data structure for the BFACF and bilocal algorithms.

2.7.1.2 Testing self-avoidance

If the walk configuration were stored only as a linear list (whether sequentially allocated or linked), then this entire list would have to be searched each time we want to add one new site to the walk, in order to check the self-avoidance constraint. This would take a time of order N, which is a disaster.

The solution is to maintain *two* (redundant) data structures to store the current walk configuration: a linear or circular list as described in the preceding subsection, and a *bit table* or *hash table*.* The latter can be defined abstractly as data structures that perform the following functions: Given a finite (but typically enormous) set K of "possible keywords", we wish to store a subset $H \subset K$ (of cardinality $\leq$ some maximum M) in such a way that, for any $x \in K$, the following operations can be carried out rapidly:

1. *Query*. Is $x \in H$?
2. *Insertion*. Insert x into H (if it is not in H already).
3. *Deletion*. Delete x from H (if it is in H currently).

Here "rapidly" means "in a time of order 1, on the average". In our application, the set K of possible keywords will be the set of all points in some box $B \subset \mathbb{Z}^d$ which is large enough to contain all possible points in the walk ω: e.g., a cube of side $\geq 2N$ centered at the origin for a fixed-N simulation, or a cube of size $\gtrsim 140\langle N \rangle$ for a variable-N simulation.

A *bit table* is simple to describe: it is a large block of memory in which each possible keyword $x \in K$ (i.e. each site of the large box B) is assigned one distinct bit. That bit is set to 1 if $x \in H$ (i.e., if the site in question currently belongs to the walk), and 0 otherwise. Clearly, the three operations of query, insertion and deletion can each be carried out in a time of order 1. The trouble is that for large N, the memory requirements become prohibitive, especially in dimension $d > 2$: e.g., at $N = 1000$ the memory needed is 0.5 megabyte for $d = 2$, and 1 gigabyte for $d = 3$!

The memory requirements can be reduced by a clever trick called the *sliding bit table*.[196,197] This approach is adequate in algorithms in which the new sites are all being added (or old sites being deleted) in the same small vicinity. In practice this means the slithering-tortoise algorithm.

*To my knowledge, the first use of a bit table for self-avoidance checking was by McCrackin,[194] and the first such uses of a hash table were by Alexandrowicz[97] and Jurs and Reissner.[195]

A more general approach is provided by the *hash table* (Ref. 94, Chapter 12; Ref. 198, Section 6.4): an array of M words is assigned, and each keyword $x \in K$ is assigned a *primary address* $h(x)$ in this array. Since in general $M \ll |K|$, the "hash function" h is necessarily many-to-one, i.e., many distinct keywords may share the same primary address, leading to the possibility of *collisions*. The various hash-coding algorithms are distinguished by the method they use to resolve collisions, i.e. to decide where to store a keyword if its primary address happens to be occupied by some other keyword. One of the simplest collision-resolution schemes is *linear probing*: if the primary address $h(x)$ is occupied, the algorithm searches successively in addresses $h(x)+1$, $h(x)+2, \ldots$ (modulo M) until it finds either the keyword x or an empty slot. Other approaches involve *chaining* via linked lists.

In the worst possible case, a single query or insertion into a hash table containing N entries could take a time of order N. However, it can be shown (Ref. 198, Section 6.4) that as long as the hash table does not get close to full (i.e., N does not get near M), then the *average* time for a single query or insertion (assuming random distribution of the points) is of order 1. So the hash-table method is nearly as fast as the bit-table method, and far more space-effective.

We remark that *deletion* from a linear-probing hash table requires some care: if done naively, entries can get "lost" (Ref. 198, pp. 526–7). However, these subtleties are irrelevant if deletions always occur in a last-in-first-out manner (as in the slithering-tortoise algorithm), or occur only when cleaning up the table at the end (as in the pivot and cut-and-paste algorithms). In the latter case it suffices to keep a linear list of the memory locations in which elements have been inserted; these locations can then be cleared at the end.

Depending on the application, it may be desirable to maintain the bit table/hash table either as a *permanent* data structure (containing the current walk ω) or as a *scratch* data structure (for checking self-intersection of proposed new walks ω'); or it may be desirable to maintain one of each, together with a flag saying which is which. See Ref. 132, Section 3.3 and Ref. 165, Section 3.2 for details.

2.7.2 Measuring virial coefficients

The virial coefficients B_k play a central role in the theory of dilute polymer solutions. But to measure them it is *not* necessary (or even desirable) to simulate a many-chain system; rather, it suffices to simulate k *independent* polymer chains and then measure a suitable overlap observable. Consider, for concreteness, the second virial coefficient $B_2^{(N_1,N_2)}$, defined by eqs (2.11)–(2.13). We have

$$B_2^{(N_1,N_2)} = \frac{1}{2}\left\langle T(\omega^{(1)}, \omega^{(2)})\right\rangle_{N_1,N_2}, \tag{2.94}$$

where $\omega^{(1)}$ and $\omega^{(2)}$ are *independent* SAWs of N_1 and N_2 steps, respectively, and $T(\omega^{(1)}, \omega^{(2)})$ is the number of translates of $\omega^{(2)}$ which somewhere intersect $\omega^{(1)}$:

$$T(\omega^{(1)}, \omega^{(2)}) = \#\{x \in \mathbb{Z}^d : \omega^{(1)} \cap (\omega^{(2)} + x) \neq \varnothing\}. \tag{2.95}$$

So we can run in parallel two independent simulations (using for example the pivot algorithm), and then every once in a while measure the observable $T(\omega^{(1)}, \omega^{(2)})$.

The straightforward method for determining $T(\omega^{(1)}, \omega^{(2)})$ is to compute $x = \omega_i^{(1)} - \omega_j^{(2)}$ for each of the $(N_1 + 1)(N_2 + 1)$ pairs i, j, write these points into a hash table (see Section 2.7.1.2), and count how many distinct values of x are obtained. Unfortunately, this takes a CPU time of order $N_1 N_2$, i.e., order N^2 if $N_1 = N_2 = N$. By contrast, we expect that one "effectively independent" sample of the pair $(\omega^{(1)}, \omega^{(2)})$ can be produced by the pivot algorithm in a CPU time of order N. So this approach would spend more time analyzing the data than generating it!

Fortunately, there exist *Monte Carlo* algorithms which can produce an *unbiased estimate* of $T(\omega^{(1)}, \omega^{(2)})$, with statistical errors comparable to or smaller than those already intrinsic in the observable $T(\omega^{(1)}, \omega^{(2)})$, in a mean CPU time of order N. So the idea is to perform a "Monte Carlo within a Monte Carlo". At least two such algorithms are known: the "hit-or-miss" algorithm,[39] and the Karp–Luby algorithm.[199,200] See Ref. 96, Section 5.3 for a preliminary discussion, and Ref. 39 for a fuller account.

The "hit-or-miss" algorithm can easily be generalized to compute higher virial coefficients. I do not know whether the Karp–Luby algorithm can be generalized in this way.

2.7.3 Statistical analysis

For the most part, the statistical analysis of SAW Monte Carlo data uses the same methods as are employed in other types of Monte Carlo simulation. In particular, with dynamic Monte Carlo data it is essential to carry out a thorough *autocorrelation analysis*: only in this way can one test the adequacy of the thermalization interval and the run length and properly assess the statistical error bars. For details, see e.g., Ref. 36, Section 3; Ref. 37, Section 2; Ref. 96, Appendix C; and Ref. 11, Sections 9.2.2 and 9.2.3.

Typically one carries out fixed-N simulations at a (wide) range of values of N, and then uses *weighted least-squares estimation*[201] to extract the critical exponent ν and the various critical amplitudes. However, for high-precision work it is important to take account of *corrections to scaling*. According to renormalization-group theory,[202,203] the mean value of any global observable $\mathcal{O}$ behaves as $N \to \infty$ as

$$\langle \mathcal{O} \rangle_N = AN^p \left[1 + \frac{a_1}{N} + \frac{a_2}{N_2} + \ldots + \frac{b_0}{N^{\Delta_1}} + \frac{b_1}{N^{\Delta_1+1}} + \frac{b_2}{N^{\Delta_1+2}} + \ldots \right.$$
$$\left. + \frac{c_0}{N^{\Delta_2}} + \frac{c_1}{N^{\Delta_2+1}} + \frac{c_2}{N^{\Delta_2+2}} + \ldots \right]. \qquad (2.96)$$

Thus, in addition to "analytic" corrections to scaling of the form a_k/N^k, there are "non-analytic" corrections to scaling of the form b_k/N^{Δ_1+k}, c_k/N^{Δ_2+k} and so forth, as well as more complicated terms not shown in eq. (2.96). The leading exponent p and the correction-to-scaling exponents $\Delta_1 < \Delta_2 < \ldots$ are universal; p of course depends on the observable in question, but the Δ_i do not. (Please note that the exponents $\Delta_1 < \Delta_2 < \ldots$ have no relation whatsoever to the gap exponent Δ_4 defined in eq. (2.10). The notation used here is standard but unfortunate.) The various amplitudes (both leading and subleading) are all nonuniversal. However, *ratios* of the corresponding amplitudes A, b_0 and c_0 (but not a_k or the higher b_k, c_k) for different observables are universal.[52,203]

Obviously it is hopeless to try to estimate from noisy Monte Carlo data more than the first one or two terms in eq. (2.96), i.e.,

$$\langle \mathcal{O} \rangle_N = AN^p \left[1 + \frac{a}{N^{\Delta}} \right] \qquad (2.97)$$

where $\Delta \equiv \min(\Delta_1, 1)$. A reasonable approach is as follows: First truncate the series at zeroth order ($\langle \mathcal{O} \rangle_N = AN^p$) and perform a weighted least-squares fit using only the data at $N \geq N_{min}$, for a sequence of successively larger values N_{min}. For each such fit, the χ^2 value indicates whether the hypothesis $\langle \mathcal{O} \rangle_N = AN^p$ for $N \geq N_{min}$ is consistent with the data—or in other words, whether the corrections to scaling for $N \geq N_{min}$ (which surely exist) are statistically significant.* If they are not, then one is done; the estimates of p and A ought to be roughly independent of N_{min}, within statistical error bars. If the corrections are significant, then one can insert the first correction-to-scaling term and redo the least-squares fit and χ^2 test.† However, one must keep in mind that the estimate of Δ (and the correction amplitude) produced by such a fit is merely an *effective exponent* which may be imitating the combined effect of *several* correction-to-scaling terms over some particular range of N. Such an effective exponent is of no intrinsic physical interest; this approach should simply be thought of as a

*Statistical significance (resp. insignificance) of the corrections for $N \geq N_{min}$ means only that these corrections are larger than (resp. comparable to or smaller than) the statistical errors *in this particular simulation*. By making sufficiently long runs, the statistical error bars can in principle be made arbitrarily small, and so the corrections to scaling will *always* eventually be found to be statistically significant.

†One may either make a guess for Δ and then estimate the amplitude via linear least-squares, or estimate simultaneously both Δ and the amplitude by nonlinear least-squares.

semi-empirical method for extrapolating the data to $N \to \infty$. In principle the true Δ can be found by taking N_{min} very large—large enough so that the second correction-to-scaling term is negligible compared to the first—but this means that the first correction-to-scaling term will also be negligible compared to the leading term, and the *statistical* errors in Δ and the corresponding amplitude will therefore be enormous. At present it seems feasible to obtain only *rough* estimates of Δ by direct Monte Carlo simulation.[39,204] However, the situation may improve in the future, as more powerful computers become available.

A novel point arises when estimating μ and γ (or μ and α_{sing}) from a variable-N simulation. Thanks to eqs (2.2)/(2.18) [or (2.3)/(2.15)], the probability distribution of chain lenghts N is known *exactly* except for the two unknown parameters (μ, γ) [or (μ, α_{sing})], provided that we temporarily neglect corrections to scaling. This fact allows us to use *maximum-likelihood estimation*[201] to obtain estimates of μ and γ which are not only demonstrably optimal in a rigorous statistical sense—that is, they achieve the minimum possible mean-square error for a given quantity of Monte Carlo data—but which also provide *a priori* error estimates. This means that the statistical errors can be computed reliably, *in advance* of performing the Monte Carlo simulation. Or to put it more strikingly: before performing the simulation, one cannot know what the final central estimates will be, but one *can* know the error bars! See Ref. 182, Section 4.2 and Ref. 165, Section 4 for details.

2.8 Some applications of the algorithms

The development over the past decade of efficient Monte Carlo algorithms for the SAW (Section 2.6) has combined with recent advances in computer hardware (notably clusters of high-speed RISC workstations) to make possible high-precision studies of SAWs that would have been unimaginable only a few years ago. For example, a recent study[39] of SAWs in $d = 2$ and $d = 3$ has employed chain lengths up to $N = 80\,000$, obtaining error bars of order 0.1–0.3%. (To be sure, this work took several *years* of CPU time!)

In this section I cannot hope to do justice to all the applications of the new algorithms. Rather, I shall limit myself to giving an informal account of a few illustrative examples drawn from my own work (most of which is joint work with Sergio Caracciolo, Bin Li, Neal Madras, and Andrea Pelissetto).

2.8.1 Linear polymers in dimension $d = 3$

Probably our most important work is a high-precision study of the critical exponents ν and $2\Delta_4 - \gamma$ (and in particular the hyperscaling law $d\nu = 2\Delta_4 - \gamma$) and universal amplitude ratios for SAWs in both two and three dimensions, using the pivot algorithm.[39] In dimension $d = 3$, the renormalization-group prediction for the exponent ν is 0.5880 ± 0.0015[16–20] or 0.5872 ± 0.0014.[21] However, this method is susceptible to serious (and quite possibly undetectable) systematic errors arising from a confluent singularity at the RG fixed point;[25,26] this led me to be rather skeptical of the claimed error bars. Indeed, when we began this work five years ago, the series-extrapolation[205] and Monte Carlo[96,206] estimates of ν were significantly higher, around 0.592 ± 0.0015 (one standard deviation). I was therefore looking forward to improving the statistics on the Monte Carlo studies by a factor of 10 or so (i.e., reducing the error bar by a factor of about three), so as to definitively rule out the RG prediction (or at least its claimed error bar). But what actually happened is rather different from what I had envisaged!

The original Monte Carlo studies[96,206] predicting $\nu \approx 0.592$ were based on walks of length $N \lesssim 3000$. But as we studied longer and longer walks, the apparent exponent fell.* Eventually things stabilized, but we had to use walks up to $N = 80\,000$ and (if we fit to a pure power law) to *throw away* all walks with $N \lesssim 5000$! And the value at which things stabilized was—surprise!—almost exactly the RG value: our estimate is $\nu = 0.5877 \pm 0.0006$ (68% confidence limits), taking account of both statistical errors and corrections to scaling.[39] The moral is that corrections to scaling are an extremely serious effect in high-precision Monte Carlo studies; it is necessary to be very careful, and sometimes to go to enormous chain lengths, to escape from their effects. The extraordinary accuracy of the RG prediction remains a mystery (at least to me).†

Another startling conclusion from this study concerns the interpenetration ratio Ψ (cf. eq. (2.14)), and goes to the heart of polymer theory. But this requires a brief historical digression.

For several decades, most work on the behavior of long-chain polymer molecules in dilute solution[5,12,15,208,209,210] has been based on the so-called "two-parameter theory" in one or another of its variants: traditional (Flory-type),‡ pseudo-traditional (modified Flory-type)§ or modern (continuous-chain-type).¶ All two-parameter theories predict that in the limit of zero concentration, the mean-square end-to-end distance $\langle R_e^2 \rangle$, the mean-square

*For example, a recent study of Eizenberg and Klafter[178] using walks of length $N \lesssim 7200$ found $\nu \approx 0.5909 \pm 0.0003$.

†It may be related to the apparent fact[26,207] that the confluent exponent Δ_2/Δ_1 is very close to an integer (namely, it is ≈ 2).

‡See Yamakawa, Ref. 12, Sections 11 and 16 (pp. 69–73 and 110–118) and parts of Sections 15, 20b and 21b (pp. 94–110, 153–164 and 167–169). See also DesCloizeaux and Jannink, Ref. 15, Section 8.1 (pp. 289–313).

§See Yamakawa, Ref. 12, most of Section 15 (pp. 94–110) and parts of Sections 20b and 21b (pp. 153–164 and 167–169). See also Domb and Barrett.[211]

¶These theories take as their starting point the Edwards model of a weakly self-avoiding continuous chain.[15,209,212–216] (The Edwards model is also equivalent to the continuum φ^4 field theory with $n = 0$ components.) See DesCloizeaux and Jannink[15] for a detailed treatment of the Edwards model.

radius of gyration $\langle R_g^2 \rangle$ and the interpenetration ratio Ψ depend on the degree of polymerization N (or equivalently on the molecular weight $M = N M_{monomer}$) according to

$$\langle R_e^2 \rangle = A N F_{R_e}(bN) \tag{2.98a}$$

$$\langle R_g^2 \rangle = A N F_{R_g}(bN) \tag{2.98b}$$

$$\Psi = F_\Psi(bN) \tag{2.98c}$$

where F_{R_e}, F_{R_g}, F_Ψ are claimed to be *universal* functions (which each specific two-parameter theory should predict), and A and b are *non-universal* scale factors depending on the polymer, solvent and temperature but independent of N. (The conventional notation is $\alpha_R^2 = F_{R_e}, \alpha_S^2 = 6F_{R_g}, h = \alpha_S^d F_\Psi / z$ and $z = (bN)^{2-d/2}$ in spatial dimension d.) Moreover, virtually all the theories—and in particular the modern continuous-chain-based theories—predict that F_Ψ is a monotone increasing and concave function of its argument bN, which approaches a limiting value $\Psi^* \approx 0.2 - 0.3$ as $bN \to \infty$.

But our Monte Carlo data show precisely the opposite behavior: Ψ is a *decreasing* and *convex* function of N, which approaches a limiting value $\Psi^* \approx 0.247$ as $N \to \infty$ (Fig. 2.15). The same behavior was found by Nickel.[203] Indeed, there is *experimental* evidence that for real polymers in a sufficiently good solvent, the approach to Ψ^* is also from above, contrary to the two-parameter theory.[210,217–219] This behavior was considered to be a perplexing "anomalous effect", and various explanations were advanced.[218,220,221] What is going on here?

The correct explanation, in my opinion, was given two years ago by Nickel[203] (see also Refs 222, 223): theories of two-parameter type are simply wrong. Indeed, they are wrong not merely because they make incorrect predictions, but for a more fundamental reason: they purport to make universal predictions for quantities that are not in fact universal. Two-parameter theories predict, among other things, that Ψ is a universal function of the expansion factor $\alpha_S^2 \equiv \langle R_g^2 \rangle / \langle R_g^2 \rangle_{T_\theta}$; in particular, Ψ is claimed to depend on molecular weight and temperature only through the particular combination $\alpha_S^2(M, T)$. This prediction is quite simply incorrect, both for model systems and for real polymers. Indeed, even the *sign* of the deviation from the limiting value Ψ^* is not universal.

All this has a very simple renormalization-group explanation,[203] so it is surprising that it was not noticed earlier. As mentioned already in Section 2.7.3, standard RG arguments predict, for any real or model polymer chain, the asymptotic behavior

$$\langle R_e^2 \rangle = A_{R_e} N^{2\nu}(1 + b_{R_e} N^{-\Delta_1} + \ldots) \tag{2.99a}$$

$$\langle R_g^2 \rangle = A_{R_g} N^{2\nu}(1 + b_{R_g} N^{-\Delta_1} + \ldots) \tag{2.99b}$$

$$\Psi = \Psi^*(1 + b_\Psi N^{-\Delta_1} + \ldots) \tag{2.99c}$$

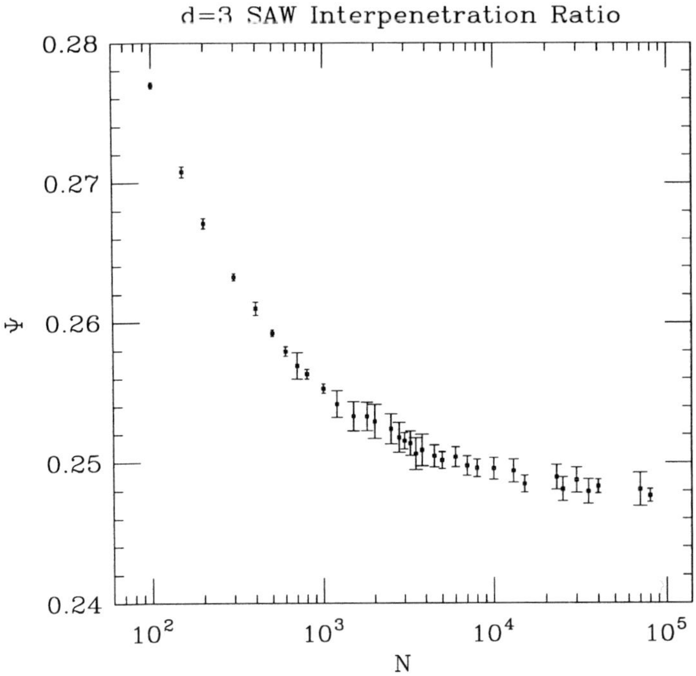

Fig. 2.15 Interpenetration ratio Ψ versus chain length N, for SAWs in $d = 3$. Error bar is one standard deviation. Data from Ref. 39.

as $N \to \infty$ at fixed temperature $T > T_\theta$. The critical exponents ν and Δ_1 are universal. The amplitudes A_{R_e}, A_{R_g}, b_{R_e}, b_{R_g}, b_Ψ are nonuniversal; in fact, even the *signs* of the correction-to-scaling amplitudes b_{R_e}, b_{R_g}, b_Ψ are non-universal. However, the RG theory also predicts that the dimensionless amplitude *ratios* A_{R_g}/A_{R_e}, Ψ^*, b_{R_g}/b_{R_e} and b_Ψ/b_{R_e} are universal.[52,203]

So there is no reason why the correction-to-scaling amplitudes should have any particular sign. In the continuum Edwards model, the effective exponents $\nu_{eff,R_e} \equiv \frac{1}{2} d\log\langle R_e^2\rangle / d\log N$ and $\nu_{eff,R_g} \equiv \frac{1}{2} d\log\langle R_g^2\rangle / d\log N$ and the interpenetration ratio Ψ all approach their asymptotic values *from below*[15,22–24,209]: that is, b_{R_e}, $b_{R_g} > 0$ and $b_\Psi < 0$. On the other hand, in lattice self-avoiding walks, these quantities approach their asymptotic values *from above*[39,203]; and the same occurs in the bead-rod model with sufficiently large bead diameter.[224] Indeed, this latter behavior is almost obvious qualitatively: short self-avoiding walks behave roughly like hard spheres; only at larger N does one see the softer excluded volume (smaller Ψ) characteristic of a fractal object. All these models agree closely, as they should, for the leading *universal* quantities ν, A_{R_g}/A_{R_e} and Ψ^*; and they agree reasonably well for the universal correction-to-scaling quantities Δ_1, b_{R_g}/b_{R_e} and b_Ψ/b_{R_e}.

In summary, the error of all two-parameter theories is to fail to distinguish correctly which quantities are universal and which are non-universal. In particular, the modern two-parameter theory begins from one special model—the continuum Edwards model—and assumes (incorrectly) that it can describe certain aspects of polymer behavior (e.g., the sign of approach to Ψ^*) which in reality are non-universal.

However, this is not the end of the story: The continuum Edwards model *does* in fact describe universal properties of polymer molecules, albeit not the behavior as $N \to \infty$ at fixed temperature $T > T_\theta$. Rather, this theory describes the *universal crossover scaling behavior in an infinitesimal region just above the theta temperature*, namely the limit $N \to \infty$, $T \to T_\theta$ with $x \equiv N^\phi(T - T_\theta)$ fixed, where ϕ is a suitable *crossover exponent*. More precisely, for suitably chosen exponents ϕ and ν_θ, the following limits are expected to exist:

$$f_{R_e}(x) \equiv \lim_{\substack{N\to\infty \\ T\to T_\theta \\ x\equiv N^\phi(T-T_\theta)\ \text{fixed}}} \frac{\langle R_e^2\rangle_{N,T}}{N^{2\nu_\theta}} \tag{2.100a}$$

$$f_{R_g}(x) \equiv \lim_{\substack{N\to\infty \\ T\to T_\theta \\ x\equiv N^\phi(T-T_\theta)\ \text{fixed}}} \frac{\langle R_g^2\rangle_{N,T}}{N^{2\nu_\theta}} \tag{2.100b}$$

$$f_{\Psi}(x) \equiv \lim_{\substack{N\to\infty \\ T\to T_\theta \\ x\equiv N^\phi(T-T_\theta)\ \text{fixed}}} \Psi(N,T) \tag{2.100c}$$

The exponents ϕ and ν_θ are universal, and the *crossover scaling functions* f_{R_e}, f_{R_g} and f_Ψ are universal modulo a rescaling of abscissa and ordinate. The exponents are believed to take the values

$$\phi = \begin{cases} 2 - \frac{d}{2} & \text{for } 3 < d < 4 \\ \frac{1}{2} \times \log^{-4/11} & \text{for } d = 3\ [225,226] \\ \frac{1}{2} + \frac{3\epsilon}{22} + \ldots & \text{for } d = 3 - \epsilon\ [44\text{-}46,225,226] \\ \frac{3}{7} & \text{for } d = 2\ [47] \end{cases} \tag{2.101}$$

$$\nu_\theta = \begin{cases} \frac{1}{2} & \text{for } d \geq 3 \\ \frac{1}{2} + \frac{2\epsilon^2}{363} + \ldots & \text{for } d = 3 - \epsilon\ [44\text{-}46,225,226] \\ \frac{4}{7} & \text{for } d = 2\ [47] \end{cases} \tag{2.102}$$

The functions f_{R_e} and f_{R_g} (and f_Ψ at least for $x \geq 0$) are monotonically increasing functions of their argument $x \equiv N^\phi(T - T_\theta)$, with the asymptotic behavior

$$f_{R_e}(x), f_{R_g}(x) \sim \begin{cases} x^{2(\nu-\nu_\theta)/\phi} & \text{as } x \to +\infty \\ (-x)^{2(\nu_{coll}-\nu_\phi)/\phi} & \text{as } x \to -\infty \end{cases} \tag{2.103a}$$

$$f_\Psi(x) \sim \begin{cases} \Psi^* \ (>0) & \text{as } x \to +\infty \\ \text{unknown } (<0) & \text{as } x \to -\infty \end{cases} \tag{2.103b}$$

where $\nu_{coll} = 1/d$. Then the claim[222] is that, for $3 \leq d < 4$, the functions $f_{R_e}(x)$, $f_{R_g}(x)$ and $f_\Psi(x)$ for $x \geq 0$ are given precisely by the continuum Edwards model, modulo the nonuniversal rescaling of abscissa and ordinate:

$$f_{R_e}(x) = K_1 \alpha_R^2(K_2 x) \tag{2.104a}$$

$$f_{R_g}(x) = (K_1/6)\alpha_S^2(K_2 x) \tag{2.104b}$$

$$f_\Psi(x) = \tilde{h}(K_2 x)/\alpha_S^d(K_2 x) \tag{2.104c}$$

Here $\alpha_R^2(z), \alpha_S^2(z)$ and $\tilde{h}(z) \equiv zh(z)$ are the conventional expansion and second virial factors of the continuum Edwards model,[12,15,22–24,210] and K_1 and K_2 are nonuniversal scale factors. Thus, the continuum Edwards model *is* a correct theory for a certain limiting regime in the molecular-weight/temperature plane—but this regime is *not* the one previously thought. The explanation of eq. (2.104) relies on a Wilson–deGennes-type renormalization group[13,227]; see Ref. 222 for details, and Ref. 223 for further discussion.

It will be very interesting to test the predictions (2.104) numerically. But this will require better algorithms for simulating SAWs near the theta point (see Section 2.9.2).

2.8.2 Linear polymers in dimension $d = 2$

Dimension $d = 2$ is very special: for many statistical–mechanical systems, the critical exponents can be determined *exactly* (though non-rigorously) by Coulomb-gas[228,229] and/or conformal-invariance[230–232] arguments. This is the case for the two-dimensional SAW, for which we know the exact exponents $\nu = 3/4$ and $\gamma = 43/32$.[228] The unknown quantities in this model are the various universal amplitude ratios, which together determine the typical shape of a SAW and the strength of its interactions with other SAWs. The pivot algorithm has been employed[164] to obtain extremely acurate values for the limiting ratios

$$A_\infty = \lim_{N\to\infty} \frac{\langle R_g^2 \rangle_N}{\langle R_e^2 \rangle_N} = 0.140\,26 \pm 0.000\,11 \tag{2.105}$$

$$B_\infty = \lim_{N\to\infty} \frac{\langle R_m^2 \rangle_N}{\langle R_e^2 \rangle_N} = 0.439\,62 \pm 0.000\,33 \tag{2.106}$$

In particular, this confirms the beautiful conformal-invariance prediction[164,233]

$$\frac{246}{91}A_\infty - 2B_\infty + \frac{1}{2} = 0. \tag{2.107}$$

Another open question, which has attracted a lot of work, concerns the correction-to-scaling exponents in the two-dimensional SAW. We are currently using the pivot algorithm to investigate this question.[204]

2.9 Conclusions

2.9.1 *Practical recommendations*

What is the upshot of all this for the practicing polymer scientist, who wants to know which algorithm to use when? Here are a few recommendations:

(1) When simulating linear polymers for the purpose of studying *global* observables (e.g., the critical exponent ν, universal amplitude ratios, etc.), use the pivot algorithm. For initialization, use dimerization if this can be done in a CPU time less than half of your total planned run length; otherwise use the method proposed in Ref. 168, but be careful to discard at least $\sim N/f$ iterations at the beginning of the run.

(2) When simulating linear polymers for the purpose of studying *local* observables (e.g., number of bends, nearest-neighbor contacts, etc.), use the pivot algorithm as described above, but make sure that the run length is $\gtrsim 1000N/f$ iterations. Alternatively, use the slithering-tortoise algorithm or the incomplete-enumeration algorithm.

(3) To obtain the critical exponent γ, use the join-and-cut algorithm (together with the pivot algorithm for the N-conserving moves).

(4) To obtain the critical exponent α_{sing}, use the BFACF + cut-and-paste algorithm.

(5) To obtain the connective constant μ, use the slithering-tortoise algorithm or the incomplete-enumeration algorithm.

Of course, these recommendations are not engraved in stone; other algorithms could be useful in certain situations.

2.9.2 *Open problems*

There are numerous open problems concerning the behavior of the algorithms discussed in this article. Among the most important ones are:

(1) For $d \leq 4$, does there exist any static Monte Carlo algorithm for generating a random N-step SAW (with exactly uniform distribution) in a mean CPU time that is bounded by a polynomial in N? (Section 2.4.3)

(2) What is the precise behavior of the enrichment algorithm as $N \to \infty$? (Section 2.5.2)

(3) What is the precise behavior of the incomplete-enumeration algorithm as $N \to \infty$? (Section 2.5.3)
(4) Are local N-conserving algorithms necessarily nonergodic also in dimension $d \geq 4$? (Section 2.6.4.1)
(5) What is the dynamic critical exponent of the various local N-conserving algorithms (restricted to the ergodic class of a straight rod)? Is it exactly $2 + 2\nu$ for algorithms not having special conservation laws? What about for algorithms having special conservation laws, such as Verdier–Stockmayer? (Section 2.6.4.1)
(6) What is the dynamic critical exponent for the various bilocal (or mixed local/bilocal) algorithms? Is the conjecture $\tau \sim N^2$ exact, approximate or wrong? (Section 2.6.4.2)
(7) Can we improve our theoretical understanding of the acceptance fraction and dynamic critical behavior of the pivot algorithm? (Section 2.6.4.3)
(8) Are there any bilocal (or mixed local/bilocal) algorithms that are ergodic for the fixed-N, fixed-x ensemble? If so, what is their dynamic critical behavior? (Section 2.6.5.1)
(9) What is the dynamic critical exponent of the fixed-N cut-and-paste algorithm? (Section 2.6.5.2)
(10) What is the precise dynamic critical exponent of the slithering-tortoise algorithm? Is it strictly between 2 and $1 + \gamma$? (Section 2.6.6.1)
(11) What is the precise dynamic critical exponent of the BFACF algorithm? Is it exactly 4ν? (Section 2.6.7.1)
(12) What is the precise dynamic critical behavior of the BFACF + cut-and-paste algorithm, as a function of $\langle N \rangle$ and p_{nl}? (Section 2.6.7.2)
(13) Can the Karp–Luby algorithm be generalized to compute virial coefficients B_k for $k \geq 3$? (Section 2.7.2)

In addition, there are many interesting open problems concerning the adaptation of these algorithms—or the invention of new algorithms—for polymeric systems more complicated than a single SAW (athermal linear polymer) in infinite space. For lack of space, I can merely list these problems and give a few (grossly incomplete) bibliographical references.

SAWs in confined geometries. In recent years there has been much interest in studying SAWs attached to surfaces or confined to specified regions (wedges, slabs, tubes, etc.).[234,235] Most of the algorithms described here do generalize to such situations; but it is no longer guaranteed that they are even ergodic, much less efficient. This requires a case-by-case study. See Ref. 40, Section 2.4.2 for some references.

SAWs with nearest-neighbor attraction ($\to$ *theta point*). The transition of polymer conformation from the high-temperature (good-solvent) regime to the theta point to the collapsed regime is well modeled by the self-avoiding walk with nearest-neighbor attraction. One of the most fundamental problems in polymer physics is to understand quantitatively the details of this crossover, both in $d = 3$[222,225,226] and in $d = 2$.[236] The dynamic algorithms described here can easily be modified to handle a nearest-neighbor interaction, by inserting a Metropolis accept/reject step. But their efficiency may deteriorate markedly in the neighborhood of the theta point and even more drastically in the collapsed regime; this requires a detailed study for each algorithm. See Refs 171, 172, 237, 238 for some recent preliminary work on the pivot algorithm; and see Ref. 40 for citations of older work.

Branched polymers. In recent years much attention has been devoted to the theoretical and experimental study of branched polymers, whose behavior is quite different from that of linear or ring polymers. The simplest case is that of branched polymers with *fixed* topology, such as star or comb polymers. Many of the algorithms for linear polymers can be adapted to this case, although both the ergodicity and the efficiency are nontrivial problems. See Ref. 40, Section 2.4.3 for some references, and see Refs 169–172, 174, 176 for recent work using the pivot algorithm.

A more difficult problem is that of branched polymers with *variable* topology, such as arbitrarily-branched lattice trees. Early works used simple sampling,[239] while more recent works have used variants of the slithering-tortoise or incomplete-enumeration algorithms.[240] However, in all these cases it has been difficult to generate branched polymers of more than ≈ 50 segments, because of the critical slowing-down. On the other hand, the need for large polymers is even more acute here than for linear or ring polymers, since one needs a rather large number of segments in order to feel the full effects of the branching. (That is, one expects large corrections to scaling in which the effective exponents for small N are biased toward the unbranched-polymer values.) A very promising non-local algorithm was devised recently by Janse van Rensburg and Madras.[241]

Multi-chain systems. Much current work in polymer science, both theoretical and experimental, focuses on semidilute and concentrated solutions and on melts. The simulation of multi-chain systems poses very difficult problems: for example, it is an open question whether *any* known algorithm is ergodic at *any* nonzero density! In the semidilute case, many of the algorithms described here can be applied with minor modification, and their performance in practice (if one disregards the ergodicity problem) will probably be similar to that for single chains. On the other hand, dense solutions and melts constitute a much more difficult problem, due to the possibility of "gridlock". Several algorithms have been proposed,[242] involving both local-deformation and subunit-exchange moves, but the relaxation seems in general to be very slow. Progress in this area will probably require new physical and algorithmic ideas. See Ref. 40, Section 3 for some references.

Acknowledgments

Many of the ideas reported here have grown out of joint work with my colleagues Alberto Berretti, Sergio Caracciolo, Tony Guttmann, Greg Lawler, Bin Li, Neal Madras, Andrea Pelissetto, and Larry Thomas. I thank them for many pleasant and fruitful collaborations. I have also learned much from discussions with Carlos Aragão de Carvalho, Jim Barrett, Marvin Bishop, Bertrand Duplantier, Roberto Fernández, Michael Fisher, Jürg Fröhlich, Paul Gans, Takashi Hara, Mal Kalos, Bernie Nickel, Gordon Slade, Stu Whittington, and too many others to thank here. Finally, I wish to thank Kurt Binder and Neal Madras for many helpful comments on early drafts of this manuscript.

The author's research is supported in part by U.S. National Science Foundation grant DMS-9200719, U.S. Department of Energy contract DE-FG02-90ER40581 and NATO Collaborative Research Grant CRG 910251, as well as by a New York University Research Challenge Fund grant. Acknowledgment is also made to the donors of the Petroleum Research Fund, administered by the American Chemical Society, for partial support of this research.

References

1. W. J. C. Orr, *Trans. Faraday Soc.* **43**, 12 (1947).
2. E. W. Montroll, *J. Chem. Phys.* **18**, 734 (1950).
3. M. E. Fisher, *Rev. Mod. Phys.* **46**, 597 (1974).
4. M. E. Fisher, in *Critical Phenomena (Stellenbosch 1982)*, Lecture Notes in Physics #186, edited by F. J. W. Hahne (Springer-Verlag, Berlin-Heidelberg-New York, 1983), pp. 1–139.
5. P. G. DeGennes, *Scaling Concepts in Polymer Physics* (Cornell University Press, Ithaca, NY, 1979).
6. A. Yamamoto, M. Fujii, G. Tanaka, and H. Yamakawa, *Polym. J.* **2**, 799 (1971).
7. M. Fukuda, M. Fukutomi, Y. Kato, and T. Hashimoto, *J. Polym. Sci., Polym. Phys. Ed.* **12**, 871 (1974).
8. Y. Miyaki, Y. Einaga, and H. Fujita, *Macromolecules* **11**, 1180 (1978).
9. J. P. Cotton, *J. Physique Lett.* **41**, L231 (1980).
10. A. D. Sokal, Fundamental problems in the static scaling behavior of high-molecular-weight polymers in dilute solution II. Critical review of the experimental literature, in preparation.
11. N. Madras and G. Slade, *The Self-Avoiding Walk* (Birkhäuser, Boston-Basel-Berlin, 1993).
12. H. Yamakawa, *Modern Theory of Polymer Solutions* (Harper and Row, New York, 1971).
13. K. G. Wilson and J. Kogut, *Phys. Reports* **12C**, 75 (1974).

14. E. Brézin, J.-C. LeGuillou, and J. Zinn-Justin, in *Phase Transitions and Critical Phenomena*, Vol. 6, edited by C. Domb and M. S. Green (Academic Press, London, 1976).
15. J. des Cloizeaux and G. Jannink, *Polymers in Solution: Their Modelling and Structure* (Oxford University Press, New York, 1990).
16. G. A. Baker, Jr., B. G. Nickel, and D. I. Meiron, *Phys. Rev.* **B17**, 1365 (1978).
17. J. C. LeGuillou and J. Zinn-Justin, *Phys. Rev.* **B21**, 3976 (1980).
18. D. Z. Albert, *Phys. Rev.* **B25**, 4810 (1982).
19. J. C. LeGuillou and J. Zinn-Justin, *J. Physique Lett.* **46**, L137 (1985).
20. J. C. LeGuillou and J. Zinn-Justin, *J. Physique* **50**, 1365 (1989).
21. D. B. Murray and B. G. Nickel, Revised estimates for critical exponents for the continuum *n*-vector model in 3 dimensions, preprint (1991).
22. M. Muthukumar and B. G. Nickel, *J. Chem. Phys.* **80**, 5839 (1984).
23. M. Muthukumar and B. G. Nickel, *J. Chem. Phys.* **86**, 460 (1987).
24. F. C. Shanes and B. G. Nickel, Calculation of the radius of gyration for a linear flexible polymer chain with excluded volume interaction, *J. Chem. Phys.* (to appear).
25. B. G. Nickel, in *Phase Transitions* (1980 Cargèse lectures), edited by M. Lévy, J.-C. LeGuillou and J. Zinn-Justin (Plenum, New York-London, 1982).
26. B. G. Nickel, *Physica* **A177**, 189 (1991).
27. A. R. Conway, I. G. Enting, and A. J. Guttmann, *J. Phys.* **A26**, 1519 (1993).
28. D. S. Gaunt and A. J. Guttmann, in *Phase Transitions and Critical Phenomena*, Vol. 3, edited by C. Domb and M. S. Green (Academic Press, London, 1974).
29. G. A. Baker Jr. and P. Graves-Morris, *Padé Approximants*, 2 vols (Addison-Wesley, Reading, MA, 1981).
30. A. J. Guttmann, in *Phase Transitions and Critical Phenomena*, Vol. 13, edited by C. Domb and J. L. Lebowitz (Academic Press, London, 1989).
31. G. W. King, U. S. National Bureau of Standards, Applied Mathematics Series, No. 12 (1951).
32. F. T. Wall, L. A. Hiller Jr., and D. J. Wheeler, *J. Chem. Phys.* **22**, 1036 (1954).
33. G. H. Hardy, *A Mathematician's Apology* (Cambridge University Press, 1967).
34. A. D. Sokal, in *VIIIth International Congress on Mathematical Physics*, edited by R. Sénéor and M. Mebkhout (World Scientific, Singapore, 1987).
35. A. D. Sokal, in *Computer Simulation Studies in Condensed Matter Physics: Recent Developments*, edited by D. P. Landau and H.-B. Schüttler (Springer-Verlag, Berlin-Heidelberg, 1988).
36. A. D. Sokal, *Monte Carlo Methods in Statistical Mechanics: Foundations and New Algorithms*, Cours de Troisième Cycle de la Physique en Suisse Romande (Lausanne, June 1989).
37. A. D. Sokal, in *Quantum Fields on the Computer*, edited by M. Creutz (World Scientific, Singapore, 1992).
38. H. Müller-Krumbhaar and K. Binder, *J. Stat. Phys.* **8**, 1 (1973).
39. B. Li, N. Madras, and A. D. Sokal, Critical exponents, hyperscaling and universal amplitude ratios for two- and three-dimensional self-avoiding walks, *J. Stat. Phys.* (to appear August 1995).
40. K. Kremer and K. Binder, *Comput. Phys. Reports* **7**, 259 (1988).

41. R. Vilanove, D. Poupinct, and F. Rondelez, *Macromolecules* **21**, 2880 (1988).
42. D. Poupinet, R. Vilanove, and F. Rondelez, *Macromolecules* **22**, 2491 (1989).
43. M. J. Stephen and J. McCauley, *Phys. Lett.* **A44**, 89 (1973).
44. M. J. Stephen, *Phys. Lett.* **A53**, 363 (1975).
45. A. L. Lewis and F. W. Adams, *Phys. Rev.* **B18**, 5099 (1978).
46. B. Duplantier, *Europhys. Lett.* **1**, 491 (1986). Equations (3.5), (3.6) and (4.7) of this paper have erroneous coefficients; the corrected formulae can be found in B. Duplantier, *J. Chem. Phys.* **86**, 4233 (1987) and *Phys. Rev.* **A38**, 3647 (1988).
47. B. Duplantier and H. Saleur, *Phys. Rev. Lett.* **59**, 539 (1987).
48. T. Hara and G. Slade, *Commun. Math. Phys.* **147**, 101 (1992).
49. T. Hara and G. Slade, *Rev. Math. Phys.* **4**, 235 (1992).
50. T. Hara, G. Slade, and A. D. Sokal, *J. Stat. Phys.* **72**, 479 (1993); **78**, 1187(E) (1995).
51. T. Hara and G. Slade, The self-avoiding-walk and percolation critical points in high dimensions, Isaac Newton Institute preprint NI93011 (October 1993), mp_arc@math.utexas.edu #93-286—to appear in *Combinatorics, Probability and Computing.*
52. V. Privman, P. C. Hohenberg, and A. Aharony, in *Phase Transitions and Critical Phenomena*, Vol. 14, edited by C. Domb and J. L. Lebowitz (Academic Press, London-San Diego, 1991).
53. G. Slade, *Commun. Math. Phys.* **110**, 661 (1987).
54. G. Slade, *Ann. Probab.* **17**, 91 (1989).
55. G. Slade, *J. Phys.* **A21**, L417 (1988).
56. J. des Cloizeaux, private communication cited in E. Brézin, in *Order and Fluctuation in Equilibrium and Nonequilibrium Statistical Mechanics* (17th Solvay Conference, 1978), edited by G. Nichols, G. Dewel, and J. W. Turner (Wiley-Interscience, New York, 1981).
57. M. E. Fisher, *Rep. Prog. Phys.* **30**, 615 (1967).
58. H. E. Stanley, *Introduction to Phase Transitions and Critical Phenomena* (Oxford University Press, Oxford, 1971).
59. M. Daoud, J. P. Cotton, B. Farnoux et al., *Macromolecules* **8**, 804 (1975).
60. V. J. Emery, *Phys. Rev.* **B11**, 239 (1975).
61. C. Aragão de Carvalho, S. Caracciolo, and J. Fröhlich, *Nucl. Phys.* **B215 [FS7]**, 209 (1983).
62. R. Fernández, J. Fröhlich, and A. D. Sokal, *Random Walks, Critical Phenomena, and Triviality in Quantum Field Theory* (Springer-Verlag, Berlin-Heidelberg-New York, 1992).
63. J. M. Hammersley and D. C. Handscomb, *Monte Carlo Methods* (Methuen, London, 1964).
64. J. G. Kemeny and J. L. Snell, *Finite Markov Chains* (Springer-Verlag, New York, 1976).
65. M. Iosifescu, *Finite Markov Processes and Their Applications* (Wiley, Chichester, 1980).
66. K. L. Chung, *Markov Chains with Stationary Transition Probabilities*, 2nd edition (Springer-Verlag, New York, 1967).

67. E. Nummelin, *General Irreducible Markov Chains and Non-Negative Operators* (Cambridge University Press, Cambridge, UK, 1984).
68. Z. Šidak, *Czechoslovak Math. J.* **14**, 438 (1964).
69. A. D. Sokal and L. E. Thomas, *J. Stat. Phys.* **54**, 797 (1989).
70. P. C. Hohenberg and B. I. Halperin, *Rev. Mod. Phys.* **49**, 435 (1977).
71. S. Caracciolo and A. D. Sokal, *J. Phys.* **A19**, L797 (1986).
72. A. D. Sokal, *Nucl. Phys. B* (Proc. Suppl.) **20**, 55 (1991).
73. A. J. Guttmann and I. G. Enting, *J. Phys.* **A21**, L165 (1988).
74. A. J. Guttmann, private communication, cited in Ref. 50.
75. A. J. Guttmann, *J. Phys.* **A11**, L103 (1978).
76. A. J. Guttmann, *J. Phys.* **A14**, 233 (1981).
77. M. E. Fisher and D. S. Gaunt, *Phys. Rev.* **A133**, 224 (1964).
78. H. Kesten, *J. Math. Phys.* **5**, 1128 (1964).
79. A. M. Nemirovsky, K. F. Freed, T. Ishinabe, and J. F. Douglas, *J. Stat. Phys.* **67**, 1083 (1992).
80. F. T. Wall, R. J. Rubin, and L. M. Isaacson, *J. Chem. Phys.* **27**, 186 (1957).
81. J. M. Hammersley and K. W. Morton, *J. R. Statist. Soc.* **B16**, 23 (1954).
82. M. N. Rosenbluth and A. W. Rosenbluth, *J. Chem. Phys.* **23**, 356 (1955).
83. S. Hemmer and P. C. Hemmer, *J. Chem. Phys.* **81**, 584 (1984).
84. P. C. Hemmer and S. Hemmer, *Phys. Rev.* **A34**, 3304 (1986).
85. J. W. Lyklema and K. Kremer, *J. Phys.* **A19**, 279 (1986).
86. J. Batoulis and K. Kremer, *J. Phys.* **A21**, 127 (1988).
87. F. L. McCrackin, *J. Res. Nat. Bur. Stand.* **B76**, 193 (1972).
88. S. J. Fraser and M. A. Winnik, *J. Chem. Phys.* **70**, 575 (1979).
89. H. Meirovitch, *J. Phys.* **A15**, L735 (1982).
90. H. Meirovitch, *J. Chem. Phys.* **79**, 502 (1983); **81**, 1053 (E) (1984).
91. H. Meirovitch, *Macromolecules* **18**, 563, 569 (1985).
92. H. Meirovitch, *J. Chem. Phys.* **89**, 2514 (1988).
93. K. Suzuki, *Bull. Chem. Soc. Japan* **41**, 538 (1968).
94. T. H. Cormen, C. E. Leiserson, and R. L. Rivest, *Introduction to Algorithms* (MIT Press/McGraw-Hill, Cambridge MA/New York, 1990).
95. E. Horowitz and S. Sahni, *Fundamentals of Computer Algorithms* (Computer Science Press, Rockville, MD, 1978), Chapter 3.
96. N. Madras and A. D. Sokal, *J. Stat. Phys.* **50**, 109 (1988).
97. Z. Alexandrowicz, *J. Chem. Phys.* **51**, 561 (1969).
98. Z. Alexandrowicz and Y. Accad, *J. Chem. Phys.* **54**, 5338 (1971).
99. F. T. Wall and J. J. Erpenbeck, *J. Chem. Phys.* **30**, 634 (1959).
100. A. D. Sokal, unpublished manuscript (1984).
101. H. Kesten, *J. Math. Phys.* **4**, 960 (1963).
102. T. E. Harris, *The Theory of Branching Processes* (Springer, Berlin, 1963 and Dover, New York, 1989).
103. R. Grishman, *J. Chem. Phys.* **58**, 220 (1973).
104. S. Redner and P. J. Reynolds, *J. Phys.* **A14**, 2679 (1981).
105. D. Dhar and P. M. Lam, *J. Phys.* **A19**, L1057 (1986).
106. P. H. Verdier and W. H. Stockmayer, *J. Chem. Phys.* **36**, 227 (1962).
107. O. J. Heilmann, *Mat. Fys. Medd. Dan. Vid. Selsk.* **37**, No. 2 (1968).
108. O. J. Heilmann and J. Rotne, *J. Stat. Phys.* **27**, 19 (1982).

109. M. Lax and C. Brender, *J. Chem. Phys.* **67**, 1785 (1977).
110. D. E. Kranbuehl and P. H. Verdier, *J. Chem. Phys.* **71**, 2662 (1979).
111. M. E. Fisher and Z. Rácz, *Phys. Rev.* **B13**, 5039 (1976).
112. M. Suzuki, *Int. J. Magnetism* **1**, 123 (1971).
113. K. Binder, *Phys. Rev.* **B8**, 3423 (1973).
114. M. Delbrück, in *Mathematical Problems in the Biological Sciences* (Proc. Symp. Appl. Math., Vol. 14), edited by R. E. Bellman (American Mathematical Society, Providence, 1962), pp. 55–63.
115. P. H. Verdier, *J. Comput. Phys.* **4**, 204 (1969).
116. H. J. Hilhorst and J. M. Deutch, *J. Chem. Phys.* **63**, 5153 (1975).
117. M. T. Gurler, C. C. Crabb, D. M. Dahlin, and J. Kovac, *Macromolecules* **16**, 398 (1983).
118. T. M. Birshtein, V. N. Gridnev, Yu. Ya. Gotlib, and A. M. Skvortsov, *Vysokomol. Soyed.* **A19**, 1398 (1977) (*Polym. Sci. USSR* **19**, 1612 (1977)).
119. Yu. A. Taran and L. B. Stroganov, *Vysokomol. Soyed.* **A20**, 1787 (1978) (*Polym. Sci. USSR* **20**, 2007 (1978)).
120. P. H. Verdier and D. E. Kranbuehl, *Polym. Preprints* (ACS) **17**, 148 (1976).
121. H. Boots and J. M. Deutch, *J. Chem. Phys.* **67**, 4608 (1977).
122. D. E. Kranbuehl and P. H. Verdier, *Polym. Preprints* (ACS) **21**, 195 (1980).
123. P. Romiszowski and W. H. Stockmayer, *J. Chem. Phys.* **80**, 485 (1984).
124. H. Meirovitch, *Macromolecules* **17**, 2038 (1984).
125. N. Madras and A. D. Sokal, *J. Stat. Phys.* **47**, 573 (1987).
126. A. T. Clark, M. Lal, M. A. Turpin, and K. A. Richardson, *Faraday Discuss., Chem. Soc.* **59**, 189 (1975).
127. M. Lal and R. F. T. Stepto, *J. Polym. Sci.: Polym. Symp.* **61**, 401 (1977).
128. L. Monnerie and F. Gény, *J. Chim. Phys.* **66**, 1691 (1969).
129. L. Monnerie, F. Gény, and J. Fouquet, *J. Chim. Phys.* **66**, 1698 (1969).
130. F. Gény and L. Monnerie, *J. Chim. Phys.* **66**, 1708 (1969).
131. K. Kremer, A. Baumgärtner, and K. Binder, *J. Phys.* **A15**, 2879 (1981).
132. S. Caracciolo, A. Pelissetto, and A. D. Sokal, *J. Stat. Phys.* **60**, 1 (1990).
133. P. H. Verdier, *J. Chem. Phys.* **45**, 2118 (1966).
134. P. H. Verdier, *J. Chem. Phys.* **45**, 2122 (1966).
135. P. H. Verdier, *J. Chem. Phys.* **52**, 5512 (1970).
136. P. H. Verdier, *J. Chem. Phys.* **59**, 6119 (1973).
137. K. Iwata and M. Kurata, *J. Chem. Phys.* **50**, 4008 (1969).
138. K. Iwata, *J. Chem. Phys.* **54**, 12 (1971).
139. R. A. Orwoll and W. H. Stockmayer, *Adv. Chem. Phys.* **15**, 305 (1969).
140. W. H. Stockmayer, W. Gobush, and R. Novich, *Pure Appl. Chem.* **26**, 537 (1971).
141. T. E. Harris, *J. Appl. Prob.* **2**, 323 (1965).
142. R. Gisselquist, *Ann. Prob.* **1**, 231 (1973).
143. R. Arratia, *Ann. Prob.* **11**, 362 (1983).
144. A. K. Kron, *Vysokomol. Soyed.* **7**, 1228 (1965) (*Polym. Sci. USSR* **7**, 1361 (1965)).
145. A. K. Kron, O. B. Ptitsyn, A. M. Skvortsov, and A. K. Fedorov, *Molek. Biol.* **1**, 576 (1967) (*Molec. Biol.* **1**, 487 (1967)).
146. F. T. Wall and F. Mandel, *J. Chem. Phys.* **63**, 4592 (1975).

147. F. Mandel, *J. Chem. Phys.* **70**, 2984 (1979).
148. A. D. Sokal, unpublished (1993).
149. I. Webman, J. L. Lebowitz, and M. H. Kalos, *J. Physique* **41**, 579 (1980).
150. J. Reiter, *Macromolecules* **23**, 3811 (1990).
151. A. Pelissetto, unpublished (1992).
152. M. Lal, *Molec. Phys.* **17**, 57 (1969).
153. O. F. Olaj and K. H. Pelinka, *Makromol. Chem.* **177**, 3413 (1976).
154. A. T. Clark and M. Lal, *Br. Polym. J.* **9**, 92 (1977).
155. S. D. Stellman and P. J. Gans, *Macromolecules* **5**, 516 (1972).
156. S. D. Stellman and P. J. Gans, *Macromolecules* **5**, 720 (1972).
157. J. J. Freire and A. Horta, *J. Chem. Phys.* **65**, 4049 (1976).
158. J. G. Curro, *J. Chem. Phys.* **61**, 1203 (1974).
159. J. G. Curro, *J. Chem. Phys.* **64**, 2496 (1976).
160. H. L. Scott, Jr., *Biochem. Biophys. Acta* **469**, 264 (1977).
161. B. MacDonald, N. Jan, D. L. Hunter, and M. O. Steinitz, *J. Phys.* **A18**, 2627 (1985).
162. D. L. Hunter, N. Jan, and B. MacDonald, *J. Phys.* **A19**, L543 (1986).
163 K. Kelly, D. L. Hunter, and N. Jan, *J. Phys.* **A20**, 5029 (1987).
164. S. Caracciolo, A. Pelissetto, and A. D. Sokal, *J. Phys.* **A23**, L969 (1990).
165. S. Caracciolo, A. Pelissetto, and A. D. Sokal, *J. Stat. Phys.* **67**, 65 (1992).
166. S. Caracciolo, G. Ferraro, and A. Pelissetto, *J. Phys.* **A24**, 3625 (1991).
167. G. Zifferer, *Macromolecules* **23**, 3166 (1990).
168. G. Zifferer, *Molec. Simul.* **6**, 103 (1991).
169. G. Zifferer, *Makromol. Chem.* **191**, 2717 (1990).
170. G. Zifferer, *Makromol. Chem.* **192**, 1555 (1991).
171. G. Zifferer, *Makromol. Chem., Theory Simul.* **1**, 55 (1992).
172. G. Zifferer, *Makromol. Chem., Theory Simul.* **2**, 653 (1993).
173. M. Bishop and J. H. R. Clarke, *J. Chem. Phys.* **94**, 3936 (1991).
174. M. Bishop, J. H. R. Clarke, A. Rey and J. J. Freire, *J. Chem. Phys.* **94**, 4009 (1991).
175. M. Bishop and C. J. Saltiel, *J. Chem. Phys.* **95**, 606 (1991).
176. M. Bishop, J. H. R. Clarke, A. Rey, and J. J. Freire, *J. Chem. Phys.* **95**, 608 (1991).
177. A. J. Chorin, *Commun. Math. Phys.* **132**, 519 (1990).
178. N. Eizenberg and J. Klafter, *J. Chem. Phys.* **99**, 3976 (1993).
179. N. Madras, A. Orlitsky, and L. A. Shepp, *J. Stat. Phys.* **58**, 159 (1990).
180. L. E. Dubins, A. Orlitsky, J. A. Reeds, and L. A. Shepp, *IEEE Trans. Inform. Theory* **34**, 1509 (1988).
181. E. J. Janse van Rensburg, S. G. Whittington, and N. Madras, *J. Phys.* **A23**, 1589 (1990).
182. A. Berretti and A. D. Sokal, *J. Stat. Phys.* **40**, 483 (1985).
183. G. F. Lawler and A. D. Sokal, *Trans. Am. Math. Soc.* **309**, 557 (1988).
184. D. Randall and A. Sinclair, Testable algorithms for self-avoiding walks, Berkeley/ICSI preprint (1993).
185. G. F. Lawler and A. D. Sokal (unpublished work).
186. B. Berg and D. Foerster, *Phys. Lett.* **106B**, 323 (1981).
187. C. Aragão de Carvalho and S. Caracciolo, *J. Physique* **44**, 323 (1983).

188. E. J. Janse van Rensburg and S. G. Whittington, *J. Phys.* **A24**, 5553 (1991).
189. E. J. Janse van Rensburg, *J. Phys.* **A25**, 1031 (1992).
190. A. D. Sokal and L. E. Thomas, *J. Stat. Phys.* **51**, 907 (1988).
191. S. Caracciolo, A. Pelissetto, and A. D. Sokal, *J. Stat. Phys.* **63**, 857 (1991).
192. G. Burde and H. Zieschang, *Knots* (de Gruyter, Berlin, 1985).
193. E. J. Janse van Rensburg and S. G. Whittington, *J. Phys.* **A23**, 3573 (1990).
194. F. L. McCrackin, *J. Chem. Phys.* **47**, 1980 (1967).
195. P. C. Jurs and J. E. Reissner, *J. Chem. Phys.* **55**, 4948 (1971).
196. M. D'Onofrio, Thesis, Università di Roma I (June 1986).
197. A. Berretti and A. D. Sokal, *Comput. Phys. Commun.* **58**, 1 (1990).
198. D. E. Knuth, *The Art of Computer Programming*, Vol. 3 (Addison-Wesley, Reading, Massachusetts, 1973).
199. R. M. Karp and M. Luby, in *24th IEEE Symposium on Foundations of Computer Science* (IEEE, New York, 1983), pp. 56–64.
200. R. M. Karp, M. Luby, and N. Madras, *J. Algorithms* **10**, 429 (1989).
201. S. D. Silvey, *Statistical Inference* (Chapman and Hall, London, 1975), Chapter 4.
202. F. J. Wegner, *Phys. Rev.* **B5**, 4529 (1972).
203. B. G. Nickel, *Macromolecules* **24**, 1358 (1991).
204. S. Caracciolo, A. J. Guttmann, B. Li, A. Pelissetto, and A. D. Sokal (in preparation).
205. A. J. Guttmann, *J. Phys.* **A20**, 1839 (1987).
206. D. C. Rapaport, *J. Phys.* **A18**, 113 (1985).
207. K. E. Newman and E. K. Riedel, *Phys. Rev.* **B30**, 6615 (1984).
208. P. J. Flory, *Principles of Polymer Chemistry* (Cornell University Press, Ithaca, NY, 1953).
209. K. F. Freed, *Renormalization Group Theory of Macromolecules* (Wiley, New York, 1987).
210. H. Fujita, *Polymer Solutions* (Elsevier, Amsterdam, 1990).
211. C. Domb and A. J. Barrett, *Polymer* **17**, 179 (1976).
212. S. F. Edwards, *Proc. Phys. Soc. Lond.* **85**, 613 (1965).
213. S. R. S. Varadhan, appendix to article of K. Symanzik, in *Local Quantum Theory*, edited by R. Jost (Academic Press, New York-London, 1969).
214. J. Westwater, *Commun. Math. Phys.* **72**, 131 (1980).
215. J. Westwater, *Commun. Math. Phys.* **84**, 459 (1982).
216. A. Bovier, G. Felder and J. Fröhlich, *Nucl. Phys.* **B230**[FS10], 119 (1984).
217. H. Fujita and T. Norisuye, *Macromolecules* **18**, 1637 (1985).
218. K. Huber and W. H. Stockmayer, *Macromolecules* **20**, 1400 (1987).
219. H. Fujita, *Macromolecules* **21**, 179 (1988).
220. J. F. Douglas and K. F. Freed, *Macromolecules* **18**, 201 (1985).
221. J. F. Douglas and K. F. Freed, *J. Phys. Chem.* **88**, 6613 (1984).
222. A. D. Sokal, Static scaling behavior of high-molecular-weight polymers in dilute solution: A reexamination, NYU preprint NYU-TH-93/05/01 (May 1993), hep-lat@ftp.scri.fsu.edu #9305009 (rejected five times by *Phys. Rev. Lett.*). A slightly abridged and revised version appeared in *Europhys. Lett.* **27**, 661 (1994).

223. A. D. Sokal, Fundamental problems in the static scaling behavior of high-molecular-weight polymers in dilute solution I. Critique of two-parameter theories (in preparation).
224. A. J. Barrett and B. G. Nickel (private communication).
225. B. Duplantier, *J. Chem. Phys.* **86**, 4233 (1987).
226. B. Duplantier, *Phys. Rev.* **A38**, 3647 (1988).
227. P. G. De Gennes, *J. Physique Lett.* **39**, L-299 (1978).
228. B. Nienhuis, *J. Stat. Phys.* **34**, 731 (1984).
229. B. Nienhuis, in *Phase Transitions and Critical Phenomena*, Vol. 11, edited by C. Domb and J. L. Lebowitz (Academic Press, London, 1987).
230. A. A. Belavin, A. M. Polyakov, and A. B. Zamolodchikov, *Nucl. Phys.* **B241**, 333 (1984).
231. J. L. Cardy, in *Phase Transitions and Critical Phenomena*, Vol. 11, edited by C. Domb and J. L. Lebowitz (Academic Press, London, 1987).
232. P. Ginsparg, in *Fields, Strings and Critical Phenomena* (Les Houches 1988), edited by E. Brézin and J. Zinn-Justin (North-Holland, Amsterdam-New York, 1989).
233. J. L. Cardy and H. Saleur, *J. Phys.* **A22**, L601 (1989).
234. P. G. DeGennes, *Adv. Colloid Interface Sci.* **27**, 189 (1987).
235. K. De'Bell and T. Lookman, *Rev. Mod. Phys.* **65**, 87 (1993).
236. A. Coniglio, N. Jan, I. Majid and H. E. Stanley, *Phys. Rev.* **B35**, 3617 (1987); B. Duplantier and H. Saleur, *Phys. Rev. Lett.* **59**, 539 (1987); P. H. Poole, A. Coniglio, N. Jan and H. E. Stanley, *Phys. Rev. Lett.* **60**, 1203 (1988); B. Duplantier and H. Saleur, *Phys. Rev. Lett.* **60**, 1204 (1988); H. Meirovitch and H. A. Lim, *Phys. Rev.* **A39**, 4186 (1989); B. Duplantier and H. Saleur, *Phys. Rev. Lett.* **62**, 1368 (1989); H. Meirovitch and H. A. Lim, *Phys. Rev. Lett.* **62**, 2640 (1989); B. Duplantier and H. Saleur, *Phys. Rev. Lett.* **62**, 2641 (1989); I. S. Chang, Y. Shapir and H. Meirovitch, *J. Phys.* **A23**, L537 (1990); C. Vanderzande, A. L. Stella and F. Seno, *Phys. Rev. Lett.* **67**, 2757 (1991); I. S. Chang and H. Meirovitch, *Phys. Rev. Lett.* **69**, 2232 (1992).
237. S. Caracciolo, G. Ferraro, A. Pelissetto, and A. D. Sokal (work in progress).
238. S. Caracciolo, A. Pelissetto, and A. D. Sokal, *Phys. Rev. Lett.* **72**, 179 (1994).
239. S. Redner, *J. Phys.* **A12**, L239 (1979).
240. W. A. Seitz and D. J. Klein, *J. Chem. Phys.* **75**, 5190 (1981); S. Caracciolo and U. Glaus, *J. Stat. Phys.* **41**, 41 (1985); U. Glaus, *J. Phys.* **A18**, L789 (1985); J. A. M. S. Duarte, *J. Phys.* **A19**, 1979 (1986); P. M. Lam, *Phys. Rev.* **A34**, 2339 (1986).
241. E. J. Janse van Rensburg and N. Madras, *J. Phys.* **A25**, 303 (1992).
242. O. F. Olaj, W. Lantschbauer and K. H. Pelinka, *Chemie-Kunststoffe Aktuell* **32**, 199 (1978); M. L. Mansfield, *J. Chem. Phys.* **77**, 1554 (1982); O. F. Olaj and W. Lantschbauer, *Makromol. Chem. Rapid Commun.* **3**, 847 (1982); W. G. Madden, *J. Chem. Phys.* **87**, 1405 (1987) and **88**, 3934 (1988); J. Reiter, G. Zifferer and O. F. Olaj, *Macromolecules* **22**, 3120 (1989).

3

STRUCTURE AND DYNAMICS OF NEUTRAL AND CHARGED POLYMER SOLUTIONS: EFFECTS OF LONG-RANGE INTERACTIONS

Burkhard Dünweg, Mark Stevens, and Kurt Kremer

3.1 Introduction

The physics of polymer solutions has been subject to vast development over the last 20 years. The introduction of the scaling concept by DeGennes marked the beginning of a new era of polymer theory.[1] In contrast to the earlier mean field methods, the scaling theory allowed one to successfully treat the spacially short-ranged interactions which are mediated by a long-range chemical interaction, the backbone of the chain. Within the $n \to 0$ limit of the n-vector model analogy it was possible to use the sophisticated theoretical methods developed for renormalization group investigations of second order phase transitions to study solution properties of polymers.[2] Computer simulations successfully participated in this research, since the scaling concept allowed one to use simple and highly efficient models and methods (see, e.g., the chapter by Sokal in this volume).[3] Characteristic to all the polymer problems now so well understood however is the spatially short range nature of the interaction potential mediated by eventually very long chemical paths. As soon as the spatial interaction range is longer than just a few monomer diameters, effecting the chain extension at even much larger distances, the above-mentioned story of success looks very different. Standard methods, as used for the above-mentioned problems, fail or have to be taken with many reservations.[4] Such long-range interactions can be relevant for dynamic properties, as for the dynamics of a neutral polymer in solution, or for both dynamic and static properties, namely in the case of polyelectrolyte or polyamphilyte solutions. While the first problem has been subject to investigation and discussion for many years, the second problem has been only very recently studied sensibly by computer simulations.

Such long-range interactions pose special challenges to analytic theory and experiment as well as to computer simulations. In the case of the dynamics of neutral chains in solution, the hydrodynamic interaction, mediated by momentum transport through the solvent, falls off with a $1/r$ power law. This introduces significant changes in the dynamical properties of the chains and consequently the solutions. It does not, however, affect the static properties.[1,5] Theoretically even the "simplest" problem of a single chain in a good solvent can only be solved within severe approximations. The case of Θ-solvents or poor solvents is even more difficult due to the additional short-range attractive contacts along the chains. On the other hand, experimentally it is difficult to reach the limit of sufficient dilution to test the basic concepts. For example, in scattering experiments it is very difficult to work at sufficiently small concentration and maintain good contrast. Computer simulations give the chance of studying such questions under idealized and thus cleanly controlled conditions.

For polyelectrolytes, or, more generally, charged chains, the situation is even more severe. Although of very high technical and biological importance, all three approaches, analytic theory, experiment and computer simulations, are in a state of infancy. The electrostatic interactions dominate the chain conformation statistics and pose severe difficulties to the extent that not even the static properties are presently understood. Well-controlled experiments are extremely difficult to perform because of preparation problems and the fact that both chains and solvent scatter light weakly and of comparable strength.[6] Theoretically the long-range Coulomb interaction as well as the influence of charge fluctuations (due to the discrete nature of the ions) are not understood at all but are thought to be crucial, for example, in the stabilization of biological bilayers. The use of computer simulations will, for the first time for charged chains, give direct insight into the conformations of polyelectrolyte chains under realistic but controlled conditions. Considering the huge variety of charged polymer systems, such as stiff polyelectrolytes, e.g., DNA, or flexible polyelectrolytes, such as RNA or NaPSS (sulfonated polystyrene), to weakly charged systems, with a charge only every now and then and long neutral pieces in between, the recent simulations, which will be discussed here, are only very first attempts seriously to investigate the systems.

These are two typical cases where computer simulations slowly mature to play an important role in our effort to gain a better understanding. The simulations allow an investigation under controlled and partially idealized conditions, beyond the experimental limits. The price for this is still relatively small total system sizes and relatively short chains. This chapter gives a short overview of the work done for polymer systems governed by long-range interactions. Special attention is given to the most recent developments.

The following section deals with the problem of polymer chain dynamics in solutions. Several methods are discussed, which use different levels of model description as a starting point. The third part reviews some work on charged polymers. There again, a variety of different approaches has been used throughout the recent years, ranging from simple Debye–Hückel chains to systems with explicit counterions. Finally we will comment briefly on the future perspectives considering the modern parallel computers, which will be generally available in a few years.

3.2 Dynamics of neutral polymer chains in dilute solution

Flexible polymer chains immersed in solvent constitute a complicated many-body system, which today is still not satisfactorily understood. This is true even for simple neutral solutions. Ideas of universality and scaling[1] as well as simplified descriptions of the polymer–solvent dynamics[5,7,10] have yielded significant insight into the asymptotic behavior (both static and dynamic) in various limiting cases, such as high dilution, long chains, long time-scales or large length-scales. However, our knowledge in the nonasymptotic or crossover regimes is still rather poor, the most important of these crossovers being the dramatic change in static and dynamic behavior when the chain density is increased. Typically in these regimes analytical theories become cumbersome or even unfeasible, while experiments are often limited to a few quantities they can measure. It is therefore worthwhile to explore the possibilities and limitations of computer simulations.

The statics of neutral polymer solutions is roughly understood in terms of static scaling,[1] and the influence of density and interactions is taken into account in the Flory–Huggins theory[1,11] or similar, much more sophisticated approaches. Computer simulations, in particular extensive Monte Carlo (MC) simulations on a lattice, have been able to deepen this picture, and to verify the phenomenological crossover scaling for the crossover from dilute to dense systems.[12] These methods are also able to describe the crossover from Rouse dynamics[5,9] in the semidilute regime to reptation dynamics[5] in dense melts.[13] For these problems, molecular dynamics (MD) simulations (Refs 14, 15; see contribution by Kremer and Grest in this volume) have been successfully employed as well, giving consistent results. In the dilute regime, however, dynamic MC and related methods are unable to provide a realistic description of the dynamics: here the momentum transport through the solvent becomes very important, and has, in some way, to be included in the simulation. The phenomenon is usually referred to as "hydrodynamic interaction", and amounts to long-range correlations in the stochastic displacements of the chain monomers, mediated by the solvent flow. The standard model of polymer dynamics in the dilute limit, which does incorporate this effect, is called the Zimm model*.[10]

*In the original papers by Rouse[9] and Zimm[10] a Smoluchowski equation for the polymer motion was solved for the special case of a random walk chain. In the Zimm paper the additional approximation of a pre-averaged diffusion tensor was introduced. Here we use the terms "Rouse model" and "Zimm model" in the more modern and generalized sense of dynamic universality classes: the dynamic scaling relations are determined only by the static scaling (random walk versus self-avoiding walk) and by the type of Smoluchowski equation (Rouse model = a single chain coupled to a viscous background with uncorrelated stochastic displacements; Zimm model = Rouse model plus hydrodynamic interactions = viscous background with correlated displacements).

The concept of hydrodynamic interaction is certainly not particularly intuitive. Even more difficult is the phenomenon of "hydrodynamic screening": On increasing the polymer concentration, the range of these correlations decreases more and more, and the dynamics crosses over from Zimm to Rouse behavior and eventually reptation.[5,16–21] There is hope that computer simulations might provide additional insight. However, the status of dynamics simulations for solutions is still rather poor. Quite generally, simulations which study dynamics tend to be more demanding than those which study statics. This is because measuring time-displaced correlation functions needs an increased statistical effort, and because the correct reproduction of the dynamical behavior puts more stringent requirements on the simulation algorithm. For these reasons, computer simulations of the dynamics of polymer solutions are less developed than those for the statics. The field has just matured enough to have provided us with a partial understanding of dynamics data in the dilute limit which as the theoretically easiest case has been attacked first. In the semidilute regime, the dynamics has not yet been treated by satisfactory simulational approaches. As will become clear below, there are still problems left even in the simple case of a single chain in a solvent of its own monomers, which is the optimal good solvent. Most of this section will deal with MD simulations of such a system, where the solvent particles are taken into account explicitly. Besides such a "first principles" approach there are several levels of investigation. The simplest is certainly what can be called static dynamics. For this the validity of the Zimm model is taken for granted. Then one can use any numerical method to generate conformations and use these to calculate, e.g., the hydrodynamic radius of the chains $\langle 1/R_H \rangle$ or the initial decay rate of the dynamic structure factor of the polymer. What is implicitly assumed there is that the most interesting features of the dynamics show up in the short-time behavior, which is accessible as a static average over a dynamical operator. While this approach does not necessarily yield insight into the basic theoretical problems, it is very helpful for more complicated systems such as star polymers or chains in confined geometries. To test the very basic concepts one has to investigate on a fundamental level. The next step would be to accept the usual modeling of the hydrodynamic interaction via a diffusion tensor but actually perform a Brownian dynamics (BD) simulation in order to obtain the dynamical implications of the model beyond its short-time limit. This was done by several investigators, but, as will be seen later, has severe numerical difficulties for longer chains, since the square root of a $3N \times 3N$ matrix has to be calculated every time step, where N is the number of monomers in the system. The most basic approach employed so far is a direct MD simulation of a single chain in a solvent. Though this was attempted by several groups over the last years, it has only recently been possible by fast computers and optimized algorithms that any significant results were obtained. Before we report results pertaining to all three approaches, we first give a short overview of the theory.

3.2.1 *Theoretical background*

The theoretical treatment of the dynamics of a polymer chain in solution usually starts by writing down Kirkwood's diffusion equation (Smoluchowski equation)[5]

$$\frac{\partial}{\partial t}P(\{\vec{r}_i\},t|\{\vec{r}_i^{\,0}\},0)=\sum_{ij}\frac{\partial}{\partial\vec{r}_i}\cdot\overleftrightarrow{D}_{ij}(\{\vec{r}_i\})\cdot\left(\frac{\partial}{\partial\vec{r}_j}-\frac{\vec{F}_j}{k_BT}\right)P(\{\vec{r}_i\},t|\{\vec{r}_i^{\,0}\},0). \tag{3.1}$$

Here the $\vec{r}_i$ denote the positions of the monomers, so that $\{\vec{r}_i\}$ specifies the conformation of the polymer chain. $P(\{\vec{r}_i\},t|\{\vec{r}_i^{\,0}\},0)$ is the conditional probability density for a transition from $\{\vec{r}_i^{\,0}\}$ at time 0 to $\{\vec{r}_i\}$ at time t. The forces $\vec{F}_j$ are defined thermodynamically via the configurational equilibrium distribution function $\rho(\{\vec{r}_i\})$ at absolute temperature T (k_B is Boltzmann's constant):

$$\frac{\vec{F}_j}{k_BT}=\frac{\partial}{\partial\vec{r}_j}\ln\rho. \tag{3.2}$$

Eq. (3.1) describes the diffusive behavior of a chain (i.e., the movement of the center of mass) as well as its conformational rearrangements as a function of time. The equation is stochastic because the chain performs Brownian motion, and it has many different conformations which all have the same probability. The monomer–monomer interactions are described by the $\vec{F}_j$. We will assume here that there are no long-range interactions present (in marked contrast to the case of polyelectrolytes!) and that hence the chain's structure is a random or self-avoiding walk. Motion in three-dimensional space is assumed throughout. The diffusion tensor $\overleftrightarrow{D}_{ij}$ specifies the dynamics. Mathematical consistency of eq. (3.1) requires that $\overleftrightarrow{D}_{ij}$ is symmetric and positive–definite *for all possible polymer conformations* (no other property is necessary). In the Rouse case, $\overleftrightarrow{D}_{ij}$ is simply diagonal,

$$\overleftrightarrow{D}_{ij}=D_0\delta_{ij}\overleftrightarrow{1}, \tag{3.3}$$

where D_0 is the diffusion coefficient of a single monomer. In the Zimm case the hydrodynamic interaction is described by nonvanishing off-diagonal elements. Many different forms have been given in the literature,[5,22–26] all of which give the same decay for long distances, however significantly differ for shorter distances. The simplest is the Oseen tensor:

$$\overleftrightarrow{D}_{ij} = D_0 \delta_{ij} \overleftrightarrow{1} + (1 - \delta_{ij}) \frac{k_B T}{8\pi\eta r_{ij}} (\overleftrightarrow{1} + \hat{r}_{ij} \otimes \hat{r}_{ij}), \tag{3.4}$$

where $\hat{r}_{ij} \otimes \hat{r}_{ij}$ denotes the tensor product of the unit vector in the $\vec{r}_{ij}(= \vec{r}_i - \vec{r}_j)$ direction with itself. η is the solvent shear viscosity. An easy way to derive this is based on the Navier–Stokes equation for incompressible, low Reynolds number flow: apart from a factor $k_B T$, the diffusion tensor $\overleftrightarrow{D}_{ij}$ is nothing but the mobility tensor which describes the linear velocity response of particle i to an external force applied at the location of particle j (generalized Einstein relation). However, this velocity response can be found by just calculating the flow field which is generated by this force (which is modeled as a point force), because particle i is assumed to be "embedded" in the surrounding flow (stick boundary conditions).

Zimm[10] solved the chain dynamics for random walks, using the so-called preaveraged version of the Oseen tensor. In his original treatment, $\overleftrightarrow{D}_{ij}$ was averaged over the chain conformations before solving the Smoluchowski equation. Thus $\hat{r}_{ij} \otimes \hat{r}_{ij}$ in eq. (3.4) was replaced by $1/3 \overleftrightarrow{1}$ and the inverse distance between the particles by the average inverse distance. This gave a diagonal tensor, for which the Smoluchowski equation can be solved in much the same way as for the Rouse model. However, at this time it was rather unclear how severe the pre-averaging approximation is. It was only recently that some light has been shed on this question, using Brownian dynamics simulations (see below). Strictly speaking the Oseen interaction results in an ill-defined problem, since $\overleftrightarrow{D}_{ij}$ is *not* positive–definite for *all* conformations. Hence, the numerical approach to solving the Smoluchowski equation (Brownian dynamics) has to use different interaction tensors, as will be discussed below.

For the more microscopic approach of an MD simulation of a chain in solvent particles, it is useful to also look at the theory from a more microscopic point of view, in particular in order to assess its limitations. The derivation of equations of motion of the Smoluchowski type and the discussion of the involved errors is a standard problem in modern transport theory. In the present case, the form of the hydrodynamic interaction tensor has to be derived from the microscopics, too.[24,27–30] However, analytical considerations have not yet given a satisfactory answer to the question for which time and length scales one should expect the asymptotic Brownian and Oseen behavior, respectively.

A treatment that studies the case of *unphysical* dynamics of the solvent particles has been carried out in Ref. 31, also using a microscopic approach. This was done in order to discuss if it is justified to replace an MD algorithm in which the solvent particles obey strictly Newton's equations of motion by a Langevin simulation in which every solvent particle is artificially coupled to a weak friction and a weak heat bath[32] (this latter method has some technical advantages, as discussed below). The main result is that the Navier–Stokes equation

$$\rho \frac{\partial}{\partial t} \vec{u} = \eta \Delta \vec{u} \tag{3.5}$$

(ρ mass density, $\vec{u}$ velocity flow field) is modified to

$$\rho \frac{\partial}{\partial t} \vec{u} = \eta \Delta \vec{u} - \frac{1}{m} \rho \zeta \vec{u} = \eta (\Delta - \kappa^2) \vec{u} \tag{3.6}$$

where m is the mass of a solvent particle and ζ the friction constant that is put artificially into the algorithm. While the Green's function of eq. (3.5) is the Coulomb-like Oseen tensor, eq. (3.4), the Green's function of eq. (3.6) exhibits a Yukawa-like decay $\propto \exp(-\kappa r)/r$. Hence, the algorithm introduces a *screening* into the hydrodynamic interaction, with a screening length of

$$l = \kappa^{-1} = \sqrt{\frac{m\eta}{\rho\zeta}}. \tag{3.7}$$

That is, such a modification of the algorithm would result in a loss of the most decisive property of the hydrodynamic interaction, which is its long-range nature. An interaction which has a typical length-scale built in can be renormalized to a localized interaction, i.e., on a sufficiently coarse-grained scale the displacements of the monomers could be considered as independent. In this case the dynamics would be asymptotically Rouse-like on scales large compared to l.

However, the inclusion of the long-range correlations places the Zimm model into a different dynamic universality class than the Rouse model.[1,5,33–36] Although there exist sophisticated renormalization-group treatments,[4,20] the essential physics can be understood rather easily on the phenomenological level: we first consider the center of mass diffusion constant D which within the framework of the Smoluchowski equation differs only weakly from the Kirkwood value $D^{(K)}$[37,38] (the latter describing the short-time behavior):

$$D^{(K)} = \frac{1}{3N_{ch}^2} \sum_{ij} \mathrm{Tr} \left\langle \overleftrightarrow{D}_{ij} \right\rangle, \tag{3.8}$$

where N_{ch} is the number of monomers. For the Rouse model, one has $D^{(K)} = D_0/N_{ch}$, or $D \propto R_G^{-1/\nu}$, where the gyration radius $\langle R_G^2 \rangle = 1/(2N_{ch}^2) \sum_{ij} \langle r_{ij}^2 \rangle$ describes the typical dimensions of the chain, while ν is the exponent which describes the chain's fractal structure via the scaling law $R_G \propto N_{ch}^{\nu}$. For a random walk, ν is 1/2, while $\nu \approx 0.59$ for a self-avoiding walk. For the Zimm model, eqs (3.4) and (3.8) yield

$$D = \frac{D_0}{N_{ch}} + \frac{k_B T}{6\pi\eta} \left\langle \frac{1}{R_H} \right\rangle, \tag{3.9}$$

where the hydrodynamic radius is defined as

$$\left\langle \frac{1}{R_H} \right\rangle = \frac{1}{N_{ch}^2} \sum_{i \neq j} \left\langle \frac{1}{r_{ij}} \right\rangle. \tag{3.10}$$

Hence, in Zimm scaling, $D \propto R_H^{-1} \propto R_G^{-1}$ (asymptotically). From this one concludes the dynamic exponent z, which relates the longest relaxation time τ_R or τ_Z (Rouse or Zimm case, respectively) with the radius of gyration ($\tau \propto R_G^z$): Since τ is, by order of magnitude, given by the time the chain needs to move its own size ($\tau \approx R_G^2/(6D)$), one finds $z = 2 + 1/\nu$ in the Rouse case and $z = 3$ for the Zimm model (i.e., faster dynamics due to the hydrodynamic correlations, which speed up the diffusion). This exponent also describes the relation between time scales and length scales in the scaling regime (i.e., length scales intermediate between microscopic lengths of the order of a bond length and R_G, and time scales intermediate between microscopic times and τ): here the mean-square displacement of a single monomer scales as

$$\langle (\Delta r)^2 \rangle \propto t^{2/z}, \tag{3.11}$$

while the dynamic structure factor

$$\begin{aligned} S(k,t) &= N_{ch}^{-1} \sum_{ij} \langle \exp(i\vec{k} \cdot (\vec{r}_i(t) - \vec{r}_j(0))) \rangle \\ &= N_{ch}^{-1} \sum_{ij} \left\langle \frac{\sin(k|\vec{r}_i(t) - \vec{r}_j(0)|)}{k|\vec{r}_i(t) - \vec{r}_j(0)|} \right\rangle, \end{aligned} \tag{3.12}$$

which is measured in dynamic light scattering experiments,[39–44] is supposed to obey the relation

$$S(k,t) = k^{-1/\nu} f(k^z t). \tag{3.13}$$

The theory as sketeched above is well established and has, in part, been confirmed by light scattering experiments[39–44] and some neutron scattering data.[45] However, there are a number of problems with the theory. First of all, the Smoluchowski equation is only valid on time scales on which the motion of the monomers can be viewed as Brownian. This requires some care in the interpretation of MD results where the ballistic short time regime should be cut off. Furthermore, the theory totally disregards retardation effects: the hydrodynamic interaction in reality does not spread infinitely fast, but rather diffusively with a finite diffusion constant which is nothing but the kinematic viscosity $\eta_{kin} = \eta/\rho$ where ρ is the solvent's mass density. The approximation to regard the flow field as instantaneously generated by the current polymer conformation can only be valid if $D_0 \ll \eta_{kin}$, i.e., if momentum transport is much faster than the monomer motions. It is not immediately obvious that any polymer–solvent system should satisfy that condition, i.e., considering that the typical solvent molecule is roughly of the same size as the typical monomer.

Similar caveats also concern length scales: from the hydrodynamic derivation of the Oseen tensor (eq. [3.4]) it is completely obvious that it can describe the correlations in the stochastic displacements only on sufficiently long length scales. For short interparticle distances, the $1/r$-like interaction becomes completely unphysical, and, even worse, mathematically ill-defined: there are polymer conformations in which the monomers are too close to each other and which hence violate the condition of positive-definiteness of $\overleftrightarrow{D}_{ij}$ (the reader can easily verify this for the case of a dumbbell). Mainly for this reason many authors have introduced modified interaction tensors which all cross over to the leading Oseen behavior for large distances, but which are supposedly more accurate for short distances: one approach is to take the finiteness of the monomers seriously and try to solve the flow problem around an array of spheres,[22,23,25,26] while other authors have introduced a phenomenological length as a lower boundary for the validity of hydrodynamics[24] or purely empirical regularization procedures.[46] All of these attempts have in common that they introduce an additional microscopic length scale into the problem which is not *a priori* built into the original Hamiltonian of monomers and solvent particles, and is hence unknown. MD simulations with explicit solvent particles, as the most basic tool, have the unique advantage of not assuming the validity of the Smoluchowski equation from the outset, and of being able to address such questions quantitatively. The other approaches take the validity of the Smoluchowski equation on the typical polymeric length scales for granted.

The typical limitation of light scattering experiments[39–44] is that they cannot measure all quantities independently in order thoroughly to test the theory. For example, eq. (3.9) relates a dynamic quantity, the diffusion constant D, to a static quantity, the hydrodynamic radius R_H. D is usually measured by determining the initial decay rate of the dynamic structure factor $S(k, t)$ in the low-k limit.[5] However, R_H is not measured independently, but rather determined indirectly using eq. (3.9). Of course, an independent measurement of R_H is easily possible in a simulation.

A similar problem occurs for the initial decay rate at the higher k values (scaling regime): writing

$$S(k, t) = S(k, 0) \exp(-D(k)k^2 t) \qquad (3.14)$$

in the short-time regime, Akcasu *et al.*[41] have shown that within the framework of the Smoluchowski equation one has

$$D(k) = \frac{\sum_{ij} \left\langle \hat{k} \cdot \overleftrightarrow{D}_{ij} \cdot \hat{k} \exp(i\vec{k} \cdot \vec{r}_{ij}) \right\rangle}{\sum_{ij} \left\langle \exp(i\vec{k} \cdot \vec{r}_{ij}) \right\rangle}. \tag{3.15}$$

(This reduces to eq. [3.9] in the limit $k \to 0$.) Analytically, this can be calculated only approximately for finite chains because of the problem of polymer conformational statistics. In the asymptotic limit of an infinite chain with vanishing bead size, one has (for k in the scaling regime)[47]

$$D(k) = C_0 \frac{k_B T}{\eta} k, \tag{3.16}$$

where C_0 is a constant dependent on the exponent ν: $C_0 = 0.078\,8$ for $\nu = 3/5$ and $C_0 = 0.062\,5$ for $\nu = 1/2$. However, experiments usually see a prefactor which is smaller.[42–44] We believe that this is due to both finite chain length and, more importantly, finite bead size, the latter usually being neglected in the data analysis.[48] Again, the right-hand side of eq. (3.15) is not measured independently in experiments, while in a computer simulation this is possible.

3.2.2 Simulations

The above discussion now clearly defines the various levels one can employ to attack the problem. As mentioned throughout the discussion, the first and most ambitious way is to perform a microcanonical MD simulation. The typical problems will be discussed in detail later. The next level would be to simulate the Smoluchowski equation directly via Brownian dynamics. This assumes its validity on the relevant time-scales. However, because of the necessary calculation of the square root of the hydrodynamic interaction tensor the method easily runs into numerical problems: if the tensor is not properly regularized, negative eigenvalues are bound to occur sooner or later in the course of the simulation. Moreover, due to the computational complexity of the square root (the number of operations is proportional to the third power of the number of the Brownian particles[49] the method is limited to relatively short chains. This approach has been used by several investigators throughout the last 15 years. The simplest and most widely used method is to assume the validity of the Zimm description and calculate the hydrodynamic radius etc. from standard MC simulations. Within the Zimm model then the consequences for the dynamics are analyzed. In the present paper we give an overview for all three different approaches. We follow the discussion of the theory, namely first to discuss the more fundamental attempts which will not necessarily be able to reach the asymptotic regime in every aspect, and then move towards the "static dynamics" which one can call the standard MC procedures. The main emphasis will be on the discussion of the MD simulations.

3.2.2.1 *MD simulations of a single polymer chain in a bath of solvent particles*

(a) *Overview* The most obvious simulational approach attacking the problem of Brownian motion in complex systems is to perform MD on a system of both solvent particles as well as the Brownian particles. For a single polymer chain in solution, however, this is not an easy task, since the chain relaxes on a much slower time-scale than the solvent. Moreover, long chains are required to clearly see scaling behavior, which means large systems in order to keep the solution dilute.

Such simulations have been done since the late Seventies.[50–61] However, due to the large computational effort necessary, the older work was limited to either studies of static quantities, or analysis of dynamic quantities for short chains with rather inaccurate statistics. It was only recently that some progress was made by two independent studies,[58–61] which were able for the first time to attain enough statistics to make a detailed analysis of the dynamics worthwhile. Pierleoni and Ryckaert (PR)[58,59] studied chains of length $N_{ch} = 6, 9, 20, 30$, while Dünweg and Kremer (DK)[60,61] explored the range $N_{ch} = 30, 40, 60$. Dünweg and Kremer did not study shorter chains more carefully, because preliminary test runs indicated that these do not exhibit a sufficiently large time window between microscopic times and the Zimm time to make the $t^{2/3}$ behavior of the mean square displacement clearly visible. These shorter chains seemed rather to exhibit an effective exponent which decreased with chain length.[62] For the longer chains, however, DK observed a mean square displacement that was independent of chain length and exhibited an effective exponent of 0.7 ± 0.05 (see Figs 3.1/3.2). Pierleoni and Ryckaert did not calculate $\langle \Delta r^2 \rangle$, but rather calculated the dynamic structure factor $S(k, t)$, for which they did observe a $k^3 t$ decay behavior. On the other hand, DK analyzed $S(k, t)$ only for the longer chains ($N_{ch} \geq 30$), where the same dynamic scaling was found[61] (the statistics were not sufficient for the shorter chains). A comparison of Rouse scaling and Zimm scaling of $S(k, t)$ for $N_{ch} = 60$ is shown in Figs 3.3 and 3.4.

At first glance, the combination of these results look somewhat surprising: short chains apparently exhibit scaling only for $S(k, t)$ but not for $\langle \Delta r^2 \rangle$, while longer chains obey scaling for both quantities. Of course, the models of PR and DK are slightly different (see below), but we still think such a combination is valid. However, there are reasons to believe that neither of the studies are fully within the asymptotic hydrodynamic regime, but rather in some sort of crossover regime between atomistic dynamics and hydrodynamics. It seems that the dynamic exponent $z = 3$, which was derived from hydrodynamic arguments in the limit of long length and time scales, is valid well beyond those scales, down into a regime where practically no theory exists.[61] Then it is conceivable that some quantities

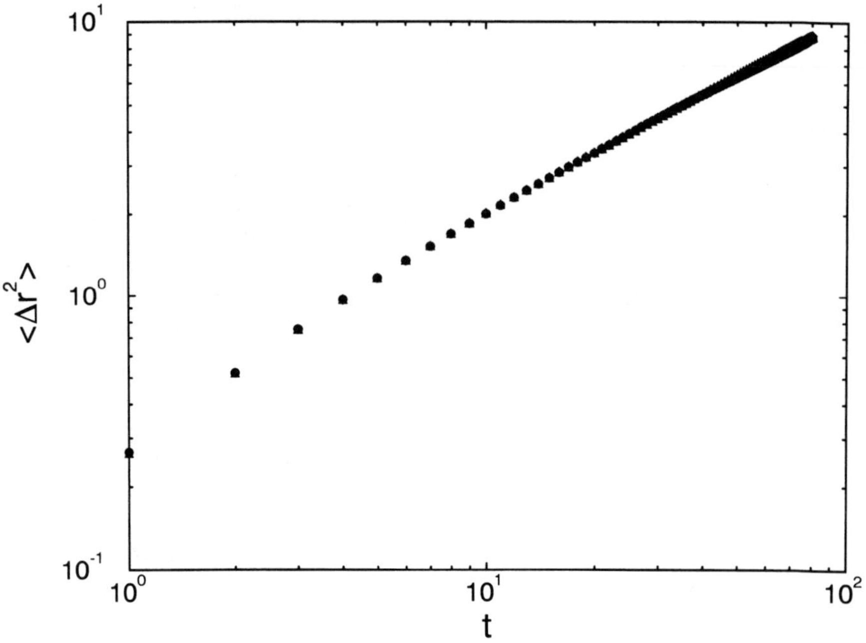

Fig. 3.1 Log–log plot of the time dependence of the mean square displacement of the single monomer in an MD simulation of a polymer chain in good solvent. Only the four inner monomers are taken into account in order to suppress end effects. Chain-lengths are $N_{ch} = 30$ (points), 40 (triangles) and 60 (diamonds) (from Ref. 61).

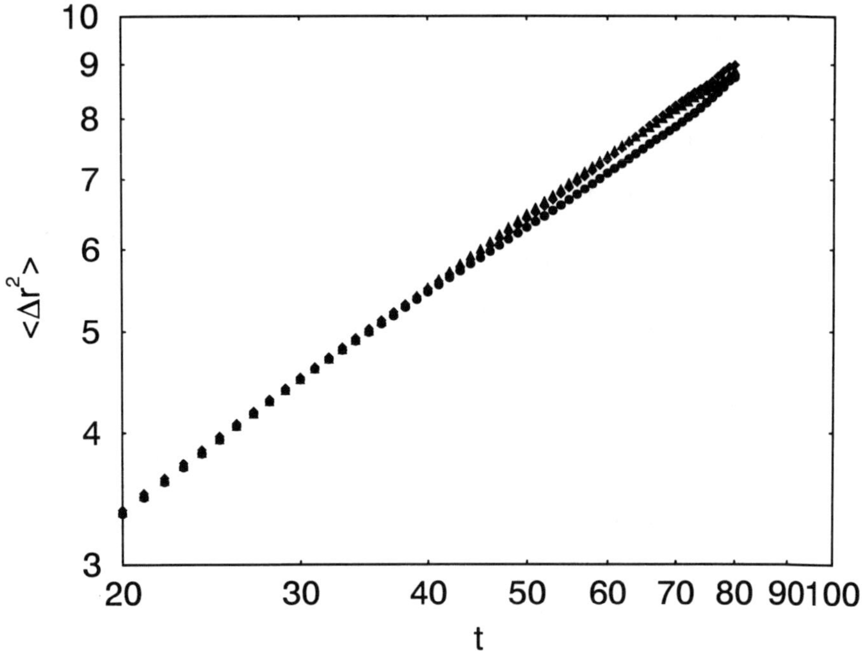

Fig. 3.2 Same as Fig. 3.1, for the time interval $20 \leq t \leq 80$ (scaling regime). The exponent was estimated as 0.70 ± 0.05 (from Ref. 61).

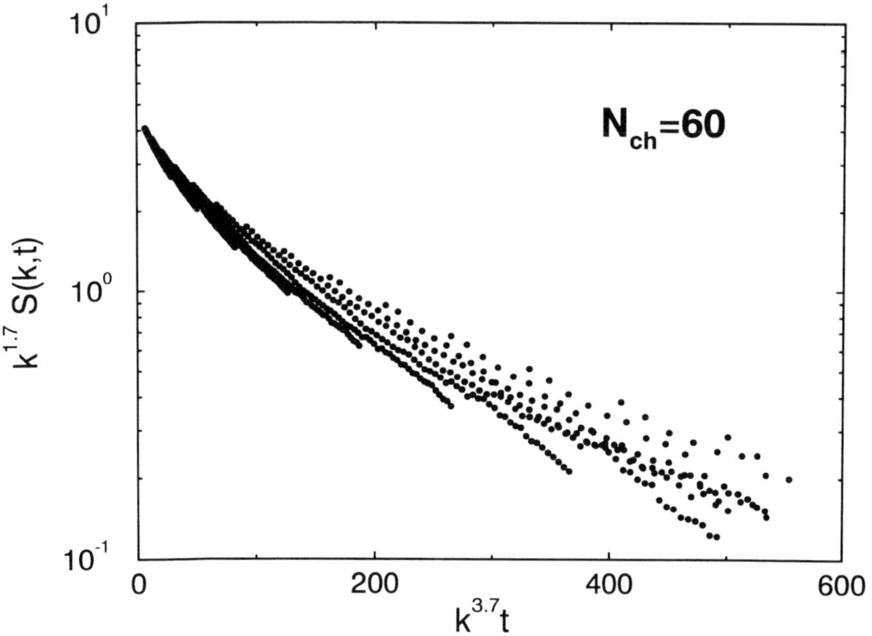

Fig. 3.3 Log–linear "data collapsing" plot of the decay of the dynamic structure factor $S(k, t)$ for $N_{ch} = 60$, in Rouse scaling form, $k^{1/\nu}S(k, t)$ vs. $k^{3.7}t$, using $\nu = 0.59$, and restricting the data to the scaling regimes $0.7 \leq k \leq 3$ and $20 \leq t \leq 80$. These scaling regimes were determined by Figs 3.1 (mean-square displacement of the single monomer) and 3.5 ($S(k, t = 0)$). Moreover, data with $S \leq 0.05$ were eliminated for reasons of statistical accuracy (from Ref. 61).

exhibit the asymptotic power law earlier than others. The arguments[61] can be summarized as follows: in contrast to phenomenological scaling theories, the Kirkwood theory (summarized in eqs [3.1] and [3.4]) has the additional predictive power to give definitive statements about the prefactors of the scaling laws. Therefore, a much more detailed comparison is possible. The main example is eq. (3.15) and its $k \to 0$ limit, eq. (3.9). While eq. (3.9) can actually be confirmed,[59,61] $D(k)$ according to eq. (3.15) is no longer in agreement with the measured decay rate if k is sufficiently large.[61,63] This is not too surprising since these larger k-values correspond to length scales on which the liquid still has structure, as measured by the solvent's pair correlation function (note that even for the longest chain, $N_{ch} = 60$, the radius of gyration is of comparable size as the typical correlation length of fluid structure!).[61] Furthermore, one can also measure the time-displaced autocorrelation function of transversal modes of the solvent's velocity flow field. The corresponding Green–Kubo integral should provide the Fourier transform of the Oseen tensor in the hydrodynamic regime.[31,61] Indeed, one sees a nice k^{-2} decay, with even the prefactor being described accurately by

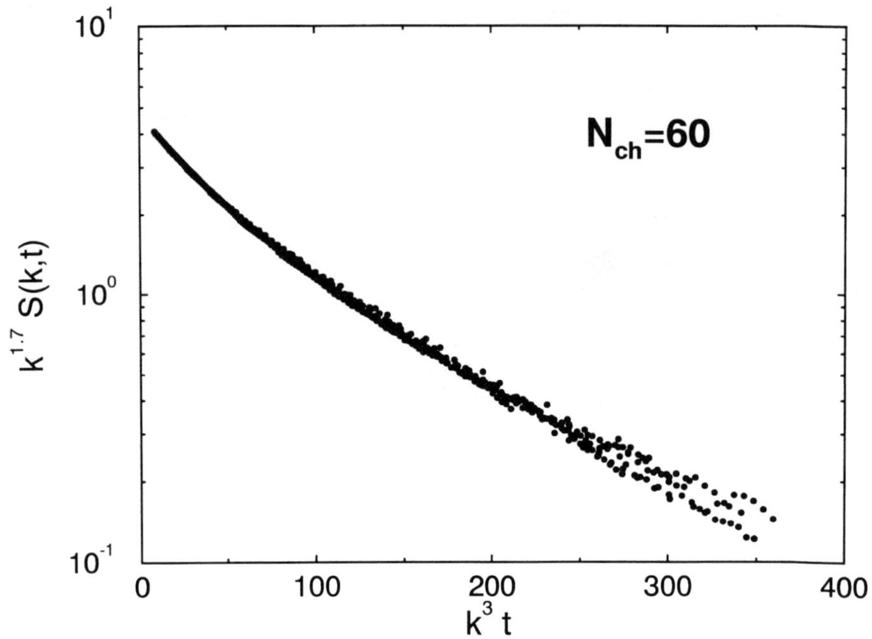

Fig. 3.4 Same as Fig. 3.3 for Zimm scaling: $k^{1/\nu}S(k,t)$ vs. k^3t (from Ref. 61).

the hydrodynamic theory.[61] However, this only works down to some length scale ($k \approx 1 \ldots 2$ [all quantities are expressed in Lennard-Jones units[64]]), where deviations occur and, apparently, the atomic motions come into play.[61] But on the other hand, DK observed the k^3t decay of $S(k,t)$ in the regime $0.7 \leq k \leq 3$! Even more dramatic seems to be the extension of dynamic scaling into short, non-hydrodynamic time scales: Pierleoni and Ryckaert[59] saw the k^3t decay down to very short times $(0.5\tau_{LJ})$, which are probably significantly shorter than the time needed to cross over from ballistic to Brownian particle motion (for comparison: DK cut off all data below $20\tau_{LJ}$ as ballistic short-time regime).

This slow crossover into asymptotic hydrodynamic behavior stands, interestingly enough (also for experiments), in marked contrast to the dynamic behavior of melts of short chains:[15] in this latter case, the corresponding length and time scales are sufficient to observe Rouse dynamics. For the case of solutions however, the chains have to be longer than in the melt case in order to observe the typical "polymer behavior". Moreover, these results indicate that the observation of a k^3t decay of $S(k,t)$ *alone* is not sufficient to prove asymptotic Zimm behavior.

(b) *Finite size effects and prefactors of scaling laws* The minimum linear size of the simulation cell, L, is determined by fitting the chain within the cell

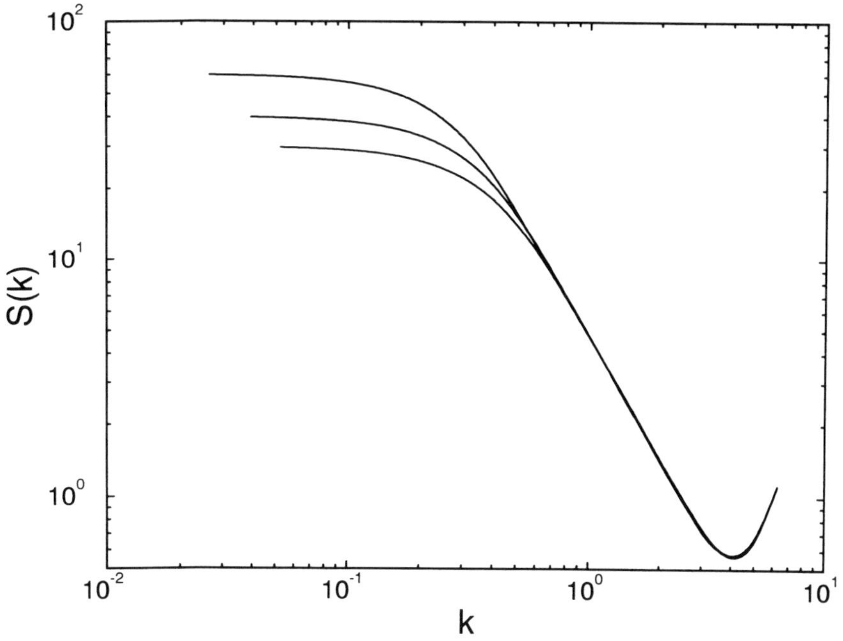

Fig. 3.5 Log–log plot of the static structure factor of the chain, for $N_{ch} = 30$ (lower curve), 40 (middle curve) and 60 (upper curve) (from Ref. 61).

(i.e., it does not overlap with its periodic images). Pierleoni and Ryckaert[59] varied R_G/L between 0.1 and 0.3 and in Dünweg and Kremer's work[61] this ratio was ≈ 0.2 for all systems. No effect of self-overlap is visible in the statics which is most readily analyzed in terms of the static structure factor $S(k, t = 0)$ (see Fig. 3.5). The $k^{-1/\nu}$ decay in the scaling regime can be nicely verified: from Fig. 3.5 one extracts an exponent $\nu = 0.58 \pm 0.01$, as expected in good solvents (similar results are found in Ref. 59, however with a less pronounced scaling regime, due to the shortness of the chains).

However, the $1/r$-dependence of the Oseen tensor indicates that the dynamical properties should be strongly system-size dependent, because, in the Oseen picture, the chain hydrodynamically interacts with its own periodic images. In Ref. 60 Dünweg and Kremer pointed out that, for this reason, one should compare the data to a modified Kirkwood theory where the Oseen tensor is replaced by the corresponding Ewald sum which takes the periodic images into account.[65,66] This has to be done both in eq. (3.8), yielding a system-size dependent hydrodynamic radius, as well as in eq. (3.15), defining a system-size dependent initial decay rate.

Perhaps a more intuitive picture of the effect is obtained if, by Fourier transform, one views the Oseen tensor as the superposition of hydrodynamic modes. The main contribution arises at the origin $k = 0$ due to the k^{-2}

divergence. However, it is just those long-wavelength modes which are cut off by the finite system size, reducing the diffusion constant significantly. Similar finite size effects, which scale in leading order like L^{-1}, have been observed in completely different contexts,[67,68] while the basic physics is very similar to the present case.

For the center of mass diffusion constant, this analysis was carried through in both studies.[59,60,61] The result is that the Kirkwood prediction for the diffusion constant is in very good agreement with the MD data *if and only if the Ewald correction is taken into account.* A glance at eq. (3.9) shows that five quantities are involved in this comparison: D, the actual diffusion constant, which can, e.g., be measured by the mean square displacement of the center of mass (see Refs 59 and 61); the temperature, which is measured via the kinetic energy; the solvent viscosity, which can be obtained via Green–Kubo integration[59,61,69] with reasonable accuracy; the Ewald generalized hydrodynamic radius, which is obtained from static averages over the chain conformations; and the monomeric diffusion coefficient D_0, which is not easily accessible. Pierleoni and Ryckaert, for their comparison, therefore replaced D_0 by the diffusion coefficient of a solvent particle. Dünweg and Kremer allowed for D_0 being different from the solvent particle value, since their monomers were twice as heavy as the solvent particles, and hence used eq. (3.9) to solve for D_0, finding a value practically independent of the chain length and rather close to the solvent particle diffusion constant. This agreement is very reasonable since in both models the chain monomers are "chemically" identical to the solvent particles (i.e., apart from chain connectivity, the interactions are the same). Moreover, the main contribution to the particle diffusion constant seems to come from the particle "size": a comparison with Stokes' law for slip boundary conditions ($D_0 = k_B T/(4\pi\eta a)$) shows good agreement if one extracts the particle radius a from the first peak of the pair correlation function.[61]

By putting the chain's conformations into hypothetical boxes of varying size, the Ewald theory also allows one to study the L-dependence of D. The finite size effect is very strong ($\propto L^{-1}$ in leading order) and an MD box which would be large enough to neglect it is computationally not feasible today.[60,61] From the structure of the theory it follows that the dimensionless parameter which controls the finite size effect on D is R_G/L. This dependence has been confirmed by Pierleoni and Ryckaert[59] who actually ran an MD simulation of a 9-monomer chain in boxes of different sizes, L varying as much as by a factor of three. They found that D increased with L, as predicted by the theory, and obtained quantitative agreement. It is interesting to note that the effect is even observable on the single-particle level: in DK's system, the diffusion constant of the solvent particles increased with the system size, in reasonable agreement with the prediction resulting from the Ewald sum.[61] Of course, this finite size effect is much weaker than that of the chain, since the solvent particles are much smaller.

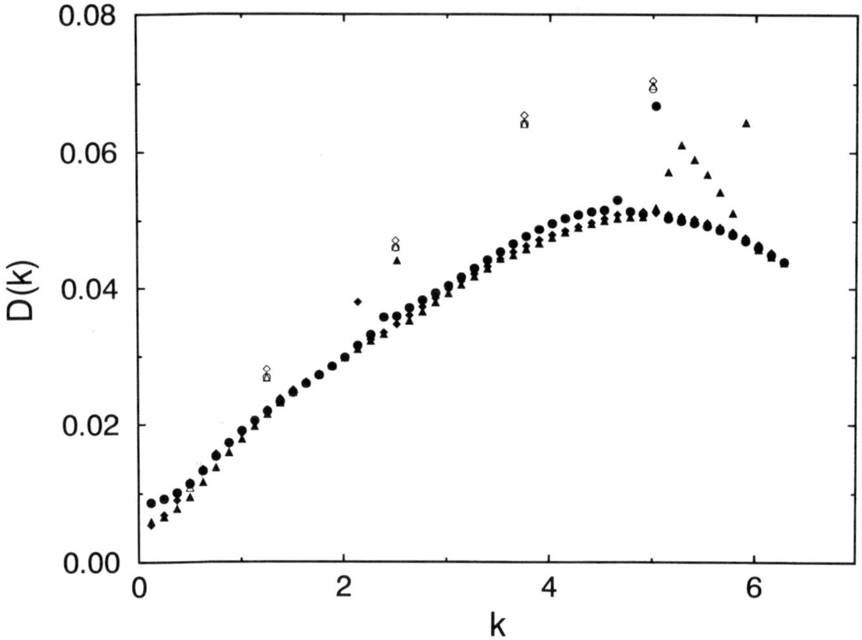

Fig. 3.6 $D(k)$ (cf. eq. [3.14]) obtained from the dynamical data for $N_{ch} = 30$ (filled circles), 40 (filled triangles) and 60 (filled diamonds). Instead of trying to perform the limit $t \rightarrow 0$ the maximum value of $D(k, t)$ (see eq. [3.17]) was taken. Hence, the data should be viewed as an upper limit to the actual initial decay rate. For comparison, the data resulting from the static evaluation with Ewald sums are also included with corresponding open symbols (from Ref. 61).

In Ref. 61 the analysis of the initial decay rate of $S(k, t)$ was also done for $k \neq 0$, studying the Ewald generalization of eq. (3.15). For the monomeric diffusion coefficient D_0 Dünweg and Kremer used the value which was obtained from the diffusion constant D. In that sense the procedure can be viewed as "fitted" to $k = 0$; however, it should be noted that the value is physically very reasonable. In Fig. 3.6 these static values are compared to the data extracted from the actual dynamics. An accurate determination of $D(k)$ is not easy. In Ref. 61 the following procedure was adopted: inspired by eq. 3.14, one can define

$$D(k, t) = -\frac{1}{k^2 t} \ln\left(\frac{S(k, t)}{S(k, 0)}\right), \tag{3.17}$$

whose "initial" value defines $D(k)$. However, what is actually meant is a typical value at the initial stages of Brownian motion. Since there is a smooth crossover from ballistic into Brownian dynamics, it is not obvious how to define the relevant time. Dünweg and Kremer just took the *maximum* value of $D(k, t)$, maximized over the observation time. Therefore, what is plotted in Fig. 3.6 is an upper limit to the actual decay rate, and the data is somewhat noisy.[61] However, it seems that at the higher k values the actual dynamics is slower than predicted by the hydrodynamic theory, which we again see as indication for a crossover to more atomistic dynamics: the discrepancies start to occur roughly at the same length scales where the other indicators also point to deviations from hydrodynamic behavior (see above). It should also be mentioned that all data (the static evaluation for $L = \infty$ but $N_{ch} < \infty$ according to eq. (3.15), the Ewald sum modification of eq. (3.15), and the actual MD) lie *below* the asymptotic Benmouna–Akcasu value (Ref. 47, see eq. [3.16]), as do the light scattering data.[42–44]

The finite size effect is most severe for $k = 0$ (i.e., the center of mass diffusion constant), while its relative contribution gets less and less important for the higher k modes. This has been seen in the simulation by Pierleoni and Ryckaert,[59] who tried to explain this observation by a retardation argument.[59] It should be noted, however, that the same behavior is also to be expected from the pure retardation-free theory. This is borne out both by the numerical evaluation of the Ewald sum[61] as well as by an approximate analytical calculation[61] which indicates that in the scaling regime the finite size effect is no longer controlled by R_G/L, but rather by $1/(kL)$. (The finite size scaling theory which was invented in Ref. 60 is apparently wrong.)

(c) *Methodological considerations* In this section, we want to give some details about modeling and simulation technique for the two studies which had been the main focus of the preceding parts.[58–61] The models are rather similar, which, in spite of the independence of the studies, is not very surprising since most of the modeling is driven by considerations of simplicity, computational efficiency and historical development of MD simulations. First, the solvent particles mainly fulfill the function of transporting momentum. Therefore one needs particles with a strong repulsive interaction at short distances, while an attractive part for longer distances may be included but would only waste computer time. The "ideal" solvent would be a system of hard spheres. However, there is no known fast vectorizing algorithm for hard spheres (efficient algorithms for the scheduling of collision events[70] are inherently scalar), and since both studies[58–61] were done on Cray vector computers, it was practically imperative to use continuous potentials for which very efficient vectorizing MD algorithms have been developed.[71,72] Soft spheres are conveniently modeled by a truncated Lennard-Jones (WCA[73]) potential:

$$U_{LJ}(r) = \begin{cases} 4\epsilon\left(\left(\frac{\sigma}{r}\right)^{12} - \left(\frac{\sigma}{r}\right)^{6} + \frac{1}{4}\right) & r \leq 2^{1/6}\sigma \\ 0 & r \geq 2^{1/6}\sigma. \end{cases} \tag{3.18}$$

This type of interaction has been used in both studies. Lengths are measured in units of σ, energies in units of ϵ, and times in units of $\tau_{LJ} - (m\sigma^2/\epsilon)^{1/2}$, where m is the mass of a solvent particle. The solvent state point was slightly different in the two studies: the density was $\rho = 0.8$ (Pierleoni and Ryckaert) and $\rho = 0.864$ (Dünweg and Kremer), while for the temperature the values $k_B T = 1.5$ (Pierleoni and Ryckaert) and $k_B T = 1.2$ (Dünweg and Kremer) were chosen. Note that the density is rather high, which we believe is necessary to ensure quasi-incompressible flow, and to make sure that momentum transport is much faster than mass transport (i.e., that the kinematic viscosity is much larger than the particle diffusion constant). Both conditions are tacitly built into the theory, and are reasonably met in both studies.[59,61]

A small fraction (slightly less than 1% in Refs 60 and 61, and about 4% for the longest chain of Refs 58 and 59) of the solvent particles is then relabeled as chain monomers and connected via an attractive backbone potential. Pierleoni and Ryckaert connected their monomers by hard rods of length $d = 1.075$. Dünweg and Kremer used the FENE potential

$$U_{ch}(r) = -\frac{k}{2} R_0^2 \ln\left(1 - \frac{r^2}{R_0^2}\right) \tag{3.19}$$

with parameters $k = 7$ and $R_0 = 2$ as attractive backbone. Note that these potentials do not introduce any angle dependence in order to reach the flexible chain limit as easily as possible. Moreover, the repulsive interaction acts between all particles in the same way. This makes the solvent an ideally good one, in which the Θ collapse never occurs. Hence, the statics of the chain is that of a pure self-avoiding walk, without any obscuring effects due to crossover to a Θ solvent. Solvents like the present one are sometimes called "athermal"[1] and are the simplest for a computer simulation.

Given the very long relaxation times and the moderate to large system sizes, it is crucial to have a well-optimized program. The "layered link cell" algorithm employed by Dünweg and Kremer[72] attained a speed of roughly 3×10^5 particle updates per second on one Cray-YMP processor, for the WCA potential at the state point in consideration.

In both studies, the runs were long enough to obtain typically several hundred statistically independent conformations.[59,61] On these timescales, purely Newtonian dynamics easily leads to numerical instabilities so that the total energy of the system drifts by an unacceptable amount. Pierleoni and Ryckaert stabilized their simulation by a Nosé–Hoover thermostat[74,75] and used the data to analyze both statics as well as dynamics. In principle, we view it as always dangerous to use non-Newtonian dynamics in order to obtain time-displaced correlation functions, in particular in problems involving hydrodynamics. However, it seems that the procedure is sound and no hydrodynamic screening is introduced artificially. This is intuitively understandable since the Nosé–Hoover thermostat, in contrast to a Langevin simulation,[32] conserves the global momentum of the system, and confirmed by numerical evidence.[76] However, a more rigorous analytical investigation along the lines of Ref. 31 has not yet been done, and seems to be more difficult than for the case of Langevin dynamics. The procedure adopted by Dünweg and Kremer was to first run the system with Langevin dynamics[32] in order to generate a statistical ensemble of conformations, from which short runs with purely Newtonian dynamics were started. The dynamical information was then obtained by averaging over these runs, which were long enough to reach well into the scaling regime, but not long enough to directly measure the Zimm time which was rather estimated by $\tau_Z = R_G^2/(6D)$. For further details, see the original papers.[59,61]

3.2.2.2 Lattice gas cellular automata (LGCA)

This is a new and potentially powerful method to simulate hydrodynamic phenomena.[77] It is still a developing field which might have a big impact on polymer solutions simulations. An application to a polymer chain in solution (in *two* dimensions) is found in Ref. 78. A hybrid scheme between LGCA and MD was developed in Ref. 79. We feel however that the field has to mature further (and the relation of LCGA dynamics to the atomistic particle dynamics has to be further clarified) before they can be used as a standard tool in polymer dynamics. Once the relation between LGCA the local microscopic dynamics is established, one is tempted to expect a next "quantum jump" for the hydrodynamic simulations of complex fluids, such as polymers.

3.2.2.3 Brownian dynamics

The previous sections described in some detail the very microscopic point of view. The characteristic length scales are typically relevant for shorter chains, as they occur often in experiment. However most theories for polymers are set up to describe the very long time and length scales. While the MD simulation can test the short length scale limit of this approach, it is presently not able to properly investigate the other limit. Here "long" not necessarily means very many model monomers, but large gyration radius compared to the characteristic length scales of the solvent. On that time and length scale, with all the reservations from the previous MD simulations, it is necessary and useful to consider the solvent as a continuous background and assume the general applicability of the Smoluchowski equation. This leads to Brownian dynamics (BD) simulations of single, still relatively short chains.[80] Historically this approach of course was used much earlier than the MD simulations.

Brownian dynamics is nothing but the numerical solution of the Smoluchowski equation. The method exploits the mathematical equivalence between a Fokker–Planck type of equation and the corresponding Langevin

equation.[81,82] For systems with hydrodynamic interactions, this equivalence is worked out in more detail in Ref. 83. The system is propagated for a small timestep Δt, in which every monomer is displaced by a small amount. This displacement is the sum of a deterministic part, governed by the forces $\vec{F}_j$ in eq. (3.1), and a stochastic part. The equation of motion, for the simplest case of no external flow and force field[80] reads

$$\vec{r}_i(t+\Delta t) = \vec{r}_i(t) + (\Delta t/k_B T)\sum_j \overleftrightarrow{D}{}^0_{ij}\vec{F}{}^0_j + \Delta t \sum_j \overleftrightarrow{D}{}^0_{ij}/\partial \vec{r}_j + \vec{\rho}{}^0_i(\Delta t). \quad (3.20)$$

The superscript ^{0} identifies the conformation at the beginning of the timestep. For small timesteps Δt this should be reasonable to do. $\vec{F}_j$ in the above equation is the force exerted on particle j. The so-called "spurious drift", i.e., the third term in the r.h.s. of eq. (3.20) usually vanishes, since most diffusion tensors which have been used in the literature have zero divergence (this is directly related to the assumption of incompressible flow). $\vec{\rho}{}^0_i(\Delta t)$ is the random displacement by the coupling to the heat bath. The crucial difficulty comes from the connection of the displacement by the heat bath and the hydrodynamic interaction tensor $\overleftrightarrow{D}_{ij}$ via the fluctuation dissipation theorem. This fixes the first two moments to be

$$\langle \vec{\rho}{}^0_i(\Delta t)\rangle = \vec{0} \quad (3.21)$$

and

$$\langle \vec{\rho}{}^0_i(\Delta t) \otimes \vec{\rho}{}^0_j(\Delta t)\rangle = 2\Delta t\, \overleftrightarrow{D}_{ij}. \quad (3.22)$$

That is, the correlation function of the stochastic displacements is proportional to the diffusion tensor, which hence has to be evaluated every time step. Calculation of the random displacements then requires finding a root of the diffusion tensor. For this reason, it is absolutely crucial to have a diffusion tensor which is positive–definite. This is not guaranteed for the Oseen tensor which deals with point particles. Thus in most cases the Oseen tensor is replaced by a tensor which has no singularity for strongly approaching particles. Mostly the so called Rotne–Prager–Yamakawa (RPY) tensor[22,23] is used, which avoids this problem. For short distances the singularity is removed and the tensor reads

$$\overleftrightarrow{D}{}^{RPY}_{ij} = D_0\delta_{ij}\overleftrightarrow{1} + (1-\delta_{ij})\frac{k_B T}{8\pi\eta r_{ij}}\begin{cases}\left(1+\dfrac{2a^2}{3r_{ij}^2}\right)\overleftrightarrow{1} + \left(1-\dfrac{2a^2}{r_{ij}^2}\right)\hat{r}_{ij}\otimes\hat{r}_{ij} & r_{ij}\geq 2a\\[2ex] \dfrac{r_{ij}}{2a}\left(\dfrac{8}{3}-\dfrac{3r_{ij}}{4a}\right)\overleftrightarrow{1} + \dfrac{r_{ij}}{4a}\hat{r}_{ij}\otimes\hat{r}_{ij} & r_{ij}<2a\end{cases} \quad (3.23)$$

where a is an effective bead radius which can be interpreted as the Stokes friction radius of the monomers. As for the Oseen case, the RPY tensor has zero divergence, but now is also positive definite, since it phenomenologically takes the finite size of the beads into account. This however is a technical trick, in order to make the simulations possible, even if the beads are allowed to approach each other arbitrarily.

A rather efficient method to calculate the root of the hydrodynamic interaction tensor is Cholesky decomposition.[49] The random displacements are then obtained via multiplying the root matrix with a vector of random numbers. The root is usually not unique, i.e., there are several matrices whose square is the diffusion tensor, but since any of these matrices yields random displacements which satisfy the condition eq. (3.22), this non-uniqueness averages out in the course of the simulation. These matrix operations become numerically rather intensive if the number of monomers becomes large (the number of operations is proportional to the third power of the number of monomers). The numerical algorithms for Langevin equations are well established,[80,81] however, some details are still discussed today.[84–86]

Ermak and McCammon[80] were the first to simulate a Smoluchowski equation with hydrodynamic interactions. Since then, the approach has been applied to polymer chains by several investigators.[37,38,46,87–89]. One advantage is that it yields both dynamic and static information, studying precisely the implications resulting from the Smoluchowski equation (which is also its main disadvantage, since it takes it for granted). In particular one has a well defined "short-time" behavior (i.e., short compared to the Zimm time, but long compared to the microscopic timescales of the solvent). As discussed before, this separation is not at all trivial in MD simulations.

It is still an interesting, important and active field to look at the consequences of various approximations below the full Smoluchowski equation. Fixman[37] and Rey *et al.*[38,89] used Brownian dynamics, amongst others, in order to study the difference between the short-time Kirkwood diffusion constant and the long-time limit (for time scales long compared to the Zimm time). Similarly, one can also investigate the difference between "fluctuating hydrodynamics" (i.e., the solution of the full Smoluchowski equation) and "preaveraged hydrodynamics" (where $\overleftrightarrow{D}_{ij}$ is replaced by its static average). These studies are of special interest, since they give some clear hints on the validity of the pre-averaging, which is the basis for the investigations discussed in the following section. Slight differences between the short and long time behavior of the diffusion are up to now certainly beyond the limits of MD simulations, mainly because there is no well-defined short-time limit.

Fixman[37,87] studied in some detail the properties of Gaussian chains of up to 56 monomers, comparing fluctuating hydrodynamics to pre-averaged hydrodynamics (this latter case can be done analytically). It was observed that the effect of fluctuating hydrodynamic interactions seems to *decrease* the diffusion constant in the long time, long chain limit only weakly ($\approx 5\%$).

This observation is in accordance with a supposedly rigorous variational bound for the diffusion constant, which states that the Kirkwood value is an *upper* bound.[5] A much stronger effect ($\approx 30\%$) was found for the intrinsic viscosity $[\eta]$, which measures the polymer contribution to the viscosity η of the solution, relative to the viscosity η_s of the pure solvent. It is experimentally defined as

$$[\eta] = \lim_{\rho \to 0} \frac{\eta - \eta_s}{\rho \eta_s}, \tag{3.24}$$

where ρ is the polymer density (number or mass density). For the simulations of single chains this relation changes to the time integral over the intrinsic stress autocorrelation function of the chain.

A severe problem, however, occurred in the extrapolation to long chains: it was found that the bead size a has a strong influence on the diffusion constant, and an even stronger one on the intrinsic viscosity. This suggested the conclusion in Ref. 37 that even in the long chain limit the transport coefficients might be sensitive to microscopic details, in clear contradiction to renormalization-group arguments.[4] It should however be noted that the extrapolation to the infinite chain length limit is probably not trivial: for hydrodynamic properties like the hydrodynamic radius (and hence the diffusion constant) it is known that corrections to scaling are extremely important (see below). We should also mention that in some MD studies transport coefficients such as the diffusion constant of a single particle were investigated as the function of some presumably unimportant parameters as, e.g., its mass,[62,90] where a nontrivial dependence was found. Here certainly more work has to be done in order to resolve these uncertainties.

While Fixman's investigations were restricted to Gaussian chains, Rey *et al.*[38,89,91] studied both Gaussian and self-avoiding walks, combining Monte Carlo simulations for the statics with BD using the RPY tensor for the dynamics. Beyond the diffusion constant, they also measured the dynamic structure factor and the relaxation of Rouse modes.[5] This allows a direct comparison to scattering experiments. For the random walk case[89] their results are, within the error bars, in agreement with the earlier data of Fixman. They confined themselves to significantly shorter chains (up to $N_{ch} = 20$), probably in order to be able to attain the necessary statistics. For the self-avoiding walk (SAW) chains of length of up to $N_{ch} = 25$, a rather soft repulsive potential for the excluded volume interaction was introduced. This potential still allowed the chains to cut through themselves, however, since entanglement effects are not important for such a system this artifact is of course of no relevance here. The potential used was $V(r) = A\exp(-\alpha r)$ with $A = 75k_BT, \alpha = 4/b$ and an interaction cutoff $r_c = 0.512b$, where b is the root mean square distance of neighboring beads in the absence of excluded volume interactions, resulting from a harmonic backbone potential. This soft excluded volume interaction allowed a BD time-step about a factor 100 times larger than the conventional short range repulsive LJ potential.* For those chains both fluctuating as well as pre-averaged hydrodynamics was studied. In the latter case, the pre-averaged diffusion tensor was obtained from actual simulation data, i.e., apart from some statistical error, no additional approximation beyond pre-averaging was introduced. Fig. 3.7 gives an example for a chain of length $N_{ch} = 25$, showing the averaged inverse distances.

The first results concern the comparison between the Kirkwood formula for the diffusion constant and the diffusion constant obtained from pre-averaged BD simulations. Within the error bars, the numbers are identical. However, for fluctuating hydrodynamics a diffusion constant systematically *above* the Kirkwood value was found, at variance with the variational bound.[5] Since the data seem to be rather accurate, this might be an indication that something is wrong with the rigorous bound. For a discussion, see Ref. 38. These questions must be regarded as completely unresolved today.

Rey *et al.* also evaluated the theoretical prediction for the correction term D_1, by which the true diffusion constant should be smaller than the Kirkwood value[38]:

$$D_1 = \frac{1}{3N} \int_0^{\infty} C_A(\tau) d\tau \tag{3.25}$$

where $C_A(\tau) = \langle \vec{A}(t)\vec{A}(t+\tau) \rangle$ is the autocorrelation function of the deterministic part of the displacement vector per time unit, $\vec{A} = (1/k_B T) \sum_{i=1}^{N_{ch}} \sum_{j=1}^{N_{ch}} \overleftrightarrow{D}_{ij} \vec{F}_j$. Of course, taking this correction into account makes the above-mentioned discrepancy even worse (D_1 is positive). However, the behavior of $C_A(\tau)$ shows very nicely that apparently fluctuating hydrodynamics can be approximated better and better by pre-averaged hydrodynamics when the chain length is increased: Fig. 3.8 shows the time dependence of the ratio $C_A^F(\tau)/C_A^P(\tau)$, where C_A^F and C_A^P were obtained from simulations with fluctuating and pre-averaged hydrodynamics, respectively. The ratio is always larger than 1, and very soon it approaches 1. In addition the deviation is reduced strongly with increasing chain length.

Moreover, it was found that with increasing chain length the relative deviation of the true diffusion constant from the Kirkwood value seemingly tends to zero. In view of these results it seems conceivable that in the limit of infinitely long chains there is no difference between pre-averaged and fluctuating hydrodynamics and that hence the "static dynamics" approach gives good results even in the long time limit.

*Note that this idea does not apply for the previously considered MD simulations. Such a soft potential would change the solvent's compressibility dramatically. The assumption of an effectively incompressible solvent then would certainly be in error.

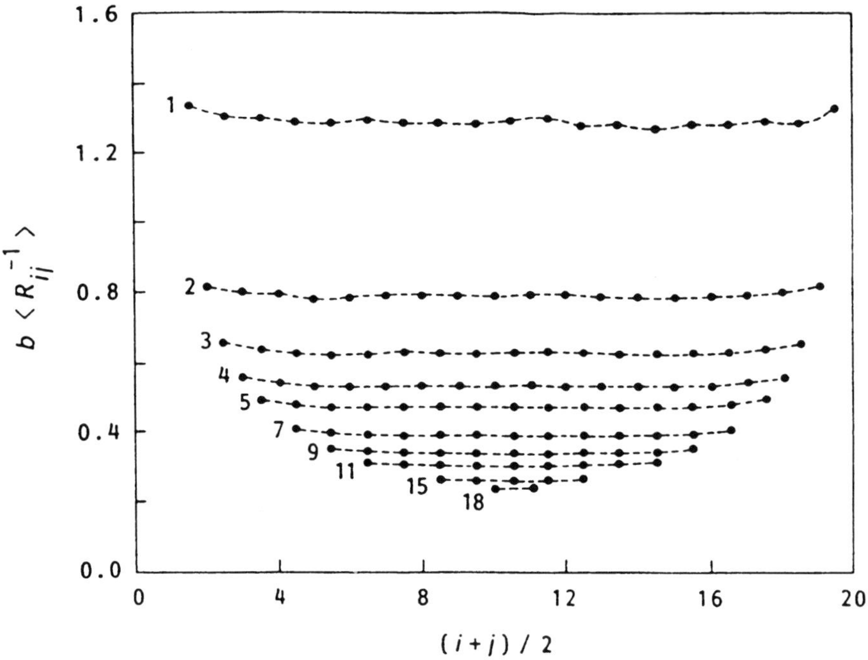

Fig. 3.7 Average values of $\langle r_{ij}^{-1} \rangle$ versus the position of the center of pair ij for a linear chain with $N_{ch} = 20$ units. The lines connect points with a fixed length of the subchain as indicated (from Ref. 91).

Besides these investigations first attempts to study shear flow have been made, using BD simulations.[92,93] We should also finally mention that the BD approach can be applied to suspensions as well. For these systems, there exist highly developed simulation schemes, which employ rather accurate diffusion tensors taking into account both the finiteness of the Brownian particles[25,26] as well as the hydrodynamic interactions with the periodic images via Ewald sums.[65,66] For more information on these so-called "Stokesian Dynamics", see, e.g., Ref. 94. Due to the computational complexity, only moderate system sizes can be studied.

3.2.2.4 "Static dynamics"

As was mentioned before, a static average over a dynamical operator yields the short-time behavior (i.e., the initial decay rate of the correlation functions). Eqs (3.9) and (3.15) are based on this reasoning. Moreover, as discussed in the previous section, there are reasons to believe that in the long chain limit the approach gives good results even for long times.

The application of this idea originates with Zimm.[95] Since then it has been used to study single isolated objects, such as chains, stars and combs etc.

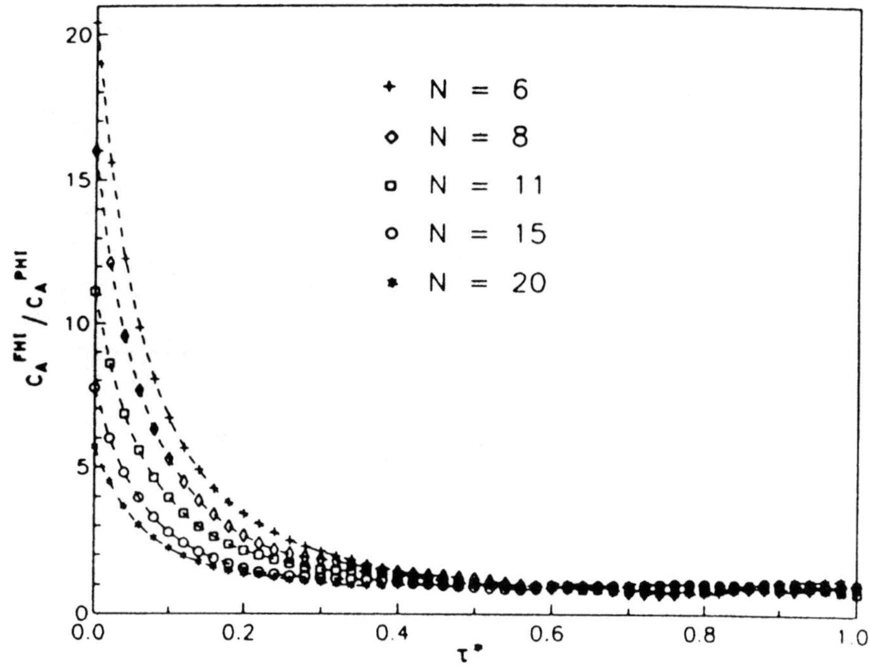

Fig. 3.8 Ratio of the time correlation function $C_A(\tau)$ as defined in the text (cf. eq. [3.25]) for the full hydrodynamic interaction compared to the pre-averaged one for different chain-lengths as indicated (from Ref. 38).

under various physical conditions. Since only conformational properties are of interest, the most effective method is usually a static MC simulation (see e.g., Chapter 2 by Sokal or Ref. 96). We will here discuss the implications on two levels. First a few examples dealing with purely static quantities which are relevant for the hydrodynamics will be given. Then we review some attempts in which the strength of the hydrodynamic interaction is taken into account, e.g., by varying the Stokes friction radius of the beads. Since the main aspect of the present chapter is the chain hydrodynamics, we mention the simulation algorithms only in passing. In addition, we will confine ourselves mainly to recent developments.

The first and simplest quantity to analyze is the hydrodynamic radius $\langle R_H^{-1} \rangle^{-1}$ as a purely geometric quantity. From light-scattering experiments assuming proper separation of the hydrodynamic and the bare friction contributions, it is known[97] that the corrections to scaling for the averages are very strong. Some experiments even conclude a different, somewhat smaller asymptotic exponent for the hydrodynamic radius $\langle R_H^{-1} \rangle^{-1}$ than for the end-to-end distance or the radius of gyration.[40,98–100] This is of course wrong,[99] but since for R_H the inverse distances are averaged, corrections due to close approach in space or along the chain are very important. Especially when some effective bead radius is incorporated this can obscure the results for shorter chains significantly. Within an extremely simplified model, it can actually be shown that for the hydrodynamic radius the exponent of the leading correction to scaling is $\delta = 1 - \nu$.[48] Thus $\langle R_H^{-1} \rangle^{-1}$ reads

$$\langle R_H^{-1} \rangle^{-1} = N_{ch}^{\nu}(A + BN_{ch}^{\nu-1}). \tag{3.26}$$

In most cases $\langle R_H^{-1} \rangle^{-1} N_{ch}^{-\nu}$ is plotted vs. either N_{ch}^{-1} or $N_{ch}^{-1/2}$. For the latter the deviation from the correct power is presently too small to be detected in a simulation. The calculation of the purely geometrical hydrodynamic radius is unambiguous while the subsequent use for the intrinsic viscosity or the sedimentation velocity needs further assumptions. Fig. 3.9 gives an example from a simulation of athermal walks and stars on the fcc lattice, using a modified dimerization method.[101] The fcc lattice has the advantage of being closer to the SAW fixed point than other $3d$ lattices. This means that the amplitudes of the corrections to scaling are very small compared to all other $3d$ lattices and the persistence length is not significantly larger than a bond length. The figure shows the results for a linear polymer and for stars of $f = 3, 4, 5, 6$ arms. Corrections to scaling for $\langle R_G^2 \rangle$ are almost undetectable for the chain lengths presented in the figure, while for R_H they are very strong. The difference, for example, between the asymptotic amplitude A and the effective amplitude for $N_{ch} = 100$ is about 20% with a slight increase for higher f. Taking the length of the chains into account, the rather disturbing results of some experiments are not too surprising. To eliminate model specific aspects an important quantity is the ratio

$$\rho(f) = \frac{\langle R_G^2(N_{ch} \to \infty, f) \rangle^{1/2}}{\langle R_H^{-1}(N_{ch} \to \infty, f) \rangle^{-1}}, \tag{3.27}$$

which for random walk stars[102] can be evaluated exactly to be

$$\rho(f) = \left(\frac{3f - 2}{f\pi} \right)^{1/2} \frac{8(2 - f + 2^{1/2}(f-1))}{3f}. \tag{3.28}$$

Note that of course $\rho(f = 1) = \rho(f = 2)$. Fig. 3.10 shows a comparison between the simulation of the athermal lattice SAW polymers and the random walk (RW) calculation and the $f = 3$ experiment by Huber *et al.*.[103] The strong deviation between simulation and experiment shows that current computer simulations are by far superior to experiment for studying such fundamental relations. The data show that for a small number of arms the SAW polymers obviously display fluctuations which increase $\rho(f)$ compared to the RW result. This changes with increasing f. It would certainly be of interest to see $\rho(f \gg 1)$.

To go further and to estimate the sedimentation velocity from such a calculation, the coupling to the solvent must explicitly be incorporated, as was done, e.g., in Ref. 95. In order to remove such problems in experiments

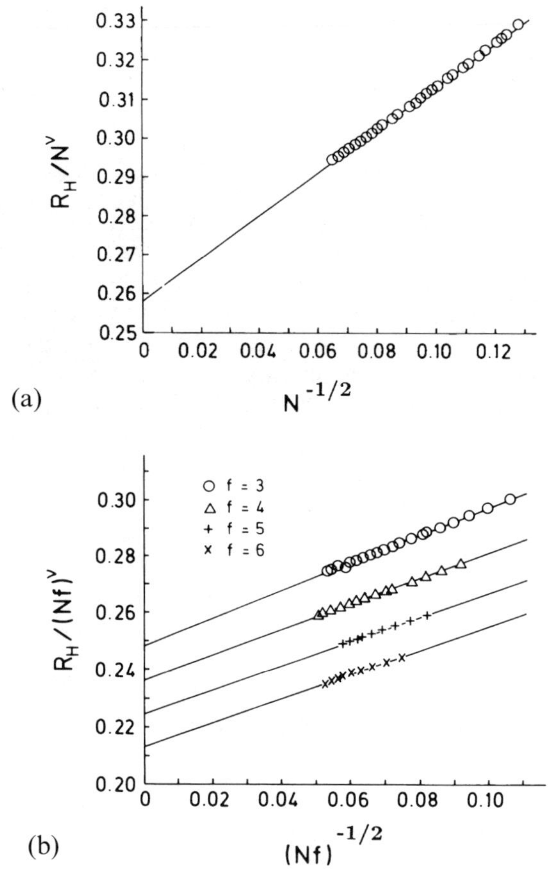

Fig. 3.9 $R_H(N_{ch})/N_{ch}^{\nu}$ versus $1/N_{ch}^{1/2}$ for chains (part [*a*]) and the ratio $R_H(f, N_{ch})/(fN_{ch})^{\nu}$ versus $1/(N_{ch}f)^{1/2}$ for stars with f arms (part [*b*]). The straight lines give the amplitude of the corrections to scaling and show the very slow crossover towards the asymptotic behavior (from Ref. 101).

and to compare to theory, characteristic ratios such as ρ need to be calculated/measured. Thus in most simulations point monomers sitting on lattice positions are considered, since the ratio ρ is expected to be universal, i.e., not to depend on the details of the model.

Another interesting quantity is the initial decay rate of the dynamic light scattering function. Again, in order to provide a direct basis for comparison to experiment, the local coupling to the solvent has to be inserted. As was discussed before, the initial decay rate $\Gamma^{(0)}(k) = -(d/dt)(\ln S(k,t))_{t=0}$ can be written as $\Gamma^{(0)}(k) = D(k)k^2$ where $D(k)$ is given by eq. (3.15). For $kR_G \ll 1$, one expects $\Gamma^{(0)} = D^{(K)}k^2$, where $D^{(K)}$ is the short-time (i.e., Kirkwood) value of the center of mass diffusion constant, eq. (3.9). For $kR_G \gg 1$

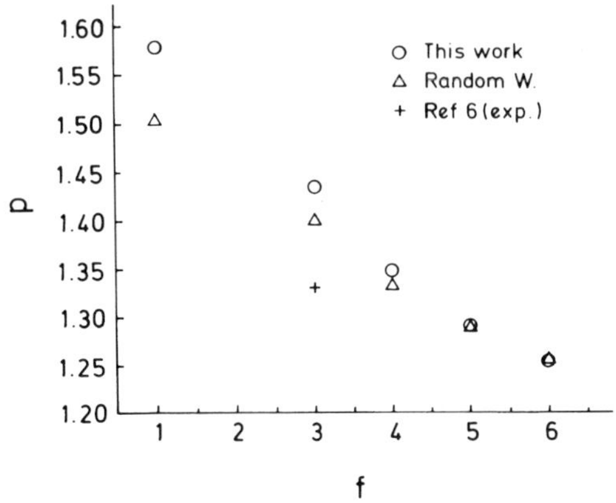

Fig. 3.10 The ratio $\rho = R_G(N_{ch},f)/R_H(N_{ch},f)$ versus f, comparing the MC data to the exact random walk results. One experimental point is included (cf. text) which also illustrates the experimental difficulties (from Ref. 101).

one should observe $\Gamma^{(0)} = (C_0 k_B T/\eta_S)k^3$ with $C_0 = 0.0788$ for SAW statistics.[47] Fig. 3.11 shows the results for a linear chain and star polymers on a fcc lattice. Since the prefactors are taken to be unity ($D_0 = 1$ and $(k_B T)/(8\pi\eta_S) = 1$) this can only be used to compare to experiment on a rather qualitative level, especially for smaller systems. First one observes the expected divergence of $\Gamma^{(0)}/k^3$ for small k. With increasing number of arms the de Gennes correlation hole becomes more and more pronounced. It

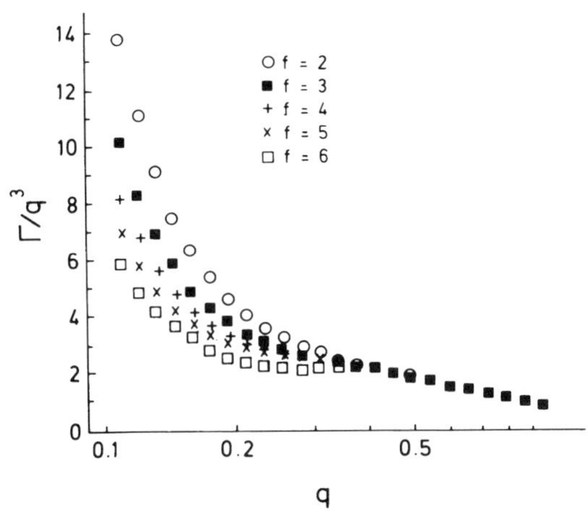

Fig. 3.11 Initial decay rate $\Gamma^{(0)}$ for star polymers with a varying number f of arms, as indicated. In all cases the arm length is $N_{ch} = 40$. In the high wave number regime, where the scatterer only feels the single arm, all curves coincide (from Ref. 101).

simply means that the space which is occupied due to the presence of other strands is not available for conformational relaxation. For larger k with $2\pi/k$ smaller than the diameter of the largest blob[1] in the star, one essentially observes the effect of the individual strands of the stars. There the data have to become independent of f, displaying the same result as for the isolated chain.

Many authors, and such was the original attempt by Zimm,[95] try to go beyond the above calculations. They want to get information on the sedimentation velocity (which is directly related to the diffusion constant), or the intrinsic viscosity. According to Fixman[104] and Wilemski and Tanaka[105] the "static dynamics" (i.e., zero time limit) approach should give an upper bound for transport properties such as the diffusion constant or the zero shear viscosity. (As was discussed in the previous section, this was not confirmed by the BD simulations of Rey *et al.*[38]) However, in order to conclude these short-time approximations for the transport coefficients from the geometric properties of the chain, further input data are needed. Of these, one should distinguish between solvent properties (in particular the solvent viscosity) and microscopic bead properties like the monomeric diffusion constant D_0 (or the related friction coefficient ζ or the corresponding Stokes radius a). Of course, from renormalization-group theory[4] one expects that these latter properties should have no influence on the transport coefficients in the asymptotic limit (diverging chain length and vanishing density), but in both simulations as well as experiments one deals with finite chains. In this context we again mention the BD results of Fixman,[37] where severe doubts regarding the independence on microscopic model parameters were put in place. The strength of the coupling is usually, in addition to the solvent viscosity, introduced via an effective bead size or Stokes radius. This however is ambiguous. Since rotations and translations enter in Zimm's approach with different friction constants, as they might be of importance for the sedimentation problem,[95] there is another numerical parameter of order unity. The finally resulting numerical factor usually is denoted as h^* and is called strength of the hydrodynamic interaction. This altogether leads to a relation between microscopic friction and viscosity. For the assumption of a simple spherical molecule one expects the standard Stokes law $\zeta = 6\pi\eta_S a$. This would be the ideal case of stick boundary conditions. However, it is assumed that because of the microscopic nature of the beads there is a situation in between stick and slip boundary conditions. Thus, as this short and rather incomplete discussion shows, the direct quantitative use of essentially both BD and "static dynamics" bears many unresolved questions.

The influence of the microscopic parameters on the predicted diffusion or sedimentation constants is shown in Fig. 3.12 from Ref. 106. There for short linear chains and small star molecules the calculated diffusion constant within Zimm's scheme is given for two different effective bead sizes. The

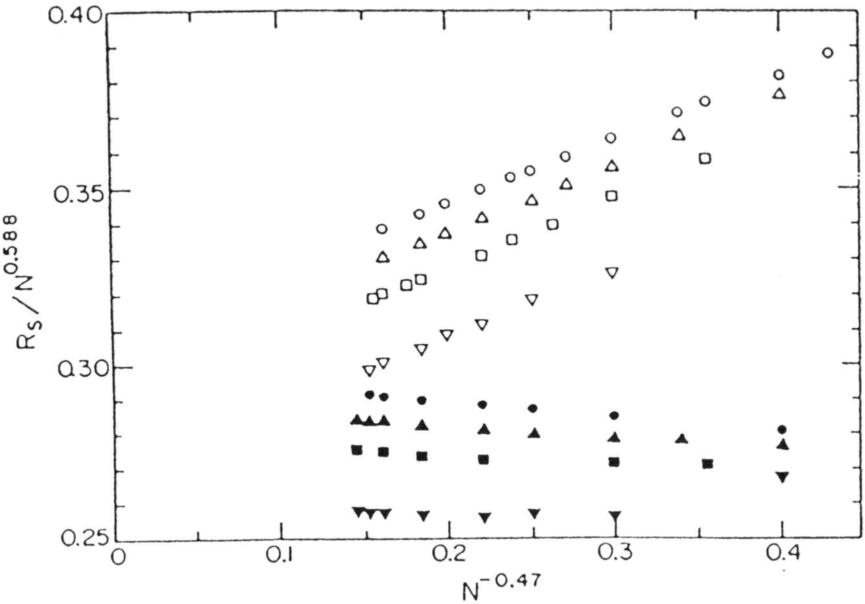

Fig. 3.12 Ratio of the hydrodynamic radius R_H to N_{ch}^{ν} (in our notation), plotted versus $N_{ch}^{-0.47}$. Here the hydrodynamic radius is not defined as a purely geometric quantity, but rather as the Stokes radius of a sphere which would have the same sedimentation velocity as the polymer. The latter is obtained via "static dynamics", also taking rotational diffusion into account (Zimm's approach.[95]). Data are shown for star polymers on the cubic lattice and $f = 1, 3, 4, 6$ (from top to bottom) for different effective monomeric Stokes radii $a = 1/4$ (full symbols) and $a = 1/2$ (open symbols) (from Ref. 106).

results differ dramatically and also the prefactor of the finite size corrections shows different signs, which is a clear indication that the data are by far not asymptotic. The questions arise again for how far with increasing N_{ch} such problems remain.

Nevertheless, one can use such simulations to determine, e.g., the relative contributions of different parts of the conformational properties to the diffusion. Fig. 3.13 shows an example in which different approximations are compared to the actual simulations.[106]

To avoid the above-discussed finite size effects and problems with the microscopic form of the local hydrodynamic interaction, at least to some extent, typical ratios are calculated. For this the intrinsic viscosity is the most widely used example. This viscosity is expected to be independent of microscopic aspects. It is also used experimentally to determine molecular weights.[11] Here we want in a little more detail to review some recent studies of chains under Poiseuille or Couette flow.

Van Vliet *et al.*[107,108] recently used the approach of Wilkinson *et al.* in order to study the influence of confined geometry on the flow problem. The

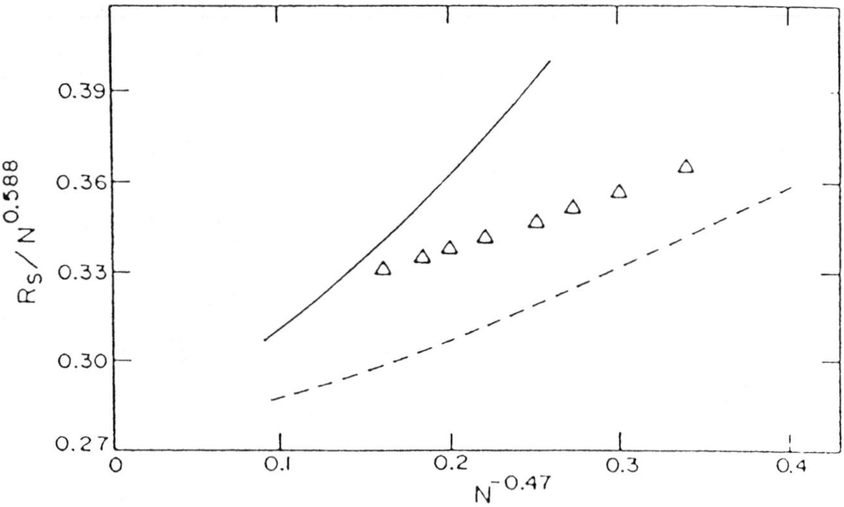

Fig. 3.13 Scaled hydrodynamic radius as in Fig. 3.12 for an $f = 3$ arm star polymer for $a = 1/2$ versus $N_{ch}^{-0.47}$. The three sets of data correspond to different approximations. The full line gives the hydrodynamic radius as defined in eq. (3.10). The broken curve also takes into account the "free-draining part" (i.e., the term D_0/N_{ch} in eq. 3.9), while the triangles give the results from the Zimm approach[95] (from Ref. 106).

problems investigated were planar Couette (linear velocity profile) and planar Poiseuille (parabolic velocity profile) flow. The first is important for studies of chains under shear and chains near surfaces while the latter certainly is of high practical importance, e.g., for chains flowing in very narrow tubes or in viscometers. In these studies the change in the intrinsic viscosity compared to the unconfined situation is studied for both good solvent and Θ-solvent chains on the simple cubic lattice. The chains are mainly generated using the Wall–Mandel slithering snake algorithm.[96] They study the zero shear intrinsic viscosity as a function of the ratio of plate distance L over unconfined chain dimension which scales like N_{ch}^{ν}. The main difference to the bulk studies mentioned earlier is the inclusion of the angular/rotational contribution to the intrinsic viscosity and the RPY tensor. The RPY tensor had to be modified in order to take the hydrodynamic coupling to the walls properly into account. In Zimm's original treatment, as adopted by others, the rotational contribution is included via essentially the average asphericity of the chains. This has to be changed here, since the chains near the surface are not free to rotate. The Zimm angular velocity is given by

$$\Omega_L = \dot{\gamma} \frac{\sum_{i=1}^{N_{ch}} \langle z_i^2 \rangle}{\sum_{i=1}^{N_{ch}} \langle x_i^2 + z_i^2 \rangle}, \tag{3.29}$$

where $\dot{\gamma}$ is the shear rate, which is a constant for Couette flow, but dependent on z for Poiseuille flow. The flow is in the xz plane with flow lines parallel to the x-axis. For the Poiseuille flow $\dot{\gamma}$ is taken as the shear rate at the z-coordinate of the center of mass of the chain. The Zimm angular velocity is approximated either by the average angular velocity under these conditions in a bulk solution or by the actual conformation dependent velocity. The differences turn out to be negligible within the error bars of the simulation. Within this scheme the intrinsic viscosities are then given for the Couette case by

$$[\eta]_L = -\frac{1}{\eta_S \dot{\gamma}} \sum_i z_i F_{x,i}, \tag{3.30}$$

where z_i is measured from the center-of-mass of the chain. (In many applications a prefactor N_A/M, where N_A is Avogadro's number and M the molecular mass of the species under consideration, is added in order to get experimental units.) For Poiseuille flow the situation is more complicated. There the position of the center of mass is fixed and the contribution to $[\eta]_L$ calculated. Then these individual numbers are averaged over the probabilities of the z-coordinates of the center of mass.

Figs 3.14 and 3.15 give two examples of the change of the intrinsic viscosity relative to the corresponding bulk solution. The first one shows the different contributions of the angular and translational parts for Couette flow as a function of the scaled thickness of the slit. For very small slits only the translational part of the viscosity remains. As soon as the layer thickness reaches about 1.5 times the unperturbed chain dimensions the rotational part dominates the behavior. The main conclusion is that the characteristic layer thickness for this to happen is clearly smaller than predicted by dumbbell models. Secondly, the sliding conformations which occur in the very narrow slits contribute significantly to the intrinsic viscosity. The second figure shows the difference between the Couette geometry and the Poiseuille geometry for the case where the angular contribution is analyzed for each conformation separately. This explains the larger scatter of the data. Because of the stronger variation of the contribution to the intrinsic viscosity with the position of the center of mass of the chain in the Poiseuille case, the intrinsic viscosity is reduced over a much larger range of layer thickness compared to before.

3.2.2.5 *First conclusion*

The various simulation approaches described in the previous sections show the possibilities of modern computer simulations for hydrodynamic properties of polymer solutions. Unfortunately, there are very severe limitations. For the MD simulations still faster computers are needed e.g., to make a serious attempt to look at semidilute solutions beyond some very

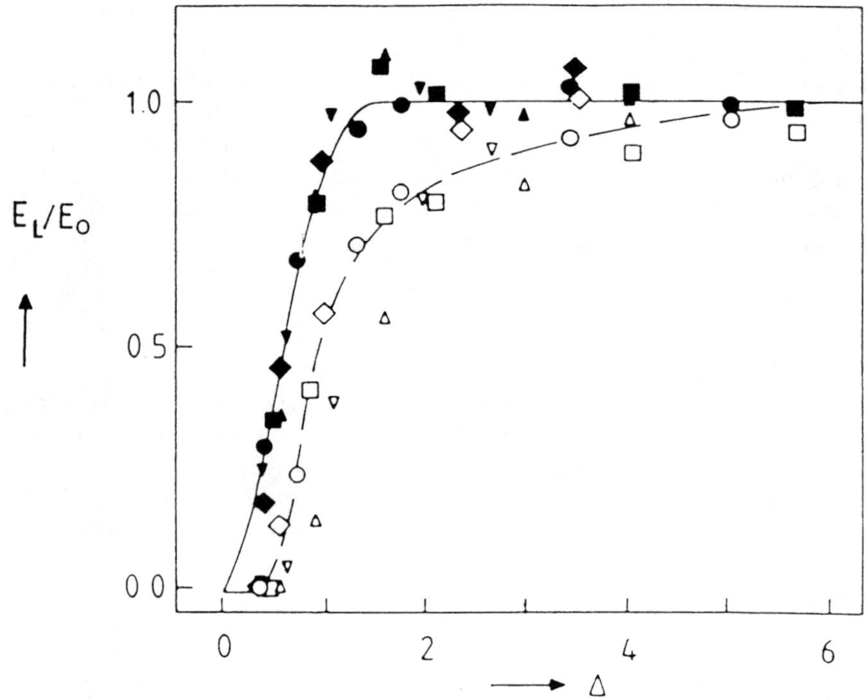

Fig. 3.14 Ratio of the intrinsic viscosity in a slit of width L to the bulk value versus the normalized slit width $\Delta = L/N_{ch}^{\nu}$ for an athermal solvent and chain-lengths between $N_{ch} = 12$ and $N_{ch} = 79$. The open symbols give the contribution from free rotation (from Ref. 107).

limiting cases. This is a problem where modern parallel computers might help. For the BD simulations the calculation of the square root of the diffusion tensor leads to severe problems, especially if one would like to go to semidilute systems or even simple systems of several chains where periodic boundary conditions have to be included. Though very promising to some extent, as far as we can see the application of BD is rather limited in the long run. More complicated situations might be studied extensively in the future by "static dynamics" methods. There the big advantage is the high accuracy of the available conformation distributions. On the other hand, this method requires by far the most serious assumptions and approximations which consequently to our impression give it only qualitative character. It was used, for example, by the authors of the above-discussed papers to test scaling laws. In general, in spite of considerable effort by various groups over the last years, the situation is not very satisfying. There will be some progress by the advent of more powerful computers, but for the really

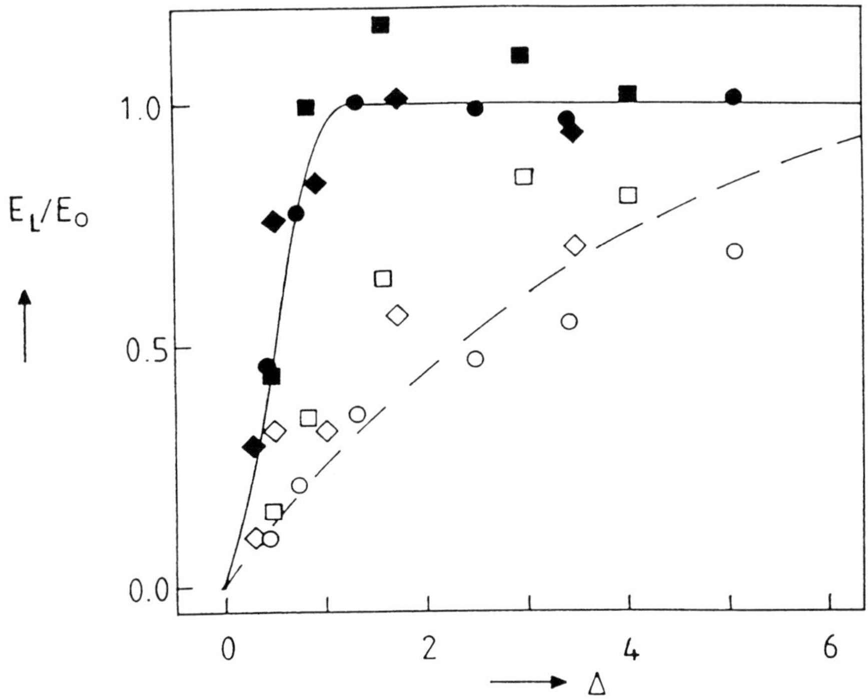

Fig. 3.15 Ratio of the intrinsic viscosity in a slit of size L to the bulk value versus the normalized slit width $\Delta = L/N_{ch}^{\nu}$ for the Couette (solid symbols) and the Poiseuille (open symbols) flow for chain-lengths between $N_{ch} = 59$ and $N_{ch} = 150$. The typical frequency is set to $\omega = \omega_L$ (from Ref. 108). For details see text.

delicate and important problems, such as hydrodynamic screening, nonzero shear etc. new algorithmic ideas are essential as well.

3.3 Structure of charged polymer solutions

So far we have reviewed the special case where the long-range nature of the interaction affected only the dynamic properties of the polymers leaving the static properties unaltered. This is the case for neutral, nonpolar polymers. As soon as the chains contain highly polar units (e.g., polyethylene oxide) or even charged units (polyelectrolytes, polyamphilytes), the conformations are also strongly affected by the long-range nature of the interaction. This modification of the average conformation certainly will be most strongly pronounced for charged chains in solution.

Polyelectrolytes in solution represent an important class of polymers. Much interest in polyelectrolytes occurs because many biological polymers are polyelectrolytes.[109] In particular DNA and RNA are prototypical examples of a polyelectrolyte. Besides the many biological polyelectrolytes, other common examples include polyacrylic acid and polystyrene sulphonate. The interest in polyelectrolytes is in addition supported by the huge potential of applications.[110] Already today polyelectrolytes are used for thickening agents, rheology, metal extraction, food stabilization, and water absorbants (diapers).[111,112] Interest in polyelectrolytes is presently rising particularly because they are water soluble and thus expected to be environmentally safer.

A large variety of charged polymers exists because of the numerous ways of placing charges on the polymer chain. From every class of neutral polymers several charged polymer classes can be created. This structural variety in part leads to a rich behavioral variety. Yet, even a single class of charged polymers such as linear polyelectrolytes in solution exhibits a wide range of physical behavior. These polyelectrolytes possess characteristics dramatically different from neutral polymers. In this review we will consider only linear polyelectrolytes in solution, because they comprise the basic type of polyelectrolyte systems. Both the current simulation studies as well as the attempts for analytical theories essentially only deal with this prototypical class of systems.

The long-range nature of the Coulomb interactions significantly alters the polymer structure in comparison with neutral polymers. Furthermore, the charges introduce more than just a long-ranged interaction into the system. Because the total system is neutral, polyelectrolytes in solution are, at their simplest, a ternary system composed of the charged polymer, counterions and solvent. In general, there is added salt which makes the system a quaternary system. Thus, polyelectrolytes in solution are more complex than their neutral counterpart.

Five parameters classify the main types of linear polyelectrolytes: the fraction f of charged monomers, the strength of the Coulomb interaction normalized to k_BT, λ, the added salt concentration c_s, the polymer concentration c and of course the chain length L. One other important quantity is the valence of the counterions or salt ions. For the most part only monovalent ions have been considered to date in simulations and theory. To completely understand polyelectrolytes requires studying the variation of each of these quantities which is a formidable task. The understanding of polyelectrolytes in solution is just beginning. While much work has been done on these systems, by no means is there a definitive understanding of their properties and structure.

Because the Coulomb interaction is long-ranged, its introduction into the system is often nonperturbative. The structure of polyelectrolytes in solution depends strongly on the strength of the Coulomb interaction. Polyelectrolytes are classified as either weak or strong depending on the Coulomb strength. The convenient dimensionless coupling strength is the ratio of the Bjerrum length, $\lambda_B = e^2/\epsilon k_BT$, and the distance separating charges along the chain backbone, a:

$$\lambda = \lambda_B / a. \tag{3.31}$$

λ is called the Manning ratio. The dimensionless Coulomb pair interaction is then

$$\beta u(r_{ij}) = z_i z_j \frac{\lambda_B}{r_{ij}}, \tag{3.32}$$

where $\beta = 1/k_B T$ is the inverse temperature, z_i is the valence of the ith particle, and r_{ij} is the pair separation distance. A strong polyelectrolyte is defined as having the coupling strength $\lambda \gtrsim 1$ which physically corresponds to the Coulomb interaction between bonded charged monomers being stronger than the thermal interaction. In general, strong polyelectrolytes are fully charged ($f = 1$), but the degree of charging and the bond length obviously combine to define f. In weak polyelectrolytes, $\lambda \ll 1$, the Coulomb interaction is dominated by thermal interactions at short distances. But, at long length scales the Coulomb interactions can add up and dominate the thermal interactions unless screened by counterions. Thus, the Coulomb interaction will tend to dominate the structure of the charged polymer at some length scale.

This book, including the first part of this chapter, is a testimony to the strong influence that simulations have had on our understanding of neutral polymers. The basic need for similar simulations on charged polymers is even stronger. In great contrast to our knowledge of neutral polymers, our knowledge of polyelectrolytes is weak. As pointed out above, polyelectrolytes are much more complex than neutral polymers. Consequently they are more difficult to characterize both experimentally and theoretically. Other "technical" difficulties are also important. Polyelectrolytes are elongated in comparison with neutral polymers due to the Coulomb repulsion between charged monomers. As a result the overlap concentration is at much lower values than for neutral polymers where the resolution of neutron scattering is poor. Consequently, no measurement of the single chain structure factor has been done. Theoretically, the long-ranged nature of the Coulomb interaction is intractable. Most theoretical work presumes a Debye–Hückel interaction which is only valid in the weak interaction limit, generally taken to be at very dilute concentrations. Furthermore, neutral polymer theory has been advanced through scaling arguments,[1] but since polyelectrolytes in solution have more than just the added length scale of the Coulomb interaction, such scaling theories are much more difficult and speculative. We thus have the classic case of theory and experiment only being best done in two different regimes. Polyelectrolytes in solutions appear to be a great opportunity for simulations to resolve the fundamental issues. However, the complexity of polyelectrolytes and the long-range Coulomb interaction affect simulations as well making the computer time much longer than for neutral chains.

Below we discuss the present status of the theory of polyelectrolytes. Some of the important experimental results with which there is simulation data to compare are also discussed. We do not mention experiments such as viscosity measurements since no simulation has calculated such quantities.

3.3.1 *Theoretical models*

Theory presently treats only limiting cases; no general theoretical scheme exists which covers the whole range of systems. The two main limiting cases are strong very stiff polyelectrolytes in solution or weak flexible polyelectrolytes in solution. For both regimes the discrete nature of the counterions and the salt in solution is only treated in a mean-field manner.

3.3.1.1 *Strong polyelectrolytes*

The structure of charged polymers is very different from neutral polymers. For polyelectrolytes in solution the like charged monomers repel each other tending to elongate the chain unless their interaction is screened by counterions or salt ions. The long-range nature of the Coulomb interaction can extend this elongation to all length scales. This can be seen from the Flory calculation below which shows that the isolated chain is rod-like. Polyelectrolyte structure can be varied greatly depending upon the amount of screening due to counterions and/or coions.

One can apply Flory arguments to calculate the end-to-end distance, R, as a function of the number of monomers for the case of a completely isolated charged chain (i.e., no counterions or salt ions). In the Flory argument the free energy is

$$F = k_B T \frac{R^2}{Na^2} + k_B T \lambda_B \frac{N^2}{R}, \tag{3.33}$$

where the first term is the usual elastic energy and the second term represents the electrostatic energy of the charged monomers. The above equation for the free energy omits all the logarithmic corrections, even though they probably are very important for polyelectrolytes in solution. Minimizing the free energy with respect to the end-to-end distance, R, yields

$$R \sim \lambda_B^{1/3} a^{2/3} N. \tag{3.34}$$

The dilute limit is a rod-like chain—the most extended a chain can be. We note that already within this scheme the case $\lambda_B > a$ leads to unphysical stretching! In comparison, $R \sim N^\nu$, where $\nu = 1/2$ for Gaussian chains and $\nu = 0.588 \approx 3/5$ for good solvent neutral chains. The more accurate renormalization group (RG) calculation also gives $\nu = 1$ for the three-dimensional isolated charged chain. More generally, in d dimensions renormalization group calculations[113] have found that $\nu_{RG} = 2/(d-2)$ for

$4 < d < 6$. Flory-type calculations give $\nu_F = 3/d$. For $d = 4$ and 5 the two calculations disagree.

Since polyelectrolytes are stretched at dilute concentrations, much theoretical work has focused on the perturbation from the rod-like conformation.[114–120] In the infinitely dilute limit the Debye–Hückel approximation to the Coulomb potential is valid:

$$\beta u(r) = \frac{\lambda_B}{r} e^{-\kappa r}, \tag{3.35}$$

where

$$\kappa^2 = 4\pi\lambda_B \sum_i c_i z_i^2 \tag{3.36}$$

defines the Debye length, $\Lambda = \kappa^{-1}$ with c_i the concentration of the ith ionic species with valence z_i. The validity of this approximation is questionable as the concentration increases, but no other means of treating the counterions (and salt ions) is amenable to calculation. The departure from completely rod-like structure is described in terms of the persistence length, L_p, and the basis of most theoretical work on polyelectrolytes in solution are persistence length calculations. The original calculations by Odijk[114] and Skolnick and Fixman[115] were done for a worm-like chain which models semiflexible polyelectrolytes like DNA. The total persistence length, L_p, is taken to be just the sum of the intrinsic persistence length of the uncharged chain, L_i, and the electrostatic persistence length, L_e. The free energy is taken to be the sum of the elastic and electrostatic interaction as in the Flory argument with the electrostatic interactions being the sum of the Debye–Hückel pair potentials. Entropy is neglected and the chain is assumed to be locally stiff (i.e., a worm-like chain).[114,115] The complete form of the electrostatic persistence length is

$$L_e = \frac{1}{12} z^2 N^2 \lambda_B [3x^2 - 8x^3 + e^{-1/x}(x + 5x^2 + 8x^3)], \tag{3.37}$$

where $x = 1/\kappa L$.[118] This expression simplifies when $x \ll 1$ to

$$L_e = \frac{z^2 \lambda_B}{4(\kappa a)^2} \tag{3.38}$$

which scales as $c^{-1/2}$.

The above expressions have been modified to include excluded volume effects by assuming the same recipe as used for neutral polymers. The radius of gyration is written in terms of an expansion factor α and the value of R_G at the Θ-temperature where the excluded volume repulsion is cancelled by the van der Waals attraction:

$$R_G^2 = \alpha^2 \langle R_G^2 \rangle_0. \tag{3.39}$$

The chain is viewed as composed of N_K Kuhn segments of length L_K. The expansion factor α is a function of these two parameters and the excluded volume, V_{ex}, only depending on the dimensionless combination

$$z = \frac{3}{2\pi} \frac{V_{ex}}{L_K^3} N_K^{1/2}. \tag{3.40}$$

Various forms of α expanded in terms of z have been given.[2,118,121] The work of Odijk and Houwaart[118] used the Yamakawa–Tanaka expression

$$\alpha^2 = 0.541 + 0.459(1 + 6.04z)^{0.46} \tag{3.41}$$

The connection with polyelectrolyte theory is through the persistence length. The persistence length of a rod is $L/2$. Thus, $L_K = 2L$, for a rod. The calculation of V_{ex} for a rod yields

$$V_{ex} = 8\pi L_p^2 \Lambda \tag{3.42}$$

which gives

$$z = \frac{3^{3/2}}{4\pi^{1/2}} \frac{L^{1/2}\Lambda}{L_p^{3/2}}. \tag{3.43}$$

Other expressions for the excluded volume have been calculated.[122] All the expressions assume a rigid rod structure.

Based on these persistence length calculations Odijk has developed for $L \gg \Lambda$ a scaling theory of polyelectrolytes for both dilute and semidilute regimes.[123] He proposed a sequence of structural changes as a function of increasing polymer concentration. The chains remain rigid rods until $L_e \approx L$ at monomer concentration $c_b^* = 1/16\pi a^2 L$. In this picture the chains can overlap before they begin to bend. To see this, consider the overlap concentration for rigid cylinders: $c^* = 6/(\pi a L^2)$. Then, $c_b^* > c^*$ when $L > 96a$, or equivalently, when $N > 96$ if each monomer is charged. Between c^* and c_b^* one has rods packed at densities which may form a nematic phase. A second transition (above c^*) occurs when the strand–strand distance equals L_e at $c_e^* = 0.04/4\pi\lambda_B^3$. By this concentration the nematic phase disappears. At higher concentrations the chains are Gaussian chains of blobs. Following the usual scaling theory recipe and using the asymptotic expressions for the persistence length given above, the following semidilute scaling results:

$$R \sim \begin{cases} c^{-5/16} & L_e \gg L_i \\ c^{-3/16} & L_e \ll L_i. \end{cases} \tag{3.44}$$

De Gennes *et al.* have proposed an alternative picture for the structure of polyelectrolytes as a function of concentration.[124,125] As with Odijk at dilute concentrations the chains are rigid rods. They point out the possibility of a crystalline phase at very dilute concentrations and long chains. In the semidilute regime, the chain is viewed as composed of straight segments each with g charged monomers which form an ideal chain (in the limit of large N). The strand–strand distance is equal to the length of a segment: $\xi = ga$. In other words, the persistence length is pinned at the strand–strand distance. In contrast to the Odijk theory the crossover density and the overlap density are the same. The end-to-end distance is given by

$$R^2 \sim \frac{N}{g}\xi^2. \tag{3.45}$$

At c^*, $\xi \sim N$, and $R \sim N$ as required. The concentration dependence can be obtained using the fact that $g \sim c\xi^3$ yielding

$$R(c) \sim N^{1/2}\left(\frac{a}{c}\right)^{1/4} \text{or } R(c) \sim N\left(\frac{c^*}{c}\right)^{1/4}. \tag{3.46}$$

The scaling for the chain structure factor comes rather simply from the segment picture.

$$S(q) \sim \begin{cases} \dfrac{1}{q} \; q > \dfrac{1}{\xi} \\ \dfrac{1}{q^2} \; \dfrac{1}{\xi} > q > \dfrac{1}{R}. \end{cases} \tag{3.47}$$

The same q dependence occurs in the Odijk picture, but ranges of q and their concentration dependence are different.

Recently Barrat and Joanny[126] have addressed one of the criticisms of the OSF theory. They have used variational methods to allow the chain to be flexible instead of locally stiff. A new expression for the persistence length is obtained scaling as the Debye length or $c^{-1/2}$ which agrees with experiment.[127–133] Their method is applicable when $a \gg (\lambda_B L_i)^{1/2}$ or $\lambda L_i/a \ll 1$, i.e., the chain is a weak polyelectrolyte. However, most experiments are in the strong polyelectrolyte regime. It remains to be seen whether Barrat and Joanny's results can be extended to the experimental regime.

One other relevant theoretical set of work concerns the counterion distribution particularly in the dilute limit. Manning[134–137] solved the Debye–Hückel equation for a single infinitely thin polyelectrolyte. He found that when $a < \lambda_B$ the counterions condense onto the line polymer reducing the charge density until the charge separation becomes equal to the Bjerrum length. The details are altered when the Poisson–Boltzmann approximation is used for a cylindrical polyelectrolyte,[111,119,120] but the basic point of condensation occuring for $\lambda > 1$ remains. In a similar vein, Oosawa[138] proposed a two-phase model of bound and free counterions. These results are especially relevant, since many prototypical polyelectrolytes, such as DNA and NaPSS, have $\lambda \approx 3$.

3.3.1.2 *Weak polyelectrolytes*

The scaling theory of weak polyelectrolytes[113,139–143] is more developed than for strong polyelectrolytes as the Coulomb interaction is more perturbative and a Flory-like ansatz is expected to give better results. The theory of weak polyelectrolytes is based on the picture of chain structure at very dilute concentrations as a sequence of Gaussian blobs that form a linear chain.[125] As usual the counterions are neglected and only the charged monomers considered. Since weak polyelectrolytes are usually only partially charged, quantities scale with λf^2. In general the separation between charged monomers, a, is not the same as the bond length, b. A blob consists of g charges or ga/b monomers such that the electrostatic energy of the blob equals k_BT. From this, we can write the blob diameter to be

$$D = g^2\lambda_B. \tag{3.48}$$

We then have

$$D \sim (ga/b)^{1/2}b = (gab)^{1/2}. \tag{3.49}$$

Since the blobs form a linear chain,

$$R \sim \frac{Nb}{ga}D. \tag{3.50}$$

These equations yield the following scaling relations in terms of the fundamental parameters.

$$g \sim a^{-1/3}b^{1/3}\lambda^{-2/3} \tag{3.51a}$$

$$D \sim a^{1/3}b^{2/3}\lambda^{-1/3} \tag{3.51b}$$

$$R \sim a^{-1/3}b^{4/3}\lambda^{1/3}N \tag{3.51c}$$

In strong polyelectrolytes the solvent quality is usually neglected as the Coulomb repulsion effectively creates a good solvent except when the screening is very strong. But, for weak polyelectrolytes the tendency for the solvent quality to be poor has fundamental consequences: the blob structure changes in a poor solvent altering the scaling relations. Specifically, in a poor solvent weakly below the Θ-point the following relations are expected to hold[139]

$$g \sim a^2b^{-2}\tau\lambda \tag{3.52a}$$

$$D \sim v^{1/3}a^{1/3}b^{-1/3}\lambda^{-1/3} \tag{3.52b}$$

$$R \sim v^{1/3}a^{-2/3}b^{2/3}\lambda^{2/3}N, \tag{3.52c}$$

where $\tau = (T - \Theta)/\Theta$ and the second virial coefficient, $B \sim v\tau$. All these relations are for single chains in dilute solution.

Phase diagrams have been proposed for weak polyelectrolytes[140–142] which exhibit a wide variety of phases. Present simulation methods have not been able to address the nature of the phase diagram, and thus we will not discuss the theories. These phases are of course of great interest for future simulations. The double structure of weak polyelectrolytes requires rather long chains to clearly observe the scaling predicted. The chain length must be at least the product of the minimum blob size that exhibits scaling and of the minimum number of blobs that together exhibit scaling. To calculate the dependence of the polyelectrolyte structure on the above parameters requires longer chains than the minimum in order to allow the chain structure to adjust to the varying parameters. While this is extremely important for many applications a severe test is still not available.

3.3.2 Experiment

Bulk quantities such as the osmotic pressure are relatively easily measured while the microscopic structure of polyelectrolytes in solution has been difficult to determine. Measurements of the osmotic pressure on several systems show two scaling regimes[144] (Fig. 3.16). At low densities, $\Pi \sim c^{9/8}$ which is the dependence predicted by Odijk's scaling theory.[123] In the high concentration regime the scaling exponent changes to the scaling for semidilute neutral polymers: $c^{9/4}$.[1] These results suggest that the polyelectrolyte chains are elongated at dilute concentrations and are similar to neutral chains in good solvents at semidilute concentrations. According to Odijk, the crossover should occur at c_2^* which is incorrect since the crossover of Π is molecular weight independent. However, Odijk's scaling relation can be modified to crossover to neutral scaling.[144]

The peak position, q_m, in the interpolymer structure factor exhibits a similar variation with concentration, c.[145–149] At low concentrations, the peak position scales as $c^{-1/3}$ which corresponds to the variation of an average interpolymer separation. At high concentrations, the peak position scales as $c^{-1/2}$ which corresponds to a correlation length that scales as the Debye length. Both the Odijk picture and the de Gennes picture predict such a dependence. Thus, this experimental data does not distinguish between the two theories and the structure to which this dependence corresponds has not been determined. A definitive understanding of this dependence remains an open problem.

Much experimental effort has been expended measuring the persistence length of polyelectrolytes in solution.[127–133] A significant discrepancy has been found between the OSF prediction and the measured dependence of L_p on the ionic strength, I. Experiments (Fig. 3.17)[127–133] find $L_p \sim I^{-1/2}$, but theory gives $L_p \sim I^{-1}$. This had led to a lot of effort in recalculating the electrostatic contribution to the persistence length.[119,120,150–153] One aspect seems clear. The original calculations are for stiff chains like DNA which

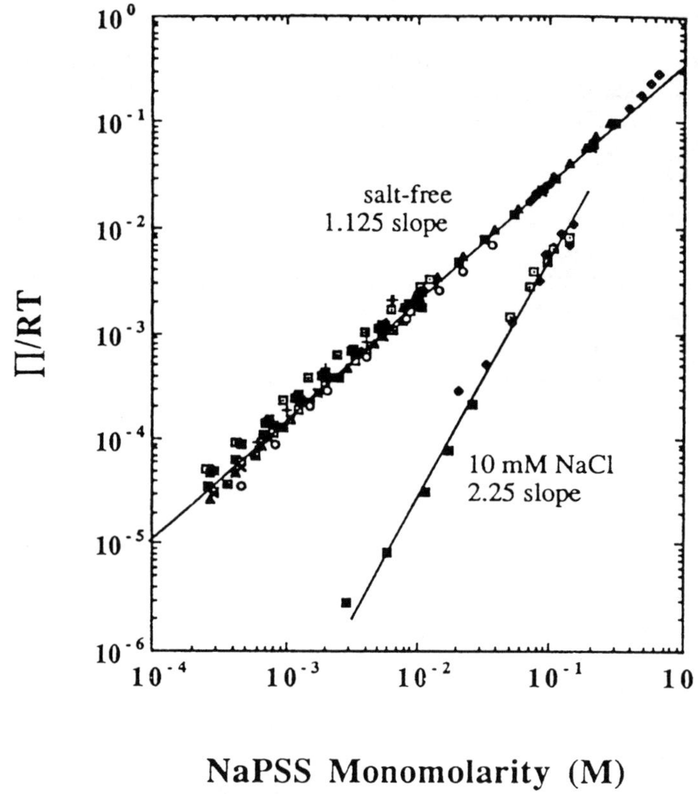

Fig. 3.16 Plot of the measured osmotic pressure at various molecular weights (from Ref. 144).

has a long intrinsic persistence length ($L_i = 500$Å). But, the experiments on flexible polyelectrolytes like polystyrene sulphonate ($L_i \approx 10$Å) do not satisfy the stiffness assumption and do not agree with the OSF predictions.

3.3.3 Simulation methods

The simulation techniques used for polyelectrolytes in solution are extensions of the standard methods used for neutral polymers.[3,96] The polymer chain is modeled as a set of connected beads. The beads are charged depending on the charge fraction, but otherwise the details of the monomer structure are neglected. Various means of connecting the bonded monomers are used. In lattice Monte Carlo the bonds are of course fixed. Two sets of simulations have used the rotational isomeric state model.[121,152,154–156] Other simulations have used Hookean springs[153] or the finite-extendable-nonlinear-elastic (FENE) potential.[157,158] No important dependence on the nature of the bonds is expected at this level of modeling the polymer chain.

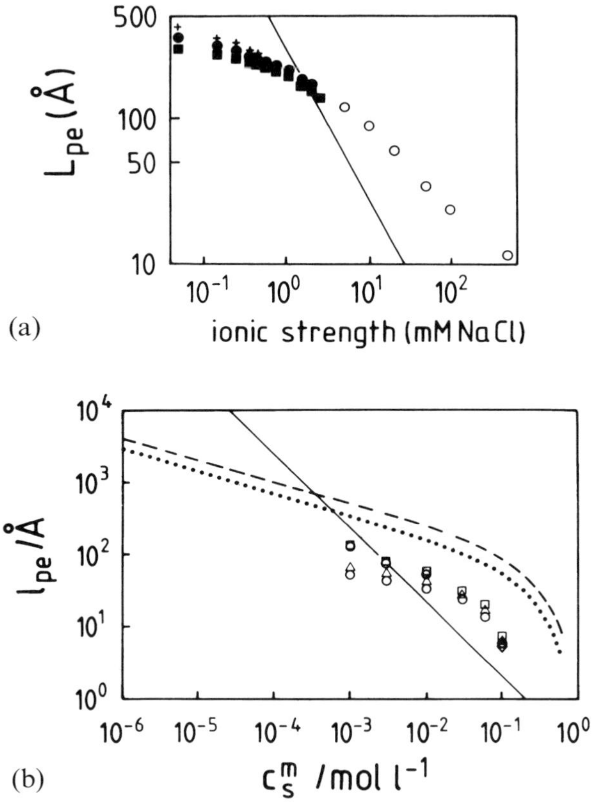

Fig. 3.17 Measured persistence length (a) by birefringence for NaPSS for molecular weight of up to $M = 200000$ (from Ref. 132), and (b) by static light scattering on PVP (from Ref. 133). In both cases the Odijk prediction is included as a straight line. (This line was wrong in the original figure in Ref. 133 (M. Schmidt, private communication); this has been corrected here.)

Most works have performed Monte Carlo (MC) simulations.[121,143,152,154–156,159–175] In these the monomers are moved via the standard reptation, crankshaft, flip, etc. moves. The only special consideration is for the case of a partially ionized chain. In this case, a reptation move can change the identity of the chain by altering the sequence of charges along the chain. Since there exists a set of sequences for a given total charge, the changing identity is viewed as sampling the ensemble of possible sequences.[156] Other MC schemes used include a grand canonical method[173] and an ensemble growth method.[143] A few recent works[157,158,176] used molecular dynamics.

Treating the solvent is more complex for charged polymers in solution than for neutral polymers in solution. In neutral polymer simulations the solvent quality can be treated by effectively altering the monomer–monomer interaction. For polyelectrolytes in solution other considerations are also important. Foremost is the presence of counterions and salt ions in the solution. Most simulations to date have not explicitly treated the ions in solution.[121,143,152,154,155,161,162,164–166,168,169,171,172,176] Instead, the Debye–Hückel approximation is used for the charged monomer interactions so that the solvent ions enter only through the screening length. This approximation is discussed below. A few simulations[156–158,163,173,174] have explicitly treated the solvent ions. One of the great interests in polyelectrolyte simulations is calculations with explicit solvent ions, because theoretically this problem is presently untenable. Such simulations should offer insight into more realistic aspects of counterion condensation and the nature of the ionic screening. Two further complications have yet to be treated: the differing dielectric constants of the solvent and the polymer chains, and the polar nature of the solvent. All works to date treat the uncharged solvent particles as a continuous background providing the bulk dielectric constant. The dielectric constant of the solvent is usually taken to be that of water at room temperature (298 K), $\epsilon = 78$. At dilute concentrations the bulk dielectric constant should pose no problem, but this difference may require more sophisticated treatment at high polyelectrolyte concentrations. With respect to the polar nature of the solvent, the potentially important hydration shells may need to be considered. These aspects are presently beyond the available computational speed.

The long-range Coulomb interaction requires special treatment.[177–180] The history of simulating Coulomb systems is full of incorrect results due to not using the Ewald method to include interactions with the image cells.[178–180] Many people have been deceived by the fact that one can accurately calculate the system energy without summing over the periodic images. But the phase diagram critically depends on the long range interactions.[181,182] Most of the simulations of polyelectrolytes in solution[156,160,163,174] have used the minimum image (MI) method which only includes all Coulomb interactions within the minimum image. These works simulated only a single chain ostensibly at infinite dilution, although the use of periodic boundary conditions yields a finite (dilute) concentration. In the infinitely dilute case, the long range interactions are not relevant though simulating even a simple electrolyte should be done with care.[182] The Ewald method is necessary when the Coulomb interactions are stronger than the thermal interactions. For most polyelectrolytes in solution only a small subset of the Coulomb pair interactions have energy greater than $k_B T$ and these tend to be the bonded monomers for which the bond interaction dominates. Thus, one expects that the full Ewald method may not be necessary, but given the history of simulations it should be checked. Only the simulations of Stevens and Kremer[157,158] have checked their results by performing Ewald summations for a few cases. For most of their simulations they used a radially symmetric form given by Adams and Dubey[177] that is an order of magnitude better in the energy calculation than the MI method. This form is found to be sufficient for their parameters.

The MI image has been chosen over the Ewald method, because for the system sizes studied the MI method is faster than the Ewald method. Even so, the MI method is slow compared to neutral system simulations. The main obstacle with charged systems is poor scaling with the number of particles. Since all particle pairs interact the computation time scales as N^2 for the MI method. (For large enough N the Ewald method changes from scaling as N^2 to $N^{3/2}$.[183]) Thus for even rather small system sizes the computational time is large. For a sufficiently large system a method based on a multipole expansion which scales linearly in N will resolve this problem.[184,185]

The computational times have been so long that most simulations of polyelectrolytes in solution have used the approximate Debye–Hückel (DH) form (eq. 3.35) for the Coulomb interactions. The advantage of this approximation is that the potential is finite ranged and the counterions and salt ions are not simulated. Thus, DH simulations are much faster than ones which explicitly treat each Coulomb interaction. This approximation is done primarily for the sake of saving CPU time and is generally viewed best at very dilute polymer concentrations. Although at concentrations where $\Lambda \approx a$, the Debye–Hückel interactions become no more than an excluded volume interaction and the chain mimics a neutral chain. Thus when the ionic screening is very strong, the DH interaction may also be reliable. But in general the disadvantage of the DH approximation is that it is not expected to be accurate for much of the relevant range of parameters. Furthermore, important phenomena like Manning condensation and fluctuations cannot be examined.

The number of simulations on polyelectrolytes in solution is rather small. We give a brief summary of the work here. The initial simulations on polyelectrolytes in solution were done by Brender *et al.*[160] Subsequently, a series of papers[162–165] examined the dependence of the polyelectrolyte structure on the Bjerrum length. The dilute limit dependence of the end-to-end distance on N has been examined by Baumgärtner[159] and Kantor.[186] Reed and Reed[121,152,171] have examined excluded volume effects on the persistence length. Valleau[173] and Woodward and Jönsson[174] have studied the accuracy of the DH approximation. Some of the other works have done calculations for both Coulomb and Debye–Hückel interactions. Only one set of works has gone beyond single chain polyelectrolyte simulations and performed simulations on systems of polyelectrolyte chains.[157,158] These simulations are for salt-free strong polyelectrolytes ($\lambda_B \approx a$) with explicitly treated counterions. Simulations in two dimensions for a system of chains have been done.[167,172,176] However, these simulations do not really treat polyelectrolytes as the charge interaction is cut off in the lattice MC at either the first or second lattice spacing. This cut-off is counter to the basic long-ranged nature of the Coulomb interaction and fundamentally alters the nature of the system. For weak polyelectrolytes various single chain simulations have been done. Barrat and Boyer[153] have examined the persistence length. Higgs and Orland[143] have tested scaling theory predictions. Hooper *et al.*[168,169] have studied the dependence on the charge fraction.

3.3.4 Simulation results

3.3.4.1 Strong polyelectrolytes

(a) Scaling exponent One of the defining characteristics of polyelectrolytes is the rod-like scaling of R with N at infinite dilution. One of the early simulations confirmed this result for three-dimensional chains.[159] Yet, for $d = 4$ and 5 the calculated exponent, ν, agreed with the Flory theory predictions. This result contradicts the experience that RG calculations are usually more accurate than Flory calculations. Recent simulations[186] (Fig. 3.18) have resolved this dilemma, finding agreement with the renormalization group calculation. In the first work the Coulomb coupling strength was $\lambda = 1/10$ which makes the chain almost a weak polyelectrolyte. The weaker the polyelectrolyte the larger the chain needs to be to observe the crossover to asymptotic scaling. Evidently, for $d = 4$ and 5 the chains in Ref. 159 were not sufficiently long and produced values close to the Flory exponent. In the later simulations $\lambda = 1$ which is a strong polyelectrolyte, and the RG scaling is seen with even shorter chains.

While the scaling of the isolated polyelectrolyte chain has been resolved, at finite concentrations the presence of counterions appears to have important consequences. Recent simulations[157,158] have shown that the effects of counterions even at very dilute concentrations are strong. The effective value of ν has been found to be 0.90, not 1 for up to 128 beads chains at monomer densities to $5 \cdot 10^{-7} a^{-3}$ for $\lambda = 0.9$. Taking $\lambda_B = 7.1\text{Å}$, the value in water, the above density corresponds to $2.4 \cdot 10^{-6}$M (molar) which is just about the ionic concentration of water. The N-dependence of the chains simulated suggest that longer chains at the same concentrations will have the same maximum value of ν.[157] The proper scaling limit is a double limit of the polymer concentration going to zero along with the chain length going to infinity. To find a value of ν closer to 1 will require both lower concentrations and longer chains, but the intrinsic salt content of the solvent may prevent the limiting value being reached. The next obvious question is how dependent is this result on the value of λ. It is possible that a larger value of λ would achieve the limiting value, but Manning condensation or similar condensation effects tend to limit the charging of the polymer to $\lambda \approx 1$. Thus, for *flexible* polyelectrolytes it is quite possible that the "ideal" state of a rod-like chain occurs only at rather unphysical parameters. Recent

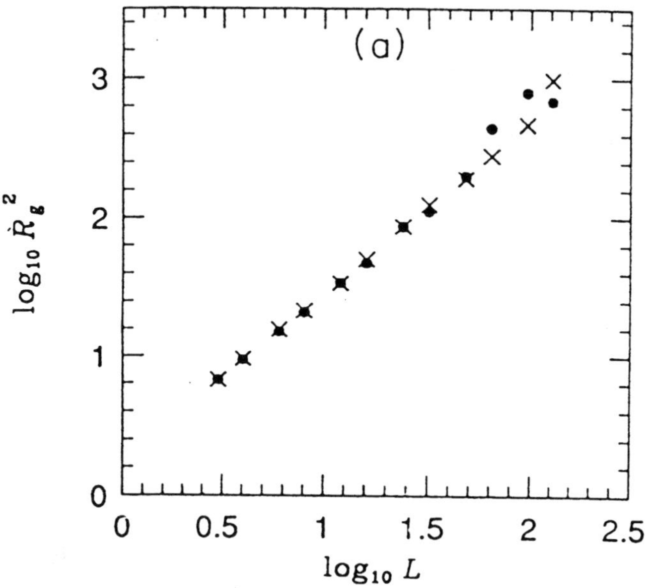

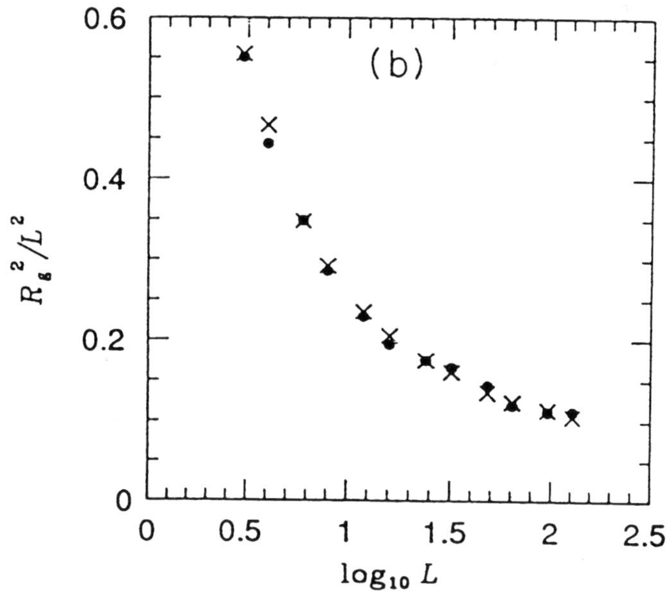

Fig. 3.18 Plot of R_G as a function of chain length, L, for (a) $d = 5$ and (b) $d = 4$ dimensions. The crosses represent fixed charges while the dots are for a mobile model. (Figure taken from Ref. 186).

experiments support this result that the flexible polyelectrolyte chain is not rod-like.[132,133,187,188]

(b) Validity of Debye–Hückel approximation Most simulations to date and just about all theories use Debye–Hückel interactions. This simplifies calculations greatly by reducing the number of species to one and by reducing the range of interaction to a finite value. However, this approximation has serious dangers and disadvantages. The approximation is rather strong and probably not realistic except at very dilute concentrations or very short Debye lengths. The solvent ions are represented as an averaged density. Thus, effects of individual counterions and fluctuations are not at all present. As mentioned above these effects can be dramatic in altering the traditional view of the dilute chain structure and the scaling exponent ν.[157,158]

The DH approximation has only been partially tested. Christos and Carnie[156] compare their Coulomb simulations with their earlier DH simulations.[154,155] While they have only a small amount of data, there is a large difference between the values for the end-to-end distance when the screening length is large. When the charge fraction is small ($f \sim 0.2$) and the salt concentration large ($c_s \gtrsim 0.1$ M) the agreement between DH and Coulomb simulations for chains with $N \leq 80$ is good. These results indicate that the DH approximation is best when Coulomb interactions are weak due to strong screening. This is consistent with the DH approximation being the weak interaction limit for the Coulomb potential. Brender has examined the λ_B dependence for both DH[162] and Coulomb[163] simulations. She finds that the ionic screening is substantially stronger than the DH approximation yields. Even small salt concentrations ($\sim 10^{-6}$ M) have a noticeable screening effect.

The validity of the Debye–Hückel interactions has been directly examined by two sets of simulations.[173,174] Valleau[173] performed grand canonical simulations on a short chain ($N = 10$) at three salt concentrations. He came to the conclusion that there is no reason to believe that the DH approximation has any utility. However, only small differences in the end-to-end distance between the DH and Coulomb simulations were observed. Subsequently, Woodward and Jönsson did similar (except canonical ensemble) simulations pointing out that Valleau had incorrectly calculated the total charge density for the DH simulations. The DH interaction is an effective monomer–monomer potential which includes the electrostatic interaction of *all* charged species. Valleau had failed to include the contribution of other monomers and the charge densities around them in the total charge density. Using the correct DH densities Woodward and Jönsson find that the DH and Coulomb total charge densities were not drastically different. Unfortunately, they compare DH and Coulomb simulations only for 10-bead chains which are short and not representative. At this chain length the total charge density about a monomer is consistently too large near the monomer for the DH simulations. In terms of the number density they find that at *high* salt concentrations (1.0 M) the screened Coulomb simulations strongly underestimate the peak value. At lower concentrations (< 0.1 M) only a very small difference between DH and Coulomb number densities is seen. Like Valleau they find that the end-to-end distance of the short chains depends very little on the interaction type. These results indicate that in the dilute limit the DH approximation is reasonable as has been presumed. Yet, for N up to 80, the simulations of Carnie *et al.*[154,155] and Brender[162,163] show that DH chains are much more stretched and have a different scaling with N than Coulomb chains, suggesting that the short chain comparison by Woodward and Jönsson is not representative. In general, the results to date suggest that DH simulations are best when the Coulomb interactions are weak due to strong screening (short Λ). In this case, the monomer–monomer potential effectively becomes just an altered good solvent interaction and the chains are not much different from neutral chains. At low salt concentrations where Debye–Hückel is supposed to be valid, the difference is the greatest. Evidently, the individual character of the counterions is rather important.

(c) Persistence length Experiments[132,151] find that the persistence length scales differently than the OSF theory predicts. In terms of the ionic strength, I, the OSF theory predicts $L_e \sim I^{-1}$. The persistence length is not measured directly, but calculated from the measured end-to-end distance or radius of gyration. For this a worm-like chain structure is assumed, which for many systems such as NaPSS is not applicable.[132] Not surprisingly, one of the results of simulations on flexible chains[158] is that a flexible chain has a different persistence length than calculated using the worm-like formula. The difference is greatest when $L \approx L_p$. On the other hand, in the limit $L \gg L_p$, the chain is Gaussian and the worm-like model is satisfied. The simulation results also show that the discrepancy between the OSF theory and simulation is only partly due to the worm-like assumption. A larger problem appears to be the neglect of the chain entropy.

Reed and Reed[121,152] have performed simulations which closely match the theoretical model. A worm-like chain is obtained by using the rotational isomeric state model chain, and DH interactions without excluded volume are used. They compare their simulation results with the excluded volume corrections to the OSF theory and find a good fit for the salt concentration dependence. They also examine the dependence on λ and find the fits to be not as linear as the simulation data. It is not clear that the excluded volume is the source of the discrepancy between theory and experiment. Other calculations have produced the appropriate scaling. Calculations of L_e based on Poisson–Boltzmann theory[119,120,189] and based on a flexible chain[126] instead of a worm-like chain all give the $I^{-1/2}$ scaling. As mentioned above another possible candidate is the neglect of entropy in the OSF theory which *is* included by simulations. Thus, given all these different sources of the scaling it is not clear which of the calculations is best.

(d) Characterization of chain size In order to characterize the polyelectrolyte chain size, we discuss the dependence of the end-to-end distance on the various parameters. Unfortunately most of the simulations have poorly analyzed R. The N dependence of R was discussed above. The only simulations that can discuss systematically the polymer concentration dependence is the recent work by Stevens and Kremer.[158] Brender has examined the Bjerrum length dependence.[162–165]

As the screening of the Coulomb interactions decreases, or equivalently for salt-free polyelectrolytes, as the concentration decreases, the chain becomes elongated by the Coulomb repulsion. The simulations (Fig. 3.19) find that the chains clearly become more extended with decreasing concentration until a saturation density is reached.[158] In the Odijk scaling theory, this saturation should occur at the concentration c_b^*. However, the simulation results show that the saturation occurs at a much lower concentration. Stevens and Kremer give an alternative picture of the saturation based on the counterion screening of the monomer interactions. The chain extension should saturate with decreasing concentration when the density of counterions within the polymer volume reaches zero. Further reduction of the concentration does not reduce the counterion screening and consequently does not extend the chain. A rough estimate of this density is obtained by assuming a uniform counterion density. Then there is one counterion in the polymer volume when the average counterion separation is equal to the end-to-end distance:

$$c_b^* \to 1/(N\pi R^3/6). \tag{3.53}$$

The screening due to the counterions should saturate at approximately this density since at lower densities there are no counterions present to screen the monomer interactions. The values obtained with this calculation are more accurate then Odijk's c_b^* and their chain-length dependence is much better. However, these values tend to be high in comparison with the simulation data. This is easily understood by realizing that eq. (3.53) assumes a uniform counterion distribution when in fact the distribution is peaked near the polymer chain. Thus, it is at lower densities where the counterion density near the chain saturates. Furthermore, given that the chain structure is not completely rod-like these results suggest that there is always some number of counterions near the polymer chain.

At densities above the saturation density, one can compare with the scaling predictions of Odijk[123] and de Gennes *et al.*,[125] In the semidilute regime where $L_e \ll L_i$, Odijk's scaling theory predicts $R \sim \rho_m^{-3/16}$,[123] where ρ_m is the monomer density. The solid lines in Fig. 3.19 have slope $-3/16$. Even though $L_i = 1.2b$ for the chains so that the condition given above is not strictly satisfied, the scaling relation agrees with the data. The data are not sufficiently accurate to determine strictly that scaling holds. In fact, de Gennes *et al.*[125] predict an exponent of $-1/8$ which is close to Odijk's

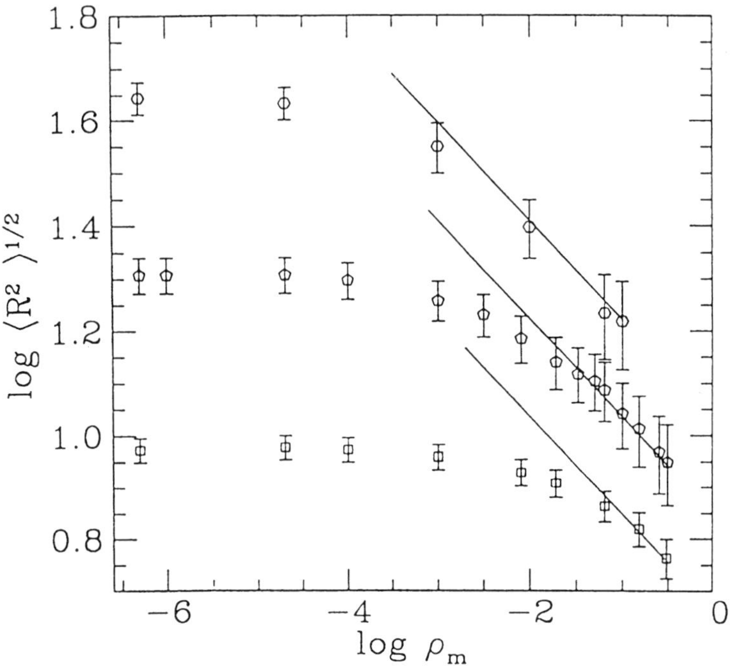

Fig. 3.19 End-to-end distance as a function of density for charged polymers. The counterions are included explicitly in the simulation. (From Ref. 190.)

value and which also fits the data. This exponent is the same as for neutral polymers. At high polymer densities the chain is not far removed from the neutral state due to screening. In fact, the good solvent limit of the ratio $\langle R^2 \rangle / \langle R_g^2 \rangle$ is reached at the highest densities simulated. It appears reasonable to think of the effect of the Coulomb interactions as just an added excluded volume for densities with a small Debye length. At lower densities our results disagree with Odijk's scaling theory which predicts a steeper curve with an exponent of $-5/16$. This disagreement is consistent with the failure of the OSF persistence length to have the proper scaling with density.

Brender[162–165] has examined the Bjerrum length dependence as a function of the salt concentration. She interprets the variation of λ_B as the same as varying the temperature, but this is not really justified. For a polyelectrolyte in water at room temperature $\lambda_B = 7.1$Å. Thus, as the Bjerrum length varies from 1 Å to 20 Å the temperature varies from 2100 K to 97 K! The Bjerrum length can be dramatically changed, but that occurs usually through the valence of the ions or the dielectric constant. For example, for divalent ion pairs the Bjerrum length in water is 28 Å. It is best to interpret these simulations as varying the coupling strength, λ. These simulations are lattice MC with a bead diameter and thus bond length of 4 Å. Then, λ varies from 1/4 to 5.

Another criticism of Brender's simulations is that she primarily employs the DH potential[162,164,165] instead of the Coulomb potential.[163] One of the most important phenomena to *directly* examine when λ varies is counterion condensation, but this can only be done by simulating the counterions. One of the papers[163] does use the Coulomb potential, but does not directly calculate the counterion distributions. It is only inferred that some condensation occurs because of great differences between the DH and Coulomb simulations. These differences are very interesting as they show that even very small salt concentrations have a strong screening effect.[163] This result is consistent with the effect of counterions seen in the simulations of Stevens and Kremer.[157,158]

The main point of Brender's series of papers is to show that the chain expands when the temperature decreases. However, this is not surprising since a decreasing temperature corresponds to an increasing λ in these simulations.

Before continuing it is important to point out a problem that exists for many of the single chain simulations. Most of these works, in particular the DH simulations, have examined the salt concentration dependence. For single chain simulations this naturally results in varying the simulation cell size. The implicit polymer density is one per cell volume which varies with the salt concentration. This coupling of c_s and c poses no problem as long as c is sufficiently dilute. However, in some cases the calculated end-to-end distance is longer than the cell length[154,156,161] effectively implying a semidilute concentration. This can be rectified by performing the simulations in a larger cell with more particles, but obviously this has not always been done. Even when the cell length is longer than R, the image chains may interact (even when just the MI method is used) because of the long-range potential. One needs to be very careful about the system size relative to the chain size.

(e) Comparison with experiment Comparison to experiment has predominantly been done for the persistence length which was discussed above. The multichain simulations of Stevens and Kremer allowed an examination of two well measured quantities—the osmotic pressure, Π,[158] and the peak position of the monomer–monomer structure factor.[190] Both of these quantities show an interesting dependence as a function of polymer concentration as discussed above.

Like the experimental data (Fig. 3.16) the simulation data (Fig. 3.20) show two scaling regimes for the osmotic pressure. The simulations were able to reach the dilute noninteraction limit, where $\Pi = ck_BT(1 + 1/N)$. Thus, at the lowest concentrations there is a deviation from the experimentally observed $c^{9/8}$ dependence. Also, at these very dilute concentrations the osmotic pressure does not exhibit a chain-length dependence. At the intermediate concentrations, the $c^{9/8}$ behavior fits the simulation data very well in agreement with experiment, and there appears some chain-length dependence.[144]

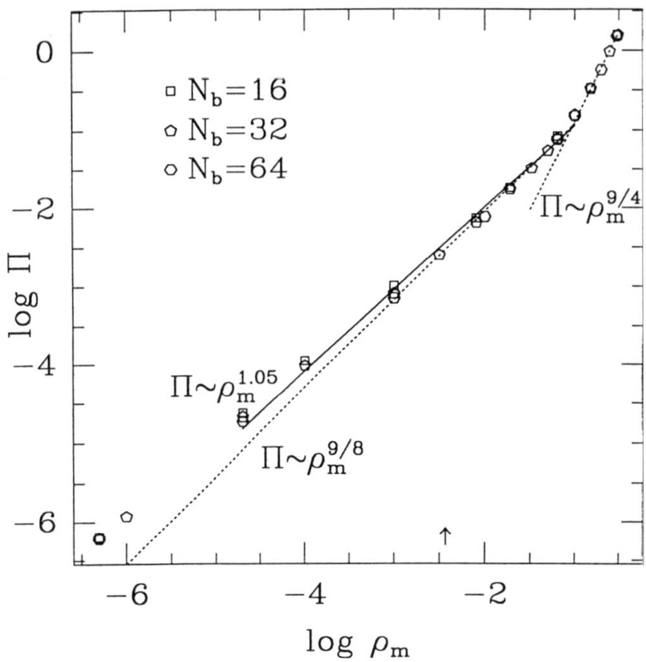

Fig. 3.20 The osmotic pressure is plotted as a function of the monomer density on a log–log scale for various chain-lengths. The 16-, 32- and 64-bead chains are represented by squares, pentagons and hexagons, respectively. For low densities in the experimental range, the pressure scales with an exponent 1.05, slightly below the Odijk prediction marked with an arrow. At high densities the exponent is 9/4 which is that of neutral chains. The errors are smaller than the points. The data show a very weak chain-length-dependence as in the experimental studies.[144] (Taken from Ref. 158).

At high concentrations the simulations find the same scaling exponent, 9/4, as in experiment and no chain-length dependence. The lack of any chain-length dependence in the crossover density of Π, $\rho_m^{(\Pi)} = 0.07 \pm 0.04b^{-3}$ is significant. Since the overlap density is chain-length-dependent, the crossover in Π cannot be associated with the dilute–semidilute transition. In fact for the 64-bead data the overlap density is below the Π crossover density and for the 16-bead data the opposite is true. The fact that for $\rho_m > \rho_m^{(\Pi)}$ the Π-dependence is that of neutral polymers suggests that the Coulomb interactions are completely screened at $\rho_m = \rho_m^{(\Pi)}$. For complete screening, the Debye screening length should be less than all interparticle distances. The monomer separation is the shortest particle separation distance. One finds $\Lambda = b$ at $\rho_m = 1/(4\pi\lambda_B b^2) = 0.066b^{-3}$ for their set of parameters. This value agrees perfectly well with the simulation data.

The calculated peak position in the structure factor exhibits a two-regime behavior as in experiments. In Fig. 3.21, the calculated q_m for $N = 32$[190] is

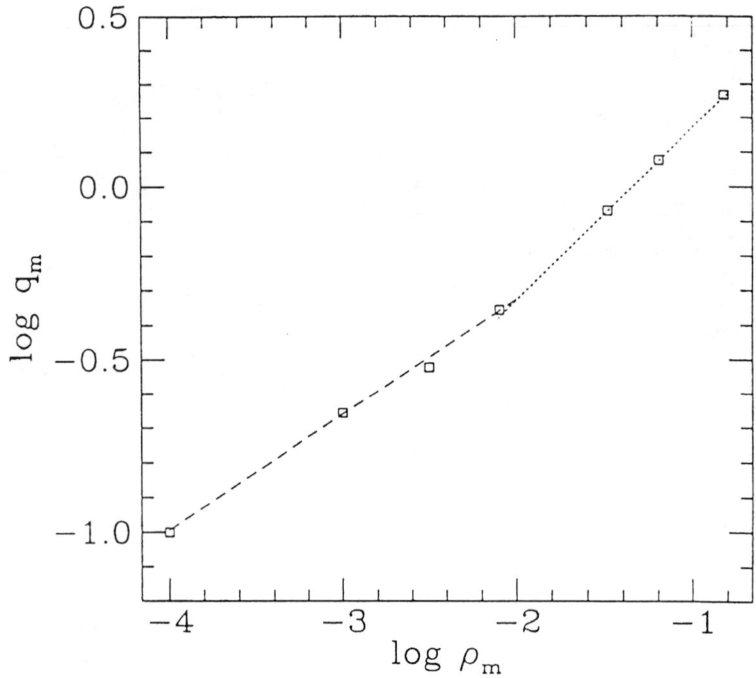

Fig. 3.21 The position of the maximum of the inter-polymer structure factor, q_m, is plotted as a function of monomer density. The same scaling dependence as found in experiments is exhibited. At low densities q_m scales as $\rho_m^{1/3}$ (dashed line) and at high densities q_m scales as $\rho_m^{1/2}$ (dotted line). (Taken from Ref. 190).

shown. The data clearly exhibit the same two scaling regimes seen in experiments. At high concentrations the peak position is chain-length-independent and scales as $\rho_m^{1/2}$. Thus, the crossover occurs *before* the chains overlap. This is consistent with the end-to-end data (see Fig. 3.19) which shows that the chains start to contract at densities below the overlap density. Consequently, at densities below ρ_m^* the interchain separation is no longer solely dependent on the average chain separation distance, but also on the intrachain contraction. Thus, we find $\rho_m^{(q)} < \rho_m^*$.

These results show that the rather simple models can reproduce experimental data quite well. Further work should be able to resolve some of the many significant questions about various experimental results.

(f) Structure factor A convenient quantity that describes the structure of the chains at all length scales is the chain structure factor or form factor. Thus, the structure factor is a particularly useful quantity to calculate in order to quantify the structure of a polymer. Moreover, since the structure factor is the experimental means to determining polymer structure, it is especially important to calculate.

From the theory of neutral polymers we know that the form factor or single chain structure factor scales as $q^{-1/\nu}$ in the range $2\pi/R < q < 2\pi/b$ for neutral polymers with $R \sim N^{\nu}$.[1] The ideal chain and good solvent chain values of ν are 1/2 and 0.588 ($\approx 3/5$), respectively. Scaling predictions for polyelectrolyte $S(q)$ have been based on the picture that the polyelectrolyte's structure is composed of rod-like segments of length L, which together form an ideal chain.[123,124,125] For $q > 2\pi/L_p$,[114,115] the structure factor should then scale as $1/q$. For $q < 2\pi/L_p$, the ideal chain would yield $\nu = 1/2$. As we have good solvent chains, we expect to see $\nu \approx 3/5$ at dilute densities and $\nu = 1/2$ for semidilute and dense solutions.

The density dependence of the structure factor, $S(q)$, for the 64-bead chains from Stevens and Kremer[157] is shown in Fig. 3.22. The inset exhibits the chain-length independence which holds up to $q = 2\pi/R(N)$. This L independence is important because within the above range of length scales "short chains" can be used to determine the structure of long chains. The slope of the curves in Fig. 3.22 gives $-1/\nu$, and the dotted lines starting from the top left going to the right of the figure have slopes of -1, $-5/3$ and -2, respectively. Clearly, the structure factor is more complex than the predictions of scaling theory. The density dependence for high and low q is distinct and we will discuss them separately.

Rather surprisingly, a density-independent regime occurs at relatively high wavevectors ($1 \lesssim qb \lesssim 2\pi/2$) which corresponds to short length scales up to about six bond lengths. The density independent behavior above $q_0 \approx 1b^{-1}$ and the change in slope at q_0 suggest the existence of blobs[125] whose length is about $6b$ for their set of parameters. The slope in this region corresponds to $\nu = 0.80$. Thus, the chains are stretched beyond the neutral good solvent value, but well below the rod value. The chain should become more elongated as the screening decreases; thus, the length of the elongated blob, $2\pi/q_0$, would increase and the slope of $S(q)$ would approach -1. Surprisingly, the slope and q_0 are *independent of density and chain length* (At high concentrations where the strand–strand distance is less than $2\pi/6a$, the position of q_0 shifts to ξ.) The structure on short length-scales certainly does not correspond to any of the theoretical pictures.[114,115,125]

At the lowest densities the chains are more stretched on long length scales than on short ones. But, the largest value of ν is 0.90 which is not rod-like. For $q < q_0$, Fig. 3.22 shows that the structure factor shows a pronounced density dependence with ν varying continuously from 0.90 to about the good solvent value, 0.6. At ρ_m^* it was found that $\nu = 0.80$ extends to below q_0. Thus, the transition between being stretched most at long or short length scales occurs at ρ_m^*. The continuous variation in the effective value of ν contradicts the theoretical pictures of just two values of ν. Not only the length over which the Coulomb interaction stretches the chain, but also the degree of stretching depends on the screening.

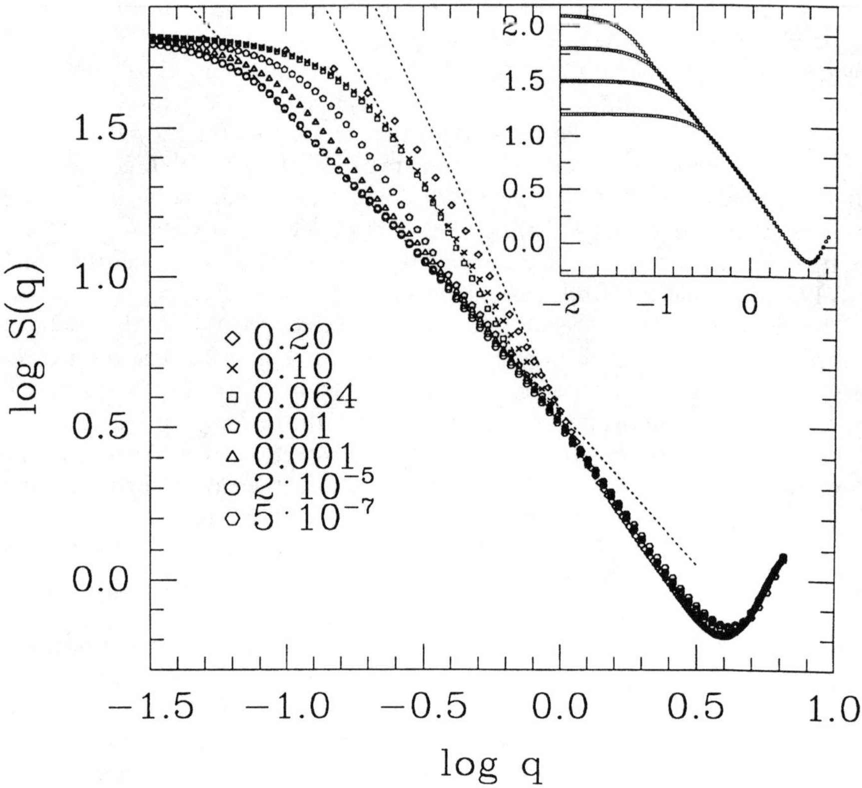

Fig. 3.22 The structure factor for 64-bead chains gradually changes from apparent rod-like to coiled form with increasing monomer density. This contradicts the theoretically expected form which is a combination of rigid rod, good solvent chain and ideal chain forms. The dotted lines give slopes of -1, $-5/3$, and -2 which correspond to the q-dependence of a rigid rod, good solvent chain and ideal chain. For $0 < \log qb < 0.2$ the slope is greater than one, implying the chains are not fully stretched in this regime. The inset shows the chain-length-independence at $q > 2\pi/R$ at $\rho_m = 0.001b^{-3}$ for $N_b = 16, 32, 64$, and 128. (Taken from Ref. 157).

3.3.4.2 Weak polyelectrolytes

Only two investigations have specifically simulated weak polyelectrolytes.[143,153] Other simulation studies have examined the dependence on the charge fraction implicitly simulating weak polyelectrolytes.[155,156,168,169] Similarly, Brender[162–166] has examined the effect of varying the Bjerrum length, but remains primarily within the strong polyelectrolyte regime. Only Higgs and Orland[143] and Hooper *et al.*[169] have considered poor solvent effects. Only Higgs and Orland and Barrat and Boyer consider the weak polyelectrolyte theory discussed above. Most of the other works offer very little comparative analysis. All that one can say is that the general trends of expansion and contraction follow what one expects from the scaling relations.

Higgs and Orland[143] specifically test the scaling of the end-to-end distance as a function of N and λ. They also do both Coulomb and Debye–Hückel simulations, although they neglect altogether the counterions in the Coulomb simulations. Their simulations covered the range $N = 10$–220. In a Θ solvent they find (Fig. 3.23) $R^2 \sim N$ at short chain length and $R^2 \sim N^2$ at long chain lengths ($N \gtrsim 20$) as expected by the Gaussian blob picture. The scaling with λ exhibits the $\lambda^{2/3}$ scaling as predicted by scaling theory.[139] The comparison with the value of 2/3 in the exponent is only rough. Given the result of Stevens and Kremer[158] which exhibits a less than linear behavior for strong polyelectrolytes, it would be interesting to put the exponent on more quantitative grounds and to explicitly simulate the counterions. For poor solvent conditions, the simulation finds that $R_G^2 \sim N^{2/3}$ at short chain lengths and $R_G^2 \sim N^2$ at long chain lengths ($N \gtrsim 50$). In terms of λ a $\lambda^{4/3}$ scaling is seen at intermediate λ. For longer chains at $\lambda \approx 0.1$ there appears to be a crossover to the Θ solvent dependence. At this point the polyelectrolyte blob is bigger than the poor solvent blob size. Thus, the electrostatic interactions are sufficiently strong to dominate and convert the solvent to a good solvent.

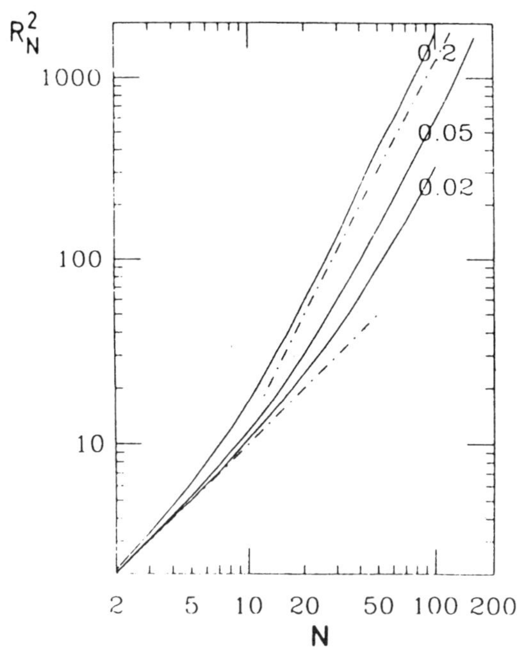

Fig. 3.23 Plot of $R^2(N)$ for a weak polyelectrolyte in a Θ solvent. The three curves are for $\lambda = 0.02$, 0.05, and 0.2. The dashed lines have slopes 1 and 2. (Taken from Ref. 143).

Barrat and Boyer's simulations[153] are similar to those of Higgs and Orland. Since the chains are totally charged so that $a = b$, the scaling relations of eqs. (3.51) and (3.52) are simplified. For example, in the Θ solvent case, $D \sim a\lambda^{-1/3}$. One of the important results of this work is a quantitative means of calculating the blob size. They compute

$$h(n) = [R^2(n)/a^2 - n]^{1/2}, \tag{3.54}$$

where $n = 1, 2, \ldots, N$. Since the chain is expected to be linear in N and thus n at large n, $h(n)$ gives the deviation of R from the linear scaling as a function of n. They compare this formula with that of the stretched chain model (SCH) in which a force F is applied to both ends stretching the chain. This problem can be solved and gives

$$R^2_{SCH}(n) = na^2 + n\frac{a^2}{D(F)^2}, \tag{3.55}$$

where $D(F)$ represents the size of the blob as a function of the applied force. In terms of $h(n)$, D is just the inverse slope. Using this method, they find for $N = 100$ at $\lambda = 0.01$ that $D = 6.3a$ and at $\lambda = 0.5$ that $D = 1.2a$. For the scaling as a function of λ they find that the ratio $D/(a\lambda^{-1/3})$ varies from 1.45 to 0.92 over the above range of $\lambda(0.01$ to $1.0)$. Thus, the scaling relation is at best approximate.

Barrat and Boyer did the only other calculation of the structure (form) factor. Since these simulations neglect the counterions, one expects the q^{-1} dependence should hold over a range of wavevectors. At $q \gtrsim 2\pi/D$, the scaling should be q^{-2} since on this length scale the chains are Gaussian. They find (Fig. 3.24) that for $\lambda = 0.1$ and $N = 200$ the intermediate q dependence is q^{-1}, although they do not state how accurate the exponent is. Fig. 3.24(b) shows the structure factor when the Debye–Hückel approximation is used for $\kappa a = 0.05$. Here it is clear that the q^{-1} dependence is only true near $qa = 10^{-2}$. Since the blob size for this case is only $2a$, the Gaussian regime would be rather small. Although in the vicinity of $qa = 1$, the DH structure factor has a slope close to -2.

3.3.4.3 Second conclusion

Simulations of polyelectrolytes in solution are still in their infancy. Nevertheless, the understanding of polyelectrolyte structure is being greatly transformed by the present results. Quite clearly, the counterion (or salt) screening is very significant and very different from the naive Debye–Hückel approximation even at dilute concentrations. The ionic screening needs to be further examined. In particular the ion distribution about the polymer chains needs to be studied. The present theoretical pictures do not describe the structure of flexible strong polyelectrolytes well excluding perhaps at high ionic concentrations where the Coulomb interactions are strongly

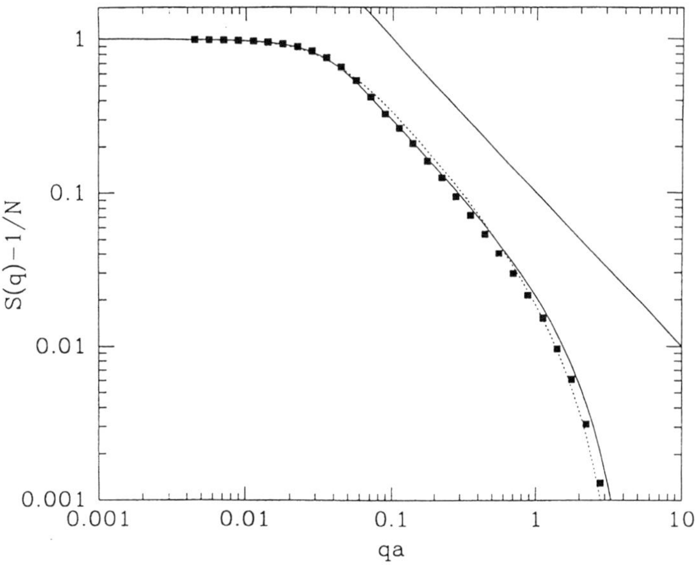

(a)

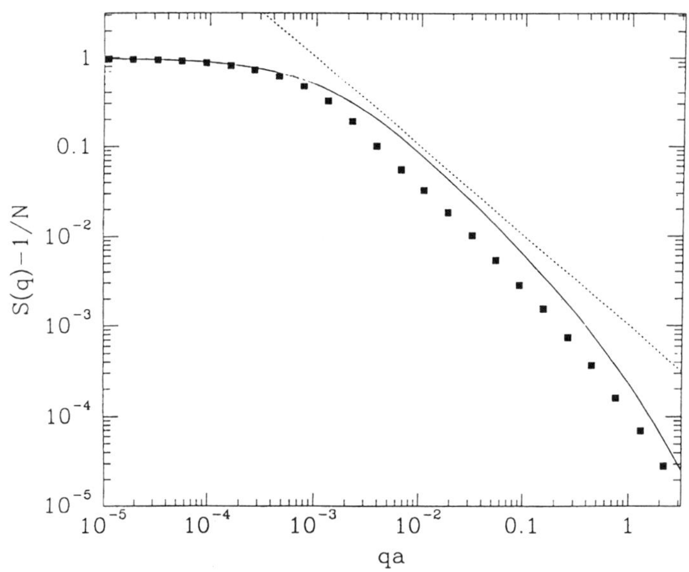

(b)

Fig. 3.24 Plot of the structure factor for a weak polyelectrolyte with $\lambda = 0.1$. Part (a) is for Coulomb interactions and part (b) is for Debye–Hückel interactions with $\kappa a = 0.05$. The dotted line has slope -1, and the points are from a variational theory. (Taken from Ref. 153).

screened. The structure of salt-free flexible strong polyelectrolytes can be summarized as follows. The chains are stretched on short length scales, but are not rod-like. The structure is more like a bent sausage link than a blob. The size of these sausages is independent of polymer concentration or chain length. The total chain composed of the sausages has a density dependent structure. At very dilute concentrations the chain is straighter than the sausages. For flexible polyelectrolytes, it appears that a rod-like chain occurs, if at all, only for *very* long chains at *very* dilute concentrations. At the overlap concentration the chains become more bent than the sausages. For weak polyelectrolytes in solution the theory appears to be much better, although many chain simulations with counterions have yet to be done.

Much interesting work remains to be done as the complexity of polyelectrolytes has only started to be touched. We have not discussed the interesting light-scattering results that show the existence of two modes.[130,131,133,146,191–195] These two modes are not yet understood.[130,131] One mode is viewed to originate from polymer diffusion, while the second is attributed to a collective mode due to clustering. The systems needed to study this effect are significantly larger than any of those which have been simulated so far. Hopefully, with parallel machines such work will be possible.

Another area of great interest is polyelectrolyte brushes or polyelectrolytes grafted to a surface.[196–202] This problem has applications to colloid stabilization, waste water treatment and oil recovery.[112] Some work has already been done in this direction.[203,204] Several theoretical works have been performed.[196–200,202] These works suggest a rich phase diagram of states. Experiments[201,205] are just beginning.

Interesting effects occur in charged systems when divalent ions are present.[206–210] Woodward and Jönsson[174] have done some simulations with divalent ions in the added salt. A serious simulation study of the effects of valence has yet to be done. Experiments along this line[211,212] have begun and suggest some significant differences between monovalent and divalent counterions.

One of the major goals of understanding polyelectrolytes is the application of the knowledge to biomolecules and thus to biochemistry. The systems studied presently are model systems which give insight in some of the fundamental aspects of the statics and dynamics of polyelectrolytes in solution.[109] However, longer chains and bigger systems are needed in the simulations. This will require advanced algorithms like the multipole methods[184,185] and advanced hardware like parallel computation.

3.4 Conclusion

The previous chapters give an impression about the numerical effort which was employed over the last 20 years in order to get some more insights into the effect of long range interactions in polymer solutions. In spite of this effort most of the calculations can only be viewed as "first attempts" to progress towards the really difficult problems. For solutions of neutral polymers the dilute, or better, single chain case, seems to be reasonably well understood, although there are many remaining problems as discussed. Qualitative aspects however seem to be reasonably under control, especially also because of the relatively large amount of work employing the "static dynamics" ansatz. The BD simulations, to our impression, as they can be viewed now, seem to be more or less approaching a natural limit. Up to now only single polymers were treated. For the bare number of monomers the N^3 operation to calculate the square root of the hydrodynamic interaction tensor certainly gives the strongest bottleneck and will naturally limit the chain length. The same holds for attempts to perform simulations for "realistic" dilute or semidilute solutions. There the interaction tensor for the whole system has to be taken, including the Ewald summation for the periodic images. We know of no attempt to do this. It is also not completely clear how the hydrodynamic interaction is screened. This statement of course holds for the corresponding "static dynamics" problem as well. Certainly the most powerful scheme to get more insight could be the full blown MD simulation. There however one faces the problem of the slow relaxation and intrinsically slow integration of the equations of motion. While the CPU time grows linearly with the number of particles in the system, the number of particles grows with the third power of the box length. Thus it is not clear which method at the end will be the one with the best perspectives. Probably the improvements of other simulation schemes, such as cellular automata, in combination with e.g., MD will provide eventually the most progress. Presently one could try to combine MD and "static dynamics" e.g., to get a better feeling for the limitations and problems turning out of both schemes. It certainly is necessary to get an improved understanding of the dynamics of semidilute polymer solutions.

For polyelectrolytes, the situation is even much more severe. Not only the dynamic properties but already the statics is only poorly understood. As the previous chapters illustrate, one currently is at the very beginning of really serious numerical investigations. Up to now mostly single charged chains with either a plain Coulomb or a screened Coulomb potential were simulated. All collective effects, which are certainly technically of extreme importance, are only included in a rather rudimentary way by now. Still simulations already at this stage turn out to be very important in many respects, since also well controlled experiments are very difficult and hardly available, and the analytic theoretical treatment is only in a very early stage as yet. Simulations will, with the use of modern multipole methods in order to properly and effectively take the long-range nature of the Coulomb interaction into account, become even more important. Then it will be possible on modern computers to treat many particle systems, which means of the order of 10^4 particles. This certainly is not enough to give very precise values of, for example, critical exponents near collective phase transitions, but the results will certainly be clearly superior to most experiments which are expected to come in the next years.

Acknowledgments

This work was supported by the Alexander-von-Humboldt Foundation, the Deutsche Forschungsgemeinschaft, and the Höchstleistungsrechenzentrum Jülich within the projects "Polymer Dynamics" and "Thermodynamics of Disordered Polymer Systems". Stimulating discussions with C. Pierleoni, J.-P. Ryckaert, K. Binder, D. C. Rapaport, G. S. Grest, P. A. Pincus, and H. van Vliet are gratefully acknowledged.

References

1. P. G. de Gennes, *Scaling Concepts in Polymer Physics* (Cornell University Press, Ithaca, 1979).
2. J. des Cloiseaux and G. Jannink, *Polymers in Solution: Their Modelling and Structure* (Clarendon Press, Oxford, 1990).
3. K. Kremer, in *Computer Simulation in Chemical Physics*, edited by M. P. Allen and D. J. Tildesley (Kluwer, Dordrecht/Boston/London, 1993).
4. Y. Oono, *Adv. Chem. Phys.* **61**, 301 (1985).
5. M. Doi and S. F. Edwards, *The Theory of Polymer Dynamics* (Clarendon Press, Oxford, 1986).
6. M. Schmidt (1993, in preparation).
7. J. G. Kirkwood and J. Riseman, *J. Chem. Phys.* **16**, 565 (1948).
8. J. J. Erpenbeck and J. G. Kirkwood, *J. Chem. Phys.* **29**, 909 (1958).
9. P. E. Rouse, *J. Chem. Phys.* **21**, 1272 (1953).
10. B. H. Zimm, *J. Chem. Phys.* **24**, 269 (1956).
11. P. J. Flory, *Principles of Polymer Chemistry* (Cornell University Press, Ithaca, NY, 1953).
12. W. Paul, K. Binder, D. W. Heermann, and K. Kremer, *J. Phys. II (Paris)* **1**, 37 (1991).
13. W. Paul, K. Binder, D. W. Heermann, and K. Kremer, *J. Chem. Phys.* **95**, 7726 (1991).
14. K. Kremer, G. S. Grest, and I. Carmesin, *Phys. Rev. Lett.* **61**, 566 (1988).
15. K. Kremer and G. S. Grest, *J. Chem. Phys.* **92**, 5057 (1990).
16. S. F. Edwards and K. F. Freed, *J. Chem. Phys.* **61**, 1189 (1974).
17. K. F. Freed and S. F. Edwards, *J. Chem. Phys.* **61**, 3626 (1974).
18. M. Muthukumar, *J. Chem. Phys.* **77**, 858 (1982).
19. M. Muthukumar and S. F. Edwards, *Polymer* **23**, 344 (1982).
20. S. Stepanow, *J. Phys. (France)* **49**, 921 (1988).
21. G. H. Fredrickson and E. Helfand, *J. Chem. Phys.* **93**, 2048 (1990).
22. J. Rotne and S. Prager, *J. Chem. Phys.* **50**, 4831 (1969).
23. H. Yamakawa, *J. Chem. Phys.* **53**, 436 (1970).
24. R. S. Adler, *J. Chem. Phys.* **69**, 2849 (1978).
25. T. Yoshizaki and H. Yamakawa, *J. Chem. Phys.* **73**, 578 (1980).
26. P. Mazur and W. van Saarloos, *Physica A* **115**, 21 (1982).
27. J. M. Deutch and I. Oppenheim, *J. Chem. Phys.* **54**, 3547 (1971).
28. T. J. Murphy and J. L. Aguirre, *J. Chem. Phys.* **57**, 2098 (1972).
29. R. Kapral, D. Ng, and S. G. Whittington, *J. Chem. Phys.* **64**, 539 (1976).

30. W. Hess and R. Klein, *Physica* **A94**, 71 (1978).
31. B. Dünweg, *J. Chem. Phys.* **99**, 6977 (1993).
32. T. Schneider and E. Stoll, *Phys. Rev. B* **17**, 1302 (1978).
33. P. G. de Gennes, *Physics* **3**, 37 (1967).
34. P. G. de Gennes, *Macromolecules* **9**, 587 (1976).
35. P. G. de Gennes, *Macromolecules* **9**, 594 (1976).
36. R. S. Adler and K. F. Freed, *J. Chem. Phys.* **70**, 3119 (1979).
37. M. Fixman, *J. Chem. Phys.* **78**, 1594 (1983).
38. A. Rey, J. J. Freire, and J. G. de la Torre, *Macromolecules* **24**, 4666 (1991).
39. B. J. Berne and R. Pecora, *Dynamic Light Scattering* (Wiley, New York, 1976).
40. M. Adam and M. Delsanti, *J. Physique* **37**, 1045 (1976).
41. A. Z. Akcasu, M. Benmouna, and C. C. Han, *Polymer* **21**, 866 (1980).
42. C. C. Han and A. Z. Akcasu, *Macromolecules* **14**, 1080 (1981).
43. Y. Tsunashima, M. Hirata, N. Nemoto, and M. Kurata, *Macromolecules* **20**, 1992 (1987).
44. M. Bhatt, A. M. Jamieson, and R. G. Petschek, *Macromolecules* **22**, 1374 (1989).
45. D. Richter, K. Binder, B. Ewen, and B. Stühn, *J. Phys. Chem.* **88**, 6618 (1984).
46. W. Zylka and H. C. Öttinger, *J. Chem. Phys.* **90**, 474 (1989).
47. M. Benmouna and A. Z. Akcasu, *Macromolecules* **13**, 409 (1980).
48. B. Dünweg and K. Kremer (to be published).
49. J. Stoer and R. Bulirsch, *Introduction to Numerical Analysis* (Springer-Verlag, Berlin/Heidelberg/New York/Tokyo, 1980).
50. M. Bishop, M. H. Kalos, and H. L. Frisch, *J. Chem. Phys.* **70**, 1299 (1979).
51. D. C. Rapaport, *J. Chem. Phys.* **71**, 3299 (1979).
52. Y. Y. Gotlib, N. K. Balabaev, A. A. Darinskii, and I. Neelov, *Macromolecules* **13**, 602 (1980).
53. W. Bruns and R. Bansal, *J. Chem. Phys.* **74**, 2064 (1981).
54. P. G. Khalatur, Y. G. Papulov, and A. S. Pavlov, *Mol. Phys.* **58**, 887 (1986).
55. B. Smit, A. van der Put, C. J. Peters, J. de Swaan Arons, and J. P. J. Michels, *J. Chem. Phys.* **88**, 3372 (1988).
56. J. Luque, J. Santamaria, and J. J. Freire, *J. Chem. Phys.* **91**, 584 (1989).
57. W. Smith and D. C. Rapaport, *Mol. Simul.* **9**, 25 (1992).
58. C. Pierleoni and J.-P. Ryckaert, *Phys. Rev. Lett.* **61**, 2992 (1991).
59. C. Pierleoni and J.-P. Ryckaert, *J. Chem. Phys.* **96**, 8539 (1992).
60. B. Dünweg and K. Kremer, *Phys. Rev. Lett.* **61**, 2996 (1991).
61. B. Dünweg and K. Kremer, *J. Chem. Phys.* **99**, 6983 (1993).
62. B. Dünweg, Ph.D. thesis, Universität Mainz, Germany, 1991.
63. C. Pierleoni (private communication).
64. M. P. Allen and D. J. Tildesley, *Computer Simulations of Liquids* (Clarendon Press, Oxford, 1989).
65. B. R. A. Nijboer and F. W. de Wette, *Physica* **23**, 309 (1957).
66. C. W. J. Beenakker, *J. Chem. Phys.* **85**, 1581 (1986).
67. M. Robbins, G. S. Grest, and K. Kremer, *Phys. Rev. B* **42**, 5579 (1990).
68. K. Chen and D. P. Landau, *Phys. Rev. B* **46**, 937 (1992).
69. J. P. Hansen and I. R. McDonald, *Theory of Simple Liquids* (Academic Press, New York, 1986).

70. D. C. Rapaport, *J. Comput. Phys.* **34**, 184 (1980).
71. D. C. Rapaport, *Comput. Phys. Rep.* **9**, 1 (1988).
72. G. S. Grest, B. Dünweg, and K. Kremer, *Comput. Phys. Comm.* **55**, 269 (1989).
73. J. D. Weeks, D. Chandler, and H. C. Andersen, *J. Chem. Phys.* **54**, 5237 (1971).
74. S. Nosé, *J. Chem. Phys.* **81**, 511 (1984).
75. W. G. Hoover, *Molecular Dynamics* (Springer-Verlag, Berlin, 1987).
76. C. Pierleoni and J.-P. Ryckaert, *Mol. Phys.* **75**, 731 (1992).
77. M. Mareschal and B. Holian (eds) *Microscopic Simulations of Complex Hydrodynamic Phenomena* (Plenum Press, New York/London, 1992).
78. J. M. V. A. Koelman, *Phys. Rev. Lett.* **64**, 1915 (1990).
79. P. J. Hoogerbrugge and J. M. V. A. Koelman, *Europhys. Lett.* **19**, 155 (1992).
80. D. L. Ermak and J. A. McCammon, *J. Chem. Phys.* **69**, 1352 (1978).
81. H. Risken, *The Fokker–Planck Equation* (Springer-Verlag, Berlin/Heidelberg/New York/Tokyo, 1984).
82. C. W. Gardiner, *Handbook of Stochastic Methods for Physics, Chemistry and Natural Science* (Springer-Verlag, Berlin, 1985).
83. R. J. A. Tough, P. N. Pusey, H. N. W. Lekkerkerker, and C. van den Broeck, *Mol. Phys.* **59**, 595 (1986).
84. A. Greiner, W. Strittmatter, and J. Honerkamp, *J. Stat. Phys.* **51**, 95 (1988).
85. B. Dünweg and W. Paul, *Int. J. Mod. Phys. C* **2**, 817 (1991).
86. M. Seeßelberg and F. Petruccione, *Comput. Phys. Comm.* **74**, 247 (1993).
87. M. Fixman, *Macromolecules* **14**, 1710 (1981).
88. M. Fixman, *Macromolecules* **19**, 1195 (1986).
89. A. Rey, J. J. Freire, and J. G. de la Torre, *J. Chem. Phys.* **90**, 2035 (1989).
90. S. Toxvaerd, *Mol. Phys.* **56**, 1017 (1985).
91. A. Rey, J. J. Freire, and J. G. de la Torre, *Polymer* **33**, 3477 (1992).
92. M. Fixman, *J. Chem. Phys.* **84**, 4085 (1986).
93. H. H. Saab and P. J. Dotson, *J. Chem. Phys.* **86**, 3039 (1987).
94. G. Bossis and J. F. Brady, in *Microscopic Simulations of Complex Hydrodynamic Phenomena*, edited by M. Mareschal and B. L. Holian (Plenum Press, New York/London, 1992).
95. B. H. Zimm, *Macromolecules* **13**, 592 (1980).
96. K. Kremer and K. Binder, *Comput. Phys. Rep.* **7**, 259 (1988).
97. M. Schmidt and W. Burchard, *Macromolecules* **14**, 210 (1981).
98. M. Adam and M. Delsanti, *Macromolecules* **10**, 1229 (1977).
99. G. Weill and J. des Cloizeaux, *J. Physique* **40**, 99 (1979).
100. N. Nemoto, Y. Makita, Y. Tsunashima, and M. Kurata, *Macromolecules* **17**, 425 (1984).
101. J. Batoulis and K. Kremer, *Macromolecules* **22**, 4277 (1989).
102. W. Burchard, M. Schmidt, and W. H. Stockmayer, *Macromolecules* **13**, 580, 1265 (1980).
103. K. H. Huber, W. Burchard, and L. J. Fetters, *Macromolecules* **17**, 541 (1984).
104. M. Fixman, *J. Chem. Phys.* **78**, 1588 (1983).
105. G. Wilemski and G. Tanaka, *Macromolecules* **14**, 1531 (1981).
106. M. K. Wilkinson, D. S. Gaunt, J. E. G. Lipson, and S. G. Whittington, *Macromolecules* **21**, 1818 (1988).

107. J. H. van Vliet and G. ten Brinke, *Macromolecules* **24**, 5351 (1991).
108. J. H. van Vliet and G. ten Brinke, *Macromolecules* **25**, 3695 (1992).
109. H. Eisenberg, *Biological Macromolecules and Polyelectrolytes in Solution* (Clarendon Press, Oxford, 1976).
110. M. Hara (ed.), *Polyelectrolytes: Science and Technology* (Marcel Dekker, New York, 1993).
111. C. Anderson and J. M. T. Record, *Ann. Rev. Phys. Chem.* **33**, 191 (1982).
112. D. Napper, *Polymetric Stabilization of Colloidal Dispersions* (Academic Press, London, 1983).
113. P. Pfeuty, R. M. Velasco, and P. de Gennes, *J. Physique, Lett.* **38**, L5 (1977).
114. T. Odijk, *J. Polym. Sci., Polym. Phys. Ed.* **15**, 477 (1977).
115. J. Skolnick and M. Fixman, *Macromolecules* **10**, 944 (1977).
116. T. Odijk and M. Mandel, *Physica A* **93**, 298 (1978).
117. T. Odijk, *Polymer* **19**, 989 (1978).
118. T. Odijk and A. Houwaart, *J. Polym. Sci., Polym. Phys. Ed.* **16**, 627 (1978).
119. M. Fixman, *J. Chem. Phys.* **76**, 6346 (1982).
120. M. L. Bret, *J. Chem. Phys.* **76**, 6242 (1982).
121. C. Reed and W. Reed, *J. Chem. Phys.* **92**, 6916 (1990).
122. M. Fixman and J. Skolnick, *Macromolecules* **11**, 863 (1978).
123. T. Odijk, *Macromolecules* **12**, 688 (1979).
124. J. Hayter, G. Jannink, F. Brochard-Wyart, and P. de Gennes, *J. Physique Lett.* **41**, 451 (1980).
125. P. de Gennes, P. Pincus, and R. Velasco, *J. Physique* **37**, 1461 (1976).
126. J.-L. Barrat and J.-F. Joanny, *Europhys. Lett.* **3**, 343 (1993).
127. T. Kitano, A. Taguchi, I. Noda, and M. Nagasawa, *Macromolecules* **13**, 57 (1980).
128. S. Miyamoto, *Makromol. Chem.* **182**, 559 (1981).
129. M. Tricot, *Macromolecules* **17**, 1698 (1984).
130. J. Schurr and K. Schmitz, *Ann. Rev. Phys. Chem.* **37**, 271 (1986).
131. W. Reed, S. Ghosh, G. Medjahdi, and J. Francois, *Macromolecules* **24**, 6189 (1991).
132. V. Degiorgio, F. Mantegazza, and R. Piazza, *Europhys. Lett.* **15**, 75 (1991).
133. S. Förster, M. Schmidt, and M. Antonietti, *J. Phys. Chem.* **96**, 4008 (1992).
134. G. Manning, *J. Chem. Phys.* **51**, 924 (1969).
135. G. Manning, *J. Chem. Phys.* **51**, 934 (1969).
136. G. Manning, *J. Chem. Phys.* **51**, 3249 (1969).
137. G. Manning, *Q. Rev. Biophys.* **11**, 179 (1978).
138. F. Oosawa, *Polyelectrolytes* (Marcel Dekker, New York, 1971).
139. A. Khokhlov, *J. Phys. A* **13**, 979 (1980).
140. A. Khokhlov and K. Khachaturian, *Polymer* **23**, 1742 (1982).
141. J. Joanny and L. Leibler, *J. Phys. (France)* **51**, 51 (1990).
142. P. Higgs and E. Raphael, *J. Phys. I* **1**, 1 (1991).
143. P. Higgs and H. Orland, *J. Chem. Phys.* **95**, 4506 (1991).
144. L. Wang and V. Bloomfield, *Macromolecules* **23**, 804 (1990).
145. M. Nierlich *et al.*, *J. Physique* **40**, 701 (1979).
146. M. Drifford and J.-P. Dalbiez, *J. Phys. Chem.* **88**, 5368 (1984).
147. G. Jannink, *Makromol. Chem., Macromol. Symp.* **1**, 67 (1986).

148. K. Kaji, J. Urakawa, T. Kanaya, and R. Kitamaru, *J. Physique* **49**, 993 (1988).
149. L. Wang and V. Bloomfield, *Macromolecules* **24**, 5791 (1991).
150. K. Schmitz, *Polymer* **31**, 1823 (1990).
151. M. Schmidt, *Macromolecules* **24**, 5361 (1991).
152. C. Reed and W. Reed, *J. Chem. Phys.* **94**, 8479 (1991).
153. J. Barrat and D. Boyer, *J. Phys. II (France)* **3**, 343 (1993).
154. S. L. Carnie and G. A. Christos, *J. Chem. Phys.* **89**, 6484 (1989).
155. G. A. Christos and S. L. Carnie, *J. Chem. Phys.* **91**, 439 (1989).
156. G. A. Christos and S. L. Carnie, *J. Chem. Phys.* **92**, 7661 (1990).
157. M. Stevens and K. Kremer, *Macromolecules* **26**, 4717 (1993).
158. M. Stevens and K. Kremer, *Phys. Rev. Lett.* **71**, 2228 (1993); submitted to *J. Chem. Phys.*
159. A. Baumgärtner, *J. Physique Lett.* **45**, L515 (1984).
160. C. Brender, M. Lax, and S. Windwer, *J. Chem. Phys.* **74**, 2576 (1981).
161. C. Brender, M. Lax, and S. Windwer, *J. Chem. Phys.* **80**, 886 (1984).
162. C. Brender, *J. Chem. Phys.* **92**, 4468 (1990).
163. C. Brender, *J. Chem. Phys.* **93**, 2736 (1990).
164. C. Brender, *J. Chem. Phys.* **94**, 3213 (1991).
165. C. Brender and M. Danino, *J. Chem. Phys.* **97**, 2119 (1992).
166. C. Brender, *J. Phys. Chem.* **96**, 5553 (1992).
167. J. Takashima, M. Takasu, and Y. Hiwatari, *Phys. Rev. A* **40**, 2706 (1989).
168. H. H. Hooper, H. W. Blanch, and J. M. Prausnitz, *Macromolecules* **23**, 4820 (1990).
169. H. H. Hooper, S. Beltran, A. P. Sassi, H. W. Blanch, and J. M. Prausnitz, *J. Chem. Phys.* **93**, 2715 (1990).
170. Y. Kantor, H. Li, and M. Kardar, *Phys. Rev. Lett.* **69**, 61 (1992).
171. C. Reed and W. Reed, *J. Chem. Phys.* **96**, 1609 (1992).
172. J. Takashima, M. Takasu, and Y. Hiwatari, *Mol. Simul.* **6**, 199 (1991).
173. J. Valleau, *Chem. Phys.* **129**, 163 (1989).
174. C. Woodward and B. Jönsson, *Chem. Phys.* **155**, 207 (1991).
175. Y. Kantor, H. Li, and M. Kardar, *Phys. Rev.* **E49**, 5103 (1994).
176. T. Hasegawa, J. Takashima, M. Takasu, and Y. Hiwatari, preprint (unpublished).
177. D. Adams and G. Dubey, *J. Comput. Phys.* **72**, 156 (1987).
178. S. de Leeuw, J. Perram, and E. Smith, *Ann. Rev. Phys. Chem.* **37**, 245 (1986).
179. B. Smit, in *Computer Simulation in Chemical Physics*, edited by M. P. Allen and D. J. Tildesley (Kluwer Academic Publishers, Dordrecht, 1993), pp. 173–209.
180. A. Chandra, D. Wei, and G. Patey, *J. Chem. Phys.* **99**, 4926 (1993).
181. A. Panagiotopoulos, *Mol. Simul.* **9**, 1 (1992).
182. A. Panagiotopoulos, *Fluid Phase Eq.* **76**, 97 (1992).
183. J. Perram, H. Petersen, and S. de Leeuw, *Mol. Phys.* **65**, 875 (1988).
184. J. Barnes and P. Hut, *Nature* **324**, 446 (1986).
185. L. Greengard, *The Rapid Evaluation of Potential Fields in Particle Systems* (MIT Press, Cambridge, MA, 1988).
186. Y. Kantor, *Phys. Rev.* **42**, 2486 (1990).
187. H. Vink, *J. Chem. Soc., Faraday Trans. I* **83**, 801 (1987).

188. J. Yamanaka, H. Matsuoka, K. Kitano, M. Hasegawa, and N. Ise, *J. Am. Chem. Soc.* **112**, 587 (1990).
189. R. Davis and W. Russel, *J. Polym. Sci.: Part B: Polym. Phys.* **24**, 511 (1986).
190. M. Stevens and K. Kremer, in *Macro-Ion Characterization: From Dilute Solutions to Complex Fluids*, edited by K. Schmitz (American Chemical Society Press, New York, 1993).
191. M. Drifford and J.-P. Dalbiez, *J. Physique Lett.* **46**, 311 (1985).
192. S. Förster, M. Schmidt, and M. Antonietti, *Polymer* **31**, 781 (1990).
193. M. Sedlak and E. Amis, *J. Chem. Phys.* **96**, 817 (1992).
194. M. Sedlak and E. Amis, *J. Chem. Phys.* **96**, 826 (1992).
195. M. Sedlak, *Macromolecules* **26**, 1158 (1993).
196. S. Miklavic and S. Marčelja, *J. Phys. Chem.* **92**, 6718 (1988).
197. S. Misra, S. Varanasi, and P. Varanasi, *Macromolecules* **22**, 4173 (1989).
198. P. Pincus, *Macromolecules* **24**, 2912 (1991).
199. O. Borisov, T. M. Birshtein, and E. Zhulina, *J. Phys. II* **1**, 521 (1991).
200. E. Zhulina, T. Birshtein, and O. Borisov, *J. Phys. II* **2**, 63 (1992).
201. H. Watanabe *et al.*, *Mat. Res. Soc. Symp. Proc.* **249**, 255 (1992).
202. R. Ross and P. Pincus, *Macromolecules* **25**, 2177 (1992).
203. M. Granfeldt, S. Miklavic, S. Marčelja, and C. Woodward, *J. Chem. Phys.* **23**, 4760 (1990).
204. L. Sjöström, T. Å. Kesson, and B. Jönsson, *J. Chem. Phys.* **99**, 4739 (1993).
205. S. Patel and M. Tirrell, *Ann. Rev. Phys. Chem.* **40**, 597 (1989).
206. L. Gulbrand, B. Jönsson, H. Wennerström, and P. Linse, *J. Chem. Phys.* **80**, 2221 (1984).
207. R. Kjellander and S. Marčelja, *Chem. Phys. Lett.* **112**, 49 (1984).
208. R. Kjellander and S. Marčelja, *Chem. Phys. Lett.* **114**, 125 (1985).
209. R. Kjellander, S. Marčelja, R. M. Pashley, and J. P. Quirk, *J. Chem. Phys.* **92**, 4399 (1990).
210. M. J. Stevens and M. O. Robbins, *Europhys. Lett.* **12**, 81 (1990).
211. V. Bloomfield, *Biopolymers* **31**, 1471 (1991).
212. M. Ferrari and V. Bloomfield, *Macromolecules* **25**, 5266 (1992).

4

ENTANGLEMENT EFFECTS IN POLYMER MELTS AND NETWORKS

Kurt Kremer and Gary S. Grest

4.1 Introduction

Dense polymer systems, such as melts, glasses, and crosslinked networks or rubber are extremely complex materials. Besides the local chemical interactions and density correlations which are common to all disordered liquids and solids, the chain conformations also play an important role. Their influence is twofold. First the intrachain entropy dominates over the positional entropy of the center of mass of the chains. This leads to the well known effect that a weak effective repulsion between different types of chains is sufficient to drive phase segregation.[1] The static and dynamic properties of mixtures of two types of chains is an important and challenging problem, which is reviewed by Binder in Chapter 7 of this volume. Here we consider dense melts of chains of the same chemical composition. In this case the entropy is at its maximum when the chains have a random walk structure.[2] Since the average end-to-end distance $\langle R^2(N)\rangle \sim N$ for a random walk, the average volume of a chain of length N monomers is

$$V \sim \langle R^2(N)\rangle^{3/2} \sim N^{3/2} \tag{4.1}$$

Assuming the density is homogeneous, there are $N/V \sim N^{1/2}$ other chains in the typical volume of a chain. These other chains effectively screen the long-range excluded volume interaction,[1,2] since a monomer cannot distinguish whether another monomer belongs to the same chain or to another. This property has been firmly established both by experiment and simulation. The second important property which must be taken into account is that the chains cannot move through each other. This rather elementary property plays a dominant role in dynamical and relaxational properties of a melt, even for fairly short chains. In contrast to small molecules, the motion of a polymer chain is more complex than simply random diffusion, at least over distance scales smaller than its own size. It is this difference with the motion of small molecule liquids that we are interested in studying here. In many respects, a dense melt of polymer chains is easier to understand than a dilute or semidilute solution in that hydrodynamic interactions are screened and do not play a role. A detailed discussion of hydrodynamic effects is given by Dünweg, Stevens and Kremer in Chapter 3 of this volume.

In a melt, the motion of a monomer is a consequence of its connectivity to other monomers on the same chain and its interaction with monomers on other chains. To a very good first approximation, the other chains can be viewed as providing a viscous background and a heat bath. While this certainly is a drastic oversimplification, which ignores all correlations due to the structure of the other chains which make up the background, it has been found to work very well. The advantage of this simplification is that the Langevin dynamics of a single chain of point masses connected by harmonic springs can be solved. This was first done in a seminal paper by Rouse[3] and about the same time in a similar fashion by Bueche.[4] In this model, which is commonly referred to as the Rouse model, the diffusion constant of the chain $D \sim N^{-1}$, the longest relaxation time $\tau_d \sim N^2$ and the viscosity $\eta \sim N$. This model has been found to describe the dynamics of a melt of relatively short chains both qualitatively and quantitatively almost perfectly,[2,5–10] though the reason is not well understood. Only very recently have some deviations, not only in the short time regime, been observed. For Rouse chains the effects are rather subtle[11–13] and would require a detailed discussion beyond the scope of this review. This model works well for relatively short chains ($M \lesssim 20\,000$ for polystyrene [PS] or $M \lesssim 2000$ for polyethylene [PE]), but definitely well beyond what one would call an oligomer. The exact value of M at which the Rouse model no longer holds is temperature- and density-dependent. For longer chains, the motion of the chains are observed to be significantly slower. Experiments show a dramatic decrease in the diffusion, $D \sim N^{-2}$, and an increase in the viscosity towards $\eta \sim N^{3.4}$. The time-dependent modulus $G(t)$ also shows a solid or rubber-like plateau at intermediate times before decaying completely.[14] Since the properties of the system starts to change at fixed chemistry-dependent chain length N_e or molecular weight M_e, one is led to the idea that the chain connectivity together with the fact that the chains cannot pass through each other are essential. If this is the case then the behavior should, as for the short Rouse chain, be universal and only the characteristic chain length N_e should depend on the chemical species.

The most well known approach to understanding these phenomena is due to Edwards and de Gennes. Edwards[15] in his work on crosslinked networks introduced the concept of obstacles produced by the other chains, resulting in a "tube" in which the monomers move. Later, de Gennes[16] realized that for long chains, a network is not needed. Figure 4.1 shows a "historical" map of the development of this concept. First consider a network. The figure shows one strand of the network in the center marked by a thick line and a rather crude sketch of the surrounding. Edwards[15] discussed how the black center chain could move around the obstacles created by all the other chains which in his case are part of the network. He noted that due to the topological constraints the chain is much more localized than just by the fact that the two ends are connected to a crosslink. Since he was interested in

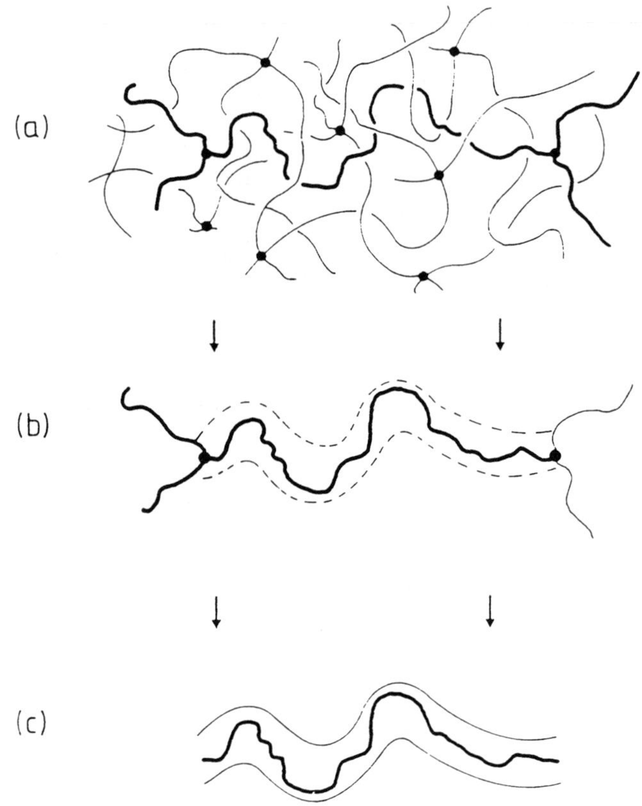

Fig. 4.1 Sketch of the historical development of the tube constraint and reptation concept. Starting from a network Edwards[15] in 1967 defined the confinement to the tube, while deGennes[16] in 1971 realized that for long chains the ends only play a small role for intermediate times.

studying networks in which the disorder was quenched, all loops and knots in the system were conserved. His idea was that the nearby chains create effective obstacles which cause the strand to be essentially bound to a tube-like structure (Fig. 4.1, middle part). This hypothetical tube, built by all the other chains, followed the coarse-grained conformation of the chain. The length scale of this coarse graining is called the entanglement length N_e and a sphere of the same diameter as the tube d_T typically contains $d_T^{1/\nu} = N_e$ monomers, where $\nu = 1/2$ is the random walk exponent. Within this picture the strand can perform a quasi one-dimensional Rouse relaxation along that tube. Later de Gennes,[16,17] realized that the dynamics of long chain melts should be governed by the same mechanism (Fig. 4.1, lower part). If the chains are very long, most of the monomers are far from the chain end. Then, on intermediate time scales, these monomers do not realize that the ends are free. Since there are only a very few chain ends, the topology of the surrounding should not change significantly on these intermediate time scales and a chain can only diffuse by reptating out of its original tube. This gives $D \sim N^{-2}, \tau_d \sim N^3$ and $\eta \sim N^3$ as well as a plateau modulus at intermediate time scales. Considering the simplicity of the concept, the model describes many experimental findings remarkably well. However, in spite of its successes, many open questions remain, including how to formulate the reptation concept on a more fundamental basis[18–20] and how to explain the discrepancy between the observed viscosity $\eta \sim N^{3.4}$ and the reptation prediction N^3.[21–28] A quantitative model of what an entanglement physically really is remains largely unsettled.[29,30] There are also a variety of non-reptation/tube phenomenological approaches which only treat the interactions between the chains either in a mean-field approximation or develop a memory functional formalism.[31–34] Microscopic statistical dynamical theories have also been recently advanced,[19,35–40] including a polymeric mode coupling theory,[36–38] in which the motion of the chains is assumed to be isotropic. In many of these models, the cooperative slithering snake motion of the chain is assumed, as in the work of Loring and coworkers[41–43] who have developed a stochastic model of polymer dynamics which treats the single molecule explicitly and the dynamics of its environment in a mean-field sense. Most of these models give results which agree with at least some experimental results but none gives a completely satisfactory explanation of polymer dynamics. A comprehensive review of the dynamics of an entangled melt has been published recently by Lodge, Rotstein and Prager.[44]

Our understanding of networks is in a rather similar situation. It has been known for a long time, as Fig. 4.1 illustrates, that the noncrossability of the chains plays an important role. Experiments on the modulus of melts, crosslinked at different initial concentrations directly prove this point. However it is not clear how this comes about and what the actual contributions of the entanglement compared to the crosslinks are. The reason for this is twofold. First, experiments which allow good control of all the microscopic parameters are very difficult to perform.[45] Theoretical descriptions are very complicated and usually contain several adjustable parameters, which are related to microscopic details, which are almost impossible to determine uniquely from experiment.[46]

While there are tremendous computational difficulties in studying the dynamics of dense melts of polymers and crosslinked networks, computer simulations have already played an important role in this field of research and will continue to do so in the future. Here we will review some of the successes of simulations in this area and discuss the outlook for the next few years. In fact, it can be argued, that the first microscopic demonstration that a new characteristic time and length scale is present in a melt of linear chains was from a computer simulation.[47] While simulations do not directly prove that the reptation concept is correct in the long time limit (that must await faster computers), they present a detailed picture of the motion of chain in a dense melt. Experimentally most measurements, such as the viscosity, diffusion constant and modulus, are for macroscopic properties and do not directly probe the microscopic motion of monomers on the chain directly. While neutron spin-echo scattering[48] is a very powerful technique which covers the appropriate length scales, the time range over which it can follow the motion of a polymer chain is rather limited. Pulsed gradient spin-echo NMR, with large field gradients, is a new technique[49–52] which is able to address the appropriate time and distance scales. However both of these methods probe only one aspect of the motion of the chains. Since most samples are never really ideal, experimentally one must often deal with polydispersity effects. Simulations, on the other hand, offer the unique opportunity to work under well controlled conditions, with either monodispersed samples or on samples with a specific amount of known polydispersity. In addition one can not only use models with full excluded volume, which is normally the case, but one can also study models in which chains are allowed to pass through each other in order to test specific aspects of the theory. From an analysis point of view, all interesting dynamic and static properties can be determined simultaneously. The mean-square displacements of the monomers and the motion of chains or parts of the chain, can be determined and visualized, offering opportunities well beyond the experimental limitations. The price to pay of course is that one has to follow the slow physical relaxation of the system, which means that we cannot use any of the smart and extremely fast methods, such as those discussed by Sokal in Chapter 2 of this volume. At present, the limitation of chain length to about $6 - 7N_e$ is simply due to the CPU time available. As computers increase in speed, the chain-lengths which can be studied will also. Simulations on chains of $15 - 20N_e$, which should be feasible within the next few years, will play a critical role in settling a number of the outstanding issues in polymer dynamics.

In the next section, we review in more detail the relevant theoretical concepts. In Section 4.3, we discuss the different simulation methods, both Monte Carlo and molecular dynamics, which have been used to study both the static and dynamic properties of a melt. We also show why it is essential to use a coarse-grained model, instead of a more realistic model, if one is to have any chance of addressing the issues related to whether reptation theory is correct or not. In Section 4.4, we review the progress which has been made recently in understanding the motion of long linear chains in a melt. In Section 4.5, we consider the properties of highly crosslinked networks or rubbers. Finally, in Section 4.6, we briefly summarize our main results and present our outlook for future simulations in this area.

4.2 Theoretical concepts

In a melt of homopolymers, the excluded volume interaction is effectively screened.[1,2] There is no tendency for a chain to swell beyond the ideal random-walk dimension. Only the prefactor, or more precisely the persistence length, is governed by the very local monomer–monomer interactions. The mean-square end-to-end distance of a chain of length N has the form

$$\langle R^2(N)\rangle = \langle(\mathbf{r}_1 - \mathbf{r}_N)^2\rangle = l^2 l_p^2 (N-1). \tag{4.2}$$

where $\mathbf{r}_1$ and $\mathbf{r}_N$ are coordinates of the chain-ends. Here l is the average bond length between two monomers on the chain and l_p is the persistence length ($l_p = \sqrt{C_\infty}$ in Flory's terminology[53]). Each polymer chain can therefore be viewed as a freely joined chain of subchains each of size l_p. The mean-squared radius of gyration,

$$\langle R_G^2(N)\rangle = \frac{1}{N}\left\langle \sum_{i=1}^{N} (\mathbf{r}_i - \mathbf{r}_{cm})^2 \right\rangle, \tag{4.3}$$

also scales linearly with N, $\langle R_G^2\rangle = l^2 l_p^2 (N-1)/6$. Here $\mathbf{r}_i$ denotes the position of monomer i and $\mathbf{r}_{cm}$ is the center of mass of the chain. Simulation results, both MD[47,54] and MC,[9,55–59] have clearly shown that both quantities scale linearly with N and the ratio $\langle R^2\rangle/\langle R_G^2\rangle = 6$, as expected. Neutron scattering experiments[60] have also demonstrated that $\langle R_G^2\rangle \sim N$. While the excluded volume interactions are screened in a melt, there is no reason that l_p for the melt should be the same as at the Θ point for a dilute chain. Simple lattice simulations implied this,[9] but there the Θ-solvent is modeled by lattice vacancies and an attractive monomer–monomer potential. This however is only "one" possible Θ-solvent. Recent simulations on different models by Rodriguez and Freire[61] and Milchev, Paul and Binder[62] found that at $T = T_\theta$, $\langle R_G^2(N)\rangle$ increases with concentration.

The dynamics of polymeric liquids are often described in terms of the Rouse and reptation models. For short chains the topological constraints do not play a dominant role and chains in a melt satisfy the Rouse model reasonably well.[5,63] However for chains of length $N > N_e$, where N_e is the entanglement length, the topological constraints begin to affect the motion of the monomers. These constraints are a consequence of the chain connectivity and the fact that the chains cannot pass through each other. Before discussing the simulation results, we first review some of the background concerning these two theoretical approaches. We restrict ourselves mostly to quantities which can be investigated directly by simulation.

4.2.1 Unentangled melt

The motion of a single monomer is governed by the connectivity of the chain and the interaction of the monomers with its surroundings. In the simplest model, all the complicated interactions are absorbed into a monomeric friction and a coupling to a heat bath. This so-called Rouse model[3] was originally proposed to model an isolated chain in solution, though it actually works better for unentangled chains in a melt. For chains in solution we refer to the chapter on solutions by Dünweg *et al.* in this volume. In this model the polymer is modeled as a freely jointed chain of N beads connected by $N-1$ springs, immersed in a Newtonian continuum. Hydrodynamic interactions are neglected. Each bead acts as a point source of friction, with friction coefficient ζ. The beads are connected with a frictionless Hookean spring with a force constant $k = 3k_BT/b^2$, where $b = ll_p$ is the root mean-square separation of adjoining beads. Each bead-spring unit is intended to model a subchain of the real molecule, not a monomer. The equation of motion of the beads can be described by either the Smoluchowski or Langevin equations.[5] Neglecting inertia forces, which is reasonable, since the motion is overdamped at low-frequency in viscous media, the Langevin equation for monomer $i(i \neq 1,\ N)$ reads,

$$\zeta\dot{\mathbf{r}}_i = -k(2\mathbf{r}_i - \mathbf{r}_{i-1} - \mathbf{r}_{i+1}) + \mathbf{f}_i, \tag{4.4}$$

Usually the model is solved for a ring with no free ends. If the chain ends are free, as for all linear chains, the first and the last monomer have to be treated differently. For $i = 1$, the first term on the right hand side is $-k(\mathbf{r}_1 - \mathbf{r}_2)$ and similarly for $i = N$. The distribution of random forces $\mathbf{f}_i$ is Gaussian with zero mean and second moment,

$$\langle \mathbf{f}_i(t) \cdot \mathbf{f}_j(t') \rangle = 6\zeta k_B T \delta_{ij} \delta(t - t'). \tag{4.5}$$

Note that this model does not contain any specific interactions between monomers except those due to the chain connectivity. Since the model neglects both hydrodynamic effects and excluded volume, it does not describe the dynamics of an isolated chain in solution. However in a dense melt, the long-range hydrodynamic interactions are screened out, much like the excluded volume interactions. For this reason, it was suggested[14] that this model could describe the motion of chains in a melt, except that ζ arises from other chains in the melt and is larger than in a dilute solution.

The Rouse model can be solved by transforming to the normal coordinates $\mathbf{X}_p(t)$ of the chain. For a discrete monomer chain these are given by[64–66]

$$\mathbf{X}_p(t) = \frac{1}{N}\sum_{i=1}^{N} \mathbf{r}_i(t)\cos\frac{p\pi(i-1)}{N-1} - \frac{b'}{2N}[\mathbf{r}_1(t) + (-1)^p\mathbf{r}_N(t)], \tag{4.6}$$

where b' is zero for free chains, while for network chains it is one,* and $p = 0, 1, 2, \ldots, N-1$. Equation (4.4) can then be rewritten as

$$\zeta_p\dot{\mathbf{X}}_p = -k_p\mathbf{X}_p + \mathbf{f}_p \tag{4.7}$$

where $\zeta_0 = N\zeta$ and $\zeta_p = 2N\zeta$ for $p \geq 1$ and

$$k_p = 8Nk\sin^2\frac{p\pi}{2(N-1)}. \tag{4.8}$$

For small p/N, one recovers the usual result[5]

$$k_p = 2\pi^2kp^2/N = \frac{6\pi^2k_BT}{Nb^2}p^2 \tag{4.9}$$

with Nb^2 being the mean-squared end-to-end distance. Since the random forces $\mathbf{f}_p$ are independent of each other, the motion of the $\mathbf{X}_p$ are independent of each other. The motion of the polymer can therefore be decomposed into independent modes.

For chains in a melt, the Rouse modes are expected to be eigenmodes of the chains. This has been verified by MD[54] for a melt of short chains by comparing the time correlation functions of the normal modes with the theoretical result for $p \geq 1$, namely

$$R_p(t) = \frac{< \mathbf{X}_p(t)\cdot\mathbf{X}_p(0) >}{< X_p^2(0) >} = \exp(-t/\tau_p), \qquad \tau_p = \frac{\zeta(ll_p)^2N}{3\pi^2k_BT}\frac{N}{p^2}, \tag{4.10}$$

where we have used the small p/N result for k_p, eq. (4.9). The longest relaxation time is $\tau_R = \tau_1 \sim N^2$. For long chains in a melt, this equation is expected to only describe the relaxation of high p modes with $N/p \lesssim N_e$. For smaller p, topological constraints of the chain play an important role and the Rouse model is no longer valid.

The case of $b' = 1$ corresponds to a coupling constant $k' = 2k$ for the end-bonds. This correction for free chains is in most cases negligible, since this results in a $1/N$ correction term. All the MD simulations presented in this review used $b' = 1$. For walks within a network $b' = 1$[13] since there the last monomer within an arbitrary walk through the net is not only coupled to the previous monomer, but also to the following one. This leads to an effectively stronger coupling, which for calculating motion quantities is not important. However for a correct estimate of the modulus this extra term is crucial. For more details see Refs 13, 162 and Section 4.5 on networks.

Within this scheme the relaxation modulus of the melt is given by[5]

$$G(t) = \frac{\rho k_B T}{N} \sum_p R_p(2t). \tag{4.11}$$

where ρ is the monomer density. This of course assumes that the single chain Rouse modes can be taken as eigenmodes of the whole melt. This certainly is an assumption, as briefly discussed in the introduction. For the Rouse model, using eq. (4.10), this gives

$$G(t) = \frac{\rho k_B T}{N} \sum_p \exp(-2tp^2/\tau_R). \tag{4.12}$$

From this, one can easily calculate the viscosity η since

$$\eta = \int_0^\infty G(t)dt = \frac{\rho k_B T}{2N} \sum_p \tau_p = \frac{\pi^2}{12} \frac{\rho k_B T}{N} \tau_R = \frac{\rho \zeta}{36} N(ll_p)^2 \sim N\zeta. \tag{4.13}$$

The self-diffusion constant D can be determined from the mean-square displacement of $\mathbf{X}_0 = \mathbf{r}_{cm}$, the center-of-mass of the chain,

$$g_3(t) = <(\mathbf{r}_{cm}(t) - \mathbf{r}_{cm}(0))^2> . \tag{4.14}$$

Within the Rouse model $g_3(t) \sim t$ for all times. The diffusion constant $D(N) = lim_{t\to\infty} g_3(t)/6t$ is expected to reach the asymptotic value

$$D = \frac{k_B T}{N\zeta}. \tag{4.15}$$

even for relatively short times.

In simulations, a direct way to study the motion of the monomers is to measure the mean-square displacement of a monomer $g_1(t)$ as a function of time t, which is given by

$$g_1(t) = \frac{1}{N} \sum_{i=1}^{N} \langle [\mathbf{r}_i(t) - \mathbf{r}_i(0)]^2 \rangle. \tag{4.16}$$

Using the fact that the chain structure is that of a random walk, it is easy to show that

$$g_1(t) \sim \begin{cases} t^1, & t < \tau_o, g_1(t) < l^2; \\ t^{1/2}, & \tau_o < t < \tau_R, g_1(t) < \langle R^2 \rangle; \\ t^1, & t > \tau_R, g_1(t) > \langle R^2 \rangle, \end{cases} \tag{4.17}$$

Thus for very early times, when a monomer has moved less than its own diameter σ, it is affected little by its neighbors along the chain. This short time regime, $t < \tau_o$, is governed by the local chemical properties of the chains or for simulations depends on the model used. Within the worm-like chain model in which the chain is a continuous flexible path, τ_o is zero.

However for intermediate times, the motion of a monomer is slowed down because it is connected to other monomers. This can be viewed as the diffusion of a particle with increasing distance dependent mass. The actual mass at time t is just the number of monomers within a sphere of diameter $\sqrt{g_1(t)}$. This also naturally leads to the correct good solvent scaling. It continues until the chain has moved a distance comparable to its size. Finally for long times, when the chain has moved a distance greater than $\langle R^2 \rangle^{1/2}$, they undergo free diffusion with a diffusion coefficient $D \sim N^{-1}$, eq. (4.15). It turns out experimentally that this extremely simple model provides an excellent description of polymer dynamics, provided that the chains are short enough. Measurements[63,67–69] of η as well as NMR[70–76] and neutron spin-echo scattering experiments[6,77–79] which probe the motion of the monomers are in agreement with eq. (4.17). Results for molecular dynamics simulations on short chains also agree with this model.[54] In the Monte Carlo and Brownian dynamics methods, the Rouse model is inherently built in so it is more difficult to decide how much of the observed behavior can be directly attributed to the interactions and how much to the algorithm. For short chains, it turns out that the noncrossability of the chains, as well as the chain nature of the surrounding of each monomer, only affects by changing the prefactors in the diffusion coefficent through the monomeric friction coefficient ζ. Why these effects turn out to average to such a simple contribution certainly is not understood.

4.2.2 Entangled melt

For chains which significantly exceed a critical entanglement length N_e, the motion of the chain is slowed down drastically. The clearest evidence for this slowing down comes from the diffusion constant D, which can be measured by following the motion of labeled chains using infrared microdensitometry,[80] forced Rayleigh scattering,[81,82] forward recoil spectrometry[83–85] or field gradient NMR.[70–76] See Tirrell[86] and Lodge *et al.*[44] for comprehensive reviews of experimental results. For $N \gtrsim N_e$,

$$D \sim N^{-2}. \tag{4.18}$$

Several forms for the prefactor of D have been discussed in the literature.[19,87] The effects of the entanglements are also clearly seen in the viscosity, which for large N, has the scaling form[63,67,68,69,88]

$$\eta \sim N^{3.4}. \tag{4.19}$$

compared to N for short chains. Direct observation of the motion of molecules across an interface between two types of partially deutrated polystyrene polymers[89] shows some evidence for entanglements, though the interpretation is not unique.

There have been several theoretical models which try to understand this behavior. However only the reptation concept of Edwards[15] and de Gennes[16,17] or a few variants[19,21,90,91] of this approach take explicitly the noncrossability of the chains into account. The first detailed treatment of this concept to the viscoelastic properties of polymer melts and solutions was carried out by Doi and Edwards.[5,92] Curtiss and Bird[18] also applied the reptation concept to study stress relaxation. For a recent review of the theories of stress relaxation see Bird and Öttinger.[93] In this model a chain moves on a coarse-grained scale mainly along its own contour, except at the chain ends. The reason for this is that the topology of the surrounding suppresses the motion transverse to its own contour. For short time scales the motion of the monomers cannot be distinguished from that of the Rouse model. Until the distance a monomer moves exceeds a critical size, namely the so-called tube diameter $d_T^2 \equiv \langle R^2(N_e) \rangle$, the motion of the monomer is isotropic and $g_1(t) \sim t^{1/2}$ as in the Rouse model. The typical time for the onset of this constrained motion is the Rouse relaxation time for a chain of length N_e,

$$\tau_e \sim N_e^2. \tag{4.20}$$

After this time the motion is no longer isotropic. The monomers only can diffuse along the chain's own coarse-grained contour, which is the backbone of the tube. By this forward and backward motion of the monomers, the chain explores new space and slowly destroys the original tube. For $t > \tau_e$, one only has Rouse relaxation along this coarse-grained random walk structure[16,17] and the $t^{1/2}$ power law for $g_1(t)$ becomes a $t^{1/4}$ power law. This is because a monomer on the chain is undergoing a random walk motion along a random walk contour as illustrated in Fig. 4.2. However at the Rouse relaxation time τ_R of the entire chain, the chain has only moved a distance of order the square root of the contour length of the tube. Following this regime the chain creeps out of its tube (see Fig. 4.2). The overall diffusion along the tube gives a second $t^{1/2}$ regime for the motion in space. In this second $t^{1/2}$ time regime the diffusive motion of the whole chain along the tube can be represented as a random walk of a single particle. An elemental step is the displacement of a polymer chain along its tube in direction of one of its ends. Only the ends explore new space. The longest relaxation time is simply the mean lifetime of the tube. The initial tube will be destroyed when one of the segments has visited N different contiguous sites. This requires a time $\tau_d \sim N^3/N_e$. In this time the chain has moved a distance comparable to its own size, therefore the diffusion constant D scales as in agreement with experimental results, eq. (4.18). The theory predicts the following general power-law sequence for the mean-square displacement in space, $g_1(t)$:

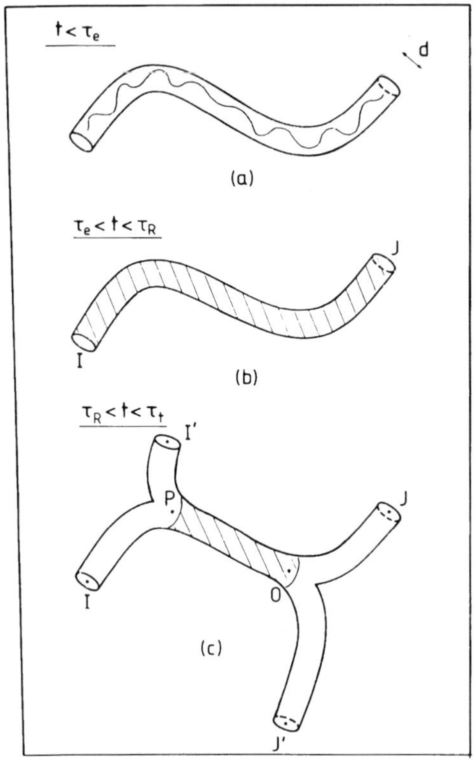

Fig. 4.2 Schematic representation of a reptating chain in different time regimes: (a) short-time unrestricted Rouse motion, (b) equilibration of density fluctuations along the chain, and (c) creep motion of chain out of the tube (from Ref. 135).

$$g_1(t) \sim \begin{cases} t^1, & t < \tau_o; \\ t^{1/2}, & \tau_o < t < \tau_e \sim N_e^2; \\ t^{1/4}, & \tau_e < t < \tau_R \sim N^2; \\ t^{1/2}, & \tau_R < t < \tau_d \sim N^3/N_e; \\ t^1, & t > \tau_d \end{cases} , \tag{4.21}$$

which is shown schematically in Fig. 4.3. For the motion of the center-of-mass $g_3(t)$ one expects

$$g_3(t) \sim \begin{cases} t^1, & t < \tau_e \sim N_e^2; \\ t^{1/2}, & \tau_e < t < \tau_R \sim N^2; \\ t^1, & \tau_R < t \end{cases} \tag{4.22}$$

Direct evidence for these intermediate time regimes has been seen by pulsed gradient spin-echo NMR,[50,94] field cycling and rotating-frame NMR[95] and diffusion of polymers at an interface.[89,96–98]

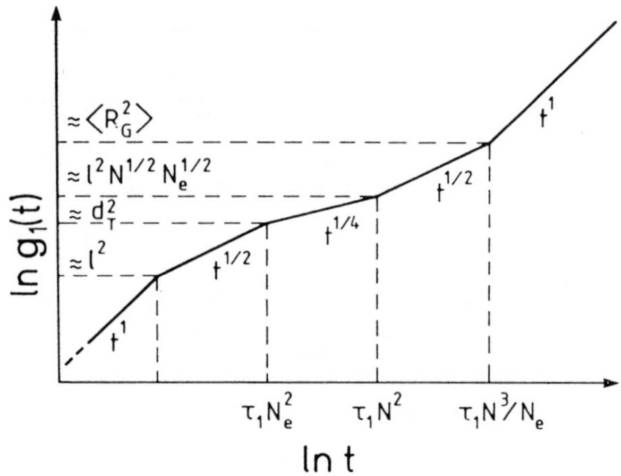

Fig. 4.3 Schematic plot of the mean–square displacement for a monomer in the reptation model.

The reduced mobility of the chains can also be seen in the relaxation of the long-wavelength modes of the long chain. In the reptation model, the relaxation time of a Rouse mode p, with $N/p > N_e$ is enlarged by a factor of N/N_e, giving[5]

$$\tau_{p,Rep} = \frac{N\langle R^2\rangle}{p^2}\frac{\zeta}{\pi^2 k_B T}\frac{N}{N_e} \sim \frac{N^3}{p^2}. \tag{4.23}$$

Hess[19] derived a similar expression from his microscopic model by explicitly considering the effective entanglement as a dynamic effect. Hess included the important many chain cooperative effects of constraint release and tube renewal,[87,90,99–108] which are necessary in order to get quantitative predictions for the stress relaxation functions. Ultimately this does not affect the N dependence of the relaxation time. He found that after an initial fast Rouse-like decay up to time τ_e, $\tau_{p,Hess} = 2/3\tau_{p,Rep}$. Both models describe essentially the same physical picture. For the generalized Rouse model, Kavassalis and Noolandi[35] found that $\tau_{p,GRM} \sim N^3/p^4$. MD simulation results of Kremer and Grest[54] could not distinguish between the standard reptation and Hess models but could rule out the generalized Rouse model.

In the reptation model, the viscosity η scales with $\tau_d, \eta \sim N^3$. The exponent 3 is slightly smaller than that found experimentally, eq. (4.19). There have been numerous models[20–28,36] to explain this difference between the reptation model and experiment. Though the question is not completely settled, it is believed by many that the higher exponent is a finite size crossover effect due to the tube length fluctuations.[21,28] Doi[21] was the first to suggest this difference was a crossover effect and that for finite chain length, contour length fluctuations enhance the relaxation of the stress. Doi suggested that the dependence on η and the longest relaxation time on N is no longer a power law for intermediate values of N and only for very large N is the reptation prediction recovered. Early computer simulations[24] of the dynamics of a single entangled chain seemed to agree with the original reptation model. However, more extensive numerical studies[28,109–112] using a discretized reptation model based on de Gennes' model of moving defects found that for finite chain length, η had an effective exponent in the range 3.2–3.5, consistent with the experiment. The analytical treatment[91] of tube fluctuations is in agreement with these numerical calculations. In this approach, tube length fluctuations lead to a very broad crossover with an apparent exponent for the relaxation time ~ 3.4 and the asymptotic value of 3 is achieved only for very large value of N. There have been several alternative explanations of the 3.4 power law which do not rely on a crossover but suggest that the true asymptotic power law is greater than 3.[20,26,27,38] All of these models incorporate in some way the fact that reptation is a single chain model and do not necessarily describe all the complicated many-chain effects which contribute to the stress relaxation.

One difficulty with the crossover explanation for the 3.4 power law for η and τ_d is that it also predicts a larger exponent for the diffusion constant D. In a single chain model, as noted above, it is assumed that a linear chain diffuses a distance of order R during its longest relaxation time τ_{relax}, so that $D \sim R^2/\tau_{relax}$. Since for chain lengths of experimental interest, the longest relaxation time $\tau_{relax} \sim N^{3.4}$, instead of N^3, one would predict that $D \sim N^{-2.4}$ instead of N^{-2}. However this does not agree with experiment. A more appropriate way to interpret this result is not the $D \sim N^{-2.4}$ but that the mean squared distance a polymer diffuses during its relaxation time grows as $N^{1.4}$ which is larger than its mean-squared size $R^2 \sim N$. For this to happen the memory of the polymer configuration has to remain in the region even after the chain leaves this region. This memory effect was included in a phenomenological way by Scher and Shlesinger but they did not explain its physical origin. Recently Rubinstein and Obukhov[113] argued that a reptating chain leaves a trace in a melt in the form of elastic distortions of an entanglement net. Neighboring chains are attracted to this trace and can relax these elastic distortions by partially reproducing the configuration of the first polymer. They showed that the difference in the longest relaxation time τ_{relax} and the time a chain takes to move its own size $\tau_d \sim N^3$ is $\tau_{relax}/\tau_d \sim N^{1/3}$ in agreement with experiment.[74] Semenov[114] argued in a similar way that the diffusion is retarded due to long-living density fluctuations in the system. These fluctuations, which allow the chains reptate out of the tube themselves only can relax via reptation itself.

Other theories which are not based on the reptation/tube model have also been developed. While some make an ansatz about how a polymer moves in a melt, others are more microscopic.[19,36–40] Schweizer[36–38] uses a mode-coupling approach. His theory predicts the emergence of a plateau shear modulus due to dynamical correlations induced by excluded volume interactions and chain connectivity, which is included to linear order in the effective equation of motion. The linear and scalar nature of the effective friction in the equation of motion implies isotropic dynamics,[115] which precludes an explicit treatment of anisotropic single chain motion as assumed in the reptation model. *A priori*, it is unclear how serious this approximation is, particularly as many of the results of the theory agree very closely with those of the reptation model. Both models predict the same N dependence for the diffusion constant D and viscosity η. The reduction in D is due to isotropic, but highly correlated, force fluctuations or motions of clusters.[33] Schweizer also finds that there is a system-specific, density-dependent length scale significantly larger than the statistical segment length, above which there is a crossover from unentangled Rouse to slowed down mode-coupling dynamics at a time τ_c.[38] The mean-square displacement for a monomer $g_1(t)$ has several regimes. For $\tau_c < t < \tau_R, g_1(t) \sim t^{9/32}$, followed by a $t^{3/8}$ regime. In the early time regime, the power law is very nearly the same as the reptation prediction, eq. (4.21), and it is difficult at present to distinguish the two approaches. Simulations on longer chains, greater than those that are presently accessible $(6 - 7N_e)$ are necessary to distinguish which model best describes the intermediate regime, $t^{1/2}$ or $t^{3/8}$. Such a determination must await access to the next generation of supercomputers. In either case, such an isotropic approach might seem reasonable, in at least an "effective" medium sense in order to describe experimental data, though the motion of the monomers appears to be anisotropic.[116] A self-consistent version of the theory in the spirit of mode-coupling theories of glasses and critical phenomena has recently been developed.[38] For chains, this theory predicts a localization transition which may signal the breakdown of the isotropic caging assumption and the onset of anisotropic motion such as reptation. Extensions of the theory to finite N, including nonuniversal corrections to the asymptotic mode-coupling theory are needed before the model can be compared directly to present simulations. However we can make a rather crucial test of this model from our simulations of polymer networks as will be discussed in Section 4.5. Since in Schweizer's theory the monomers move isotropically in space, the chain ends do not play an especially crucial role. Thus for long times the motion of monomers in the network should be isotropic, since it is expected to be confined to a tube within the reptation scheme. This will be discussed in more detail below.

The plateau modulus G_N^o is determined from rheological measurements. In the reptation model, it is related to the tube diameter $d_T = \langle R^2(N_e) \rangle$,[117]

$$G_N^o = \frac{4}{5}\frac{\rho k_B T}{N_e}. \tag{4.24}$$

The 4/5 prefactor arises from tube length fluctuations, which allow for a better relaxation under deformation, while this is not the case in networks.

While this is the fundamental relationship of the Doi–Edwards[92] theory of viscoelastic behavior of a polymer melt, the molecular origin of the entanglement distance is still not very well understood. Graessley and Edwards[118] suggested that G_N^o should be determined solely from the density of chain contour lengths ($\rho L/N$, where ρ/N is the number density of chains and L is the contour length of a chain) and the Kuhn length $l_K = C_\infty l$ of a chain. A simple dimensional argument led them to conclude that

$$G_N^o \frac{l_K^3}{k_B T} \sim \left(\frac{\rho}{N} L l_K^2\right)^a, \tag{4.25}$$

where a was not calculated explicitly by Graessley and Edwards.[118] In terms of measured quantities, C_∞, ρ and l,

$$\frac{G_N^o}{k_B T} \sim C_\infty^{2a-3} l^{3a-3} \rho^a. \tag{4.26}$$

This leads to an entanglement length $N_e \sim C_\infty^{3-2a} \rho^{1-a}$ and a tube diameter $d_T \sim \rho^{(1-a)/2}$.

A number of models have been developed which predict the exponent a. The local packing models of Ronca,[31] Kavassalis and Noolandi[35] and Lin[119] suggest that entanglements are determined from a fixed number of strands in an entangled volume $\langle R^2(N_e)\rangle^{3/2}$. This leads to the prediction $a = 3$. This result agrees with an earlier model by Doi.[120] An alternative scaling based on a fixed number of binary contacts[15,118,121] making an entanglement strand gives $a = 2$ while Colby and Rubinstein[122] conjectured that entanglements are determined by a constant number of binary contacts in an entangled volume, which gives $a = 7/3$. Helfand[123] has pointed out that the differences between these scaling ideas arises because there are two independent length scales, the Kuhn length l_K and the average distance between contour lines $(\rho L/N)^{-1/2}$, which appear in the analysis. This means that some additional conjecture about the entanglements is required for scaling to work. Unfortunately there are experiments to support all three pictures of an entanglement. Rheological[118] and neutron spin-echo experiments[124] on concentrated polymer solutions result in a scaling exponent $a \simeq 2.2$, midway between the two binary contact models. This result is in agreement with the variation of G_N^o with concentration in a Θ solvent which give $a = 2.33 \pm 0.13$.[125] However scaling melt data for a range of polymers, according to eq. (4.26) give $a = 2.8 \pm 0.4$[125] and 2.9 ± 0.1,[126] consistent with the packing models. Simulations, which will be discussed later,[12] give an exponent between $a = 2.53(\nu = 0.59)$ and $a = 3(\nu = 1/2)$ covering the experimental values. These later results are in disagreement with simple models[127–129] which assume that the entanglement molecular weight scales only with a positive power of C_∞ and does not depend on the density ρ. To date, simulations have not really been able to address what determines the plateau modulus since the longest chains are only about $6 - 7N_e$ and longer chains are needed to observe a clear plateau modulus. However simulations can study the dependence of C_∞ and N_e on temperature and density,[12,59] though more needs to be done.

An alternative approach to the scaling theories for the entanglements are topological methods. These are based on the assumption that entanglements are purely geometrical. If this is the case, then d_T and N_e can only be a function of the two length scales, l_K and $(\rho L/N)^{-1/2}$, as mentioned above. Using the Gaussian topological invariants, which counts the windings swept out by one curve around another, Edwards[130] calculated the dependence of d_T in two limiting cases. For a pure random flight polymer, Edwards' results agree with the packing model. His results for a Gaussian coil in a network of rods,[130] are identical to that for the binary contact model. Iwata and Edwards[29,30] extended the Gaussian integral method to what is known as localized Gaussian integrals. The theory introduces a topological interaction parameter $\bar{\gamma}$, which describes the capability of a chain to entangle. $\bar{\gamma}$ is mainly determined by the diameter of a polymer chain. The smaller the diameter, the more a polymer can entangle. Also the larger C_∞ is, the more a chain is extended making it easier to become entangled. The theory does not predict an analytic form for the plateau modulus, but for the concentration dependence of G_N^o, an effective exponent a can be determined. $a = 2$ for high concentrations and increases slightly with decreasing concentration.

As for long chains in a melt, chain entanglements are also expected to play a role for the modulus of polymer networks. By randomly crosslinking the chains at several places a system of strands is produced, in which the strand ends are in most cases connected to crosslinks. This holds for systems far above the percolation/vulcanization threshold. Thus most of the monomers are in strands which have both ends connected to crosslinks. These strands are of course highly polydisperse. As already mentioned in the introduction the static plateau modulus for such polymer networks of long strands is expected to result from the same reasons as for entangled melts. Thus the entanglement picture of networks gives a modulus[14]

$$G(N \gg N_e) = \rho k_B T / N_e, \tag{4.27}$$

which is the same as for melts up to a prefactor of 4/5. As in a melt, the monomers are supposed to be confined to a tube-like regime in space. Since there is no tube renewal due to the lack of free ends, one should expect to observe the tube confinement for even shorter chains than in a melt. This then should allow one to test the mode coupling theory[36–38] for melts from the network simulations. The mode coupling theory does not predict a confinement to a tube along the coarse-grained contour of the strands. Eventually the monomers should move isotropically in space and explore a regime which ultimately is expected to scale with the radius of gyration of the strands. This range should be somewhere in between the limits of the phantom network and the affine network model.[45] Any significant deviation from this scaling can be interpreted with respect to the differences between reptation theory and the corresponding mode coupling approach. This point will be discussed in Section 4.5 on polymer networks.

4.3 Model and method

To address the question of what is an entanglement and how it affects the properties of a dense melt or network is a formidable task for computer simulations, as it is for experiment. In a simulation, the greatest difficulty arises simply because the chain-length N must be much larger than the entanglement length N_e and the longest relaxation time increases rapidly with $N, \tau_d \sim N^3$ for $N > N_e$. Experimentally the longest relaxation time increases with even a slightly higher power, $N^{3.4}$. To study the crossover from the early Rouse regime to the entangled regime, N should be at least 4–5N_e. Within the original reptation model, much longer chains ($N \gtrsim 15N_e$) would be necessary in order to clearly observe the second $t^{1/2}$ regime and to observe the plateau modulus.[131] For polymer networks, in which the entanglements are trapped by the crosslinking, the relevant distance is the mean separation between crosslinks. In this case, the effects of the entanglements can clearly be seen even for chains of order $1 - 3N_e$. However in this case larger systems are needed in order to average over the disorder. For such long chains, simply equilibrating the sample before collecting statistics requires considerable amount of CPU time. While the effect of the entanglements can be observed directly in the mean-square displacement of a monomer or the modulus, it is also important to try to identify exactly what an entanglement is. This requires some theoretical input to decide how to identify an entanglement from the monomer coordinates.

To estimate the relevant time scales for the simulation, let us consider the simplest polymer — polyethylene (PE).[132–134] Since the relaxation time decreases as T increases, we choose a high temperature somewhat below that at which the polymer would degrade, to the estimate a lower CPU time limit. While the diffusion constant D increases and the relaxation time decreases with increasing T, this is at least partially compensated for by an increase in N_e. (For PE it turns out that it is better to do the simulation at high temperature but there is no general rule about whether it is better to carry them out at the highest possible temperature or not, particularly since the time step Δt also decreases with increasing T.) From recent neutron spin-echo measurements (which will be discussed in more detail below), Richter *et al.*[135] found that the entanglement time $\tau_e \simeq 5.0 \times 10^{-9}$ s at $T = 509$ K for saturated polybutadiene (PEB-2). PEB-2 is essentially PE with one ethyl branch every 50 main-chain bonds. Using this time and the entanglement molecular weight, which for PEB-2 at $T = 509$ K is $M_e \simeq 2000$ or about 137 monomers,[79,135] it is possible to estimate the minimal amount of CPU time which would be needed to observe the onset of entanglement effects using the published results. To simplify the model, the hydrogen atoms are not included explicitly but incorporated into a larger effective united carbon atom. The interactions include a spring potential between the monomers, a bond-bending potential and a torsional potential to account for the backbone structure of the chain. Monomers (in this case CH_2 and CH_3 units) that are not neighbors along the chain sequence of the polymer interact via a standard Lennard-Jones potential. In these models, the center for the united atom nonbonded potential was either at the center of the carbon atom[132,133,136] or moved towards the hydrogens.[134,137] The parameters are chosen to fit experimental data. Since these potentials are nonlinear, the stability of the simulation requires fairly small time-steps. Typically, one needs of order 50 integration time-steps per oscillation time of the highest-frequency mode of the system. This is the first problem for a realistic simulation. Softening the spring constant k for the C–C bond from its typical spectroscopic value reduces the frequency of the highest modes somewhat allowing a larger time-step Δt. Reducing k by a factor of 7, Rigby and Roe[132] found that they could use a time-step as large as 10^{-14} s for $T \leq 360$ K. Other authors[133,134,136,137] have used smaller Δt. For $T \sim 500$ K, a smaller Δt would probably be needed, though we use this larger value for our estimate. For a 500 monomer system, their optimized MD program ran at 50 time-steps per second on the Cray XMP. For this system size, a run of length τ_e would take about 2.8 h on the XMP or about 2 h on a YMP.

Using Rigby and Roe's timings,[132] it is straightforward to estimate the CPU time required to simulate a minimal system for studying entanglement effects. As an example consider the relaxation time for 10 chains of length $5M_e$. Such a system is significantly smaller than most of the subsequently discussed investigations. For this system, we estimate that the relaxation time within the ideal reptation model is about $\tau_d(5N_e) = \tau_e(N/N_e)^3 \simeq 6.25 \times 10^{-7}$ s. Transformed into CPU time, this would require about 3400 hours of CPU time on a single Cray YMP processor for a run of length one relaxation time. For 100 chains of $6M_e$, which is more comparable to the largest system studied by Kremer and Grest,[54] this time increases to 70 000 CPU hours on a single YMP processor. For a complete and conclusive study, even for this rather short chain length, many more chains and longer runs are needed to obtain reasonable statistics. Note that this result is for quite high T, far above the glass temperature. An earlier estimate by us[138] was for the temperature of Rigby and Roe's simulation, which though also above the glass transition temperature, gave a CPU time which was significantly larger. One should also keep in mind, that the intrachain potentials are still subject to some discussion. A recent recalculation of the energy barriers for the torsion angles for PE revealed that the barrier is significantly *too low* in most of these models.[139] Even allowing for the faster Cray C90 processors, the time for a comprehensive study is prohibitive at present. On future computers, with say, a 10^2 net speed-up in speed (probably on a massively parallel system), it will be possible to make some runs at high temperatures for simple systems such as PE. However on today's fastest computers, a conclusive investigation with more and longer chains would require hundreds of thousands of CPU hours. Thus the prospects of such an investigation are not very good. The situation becomes even more hopeless if one wants to study more relevant polymers like polystyrene (PS)[140] or more complicated ones like polydimethylsiloxane (PDMS) or polycarbonates (PC). In these cases, the potentials are considerably more complicated and the entanglement lengths are often significantly larger. Thus even for tomorrow's computers, at least for those that are likely to come in the foreseeable future, these problems are beyond the limits. Consequently, a complete, detailed simulation of chemically and industrially interesting polymers on today's fastest supercomputer would easily require many years for the chains are to move a distance comparable to their own length at temperatures of practical interest. This is not to say that simulations on more realistic models are not important. In fact, there are a number of other important properties which do depend on the exact nature of the interactions which do require realistic models. These include local properties such as short-range and short-time correlations[132–134,140,141] orientational correlations,[141] high-frequency excitations, subglass relaxation of side groups,[142] and diffusion of small penetrant molecules in a polymer melt or network.[143,144] Also many macroscopic aspects, such as the viscosity prefactor and the glass transition temperature, depend strongly on the local properties of the polymers. In order to get more information on the chemistry-dependent properties of complicated polymers, one has to work with modifications of the simple models discussed above.[145] Work in this direction is discussed by Clarke in Chapter 5 of this book.

To obtain insight into the long-time, large-distance properties of melts and networks, another approach is needed, since a microscopic approach for long-time properties is a hopeless undertaking. Consequently, the classical approach of using highly simplified models[58,146–151] is still the only feasible way to gain insight into the dynamical properties of complicated polymer systems. We refer to these models as coarse-grained models, in that the monomers or units of the polymer chain do not correspond explicitly to atoms but to coarse-grained units. However by making a proper mapping of these simple model systems onto different chemical species, it is possible to give quantitative estimates of time and length scales for a variety of different polymeric systems.[54] Since this can be done from a single set of runs for one model, such an approach naturally gives more insight into the universal properties[2] of polymeric systems.

What is the minimal amount of detail which must be included to study the long-time and large-distance properties of melts and networks? Clearly self-avoidance of the monomers is important as well as noncrossability of the chains. Both of these are essential if one wants to study the effect of entanglements. If the chains can pass through each other then the topological constraints of long chains do not play a role. While it is sometimes useful for crosslinked networks to study a model in which nonconnected monomers are allowed to pass through each other in order to test simple theories (see Section 4.5), this is usually not the case for melts. Fortunately these two requirements are sufficient for constructing a model which can be used to test the predictions of theory and model experimental systems on large distance and long time scales. The introduction of bond-bending and torsional forces, which slow down the simulation, are not necessary. There are several equivalent methods/models which satisfy these two requirements. The classic method is to embed the chains as self-avoiding and mutually avoiding walks on a lattice.[58,131,148,149,151] In this method monomers reside on the lattice sites which are connected to form chains. Typically one uses the diamond[9,10] or simple cubic lattice.[32,57,152,153] The monomers are moved using MC methods which, when chosen properly, satisfy detailed balance and reproduce the Rouse[3] dynamics for short chains.[148,151] The majority of the simulations on polymer melts prior to 1990 used this approach. Since that time, other methods have been adopted. One of them is the bond-fluctuation method, which was first introduced by Carmesin and Kremer.[66] In this method, each monomer consists of 2^d lattice sites. In addition to excluded-volume interactions, the bond length l is restricted to a maximum extension to avoid bond crossing. This model has turned out to be considerably more efficient than the standard lattice MC method for simulating dense melts.[12,59,154,155] One major reason is that the effective density for models with fluctuating bond lengths is much higher than for simple lattice models,[12,59] which results in a much shorter entanglement length N_e. Off-lattice, pearl necklace[146,147] and bead-spring models, have also been used. In the former, the bond length is fixed, while in the later it is allowed to vary. MC methods, in which subsequent configurations of the polymer are generated stochastically, have been applied to both of these models.[62,146,147]

Molecular dynamics methods[156–158] have been used for bead-spring models to study dense melts of linear chains[54,131,150] and polymer networks [159–162] as well as many-arm stars and end-grafted polymer brushes.[163] It is very effective for dense systems where strong density fluctuations occur. Each monomer i of the system moves according to an equation of motion,

$$m\frac{\mathrm{d}^2\mathbf{r}_i}{\mathrm{d}t^2} = -\nabla \sum_{j,j\neq i} U_{ij} - m\Gamma\frac{\mathrm{d}\mathbf{r}_i}{\mathrm{d}t} + \mathbf{W}_i(t). \tag{4.28}$$

where m is the monomer mass. The interaction potential has two parts. The first U^{l}_{ij} is a potential which acts between all monomers in the system. Often it is taken to be a purely repulsive Lennard-Jones potential.[54] Along the sequence of the chain or between crosslinked monomers, a strong attractive potential U_{ij}^{ch} is added. The parameters of U_{ij}^{ch} must be chosen carefully so as to avoid the chains from cutting each other yet be soft enough to allow for a large time step.[54] Here Γ is the bead friction which acts to couple the monomers to the heat bath and $\mathbf{W}_i(t)$ describes the random force acting on each bead. The strength of the random force is coupled to the bead friction by the fluctuation–dissipation theorem. The equations can be integrated with a variety of techniques.[156–158] We found that the velocity-Verlet[164] algorithm is very efficient and allows us to use a time step Δt in the range 0.012–0.0135τ, where $\tau = \sigma(m/\epsilon)^{1/2}$ is the standard time unit for a Lennard-Jones fluid. This was about twice as large as our earlier simulations[54] which used a fifth order predictor–corrector algorithm.[165] Since the potential is purely repulsive, the system is close to athermal. Most of the MD results will be for $T = 1.0\epsilon/k_B$ and $\Gamma = 0.5\tau^{-1}$ at a density $\rho\sigma^3 = 0.85$. Results in this review will be presented in reduced units in which σ, ϵ and $m = 1$. The introduction of a weak coupling to the background is very important not only to keep the temperature at the preset value but also to keep the system stable over the course of the simulations. Because we are interested in the dynamics of a melt in which the relaxation time is very long, runs of up to $20 \times 10^6 \Delta t$ after equilibration were made. We found that coupling the system weakly to the heat bath was a physically appealing way to reduce the effect of the numerical errors which accumulate during the course of any long simulation. This overall coupling to a heat bath does, however, lead to a diffusion of the entire system. This diffusion has to be removed when analyzing the motion of the chains. This is done by simply calculating all quantities in the center-of-mass coordinates of the whole system. For a more complete review of all of these methods see Chapter 9 by Grest and Murat in this volume.

Recently Schulz *et al.*[166] combined the bond fluctuation MC and MD simulation. They used the MC method to follow the large-scale motion of the chains but periodically used MD for short intervals. They claim that this method gives a significantly increase in the overall speed of the code. This is probably because the MD allows one to resolve "blocked" conformations better due to the cooperativity of the motion, while the MC is much faster in more standard environment. Heermann and Yixue[167] have also discussed hydrid MC–MD methods but so far have only presented results for very short chains.

The static conformations on scales larger than the persistence length or the average spacing between nonbonded nearest neighbors do not depend on the specific model. For shorter distances differences occur and the lattice models are somewhat less realistic than the continuous space models. The MD simulations of Kremer and Grest are described very well by the PRISM (polymer reference interaction site model) model of Curro and Schweizer.[168]

For a general discussion of the microscopic structure of the various polymer melt models see Ref. 169.

Which model or method is best to use for studying dense linear melts and networks depends on the particular system and question under consideration. Which is better depends somewhat on whether it is more appropriate to work on a lattice or in the continuum and whether it is acceptable to have stochastic dynamics on all time scales. On a lattice, we believe the best method is clearly the bond fluctuation method. For off-lattice, MD using a bead-spring model is very efficient particularly on vector supercomputers. On super scalar workstations, there are some indications that MC for the bead-spring model is also efficient.[62] However on the Cray due to the extremely fast integer routines, this is questionable. The bond fluctuation algorithm as well as the MD are highly vectorized[154,170], while the MC bead-spring simulations have not been vectorized as of yet. Comparisons between the various methods are difficult, particularly between lattice and off-lattice simulations since the models and methods are so different. Recently Everaers and Kremer[171] improved the vectorized MD due to a significantly more effective setup of the Verlet table. For linear chains in a melt or semidilute solution, lattice simulations are a good choice. However in some applications, such as constant pressure simulations[172,173] or for polymers under shear, one should use a continuum model. Similarly, if one is interested in the behavior of gels or polymer networks under swelling or elongation then a continuum simulation using MD is probably the most appropriate way to proceed. Continuum simulations, either MC or MD, are also essential for simulations on the more detailed, microscopic polymer models.[174,175] The recent approach by Schulz *et al.*[166] might point to a very interesting alternative, however the description in the paper is too short and incomplete in order to judge at this point.

There is one case where a fairly reasonable comparison between the methods can be made. That is in the study of the dynamics of a dense melt of linear chains. For the purely repulsive case, in which only excluded volume interactions are taken into account, there is sufficient data from the different methods to allow for a comparison. However since the simulations are done on different models, one needs some measure of effective speed, since MC or MD steps per second is not a useful comparison. In this case, a good measure of the effective speed of the algorithm is to measure the CPU time to reach the crossover time τ_e from the Rouse regime to the entangled regime. On a Cray YMP, Grest and Kremer[131,150] estimated that at present the bond fluctuation method is clearly faster than MD. However the recent improvement for the MD simulation[171] of chains interacting with the short range LJ potential to about 300 000 particle moves per second on a Cray YMP for the situation discussed above, reduces the difference to some extent. However the inclusion of additional interactions has a much stronger effect on slowing down the bond fluctuation method than it does for the

MD. On an RISC workstation, Milchev *et al.*[62] found that the CPU time to reach τ_e for their off-lattice MC simulations are only about 30% slower than the bond fluctuation algorithm on the same computer. Generally speaking, the off-lattice bead-spring models are the most flexible but not always the fastest. For those problems in which it is applicable, the bond fluctuation method is currently the most efficient.

4.4 Simulations of uncrosslinked polymers

The numerical/simulational investigation of polymer melts can be separated into two classes. In the first the reptation model is assumed to be correct and the system is simulated by elementary moves which directly follow the reptation concept. These simulations are mainly used nowadays to examine crossover properties within the model, such as whether the $N^{3.4}$ power law is asymptotic or just an effect of the experimentally accessable chain lengths, though they reach up to more than $1000N_e$.[88] We refer to this group of numerical investigations as "reptation simulations". However to test the validity of the concept itself, one has to go to a more basic level and use the methods such as those discussed in the previous section which incorporate only local moves which do not contain any bias towards reptative motion. This is essential to test whether the basic ansatz of the reptation model is valid or not. In this section we review several recent investigations using both MC and MD simulations which studied the dynamics of a dense melt of linear chains. First we discuss some recent results using the "reptation simulations" and then we present a more detailed account of the more basic "molecular level" simulations.

4.4.1 Reptation simulations

When simulations were first applied to the dynamics of a melt of mobile chains it was very difficult to set up a program which was capable of dealing with such a complex problem in a reasonable manner. Among the early attempts, Deutsch[176] employed a special lattice model in which interchain excluded volume interaction were included but monomers along the chain were allowed to overlap. This significantly enhanced the mobility of the chains. Deutsch, already in 1982, observed a signature of the $t^{1/4}$ regime in the mean-square displacement of the inner monomers. Though at the time the validity of such an approach compared to other simulations was highly disputed,[177] it now seems clear that this simulation should be included within the "reptation simulations". Similar ideas were also put forward by Evans and Edwards.[178] A related approach which employs dynamics similar to the moving de Gennes defects has been used extensively by Pakula and coworkers.[179] However since there is no direct interpretation of the

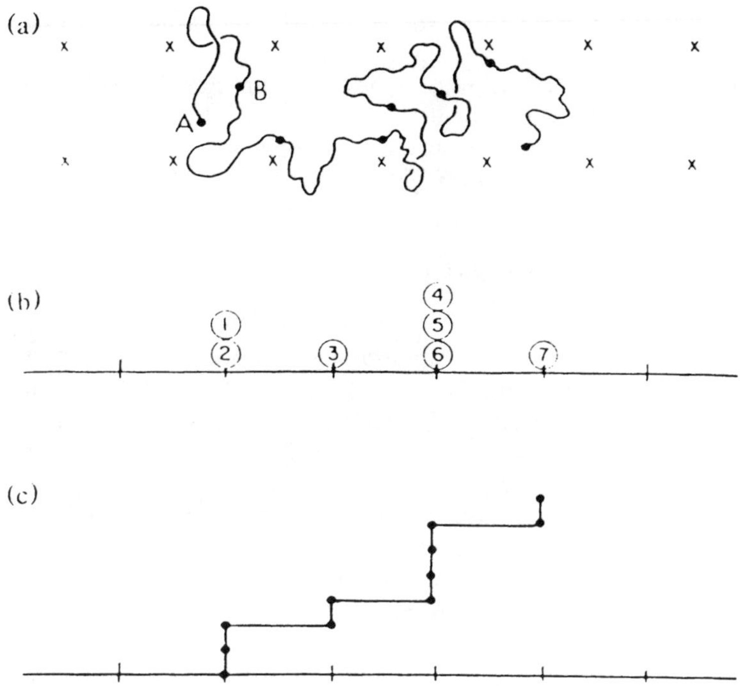

Fig. 4.4 Illustration of the repton idea, from Ref. 28. (a) A polymer in an entangled net is confined to a tube. Filled circles divide the chain into segments of stored length. (b) Repton model representation of conformation in part (a). Cells of the entanglement net along the confining tube are represented by a one-dimensional lattice. Sections of chain length stored in these cells are modeled by reptons on the lattice sites. (c) Directed walk representation of the same conformation.

dynamics in their cooperative motion algorithm,[148,151] we omit a discussion of their results here.

The reptation concept and the crossover for large N was directly analyzed recently by Rubinstein.[28] This work was motivated by the long-standing problem of the observed $N^{3.4}$ power law for the longest relaxation time τ_d. As discussed above, the reptation model predicts a N^3 power law. One of the main questions in this context was to investigate whether for large but finite N, the reptation model would give an apparent power law consistent with experiment. Rubinstein[28] employed the so-called repton model, where the topological constraints of motion are directly simulated within a single chain picture. Figure 4.4 shows his model. The repton model actually reduces the problem to a one dimensional directed walk. As illustrated in Fig. 4.4 the idea is to look at a SAW chain in the tube. Thus the order of the units, the reptons, has to be preserved. In the example illustrated in the figure, repton number 1 can move to the left but not to the right. One attempted move per repton then corresponds to the time τ_e. In the actual simulation the probability for the outer monomers is different for the direction "into" the tube and "out of" the tube. If z is the lattice coordination number for the hypothetical repton chain lattice, the probability for moves "out" of the tube is $(1 - 1/z)$ while it is $1/z$ for moves "into" the tube. For the inner parts the probabilities are direction independent. This on a very coarse-grained level is the simulation of a chain in a tube,[180] where the intrachain excluded volume is effectively taken into account by the preserved order of reptons. Rubinstein used this model to estimate both $D(N, z)$ as well as $\eta(N, z)$. The viscosity is given by the integral of the stress relaxation function, which is directly related to the conformational relaxation of the chains. Thus the repton model can provide us with an estimate of how η depends on N, though it certainly cannot provide the prefactor. Figure 4.5 shows the results for the viscosity as a function of the chain length N (here of course in units of reptons) and lattice coordination number z. The slope, giving the viscosity exponent, varies between 3.22 for $z = 2$ and 3.52 for $z = 12$.[28] Thus even within such a very simple model, where only the intrachain EV within the tube is considered beyond the original

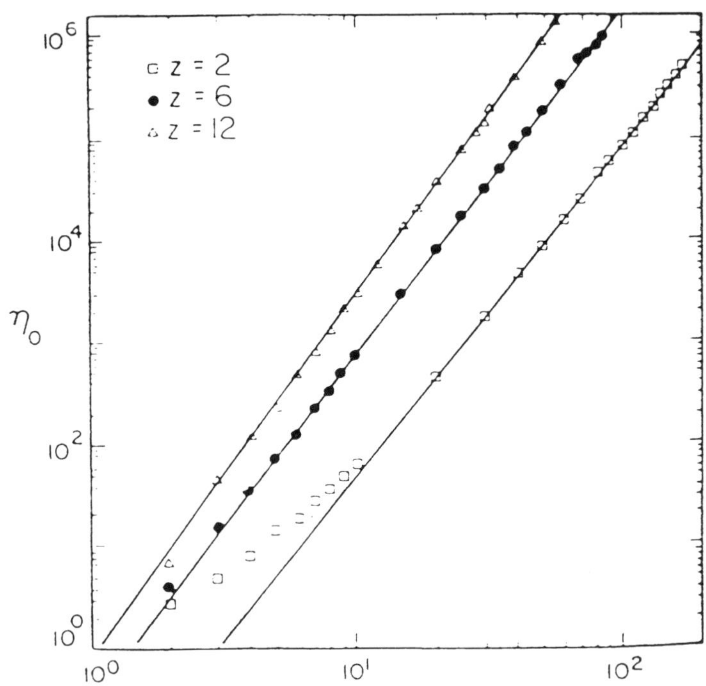

Fig. 4.5 Repton model prediction for the viscosity η_o as function of the number of reptons per chain N for different values of z, as indicated. The slope varies from 3.22 ($z = 2$) to 3.52 ($z = 12$), (from Ref. 28).

approach, one recovers the experimental results for the viscosity. Similar results were obtained by Deutsch and Madden.[110] Along such lines several authors tried to elucidate the problem of one moving chain in an environment of obstacles. Baumgärtner and Muthukumar were mainly interested in the problem of chain localization in big pores.[181] This is not of special relevance here since, for the entanglement problem, the average distance between obstacles has to be much smaller than the chain extension. Deutsch and Madden[110] as well as Reiter[111] discuss the motion of random walks on a lattice of obstacles or entanglements. While Deutsch and Madden[110] describe the motion of a random walk on such a lattice, Reiter[111] compares the Rouse, slithering snake and Evans and Edwards dynamics.[178] In general both find agreement with the reptation scheme, however the diffusion constant $D(N)$ seems to display an N-dependence which is too strong.

More recently a more detailed approach was presented by O'Connor and Ball,[112] who extended the ideas of Rubinstein[28] and Ketzmerick and Öttinger[109] by introducing Rouse dynamics for the chains in the tube. This turns out to be especially important for the end beads. In this way the contour length fluctuations are properly taken into account. This leads to an agreement with experiment, similar to that found by Rubinstein.[28] They also included tube renewal to account for the fact that the tube is only expected to be a preserved quantity, if there is no chain end in the neighborhood. In this way, they were able to obtain quantitative agreement with experiment. The only adjustable parameter in their model was the plateau modulus or the entanglement molecular weight N_e. Thus on the level of these simulations the reptation model seems to account perfectly for the rheological properties of long chain melts.

In the original reptation theory the stress is transported along the chains within the tube, the so called backbone stress. This is in contrast to simulations,[13,182–184] where it was found that under shear or elongation most of the forces are due to excluded volume contributions. Fixman[20,185] used this to argue against the reptation model. He employs a model of chains with a fixed bond angle of 90°.[185] Each bond is modeled as a long hard rod of length L and diameter d. The aspect ratio L/d varies between 15.2 and 69.4. A Langevin simulation of a semidilute solution of such rods was carried out. Typically the volume unit of L^3 (length of the stiff bonds) was occupied by six different strands. By using this rather constrained model he argued that the main contribution to the plateau modulus is related to excluded volume forces. His model however is certainly very questionable when applied for a melt of flexible polymers. The model is also unable to reproduce several experimental results as well as the previous simulation or the subsequent simulations on a more molecular level. Models like this may however be useful for semidilute systems of, e.g., block-copolymers made of long, very stiff units and small flexible spacers.

For such systems Fixman's model however would be more of a molecular level model.

4.4.2 *Melt simulations on a "molecular level"*

From the discussion in Section 4.3, it is clear that for long-time, large-scale phenomena the term molecular level still means coarse-grained models. As shown before it is impossible to perform simulations for melt dynamics for the case of a very detailed, chemically accurate model. In this subsection we report on recent MD and MC results as well as one hybrid method simulation which incorporates both MD and MC to study the dynamics of a melt of linear chains. In all cases the moves are employed in a way that the appearance of a slowed down motion is a direct consequence of the non-crossability of the chains and not of the model.

The first important question to ask about any melt simulation is whether the chain lengths and relaxation times which can be simulated are sufficient to observe the most obvious effect of the entanglements, namely the decrease in the diffusion constant. If this decrease in $ND(N)$ can be observed then the simulations can at least be used to characterize the transition from Rouse towards reptation. Eqs (4.15) and (4.18) give the expected behavior of the diffusion constant $D(N)$ for short and long chains, respectively. In order for a simulation to make a significant contribution, it must at least be able to cover this crossover regime. Following eq. (4.15), $D(N)N$ should define a plateau for small N, from which the monomeric friction or mobility can be obtained. To compare results from different simulations to experimental results for different polymers, a plot of $D(N)/D_{\mathrm{Rouse}}(N)$ versus N/N_e or M/M_e, respectively, should give one universal curve, where $D_{\mathrm{Rouse}} = k_B T/\zeta N$. This mapping is important for our understanding, since experiment and simulation use different methods to estimate M_e or N_e. In simulations N_e is mostly determined by the crossover towards the $t^{1/4}$ regime in $g_1(t)$, while experimentally the plateau modulus is usually determined from the stress relaxation function. More recent simulations, which extend the time window and the effective chain lengths[12,166] could also use the length of the $t^{1/4}$ regime to determine N_e. However the aforementioned method is generally used. A collapse of all available data onto one curve shows that not only the different methods of determining N_e really define the same length scale, but it also strongly suggests that there is a chemistry-dependent length N_e, which in all cases plays the very same role. Figure 4.6 includes data from MD simulations,[54] standard lattice MC,[58] bond fluctuation MC,[59] and a hybrid combination of bond fluctuation MC and MD,[166] as well as experimental results from NMR measurements on PE.[74]

The MD[54] simulations were performed at a density of $\rho = 0.85\sigma^{-3}$ and the bond fluctuation MC simulations[12,59] at two different volume fractions $\Phi = 0.4, 0.5$. For the hybrid method of Schulz *et al.*,[166] we used $N_e = 40$ as

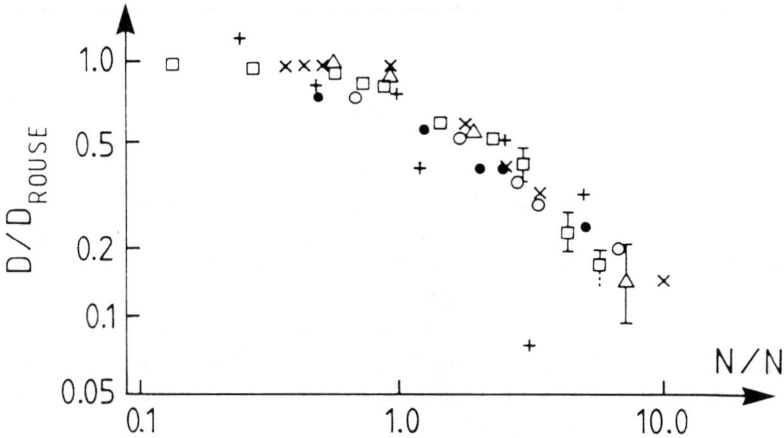

Fig. 4.6 Scaled diffusion constants from experiment and simulation as a function of chain length scaled by the entanglement lengths N_e. Diffusion constant is normalized to the Rouse diffusion constant. Data for polyethylene (×) from the NMR experiments of Pearson *et al.*[74] Simulation results are from the MD work (□) of Kremer and Grest,[54] the bond fluctuation MC work of Paul *et al.*[59] (•, ○) (two different densities $\Phi = 0.5, 0.4$) and Schulz *et al.*[166] (+) and the standard lattice MC work (△) of Skolnick *et al.*[153] (Adapted from Ref. 131).

given in their paper and $D_{\text{Rouse}} = 1.63/N$ deduced from their diffusion constant plot. The coefficient was determined by taking the mean value of $D(N = 10)$ and $D(N = 20)$ for their simulation. As the figure shows, the scatter of the data is significant, indicating that the precision is considerably smaller than for the other data. In the lattice simulations of Skolnick *et al.*,[58] the entanglement length turns out to be about $N_e = 110 - 120$ for chains on a simple cubic lattice at density $\Phi = 0.5$.[131] There the determination of the Rouse regime is rather difficult, since simple lattice chains at $\Phi = 0.5$ model a semidilute solution rather than a dense melt. Comparing their results to earlier MC simulations on the diamond lattice by Kremer[9] one expects a screening length for the EV interaction of about 10–20 monomers. Thus in order to determine D_{Rouse} only $N \geq 20$ was considered. Then one finds for the simple cubic lattice at $\Phi = 0.5$, $D_{\text{Rouse}} = 0.106/N$. The experimental data are from NMR measurements of Pearson *et al.*,[74] for PE at $T = 448$ K. The entanglement molecular weight M_e is taken from the plateau modulus compared to the Doi–Edwards theory.[5,14] They give $M_e = 1350$ or equivalently $N_e = 96$ monomers for this temperature, as indicated in the figure. Using this value for N_e the data nicely fall on one curve. All four simulation schemes as well as the experiment show a clear crossover from the Rouse to the entangled reptation regime. In all cases the simulations cover up to 6–7 entanglement lengths.

From these results it is clear that the simulations are in a position to analyze the crossover towards the reptation regime in some detail, however they do not cover the fully entangled regime. The crossover is actually better described by the formula

$$D(N)/D_{\text{Rouse}}(N) = N_e/(N_e + N) = \left(1 + \frac{N}{N_e}\right)^{-1} \tag{4.29}$$

derived by Hess[19] than by $D/D_{\text{Rouse}} = \frac{4}{15} N_e/N$ given by Graessley.[87] Hess's theory, where the motion confinement is taken into account via a projection operator formalism, eventually leads to the reptation model for long chains.

A key signature of the reptation is given by the mean-square displacements of the monomers and the center of mass of the whole chain. Since all present computer simulations still only can deal with rather short chains, it is necessary to confine the analysis to the innermost monomers in order to identify the long chain behavior. The ends show an enhanced mobility, which dominates the data. Thus we define

$$g_1(t) = \frac{1}{5} \sum_{i=N/2-2}^{N/2+2} < (\mathbf{r}_i(t) - \mathbf{r}(0))^2 > \tag{4.30}$$

Figure 4.7 gives results for $g_1(t)$ from the bead spring MD simulations of Kremer and Grest.[131] For short times all data collapse onto a single curve, indicating that the inner monomers of the chains do not feel the overall length and the resulting constraints. This is the same as de Gennes ansatz[16] for the introduction of the reptation as illustrated in the introduction. This signal of the slowed-down motion is completely smeared out if the whole chain is considered. Figure 4.8 for $g_1(t)$ for monomers at different locations

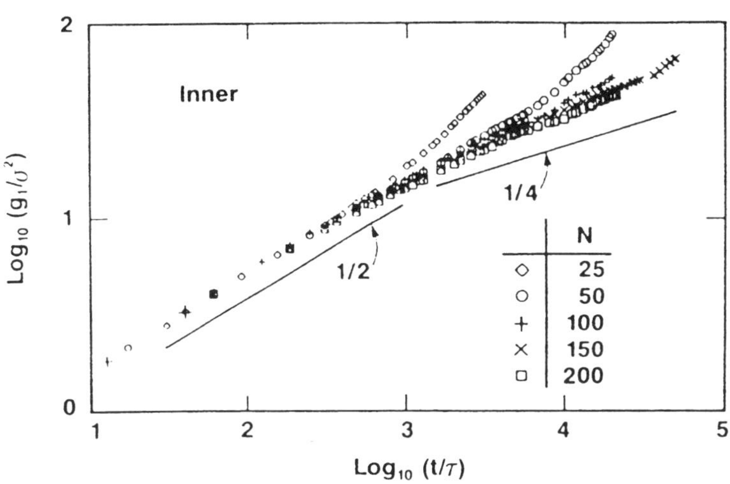

Fig. 4.7 Mean-square displacement $g_1(t)$ of the inner monomers from the MD simulations of Ref. 54 for a variety of chain lengths.

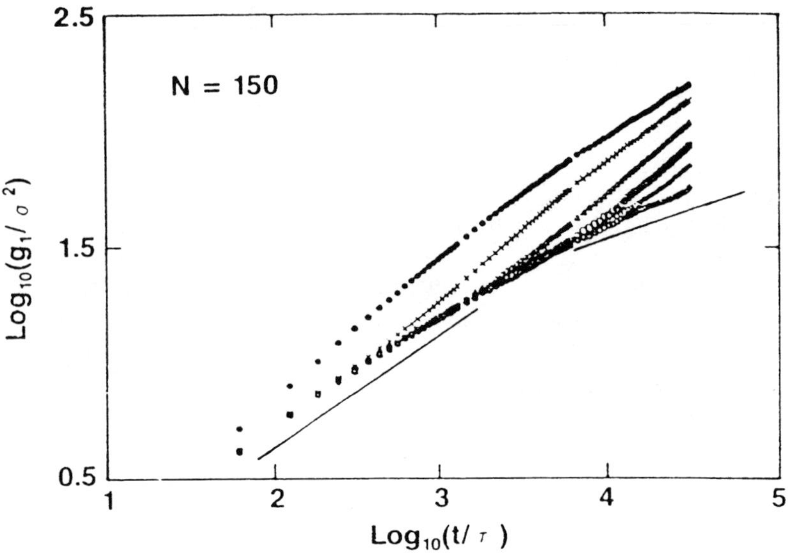

Fig. 4.8 Variation of the mean-square displacements of monomers as a function of their position along the chains for $N = 150$, from Ref. 54. The data progressively give monomers from the chain ends (upper curve) inward with an increment of 10.

along the chain clearly shows this effect. Only for larger times do we find the typical crossover to a slower motion. It is important to notice, that this crossover occurs within the error bars at the same times and monomer displacements, independent of chain length for the longer chains. This indicates a characteristic length N_e, the entanglement length, and time τ_e, the Rouse relaxation time of a subchain of length N_e. For intermediate times the slope in Fig. 4.7 is around 0.28, a little larger than the $t^{1/4}$, expected from the reptation model. Using for the crossover $g_1(\tau_e) = 2\langle R_g^2(N_e)\rangle$ the data of Fig. 4.7 give $N_e \simeq 35$ and $\tau_e \simeq 1800\tau$ for the MD data.

By the same approach one can find N_e and τ_e for the other simulations. The observed slope in the intermediate time regime for middle monomers on a chain of length 800 on the simple cubic lattice MC simulation[58] at $\Phi = 0.5$ is 0.28, while for the bond fluctuation MC simulation[12] at $\Phi = 0.5$ for $N = 200$, the slope is 0.30. Results for the hybrid method of Schulz *et al.*,[166] had a larger error due to the limited number of chains for $N = 200$. They quote a value for the slope as 0.25 ± 0.1. Thus even though the three models are very different in detail, they all are for $N/N_e \simeq 6 - 7$ and give approximately the same result for the exponent as the MD simulations. The interesting question for the future is whether for longer chains, is the power 1/4 or somewhat larger. Within the reptation model one expects the $t^{1/4}$ regime to stretch over a time range up to the Rouse time of the chains. From the MD simulations only the data for $N = 150$ are for sufficiently

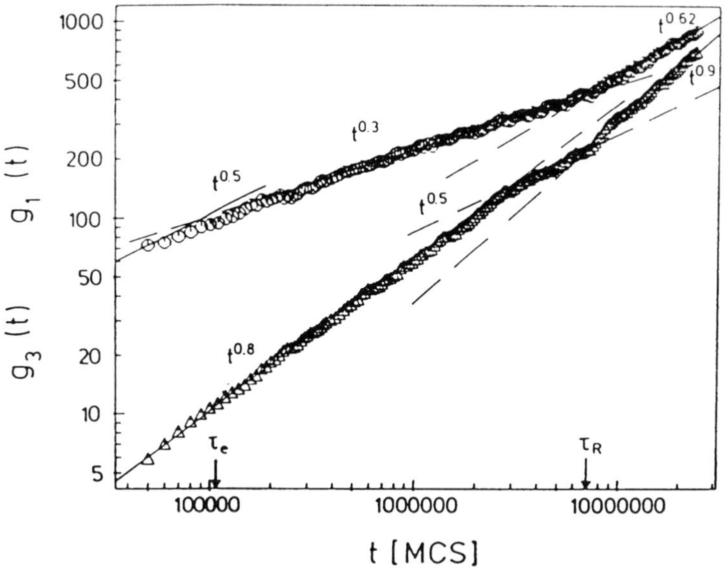

Fig. 4.9 Mean-square displacements $g_1(t)$ of the inner monomers and $g_3(t)$ of the center of mass of the chain of $N = 200$ at density $\Phi = 0.5$ from bond fluctuation MC.[12] Here the region around τ_N is shown.

long runs that one could look for this, though the data are not good enough to derive a reliable estimate of the time ratio $\tau_N/\tau_e = N^2/N_e^2$. This prediction can more clearly be tested using the bond fluctuation data as shown in Fig. 4.9. The data in Fig. 4.9 are accurate enough to clearly exhibit a crossover towards a steeper slope at the Rouse time τ_N of the chains. Using $N_e = 30$ for $\Phi = 0.5$, the ratio between the crossover times towards the slowing down and the increase of the slope of $g_1(t)$ should be given by $(N/N_e)^2 = (200/30)^2 \approx 44$, which is in excellent agreement with the data. Even more importantly the data are accurate enough to display the slowing down in the motion of the center-of-mass of the chains. After the initial effective slope of 0.8 it seems to decay to approximately 0.5 after it again increases to a value slightly below 1 at about the Rouse time of the chains, as expected by the reptation model.

These bond fluctuation data[12,59] were the first to give any numerical indication of the expected second $t^{1/2}$ regime in the motion of the middle monomers. More recently Schulz *et al.*,[166] used a combination of MD and MC for very similar model parameters as in the bond fluctuation simulations of Paul *et al.*,[12,59] They improved the statistics significantly. Figure 4.10 shows their results for $N = 200$.[166] Included are the mean-square displacements of the end monomers, the middle monomers and the center of mass. Because of the longer runs, the existence of a second $t^{1/2}$ regime is

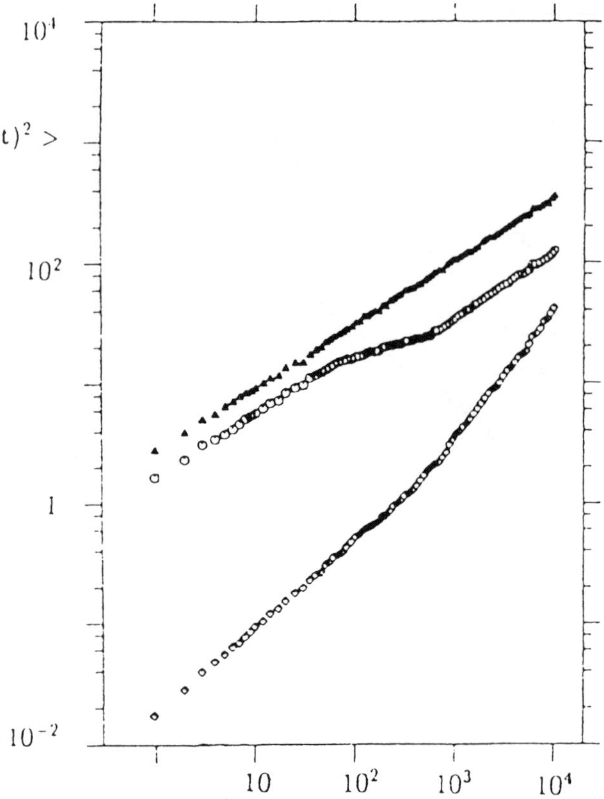

Fig. 4.10 Mean-square displacement of the outer monomers, middle monomers and the center of mass for chains of length $N = 200$ from the hybrid bond fluctuation MC–MD simulation of Schulz *et al.*[166]

clearly seen. Also the crossover of $g_3(t)$ towards the t^1 regime at the Rouse time (not the longest relaxation time of the system!) is observed as in other simulations.[12] However these results have to be taken with some reservation. As shown in Fig. 4.6, the general accuracy is fairly low. The systems sizes used were so small that chains of length $N \geq 100$ overlap with themselves, since only $NM \approx 1700$ monomers are considered. This is compared to systems of size up to $MN = 20\,000$ used by Kremer and Grest.[54] It is not clear to what extent this affects the large-scale motion. The length of the $t^{1/4}$ regime is expected to give a good indication of the entanglement chain length N_e. An interesting point however is raised by the length of the second $t^{1/2}$ regime. As discussed before this second regime is expected to last up to $\tau_d = N/N_e\tau_R \sim N^3/N_e$, which is less than a decade for $N = (6-7)N_e$. However as shown in Figs. 4.9 and 4.10 this regime is longer than expected. This increased length might be an indication for long lived density fluctuations suggested by Semenov[114] and Rubinstein and Obukhov.[113] However in both cases the number of chains was relatively small as only eight or nine chains are considered for $N = 200$.[166] Kremer and Grest[54] also studied a system of $N = 10$ chains of $N = 400$, where it was found that the motion at larger times was severely affected by the small number of chains. Since the diffusion is calculated in the center-of-mass reference frame of the whole system, this reduces to a system with a very small number of particles. Thus, although it is tempting to speculate what this data really mean, the results for longer time scales from this simulation have to be taken with some reservations.

A very important effect which has been observed in all simulations is the characteristic short time behavior of the center-of-mass displacement $g_3(t)$. All methods, including the MD,[54,131] the bond fluctuation MC,[12,59] and the hybrid method,[166] give for short times a $t^{0.8}$ power law instead of the expected Rouse result of t^1. The origin of this clear deviation from Rouse is not understood. It should be noted that this is certainly a many-chain effect, since earlier investigations on free chains or chains in a frozen environment do not show this deviation.

The data show from the mean-square displacements a clear signature of the existence of a unique length scale $d_T^2 \propto N_e$. This should also show up in the Rouse spectrum of the chains. In Ref. 54 the Rouse modes of the chains were analyzed using eq. (4.6) with $b' = 1$. Though this term is not needed for free chains, the introduced error is expected to be very small, of the order $1/N$. For $N/p < N_e$ one would expect standard Rouse behavior, as given by eq. (4.10). This is observed for short chains[54] as shown in Fig. 4.11. The observed friction coefficient $\zeta = (17 \pm 2)\tau^{-1}$ for the MD simulations is in excellent agreement with the value of $\zeta = 16\tau^{-1}$ obtained from the diffusion constant.[54] (Note that the scattering function for both experiment and simulation give a different monomeric frictions coefficient. To be consistent one has to compare the results from the modes and the diffusion.) There is, however, some deviation from the single exponential decay for the very short times. For longer chains much stronger deviations from the Rouse behavior occur.

Although the data[54] tend to agree more with the reptation model than other models, it is difficult to distinguish on this basis alone. For the reptation model the relaxation of the modes is expected to follow eq. (4.10) for times up to τ_e for $N/p > N_e$. Then the relaxation time should increase by a factor N/N_e with a plateau in between. The plateau value is expected to be around $\exp(-\tau_{\text{Rouse}}(N_e)/\tau_p)$ and last up to the disentanglement time. We should mention here that this time is dependent on the mode number p, as was shown for the case of chain in a straight tube.[180] Here one however has to deal with the problem that for the present N/N_e values, the free chain ends dominate the relaxation spectrum, as seen above for the mean-square displacements. Thus we used an approach similar to the idea employed in

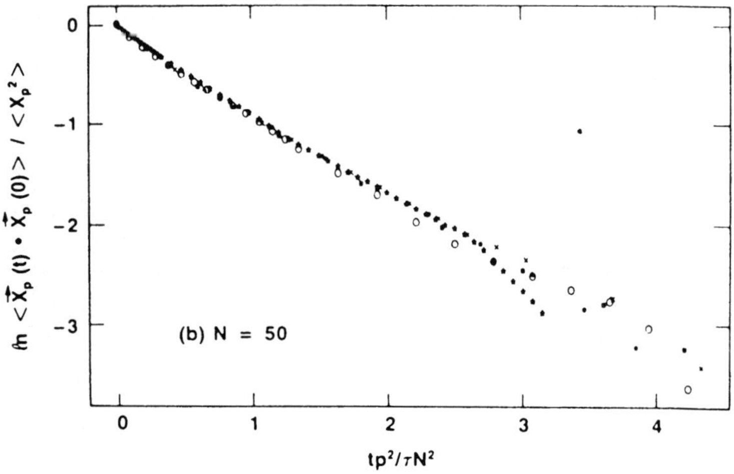

Fig. 4.11 Autocorrelation function of the first four Rouse modes of a chain with $N/N_e < 1, N = 20$ (from Ref. 54).

$g_1(t)$ for middle monomers for the Rouse modes. Since each subchain within the walk is a random walk itself, one can construct Rouse modes from the inner part of the chain. Following eq. (4.6) one can calculate

$$\mathbf{X}_{in,p}(t) = \frac{1}{N'} \sum_{i=N/2-N'/2}^{N/2+N'/2-1} \mathbf{r}_i(t) \cos\frac{p\pi(i-1)}{N'-1} - \frac{1}{2N'}[\mathbf{r}_{i_0}(t) - \mathbf{r}_{i_{end}}(t)] \qquad (4.31)$$

where $N' = N/p$ and i_0 and i_{end} are the first and the last monomer of the center section of the original chain. The $(-1)^p$ reduces to -1 in the last term of eq. (4.31) since one always deals with the first mode of the middle subchain of length $N' = N/p$. Results for the middle sections from the $M = 100$, $N = 200$ system were analyzed for $N' = 30$ to 100.[131] In Fig. 4.12, the Rouse scaling plot for the mode relaxation is shown. Although we do not find a single exponential decay in the Rouse regime, all the data collapse reasonably well onto one common curve. At a time which is approximately the same for all modes, the individual relaxation functions deviate from the common Rouse curve towards a much slower decay. This time is about a factor of 4–5 larger than τ_e, as deduced from the mean-square displacement $g_1(t)$ (Fig. 4.12). This scale factor for the time is in agreement with the ideal Rouse model and the prediction of the reptation model.[12,59] However the strong nonsingular, exponential character for shorter times is surprising and not completely understood. The fact that the crossover time for the modes is well defined and mode independent, rules out the coupling scheme put forward by Ngai *et al.*[186] They propose a smooth p-dependent crossover.

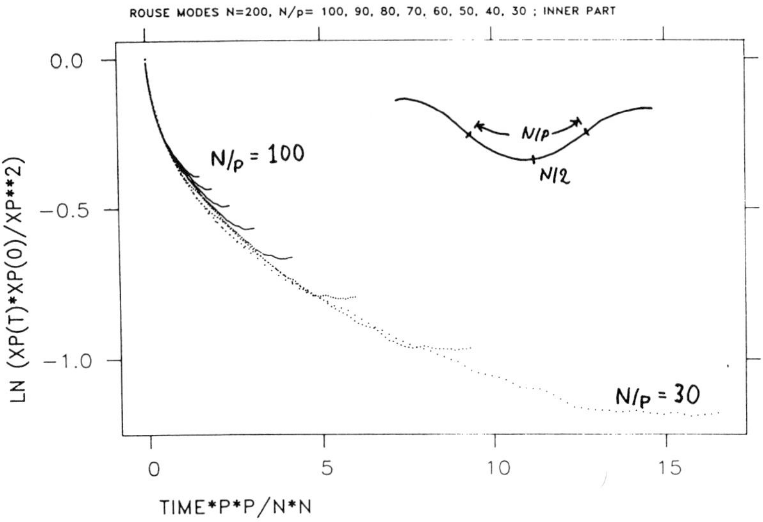

Fig. 4.12 Autocorrelation function of the Rouse modes of an inner subchain, as defined in eq. (4.10) for $N = 200$ from the MD data of Ref. 54. The construction is indicated in the figure (from Ref. 131).

So far all the evidence from simulations supports the reptation model or its variants very strongly. Although the chains are relatively short, we can still ask to what extent do the chains move along the tube given by the coarse-grained contour of the chain. It is clear that one here is only able to observe the very onset of this motion. For the original chain of $N = 200$ for the MD data the tube diameter d_T is only $(N/N_e)^{1/2} \approx 2.4$ times smaller than the mean end-to-end distance of the chain itself. To see the confinement directly, one can construct a primitive chain (PC) as[10,54,153]

$$\mathbf{R}_1 = \frac{1}{N_e}\sum_{i=1}^{N_e} \mathbf{r}_i, \quad \mathbf{R}_2 = \frac{1}{N_e}\sum_{i=3}^{N_e+2} \mathbf{r}_i, \quad \text{etc.} \tag{4.32}$$

The PC feels the constraints of the tube more strongly than the individual monomers. From $g_1(t)$, now evaluated for the PC, one finds a tube diameter $d_{T,PC} \approx 5\sigma$.[54] If one now simply plots conformations of the PC, separated by a constant time increment, we can directly observe the confinement into the tube. Figure 4.13 gives plot of 20 configurations of the PC for one chain spaced 1200 τ apart for a total elapsed time of $24\,000\tau$ (about $13\tau_e$) for $N = 200$. The other chains are not shown for clarity. For this time we expect an inner monomer of the PC to travel a distance of $(3g_1(\tau_{max}))^{1/2} \approx 13\sigma$ compared to the tube diameter $d_t \approx (3g_1(\tau_e))^{1/2} = 5\sigma$ if the chains did not reptate.[54] The figure clearly shows the confinement of the chain into a tube, showing visually that the reptation concept qualitatively gives a profound picture of the underlying physical mechanism. One can also look at a variety

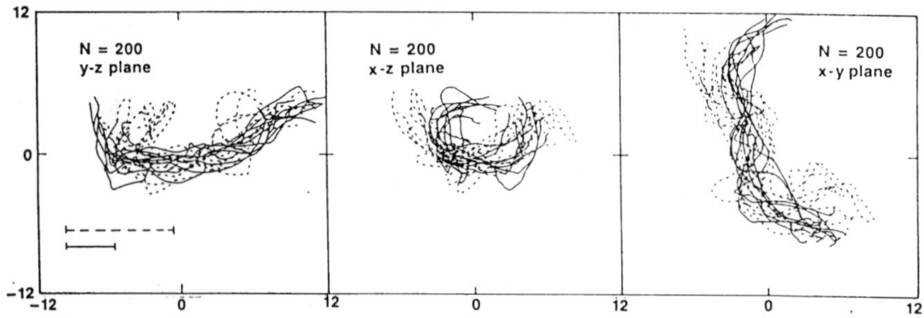

Fig. 4.13 Plot of the configurations of the PC of a chain of $N = 200$ from MD simulations. The two bars indicate the estimated tube diameter and the estimated motion distance from isotropic motion (from Ref. 54).

of other properties of the primitive chain as was done in Ref. 54. Since the motion is anisotropic, the data do not agree to the mode coupling approach of Schweizer.[36–38] Kolinski *et al.*[153] show similar plots and perform a similar analysis of the PC for their lattice MC simulation. However they claim that even for their longest chain of $N = 800$, $N/N_e \approx 7$, on the sc lattice there is no evidence for reptation. We strongly disagree with their interpretation, since they disregard the tube diameter and the enhanced mobility of the ends.

Recently the simulations discussed above have been extended by Wittmer *et al.*[155] In one part of their investigation, the dynamic structure function $S(q,t)/S(q,0)$ for bond fluctuation chains of length $N = 200$ at a density $\Phi = 0.5$ was studied. For the earlier MD results[54] on chains of $N \leq 150$ the influence of the ends was too strong to observe a clean indication of a crossover from Rouse to reptation, since $N/N_e \approx 4.3$ while for the MC chains $N/N_e \approx 6.6$. Since the enhanced mobility of the ends reaches about 1 to 1.5 entanglement lengths into the chains (cf. Figs. 4.8 and 4.13), the longest MD chains ($N = 150$) analyzed in this way were rather short, while the effective middle section in the MC case is significantly longer. In the Rouse model, de Gennes[187] showed that for long chains

$$S(q,t)/S(q,0) = f(q^2\sqrt{t}) \tag{4.33}$$

Closed form expressions for $S(q,t)$ in the Rouse model have been derived,[187,188] but they are time consuming to evaluate. Recently Grayce[189] has developed an approximate form which works quite well. For times after τ_e for $N > N_e$, there is a a q-dependent splitting towards a plateau of

$$S(q,t)/S(q,0) = 1 - \frac{q^2 d_T^2}{36}, \quad \tau \ll t \ll \tau_d. \tag{4.34}$$

This is essentially the Debye–Waller factor of the chain smearing out the tube. This asymptotic form was first derived by deGennes.[16,190] For intermediate times, early simulations of a chain in a frozen network[9] as well as recent neutron spin-echo experiments[135] showed that the crossover towards the asymptotic form is reasonably well fitted by Ronca's self-consistent approach.[31] Ronca introduced a memory kernel into the equation of motion in a self-consistent way, which led to a memory term favoring motion along the contour. In his theory he calculated the dynamic scattering function of the chains as a function of qd_T. While earlier spin-echo data[135,191] for poly (ethylene-propylene) (PEP) and polyisoprene (PI) appeared to agree with the Ronca model, more recent results for $S(q, t)$ on PE (hydrogenated polybutadiene) are incompatible with Ronca's predictions.[135] Recently des Cloizeaux[192] has introduced the concept of entanglements or stress points into the original de Gennes calculation of $S(q, t)$. This new result fits the experimental data for PE better than does Ronca's model. Chatterjee and Loring[43] have obtained semiqualitative agreement with Richter *et al.*[135] results for PE using their model which assumes a cooperative slithering snake motion for a chain.[41,42]

Results for the scattering function $S(q, t)$ for $N = 200$, $\Phi = 0.5$ from bond fluctuation MC is shown in Fig. 4.14.[155] The data were fit very well by Ronca's model with a single q-independent tube diameter $d_T = 28$, which is systematically too large. One should, however, keep in mind that the Ronca scheme describes the crossover region reasonably well. We found similar results for the MD simulation data for $N = 350$.

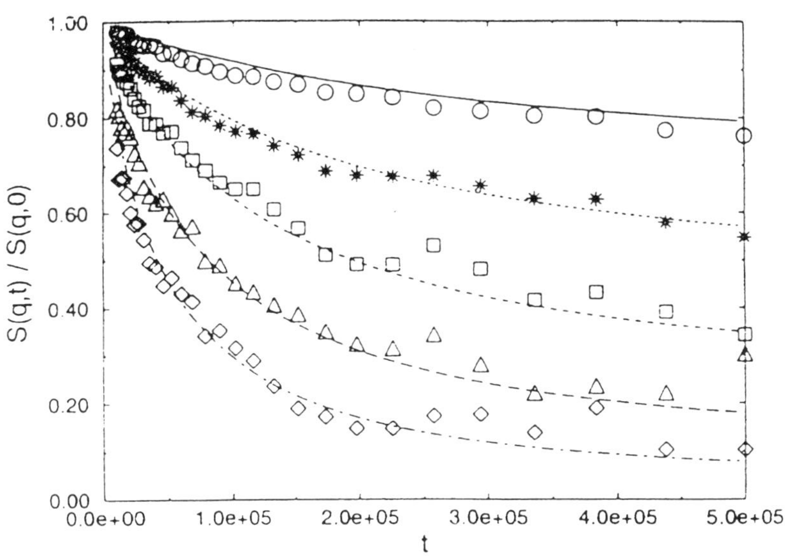

Fig. 4.14 The dynamic structure factor $S(q, t)$ of the chains for the bond fluctuation Monte Carlo data for $q = 0.15(\circ), 0.2(*), 0.25\ (\square), 0.3\ (\triangle)$, and $0.35\ (\diamond)$ (from Ref. 155). The fit is to the Ronca model.[31]

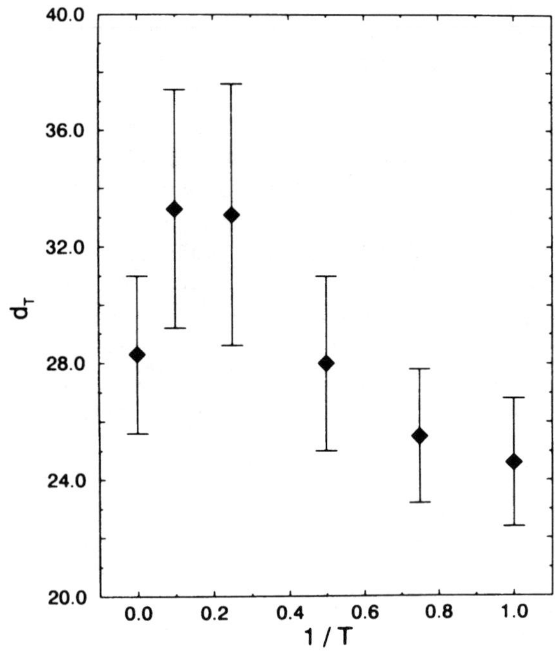

Fig. 4.15 Temperature-dependence of the tube diameter obtained from fitting the dynamic structure factor. $T = \infty$ corresponds to the standard bond fluctuation model. (From Ref. 155).

In the second part of this investigation, an inital attempt was made to calculate N_e as a function of the chain stiffness. Wittmer *et al.*[155] introduced a bond angle potential into the bond fluctuation model. They used a bond potential $U(\alpha) = \epsilon' \cos \alpha$, where α is the angle between subsequent bonds. Using $\epsilon' = 1, k_B = 1$ the interaction strength is given by the inverse temperature $1/T$. $1/T = 0$ corresponds to the standard MC model. As a function of T, C_∞ varies between 1.3 ($1/T = 0$) and 7.0 ($1/T = 5$). In order to avoid artifacts of the lattice, the dynamics were only analyzed for $0 \leq 1/T \leq 1 (1.3 \leq C_\infty \leq 2.06)$. In Fig. 4.15 the variation of the tube diameter with temperature is shown. To be consistent, d_T was always taken from a Ronca fit of $S(q, t)$. The change in stiffness also changes d_T as increasing the stiffness seems to reduce d_T. One should keep in mind, that the change in chain stiffness in this model does not affect the bond length fluctuations. Recently Wittmer *et al.*[193] reanalyzed their data in order to compare the dynamic scattering function of their chains of length 6–7N_e to the predictions of des Cloizeaux.[192] Within the limitations of the short chains compared to the asymptotic predictions of the theory, the agreement is satisfactory. Another interesting observation was made recently by Kremer,[194] who introduced a short-range attractive well potential which only affected nearest neighbors in space, into the MD model. Since the potential was set to zero at the first minimum of the radial distribution function $g(r)$, the conformational properties of the chains to did not change. He found however that $g_1(t)$ deviated from the Rouse model at earlier times and shorter distances as the well depth was increased. It is not clear at present to what extent this really changes N_e or whether this is just a short time effect due to the monomer attraction, which is absent in all of the previously discussed cases. There is also some experimental indication for such short time effects in PEO.[195] Thus in spite of considerable effort it is still not known what an entanglement means physically, although the reptation model appears to work quite well.

So far we have only discussed a variety of different equilibrium simulations, which try to follow the motion of individual chains. Kröger and coworkers[196–198] used the MD simulation model of Grest and Kremer to perform NEMD (nonequilibrium MD) simulations, where the sample is subject to a steady shear or stress. They were able to cover chains of up to $N = 100$ monomers or equivalently about $3N_e$. While this is too short to study e.g., the viscosity or the shear alignment of highly entangled chains, changes from Rouse to the entangled regime were observed. Such methods will probably become much more important with the coming faster generation of computers.

4.4.3 Comparison to experiment

The simulations discussed above demonstrate that the reptation model describes the motion of a polymer chain in a melt rather well. The entanglement length in the simulations in most cases is determined from the mean square displacement of the middle monomers. Experimentally N_e is usually determined by the plateau modulus for very long chains ($N \gg N_e$). Figure 4.6 shows that both methods give the same result. The underlying assumption is that there is no difference between the apparent short time entanglement length N_e and the long time limit, which is given by the plateau modulus and could be accounted for as the "true" N_e. For the above models as well as PE this seems to be the case. It is not absolutely clear whether this is generally true. As long as there are no strong local, attractive forces between different groups within the chains, one does not expect a significant deviation. For other cases however, e.g., for chains with strongly polarizable groups along the backbone such as poly(ethylene oxide) (PEO), or V for PS, this might not be the case. Recent simulations[194] with strong short-range potentials reveal problems in this direction as well as some experimental hints from PEO.[195] For the subsequent analysis and discussion we assume however that there is only one characteristic length N_e.

Under the above assumption one now is able to make quantitative predictions for a variety of different polymers from the simulation results. To do this, one needs to map the model chain onto chemical species. A very similar mapping is necessary to compare experimental results from one system to another.

When one discusses the static properties of polymers, the persistence lengths are usually mapped onto each other. However, this mapping is not unique and does not take into account any dynamics. The persistence length mapping only gives the limit of the smallest statistical unit one can compare. It is also known from recent experiments that simply mapping the persistence lengths is not sufficient.[124] To make the mapping unique, one needs additional information, which fixes the mapping in a way so that both static and dynamic properties can be interpreted. From the previous discussions, it should be clear that the obvious characteristic length in the system is the entanglement length, N_e, or entanglement molecular weight, M_e. A reasonable procedure is to compare chains in units of N_e. Two chains are equivalent if N/N_e is the same. This is the procedure, which was already adopted to compare all the different simulation methods. To include dynamics, one has to identify the time scale between simulation and experiment. For simulations a convenient way to do this, is to compare the extrapolated short chain diffusion constants, by plotting D/D_{Rouse} versus N/N_e as is shown in Fig. 4.6. One simply equates the Rouse diffusion constant of different experiments or simulations of the equivalent number of monomers (N_e, M_e) with each other. For systems where the diffusion constant is not available, one can compare the bead friction, which determines the viscosity in the Rouse limit. Between two simulations a simple mapping can be done by comparing the time at which $g_1(t)$ crosses over from $t^{1/2}$ to the $t^{1/4}$ regime. Once this mapping has been done, it is possible to use the simulation results to predict the crossover time τ_e, the crossover distance $g_1(\tau_e)$, or the tube diameter d_T. For polymers whose monomeric units do not contain large side groups or strongly polarizable parts we expect this mapping to be quite accurate.

In Table 4.1 we present the results of mapping the MD simulation results to other simulations (MC)*† and to experimental data for PS, PE, PDMS, PI, PEP, and polytetrahydroflurane (PTHF) at a variety of temperatures.[54] The temperatures chosen are limited due to the limited availability of experimental data to do the time mapping. The simplest way to do the mapping and estimate the time and length scales is to equate the Rouse diffusion constant for an equivalent number of monomers. With the knowledge of

*The data of Schulz *et al.*[166] are excluded from Table 4.1 since not enough details on the precise procedure are given. The bond fluctuation MC part of their simulation is at a density $\Phi = 0.5$, for which Paul *et al.*[12,59] found $N_e = 30$. However Schulz *et al.*[166] find $N_e = 40$, which suggest that they are mixing two different models with somewhat different entanglement lengths.
†For the data of Skolnick *et al.*,[58,153] the Rouse diffusion constant is the same as used for Fig. 4.6. It should be noted that the time definition, which incorporates an arbitrary prefactor for the MC simulations of Skolnick *et al.* contains about five attempted moves per monomer, while the usual definition uses one attempted move, as indicated in the table.

Table 4.1 *Estimates of d_T and τ_e based on the results from the MD,[54] bond fluctuation MC,[12,59] and standard lattice MC simulations[58,153] for a variety of polymers which have been investigated by neutron spin-echo. Either the monomer friction constant as measured by short chain viscosity or the Rouse diffusion constant and the entanglement length N_e (or better $< R^2(N_e) >^{1/2}$) are used in the mapping.*

System	T	Monomer mass	M_e	Equivalent no. of beads	Equivalent mol. mass	ll_p	$l\sigma =$	d_τ	$[g_1(\tau_e)]^{1/2}$	$1\tau =$	τ_e
MD system	$1\epsilon/K_B$	1	35	1	1	1.3σ	...	7.7σ	4.5σ	...	1800τ
PS[a]	485 K	104	18 000	4.95	515	7.4 Å	12.6 Å	97 Å	56 Å	3.1×10^{-8} s	5.5×10^{-5} s
PE[b]	448 K	14	1350	2.76	38.6	4 Å	5.1 Å	39 Å	23 Å	6.6×10^{-12} s	1.2×10^{-8} s
PDMS[c]	300 K	74	9000	3.47	257	6.2 Å	8.7 Å	68 Å	39 Å	2.3×10^{-10} s	4.1×10^{-7} s
PDMS[c]	373 K	74	$\approx$ 11 500	4.44	328	5.9 Å	8.3 Å	64 Å	37 Å	9.5×10^{-11} s	1.7×10^{-7} s
PTHF[d]	418 K	72	1440	0.57	41	8 Å	4.6 Å	35 Å	21 Å	1.8×10^{-12} s	3.2×10^{-9} s
PEP[e]	492 K	70	2950	1.20	84	7.74 Å	6.5 Å	50 Å	29 Å	0.55×10^{-11} s	1.0×10^{-8} s
PI[e]	307 K	68	4100	1.73	117	6.6 Å	6.7 Å	51 Å	30 Å	1.0×10^{-10} s	1.8×10^{-7} s
BF–MC[5]											
$\phi = 50\%$	—	1	30	0.86	—	3.16	2.65	17.3	10.0	58.1 MCS	$1.04 \cdot 10^5$ MCS
$\phi = 40\%$	—	1	40	1.14	—	3.22	2.25	20.4	11.7	60.65 MCS	$1.09 \cdot 10^5$ MCS
SC–MC[132]											
$\phi = 50\%$	—	1	110	3.14	—	1.87	1.86	14.3	8.3	6.8 MCS*5	$1.22 \cdot 10^4$ MCS*5

[a]For PS ll_p are taken from Ferry (Ref. 14) for PS in the Θ solvent. For the mapping the bead friction due to diffusion constant of Antonietti *et al.* (Ref. 82) was used with a consistency check with ξ_o from viscoelasticity gave only $\sim 10\%$ deviation! Note that $N_e = 50$ would change the time mapping, giving $\tau_e \approx 2.6 \times 10^{-5}$ s.

[b]Persistence length as given by Flory (Ref. 1) M_e and the diffusion constant taken from Pearson *et al.* (Refs 74, 75)

[c]Persistence length and bead friction due to Ferry (Ref. 14). Note that the viscoelastic bead friction is obtained from entangled polymers. Thus for the Rouse diffusion equation used for comparison, we use the long-chain bead friction (Section 4.4). The data are for $T = 274$ K. For $T = 373$ K, τ_e would change to 1.0×10^{-7} s because of the reduced bead friction. This reduction however is partially compensated for by the increase in M_e.M_e data estimated from Ref. 126.

[d]Persistence length, M_e and bead friction due (Ref. 79) and Pearson (private communication). Here also the long-chain bead friction has to be taken.

[e] ξ, ll_p, and M_e from Fetters (private communication). For $\xi(PI), N_e$ was taken to be $2.5N_e, 2N_e$ would reduce τ_e by about 30%.

$C_\infty l^2$ and N_e, M_e these scales are fixed. This procedure was used for the data of Table 4.1. A direct and simple test is to compare τ_e for the bond fluctuation simulation from Table 4.1 with Fig. 4.9. The agreement and thus the consistency of different simulations is excellent. From this mapping, one finds that the crossover time τ_e varies considerably from one polymer to another as one would expect. Some typical values for τ_e are 5.5×10^{-5} s for PS at 485 K compared to 3.2×10^{-9} s for PTHF at a comparable temperature, 500 K. For PDMS at 273 K, $\tau_e = 4.1 \times 10^{-7}$ s while it decreases to 1.7×10^{-7} s at 373 K. This sheds some light on the long-standing discussion about whether neutron spin-echo scattering could be used to observe the predicted plateaus in $S(q, t)$ or not. The first spin-echo experiments were for PDMS[6,78] and PTHF.[77] From the estimates of Ref. 54 of τ_e for the temperatures of the experiments it became clear that the neutron spin-echo experiments[77] on PTHF should have seen a deviation from the Rouse model which they did. However for PDMS this was not the case, since the times[54] were beyond the resolution of the experiment, which was around 10^{-8} seconds at the temperature used.[6] In both cases the q-range was sufficient in spite of some early concerns.[6,199] Later experiments on PEP,[124,135,191] PI,[135] and PE (PEB-2)[124,135] all showed a well defined crossover from Rouse to reptation-like behavior at times which were well described by the mapping of the simulation results. If neither the Rouse diffusion constant, nor the friction coefficient from rheology is known, one can also use the short time/small chain decay rate of the Rouse regime of $S(q, t)/S(q, 0)$, since this can directly be used to determine ζ. However one should keep in mind that using the Rouse equations, the scattering function gives a friction constant, which is too large by about a factor of two[54] compared to diffusion measurements.

One of the systems investigated by Richter *et al.*[124,135,191] were well characterized PEP samples. They used an improved spin-echo machine which could go to a maximum time of 4×10^{-8} s at $T = 492$ K. For PEP we predicted a crossover time of 1.0×10^{-8} s, which is within the resolution of the new instrument. For times longer than τ_e but shorter than τ_d (the actual longest time[180] depends on the wave vector q) the single chain scattering function $S(q, t)$ should deviate from the simple Rouse decay and approach a constant $S(q, t \gg \tau_e) = 1 - q^2 d_T^2/36$ as discussed above. If one extrapolates from this plateau value to short times, the crossing time with the Rouse relaxation curve should be τ_e. In Fig. 4.16 we show the experimental result for two scattering vectors for PEP at $T = 492$ K. The agreement to the data is excellent, displaying the overall consistency of the numerical and experimental results combined with the reptation concept.

While there are rheology and scattering data for other polymers, they are usually for different temperatures. The rheology data often end at a temperature where the neutron spin-echo data are just beginning, since the maximum time for the spin-echo is relatively short. Thus a good consistency

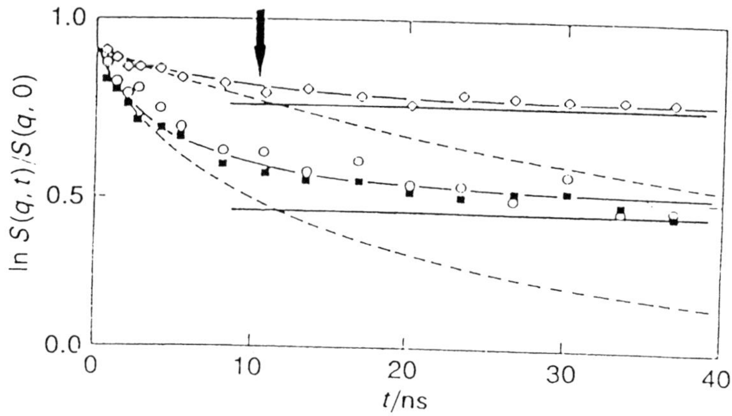

Fig. 4.16 Dynamic structure function $S(q, t)$ from neutron spin-echo scattering[193] for PEP compared to the crossover time extracted from the simulation. The crossing of the Rouse curve and the plateau level gives the approximate experimental crossover time. (From Ref. 131).

check is difficult. For Table 4.2 we used the extended set of scattering data of PEB-2[124] which gives d_T and C_∞. This was combined with the friction constant from diffusion experiments[74] and the time scaling for τ_e from the MD simulations[54] to determine τ_e for a variety of temperatures. Note that τ_e only decreases about a factor 3 from 556 K to 418 K even though the momeric friction coefficent ζ changes by a factor of nearly 6. This change in ζ is partially compensated by a reduction in C_∞ and an increase in N_e with increasing T. The small deviation between the results for PE shown in Tables 4.1 and 4.2, where d_T is determined from two different methods, rheology and scattering, respectively, illustrates the general accuracy of the mapping as well as different experimental approaches.

4.4.4 Semidilute solutions

So far in this review, we have confined our attention to dense melts, where we found good agreement to the reptation model. For short times, however, not all the data fit to the Rouse model perfectly. One way to examine this in more detail is to study crossover from solution to melt in the free draining limit as a function of density. Experimentally this certainly is not possible, because of the effects of hydrodynamics, which influence the dynamics very strongly. The bond fluctuation algorithm was used because at the relatively low densities of interest the MD is not as suitable.[151]

The absence of any hydrodynamic interaction allows one directly to ask how the entanglement length scales as a function of density. Long-range hydrodynamic interactions in real solutions make this problem more complicated. One would like to know how N_e scales with the static excluded

Table 4.2 *Temperature variation of τ_e for PEB-2.*

T (K)	C_∞	$\zeta[\frac{dyns}{cm}] \cdot 10^9$	N_e	1σ[Å]	$\tau_e \cdot 10^8$ s	d_T[Å]
418	6.52	4.70	98	5.0	6.4	38.5
446	6.32	2.84	109	5.17	4.3	39.8
463	6.19	2.20	125	5.48	4.15	42.2
484	6.05	1.67	120	5.31	2.7	40.9
509	5.87	1.25	140	5.65	2.5	43.5
529	5.74	1.03	147	5.73	2.2	44.1
556	5.57	0.81	164	5.95	2.0	45.8

The entanglement length, tube diameter d_T and C_∞ are taken from Ref. 124. The monomeric friction coefficient ζ is taken from the Vogel–Fulcher form of PE from Ref. 74, since there is little difference between PEB-2 and PE. Pearson *et al.*[74] fit their data for PE to the form $\zeta = \zeta^\infty \exp(B/\alpha(T - T_o))$ with $\zeta^\infty = 3.7 \times 10^{-11}$ dyn-s/cm, $B = 0.6, \alpha = 5.1 \times 10^{-4}/K$ and $T_o = 175$ K, which are the asymptotic values for large N. A bond length $l = 1.52$Å was used. For the determination of τ_e, the results from the MD simulations, $\tau_e = 1800\,\tau$ and $d_T =< R^2(N_e) >^{1/2}= 7.7\,\sigma$, are used, as described in the text. Using the short chain results for ξ, as for Table 4.1, would reduce our estimate of τ_e by approximately a factor of 1.5–2.

volume screening length ξ. In a good solvent semidilute solution ξ is given by the average strand–strand distance. At a monomer density Φ, $\xi \propto \Phi^{-\nu/(3\nu-1)}$, where ν is the good solvent exponent of the end-to-end distance $R \propto R_G \propto N^{\nu}$. On length scales up to ξ the chains show the single chain good solvent behavior with an exponent $\nu = 0.588$.[200] For scales larger than ξ one has a melt of blobs of diameter ξ and the chains behave like random walks with $\nu = 0.5$. At the density $\Phi^* \propto N^{-(3\nu-1)}$, where R_G and ξ are comparable, the coils just start to overlap.

In order to analyze the crossover scaling properly one has to take into account the proper reference length. For a fluctuating bond length the unit of length is the average bond length and not one lattice constant as in the standard lattice MC simulations. Physically the variation of ll_p shows that the number of monomers in each correlation volume depends not only on Φ but also on $ll_p(\Phi)$. This shifts Φ^* in addition to the standard scaling for fixed ll_p. Identifying $< l^2(\Phi) >^{1/2}$ as the basic unit of length, the crossover scaling function for the radius of gyration, and similarly for the end-to-end distance, can be written as

$$\frac{< R_G^2(N) >}{(N-1)^{2\nu} < l^2 >_{\Phi=0}} =< l^2(\Phi) > f_{RG}[(N-1)(\Phi l^3)^{1/(3\nu-1)}] \tag{4.35}$$

This scaling approach gives a satisfactory scaling, while the naive approach, which does not take the bond length variation into account, does not, even though $< l^2(\Phi) >$ only varied from 6.94 ($\Phi = 0.5$) to 7.47 ($\Phi = 0.025$).[12] In these simulations, the entire crossover from dilute solution to dense melt was covered. The screening lengths (as derived from the chain structure function) ranged from $\xi(\Phi = 0.1) = 21.2$ to $\xi(\Phi = 0.5) = 6.46$ at the highest density. At $\Phi = 0.5\,\xi$ is only equal to the distance between next-nearest neighbor along the chain. Such data should display the crossover from dilute Rouse behavior to dense melt reptation behavior. One assumption in most theories for the dynamics of semidilute systems in the free draining limit is that $N_e(\Phi)$ is just a multiple of the number of monomers of a single strand in a correlation volume (blob volume) ξ^3, namely proportional to $\xi^{-1/\nu}(\Phi)$.

One can then plot the diffusion constant versus N/N_e for different Φ as was done in Fig. 4.6. However the data were found not to scale at all. Similar to the problem with the bond length, the monomeric friction ζ and the mobility $W(\zeta)$ are Φ dependent. Paul *et al.*[12] used the mobility W, determined from the mean-square displacements of the monomers

$$g_1(t) =< l^2 > Wt^{0.54}, \tag{4.36}$$

to set the time scale. The power law for $g_1(t)$ takes the difference between $\nu = 0.59$ for the dilute solution in good solvent and $\nu = 0.5$ for the melt into account.[148] From $g_1(t)$ one can then determine $W(\zeta)$. Increasing density

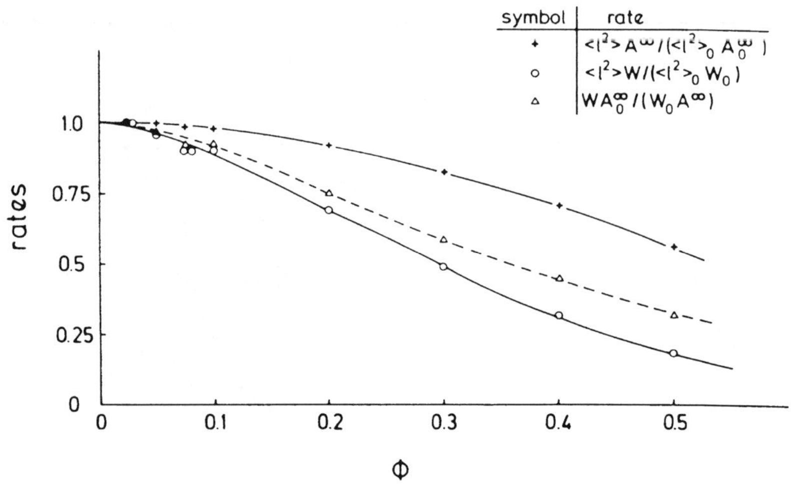

Fig. 4.17 Density dependence of the acceptance rate of the moves $A(\Phi)/A(\Phi = 0)$ and mobility $W(\Phi)/W(\Phi = 0)$ extrapolated to large N (from Ref. 12).

certainly causes a reduction of accepted moves. Thus one might try to interpret W as a simple function of the acceptance rate A for attempted moves. This however is not the case, as Fig. 4.17 shows. The mobility W clearly decays more strongly than the acceptance rate A. In a recent simulation of glassy polymers it was found that while the mobility became zero, the acceptance rate was finite.[201]

The effect of the surrounding reduces the mobility more strongly than the bare acceptance rate, since correlated forward–backward moves are not hindered significantly. Consequently, in a scaling plot of the diffusion constant $D(N, \Phi)$, both the proper length scale normalization, as defined by the variable bond length, and the proper time scale normalization, as given by the density dependent mobility, has to be taken into account. For such a general case, assuming that $N_e = \text{const} \cdot \xi^{-1/\nu}$, namely a constant times the number of monomers per excluded volume screening blob, one can plot (with $l = l(\Phi))D(N, \Phi)N/Wl^2$ versus $N(\Phi l^3)^{1/(3\nu-1)}$ as shown in Fig. 4.18. Again one finds, as for the melts, that Hess's form[19] of the diffusion crossover, which considers an enhanced constraint release mechanism, fits the data well, giving $N_e(\Phi) \approx 500(\Phi l^3)^{-1/(3\nu-1)}$.

At this point we come back to the discussion after eq. (4.26). The data from the MC simulations reveal that $\xi \propto d_T$ which gives $N_e^\nu \propto \xi$. Using a hypothetical value of $\nu = 1/2$, which should be the correct one for e.g., a mixture of short and long chains, one gets $a = 3$, just as for the packing arguments discussed in Section 4.2. For the present athermal simulations, $\nu = 0.588$ since the smeared-out background acts as a good solvent. This gives $a = 2.53$. For experiments one should expect an effective exponent

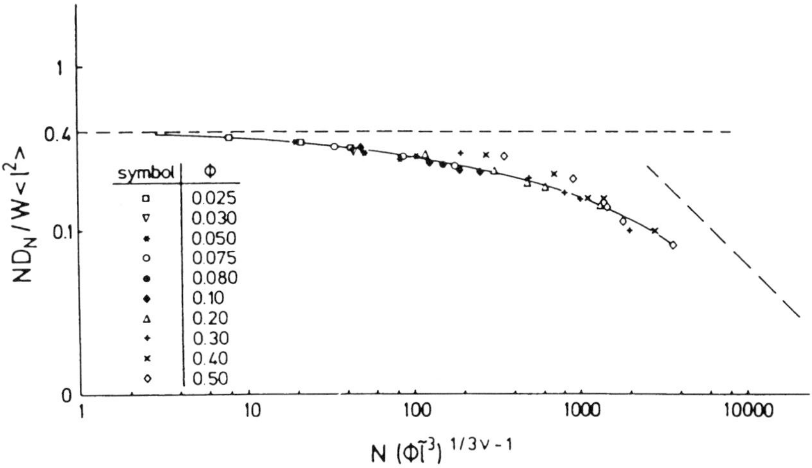

Fig. 4.18 Scaling of the density- and N-dependent diffusion constant with the correct scaling of the lengths and the mobilities (from Ref. 12).

somewhere in between these two values. In those cases in which hydrodynamics can be neglected, the screening length ξ is small and the strand length in a volume ξ^3 exhibits an effective exponent between the two asymptotic limits of ν. An example of such a system is a melt of very short and long chains. There the relevant correlation length is not the excluded volume screening length, but the strand–strand distance of the long chains. From this we can conclude that one cannot expect a unique exponent a from experiments and can give bounds from the simulation, namely $2.53 \geq a \geq 3$. This result agrees very nicely with experiment.

Paul *et al.*[12] extended the above discussion to a variety of different relaxation times τ_x. Here we confine our discussion to the time they call τ_D, which is the time when $g_3(t)/t \approx$ const. This time is determined by the crossover from the slowed-down displacement at shorter times to the asymptotic free diffusion $g_3(t) \propto t$. As seen above $g_3(t)$ starts out as $g_3(t) \propto t^{0.8}$. Compared to the almost perfect Rouse scaling of the mode spectrum, this is a significant difference. The mode spectrum however only contains fluctuations inside the chains but not the overall motion. This effect on $g_3(t)$ probably reflects the difference between the local hopping rates and the mobility, however now on a more global scale than for the monomer motion. Following the Rouse and reptation concept $\tau_D \propto N^2$ independent whether $N < N_e$ or $N > N_e$.

The physical picture, giving $g_3(t > \tau_N) = 6D(N)t$ comes from the diffusion of the chain along the tube. This motion is the same as used in the standard slithering snake algorithm.[202] If we follow Semenov[114] and Rubinstein and Obukhov,[113] this diffusion is retarded due to the long lifetime of density fluctuations. These fluctuations, which allow the chain to reptate out of the tube, should only relax via reptation itself. The data from the present microscopic simulations are not sufficient to clarify this point. However what one can do is to look at the time τ_D, at which $g_3(t)/t$ becomes constant. While τ_D cannot be obtained with very high precision, one can do well enough for the present discussion. The interesting question is whether $\tau_D \approx \tau_N \propto N^2$ or $\tau_D \approx \tau_d \propto N^3 - N^{3.4}$. To test this one can plot the scaled relaxation time $\tau_D W(\Phi)/N^{1+2\nu}$ versus the same scaling variable, x, as for the diffusion constant, $x = (N-1)(\Phi l^3)^{1/(3\nu-1)}$. In the dilute good solvent limit, $\tau_N \propto N^{1+2\nu}$, leading to a plateau for $x \to 0$. In the reptation picture one expects for both the semidilute Rouse regime and the reptation regime a slope of $N^2/N^{1+2\nu} = N^{-0.18}$. If however τ_D shows the same behavior as the reptation relaxation time τ_d one expects for $x \approx 500$ (for the present MC model one has $N_e = 500x^{-1}$)[12] again a turn upwards to the asymptotic $N^3/N^{1+2\nu}$ power. Figure 4.19 indicates the latter.

While there are still many interesting, unsolved problems within the Rouse/reptation regime for polymers, the reptation model clearly gives a good starting point. Improvements to the basic model have been developed and are important for obtaining a better quantitative agreement to the data. Chain entanglements are proven to play a key role. The nature of an entanglement however has not yet been resolved by any these investigations.

4.5 Polymer networks

So far we have discussed the striking effects of the noncrossability of polymers for uncrosslinked polymer melts and for solutions in the free draining

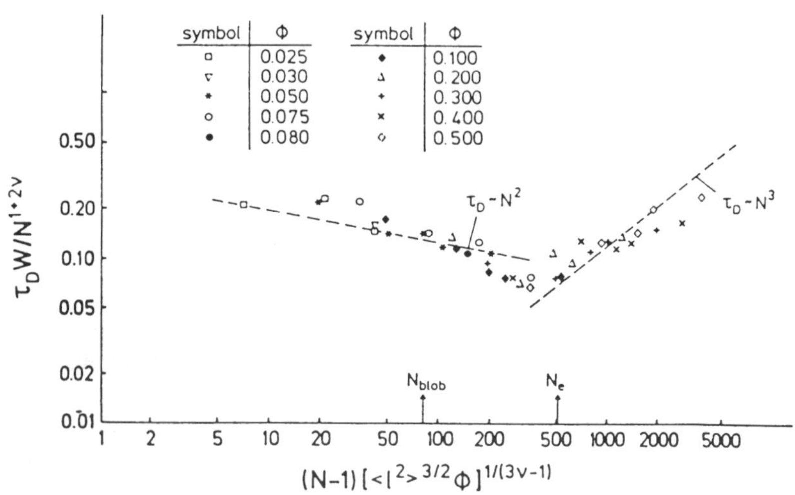

Fig. 4.19 Scaling of the diffusion relaxation time τ_D. (From Ref. 12).

limit. The effects are especially strong for the inner monomers of the chains. When such systems are crosslinked, all of the monomers are "inner", except for some dangling ends. Thus one expects the effects of conserved topology to be even stronger in crosslinked systems than in uncrosslinked ones. This should of course hold for polymer gels as well as for dense networks. However, due to the importance of hydrodynamic effects, gels are even more complicated to simulate than crosslinked melts and there are at present no simulations for high crosslinked gels in which the effects of the topology are treated explicitly. However there have been a few simulations on the properties of dense polymer networks. These simulations range from rather rigid model systems, where the crosslinks cannot move,[182,183] to highly complicated, fully mobile randomly crosslinked polymer melts.[159,203] The simulation techniques used are both MD[159] as well as MC (lattice).[203]

Here we will discuss three groups of simulation. In all cases we consider only systems in which the role of the conserved topology is explicitly taken into account. First we discuss some work on short chain networks, where the crosslinks are fixed in space.[182,183] These simulations were used to investigate to what extent the entropic spring concept, which is the basis of the theory of elasticity of networks, is valid in systems in which the excluded volume interactions are present. We then review some MD[159] and MC[203] simulations on randomly crosslinked networks, end crosslinked melts,[161,162] and interpenetrating model systems with diamond lattice topology.[13] The later simulations allows one to study systems in which the only source of disorder comes from knotted loops. Before going into the details of the simulations, we briefly review a few theoretical concepts which were not discussed in Section 4.2 For a more complete summary, there have been several articles which review the theoretical concepts[87,204–206] and some of the experimental issues[207–209] of dense polymer networks. In addition, a nice collection of articles[45] dedicated to the late Eugene Guth appeared recently.

4.5.1 Network elasticity

As discussed briefly in the introduction the elastic and relaxational properties of polymer networks are also expected to be influenced significantly by the presence of entanglements. The classical theories, the phantom network model[204,210–213] and the affine deformation model,[207] describe the two extreme points of view. In the first, at least in its original form,[210] the network strands and the crosslinks are not subject to any constraint besides connectivity and functionality. The other extreme considers the crosslinks to be fixed in space and deform affinely under deformation. A number of modifications of these theories have been proposed in which the junction fluctuations are partially suppressed.[211,214–217] All of these models however consider the network strands as entropic springs. The entropic force, as resulting from the conformational entropy of the strands, and the excluded volume interaction, with its various consequences, are completely decoupled. The excluded volume is responsible only for the constant volume of the incompressible, isotropic monomer liquid. As we will see below this is certainly not the case. Within these models the elasticity of a network is given by the density of elastic springs (affine case) or the density of elementary cycles (phantom case). A monomer in the network is elastically active if at least two independent paths lead to the backbone of the network though monomers in loops formed by a single primary chain are usually excluded.[218–222] Crosslinks in this elastically active backbone are referred to as elastically active. The number of elementary cycles then can be shown to be the difference between the number of active strands μ and active crosslinks ν .[222] Thus the modulus G^o of both models is given by

$$G^o = \frac{k_B T}{V}(\mu - h\nu) \tag{4.37}$$

V is the total volume of the system and $0 \leq h \leq 1$, is a constant which interpolates between the two extreme cases. $h = 1$ for the phantom model and $h = 0$ for the affine model.[222] For short strands, $N_s \ll N_e$, this equation is expected to reasonably describe the behavior in the linear stress regime. It is important to note here that there is no spring–spring interaction. This picture can be extended to take into account a distribution of strand lengths, but this does not give significant differences.[223]

It should be obvious that the above picture cannot hold for arbitrary chain lengths. When the strands between two subsequent crosslinks along a chain are longer than N_e, additional contributions, not included in either of these two simple models, due to the confinement of the strands to the reptation tube are also expected to play a role in the motion of the crosslinks and the modulus of the networks. The introduction of this extra constraint has been included in a variety of ways.[15,46,68,206,213,224–233] In a long chain melt the elastic time-dependent modulus shows a chain-length independent plateau for $N_s \gg N_e$. This so-called rubber-like plateau reveals that one has to expect strong similarities between networks and melts. One extension to eq. (4.37), which tries to take the entanglements into account is[220–222]

$$G^o = \frac{k_B T}{V}(\mu - h\nu) + T_e G_N^o \tag{4.38}$$

where G_N^o is the polymer melt plateau modulus for large N, and T_e is the so-called trapping factor, which can be estimated from the gel fraction of the network. T_e phenomenologically incorporates the influence of free chains and dangling ends on the modulus. There have been several different estimates of T_e.[222,234]

So far the description is essentially phenomenological. There are many extensions, such as the Mooney–Rivlin expansion[235,236] to describe the properties in the nonlinear regime,[206] but a discussion of these effects is beyond the scope of the present review. We simply note that for $N_s \to \infty$ the first term in eq. (4.38) vanishes and $G^o = T_e G_N^o$. Since $T_e = 1$ in a perfect network, the modulus of a rubber and the plateau modulus of a melt should be the same, besides a small reduction by a factor 4/5 due to the tube length fluctuations for melts. This is because under shear/elongation in a melt the tube length is conserved, while this is not the case for a network. Experimentally it is difficult to test these predictions, since it is extremely difficult to prepare well controlled networks. The experimental problem is that the final networks are often inhomogeneously crosslinked and contain additional disorder which is not easily accounted for within the models. In the vulcanization process, for example, since the crosslinking is reaction limited, strong fluctuations in the local density are often present. These types of effects are difficult to include in any theoretical model.[237,238] This is one reason why computer simulations can be very important in trying to understand the properties of networks.

Here the focus of the interest is on the second term and what role the conservation of the network topology plays. All theories consider this only for $N_s > N_e$. While there have been many approaches, the most prominent and to our knowledge the first microscopic ansatz is the Edwards tube model.[15] Within this model, for long chains the modulus is given by[46]

$$G^o(N_s \gg N_e) = \frac{\rho kT}{N_e} \frac{1}{(1+y)^2} \tag{4.39}$$

where y is a parameter to be expected to lie between zero and one. For $y = 0$ the entanglement network would act like a perfect affine network of strand length N_e, while for larger y the elastic effectiveness of the entanglements is reduced. This again points to the important question of the physical nature of the chain entanglements. There are many other extensions and modifications of the original network theories given in the literature, which we cannot discuss here. For details the interested reader is referred to the review by Edwards and Vilgis.[46]

4.5.2 Networks with fixed crosslinks

The classical models of rubber elasticity reduce the elastic properties to a study of entropic springs. In the phantom model the EV interaction only gives the volume conservation, while in the affine deformation model it also is responsible for the affine position transformation of the crosslinks. Otherwise the entropic forces and the excluded volume interactions and consequently the entanglements, as they are a result of the EV interaction, completely decouple from the chain elasticity. The elasticity is entirely determined by the strand entropy. It is obvious that this is a severe oversimplification; however there are not many alternatives to these models available, as discussed above. Gao and Weiner have set up a series of simulations to shed some light on these problems for both networks[182,183] and melts.[184] They focus especially on the differences and relative contributions of the covalent and noncovalent interactions.

Their model consists of a bond potential, a hard linear spring, which only allows for very small bond length variations around the middle bond length l_o. The EV interaction is, just as in the previously discussed melt simulations, represented by a purely repulsive Lennard-Jones potential with a cut-off at $r_c = 2^{1/6}\sigma$. Typically $l_o = \sigma$ and the force constant of the bond force is 200ϵ. These parameters are sufficient to prevent the chains from passing through each other, while for the runs with larger l_o this is not the case. Though it is known that the fluctuations of the crosslinks, or network junctions, are important, Gao and Weiner idealize to the case of fixed crosslink positions, which results in the affine deformation scheme. The system is then characterized by the set of strand end-to-end vectors $\mathbf{R_i}$, $i = 1, ..., \mu_s$, where μ_s is the number of strands, which deform affinely under deformation. This, within the classical models completely describes the system, since only the **R**s are relevant. Their ansatz, of course, is computationally much less CPU intensive than those that also take into account the motion of the crosslinks. Since the chain-ends are fixed, these authors introduce another simplification, namely, they only take chains placed randomly into a melt into account. Since the end-to-end length and direction is fixed, Gao and Weiner[182] argue that there should be no severe difference between a network and their melt. Such a system then is deformed, transforming every end-to-end vector $\mathbf{R_i}$ to $\mathbf{R_i'}$ within the affine deformation scheme. Otherwise the monomers of the chains can move freely. The deformations/transformations are either in an oriented melt model or in what they call a "Four Chain Model".[182] In the former, the chain-end vectors are fixed and are chosen randomly from a unit sphere while in the latter, the deformation cell is constructed by $\mathbf{R_i}$, and three other corresponding tetrahedral vectors. The typical chain length was $N = 20$ and the number density $\rho = 0.8\sigma^{-3}$. From the simulations discussed in Section 4.4, it is clear that the chains used in this study are well below the entanglement threshold.

One of the main aims of this simulation was to determine the relative contributions of the EV stress and the covalent stress along the backbone. Gao and Weiner vary the density $\rho = n\sigma^3/V$ by varying σ at fixed l, where n is the number of monomers and V is the volume. This covers the range from the purely decoupled case for $\sigma = 0$, $n/V = 0.8$ to typical melt densities $\sigma = 1$, $\rho = 0.8$. They find that under elongational deformation λ, at fixed λ the relative contribution to the stress originating from the covalent bonds and the noncovalent EV interaction changes dramatically with σ/l. There is a strong coupling of the stress to the EV forces at realistic melt densities. In Fig. 4.20, we show an example of their results for short chains at fixed

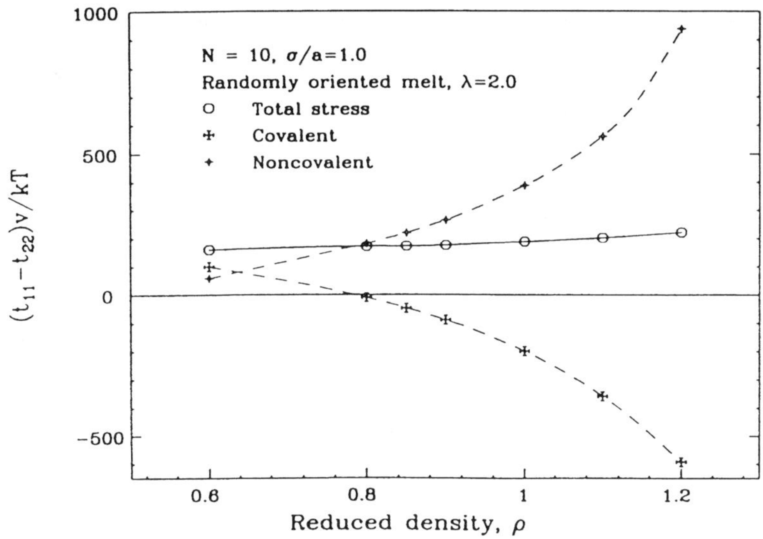

Fig. 4.20 Covalent and noncovalent contribution to the stress $t_{11} - t_{22}$ in a randomly oriented melt subject to a uniaxial extension of $\lambda = 2$ as a function of density. (From Refs 182, 239).

elongation $\lambda = 2$ as a function of density for the oriented melt model.[182,239] For comparison, the previous melt simulations of Kremer and Grest[54] were made, using the same units at a reduced density ρ slightly above 0.85 (because the bond length in the MD was $l = 0.97\sigma$). While the total stress depends only relatively weakly on the effective density, the different contributions depend very strongly. The covalent part for their conditions/model even turns out to be negative value. The main conclusion from this work is, that the dominant contribution to the restoring force, which only exists due to the connectivity of the chains, is mediated by the excluded volume interaction.

4.5.3 Fully mobile systems

The previous section described a rather artificial situation, namely a system of chains with fixed end positions. While this is a very interesting ansatz to study which forces are important, it certainly cannot describe the relaxational properties in a realistic way. Also the question of the role of conserved topology in a network can only be addressed indirectly. In order to shed more light on such experimentally directly accessible or relevant questions, Duering *et al.*[159–162] and Everaers and Kremer[13] performed a series of MD simulations on a variety of different model networks with two different interaction potentials. Both randomly crosslinked[159,240–242] and end-linked polymer melts[160–162] were studied. Everaers and Kremer[13] investigated a system of interpenetrating networks, where each subnet was topologically a diamond lattice. These different simulations were used to study the influence of the possibility of chains crossing each other, the influence of different kinds of disorder (strand length, connectivity, topology) as well as the elasticity and relaxation as a function of the strand length.

Figure 4.21 illustrates the three different classes of networks: randomly crosslinked, end-crosslinked and diamond networks. The first system, an equilibrated melt with randomly placed crosslinks, certainly resembles most closely many experimental systems such as those produced by radiation crosslinking or vulcanization.[243] These systems have disorder on several levels. The strand length between subsequent crosslinks are extremely polydisperse, as the distribution of chain lengths follow an exponential distribution.[241,244–246] Only a few crosslinks per chain are needed to give a gel-fraction which is nearly 100%, though the elastically active part of the network is significantly smaller. In this case there are many dangling chain ends or even some dangling clusters, like little trees which are connected to the network by only one strand. Those dangling pieces are expected to be responsible for the extremely slow decay in most experiments.[159,247] In addition loops and knots may play an important role in both the quasi-linear elasticity and in nonlinear effects. For this the shortest connected piece going through the cluster is expected to be of special relevance. Such disordered systems, although it is possible to identify the elastically active part of the cluster unambiguously,[160,242] are extremely difficult to analyze in a systematic manner. The relaxation of highly crosslinked melts[203,248] as well as the microphase separation of two interpenetrating nets of different polymer species[249] have been investigated using the bond fluctuation model.

A somewhat more idealized case is an end crosslinked polymer melt. Such systems have been considered experimentally as well.[208,209,250–255] There one starts with a monodisperse melt. After equilibrating the chains are kinetically crosslinked at the ends. By this all strands in the network have the

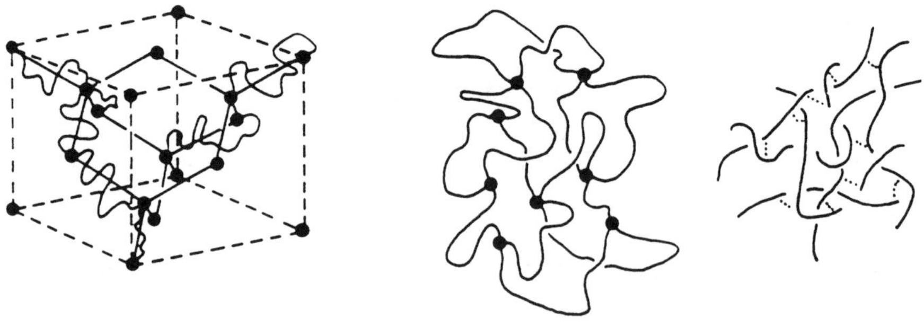

Fig. 4.21 Illustration of different classes of polymer networks used in computer simulations. While analytic theory becomes more and more difficult with increasing disorder, the experimental situation is just the opposite. Simulations can interpolate between the two extremes.

same length. Thus there are only two sources of quenched disorder in the system. One is knots. The second is chemical paths through the system, which again might be polydisperse but now on the level of number of strands, which are needed to percolate through the system.

Finally one can consider systems, which certainly are somewhat remote from experiments, where the only source of disorder is due to the occurrence of knotted loops. Such are the interpenetrating lattice networks.[13] They can be analyzed and investigated in a variety of ways in order to shed some light on the question of entanglements and conserved topology.

Although all of these simulations are very recent, in the following we follow the historical development in the order given above.

4.5.3.1 Randomly crosslinked melts

To our knowledge there are only two sets of simulations on completely mobile randomly crosslinked polymer melts, the MD simulation of Duering *et al.*[159] and the bond fluctuation MC simulation of Schulz and Sommer.[203,248] Note that standard lattice models certainly are not applicable, since the crosslinks would be immobile. (One might consider certain tricks of double occupancy near the crosslinks, as sometimes used for star polymers, though problems may occur with the conservation of the network topology.)

The bond fluctuation lattice simulations investigate the relaxational properties of randomly crosslinked polymer melts, starting out from an equilibrated melt and then add additional bonds between monomers.[248] Besides studying the gelation threshold they are concerned with two cases. One is the slip link problem, the other the collective relaxation of the whole network.

Sommer[256] first studied two knotted chains which are strongly connected since they cannot pass through each other. However such knots, as they occur in networks as well, cannot be regarded as fixed crosslinks, since the chains can slip along each other. Sommer tried to elucidate this point. While Baumgärtner and Muthukumar[257] in an earlier attempt simulated the unwinding of two walks, Sommer tried to estimate the entropy of knotted paths. These knots are modeled by the classical slip links.[46,224] Since only the entropy of an ideal walk is included, Sommer finds an entropically driven tendency for the crosslinks to cluster. This is not surprising. This clustering in a network provides the maximal length of the undisturbed parts of the strands, which certainly maximizes the total entropy. Though this is an interesting point for swollen networks, it is questionable whether in a dense network the knots can be represented by freely movable slip links, since on the scale of a blob of the tube diameter, many chains pass through this volume.

More recently Schulz and Sommer[203,248] studied randomly crosslinked polymer networks by the bond fluctuation method and compared their results to the generalized Rouse model. In the generalized Rouse model (GRM) the connectivity of the whole system is considered, but there is no excluded volume or topology conservation. With the periodic boundary conditions used in the simulations, such a calculation is expected to describe the corresponding phantom network. They studied samples containing 84 chains of length $N = 50$ at $\Phi = 03$. Following Paul *et al.*,[59] at this density the entanglement length is expected to be around $N_e \approx 51$. Thus the primary chains are too short to be entangled from a reptation theory point of view. The samples were then randomly crosslinked. The smallest crosslink density used was still very high, as on average 20% of the monomers were crosslinked, while for the highest crosslinking state 67% of the monomers were crosslinked. Thus in their samples the average strand length between crosslinks was at most $N_s = 5$ bonds, or about $N_e/5$. Thus any significant effect from the noncrossability of the chains can not be expected. For these model systems the Rouse matrix was constructed by replacing the bond fluctuation by a harmonic potential and solved numerically. Except for the highest crosslink density the effect of the discrete lattice does not seem to be significant. For all other cases, the GRE seems to describe the time-dependent fluctuations reasonably well for the short chain lengths studied. This is of interest also with respect to the subsequently discussed end-linked networks. There the mean square displacements follow the phantom model, while the modulus is much higher.

Duering *et al.*,[159,242] somewhat earlier, studied randomly crosslinked networks using the same MD simulation technique used for melts. They considered systems of chain length $N = 50$ and up to 400 chains. After equilibrating a melt of linear chains, crosslinks are introduced randomly. In Ref. 159 $p = 2$ crosslinks per chain are introduced by randomly selecting one monomer in space and then adding a bond to one monomer at random which is within a distance 1.3σ of the first one. Nearest and next nearest neighbors along the chains are excluded. The capture radius 1.3σ corresponds to a distance $l\sqrt{C_\infty}$. With $p = 2$, since each crosslink connects two strands, each chain is linked in the average by four bonds to the network. For $N = 50$ about 96% of all chains belonged to the infinite cluster. They found that the typical distance between crosslinks along the primary chains is slightly less than nine. This is because the local density near the ends of the primary chains is slightly reduced. Considering that a chain on average consists of three inner pieces and two dangling ends, almost 50% of the monomers are in dangling chain-ends. A detailed cluster analysis actually revealed that for this case only about 44% of the monomers are considered to be elastically active. The main purpose of the investigation was to check whether the conserved topology, even for such short strands, $\langle N_s \rangle \approx N_e/4$, is relevant. It is obvious that there should be a significant difference between a random walk system and the full SAW system, following the work of Gao and Weiner,[182] who found that with increasing density, the EV interaction becomes increasingly important. Here a different approach is taken. Besides the standard LJ case a second interaction potential $U^{\cos}$ is considered. There the repulsive Lennard-Jones potential between nonbonded monomers is replaced by[159]

$$U^{\cos}(r) = \begin{cases} 2\epsilon_a & r < r_a; \\ \epsilon_a(\cos\dfrac{\pi(r - r_a)}{r_c - r_a} + 1) & r_a \leq r \leq r_c, \\ 0 & r > r_c. \end{cases} \tag{4.40}$$

Then the parameters of the FENE potential[258] of the bonded monomers and the parameters ϵ_a and r_a are chosen in a way that a melt of such chains had the same $\langle R^2 \rangle, \langle R_G^2 \rangle$, pressure and Rouse friction as the LJ melt. For $\rho = 0.85\sigma^{-3}$ the parameters are $\epsilon_a = 2.22\epsilon, r_c = 2^{1/6}\sigma, r_a = 1.0\sigma$ and $R_o = 1.75\sigma$ (in the FENE Potential). The amplitude of the FENE was also reduced by a factor of 0.175.[159] Monomers can cross each other with a penalty of a few k_BT. Since the macroscopic properties are not altered, these two potentials can be used to identify the contributions from the noncrossability of the chains. The general monomer packing constraints are only weakly affected, as the pressure and the temperature remain the same.

Both models were simulated with the same crosslinked configurations. In both cases the relaxation of the systems was extremely slow. Considering $\tau_N \approx 180\tau$ for $N = \langle N_s \rangle$ and $\approx 3700\tau$ for $N = 50$ the primary chain length, it was observed that the characteristic relaxation time increased by about a factor of 10 compared to $\tau_{N=50}$ or more than 100 compared to the Rouse time of the average inner strand for the LJ case. The soft potential model displayed an even slower relaxation.

One effect of the conserved topology is seen in the relaxation. The full LJ interaction reduces the available conformations dramatically, as they are quenched, and consequently relaxed faster than the system with non-quenched topology. In both cases the displacements of the middle monomers of the primary chains and the crosslinks exceed the radius of gyration of the strands significantly. For example, the average mean-square displacement $g_2^x(t \to \infty)$ of the crosslinks in the infinite cluster reach about $10\langle R_G^2(N_s) \rangle$ for the LJ system, much greater than expected by the phantom network theory. The most striking effect of the conserved/nonconserved topology however is shown by the mode relaxation, which is directly related to the elastic modulus. Since the samples were not deformed in this study, the Rouse modes of strands were analyzed in order to measure the modulus. Following Doi and Edwards[5] the time dependent modulus $G(t)$ of the melt is given by the sum of the autocorrelation function of the Rouse modes, namely

$$G(t) = \frac{\rho k_B T}{N} \sum_{p=1}^{\infty} \frac{\langle \mathbf{X}_p(t) \cdot \mathbf{X}_p(0) \rangle}{\langle X_p^2(0) \rangle} \overset{t \to \infty}{\longrightarrow} \mathrm{C}_{\mathrm{mod}}\, G_N^o. \tag{4.41}$$

Within the reptation theory formulation of the time dependent modulus of a melt, C_{mod} is expected to be one. Since the primary chains in the infinite cluster lose their identity, one can construct a long path through the active part of the network. These paths are random walks. The Rouse modes of these paths can be calculated and the modulus determined from eq. (4.41) by calculating the mode auto correlation functions of these paths. In Ref. 162 different path lengths were employed in order to check the dependency of the result on the lengths of these so-constructed chains. Following eq. (4.38) the strand lengths are expected to be far too short to observe any trapping contribution. Considering the low density of elastically active monomers $\langle N_s \rangle$ is far below N_e. In Fig. 4.22 we show $G(t)$ for the original primary chains and constructed walks for both interaction potentials. The modulus as derived from the primary chains is too small, since the free dangling ends are included. Striking, however, is the result of the mode relaxation from the long constructed walks. $G(t)$ shows a completely different decay for times

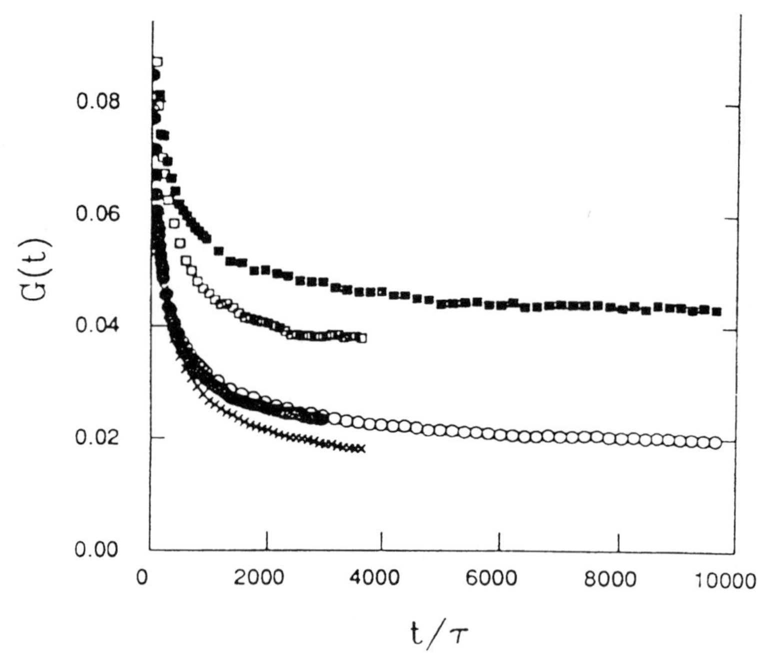

Fig. 4.22 Time-dependent modulus $G(t)$ for randomly crosslinked melts. Open squares and crosses show results for $U^{\cos}$, while full squares and circles give results for LJ interaction. The lower two curves correspond to the original chains, while the upper give the results of constructed walks of length $N = 100$. (From Ref. 159).

larger than the typical time which defines the monomer friction. For large times the extrapolation of $G(t)$ for systems interacting with the Lennard-Jones EV interaction are significantly larger than $G(t)$ for the soft cosine potential system by about 30%. This can be attributed directly to the conservation of topology in clear contradiction to the classical models and the expectation from the entanglement trapping approach. The question now is how does this affect the elastic modulus. From eq. (4.41) $G(t \to \infty) = C_{mod} G_N^o$. Recently it was found that $C_{mod} \approx 1.5$ for the LJ case and about 1.2 for the cosine potential.[13] This was determined by measuring the restoring force and comparing the resulting modulus to the prediction of eq. (4.41). For a melt this means that the Rouse modes of a chain are not eigenmodes of the system. They actually over estimate the modulus. The consequences for networks as well as viscosity measurements are discussed elsewhere.[13]

Taking this value for C_{mod} in account reduces the difference with the standard model somewhat. The actual value of the modulus then can reasonably well be described by the affine network model for the LJ system. Including $C_{mod} = 1.5$ one gets $G_N^o \approx 0.027$. Using $\rho = 0.37\sigma^{-3}$ (the density of the elastically active part) one gets within the affine model an effective strand length of $N_{eff} \approx 13$–14, while the cluster analysis yields $N_{eff} \approx 11$. The soft core model is somewhere in between the affine and the phantom model.

This simulation showed, in nice agreement with the Gao and Weiner [182–184] simulations, the relevance of the conserved topology and excluded volume. The question however remains whether these effects have anything to do with the entanglements that are believed to be important in a melt of long chains. To investigate this in more detail one has to go to longer chains and more idealized networks.

4.5.3.2 End-linked networks

Duering *et al.*[161,162] simulated end-linked model networks for chain lengths between $N = 12$ and 100, employing the MD model of Kremer and Grest.[54] Again they started out from an equilibrated melt and then defined 1/4 of the chain-ends randomly to be crosslinking sites, which can connect $f = 4$ ends. Each time a free end came within the capture radius, which was taken to be the persistence length (measured in units of the bond length l) $l\sqrt{C_\infty} = 1.3\sigma$, the two ends are connected.[160,162] In this way the system was crosslinked kinetically until almost all ends were connected. The gel fraction for the different systems varied between 96% and 100%. The elastically active part of the systems was at least 93% of the monomers, giving an effective density of at least $0.93 \times 0.85\sigma^{-3} \approx 0.79\sigma^{-3}$. The strand length varied between about $N_e/3$ to $3N_e$. Since there are no dangling clusters or only a few dangling ends, the relaxation times of these systems are expected to be considerably shorter than in the case of the randomly crosslinked networks. Another important aspect is, because there are few free ends, contributions due to the more mobile ends, which shift the crossover from Rouse to reptation to longer chains, are not expected to show up here. Duering *et al.*,[162] as in the previous case, measured the modulus via the mode relaxation, the mean-square displacements of the crosslinks and monomers and scattering functions, as well as other related quantities. Both the scattering function and the moduli can directly be related to recent experiments.[253,254] From the extrapolation of the modulus to chain length $N_s \to \infty$ one can directly estimate the trapping contribution and compare this to the reptation approach. The first interesting quantity is the average displacement of the crosslinks, with respect to their equilibrium position. Since the network is a solid, the fluctuations are finite. Vilgis and Boué[226] propose for the asymptotic mean-square displacement $g_2^x(t \to \infty)$ for the crosslinks the form

$$\frac{1}{g_2^x(t \to \infty)} = \frac{a}{N_s} + \Delta^{-2} \tag{4.42}$$

where Δ^2 is of the order of the tube diameter as in a melt. Both constants are expected to be of order unity. One can either directly determine g_2^x or investigate the dynamic incoherent structure factor of the crosslinks. (Experiments typically measure the coherent structure factor, but since the typical distance $\Delta \gg 2\pi/q$ in practical cases there is no difference between the two.) Figure 4.23 shows the dynamic scattering of the crosslinks in the convenient Rouse scaling, $S(q,t)/S(q,0)$ versus $q^2\sqrt{t}$. The data clearly show a q-dependent deviation from the common curve towards a plateau value. This indicates that the crosslinks only move a finite distance. It should be noted that the experimental situation is much less satisfying. A replot of the neutron spin-echo scattering data for $S(q,t)$ of Oeser *et al.*[253] in Ref. 162 in the Rouse scaling form shows no crossover towards a plateau. All their data are still in the initial Rouse decay regime. The simulation data now can be interpreted along the lines of the Debye–Waller factor in which the plateau describes the average smeared-out position of the crosslink. If one assumes that a crosslink is located in a harmonic potential well, then one can directly relate the plateau $\ln\{S(q,t \to \infty)/S(q,0)\}$ to $g_2^x(t \to \infty)$ by

$$-\ln S_{inc}^x(q,t \to \infty)/S(q,0) = \frac{q^2}{6} g_2^x(t \to \infty) \tag{4.43}$$

This relation is fulfilled very well by the simulation results. Thus the complicated interactions of a crosslink with the environment to a very good approximation can be viewed as a harmonic potential. The next question goes back to eq. (4.42), namely, is the width determined only by the strand length N_s or does the melt entanglement length play a crucial role for large N_s. Fig. 4.24 shows a plot of $1/g_2^x(t \to \infty)$ versus $1/N_s$ along the description of eq. (4.42). First consider the data for the shortest chain, $N_s = 12$. One

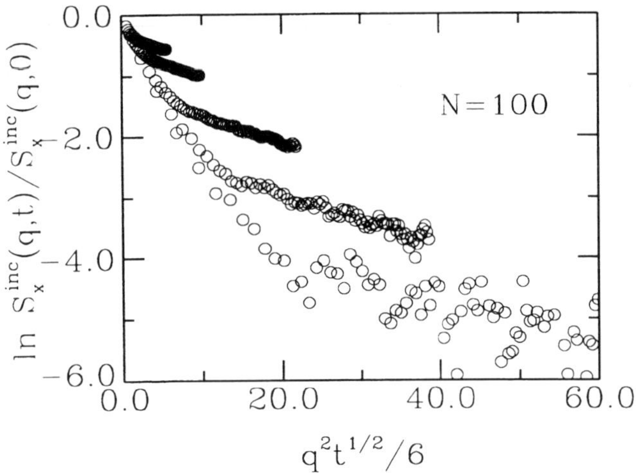

Fig. 4.23 Incoherent dynamic scattering of the crosslinks for $N_s = 100$ and various wave vectors q, from top to bottom, $q\sigma = 0.49, 0.65, 0.97, 1.29$, and 1.62. (From Ref. 162).

finds $g_2^x(N_s = 12, t \to \infty) = 7.3\sigma^2$, while $\langle R^2(N_s = 12)\rangle = 19.1\sigma^2$, which gives a ratio of almost exactly 3/8 as expected by the phantom model.[259] For larger N_s this ratio decreases significantly. This extrapolation to $N_s \to \infty$ yields $\Delta^2 \approx (18 \pm 3)\sigma^2$. This is to be compared to the characteristic distance monomers move when they crossover into the $t^{1/4}$ regime for the reptation problem, which was found to be $20\sigma^2$, namely $2\langle R_G^2(N_e)\rangle$.[54] Thus asymptotically the crosslinks are confined to a spatial area, which is given by a sphere of diameter d_T. This gives direct evidence for the relevance of the entanglement length N_e for the case of polymer networks. In a very similar

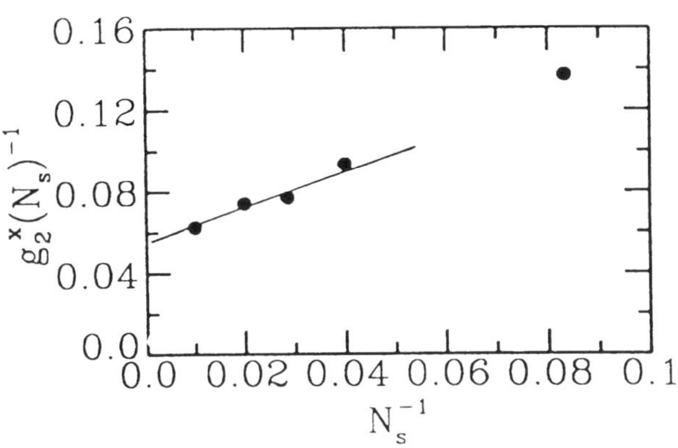

Fig. 4.24 Extrapolation of the asymptotic displacements of the crosslinks as determined from $g_2^x(t) \to \infty$ (for $N_s = 100$). (From Ref. 162).

fashion one can study the motion of inner monomers of the strands.[162] By plotting the chains, one can directly check whether chains of length $N = 100$ are long enough to show some confinement to a tube. In Section 4.4, plots of the primitive chain for the melt for $N = 200$ were given. There the dissolution of the tube from the ends was observed. For chains of length $N = 100$, the melt showed little sign of a tube. For the end-crosslinked networks, the ends are localized as part of the crosslink. Figure 4.25 shows a scatter plot of five different examples of strands in the system of $M = 500$ chains of length $N = 100$. For two strands the ends are relatively close together and resolution of the tube is not possible. In the three other cases however the tube confinement is obvious. The actual width of the scattering of the points is about the expected tube diameter, as can be seen from the diameter of the big spheres in Fig. 4.25, which is 1.2σ. The data clearly show that the entanglement length N_e plays a significant role for the fluctuation properties of the network.

At this point we can come back to the original discussion of Schweizer's approach for melt dynamics. In the case of the networks we have the ideal situation that the original reptation tube is conserved. Within the mode coupling scheme the scatter of the points in Fig. 4.25 should cover an

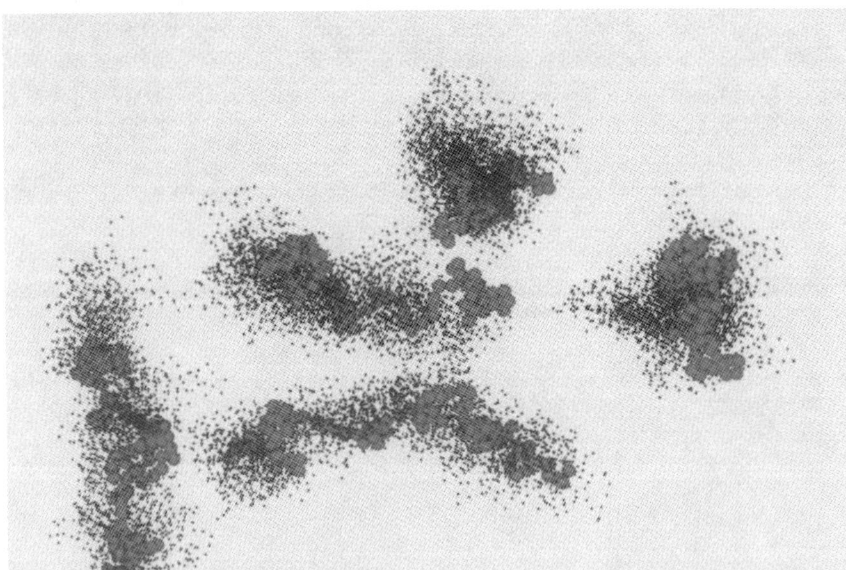

Fig. 4.25 Sample conformations of five randomly taken strands from the end-linked network with $N = 100$. The first conformations are large spheres, while the subsequent conformations are only shown as points. The total time elapsed between the first and last conformation is $15\,000\tau$. (From Ref. 162).

area of a diameter of at least $\langle R^2(N_s = 100)\rangle^{1/2} \approx 13\sigma$. Although the chains are still relatively short, this is not seen for the present simulations. Applied to the melt, this means that for the time the constraints, as they occur in the network, are conserved, the monomers are confined to a tube. The only way however to release these constraints is the motion out of the tube, which is nothing else than the reptation scheme.

Unfortunately there is not yet enough computer time available in order to provide a detailed test on the elastic properties under elongational stress. This would yield the modulus directly. Duering *et al.*[162] estimated the modulus according to eq. (4.41), which should be good to within about 10%. Here we do not review in detail their approach to calculate G_N^o; rather we show a comparison of the results to experiment, which directly points to the relevance of entanglements for network elasticity. As was done in Fig. 4.6 for the normalized diffusion constant D/D_{Rouse}, one can also scale the modulus by its value at N_e. Again one assumes that the entanglement length N_e plays the crucial role in the behavior of the system. Then one can plot $G(N_e, t \to \infty)/G(N_s, t \to \infty)$ versus N_e/N_s. $G(N_e, t \to \infty)$ is the modulus of a network of strand length N_e. In so doing, simulation and experiment should fall onto one common curve for large N_s under two conditions. First, the constant in eq. (4.41) is independent of N_s and, second, the experimental and simulational N_e describe the same physical quantity, although extracted by different methods. Duering *et al.*[162] compared their results to recent experiment on narrow molecular weight distribution end linked PDMS networks by Patel *et al.*[254] Figure 4.26 shows that the agreement is almost perfect. These results can now be compared to the predictions of entanglement theories for polymer networks. For $N_s \to \infty$ only the trapping contribution survives, while for N_s up to $\approx N_e$ the second term in eq. (4.38) is expected to be negligible. From these results, Duering *et al.*[162] concluded that roughly 2.2 entanglement lengths are needed in order to effectively act as one crosslink within a generalized affine network picture. For more details see Ref. 162.

So far only fluctuation quantities have been discussed. The stretching ratio λ for the simulations of Gao and Weiner[182] however is large enough in order to expect a strong influence of nonlinearity on the restoring force. Recently Everaers and Kremer[13] started a study of model networks, where the network connectivity is given by a diamond lattice. In order to have the situation that the average diamond lattice nearest neighbor distance corresponds to the average chain-end-to-end distance $\langle R^2(N_s)\rangle^{1/2}$ they had to use interpenetrating networks. For the same density as above ($\rho = 0.85\sigma^{-3}$) five internested networks are needed for $N_s = 12$. This class of systems allows one to study in detail the influence of random knots. Since in the diamond lattice topology, all percolating chemical paths through the system are of the same length, the only source of disorder are knots. In the end-linked case, knots and a distribution of "short paths" occurs. During the simulation the

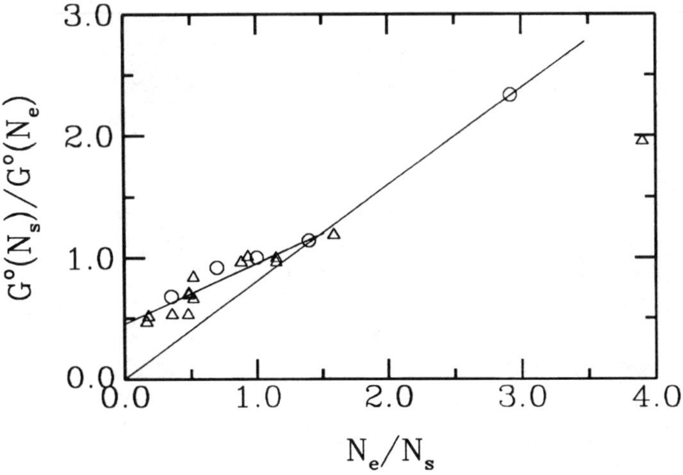

Fig. 4.26 Scaling of the experimental modulus[258] (△) for end-linked PDMS networks and the MD simulations (○) as discussed in the text. (From Ref. 162).

whole system was mobile and the lattice structure only remains for the connectivity. Two cases were studied. First the nets are regularly interpenetrating. The simulation is set up with regularly intercalating networks of fully stretched chains with full excluded volume. The density of the system is then decreased to the desired melt density $\rho = 0.85\sigma^{-3}$. Such a system has no quenched disorder. The other approach is to randomly interpenetrate networks where the strands have the correct persistence length, but no excluded volume. Then the excluded volume interaction can be switched on slowly until its full value is reached. This second system contains many random knots, in contrast to the first case. It was found that the zero shear modulus was different in the two cases, however more interesting is the behavior of the systems under strong elongational deformation. Figure 4.27 shows the results of the restoring force as a function of λ for systems with $N_s = 12$. For the system without disorder the linear regime extends about twice that of the disordered system. The reason is that the random knots in the system define topological shortest paths through the system, which are strongly elongated while the remainder of the strands are still within the Gaussian coil regime. This is shown by a detailed analysis of the forces along the chains. Here however we only reproduce one figure from their work, Fig. 4.28. The shade code is that thick intense chains carry more stress than pale thin chains. The figure clearly shows that a significant amount of stress in the case of the disordered system is concentrated on a very few strands, while it is equally distributed in the case of the regularly interpenetrating nets. Close inspection actually reveals that the knots behave as slipping links.[224] The strands seem to slip along each other until they get caught at a crosslink.

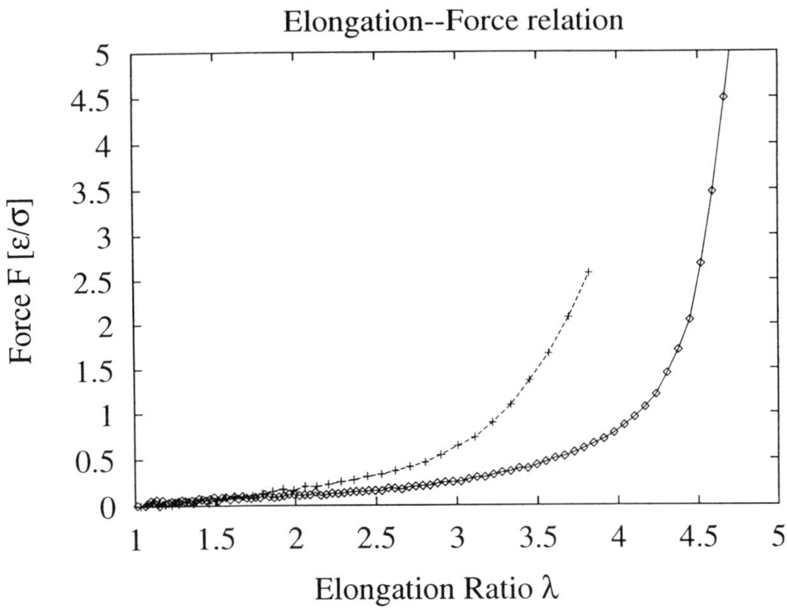

Fig. 4.27 Restoring force of two different diamond topology interpenetrating networks, with and without random knots, as indicated for $N_s = 12$. (From Ref. 13).

In more realistic networks, of course, there is another source for the onset of the nonlinear elastic response, namely chemical short paths. Nevertheless such investigations show the relevance of disorder on the actual elastic properties.

4.6 Conclusions

The initial question addressed in this review is what role do the constraints of entanglements play for the dynamics of dense polymers and what *are* entanglements. For the first problem, computer simulations in connection with theoretical models and experiments provide a rather detailed answer. The second question still remains largely unsolved.

The very nature of an entanglement is still not understood. For the most simple systems, such as the bond fluctuation MC model, results from different densities suggest that the generalized packing criterion, which takes the proper effective exponent ν on the length scale of the excluded volume blobs into account, describes the scaling of N_e rather well. This suggests that the physical origin is the overall coupling of the chains to the surrounding sea of other chains. At present little else can be said. This is especially true when the interactions become much more complicated, as is the case for systems with highly polarizable groups along the chains. In this case the

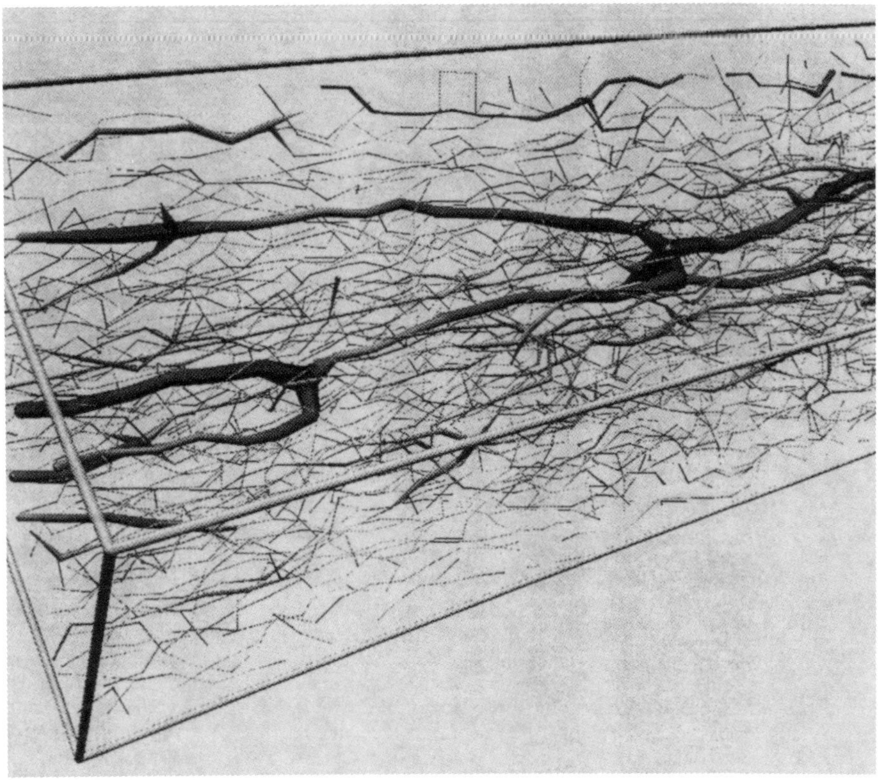

Fig. 4.28 Conformation plot of a stretched network with random knots. In the nonlinear regime there are a few strongly stretched paths, which are shown dark and thick. The less the stretching, the finer the bonds are shown. (From Ref. 13).

physical parameters determining the value of N_e are not understood. The results from the work on networks suggest that ultimately the topology conservation of the strands remains the only relevant parameter. However a lot of research still remains to be done in this case. This problem is not only an important theoretical one but one with direct technological relevance, since a controlled manipulation of flow properties is of great interest.

Significantly better is the situation if one discusses the consequences of the chain connectivity and the occurrence of entanglements. The simulations clearly show that the entanglement length is the unique length scale, which governs the slowing down of the overall motion of the polymers. For short times the middle monomers do not feel whether they are situated in long or short chains. For longer times, $t > \tau_e$, the monomer mean square displacement for $N > N_e$ is independent of N up to the Rouse time. There is even some evidence for the second $t^{1/2}$ regime for the mean square displacement.[166] However the data are not accurate enough to distinguish between the reptation scheme[16,17] and mode coupling approaches.[36–38] The mode coupling assumes that the motion is isotropic. This has been checked directly by simulations and evidence for anisotropic motion was found in all of them. This is particularly striking when one examines the motion of a chain in a network. To test whether it is a good approximation or not to ignore this anisotropic motion in a mean field sense,[116] it is important to compare the time-dependent scattering function $S(q, t)$ for the mode coupling theory and experiment. Unfortunately $S(q, t)$ for the mode coupling theories is presently not available. The time and length scales predicted by the simulations agree nicely with the recent neutron spin-echo experiments. So far one arrives at a rather complete and consistent general picture for the role of the chain connectivity and thus the entanglement length for the melt dynamics. The long time, or so called terminal regime, however is still beyond the present capabilities of even the fastest supercomputers. Many interesting questions still remain unresolved. One of these is to understand the difference between the N^3 power law for the terminal relaxation predicted by reptation theory and the observed $N^{3.4}$ law. Though a variety of different theoretical models have been proposed, this difference remains a puzzle. It is important to determine whether any of the anomalies in the relaxation times as observed for the simulations of semidilute but entangled systems can be related to this difference in the longest relaxation time. Induced bond orientation by uniaxial stretching is also another interesting area where simulations can play an important role. Recent studies on short guest chains[260] $(N < N_e)$ in a matrix of long host chains show induced orientational order comparable to that observed experimentally.[261] Additional studies for guest chains longer than N_e will be very valuable in clarifying the origin of this effect. Another potentially very fruitful path of research is pointed out by the first NEMD simulations on very weakly entangled melts. With the coming better computers this will be very interesting tool, not only to complement the equilibrium simulations.

Even more complex in many respects but somewhat clearer in others is the situation for polymer networks. It is obvious from the results reviewed here that the chain connectivity and the noncrossability of the chains play a very important role for the relaxation and the elastic modulus. Even for short chains the conservation of the network topology is crucial. This result should be seen as complementary to the investigations, in which the stress along the backbone of the strands and via the EV interaction is analyzed separately. As in the melt of uncrosslinked chains, the melt entanglement length turns out to be the crucial length scale for long chains. At this point the simulations are clearly in a much better state than any microscopic experiment, as they allow a comparison of macroscopic properties, such as the modulus, to microscopic properties, such as mean square displacements for well characterized "samples". The confinement to the tube for the longer chains is clearly demonstrated, also giving additional support to the reptation picture in comparison to mode coupling schemes. In this context it is especially interesting that the modulus in the asymptotic limit of long chains corresponds to about $1/(2.2 \pm 0.2)$ for a network of strand length $N_s = N_e$. This ratio is the same as the typical ratio between M_e and M_c for the crossover in the viscosity from Rouse to reptation. Here too many unresolved questions remain. The investigation of networks via computer modeling is in its infancy compared to the work on melts of free chains. At this point, only highly idealized systems have been investigated which in most cases have been confined to the linear regime. The range of interesting and highly relevant questions is wide, including nonlinear effects for idealized model networks to swelling/stretching of random systems. To our knowledge there is no study of systems which were crosslinked in the semidilute state and then deswelled. All these questions are of theoretical (cf. butterfly effect[262]), experimental and also technological interest.

The coming of new generations of faster computers will allow more research along the above-discussed lines and should also provide us with significantly better data on these highly disordered systems. This will improve the predictive power of the simulations, but also allow one to build even better bridges between experiments and analytic theory.

Acknowledgment

This review benefited from many helpful discussions and fruitful collaboration with K. Binder, B. Dünweg, E. R. Duering, R. Everaers, L. J. Fetters, W. W. Graessley, J. S. Huang, W. Paul, and D. Richter. Those aspects of research reviewed here which were conducted by authors was supported in part by Sonderforschungsbereich 262 of the German Science Foundation and by a NATO travel grant. Part of this work resulted from a generous CPU time grant for the Cray YMP8/32 of the German Supercomputer Centre, HLRZ, in Jülich within the "Disordered Polymers" project.

References

1. P. Flory, *Principles of Polymer Chemistry* (Cornell University Press, Ithaca, NY, 1953).
2. P. G. de Gennes, *Scaling Concepts in Polymer Physics* (Cornell University Press, Ithaca, NY, 1979).
3. P. E. Rouse, *J. Chem. Phys.* **21**, 1272 (1953).
4. F. Bueche, *J. Chem. Phys.* **22**, 603 (1954).
5. M. Doi and S. F. Edwards, *The Theory of Polymer Dynamics* (Clarendon Press, Oxford, 1986).
6. D. Richter, A. Baumgärtner, K. Binder, B. Ewen, and J. B. Hayter, *Phys. Rev. Lett.* **47**, 109 (1981); **48**, 1695 (1982).
7. A. Baumgärtner and K. Binder, *J. Chem. Phys.* **75**, 2994 (1981).

8. D. Ceperley, M. H. Kalos, and J. L. Lebowitz, *Phys. Rev. Lett.* **41**, 313 (1978); M. Bishop, D. Ceperley, H. L. Frisch, and M. H. Kalos, *J. Chem. Phys.* **76**, 1557 (1982).
9. K. Kremer, *Macromolecules* **16**, 1632 (1983).
10. A. Kolinski, J. Skolnick, and R. Yaris, *J. Chem. Phys.* **84**, 1922 (1987); *J. Chem. Phys.* **86**, 1567 (1987).
11. D. Boese, B. Momper, G. Meier, F. Kremer, J. U. Hagenah, and E. W. Fischer, *Macromolecules* **22**, 4416 (1989).
12. W. Paul, K. Binder, D. W. Heermann, and K. Kremer, *J. Chem. Phys.* **95**, 7726 (1991).
13. R. Everaers and K. Kremer (preprints), and R. Everaers, Ph.D. thesis, Universität Bonn, Germany, 1994.
14. J. D. Ferry, *Viscoelastic Properties of Polymers* (Wiley, New York, 1980).
15. S. F. Edwards, *Proc. Phys. Soc.* **92**, 9 (1967).
16. P. G. de Gennes, *J. Chem. Phys.* **55**, 572 (1971).
17. P. G. de Gennes, *J. Chem. Phys.* **72**, 4756 (1980).
18. C. F. Curtiss and R. B. Bird, *J. Chem. Phys.* **74**, 2016 (1981); *J. Chem. Phys.* **74**, 2026 (1981).
19. W. Hess, *Macromolecules* **19**, 1395 (1986); *Macromolecules* **20**, 2589 (1987); *Macromolecules* **21**, 2670 (1988).
20. M. Fixman, *J. Chem. Phys.* **89**, 3892 (1988); *J. Chem. Phys.* **89**, 3912 (1988).
21. M. Doi, *J. Polym. Sci., Polym. Phys. Ed.* **19**, 265 (1981); *J. Polym. Sci., Polym. Phys. Ed.* **21**, 667 (1983).
22. H. Wendel and J. Noolandi, *Macromolecules* **15**, 1313 (1982).
23. J. des Cloizeaux, *J. Phys. Lett. France* **45**, L17 (1984).
24. R. J. Needs, *Macromolecules* **17**, 437 (1984).
25. J. M. Deutsch, *Phys. Rev. Lett.* **54**, 56 (1985).
26. H. Scher and M. F. Schesinger, *J. Chem. Phys.* **84**, 5922 (1986).
27. G. H. Weiss, J. T. Bendler, and M. F. Schlesinger, *Macromolecules* **21**, 521 (1988).
28. M. Rubinstein, *Phys. Rev. Lett.* **59**, 1946 (1987).
29. K. Iwata and S. F. Edwards, *Macromolecules* **21**, 2901 (1988); *J. Chem. Phys.* **90**, 4567 (1989).
30. K. Iwata, *Macromolecules* **24**, 1107 (1991).
31. G. Ronca, *J. Chem. Phys.* **79**, 1031 (1983).
32. J. Skolnick, R. Yaris, and A. Kolinski, *J. Chem. Phys.* **88**, 1407 (1988); J. Skolnick and R. Yaris, **88**, 1418 (1988).
33. J. F. Douglas and J. B. Hubbard, *Macromolecules* **24**, 3163 (1991).
34. M. F. Herman, *J. Chem. Phys.* **92**, 2043 (1990); *Macromolecules* **25**, 4925 (1992); *Macromolecules* **25**, 4391 (1992); M. F. Herman and P. Tong, *Macromolecules* **25**, 6638 (1992).
35. T. A. Kavassalis and J. Noolandi, *Phys. Rev. Lett.* **59**, 2674 (1987); *Macromolecules* **21**, 2869 (1988); *Macromolecules* **22**, 2709 (1989).
36. K. S. Schweizer, *J. Chem. Phys.* **91**, 5802 (1989); *J. Chem. Phys.* **91**, 5822 (1989).
37. K. S. Schweizer, *J. Non-Cryst. Solids* **131–133**, 643 (1991).
38. K. S. Schweizer, *Physica Scripta* **T49**, 99 (1993).

39. V. G. Rostiashvili, *Zh. Eksp. Teor. Fiz.* **97**, 1005 (1990) (*Sov. Phys. JETP* **70**, 563 (1990)).
40. G. Szamel, *Phys. Rev. Lett.* **70**, 3744 (1993); G. Szamel and K. S. Schweizer, *Phil. Mag. B.* (in press); *J. Chem. Phys.* **102** (in press); U. Genz, *Macromolecules* **27**, 6452 (1994).
41. I. Szleifer and R. F. Loring, *J. Chem. Phys.* **95**, 2080 (1991).
42. I. Szleifer, J. D. Wilson, and R. F. Loring, *J. Chem. Phys.* **95**, 8474 (1991); **97**, 3875 (1992); J. D. Wilson and R. F. Loring, **97**, 3710 (1992); **99**, 7150 (1993).
43. A. P. Chatterjee and R. F. Loring, *J. Chem. Phys.* **101**, 1595 (1994).
44. T. P. Lodge, N. A. Rotstein, and S. Prager, *Advances in Chemical Physics*, Vol. 79, edited by I. Prigogine and S. A. Rice (Wiley, New York, 1990), p. 1.
45. J. E. Mark and B. Erman (eds), *Elastomeric Polymer Networks* (Prentice Hall, Englewood Cliffs, NJ, 1992).
46. S. F. Edwards, T. A. Vilgis, *Rep. Prog. Phys.* **51**, 243 (1988).
47. K. Kremer, G. S. Grest, and I. Carmesin, *Phys. Rev. Lett.* **61**, 566 (1988).
48. F. Mezei, in *Neutron Spin Echo*, edited by F. Mezei (Springer-Verlag, Heidelberg, New York, 1980).
49. R. Kimmich, W. Unrath, G. Schnur, and E. Rommel, *J. Magn. Reson.* **91**, 136 (1991).
50. P. T. Callaghan and A. Coy, *Phys. Rev. Lett.* **68**, 3176 (1992).
51. G. Fleischer and F. Fujara, *Macromolecules* **25**, 4210 (1992).
52. G. Fleischer and F. Fujara, in *NMR Basic Principles and Progress*, edited by R. Koesfeld and B. Blümich (Springer-Verlag, Heidelberg, 1994).
53. P. Flory, *Statistical Mechanics of Chain Molecules* (Interscience, New York, 1969).
54. K. Kremer and G. S. Grest, *J. Chem. Phys.* **92**, 5057 (1990); **94**, 4103 (1991) (erratum).
55. M. Bishop, M. H. Kalos, and H. L. Frisch, *J. Chem. Phys.* **72**, 3328 (1980); D. Ceperley, M. H. Kalos, and J. L. Lebowitz, *Macromolecules* **14**, 1472 (1981).
56. O. F. Olaj and W. Lantschbauer, *Makromol. Chem. Rapid Comm.* **3**, 847 (1982).
57. C. C. Crabb and J. Kovac, *Macromolecules* **18**, 1430 (1985); C. C. Crabb, D. F. Hoffmann, M. Dial, and J. Kovac, *Macromolecules* **21**, 2230 (1988).
58. J. Skolnick and A. Kolinski, in *Advances in Chemical Physics*, Vol. 78, edited by I. Prigogine and S. A. Rice (Wiley, New York, 1990), p. 223.
59. W. Paul, K. Binder, D. W. Heermann, and K. Kremer, *J. Phys. II France* **1**, 37 (1991).
60. R. G. Kirste, W. A. Kruse, and J. Schelten, *Makromol. Chem.* **162**, 299 (1973); D. G. Ballard, J. Schelten, and G. D. Wignall, *Euro. Polym. J.* **9**, 965 (1973); J. P. Cotton, D. Decker, H. Benoit, *et al.*, *Macromolecules* **7**, 863 (1974).
61. A. L. Rodriguez and J. J. Freire, *Macromolecules* **24**, 3578 (1991).
62. A. Milchev, W. Paul, and K. Binder, *J. Chem. Phys.* **99**, 4786 (1993); I. Gerroff, A. Milchev, W. Paul, and K. Binder, *J. Chem. Phys.* **98**, 6526 (1993).
63. J. D. Ferry, *Viscoelastic Properties of Polymers* (Wiley, New York, 1980).
64. P. H. Verdier, *J. Chem. Phys.* **45**, 2118 (1966).
65. M. Dial, K. S. Crabb, C. C. Crabb, and J. Kovac, *Macromolecules* **18**, 2215 (1985).

66. I. Carmesin and K. Kremer, *Macromolecules* **21**, 2819 (1989); *J. Phys. France* **51**, 915 (1990).
67. G. C. Berry and T. G. Fox, *Adv. Polym. Sci.* **5**, 261 (1968).
68. W. W. Graessley, *Adv. Polym. Sci.* **16**, 1 (1974).
69. D. S. Pearson, *Rubber Chem. Tech.* **60**, 439 (1987).
70. J. E. Tanner, *Macromolecules* **4**, 748 (1971); J. E. Tanner, K. J. Liu, and J. E. Anderson, **4**, 586 (1971).
71. R. Bachus and R. Kimmich, *Polymer* **24**, 864 (1983).
72. R. Kimmich, G. Schnur, and M. Koerf, *Prog. NMR Spectrosoc.* **20**, 385 (1988).
73. J. Kärger, H. Pfeifer and W. Heink, in *Advances in Magnetic Resonance*, Vol. 12 (Academic, New York, 1988), p. 1.
74. D. S. Pearson, G. ver Strate, E. von Meerwall, and F. C. Shilling, *Macromolecules* **20**, 1133 (1987).
75. D. S. Pearson, L. J. Fetters, W. W. Graessley, G. ver Strate, and E. von Meerwall, *Macromolecules* **27**, 711 (1994).
76. M. Appel and G. Fleischer, *Macromolecules* **26**, 5520 (1993).
77. J. S. Higgins and J. E. Roots, *J. Chem. Soc. Faraday Trans. 2* **81**, 757 (1985); J. S. Higgins, *Physica B* **136**, 201 (1986).
78. D. Richter, B. Ewen, B. Farago, and T. Wagner, *Phys. Rev. Lett.* **62**, 2140 (1989).
79. D. Richter, L. Willner, A. Zirkel, B. Farago, L. J. Fetters, and J. S. Huang, *Phys. Rev. Lett.* **71**, 4158 (1993).
80. J. Klein, *Nature* **271**, 143 (1978); *Phil. Mag. A* **43**, 771 (1981); J. Klein and B. J. Briscoe, *Proc. R. Soc. Lond. A* **365**, 53 (1979).
81. H. Hervet, L. Léger, and F. Rondelez, *Phys. Rev. Lett.* **42**, 1681 (1979); L. Léger, H. Hervet, and F. Rondelez, *Macromolecules* **14**, 1732 (1981).
82. M. Antonietti, K. J. Fölsch, and H. Sillescu, *Makrom. Chem.* **188**, 2317 (1987); M. Antonietti, J. Coutandin, and H. Sillescu, *Macromolecules* **19**, 793 (1986).
83. C. R. Bartels, B. Crist, and W. W. Graessley, *Macromolecules* **17**, 2702 (1984).
84. P. F. Green, C. J. Palmström, J. W. Mayer, and E. J. Kramer, *Macromolecules* **18**, 501 (1985).
85. P. F. Green and E. J. Kramer, *Macromolecules* **19**, 1108 (1986).
86. M. Tirrell, *Rubber Chem. Tech.* **57**, 523 (1984).
87. W. W. Graessley, *Adv. Polym. Sci.* **47**, 67 (1982).
88. R. Colby, L. J. Fetters, and W. W. Graessley, *Macromolecules* **20**, 2226 (1987).
89. T. P. Russell, V. R. Deline, W. D. Dozier *et al.*, *Nature* **365**, 235 (1993); G. Agrawal, R. P. Wool, W. D. Dozier *et al.*, *Macromolecules* **27**, 4407 (1994).
90. J. des Cloizeaux, *Europhys. Lett.* **5**, 437 (1988); *Europhys. Lett.* **6**, 475 (1988) erratum; *Macromolecules* **23**, 3992 (1990); *J. Phys. I France* **3**, 61 (1993).
91. J. des Cloizeaux, *Macromolecules* **23**, 4678 (1990); *Macromolecules* **25**, 835 (1992).
92. M. Doi and S. F. Edwards, *J. Chem. Soc. Faraday Trans. 2* **74**, 1789 (1978); *J. Chem. Soc. Faraday Trans. 2* **74**, 1802 (1978); *J. Chem. Soc. Faraday Trans. 2* **74**, 1818 (1978).
93. R. B. Bird and H. C. Öttinger, *Am. Rev. Phys. Chem.* **43**, 371 (1992).
94. M. Appel, G. Fleischer, J. Kärger, F. Fujara, and I. Chang, *Macromolecules* **27**, 4724 (1994).

95. R. Kimmich and H. W. Weber, *J. Chem. Phys.* **98**, 5847 (1993); N. Fatkulin, R. Kimmich, and H. W. Weber, *Phys. Rev. E* **47**, 4600 (1993); H. W. Weber, R, Kimmich, M. Köpf, T. Ramik, and R. Oeser, *Prog. Colloid Polym. Sci.* **90**, 104 (1992).
96. G. Reiter and U. Steiner, *J. Phys. II France* **1**, 659 (1991).
97. A. Karim, A. Monsour, G. P. Felcher, and T. P. Russell, *Mat. Res. Soc. Symp.* **171**, 329 (1990); A. Karim, G. P. Felder, T. P. Russel, *Macromolecules* **27**, 6973 (1994).
98. S. J. Whitlow and R. P. Wool, *Macromolecules* **24**, 5926 (1991).
99. J. Klein, *Macromolecules* **11**, 852 (1978).
100. M. Daoud and P. G. de Gennes *J. Polym. Sci, Polym. Phys. Ed.* **17**, 1971 (1979).
101. Y.-H. Lin, *Macromolecules* **17**, 2846 (1984); *Macromolecules* **19**, 159 (1986); *Macromolecules* **22**, 1437 (1989).
102. G. Marrucci, *J. Polym. Sci, Polym. Phys. Ed.* **23**, 159 (1985).
103. C. Tsenoglou, *J. Polym. Sci., Polym. Phys. Ed.* **26**, 2329 (1988).
104. J. L. Viovy, *J. Phys. France* **46**, 847 (1985).
105. W. W. Graessley and M. Struglinski, *Macromolecules* **19**, 1754 (1986).
106. M. Rubinstein, E. Helfand, and D. Pearson, *Macromolecules* **20**, 822 (1987).
107. M. Rubinstein and R. H. Colby, *J. Chem. Phys.* **89**, 5291 (1988).
108. J. L. Viovy, M. Rubinstein, and R. H. Colby, *Macromolecules* **24**, 3587 (1991).
109. R. Ketzmerick and H. C. Öttinger, *Continuum Mech. Thermodyn.* **1**, 113 (1989).
110. J. M. Deutsch and T. L. Madden, *J. Chem. Phys.* **91**, 3252 (1989).
111. J. Reiter, *J. Chem. Phys.* **94**, 3222 (1991).
112. N. P. T. O'Connor and R. C. Ball, *Macromolecules* **25**, 5677 (1992).
113. M. Rubinstein and S. P. Obukhov, *Phys. Rev. Lett.* **71**, 1856 (1993).
114. A. N. Semenov, *Physica A* **166**, 263 (1990).
115. K. Kawasaki, *Mod. Phys. Lett. B* **4**, 913 (1990); K. Kawasaki, T. Kawakatsu, and W. Zimmermann (to be published)
116. K. S. Schweizer (private communication).
117. W. W. Graessley, *J. Polym. Sci., Polym. Phys. Ed.* **18**, 27 (1980).
118. W. W. Graessley and S. F. Edwards, *Polymer* **22**, 1329 (1981).
119. Y.-H. Lin, *Macromolecules* **20**, 3080 (1987).
120. M. Doi, *J. Phys. A* **8**, 959 (1988).
121. F. Brochard and P. G. de Gennes, *Macromolecules* **10**, 1157 (1977); F. Brochard, *J. Phys. Lett. France* **44**, 39 (1983).
122. R. H. Colby and M. Rubinstein, *Macromolecules* **23**, 2753 (1990).
123. E. Helfand, in *Photophysical and Photochemical Tools in Polymer Science*, edited by M. A. Winnik (Reidel, New York, 1986).
124. D. Richter, B. Farago, R. Butera, L. J. Fetters, J. S. Huang, and B. Ewen, *Macromolecules* **26**, 795 (1993).
125. R. H. Colby, M. Rubinstein, and J. L. Viovy, *Macromolecules* **25**, 996 (1992).
126. L. J. Fetters, D. J. Lohse, D. Richter, T. A. Witten and A. Zirkel, *Macromolecules* **27** (1994).
127. S. M. Aharoni, *Macromolecules* **16**, 1722 (1983).
128. S. Wu, *Polym. Eng. Sci.* **30**, 753 (1990); *Polym. Eng. Sci.* **32**, 823 (1992).

129. R. P. Wool, *Macromolecules* **26**, 1564 (1993).
130. S. F. Edwards, *Proc. R. Soc. Lond. A* **419**, 221 (1988).
131. K. Kremer and G. S. Grest, *J. Chem. Soc. Faraday Trans.* **88**, 1707 (1992).
132. D. Rigby and R.-J. Roe, *J. Chem. Phys.* **87**, 7285 (1987); *J. Chem. Phys.* **89**, 5280 (1988); *Macromolecules* **22**, 2259 (1989); *Macromolecules* **23**, 5312 (1990); R.-J. Roe, D. Rigby, H. Furuya, and H. Takeuchi, *Comp. Polym. Sci.* **2**, 32 (1992); R.-J. Roe, *J. Chem. Phys.* **100**, 1610 (1994).
133. J. H. R. Clarke and D. Brown, *Mol. Sim.* **3**, 27 (1989); *Macromolecules* **24**, 2075 (1991); J. I. McKechnie, R. W. Haward, D. Brown, and J. H. R. Clarke, *Macromolecules* **26**, 198 (1993); D. Brown, J. H. R. Clarke, M. Okuda, and T. Yamazaki, *J. Chem. Phys.* **100**, 1684 (1994).
134. P. V. K. Pant, J. Han, G. D. Smith, and R. H. Boyd, *J. Chem. Phys.* **99**, 597 (1993).
135. D. Richter, R. Butera, L. J. Fetters, J. S. Huang, B. Farago, and B. Ewen, *Macromolecules* **25**, 6156 (1992).
136. J.-P. Ryckaert and A. Bellemans, *Discuss. Faraday Soc.* **66**, 95 (1978).
137. S. Toxvaerd, *J. Chem. Phys.* **93**, 4290 (1990); P. Padilla and S. Toxvaerd, *J. Chem. Phys.* **94**, 5650 (1991); *J. Chem. Phys.* **95**, 509 (1991).
138. K. Kremer and G. S. Grest, in *Computer Simulation of Polymers*, edited by R. J. Roe (Prentice-Hall, New York, 1990).
139. J. Batoulis (private communication).
140. M. Mondello, H.-J. Yang, H. Furuya, and R.-J. Roe, *Macromolecules* **27**, 3566 (1994); H. Furuya, M. Mondello, H.-J. Yang *et al., Macromolecules* **27**, 5674 (1994).
141. D. Y. Yoon, G. D. Smith, and T. Masuda, *J. Chem. Phys.* **98**, 10037 (1993).
142. G. D. Smith and R. H. Boyd, *Macromolecules* **25**, 1326 (1992); G. D. Smith, F. Liu, R. W. Devereaux, and R. H. Boyd, *Macromolecules* **25**, 703 (1992).
143. H. Takeuchi and K. Okazaki, *J. Chem. Phys.* **92**, 5643 (1990); H. Takeuchi, R.-J. Roe, and J. E. Mark, *J. Chem. Phys.* **93**, 9042 (1990).
144. P. V. K. Pant and R. H. Boyd, *Macromolecules* **26**, 679 (1993); R. H. Boyd and P. V. K. Pant, *Macromolecules* **24**, 6325 (1991) and references therein; A. A. Gusev, F. Müller-Plathe, W. F. van Gunsteren *et al., Adv. Polym. Sci.* **116**, 207 (1994).
145. W. Paul, K. Binder, K. Kremer, and D. W. Heermann, *Macromolecules* **24**, 6323 (1991).
146. A. Baumgärtner, *Ann. Rev. Phys. Chem.* **35**, 419 (1984).
147. A. Baumgärtner, in *Applications of the Monte Carlo Method in Statistical Physics*, edited by K. Binder (Springer, Berlin, 1984).
148. K. Kremer and K. Binder, *Comp. Phys. Rept.* **7**, 259 (1988).
149. K. Binder, in *Molecular Level Calculations of the Structure and Properties of Non-Crystalline Polymers*, edited by J. Biscerano (Marcel Dekker, New York, 1992), p. 221.
150. G. S. Grest and K. Kremer, in *Elastomeric Polymer Networks*, edited by J. E. Mark and B. Erman (Prentice Hall, Englewood Cliffs, 1992), p. 164.
151. K. Kremer, in *Computer Simulation in Chemical Physics*, edited by M. P. Allen and D. J. Tildesley (Kluwer Academic, Amsterdam, 1993), p. 397.
152. D. E. Kranbuehl and P. H. Verdier, *Macromolecules* **17**, 749 (1984).

153. A. Kolinski, J. Skolnick, and R. Yaris, *J. Chem. Phys.* **86**, 7164 (1987); *J. Chem. Phys.* **86**, 7174 (1987).
154. H.-P. Wittmann and K. Kremer, *Comp. Phys. Comm.* **61**, 309 (1990); **71**, 343 (1992) (erratum).
155. J. Wittmer, W. Paul, and K. Binder, *Macromolecules* **25**, 7211 (1992).
156. M. P. Allen and D. J. Tildesley, *Computer Simulation of Liquids* (Clarendon Press, Oxford, 1987).
157. R. W. Hockney and J. W. Eastwood, *Computer Simulation using Particles* (Adam Hilger, Bristol, 1988).
158. G. Ciccotti, D. Frenkel, and I. R. McDonald, *Simulations of Liquids and Solids* (North Holland, Amsterdam, 1987).
159. E. R. Duering, K. Kremer, and G. S. Grest, *Phys. Rev. Lett.* **67**, 3531 (1991).
160. G. S. Grest, K. Kremer, and E. R. Duering, *Europhys. Lett.* **19**, 195 (1992); *Physica A* **194**, 330 (1993).
161. E. R. Duering, K. Kremer, and G. S. Grest, *Macromolecules* **26**, 3241 (1993).
162. E. R. Duering, K. Kremer, and G. S. Grest, *J. Chem. Phys.* **101**, 8169 (1994).
163. G. S. Grest and M. Murat, see Chapter 9 of this volume.
164. W. C. Swope, H. C. Andersen, P. H. Berens, and K. R. Wilson, *J. Chem. Phys.* **76**, 637 (1982).
165. C. W. Gear, *Numerical Initial Value Problems in Ordinary Differential Equations* (Prentice-Hall, Englewood Cliffs, NJ, 1971).
166. M. Schulz, R. G. Winkler and P. Reineker, *Phys. Rev. B* **48**, 581 (1993).
167. D. W. Heermann and L. Yixue, *Makromol. Chem. Theory Simul.* **2**, 299 (1993).
168. J. G. Curro, K. S. Schweizer, G. S. Grest, and K. Kremer, *J. Chem. Phys.* **91**, 1357 (1989).
169. K. S. Schweizer and J. G. Curro, *Adv. Polym. Sci.* **116**, 319 (1994).
170. G. S. Grest, B. Dünweg, and K. Kremer, *Comp. Phys. Comm.* **55**, 269 (1989).
171. R. Everaers and K. Kremer, *Comp. Phys. Comm.* **81**, 19 (1994).
172. H. C. Andersen, *J. Chem. Phys.* **72**, 2384 (1980).
173. M. Parrinello and A. Rahman, *Phys. Rev. Lett.* **45**, 1196 (1980).
174. J. H. R. Clarke, see Chapter 5 of this volume.
175. For a review see *Computer Simulation of Polymers*, edited by R. J. Roe (Prentice-Hall, New York, 1990).
176. J. M. Deutsch, *Phys. Rev. Lett.* **49**, 926 (1982); *Phys. Rev. Lett.* **51**, 1924 (1983).
177. K. Kremer, *Phys. Rev. Lett.* **51**, 1923 (1983).
178. K. E. Evans and S. F. Edwards, *J. Chem. Soc. Faraday Trans. 2* **77**, 1891 (1981).
179. S. Geyler, T. Pakula, and J. Reiter, *J. Chem. Phys.* **92**, 2676 (1990); J. Reiter, T. Edling, and T. Pakula, *J. Chem. Phys.* **93**, 837 (1990); A. Gauger, A. Weyersberg, and T. Pakula, *Makromol. Chem., Theory Sim.* **2**, 531 (1993).
180. K. Kremer and K. Binder, *J. Chem. Phys.* **81**, 6381 (1984).
181. A. Baumgärtner and M. Muthukumar, *J. Chem. Phys.* **87**, 3082 (1987).
182. J. Gao and J. H. Weiner, *Macromolecules* **24**, 1519 (1991).
183. J. Gao and J. H. Weiner, *Macromolecules* **24**, 5179 (1991).
184. J. Gao and J. H. Weiner, *Macromolecules* **25**, 1348 (1992); *Macromolecules* **25**, 3462 (1992); *J. Chem. Phys.* **97**, 8698 (1992).

185. M. Fixman, *J. Chem. Phys.* **95**, 1410 (1991).
186. K. G. Ngai, S. L. Peng, and J. Skolnick, *Macromolecules* **25**, 2184 (1992).
187. P. G. de Gennes, *Phys. (NY)* **3**, 97 (1967).
188. A. Z. Akcasu, M. Benmouna, and C. C. Han, *Polymer* **21**, 866 (1980).
189. C. Grayce, *J. Chem. Phys.* **98**, 9916 (1993).
190. P. G. de Gennes, *J. Phys. France* **42**, 735 (1981).
191. D. Richter, B. Farago, L. J. Fetters, J. S. Huang, B. Ewen, and C. Lartigue, *Phys. Rev. Lett.* **64**, 1389 (1990).
192. J. des Cloizeaux, *J. Phys. I France* **3**, 1523 (1993).
193. J. Wittmer, W. Paul, and K. Binder, *J. Phys. II France* **4**, 873 (1994).
194. K. Kremer (to be published).
195. D. Richter (private communication).
196. M. Kröger and S. Hess, *Physica* **A195**, 336 (1993).
197. M. Kröger, W. Loose and S. Hess, *Journal of Rheology* **37**, 1057 (1993).
198. M. Kröger, *Makr. Chemie Macr. Symp. Theory and Simulations* **3**, 639 (1994).
199. J. M. Deutsch and N. Goldenfeld, *Phys. Rev. Lett.* **48**, 1694 (1982).
200. J. C. LeGuillou and J. Zinn-Justin, *Phys. Rev. B* **21**, 3976 (1980).
201. H. P. Wittmann, K. Binder, and K. Kremer, in *Basic Features of the Glassy State*, edited by J. Colmenero and A. Alegria (World Scientific Press, Singapore, 1990).
202. F. T. Wall and F. Mandel, *J. Chem. Phys.* **63**, 4592 (1975); F. Mandel, *J. Chem. Phys.* **70**, 3984 (1979).
203. J.-U. Sommer, M. Schulz and H. L. Trautenberg, *J. Chem. Phys.* **98**, 7515 (1993).
204. P. J. Flory, *Proc. R. Soc. Lond. A* **351**, 351 (1976).
205. A. J. Stavermann, *Adv. Polym. Sci.* **44**, 73 (1982).
206. G. Heinrich, E. Straube, and G. Helmis, *Adv. Polym. Sci.* **85**, 33 (1988).
207. L. R. G. Treolar, *The Physics of Rubber Elasticity* (Clarendon, Oxford, 1975).
208. J. E. Mark, *Adv. Polym. Sci.* **44**, 1 (1982).
209. J. P. Queslel and J. E. Mark, *Adv. Polym. Sci.* **65**, 135 (1984).
210. H. James, *J. Chem. Phys.* **15**, 651 (1943); H. James and E. Guth, *J. Chem. Phys.* **15**, 669 (1947); *J. Chem. Phys.* **21**, 1039 (1953).
211. G. Ronca and G. Allegra, *J. Chem. Phys.* **63**, 4990 (1975).
212. W. W. Graessley, *Macromolecules* **8**, 186 (1975); *Macromolecules* **8**, 865 (1975).
213. R. T. Deam and S. F. Edwards, *Phil. Trans. R. Soc. Lond. A* **280**, 317 (1976).
214. P. J. Flory, *J. Chem. Phys.* **66**, 5720 (1977).
215. B. Erman and P. J. Flory, *J. Chem. Phys.* **68**, 5363 (1978).
216. P. J. Flory and B. Erman, *Macromolecules* **15**, 800 (1982).
217. S. Kästner, *Colloid Polym. Sci.* **259**, 499 (1981); *Colloid Polym. Sci.* **259**, 508 (1981).
218. J. Scanlan, *J. Polym. Sci.* **43**, 501 (1960).
219. L. C. Case, *J. Polym. Sci.* **45**, 397 (1960).
220. N. R. Langley, *Macromolecules* **1**, 348 (1968).
221. L. M. Dossin and W. W. Graessley, *Macromolecules* **12**, 123 (1979).
222. D. S. Pearson and W. W. Graessley, *Macromolecules* **11**, 528 (1978).
223. P. G. Higgs and R. C. Ball, *J. Phys. France* **49**, 1785 (1988).
224. R. C. Ball, M. Doi, S. F. Edwards and M. Warner, *Polymer* **22**, 1010 (1981).

225. T. A. Vilgis and H.-G. Kilian, *Polymer* **25**, 71 (1983).
226. T. A. Vilgis and F. Boué, *Polymer* **27**, 1154 (1983).
227. R. J. Gaylord, *J. Polym. Bull.* **8**, 325 (1982); *Polym. Eng. Sci.* **19**, 263 (1979).
228. G. Marrucci, *Rheol. Acta* **18**, 193 (1979); *Macromolecules* **14**, 434 (1981).
229. K. Iwata, *J. Chem. Phys.* **83**, 1969 (1985).
230. M. Kosc, *Colloid Polym. Sci.* **266**, 105 (1988).
231. D. Adolf, *Macromolecules* **21**, 228 (1988).
232. P. G. Higgs and R. C. Ball, *Europhys. Lett.* **8**, 357 (1989).
233. P. Goldbart and N. Goldenfeld, *Macromolecules* **22**, 948 (1989); *Phys. Rev. A* **39**, 1402 (1989); *Phys. Rev. A* **39**, 1412 (1989).
234. D. R. Miller and C. W. Macosko, *J. Polym. Sci., Polym. Phys. Ed.* **25**, 2441 (1987); *J. Polym. Sci., Polym. Phys. Ed.* **26**, 1 (1988).
235. M. Mooney, *J. Appl. Phys.* **11**, 582 (1940).
236. R. S. Rivlin, *Phil. Trans. R. Soc. A* **241**, 379 (1948); *Phil. Trans. R. Soc. A* **242**, 173 (1949).
237. J. Bastide, L. Leibler, and J. Prost, *Macromolecules* **23**, 1821 (1990).
238. T. Vilgis, in *Elastomeric Polymer Networks*, edited by J. E. Mark and B. Erman (Prentice Hall, Englewood Cliffs, 1992), p. 32.
239. J. H. Weiner and J. Gao, in *Elastomeric Polymer Networks*, edited by J. E. Mark and B. Erman (Prentice Hall, Englewood Cliffs, 1992), p. 103.
240. G. S. Grest and K. Kremer, *J. Phys. (France)* **51**, 2829 (1990).
241. G. S. Grest and K. Kremer, *Macromolecules* **23**, 4994 (1990).
242. E. R. Duering, K. Kremer, and G. S. Crest, *Prog. Colloid Polym. Sci.* **90**, 13 (1992).
243. M. Lazár, R. Rado and J. Rychly, *Adv. Polym. Sci.* **95**, 149 (1990).
244. A. E. Tonelli and E. Helfand, *Macromolecules* **7**, 59 (1974); E. Helfand and A. E. Tonelli, *Macromolecules* **7**, 832 (1974).
245. Y. K. Leung and B. E. Eichinger, *J. Chem. Phys.* **80**, 3877 (1984); *J. Chem. Phys.* **80**, 3885 (1984).
246. L. Y. Shy and B. E. Eichinger, *Macromolecules* **19**, 2787 (1986).
247. J. G. Curro and P. A. Pincus, *Macromolecules* **16**, 559 (1983); J. G. Curro, D. S. Pearson, and E. Helfand, *Macromolecules* **18**, 1157 (1985).
248. M. Schulz and J.-U. Sommer, *J. Chem. Phys.* **96**, 7102 (1992).
249. M. Schulz and K. Binder, *J. Chem. Phys.* **98**, 655 (1993).
250. D. S. Pearson and W. W. Graessley, *Macromolecules* **13**, 1001 (1980).
251. M. Gottlieb, C. W. Macosko, G. S. Benjamin, K. O. Meyers, and E. W. Merrill, *Macromolecules* **14**, 1039 (1981).
252. W. Opperman and N. Rennar, *Progr. Coll. Polym. Sci.* **75**, 49 (1987).
253. R. Oeser, B. Ewen, D. Richter, and B. Farago, *Phys. Rev. Lett.* **60**, 1041 (1988).
254. S. K. Patel, S. Malone, C. Cohen, J. R. Gillmor, and R. Colby, *Macromolecules* **25**, 5241 (1992).
255. I. Larsson and O. Kramer, *Makromol. Chem., Macromol. Symp.* **76**, 117 (1993).
256. J.-U. Sommer, *J. Chem. Phys.* **97**, 5777 (1992).
257. A. Baumgärtner and M. Muthukumar, *J. Chem. Phys.* **84**, 440 (1986).

258. R. B. Bird, R. C. Armstrong, and O. Hassager, *Dynamics of Polymeric Liquids*, Vol. 1 (Wiley, New York, 1977).

259. A. Kloczkowski, J. E. Mark, and B. Erman, *Macromolecules* **22**, 1423 (1989); B. Erman, A. Kloczkowski, and J. E. Mark, *Macromolecules* **22**, 1432 (1989).

260. A. R. C. Baljon, G. S. Grest, and T. A. Witten, *Macromolecules* **28** (1995).

261. C. M. Ylitado, J. A. Zawada, G. G. Fuller, V. Abetz, and R. Stradler, *Polymer* **33**, 2949 (1992); C. M. Ylitalo, J. A. Kornfield, G. G. Fuller, and D. S. Pearson, *Macromolecules* **24**, 749 (1991); J. A. Kornfield, G. G. Fuller, and D. S. Pearson, *Macromolecules* **22**, 1334 (1989).

262. F. Boue, J. Bastide, and M. Buzier, in *Molecular Basis of Polymer Networks*, Vol. 42, edited by A. Baumgärtner and C. E. Picot (Springer-Verlag, Heidelberg, 1989), p. 65.

5

MOLECULAR DYNAMICS OF GLASSY POLYMERS

Julian H. R. Clarke

5.1 Introduction

Over the past fifty years or so there has built up a considerable understanding of polymer behavior based on the use of highly simplified models; other chapters in this book exemplify recent progress made in this area. So why bother with this atomistic modeling? The reason is that there is an important limitation of the simplified (sometimes referred to as "coarse-grain") models in that they do not provide a direct connection between monomer level structure and bulk properties. Atomistic level simulations offer new opportunities for making such predictions *ab initio*. Using only a knowledge of interatomic forces together with the assumptions of classical mechanics it is possible, at least in principle, to give a complete atomic-level description of a polymer system. The emergence of these modeling studies is particularly timely in view of very recent developments in experimental techniques for detailed investigation of polymer dynamics, such as multidimensional NMR[1] and inelastic neutron scattering.[2]

Of course, if one is interested in large scale configurational properties, for instance, then the atomic-level detail may be unimportant, but this is not true for many properties of dense melts and polymer glasses, where long-range configurational fluctuations are largely frozen-out and properties are determined by the detailed interplay of entropy effects with intra- and intermolecular forces. One of the great attractions of computer simulation is the control that one has over the parameters defining a particular system. Coupled with the ready availability of detailed information from the simulation this means that systematic studies can be devised to show how molecular parameters are related to bulk properties. We can look forward to computer-aided molecular design as a powerful new tool for the development of polymer materials.

The past decade has seen steadily growing activity in the detailed atomistic modeling of polymer melts and glasses using energy minimisation[3] and molecular dynamics simulation.[4–7] These studies have been aimed at achieving an atomistic level understanding of a variety of physical properties such as stress–strain behavior, diffusion of small solute molecules and local chain motions. Because of its relative simplicity, polyethylene has come under a great deal of attention in these early studies. It is ironic that pure amorphous polyethylene is extremely difficult to prepare in the laboratory so comparisons with experiment have been quite limited. From the simulation point of view, however, it is quite clear that a sound understanding of the simple systems must be achieved before embarking on more ambitious simulations of more complex polymers.

There are thus many exciting prospects for atomistic-level polymer modeling but, with such a powerful technique at our fingertips however, it is necessary to be careful and critical in its application. What you get out of a simulation exercise depends entirely on what you put into it. The aim of this chapter is not to perform an exhaustive review of the literature. Rather we wish to discuss in detail some of the application areas where progress has been made and also to focus attention on some of important technical issues that arise in such molecular dynamics simulations, particularly in regard to modeling the glassy state. The hope is to provide the reader with a survey of the opportunities that have so far been exploited and also something of a critical eye when it comes to evaluating the usefulness of simulation studies.

So how do we set about designing a simulation of a particular glassy polymer? The first question is, can we model a material that in any way resembles that formed in the laboratory? This is not just a question of having access to a realistic force field; after all, with the aid of experimental data, quantum mechanical calculations and a little persistence with empirical adjustments we can usually find a force field that will reproduce at least some properties. It is much more a question of the *completely different time scales* for the formation of laboratory and simulated polymer glasses. As discussed in a later section it is important to understand the significant differences between the glass transformation and the properties of polymer glasses formed in simulation studies and those obtained in laboratory experiments.

The application of molecular dynamics to study molecular fluids is well established[8] and many of the techniques can be transferred in a straightforward way to polymer simulations. The connectivity of polymers, however, raises some special issues which require careful attention if anything like a realistic simulation is to be obtained. For instance, the spontaneous configurational fluctuations in polymers often cover a very wide range of time scales—from picoseconds for local motions to milliseconds or even longer for large-scale fluctuations in the case of polymer melts.[9] Since atomistic modeling techniques are currently limited to times of order nanoseconds there is here a potentially serious problem of ergodicity; simulations generally cannot be run long enough to sample fully all the equilibrium fluctuations. In practical terms of course, long time fluctuations may not be important for the property being examined, e.g., guest molecule diffusion in polymer glasses. In this case we can avoid non-ergodicity problems by either using a very large system ($N \sim 100\,000$)[10] or by averaging over many smaller but independent samples.[11] Insufficient averaging of this type has been the shortcoming of many polymer simulations to date.

Having raised these issues we must nevertheless always keep in mind the objectives of the simulation, and compromises are unavoidable. For instance, if we are only interested in local chain dynamics such as motions of side groups etc., or in diffusion of guest molecules then these are probably not very dependent on the accuracy of the overall chain dimensions; a much more important criterion is likely to be the packing density and there is evidence to suggest, as is noted later, that this property relaxes much faster. If however we are interested in chain diffusion or mechanical behavior—properties which involve collective motions—then it is important to have realistic overall chain dimensions.

5.2 Molecular dynamics for polymers

Molecular dynamics simulation is basically very straightforward. It is a deterministic method in which the system follows a well defined trajectory in phase space. It is the only reliable method for examining time-dependent properties. Assuming the applicability of classical mechanics, it involves the simultaneous solution of the equations of motion for a small sample of particles interacting according to a predetermined force field and fixed conditions.

In the simplest case, both the total energy E and the volume V are kept constant and we have a microcanonical, or (N, V, E) ensemble in which case the equations to be solved are Hamiltonian. In practical simulations it is often much more convenient however to control the pressure and, in particular, the temperature of the system. This can be achieved by modifying the Hamiltonian or by using more general Lagrangian equations of motion. A good deal of work has been put into developing and testing such "extended" molecular dynamics methods to simulate the (N, V, T) and isothermal–isobaric (N, P, T) ensembles.[8] In simple molecular fluids they all give very similar results[8,12] and they have all been used at various times for polymer simulations. Whatever method is chosen the equations of motion are solved numerically using discrete time steps with the aid of well established integration algorithms.

A characteristic of small sample simulations of dense polymers under constant volume conditions is that the average pressure tensor is often anisotropic with substantial differences between the on-diagonal components and, in addition, nonzero off-diagonal terms. This has led to the use of molecular dynamics algorithms which control the pressure tensor, $\mathbf{P}$, and which allow these dynamic anisotropies to be relaxed out. Once implemented such methods have the added advantage of being easily adapted to measure the response of such systems to externally applied pressure fields as we shall show later.

For the majority of atomic and "small" molecule systems at equilibrium in the $(N,\mathbf{P},T)$ ensemble ($\mathbf{P}$ is the pressure tensor) it is widely accepted that the most rigorous approach is to use the controlled pressure technique proposed by Rahman and Parrinello (RP)[13] in conjunction with the Nosé–Hoover thermostat.[14,15] However, the choice of method must take careful account of the material we wish to study, how it is modeled and any external perturbations which we wish to apply. For polymers the Berendsen loose-coupling controlled pressure MD technique[16] is a good compromise. Although the theoretical basis of this method has been criticised[17] in practice it has been found[18,19] that to within statistical uncertainties first-order properties are the same as those obtained by more rigorous approaches.

The Berendsen method utilizes weak coupling of an external tensorial pressure field, $\mathbf{P}^*$, to the system through a simple feedback loop.[18] It is assumed that provided the coupling is loose enough it will have an insignificant effect on the first-order properties of the system. A very similar scheme can be used also to control the temperature of a simulation.[16,18] For the detailed discussion of this method readers are referred to the original articles.[16,18]

The coupling is implemented by allowing the matrix $\mathbf{h}$, made up from the basis vectors, $\mathbf{a}$, $\mathbf{b}$ and $\mathbf{c}$ which determine the shape of the primary dynamics cell, to respond to imbalances between the internally measured pressure tensor and an externally applied pressure tensor. The equation for the rate of change of the $\mathbf{h}$ matrix with time is then defined as

$$\dot{\mathbf{h}} = \frac{\mathbf{P} - \mathbf{P}^*}{M} \tag{5.1}$$

where M is a coupling constant and $\mathbf{P}$ is the internally measured pressure tensor, which in this case is defined in an "atomic" frame of reference, i.e., the momentum is localized at the positions of the CH_2 sites, and hence is symmetric,

$$\mathbf{P} = \frac{1}{V}\sum_{i=1}^{N}\left[\frac{1}{m_i}\mathbf{p}_i\mathbf{p}_i + \mathbf{r}_i\mathbf{f}_i\right] \tag{5.2}$$

Although the above equation is formally correct, great care has to be taken in a simulation using periodic boundaries when calculating $\mathbf{P}$. Criteria that can be used for choosing the value of the coupling constant M have been discussed in detail elsewhere.[18]

A simple proportional scaling of coordinates is used to minimize local disturbances.[16,18] If we define a set of scaled coordinates, $\mathbf{s}$, by

$$\mathbf{s} = \mathbf{h}^{-1}\mathbf{r} \tag{5.3}$$

differentiation gives the following equation of motion for the sites

$$\dot{\mathbf{r}}_i - \mathbf{h}\dot{\mathbf{s}}_i + \dot{\mathbf{h}}\mathbf{s}_i - \frac{\mathbf{P}_i}{m_i} + \dot{\mathbf{h}}\mathbf{h}^{-1}\mathbf{r}_i \tag{5.4}$$

The motion is thus seen to be split into two contributions which are integrated separately, that due to the momenta and that resulting from the change in shape and size of the cell. The "fast" motions due to the momenta are dealt with in the usual way using a "leapfrog" algorithm incorporating an iterative scheme to maintain the constraints whereas a simple first-order Taylor expansion is considered sufficient to integrate the equation for the relatively "slow" motion of the box

$$\mathbf{h}(t + \Delta t) = \mathbf{h}(t) + \left[\frac{\mathbf{P} - \mathbf{P}^*}{M}\right]\Delta t \tag{5.5}$$

It can then simply be shown that to first order the motion of the box results in a scaling of the position of a site

$$\mathbf{r}_i(t + \Delta t) = \mathbf{h}(t + \Delta t)\mathbf{h}^{-1}(t)\mathbf{r}_i(t) = \mathbf{H}\mathbf{r}_i(t) \tag{5.6}$$

Although the method is less rigorous than the alternative Rahman–Parrinello (RP) technique[13] it does have at least one important practical advantage. This is that the pressure imbalance is coupled to the *first derivative* of the basis vectors rather than to the second derivative; this means that motions of the box are overdamped and so there is little tendency for an unphysical oscillatory response to changes in the applied pressure. For this reason this method comes into its own for the calculation of nonequilibrium properties of dense highly viscoelastic systems.

5.3 Force fields

As in any molecular level simulation one of the first decisions to make is what inter- and intramolecular force field to use. We have basically two choices. Firstly, we can set about bringing together as much information as possible from experiment and quantum mechanical calculations to develop "good" force fields and in this way to aim for quantitatively accurate modeling. With this approach there is usually little alternative but to employ a fully atomic representation with, for instance, the GROMOS force field.[20] It must be remembered however that no force field will be completely accurate and all of them have limitations.

The alternative is to aim for a more generic model which incorporates the essential features of inter- and intramolecular interactions. For any realistic polymer model these are connectivity, chain flexibility and van der Waals interactions. The connectivity may be linear (as for the models discussed here) or may involve branches and/or crosslinks. The flexibility of chains will be limited by both "chemical" forces along the chain and local van der

Waals interactions between chains which restrict torsional and bond angle motions.

The emphasis here is not on a detailed description of molecular interactions; in fact what is more important is the detail that one can leave out and still correctly reproduce the essential behavior of the system. An example of this kind of approach is the identification of functional groups which are expected to maintain a fairly rigid structure (e.g., CH_3 and CH_2 groups) and the representation of them as single interaction sites (see below). Of course, this kind of simplification is quite different from that of the coarse-grain models where beads are used to represent statistical units in a polymer chain.

Since much of this chapter will be concerned with modeling polyethylene we will briefly outline four different models that have been used in simulations of this material and which use the second of the above approaches. We identify them as PE I to PE IV and in all four cases the monomer units are treated as single interaction sites and given masses corresponding to CH_2 groups.

There are small differences between the intramolecular potentials but these are unlikely to cause strong deviations in bulk properties. In models PE I[6] and PE II[11] neighboring sites on the chain are connected together by rigid bonds of length 0.153 nm, whereas harmonic springs are employed in models PE III[5] and PE IV.[21] The use of springs usually forces a much shorter time-step on the simulation unless an artificially small force constant is used[5] in which case there is always a risk of spurious coupling between the bond vibrations and other degrees of freedom.

Flexibility of the chains is limited by incorporating a harmonic valence angle potential, $\Phi(\theta)$, and a torsional potential, $\Phi(\alpha)$, into the model. In the case of models PE I, PE II and PE III $\Phi(\theta)$ is of the form,

$$\Phi(\theta) = \frac{1}{2}k_\theta(\cos\theta - \cos\theta_0)^2 \tag{5.7}$$

where $k_\theta = 520$ kJ mol^{-1}, $\theta_0 = 112.813°$ for PE I and PE II and $k_\theta = 500$ kJ mol^{-1}, $\theta_0 = 120°$ in the case of PE III. Use of the cosine of the angle in the harmonic potential is a computational convenience but it is possible to express $\Phi(\theta)$ directly in terms of angle displacements as in the case of PE IV which uses $k_\theta = 482$ kJ mol^{-1} and $\theta_0 = 111.6°$. For small displacements there is very little difference in the two potential forms. In the absence of precise experimental data the choice of k_θ is in any case somewhat arbitrary.

The torsional potential restricting internal rotations about a bond in the chain can be parametrized in terms of the dihedral angle, α, formed by this bond and the two adjacent bonds, using experimental and *ab initio* data for short chain alkanes. The form used in PE I for instance is that due to Steele[22] and is given below

$$\Phi(\alpha)/J\text{mol}^{-1} = C_0 + C_1 \cos\alpha + C_2 \cos 2\alpha + C_3 \cos 3\alpha \tag{5.8}$$

where $C_0 = 8832, C_1 = 18\,087, C_2 = 4880$ and $C_3 = -31800$.

There are much more significant differences between the force fields used by the four models for *inter*chain interactions and non-bonded interactions (those between sites separated by at least three others). All use the Lennard-Jones (LJ) 12-6 potential form

$$\Phi(|\mathbf{r}_{ij}|) = 4\epsilon\left\{\left(\frac{\sigma}{|\mathbf{r}_{ij}|}\right)^{12} - \left(\frac{\sigma}{|\mathbf{r}_{ij}|}\right)^{6}\right\} \tag{5.9}$$

but the parameters are different. For PE I and II $\epsilon/k_B = 57\,\text{K}$ and $\sigma = 0.428\,\text{nm}$. The potential however is truncated at different distances, r_c, for these two models. In PE I $r_c = 2.5\,\sigma$ and the appropriate long-range corrections were made to the potential energy and the virial at each step according to the density and assuming $g(r) = 1$ for $r > r_c$ so the attractive interactions are fully represented in this model. This potential gave a reasonable fit to the density of real polyethylene at 500 K, as extrapolated from experimental data.[23]

In PE II the truncation is at the minimum of the potential, so there are no attractive interactions. It is widely accepted, however, that it is the repulsive part of the potential that is primarily reponsible for structural effects in dense fluids.[24] In this case the potential is also raised by the well depth

$$\begin{aligned} \Phi(|\mathbf{r}_{ij}|) &= 4\epsilon((\sigma/|\mathbf{r}_{ij}|)^{12} - (\sigma/|\mathbf{r}_{ij}|)^{6}) + \epsilon \quad &&\text{for}\,|\mathbf{r}_{ij}| \le 2^{1/6}\sigma \\ \Phi(|\mathbf{r}_{ij}|) &= 0 &&\text{for}\,|\mathbf{r}_{ij}| > 2^{1/6}\sigma \end{aligned} \tag{5.10}$$

This short-range potential is computationally highly efficient; molecular dynamics programs run about 3.5 times faster for PE II than for PE I.[11] In PE III $\epsilon/k_B = 57\,\text{K}$ and $\sigma = 0.38\,\text{nm}$ and in this case the potential is truncated and raised at $1.5\,\sigma$[5]; this is a form previously used in studies of glass formation in the monatomic Lennard-Jones fluid.[25] In PE IV $\epsilon/k_B = 57\,\text{K}$ and $\sigma = 0.38\,\text{nm}$ but the interaction centre is offset from the centre of the CH_2 unit by an amount 0.042 nm along the bisector of the obtuse C–C–C angle. This is referred to as an anisotropic united atom potential[26] and it gave an improved fit to the density of molten laboratory polyethylene over a fairly wide range of conditions.[21] No long-range corrections are made to the pressure for PE III and PE IV and, with the dominant repulsive forces, the measured values are extremely high in molecular dynamics simulations of the melt. The density has to be arbitrarily adjusted to fit either experimental data or other simulation data.

5.4 Preparation of polymer melt samples

5.4.1 Building polymer structures

With simple molecular liquids it is relatively straightforward to prepare an equilibrium sample from an arbitrary starting configuration using molecular dynamics or Monte Carlo methods. This method does not lend itself favorably to polymers mainly due to the long time scale required for structural relaxation in these materials. Roe *et al.*[5] used the method, however, in simulations of molten polyethylene using the PE III model but only by starting the system at a very high temperature using highly reduced values for the intermolecular forces during the initial stages to speed up the equilibrium before cooling to the required temperature. This method is unlikely to give configurationally equilibrated samples for long chains in the time scale of molecular dynamics simulations.

Most simulations to date have used a two-stage process for direct preparation of amorphous samples at the required temperature. The two stages are chain growth to produce the basic topologies followed by a period of Monte Carlo or molecular dynamics, sometimes preceded by energy minimization, to "equilibrate" the excluded volume interactions. Again the device is sometimes used of reducing the magnitude or range of the van der Waals interactions during the initial period of relaxation. The first stage consumes a tiny fraction of the total computing time but is extremely important since once excluded volume interactions are fully introduced then the time scale for further topological changes becomes extremely long.

Several schemes have been adopted for the direct preparation of polymer melt samples of which here we consider only two.

1. Chain growth *including* excluded volume interactions, followed by equilibration.
2. Growth of noninteracting chains *followed by* the introduction of excluded volume with subsequent equilibration.

In Method 1 a Monte Carlo algorithm was devised which included chain conformation probabilities given by the rotational isomeric states (RIS) model together with nonbonded interactions between backbone carbons separated by four or more bonds. These structures were then relaxed by energy minimization. The same approach has also been used for polycarbonate,[27] polysulfone,[28] and polyvinylchloride[29] chains.

Method 2 uses an alternative approach of continuous intramolecular potentials instead of RIS probabilities and and this has been applied to the PE I model of polyethylene chains[30] with $N = 1000$. It has been pointed out, however, that site-by-site chain growth with excluded volume samples from a non-Boltzmann distribution of end-to-end distances.[31] In addition the effective density increases during growth so this procedure also gives a *non-uniform* distribution of conformational states along the chain—with an increased probability of more compact conformers towards the end of the chain.

A considerable improvement is obtained by carrying out the self-avoiding random walk (SARW) in the absence of all long-range interactions, a method referred to as phantom chain growth (PCG).[31] This is based on the idea, first introduced by Flory,[32] that in pure melts polymer chains display the same Gaussian statistics as under theta solvent conditions where there is full screening of long-range interactions. In the Flory model chain dimensions in the melt are determined only by local intramolecular interactions along the chain; this approach is further simplified in his RIS model. There is considerable experimental evidence to support these proposals, and recent accurate studies of the dimensions of alkane-like chains using molecular dynamics[11] have shown that to within 1–2% Flory model chains have, for instance, the same mean square radii of gyration as melt chains with full interactions up to $N = 100$.

Although PCG removed the nonuniform distribution of state along the chain, it is still not completely satisfactory since the Monte Carlo growth method used samples from a biased distribution. The discrepancy can be quite large—thus the fraction of trans states after growing an $N = 1000$ polyethylene chain was about 5% below that of an equilibrated melt.[31] The problem is easily eliminated, however, by equilibrating the distribution of phantom chains using, for instance, a pivot MC algorithm.[33] Following growth the required number of chains are then randomly positioned and oriented with the periodic system and the long-range van der Waals interactions introduced.

5.4.2 Introducing excluded volume

An unavoidable side-effect of either of the growth procedures used above is that, at a typical melt density, there are bound to be a large number of overlaps between sites for this initial configuration. In principle, energy minimization could be used to remove the high energy contacts but molecular dynamics allows relaxation at a specified temperature. It is necessary however to moderate the forces in the very early stages to avoid breakdown of the MD algorithm.

Several methods have been proposed for reducing the intensity of repulsive interactions in this initial period, such as use of a cosine potential[34] or a "soft core" potential.[35] Unfortunately for both these forms the force is actually zero for $r = 0$ and there is still a finite chance of sites remaining superimposed. One method that has proved quite robust uses a "truncated force" potential[31] in which the short-range force for neighbours i and j, where $|i - j| > 5$, is constrained to be constant below a critical separation r_{tr} i.e.,

$$\frac{-\mathrm{d}\Phi_m(r)}{\mathrm{d}r} = F_{\mathrm{tr}}\left(= \frac{-\mathrm{d}\Phi(r_{\mathrm{tr}})}{\mathrm{d}r}\right) \quad \text{for } r \leq r_{\mathrm{tr}} \tag{5.11}$$

The full definition of the resulting modified potential is then

$$\Phi_m(r) = \Phi(r) \quad \text{for } r > r_{\mathrm{tr}} \tag{5.12}$$

$$\Phi_m(r) = \Phi(r_{\mathrm{tr}}) + (r_{\mathrm{tr}} - r)F_{\mathrm{tr}} \quad \text{for } r \leq r_{\mathrm{tr}} \tag{5.13}$$

r_{tr} must be sufficiently small so that only a few pairs will be within this distance in the equilibrium distribution (which at this stage is unknown), but not so small that the large magnitude of F_{tr} causes breakdown of the algorithm.

The truncated potential is only applied to the "long-range" interactions with the full potential retained for the 1–5 interactions. The procedure is to decrease continuously the value of r_{tr} from 0.9 σ to 0.7 σ in an initial run of time length τ_v using a time step of 1 fs and re-scaling of particle momenta at each step to remove the large amounts of thermal energy released. At this stage the switch is made from the modified potential to the full potential. Satisfactory results are obtained for PE using $\tau_v = 3$ ps. So far this approach has not failed at least for our n-alkane-like models. During this initial stage it is easiest to carry out the MD simulations at constant volume with the temperature kept close to 500 K using the loose-coupling method[18] with a coupling constant $\tau_T \leq 0.1$ ps. Subsequent relaxation can be carried out either under controlled pressure or constant volume.

5.4.3 *Sample relaxation*

The growth of representive chain configurations is therefore fairly straightforward. Unfortunately much of this good work is undone as soon as the full excluded volume interactions are introduced in the second phase of the preparation! In Fig. 5.1(a) we show the effect, for instance, on the radius of gyration for alkane-like chains[10]; the results are for $N = 100$ but the effect seems to be quite general. These simulations were carried out at constant volume but can equally well be performed under controlled pressure conditions.

There are two striking features of this plot. Firstly, the introduction of excluded volume causes an immediate and substantial *decrease* in the radius of gyration (the chain configurations were initially equilibrated using pivot Monte Carlo so, within the statistical error of about 1%, the $t = 0$ value of the ordinate was unity). Secondly, the relaxation back to equilibrium values is (as might be predicted) extremely slow, even for these chains of only 100 monomers! Up to $N = 100$ the relaxation times scale approximately as n^2 so under the same conditions it would require of order 0.3 ms to fully equilibrate chains with $N = 1000$. In fact entanglement effects in long chains would lengthen the time even further.

The reasons for this spontaneous perturbation from the equilibrium configuration are still the subject of debate[33] but any explanation must take into

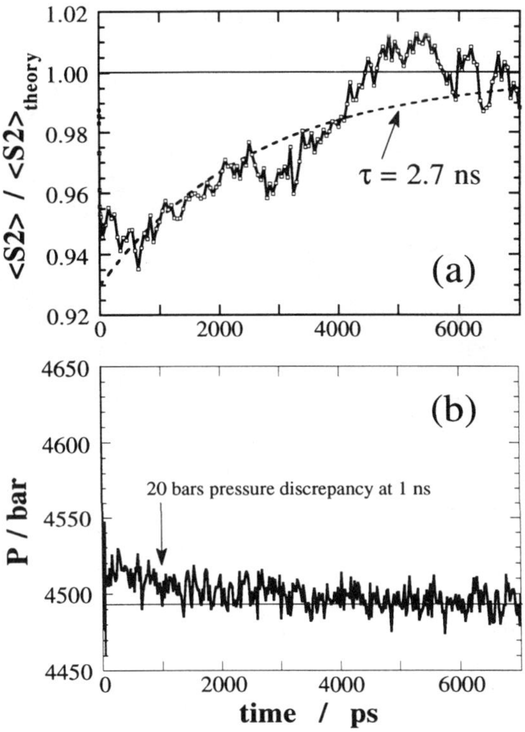

Fig. 5.1 The relaxation of (a) the square radius of gyration, S2, and (b) the hydrostatic pressure at 500 K following the introduction of excluded volume in a sample of a melt composed of 640 chains each with 100 monomers of model PE II polyethylene (see Section 5.3). Note the expanded vertical scale. The simulations were at constant volume; 20 bars pressure discrepancy would be equivalent to a density discrepancy at constant pressure of about 0.1%. S2 values are shown relative to those determined assuming complete screening of long-range interactions (see text). The density is 0.70 g cm^{-3}. These results were obtained with a Fujitsu AP1000 massively parallel computer.

account the fact that both the coordinates and the momenta in these samples are far from equilibrium during the first stages of these relaxations. Although the mean kinetic energy of the interaction sites is controlled by the loose coupling, during the first 10 ps or so there is a significant lack of equipartition in the kinetic energy between the torsional, center-of-mass and rotational motions.

All is not lost however since, as shown in Fig. 5.1(b), the relaxation of the pressure is much more rapid. If the simulations had been performed at constant pressure there would have been a similar rapid relaxation of the density. From estimates of the compressibility the calculated density discrepancy $\{\rho(\infty) - (\rho(t)\}/\rho(\infty)\}$ would be only about 0.1% after 1 ns for this model polymer with $N = 100$.[33] As indicated in the introduction the density is by far the more important property in determining local chain motions, penetrant diffusion and stress-strain behavior, so equilibration times of 0.5–1 ns may still be adequate for many simulations.

5.4.4 Sample size effects

We briefly consider here the question of choosing a sample size that will provide a good representation of bulk behavior. The traditional requirement is that the size of the primary simulation cell is large enough to prevent molecules interacting with images of themselves through the periodic boundaries and is also large enough to contain all the important characteristic structural fluctuations of the bulk system. In the case of a molecular liquid it is usually satisfactory to take N in the range 100–1000. For a polymer system with a degree of polymerization $n = 1000$ this would imply using up to 10^6 monomers in a simulation! We could relax the criterion somewhat so that, for instance, the cell dimensions are greater than the expected mean square end-to-end distance of a polymer molecule. For $n = 1000$ we would have to include around 30 chains and a total of 30 000 monomers in the simulation which, although feasible, would be a very expensive calculation.

A more radical, although controversial, approach is to use a cell which is only larger than the correlation lengths important to the phenomenon being studied. In this case we might use just one chain of 1000 monomers to form a dense amorphous polymeric system through the replicative properties of periodic boundaries. The primary chain spans many neighboring cells. The model is therefore one of a monodisperse polymer entangled with replicas of itself. A two-dimensional schematic diagram of this model is shown in Fig. 5.2. Chain ends can be eliminated from the model by arranging that the end of a chain is attached to the other end of one of its periodic images, giving "infinite" length.[36] It must be remembered, however, that the repeat length is still the number of monomers in the primary cell.

The effects of boundary conditions in this kind of model have yet to be fully evaluated but we can expect that one important condition might be the size of the unit cell in relation to the correlation length along the chain. For small N there is no doubt that the model gives a poor representation of bulk behavior, particularly for less flexible polymer chains, but as N becomes larger we expect it to be an increasingly better approximation to a dense amorphous system.

5.5 Preparation of polymer glasses

5.5.1 Glass preparation by computer simulation

Glasses are usually formed from the melt either by progressively decreasing the temperature or increasing the pressure. Cooling has been used to pro-

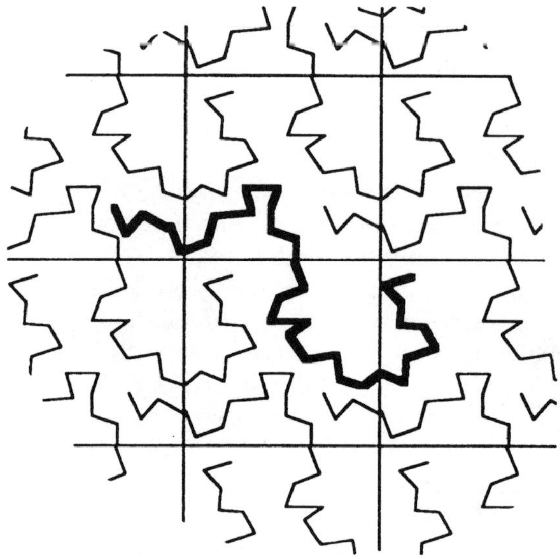

Fig. 5.2 Two-dimensional schematic diagram of the polymer model in which a single chain is replicated by the periodic boundaries.

duce glasses with both the PE I and the PE III polyethylene models.[6,37] The importance of this approach is that it provides an opportunity to characterize the transformation process by monitoring certain properties. It must be emphasized however that the cooling rate is enormous in these simulations—from 10^8 to 10^{12} K s^{-1}—so we must be careful in our comparisons with laboratory glasses (see next section). Under conditions where a material appears glassy in a simulation it might well behave as a liquid in the laboratory.

An alternative and more radical approach is to use the preparation techniques described in Section 5.3 to prepare polymer samples directly in the glassy state. This method has been used for modeling polypropylene[3] and polyethylene.[38] In the latter case careful control over the preparation procedure was used to construct amorphous samples of the *same* polymer in *different* configurational states in order to examine the relation between mechanical properties and polymer structure. Such control is only possible, of course, for nonequilibrium states where the structure can be immediately frozen in. The transformation to equilibrium liquid behavior from such samples can be observed by heating.[39]

Some results of cooling experiments on the PE I model with $N = 1000$ are shown in Fig. 5.3. Here the density is plotted as a function of decreasing temperature with the pressure controlled at 1 atm.[6] The set marked PE III is for $N = 500$ (see Section 5.3) and these data show little quantitative relation to polyethylene. For the remainder of this discussion we shall consider only the results for the PE I model.

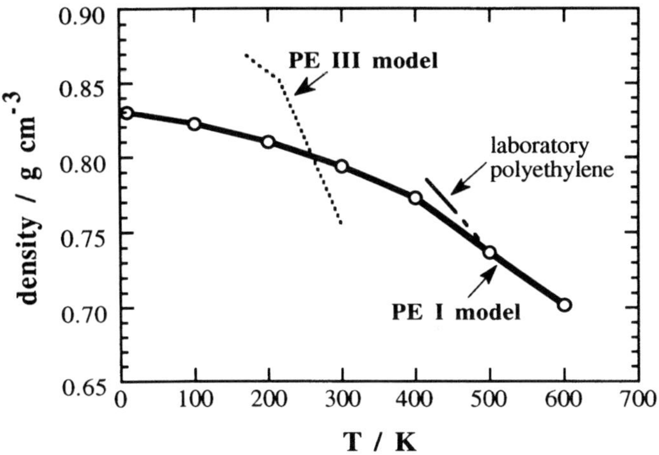

Fig. 5.3 Density as a function of temperature at 1 bar pressure, averaged over five samples of polyethylene model PE I, with a degree of polymerization of 1000. The cooling rate was about K ps^{-1} but at each temperature the samples were relaxed for about a further 1 ns. The Lennard-Jones potential parameters for the van der Waals interactions were adjusted to give a density at 500 K which agreed with a linear extrapolation of experimental data (see text for further details). Also shown are data reported for a sample of model PE III polyethylene cooled at a comparable rate.[37]

Five independent samples were generated and allowed to relax for 500 ps at 500 K at an applied isotropic pressure of 1 bar. Samples at different temperatures were then obtained by cooling or heating at a rate of 1 K ps^{-1} to the desired temperature under isotropic controlled pressure conditions (1 bar) followed by subsequent periods of relaxation of order 1 ns. Using this procedure additional samples were generated at 600, 400, 300, 200, 100, and 10 K.

As the temperature is lowered there is a gradual decrease in the thermal expansivity of the polymer (obtained from the slope of the plot) towards values typical of amorphous solids. The data in the liquid regime show a lower expansivity than the laboratory material but this may be due to shortcomings of this simplified model potential.

The decrease in expansivity as the sample is cooled is exactly as expected upon glass formation. However it would appear that any transition is rather weak and smeared-out and it seems to occur at an unexpectedly high temperature (we can only estimate T_g for real polyethylene but values found in the literature[40,41] are in the range ~ 150–300 K).

Direct observation of the freezing-out of conformational transitions at a temperature of about 400 K provides strong supporting evidence of a glass-like transformation, in accord with the notion that the torsional degrees of freedom are dominant modes of relaxation in these systems.[6] This is shown in Fig. 5.4 which gives the temperature dependence of the fraction of *trans* conformers averaged over five samples of unperturbed chains at the various temperatures. A dihedral angle is said to be *trans* if it lies between $\pm 60°$, otherwise the angle is in one of the two *gauche* states. The solid line in the plot is an extrapolation to lower temperatures made on the basis of fitting the data at 500 K and 1000 K to the form

$$\frac{\langle X_T \rangle}{\langle X_G \rangle} = A \exp\left(\frac{\Delta\Phi}{RT}\right) \tag{5.14}$$

where X denotes the fraction of trans (T) or gauche (G) states and A and $\Delta\Phi$ are the adjustable parameters. The actual values used for the curve shown[6] were $A = 1.01$ and $\Delta\Phi/R = 463\,\mathrm{K}$. The nominal difference in energies between the gauche and trans wells for the dihedral angle potential is equivalent to 530 K but this takes no account of the excluded volume effects which, for example, largely prevent sequences of the type G^+G^- from occurring.

5.5.2 *The glass transformation on different time scales*

At first sight it would appear that the characteristics of the glass transformation observed in the simulated polymer are quantitatively quite different to what is observed on macroscopic time-scales in the laboratory. Firstly, the transition is taking place at a much higher temperature than one would

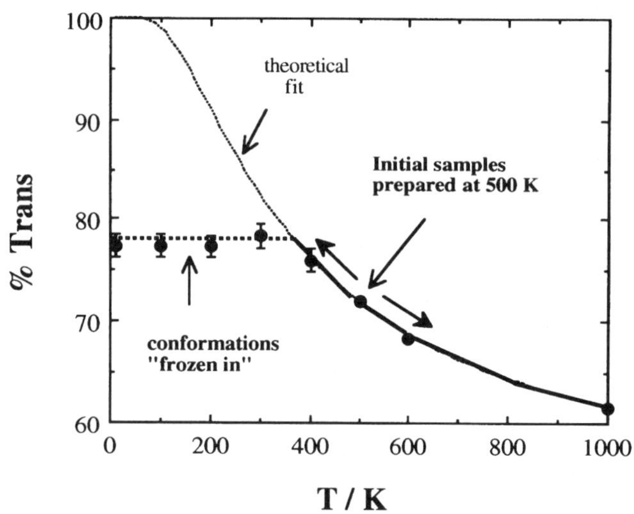

Fig. 5.4 Variation with temperature of the percentage of trans conformers in the cooling simulation of polyethylene model PE I. The theoretical equilibrium curve was calculated according to equation (5.14) in the text.

expect, and, secondly, it appears rather diffuse on the density plot (Young's modulus similarly shows a broad transition region[6]). We shall show below that *both* the above characteristics can be explained in terms of the very rapid cooling rates employed in simulations and that they have nothing to do with the details of the model.

Before proceeding to the explanation we should first note that similar behavior has been observed in simulations of glass formation for the atomic Lennard-Jones fluid[42,43] using cooling rates of the order 10^{12} s^{-1}. In addition, very similar effects have been observed for real materials in the laboratory when using "high frequency" probes of the glass transition. For instance Brillouin scattering measurements[44] have been used to probe the sound velocity at frequencies of about 10 GHz in cooling experiments on the glass forming liquid $2Ca(NO_3)_2.3KNO_3$. The high-frequency longitudinal compliance and the adiabatic compressibility derived from the measurements both show a smearing out of the transition towards high temperatures whilst the more familiar static measurements show a sharp transition at the low end of the transformation range.

The explanation of the behavior is the same for the molecular dynamics simulations as for the light scattering data.[43] Let us consider an idealized cooling experiment which, for the purposes of the present discussion, is easiest to think of as occurring in a series of discrete steps. The various molecular motions have characteristic times which are coupled to the density fluctuations in the material. As the latter slow down as a result of cooling or compression we reach a stage where full relaxation of the density is not complete before the next step in temperature or pressure so the material falls progressively out of equilibrium, and a glass transition is observed. This is shown schematically in Fig. 5.5. The faster the cooling rate the higher is the temperature at which the material falls out of equilibrium.[9] In simulation experiments where the cooling rate is enormous, the effects are extreme and the transformation occurs at a high temperature where the material may well show normal liquid behavior in the laboratory.

The smearing-out of the transformation in simulations can be understood from the temperature dependence of the relaxation times[43,44] as shown, again schematically, in Fig. 5.6. Simple linear polymers are examples of "fragile" fluids[45] which show a marked temperature dependence of the activation energies (E_a) for structural relaxation processes. At low temperatures in the supercooled regime E_a becomes extremely large and relaxation times may change by one or two orders of magnitude over a few degrees, thus producing a very sharp transition. At high temperatures a much smaller value of E_a means that the same change occurs over a much wider temperature range, thus producing a broad transformation.

Finally it is worth mentioning one additional reason why we might expect a different character to the glass transition as observed on very short time

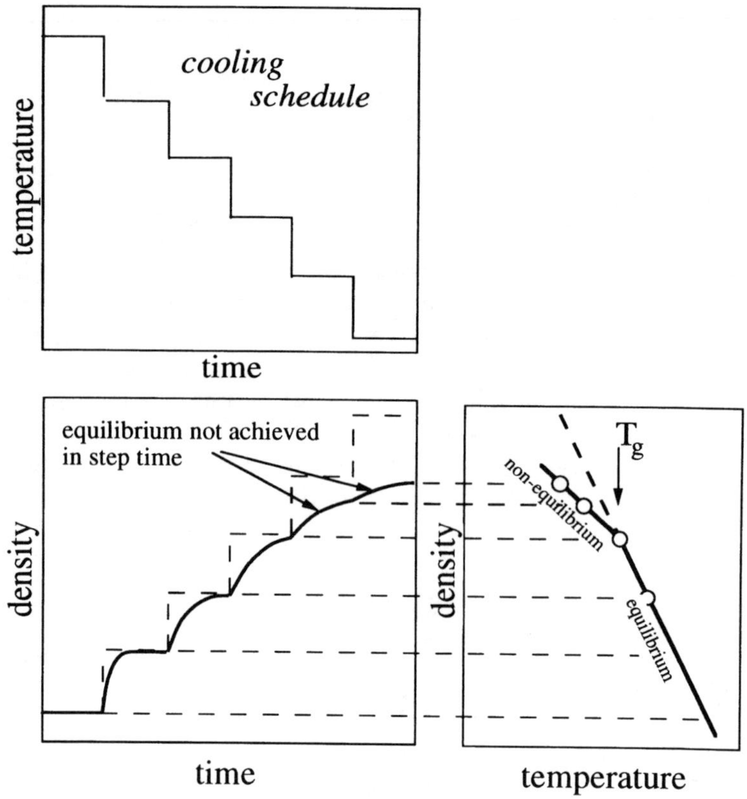

Fig. 5.5 Schematic illustration of the increasingly slower density relaxation in an idealized stepwise cooling experiment. The shorter the step times the higher the temperature at which the density falls out of equilibrium and a glass transition is observed.

scales. The above simplified discussion is based on the assumption of just one relaxation process. Of course in a supercooled polymer there are many different degrees of freedom to be frozen out. For instance Rouse-like modes appear to be a reasonable description of chain motion up to the entanglement length in dense melts[46] and it is likely that these motions have much to do with viscoelastic and mechanical properties. Each of the p Rouse modes will have a characteristic relaxation time, τ_p, and temperature dependence. They can only be "active" in the glass transformation if the cooling rate is much less than $1/\tau_p$. On the time scale of current molecular dynamics experiments only the short wavelength modes will be accessible.

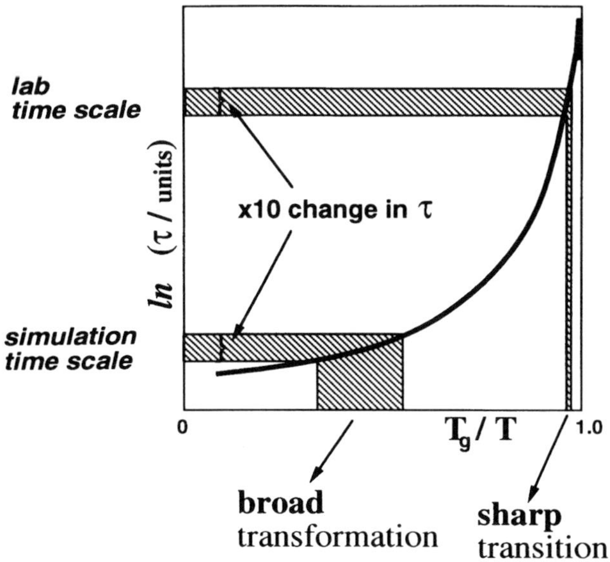

Fig. 5.6 Schematic temperature variation of a characteristic structural relaxation time in a supercooled polymer melt. It is assumed that approximately a ten times change in the relaxation time is required to give a significant change in some associated property such as expansivity or compressibility. This change takes place over a much wider range of temperature on the simulation time scale.

5.6 Stress–strain properties

5.6.1 Uniaxial tension simulations

Mention has already been made that one of the advantages of the controlled pressure molecular dynamics discussed in Section 5.2 is that the form of the applied pressure tensor **P*** can be used to impart strain to a sample as a function of time in much the same way as in laboratory experiments. Control of appropriate components of the pressure tensor can be used to produce, to take just three examples, uniaxial tension, compression or shear as illustrated in Fig. 5.7

Current limitations on simulation times have meant that very high rates of strain must be used in order to observe the system response. Nevertheless the general form of the results show striking similarities to those obtained in laboratory experiments performed on time scales many orders of magnitude slower.

To give an example of what can be achieved in such simulations we discuss below the stress–strain behavior as observed in simulations of the model PE I of polyethylene at a range of temperatures in the glass and melt.[6] The sample size was 1000 monomers formed into a single linear chain as described in Section 5.4. The cooling curve for these samples is shown in

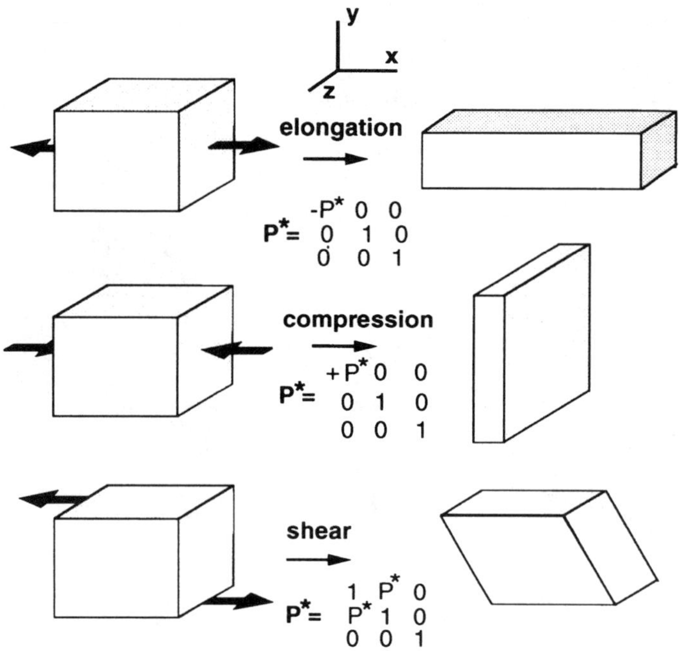

Fig. 5.7 Examples of mechanical deformations that can be induced using loose coupling molecular dynamics with an applied pressure tensor $\mathbf{P}^*$.

Fig. 5.3. In order to get a representative picture the results were averaged over only five independent samples so the statistical precision is rather limited.

The prepared samples were subjected to a gradually increasing uniaxial tension by changing the *y* component of the *applied* pressure tensor, $\mathbf{P}^*_{yy}$, at a constant rate

$$\frac{d\mathbf{P}^*_{yy}}{dt} = -\dot{\tau} \tag{5.15}$$

where the tension application rate, $\dot{\tau}$, used was either 5 bar ps^{-}1 or 1 bar ps^{-}1 and the minus sign accounts for the fact that tension is a negative pressure. Employing two different values for $\dot{\tau}$ should give some indication of the extent to which the measured properties are rate-dependent.

The strain induced in the sample is of primary interest so the applied tension is best considered as a control variable which produces a change in the strain. The response is given by the *measured* tension, i.e. $-p_{yy}$, within the sample. In these experiments then, both the strain and the measured tension are dependent variables. The method is preferable to direct control of the strain since there is no way a *priori* of predicting how the shape or density of the sample will respond to a change in the external conditions.

The simulations were continued until the sample had extended by about 50–100% of its original length. Extensions beyond about 100% were not possible without violating the truncation radius criterion for the site–site potential due to the contraction in the transverse direction. The primary information which results from these tension experiments is the response of the **h** matrix, defining the size and shape of the primary cell, and that of the measured pressure tensor, **P**. These together allow us to elucidate the stress versus strain behavior.

The average response of five independent samples at six temperatures from 10 K to 500 K are shown in Fig. 5.8 for tension application rates of 5 bar ps^{-1}. There is clearly a wide range of behavior observable in the model system; at low temperatures the material can support the tension up to strains of ~ 20% before undergoing yield and at progressively higher temperatures there is a gradual change in behavior until at 500 K it is unclear whether there is any elastic response at all.

The essential difference in character between the elastic low-temperature behavior and the viscous high-temperature response is shown by plotting the extension as a function of time (see Fig. 5.9). On a log–log scale an elastic response should have an asymptotic slope of 1 at low strains for a system with a well defined Young's modulus, E,

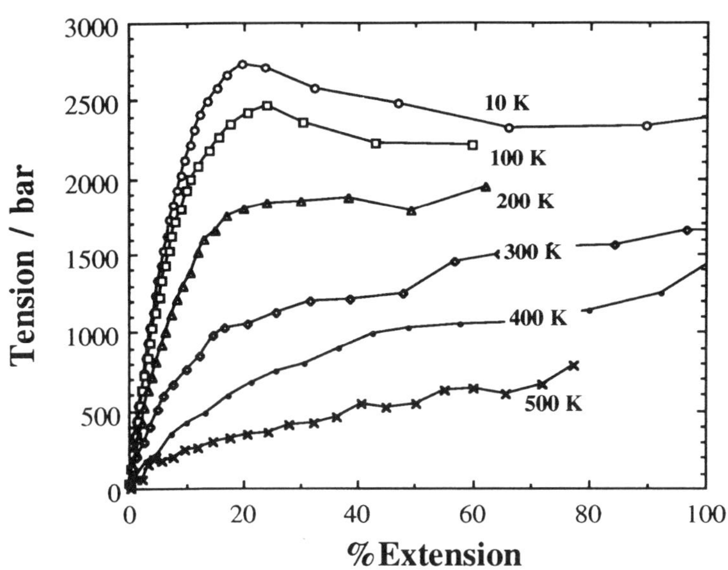

Fig. 5.8 The measured tension ($-P_{yy}$) as a function of percentage extension ($\gamma_L^* 100$) for tension applied at 5 bar ps^{-1} for polyethylene model PE I with $N = 1000$. The data at each temperature represent the average behavior over five independent samples.

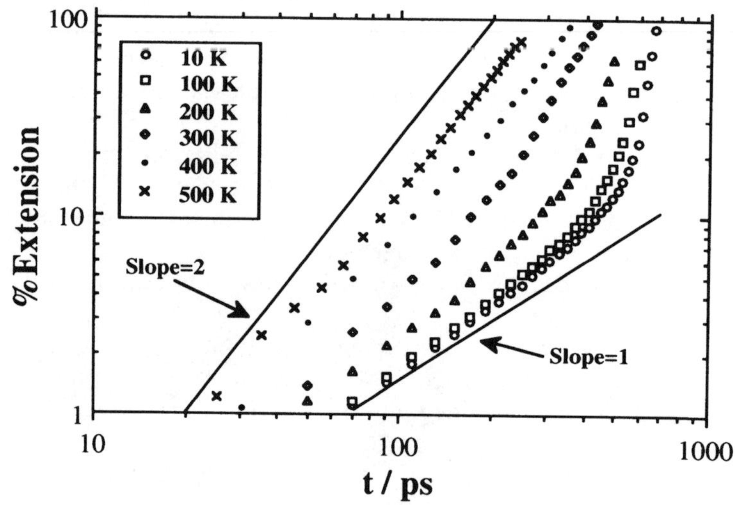

Fig. 5.9 The percentage extension as a function of time for the samples subjected to a tension application rate of 5 bar ps^{-1}. On the log–log plot a slope of 1 indicates an elastic response to the applied tension whereas a slope of 2 is that expected of a viscous material (see text for details).

$$-P_{yy} = \mathrm{E}\gamma_{\mathrm{L}} \tag{5.16}$$

Alternatively if the response to the applied tension is viscous, i.e.,

$$-P_{yy} = \eta_e \dot{\gamma}_{\mathrm{L}} \tag{5.17}$$

where η_e is an elongational viscosity coefficient, then it is easy to show that for the experiment performed here that the strain should increase quadratically in time and hence give a slope of two.

Both types of behavior are evident in confirming the trend from elastic (at low strain) to viscous response as the temperature is increased. The estimated extensional viscosity at 500 K from the simulations is of the order of 0.01 Pa s^{-1}. Although this is much lower than the equilibrium extensional viscosity of polyethylene it is known that the viscosity does decrease significantly with increasing strain rate.[47] At the extension rates used in our simulations, $10^7 \rightarrow 10^9\,s^{-1}$, the behavior is expected to be strongly non-Newtonian.

One working definition of the yield stress in the laboratory is the true stress at the observed maximum tension.[48] For convenience we have chosen to define the yield stress as the measured tension at a strain of 20% which corresponds closely to observed maxima in the load for those samples that show a maximum. The resultant values are plotted in Fig. 5.10 and the behavior is very similar to that found in real systems where the yield stress decreases approximately linearly with increasing temperature.[48]

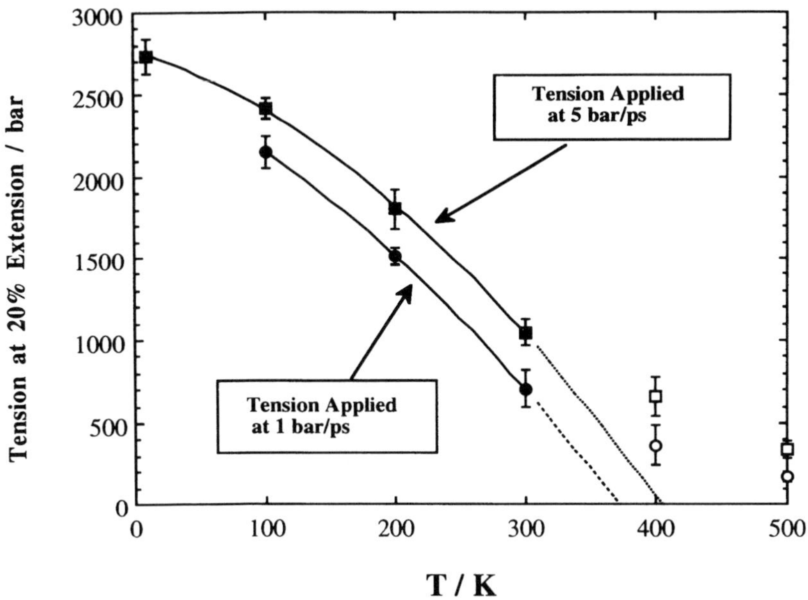

Fig. 5.10 Tension at 20% extension ("yield stress") as a function of temperature or polyethylene model PE I with $N = 1000$. Squares and circles refer to tension application rates of 5 bar ps^{-1} and 1 bar ps^{-1} respectively. Open symbols indicate data for which no discernible yield was observed; these points are excluded from the curve fits and extrapolations to zero tension. The error bars shown are the standard deviations in the results for the five independent samples.

The data covers a very wide temperature range which may account for the slight nonlinearity. It has also been shown in laboratory experiments that extrapolating the data to zero yield stress results in convergence to a temperature close to the glass transition temperature.[48] If we ignore the points above 300 K for which there is no discernible yield point, our data extrapolate to zero yield stress at around the same temperature where there is a change in expansivity (see Fig. 5.3). As for laboratory measurements there is a dependence on the rate of application of the tension with the lower rate leading to consistently lower values of the yield stress and hence a lower extrapolated temperature of zero yield stress.

It will not have gone unnoticed that the observed values of the yield stress and strain are much larger than those typically observed in the laboratory. There are several possible reasons for this. For instance, the response of the system to very high rates of strain will not include slow relaxation processes such as creep which are important in laboratory experiments. Also, simulation samples are perhaps too homogeneous; they do not contain the mesoscale heterogeneities which probably occur in laboratory samples and which may play a significant role in determining the overall relaxation behavior.

There are other subtle differences between the simulations and laboratory experience which should make us a little cautious in drawing too close a

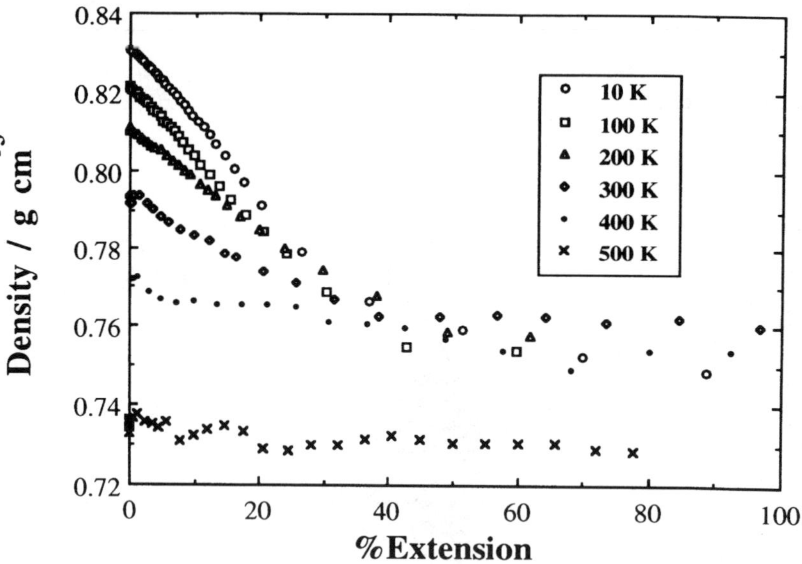

Fig. 5.11 Behavior of the density during the extension experiments for tension application rates of 5 bar ps^{-1}. Note the distinct density decrease as the extension approaches the yield point at the lower temperatures. This contrasts with post-yield flow, and the viscous flow observed at high temperatures.

parallel between the two responses. In Fig. 5.11 we show the behavior of the system densities as the strain is increased at the different temperatures. At high temperatures (400–500 K) where the flow process is predominantly viscous there is a hardly perceptible change in density during the extension of the samples.

In contrast, at low temperatures there is a noticeable dilation effect as the tension is applied and the density decrease continues until just beyond the yield point. Once the material yields the density remains relatively constant as plastic flow takes place. It appears that for a given tension application rate this apparent critical density for yield and plastic flow is *independent* of temperature over the range 10–300 K.

This behavior of the density has not been observed in laboratory experiments; in fact the density often increases slightly at yield.[48] Also it is worth recalling the well known experimental result that yield can also occur in amorphous polymers under *compression.* Simulations of model polymers under compression have yet to be reported.

One point worth making here is that the decrease in density that occurs under extension at the lower temperatures is entirely consistent with the typical values of Poisson's ratio (μ) for amorphous polymeric solids which are generally in the range 0.3 to 0.4.[49–52] Indeed our estimates of Poisson's ratio from the extensional and contractile strains

$$\mu = \lim_{\gamma_L \to 0} \frac{-\gamma_C}{\gamma_L} \tag{5.18}$$

give values of about 0.41 at the lowest temperatures.

5.6.2 *Stress–strain behavior and configurational properties*

As mentioned previously one of the great advantages of simulations is the ease with which system parameters can be controlled in order to study some physical effect. This approach was exploited in a recent study of the ways in which the phenomenon of strain hardening depends on the configurational properties of an amorphous polymer.[38] Samples of the *same* glassy polymer were prepared each with *different* configurational properties, the aim being to avoid any ambiguities introduced by comparing polymers with different chemical structure. Although it is not evident from Fig 5.8, at high strains there is nearly always an increase in the modulus which is referred to as strain hardening. This latter property is of great practical importance since the extent of strain hardening is associated with the susceptibility of the material to, for instance, necking and crazing. On a phenomenological level the large strain behavior of polymers has been rationalized in terms of mechanical models which comprise elements representing the Hookean, viscous and rubber elasticity components of the stress[53,54] but there is considerable interest in understanding strain hardening in terms of chain topology.

Necking and shear banding can also complicate the interpretation of experimental data since they make true stresses and strains difficult to determine. True stress–strain curves have been obtained for glasses composed of bis(phenol A) copolymerized to produce carbonate and phthalate chains of varying stiffness, and in this case it was noted that there was a correlation between Kuhn length (a measure of chain stiffness) and the large-strain behavior.[53] Similar results have been obtained with polyisocyanates which have extremely large Kuhn lengths and pronounced strain hardening so that they extend uniformly without necking.[55]

Let us consider the ways in which chains in an amorphous polymer can deform. Elongation can be achieved either by uncoiling the overall chain configuration as measured for instance by the persistence length, or by changing the local conformations within the "tube" formed by neighboring chains in the entangled structure. This alternative "local" mechanism corresponds in our model to segments of chains being converted from the *gauche* ("short") form to the long ("*trans*") form; the latter is the lowest energy state and has a planar zig-zag structure. For the polyethylene model used, it follows that chains which are either less coiled or contain fewer gauche states to begin with will be harder to deform. For the study in question the persistence length was used as a measure of the overall configuration. In an equilibrium polymer melt the overall configuration and the fractions of conformers are inextricably linked but this restriction does not apply for nonequilibrium glassy states where changes in the preparation procedure can provide some measure of independent control of the two properties.

The persistence length a measures the correlation in the orientation of successive monomers as we move along a polymer chain. One useful definition which is easily applied to the polyethylene model used here is[32]

$$a = b_0 \sum_{k=0}^{\infty} \langle \mathbf{e}_i \cdot \mathbf{e}_{i+k} \rangle, \tag{5.19}$$

where $\mathbf{e}_i$ is a unit bond vector. It can be further shown that

$$a = \frac{1}{2} b_0 (1 + C_\infty) \tag{5.20}$$

where the characteristic ratio C_∞ is given by

$$C_\infty = \text{limit}_{k \to \infty} \frac{\langle \mathbf{r}_k^2 \rangle}{k b_0^2} = \frac{\langle (\mathbf{r}_i - \mathbf{r}_{i+k})^2 \rangle}{k b_0^2} \tag{5.21}$$

The above equations imply that a can only be obtained accurately from an asymptotic limit. Recent work[33] suggests that this limit is only reached when $k > 1000$ for the polyethylene PE I model and an accurate result cannot be achieved with only small samples. Nevertheless correlation lengths can be deduced from averages over five configurations for values of k up to 100 (these are referred to as a_{100}) and it is reasonable to assume that to a good approximation the ratios of these values reflect the ratios of the true persistence lengths.

A set of molecular dynamics simulations was carried out on samples prepared using the direct method described in Section 5.4. In all, four different sample sets of polymer glass were produced by using different preparation procedures. All of the samples were relaxed at 200 K with $\mathbf{P}^* = 1$ bar for ~ 1 ns. The final densities were all the same to within 1% and we expect all of the samples to have glass transformations in the range 300–400 K on the simulation time scale. The associated correlation lengths and fractions of trans conformers obtained from the final 200 ps of these runs are shown in Table 5.1.

Sample set A was formed by cooling from the melt as described in Section 5.4. Set B was produced by growing chains at 200 K using PCG and relaxing them for 1 ns at the same temperature. Set C was obtained by "flash" heat treatment of set B; this involves raising the temperature instantaneously to 1000 K for 100 ps followed by rapid cooling back to 200 K. This was not sufficiently drastic to alter the overall configuration of the chain but it did allow conformational transitions to occur and the overall result of this treatment is a net decrease in the trans fraction. As shown in Fig. 5.12 the

Table 5.1 *The percentage of* trans *conformers, %* trans, *and correlation length,* a_{100}, *calculated for each of the four sample sets.*

Sample	% trans	$a_{100}/\text{Å}$
A	78	5.0
B	82	15.5
C	77	12.0
D	70	5.8

more kinked structure shows noticeable retraction of the polymer chain within the "tube" formed by its near neighbors.

Sample set D was obtained by phantom chain growth at 200 K with the torsional and 1–5 interaction potentials scaled down by a factor of 50; this has the effect of increasing the gauche fraction and produces a highly coiled chain. Van der Waals interactions were then introduced and subsequent relaxation with the full interaction potentials produced a polymer glass with a much reduced fraction of trans states.

The stress–strain behavior of all four sets of samples was obtained by subjecting them to an externally applied uniaxial tension which was increased at a rate of 5 bar ps^{-1}, exactly in the same way as described in

Fig. 5.12 Comparison of the chain structure for one sample before (thin line) and after (thick line) heat treatment. Only the coordinates of the continuous primary chain have been shown. The molecular dynamics "unit cell" which is filled by images of the primary chain, is also shown.

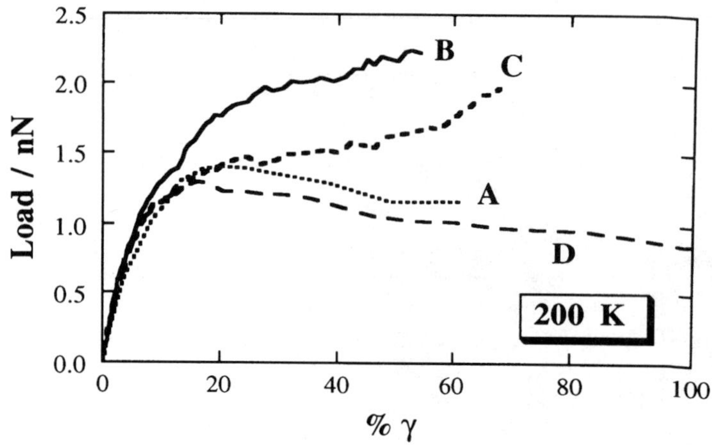

Fig. 5.13 Measured load plotted as a function of extension for the four sets of samples of the model linear polymer subjected to uniaxial tension increasing at a rate of 5 bar ps^{-1}. Data for each set are averaged over five samples. the four sets are based on polyethylene model PE I. The samples were prepared directly in the glassy state and sets A–D have different configurational properties (see Table 5.1) as a result of different preparation procedures.

the previous section. As a result of the small system size, data was obtained only up to extensions of ~ 100%. In Fig. 5.13 the load on these samples is plotted as a function of the strain, the load being determined from the product of the tension and the cross-sectional area. Since the samples are very small, the loads are also extremely small (of order 10^{-9} N).

The pattern of the stress–strain plots is similar to that discussed in the previous section—there is an initial elastic reponse followed by yield and plastic flow. In detail however the four sets of samples show rather different behavior; the differences begin to be noticeable after about 10% extension. Samples **B** and **C** both show enhanced resistance to extension beyond the point at which the **A** samples yielded (~ 20% extension). The high % trans-high correlation length samples (set **B**) show the largest extent of strain hardening and in particular produce significantly more stress than the set **C** which, within the error, has a similar correlation length but lower trans fraction. Conversely sample set **C** has practically the same % trans as set **A** so the differences here must be due to their contrasting configurational structure. Set **D** shows the lowest resistance of all to the applied tension as was expected from its highly coiled structure with a smaller fraction of trans conformations. What these results suggest is that both an increase in the fraction of trans states and an increase in the persistence length can *independently* contribute to strain hardening.

It is interesting to note the strain dependence of the fraction of trans conformers. This data is shown in Fig. 5.14 and they reveal a common feature of all uniaxial tension simulations to date, namely that there is a

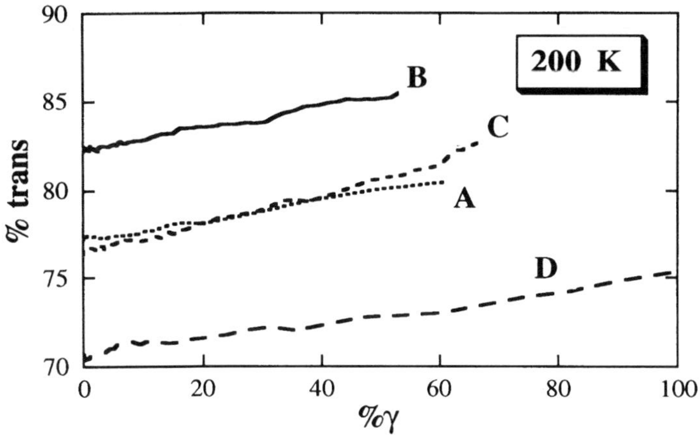

Fig. 5.14 Percentage of trans conformers plotted as a function of the extension for sample sets A–D, as in Fig. 5.13

consistently linear dependence of the % trans upon the extension. The lack of any discontinuities or breaks in the plots at the yield points confirms the conclusion that this mechanical property has nothing directly to do with the onset of transitions between different conformational states. Recent work has shown that the % trans recovers rapidly and almost completely to its initial value when the applied tension is removed, suggesting that it is to be associated with the elastic part of the overall deformation.[39]

5.7 Penetrant diffusion

The diffusion of small molecules such as He, 0_2 $C0_2$, CH_4 through glassy polymers, partially crystalline polymers and polymer membranes is of great practical interest in relation to a number of current and possible future practical applications, and it is not surprising that this phenomenon has already attracted a great deal of attention in simulation studies. This area offers many opportunities for the use of computer modeling for optimizing the molecular design of host polymers. Such small diffusing gas molecules which are present at rather low concentrations are conveniently referred to as penetrants.

The quantities of prime interest here are the solubility and diffusion coefficient of the penetrant molecules and how these quantities are related to the structure of the host polymer. Comparisons with the extensive experimental data available for a wide range of polymers provides a stiff test not only of the accuracy of potential functions but perhaps more importantly of the method of preparation techniques and methods of sampling representative polymer structures. So far only limited quantitative success has been achieved, and that is under system conditions where the diffusion rate is quite high so that the process can be characterized during the time scale of a simulation experiment. It will be much more of a challenge to compute reliable diffusion coefficients in glassy polymers and effective barrier systems where spontaneous diffusion is *required* to be extremely small. In the meantime, however, quite a lot of interesting information has already been obtained concerning transport mechanisms.

At least two research groups[56,57] are currently working on theoretical models of diffusion in polymers based on transition state theory, extending the idea first introduced more than thirty years ago[58] that diffusion in fluids was related to the redistribution of free volume. The aim here is to use molecular dynamics or Monte Carlo simulations of specific polymers to determine spatial fluctuations in the accessible volume for a particular penetrant. This information is then used to determine the adjustable parameters in a transition state theory. The advantage of this approach is that it may be possible to extend studies to glassy polymers where penetrant diffusion may be many orders of magnitude too slow for direct simulation studies. This will however require knowledge of how the parameters change as a function of temperature and density.

Self-diffusion coefficients in fluids are most commonly obtained in molecular dynamics simulations using the Einstein relation

$$D = \mathrm{Lt}_{t \to \infty} \left\{ \frac{\langle \Delta r^2 \rangle}{6t} \right\} \tag{5.22}$$

The diffusion coefficient D is then obtained from the slope of a plot of the mean squared displacement $\langle \Delta r^2 \rangle$ against time t. The problem is deciding when the limiting linear regime has been reached. Times of order 100 ps are usually sufficient to attain this condition in simple fluids. The situation may not be so straightforward in more complex systems. In viscous molecular fluids for instance the slope has been observed to change significantly over the time range 50–300 ps.[59] So-called "anomalous diffusion" has been discussed recently.[60] In this case $\langle \Delta r^2 \rangle$ is proportional to something less than the first power of t; it can occur when the simple random walk is modified by the presence of obstacles or inhomogeneities, a situation that might well exist in some polymer structures.

Molecular dynamics studies of penetrant diffusion have so far been performed for polyethylene, polyisobutylene and polydimethylsiloxane. Calculated diffusion coefficients have not been in particularly good agreement with experiment.[60–66] In the case of polyethylene, modeled using the united atom approximation, diffusion coefficients were much higher than expected and the activation energies too small although much better results have been obtained[67] using PE IV. Reliable comparisons with experimental data are made difficult by uncertainties as to the true diffusion coefficients in pure amorphous polyethylene, which have to be extracted by calculation from data on the partially crystalline material.

The sorption and diffusion of small gas molecules such as He, O_2 and H_2 in polyisobutylene has also been studied.[60] An all-atom force field with carefully chosen parameters was used but it was still found that the diffusion coefficients were up to an order of magnitude too high. Loose coupling was used to control the pressure around zero but there are no details of the system density.

Gas solubilities can also be determined by molecular dynamics simulations,[60,68] using the Widom test particle insertion method[69] to calculate the excess chemical potential μ_{ex} or free energy of the penetrant molecules. The solubility can be obtained using Henry's law. If E is the interaction energy of a virtual penetrant molecule with the polymer inserted at random within the sample (the molecule is "invisible" to the polymer) then

$$\mu_{ex} = -RT \ln\langle \exp(-E/kT) \rangle \tag{5.23}$$

In dense materials however the original implementation of this method can be quite inefficient. A "smarter" way to sample the structure is to use an accessible volume map to bias the choice of insertion points,[35,70]. Of course the accessible volume will vary according to the nature of the penetrant. For PIB at 300 K solubilities calculated in this way for He, O_2 and H_2 were typically one or two orders of magnitude greater than experimental values[60] suggesting that there is far too much spare volume in the model structures. A similar effect was noted for the solubilities of CH_4, and He in polydimethylsiloxane.[35]

It is interesting to speculate on the reasons for these discrepancies. One possibility is that the force fields have been poorly chosen. This is a controversial issue and is extremely difficult to check especially since at high polymer densities the diffusion and solubility are likely to be extremely sensitive to the fine details of, e.g., repulsive van der Waals interactions.

Anomolously high diffusion coefficients and solubilities could also result from incomplete relaxation of the polymer sample densities; one expects diffusion coefficients to be extremely sensitive to density in this regime and this again emphasizes the important issue of how model polymer samples are best prepared. In one study[60] it was pointed out that the bulk of the thermodynamic solubility of O_2 in the sample of model PIB was contributed by a single large "hole" in the structure emphasizing the need for adequate sampling of the way in which a penetrant samples the natural structural fluctuations in the polymer, either by using very large systems or by averaging over many independent simulations.[11]

There is considerable interest in the mechanism of penetrant molecule diffusion and this has been investigated in several recent studies. It is widely accepted that diffusion proceeds by a some kind of "rattle and jump mechanism" and for low temperatures and largish penetrants these two features seem to be distinguishable. For methane in PE at 300 K typical rms jump lengths appear to be about 0.5 nm with the time between jumps of order 0.5 ns.[67] At higher temperatures the hopping rate increases and the magnitude of rattling motions increases so the mechanism is less well defined. Recent simulation studies seem to suggest that the jump events occur as a result of temporary channels opening up between cavities of fluctuating size and shape[35,60] so the local mobility of the polymer chains is also important.

In the literature penetrant diffusion has been widely discussed in terms of free volume theories of polymers, and it is not surprising to see this idea taken up in simulation studies. The problem is that although geometric free volume is a physical concept that is easy to comprehend it suffers from lack of precise definition. In the simulation studies to date it has been identified with the unoccupied volume in the sample calculated as an average over many static configurations. Note that this definition is different from the accessible volume to penetrant molecules discussed earlier in this section.

This has been done for polyethylene, polyisobutylene and polypropylene models, for instance, by prescribing spheres of a chosen diameter around each interaction site, or by using Voronoi polyhedra to define the unoccupied space.[71,72] The free volume fraction, f_V, can be defined by

$$f_V = \left\{ \frac{(v_O - v_C)}{v_O} \right\} \tag{5.24}$$

where v_O is the volume of the simulation unit cell and v_C is the volume occupied by the polymer segments. In the case of united atom models with a Lennard-Jones representation of the van der Waals forces the σ-size parameter is taken as the hard core diameter in order to calculate v_C. Rather surprisingly values obtained for f_V are quite high, giving the fraction of volume unoccupied as 30–40% in PE IV and model polyisobutylene[67] although only a much smaller fraction—typically $\sim 0.1\%$—is accessible to penetrants.

It is likely that local motions of the polymer chain are also important to penetrant diffusion. Removal of the torsional potential in polyethylene considerably increases the chain mobility and gave an increase of about a factor of two in the diffusion coefficient of O_2.[73] Conversely, decreasing the magnitude of the torsion angle fluctuations in the polyethylene model PE IV drastically reduces the diffusion coefficient of methane.[67]

5.8 Local motions in amorphous polymers

Fast motions in amorphous polymers such as the reorientation of main chain bonds and side groups and conformational fluctuations can be probed by a number of spectroscopic techniques such as infrared,[74] NMR[1] and inelastic neutron scattering.[2] The relevance of computer simulation studies here is that they can be used to evaluate current theoretical models and perhaps to develop new ways of interpreting the experimental data. Detailed analyses of the orientational dynamics of the bond vectors in the PE III model of polyethylene have already been described.[37] Since molecular dynamics is particularly suitable for studying short-time dynamics then there are wide future opportunities for making useful comparisons with experimental data on, for instance, the motions of side chains and pendant groups if good representative structures can be prepared.

Most experimental techniques do not probe reorientational motions directly but measure the time-dependence of properties which themselves are sensitive to orientation.[75] Thus in far infrared and Raman spectroscopy one measures dipole or polarizability fluctuations with respect to particular normal modes of vibration. Although the interpretation can be complicated, in cases where the fluctuations arise predominantly from rotational motions of localized groups correlation functions of the first and second Legendre polynomials, $C_1(t)$ and $C_2(t)$ respectively, can be retrieved from the data.

Simulation studies are not limited in the above way and one can study the reorientational motions directly and in great detail by determining the full probability distribution function $W(\theta, t)$ for bond reorientation through an angle θ in a time t. This has been obtained for polyethylene model PE III.[37] As a result of recent developments in multidimensional NMR techniques[1] it should be possible to study this function experimentally in favorable cases although no relevant data are yet available in the case of polyethylene.

The reorientational correlation functions $C_1(t)$ and $C_2(t)$ can of course both be calculated from $W(\theta, t)$, e.g.,

$$C_1(t) = \int_0^{2\pi} \mathrm{d}\theta \, P_1(\cos\theta) \, W(\theta, t) \tag{5.25}$$

The forms of $C_1(t)$ and $C_2(t)$ can be predicted from various models for reorientational motion. One of the simplest models is small step rotational diffusion which requires that both $C_1(t)$ and $C_2(t)$ should show exponential decays and that the ratio of the respective correlation times should be three.[75] These predictions are for isotropic motion, so it would not be surprising if the data for bond vector reorientation in PE did not fit this simple model since we might guess that the motion of a single bond is likely to be highly anisotropic as a result of restriction arising from the connectivity of the chains; this has been previously observed in Brownian dynamics simulations of a polyethylene model[76] and in a dynamical rotational isomeric states model of a single chain.[77] These effects are also likely to lead to a long-time tail to the functions.

Neither $C_1(t)$ and $C_2(t)$ in amorphous polyethylene shows a simple exponential relaxation; there is a rapid initial decay over 10–20 ps followed by a much slower decay up to 100–200 ps.[37] The correlation functions fit well to the Kohlrausch–Williams–Watts form

$$C_i(t) = \exp(-t/\tau_i)^{\beta} \tag{5.26}$$

This form for the correlation function fits a wide variety of relaxation data in complex fluids and, as in the case of many other properties, β turns out to have a value close to 0.5.[37] This form for a correlation function is often interpreted in terms of a distribution of relaxation times; however in the case of model polyethylene it has been shown that the results cannot be explained in this way. Instead the result is interpreted in terms of a model for anisotropic motion in which the polymer chain is confined to a "pipe" formed by its neighbors.[78]

References

1. H. W. Spiess, *Chem. Rev.* **91**, 1321 (1991).
2. M. Bee, *Quasielastic Neutron Scattering* (Adam Hilger, Bristol, 1988).
3. D. N. Theodorou and U. W. Suter, *Macromolecules* **19**, 139 (1986); *Macromolecules,* **19**, 379 (1986).
4. D. Brown and J. H. R. Clarke, *J. Chem. Phys.* **84**, 2858 (1986).
5. D. Rigby and R. J. Roe, *J. Chem. Phys.* **87**, 7285 (1987); *J. Chem. Phys.* **89**, 5280 (1988), *Macromolecules* **22**, 2259 (1989).
6. D. Brown and J. H. R. Clarke, *Macromolecules* **24**, 2075 (1991).
7. D. Rigby and R. J. Roe, in *Computer Simulation of Polymers*, edited by R. J. Roe (Prentice-Hall, Englewood Cliffs, NJ, 1991).
8. M. P. Allen and D. J. Tildesley, *Computer Simulation of Liquids* (Clarendon Press, Oxford, 1987).
9. J. D. Ferry, *Viscoelastic Properties of Polymers*, 3rd Edition; (Wiley, New York, 1980).
10. D. Brown, J. H. R. Clarke, M. Okuda, and T. Yamazaki, *Comp. Phys. Commun.* **83**, 1 (1994).
11. D. Brown, J. H. R. Clarke, M. Okuda, and T. Yamazaki, *J. Chem. Phys.* **100**, 1684 (1994).
12. D. Brown and J. H. R. Clarke, *Mol. Phys.* **51**, 1243 (1984).
13. M. Parrinello and A. Rahman, *Phys. Rev. Lett.* **83**, 4069 (1985).
14. S. Nosé, *Mol. Phys.* **52**, 255 (1984).
15. W. G. Hoover, *Phys. Rev. A.* **31**, 1695 (1985).
16. H. J. C. Berendsen, J. P. M. Postma, W. F. van Gunsteren, A. DiNola, and J. R. Haak, *J. Chem. Phys.* **81**, 3684 (1984).
17. D. J. Evans and B. L. Holian, *J. Chem. Phys.* **83**, 4069 (1985).
18. D. Brown and J. H. R. Clarke, *Comp. Phys. Commun.* **62**, 360 (1991).
19. P. van der Ploeg, H. J. C. Berendsen, *J. Chem. Phys.* **76**, 3271 (1982).
20. H. J. C. Berendsen and W. F. van Gunsteren (eds), *GROMOS Reference Manual* (University of Groningen, Groningen, 1987).
21. P. V. K. Pant and R. H. Boyd, *Macromolecules* **26**, 679 (1993).
22. D. Steele, *J. Chem. Soc. Faraday Trans. II* **81**, 1077 (1985).

23. M. J. Richardson, P. J. Flory, and J. B. Jackson, *Polymer* **4**, 221 (1964).
24. J. D. Weeks, D. Chandler and H. C. Anderson, *J. Chem. Phys.* **54**, 5237 (1971).
25. H. Fox and H. C. Andersen, *J. Phys. Chem.* **88**, 4019 (1984).
26. S. Toxvaerd, *J. Chem. Phys.* **93**, 4290 (1990).
27. M. Hutnik, F. T. Gentile, P. J. Ludovice, U. W. Suter, and A. S. Argon, *Macromolecules* **24**, 5962 (1991).
28. C. F. Fan and S. L. Hsu, *Macromolecules* **24**, 6244 (1991).
29. P. J. Ludovice and U. W. Suter, in *Computational Modelling of Polymers*, edited by J. Bicerano (Marcel Dekker, New York, 1992), p. 401.
30. J. H. R. Clarke and D. Brown, *Mol. Simul.* **3**, 27 (1989).
31. J. I. McKechnie, D. Brown, and J. H. R. Clarke, *Macromolecules,* **25**, 1562 (1992).
32. P. J. Flory, in *Statistical Mechanics of Chain Molecules* (Hanser Publishers, New York, 1988).
33. D. Brown and J. H. R. Clarke, *J. Chem. Phys.* **100**, 6011 (1994).
34. K. Kremer and G. S. Grest, *J. Chem. Phys.* **92**, 5057 (1990).
35. R. M. Sok, H. J. C. Berendsen, and W. F. van Gunsteren, *J. Chem. Phys.* **96**, 4699 (1992).
36. T. A. Weber and E. Helfand, *J. Chem. Phys.* **71**, 4760 (1979).
37. R.-J. Roe, D. Rigby, H. Furuya, and H. Takeuchi, *Comput. Polym. Sci.* **2**, 32 (1992).
38. J. I. McKechnie, R. N. Haward, D. Brown, and J. H. R. Clarke, *Macromolecules*, **26**, 1982 (1993).
39. J. McKechnie and J. H. R. Clarke, *J. Chem. Soc. Faraday Trans. 2* (submitted).
40. P. J. Hendra, H. P. Jobic, and K. J. Holland-Moritz, *J. Polym. Sci., Polym. Lett. Ed.* **12**, 365 (1975).
41. R. Lam and P. H. Geil, *J. Macromol. Sci., Phys.* **B20**, 37 (1981).
42. J. H. R. Clarke, *J. Chem. Soc. Faraday Trans.* **2**, 1371 (1979).
43. C. A. Angell, J. H. R. Clarke, and L. V. Woodcock, *Adv. Chem. Phys.* **48**, 397 (1981).
44. C. A. Angell and L. M. Torrell, *J. Chem. Phys.* **78**, 937 (1983).
45. C. A. Angell, *J. Phys. Chem. Solids* **49**, 863 (1988).
46. M. Doi and S. F. Edwards, *The Theory of Polymer Dynamics* (Clarendon Press, Oxford, 1986).
47. R. Byron Bird, C. F. Curtiss, R. C. Armstrong, and O. Hassager, *Dynamics of Polymeric Liquids*, Vol. 1 *Fluid Mechanics*, 2nd Edition (Wiley, New York, 1987), p. 173.
48. I. M. Ward, *Mechanical Properties of Solid Polymers*, 2nd Edition (Wiley, Chichester, 1985).
49. L. E. Nielsen, *Trans. Soc. Rheol.* **9.1**, 243 (1965).
50. S. Newman and S. Strella, *J. Appl. Polym. Sci.* **9**, 2297 (1965).
51. M. H. Litt, P. J. Koch, and A. V. Tobolsky, *J. Macromol. Sci., Phys.* **B1**, 587 (1967).
52. W. Whitney and R. D. Andrews, *J. Polym. Sci. C,* **16**, 2981 (1967).
53. F. Bueche, B. J. Kinzig, and C. J. Coven, *Polym. Lett.* 399 (1965).
54. R. N. Haward and G. Thackray, *Proc. R. Soc. Lond.* **302**, 453 (1968).

55. A. A. Owadh, I. W. Parsons, J. N. Hay, and R. N. Haward, *Polymer,* **19**, 386 (1978).
56. A. A. Gusev and U. W. Suter, *J. Chem. Phys.* (in press).
57. M. L. Greenfield and D. Theodorou, *Macromolecules* (in press).
58. M. H. Cohen and D. Turnbull, *J. Chem. Phys.* **31**, 1164 (1959).
59. D. Brown and J. H. R. Clarke, *J. Chem. Phys.* **86**, 6446 (1987).
60. F. Muller-Plathe, S. C. Rogers, and W. F. van Gunsteren, *J. Chem. Phys.* **98**, 9895 (1993).
61. H. Takeuchi, *J. Chem. Phys.* **93**, 2062 (1990).
62. H. Takeuchi, *J. Chem. Phys.* **93**, 4490 (1990).
63. H. Takeuchi and K. Okazaki, *J. Chem. Phys.* **92**, 5643 (1990).
64. H. Takeuchi, R. J. Roe, and J. E. Mark, *J. Chem. Phys.* **93**, 9042 (1990).
65. F. Muller-Plathe, *J. Chem. Phys.* **94**, 3192 (1991).
66. P. V. K. Pant and R. H. Boyd, *Macromolecules* **24**, 6325 (1991).
67. P. V. K. Pant and R. H. Boyd, *Macromolecules* **26**, 679 (1993).
68. F. Muller-Plathe, *Macromolecules* **24**, 6475 (1991).
69. B. Widom, *J. Chem. Phys.* **39**, 2802 (1963).
70. G. L. Deitrick, L. E. Scriven, and H. T. Davis, *J. Chem. Phys.* **90**, 2370 (1989).
71. M. Sylvester |(private communication).
72. D. Rigby and R. Roe, *Macromolecules* **23**, 5312 (1990).
73. H. Takeuchi, R. J. Roe, and J. E. Mark, *J. Chem. Phys.* **93**, 9042 (1990).
74. G. Williams, *Chem. Soc. Rev.* **7**, 89 (1979).
75. J. H. R. Clarke, *Advances in Infrared and Raman Spectroscopy*, edited by R. J. H. Clark and R. E. Hester (Heyden, London, 1978), Vol. 4, p. 109.
76. T. A. Weber and E. Helfand, *J. Phys. Chem.* **87**, 2881 (1983).
77. I. Bahar and B. Erman, *J. Chem. Phys.* **88**, 1228 (1988).
78. H. Takeuchi and R. J. Roe, *J. Chem. Phys.* **94**, 7446 (1991).

6

MONTE CARLO SIMULATIONS OF THE GLASS TRANSITION OF POLYMERS

Wolfgang Paul and Jörg Baschnagel

6.1 Introduction

This introduction will present a very sketchy outline of physical properties of glass forming liquids pertinent to the simulations we want to present and will then focus on two main questions. Why would one want to study the glass transition in polymer melts by means of computer simulations? And if so, what can be the motivation for resorting to the dynamic Monte Carlo simulation of a lattice model as the model of choice?

The essence of the glass transition in undercooled liquids is "solidification without crystallization".[1] The former manifests itself in a steep increase in the shear viscosity over more than 10 decades (Fig. 6.1[2]) whereas the latter is clearly displayed by a look at the structure factor of the undercooled system (Fig. 6.2). The one shown is taken from the simulation discussed in Section 6.4, but it is representative of an amorphous material with small compressibility (value of $S(q)$ for $q \to 0$) and an amorphous halo at some mean particle distance (strand distance in the case of polymers) around 4–5Å. The structure at the large q-values reflects the internal structure of the polymer model under study. The way data were plotted in Fig. 6.1 led to the identification of two extreme cases, i.e., "strong glasses", which show Arrhenius behavior (straight lines), and "fragile glasses", for which the viscosity can be well described by a Vogel–Fulcher (VF) law

$$\eta = \eta_\infty \exp\left[\frac{E}{T - T_o}\right], \tag{6.1}$$

with a seemingly diverging viscosity at some temperature T_o. Close to T_g, which in this context can be defined as the temperature where the viscosity reaches a value of 10^{13} poise, the curves cross over to Arrhenius behavior again. Polymers as a rule are fragile glass formers, obeying this VF law over a wide range of viscosity values. The inset in Fig. 6.1 shows the excess specific heat of the supercooled liquid with respect to the crystal. At the glass transition part of the structural degrees of freedom of the liquid freeze, leading to a preservation of the liquid-like structure factor in the glassy regime. The shear viscosity is related to an average stress relaxation time by

$$\tau_s = \frac{\eta}{G_\infty}, \tag{6.2}$$

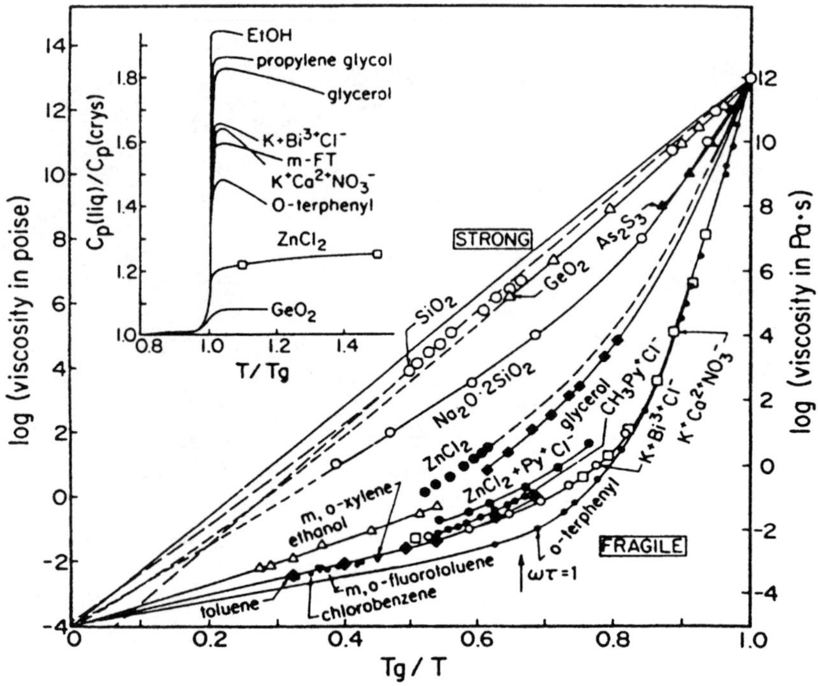

Fig. 6.1 The melt viscosity as a function of inverse temperature normalized to the T_g of the respective material. This figure is taken from Ref. 2 with permission.

where G_∞ is the high frequency elastic modulus. Typical values for G_∞ for polymers are about 10^{10}Pa,[3] so that the range of 10^{-2}–10^{12}Pa s (10^{-1} to 10^{13} poise) for the viscosity translates into 10^{-12}–10^{2} s for the typical relaxation time. Using the Stokes–Einstein relation furthermore to relate the viscosity to the self-diffusion coefficient

$$D = \frac{k_B T}{3\pi\eta a}, \tag{6.3}$$

where a is the geometric extent of the diffusing object, we see that the self diffusion essentially freezes as the temperature approaches T_g. This implies, as also seen from the specific heat behavior, that the liquid structure appears to be frozen-in on typical experimental time scales.

None of what we have discussed so far is specific to polymers. One can almost conclude that every liquid forms a glass on cooling, provided it is cooled fast enough. A cooling rate of 10^{12}K/s vitrifies liquid argon in a computer simulation,[4] and 10^5K/s produces glassy metallic alloys,[5] whereas polymers can be cooled as slowly as 10^{-4}K/s and still do not crystallize.[6] The frozen-in amorphous structure can be kept at (meta-)stable equilibrium even close to T_g and may be as close to the thermodynamic equilibrium

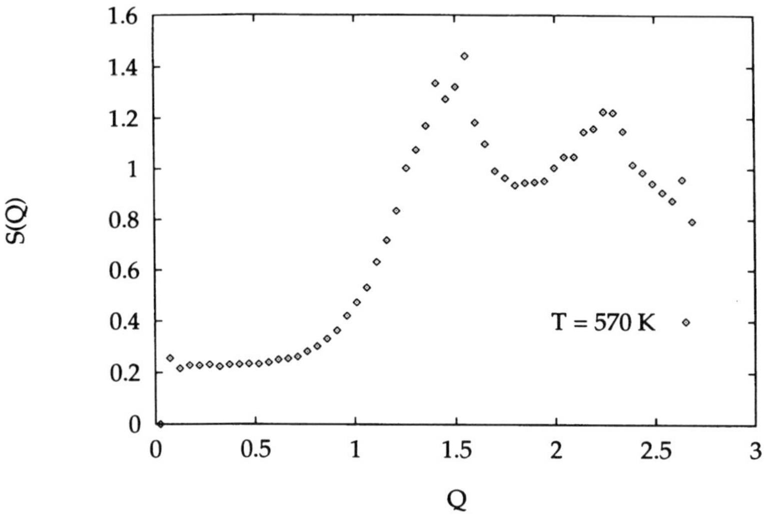

Fig. 6.2 The static structure factor for an undercooled melt in the lattice model. The temperature of the simulation is 570 K; the wave vector is given in Å^{-1}. See Section 6.4.

structure as we will ever get, especially for atactic polymers. This property makes polymers ideal systems to study the glass transition and also explains the technical relevance of such an investigation. Many of the modern polymeric materials are glasses and an understanding or even predictability of their properties is therefore of high technological importance.[7]

Great effort has been spent to characterize the liquid–glass transition and glassy regime in polymers as well as in other glass formers. But, despite all of the experimental and theoretical effort to date, there is no coherent description available covering the temperature and viscosity regime in Fig. 6.1. Due to the high degree of (meta-)stable equilibrium in a polymer glass, theories attempting to analyze the glass transition in terms of an underlying phase transition that is only blurred by the kinetic effects have been very successful in describing experimental data on the polymer glass transition. These are motivated by Kauzmann's finding[8] that the excess entropy of the supercooled liquid with respect to the crystal seems to vanish at a finite temperature T_2 when extrapolated into the glass regime. A theory dealing specifically with polymers is the one of Gibbs and DiMarzio.[9] It uses the Flory–Huggins[10] approach for calculating the configurational entropy for a melt of self-avoiding walks on a lattice. This approach reproduces the finding of a vanishing configurational entropy at a finite temperature T_2. As the entropy cannot become negative, the approximation, in which it is calculated, is assumed to break down and a second-order phase transition is postulated at that temperature. As the kinetic slowing down obviously intervenes with the phase transition, its thermodynamic signature need not be experimentally observable. Note, however, that a slight change in the calculation of the configurational entropy leads to $T_2 = 0$[11,12] and that so far no order parameter describing this speculative phase transition could be defined. This theory, as well as approaches based on the concept of a minimal free volume needed for an atom to move in the dense melt,[13] is capable of reproducing the VF law for the viscosity. None of them, however, admits for a crossover back to Arrhenius behavior around T_g, and all of them lack a first-principle derivation. At the other end of the spectrum, recent advances in the application of the mode–mode coupling formalism to the structural relaxation in supercooled liquids have led to a purely dynamic interpretation of the glass transition,[14] termed the mode coupling theory (MCT). It was developed for simple liquids and describes the onset of the need for activated processes in order for an atom to break free of the cage formed by its neighbors. The theory is applicable in the slightly supercooled region up to viscosities of around 1000 poise and produces a very detailed description of the structural relaxation not only of simple liquids but also of polymers.[15] It is not applicable around T_g and therefore falls short of reproducing the experimental phenomenology at lower temperatures.

In this situation, computer simulations have several prospects to offer. One can design simple models that reproduce the experimental phenomenology and can then be analyzed in far more microscopic detail. The models themselves can be simplified to such an extent that one can identify the molecular properties indispensable for defining a glass-forming substance. This and the complete microscopic information available allow for a guided search for possible order parameters needed to address the question of a possible underlying phase transition. Furthermore, a detailed simultaneous study of the temperature dependence of the structural and conformational relaxation in polymer melts is possible.

So, as there is good reason to study the glass transition with computer simulations, why should one want to use the Monte Carlo method for a lattice model? Polymers are complex molecules showing internal structure on length scales from 1 Å to 100 Å. The corresponding relaxation times for unentangled melts range from 10^{-13} to 10^{-7} s already at liquid-like temperatures. It is possible to study realistically detailed models of oligomer melts in a molecular dynamics or stochastic dynamics simulation[16] and to obtain equilibrated structures in the high-temperature phase. If one wants to study the glass transition in these models, however, one has to resort to very high quenching rates (10^9 K/s[17,18]). But this way one loses one of the most intriguing features of the polymer glass transition, namely the ability to perform an almost quasi-static cooling, which makes the transition very sharp. The high quenching rates introduce a strong rounding of the glass transition.[19]

If one is only interested in "universal" properties of the glass transition there is no need to deal with all the chemical details. This idea led to the very early studies of a density-driven glass transition with the use of a lattice model.[20] These simulations were done for self-avoiding walks on the diamond lattice and were intended to study the freezing of the polymer self-diffusion with increasing polymer volume fraction. The mobility of the chains in this model is mainly determined by crank-shaft type motions[21] involving the simultaneous repositioning of three adjacent monomers along the chain. With increasing density these motions, which need a lot of empty space around a chain, die out, leading to a freezing of the self-diffusion. Therefore, suitably defined lattice models should be able to capture the more universal phenomenology of the glass transition. One of these in a sense universal properties is the VF law found in a lot of different polymer systems and in many molecular glass formers, irrespective of their structure. In order to treat equilibrated samples in a computer simulation to as low temperatures as possible we will therefore resort to as simple a model as feasible. The defining properties of a polymer are connectivity and excluded volume. So one starts off with a melt of threads (Gaussian chain statistics). Such a model does not crystallize easily from the outset; but to enforce this behavior we can borrow an idea from the field of spin and orientational glasses.[22] We can choose Hamiltonians that create competition between the energetics of the chain conformation and the packing requirements of a dense melt, leading to a frustrated ground state, where there is no energetically optimal local conformation of the chains. At this point we should note further, that such a choice of Hamiltonian may be inspired by the phase transition treatment of orientational glasses, but it also has a purely kinetic effect in that it creates barriers to local structural relaxation. The chains have to assume energetically unfavorable conformations in order for the structural relaxation of their environment to proceed. By then choosing a lattice model and an imposed Monte Carlo dynamics one gets the numerically most efficient implementation of a still reasonable microscopic dynamics, which is the numerical realization of a master equation in discrete time. This stochastic dynamics cannot be expected to reproduce dynamical features on the time scale of bond length or bond angle vibrations. Conformational and structural relaxation on longer time scales, however, is brought about by jumps among the different minima in the torsional potentials along the chains. These are activated processes with a waiting time distribution depending on the Newtonian dynamics on the shorter time scales. For relaxational processes consisting of many of these jumps the dynamics can be approximated by a stochastic process, and on these time scales the Monte Carlo dynamics can be expected to be a representation of the real systems.

The choice of a lattice model also introduces a change in the ensemble as compared to the experimental situation. Experiments are usually done under constant pressure, whereas the simulations will be done at constant volume. The basic phenomenon of the glass transition is of course observable under constant volume as well, but when we model a specific polymer in Section 6.4 we will further comment on the deviations to be expected because of the different ensembles. In Section 6.2 we will discuss the bond fluctuation model used in the simulations presented further on and define the Hamiltonians chosen. Section 6.3 will then present results on general features of the glass transition in polymer melts, and Section 6.4 will show how these lattice simulations can be used to model large-scale properties of a specific polymer material. Finally, Section 6.5 will give a summary and an outlook.

6.2 Model and simulation technique

6.2.1 The definition of the bond fluctuation model

The simulation results we will be presenting were all obtained using the two-dimensional or the three-dimensional version of the bond fluctuation lattice model.[23–25] This model was proposed as an intermediate description between a highly flexible continuum treatment and the traditional lattice models, where the flexibility of the chains that are then described as self-avoiding or random walks on the underlying lattice is fully determined by the structure of this lattice. It bears on the fact that the usual high temperature persistence length in these lattice models is close to one, meaning that one repeat unit or monomer in the lattice model represents a group of atoms along the real chain and that the lattice bond should be interpreted as the vector joining the mass centers of these groups. Its length and direction will fluctuate, and the bond fluctuation model brings this idea to the square and simple cubic lattice in two and three dimensions, respectively. The repeat unit is taken to be a unit cell on the respective lattice so that the monomer occupies four lattice points in two and eight in three dimensions. The algorithmic advantage of a lattice model is that one can implement the excluded volume interaction among the monomers by simply allowing only single occupancy of each lattice vertex. The set of bond vectors defining the model in its two- and three-dimensional versions is determined by chain connectivity as the second minimal ingredient for a polymer model and the stochastic Monte Carlo dynamics we want to implement. We want to choose a monomer at random, choose a lattice direction at random, and then in the athermal case move the monomer to this new position provided it has been empty and provided no two chains intersect during the move, so that the chain connectivity is guaranteed. To implement this in a fast algorithm one can limit the set of possible bonds to maximal sets that guarantee the fulfillment of the chain connectivity requirement as long as one stays within the sets with each monomer move. In two dimensions this is the set

$$[2,0], [2,1], [2,2], [3,0], [3,1], [3,2] , \qquad (6.4)$$

where the square brackets denote the equivalence classes of all vectors that can be generated from these basic ones by applying all lattice symmetry operations, whereas in three dimensions it is

$$[2,0,0],[2,1,0],[2,1,1],[2,2,1],[3,0,0],[3,1,0]\,. \qquad (6.5)$$

Figure 6.3 shows the three-dimensional version of the model. In two dimensions there are 36 bonds with six different lengths and 41 different bond angles, whereas in three dimensions there are 108 bonds with five different lengths and 87 different bond angles. As could already be seen in Fig. 6.2, this produces a very flexible polymer model in which the underlying lattice has no discernible effect on the structure of the melt.

6.2.2 Hamiltonians and cooling procedures

If we want to study a temperature-driven glass transition with this model we have to augment it with a Hamiltonian using its natural degrees of freedom, namely the bond lengths and bond angles. The results presented in the next sections were obtained using three different Hamiltonians. The first two employed for the simulations discussed in Section 6.3 are simple model Hamiltonians designed to produce a geometric frustration effect upon cooling, as discussed in the introduction. The first Hamiltonian employed for the

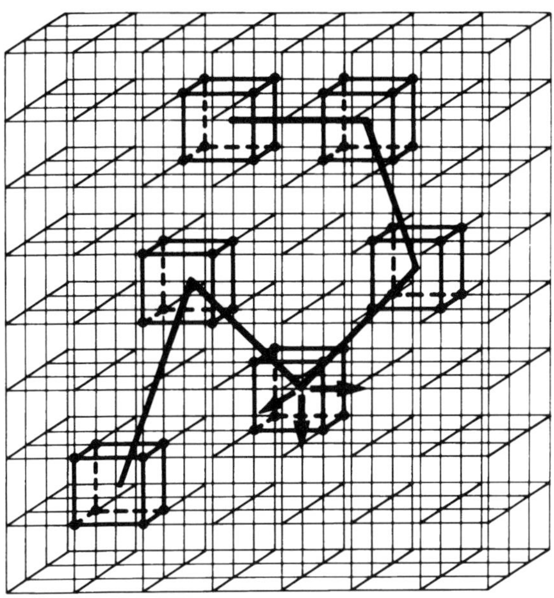

Fig. 6.3 A drawing of the three-dimensional version of the bond fluctuation model showing the unit cube repeat units and a few of the possible bond vectors.

simulation of the two-dimensional model, which we will call Hamiltonian A, is[26]:

$$\mathcal{H}(\vec{b}) = \epsilon\left(l(\vec{b}) - \sqrt{10}\right)^2 . \tag{6.6}$$

The simulations done with this Hamiltonian used a cooling scheme that is linear in the inverse temperature

$$\beta(t) = \beta_{\max}\Gamma_Q t , \tag{6.7}$$

with $\beta(t) = 1/T(t)$, $\beta_{\max} = 1/T_{\min} = 10$ and $\Gamma_Q = 4 \cdot 10^{-7}$. The temperature is measured in units of the strength of the harmonic potential in eq. (6.6). In the ground state of this Hamiltonian each bond belongs to the equivalence class [3, 1] but as will be shown in the next section it then blocks one additional lattice point besides the four vertices taken by the monomer. A volume fraction of $\phi \geq 0.8$ therefore must lead to a geometric frustration of the ground state.

The second Hamiltonian (Hamiltonian B) set up along this line of reasoning is even more simplified, using just a two-level system for the possible energies

$$\mathcal{H}(\vec{b}) = \begin{cases} 0 & \text{if } \vec{b} \in [3,0,0] \quad ([3,0] \text{ in 2d}) \\ \epsilon & \text{otherwise} \end{cases} . \tag{6.8}$$

Bonds belonging to the class [3, 0, 0] resp. [3, 0] block four additional lattice points in 3*d* and two additional ones in 2*d*, leading to a geometrically frustrated ground state for $\phi \geq 2/3$. For the quenching runs with this model the same cooling scheme is used as for Hamiltonian A (eq. [6.7]). In 2*d* one quench rate, $\Gamma_Q = 4 \cdot 10^{-5}$, is employed, whereas for the simulations in 3D the effect of the cooling rate was explicitly studied using 14 different rates.[27] We will show results for the following five representative rates: $\Gamma_Q = 4 \cdot 10^{-5}, 8 \cdot 10^{-6}, 4 \cdot 10^{-6}, 8 \cdot 10^{-7}$ and $4 \cdot 10^{-7}$.

Finally, we will use Hamiltonian C for the modeling of specific polymeric materials.[28] This Hamiltonian is constructed similar to force fields used to parametrize bond length and bond angle contributions in real polymers

$$U(l) = \epsilon_l (l - l_0)^2 ,$$
$$V(\vartheta) = \epsilon_\vartheta (\cos\vartheta - c_0)^2 . \tag{6.9}$$

The four parameters in these potentials will be optimized for the modeling of specific substances. This procedure and the application to bisphenol-A-polycarbonate will be discussed in Section 6.4. For these simulations a stepwise cooling was used, giving the system enough time to equilibrate at each temperature and cooling it down to as low temperatures as this is computationally feasible.

The thus specified Hamiltonians are then used in a dynamic Monte Carlo simulation[29] (see Chapter 1) that is a numerical realization of the discrete time master equation

$$P(x, t_{n+1}) - P(x, t_n) = \sum_{x'} W(x' \to x)P(x', t_n) - \sum_{x'} W(x \to x')P(x, t_n) \,, \tag{6.10}$$

where x denotes the configuration of the polymer melt. The transitions between configurations are implemented by the following steps:

- choose a monomer at random
- choose a lattice direction at random
- check whether the new lattice position is empty (excluded volume)
- check whether the new bonds are in the allowed set (chain connectivity)
- calculate the energy change caused by the move and accept it with probability $\min(1, \exp[-\beta\Delta E])$.

A large computational effort has gone into ensuring a high statistical accuracy of the data. The results for Hamiltonian A are averages over 12 800 chains or 128 000 monomers, those for Hamiltonian B were obtained for 28 800 chains and 288 000 monomers, and those for Hamiltonian C used 6600 chains and 132 000 monomers. The simulations were done on 64 (A), 500 (B, 2*d*), 16 or 160 (B, 3*d*) and 33 (C) replicas of a basic simulation box of size 100^2 (A), 60^2 (B, 2*d*), 30^3 (B, 3*d*) and 40^3 (C).

6.3 Results for the schematic models

If a melt of chemical polymers is cooled down, the bonds of the chains will try to adopt the energetically favorable "trans" state.[30] In order to reach this state the parts of the chains adjacent to the bonds have to be rotated. This rotation requires a certain amount of space which may or may not be available for a specific bond in a dense melt, depending upon the spatial arrangement of the polymers around it. Therefore the presence of the surrounding polymers can hinder or even prevent the tendency of a chain to stretch out. This feature is clearly apparent from various attempts to simulate the glass transition of polyethylene[17,18] rather realistically which are described in Chapter 5. It is this interplay between the expansive intrachain interaction with the effectively repulsive interchain forces, which the model Hamiltonians A and B (see eqs [6.6] and [6.8]) try to mimic in an elementary fashion and which may be summarized by the term "geometric frustration". The effects of the geometric frustration cause the melt to maintain its amorphous liquid-like structure while it is being continuously cooled down from the liquid to the frozen glassy phase according to the prescription of eq (6.7).

Figure 6.4 exemplifies this behavior by opposing a snapshot of a configuration belonging to the temperature region of the liquid (i.e., $T = 1.0$ [Fig. 6.4(a)]) with one which corresponds to the glassy phase (i.e., $T = 0.1$ [Fig. 6.4(b)]) for a two-dimensional melt that is equipped with the Hamiltonian A (eq. [6.6]) and consists of chains with $N = 25$ at a density of $\phi = 0.8$.[26] The snapshots were obtained during a cooling run with a rate $\Gamma_Q = 4 \cdot 10^{-7}$, following eq. (6.7).[26] A glance at the global features of these snapshots shows that both the high- and low-temperature configurations possess a disordered amorphous structure, proving that the chosen model Hamiltonian effectively prevents crystallization at the considered density and forces the melt to vitrify. The Hamiltonian A especially favors the large bonds of the equivalence classes [3, 0] and [3, 1]. Since the bond vectors of these classes either block one (for [3, 1]) or two (for [3, 0]) lattice sites (see Fig. 6.5) which are no longer accessible for other monomers, this loss of volume has to be compensated in order to make the density ϕ stay the same. How the melt achieves this compensation is illustrated in Fig. 6.6, which magnifies the encircled polymer of Fig. 6.4(a) together with its immediate environment. Most of the bonds of the shaded polymer belong to the equivalence class [2, 2], which allows the chain both to adopt a small energy ($\mathcal{H}([2,2]) = 0.08$) and to pack as densely as possible. This compromise between the energetic and geometric constraints leads on the one hand to an overall expansion of the chains and on the other hand to a local orientational correlation among the chains. On the length scale of the average chain's size these two aspects are the most apparent qualitative differences between the high- and the low-temperature configuration displayed in Fig. 6.4.

In order to quantify the qualitative observations from the configurations one should calculate quantities which are sensible to the structural changes

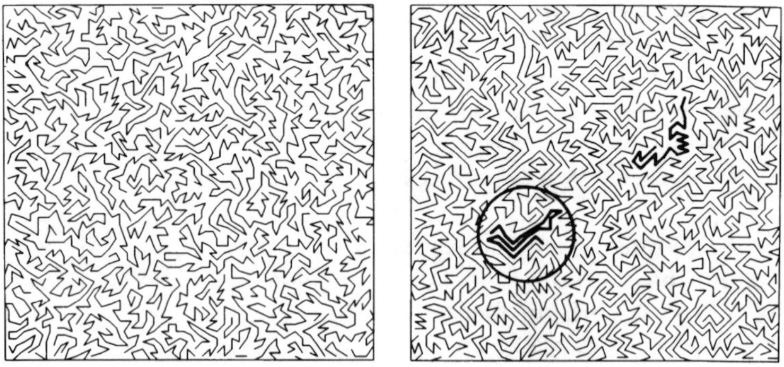

(a) (b)

Fig. 6.4 Snapshots of a polymer configuration, generated by the 2*d* version of model A at $T = 1.0$ (a) and $T = 0.1$ (b), and with the following parameters: $N = 25, P = 80$ and $\phi = 0.8$. The encircled polymer of Fig. 6.4(b) is shown in more detail in Fig. 6.6.

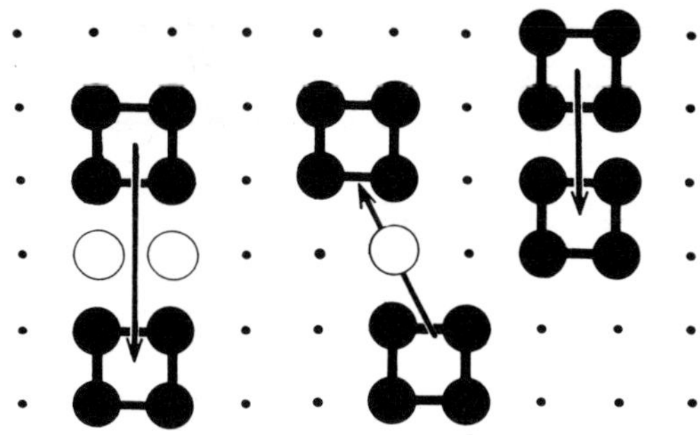

Fig. 6.5 Three possible positions of a dimer on a square lattice. The marked sites (◯) cannot be occupied by other monomers without overlapping with the dimer and are thus not available for them. In a dense melt this blocking of lattice sites leads to a competition between the energetic and geometric constraints of the chains, which prevents crystallization.

that occur on the different length scales of the melt during the cooling process. Such quantities are, for instance, the mean bond length, the radius of gyration, the chain's structure factor and the collective static structure factor, the temperature and cooling rate dependence of which will be discussed in the next subsection. From these data one can try to extract the influence of either the cooling rate or, if one additionally varies the size of the polymers, of the chain length on the glass transition temperature. The

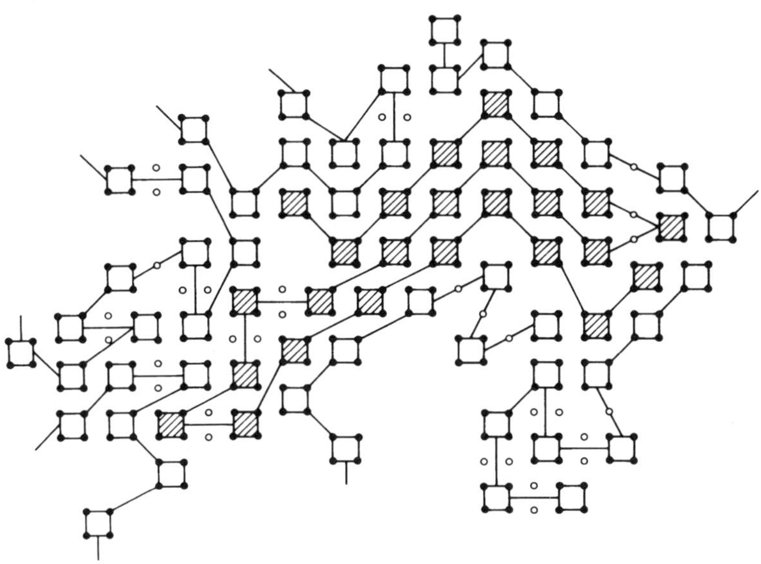

Fig. 6.6 Magnification of the encircled polymer of Fig. 6.4(b) together with parts of its environment. Most of the bonds of the shaded polymer belong to the equivalence class [2, 2] since the energy of the chain then becomes small. The open circles mark sites which are blocked by the bond vectors belonging to the classes [3, 0] and [3, 1], respectively.

results of this analysis are presented in Section 6.3.1.3. After having concentrated on the structural properties and their corollaries we address the gists of the dynamical behavior of the melt in subsection 6.3.2.

6.3.1 Structural properties of the melt

A comparison of the temperature dependence of the various structural quantities, presented in Refs 26, 27, and 31 for the model Hamiltonians A and B shows that neither the choice of the Hamiltonian nor the spatial dimension introduces qualitative differences in the freezing behavior of the melt when the cooling is done at a finite rate. Therefore we want to base the following discussion on the results obtained for the two- and three-dimensional version of model B. The cooling-rate dependence of the structural properties of the melt and of the glass transition temperature will be exemplified by the three-dimensional simulation, which was done with polymers of length $N = 10$ at a density of $\phi = 0.5\bar{3}$ (see Section 6.3.1.3), whereas the influence of the chain length on the glass transition was studied in two dimensions, using the same density as in the three-dimensional case (see Section 6.3.1.3).

6.3.1.1 Mean-squared bond length and radius of gyration

Figure 6.7[27] depicts the temperature dependence of $\langle b^2_{npc} \rangle_{npc}$ for the five representative cooling rates mentioned in Section 6.2.2. In the high-temperature region ($T \in [0.6, 2.0]$) the curves for the different cooling rates nicely collapse, indicating that the melt is in a thermally equilibrated liquid state. In this state it is mobile enough to respond easily to the speed, by which the temperature is reduced. The effect of the finite cooling rate starts to be felt below $T \approx 0.5$ and is accompanied by a strong expansion of the mean bond length. The increase of $\langle b^2_{npc} \rangle_{npc}$ continues until the temperature reaches $T \approx 0.2$, where the curves level off. In this temperature range the intrinsic relaxation times of the melt become comparable to the observation time which is determined by the cooling rate. Then the system falls out of equilibrium and $\langle b^2_{npc} \rangle_{npc}$ gets locked at a value depending on the cooling rate. The smaller the cooling rate, the more time the melt has to relax. Therefore this final value of $\langle b^2_{npc} \rangle_{npc}$ in the low temperature region ($T \leq 0.2$) increases with decreasing cooling rate. However, it is always smaller than nine, which would be the expected value for $\langle b^2_{npc} \rangle_{npc}$ if all bonds were in the ground state. This result shows that the model Hamiltonian B (as well as the Hamiltonian A [see Ref. 26]) indeed introduces strong topological constraints leading to geometric frustration. Due to the present frustration effects and due to the fact that the melt appears to be frozen on the time scale of the simulation, the above specified low temperature region $T \leq 0.2$ will be called "glassy phase".

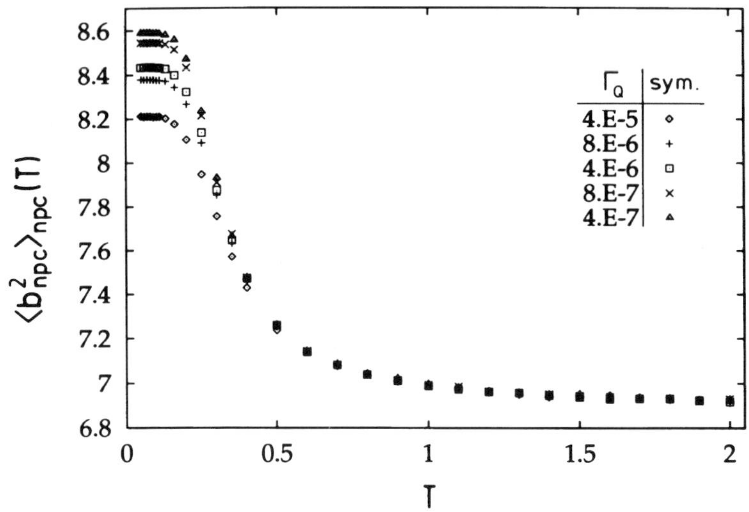

Fig. 6.7 Plot of the mean bond length vs. T for five different cooling rates: $\Gamma_Q = 4 \cdot 10^{-5}(\diamond)$, $\Gamma_Q = 8 \cdot 10^{-6}\ (+), \Gamma_Q = 4 \cdot 10^{-6}\ (\square)$, $\Gamma_Q = 8 \cdot 10^{-7}(\times)$ and $\Gamma_Q = 4 \cdot 10^{-7}(\triangle)$. This result was obtained by a simulation of the $3d$ model B with $N = 10, P = 180$, and $\phi = 0.5\bar{3}$.

In addition to the scale of a bond, another important length scale for a polymer is that of the radius of gyration. This quantity is defined by:

$$R_\mathrm{G}^2 = \frac{1}{N}\sum_{n=1}^{N}\left\langle (\vec{r}_{npc} - \vec{R}_{pc})^2 \right\rangle_{pc}, \tag{6.11}$$

where $\vec{r}_{npc}$ is the vector to monomer n in polymer p of configuration c, $\vec{R}_{pc}$ is the vector to the center of mass for the respective polymer, and the symbol $\langle \bullet \rangle_{pc}$ stands for:

$$\langle \bullet \rangle_{pc} := \frac{1}{CP}\sum_{c=1}^{C}\sum_{p=1}^{P} \bullet_{pc} \tag{6.12}$$

with the explicit values $N = 10$, $P = 180$ and $C = 160$. The result of the simulation is shown in Fig. 6.8. Qualitatively the dependence on temperature and on the cooling rate of the curves resembles very much that of the mean-squared bond length. In the high temperature range above $T = 0.5$ all five curves nicely collapse, proving that the melt is also in equilibrium on this length scale whereas R_G^2 steeply increases below $T = 0.5$ before it crosses over to a constant value in the temperature region where the melt vitrifies. Despite these qualitative similarities between the shapes of $\langle b^2_{npc} \rangle_{npc}$ and $R_\mathrm{G}^2(T)$, a closer comparison of the two quantities reveals that the R_G^2 increases by the same amount (approximately 3%) as the mean-squared bond length in the high temperature region, whereas at $T \leq 0.5$ the cooling rate crucially affects the strength of stretching on the two length scales. For

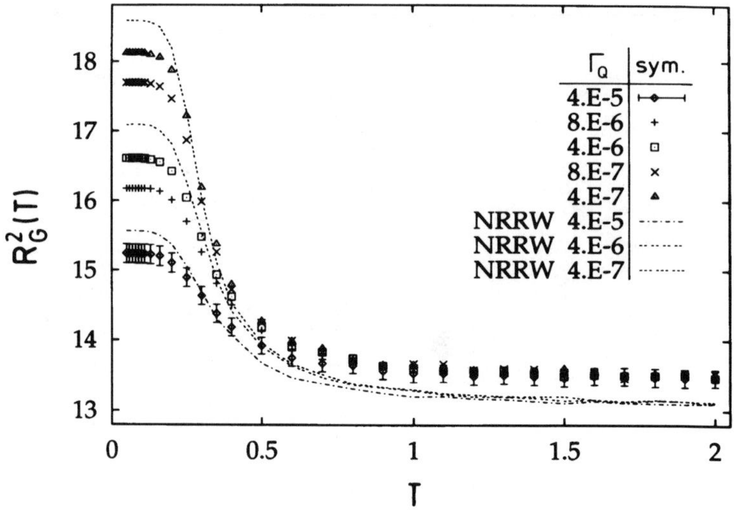

Fig. 6.8 Radius of gyration vs. temperature for the same cooling rates as in Fig. 6.7. The error bars are only included for the fastest cooling rate for the sake of clarity. The dashed–dotted and the dashed lines correspond to the generalized nonreversal random walk (NRRW) approximation.[27]

$\Gamma_Q = 4 \cdot 10^{-5}$ $\langle b^2_{npc} \rangle_{npc}$ expands faster than $R^2_G(T)$ whereas the opposite is true for $\Gamma_Q = 4 \cdot 10^{-7}$.

In order to illustrate this point further the simulation data may be compared with a generalized non-reversal-random-walk (NRRW)[21] approximation of R^2_G, which can be obtained from a prescription that Flory describes in his textbook[30] if one uses the bond vector and bond angle distribution functions from the simulation to perform the necessary averaging.[32] This analytical calculation thus depends upon the details of the cooling process. The result of the approximation is also shown in Fig. 6.8, exemplified for three of the cooling rates. Since the NRRW-approximation only partially takes into account the condition that no two monomers may overlap, the calculated R^2_G-values are systematically smaller than the simulated ones in the high-temperature region, because the self-avoidance of the monomers makes the polymer stiffer. As soon as the influence of the cooling rate starts to be felt, however, the NRRW-approximation crosses the simulation data and settles down at a higher value of R^2_G in the glassy temperature region for all of the cooling rates. This *a priori* unexpected result can be rationalized thus: since the NRRW-approximation only takes into account correlations between neighboring bond vectors, it is only affected by processes which happen on the local length scale of a bond. Changes on this length scale require a shorter amount of time than changes on a larger scale, because more and more monomers are involved in the regrouping motion the larger the length scale is. *Hence different time scales belong to different length scales.* If the temperature is lowered very fast, only the local structure of the melt may be able to adapt to this speed, whereas the larger structures essentially remain in a state which corresponds to a higher (internal) temperature.

That these retardation effects, inferred from the comparison of the NRRW-approximation and the simulation data only very indirectly, are actually present should be seen if one plots the ratio of the radius of gyration and of the mean-squared bond length, i.e., $R_G^2/\langle b_{npc}^2 \rangle_{npc}$ versus temperature for the different cooling rates. This is done in Fig. 6.9. It is clearly visible that the mean-squared bond length expands faster than the radius of gyration if the cooling rate is large. The slower the cooling rate becomes the more R_G^2 catches up with the stretching of $\langle b_{npc}^2 \rangle_{npc}$, until a cooling rate is reached, at which both length scales are affected in the same way (i.e., at $\Gamma_Q = 4 \cdot 10^{-6}$). For even smaller cooling rates the radius of gyration expands more strongly than the mean-squared bond length, due to the contribution of the bond angles to the overall size of polymer.

Presumably it would be possible to verify the result, presented in Fig. 6.9, in a neutron scattering experiment[33] if one measured the structure factor of a chain in the vicinity of the glass transition for various cooling rates. From the behavior of the structure factor at small wave numbers q one could determine $R_G^2(T, \Gamma_Q)$, whereas the end of the scaling regime, where the structure factor is proportional to q^2, would yield the temperature- and cooling-rate-dependence of the persistence length[30,34–36]. Since a monomer

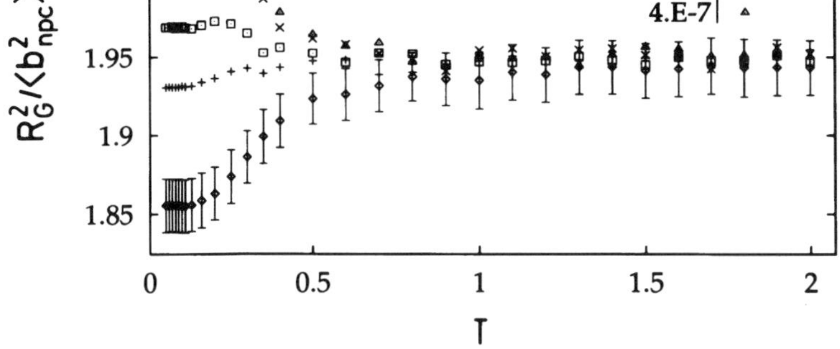

Fig. 6.9 $R_G^2/\langle b_{npc}^2 \rangle_{npc}$ vs. temperature for the same cooling rates as in Fig. 6.7. Error bars have only been included for the fastest cooling rate for the sake of clarity.

of the bond-fluctuation model stands for a group of chemical monomers (see Section 6.2.1) the persistence length of the chemical polymer chain should correspond to the bond length of the lattice model. With such a scattering experiment it would therefore be possible to decide whether the observed result is a peculiarity of the used model or a characteristic feature of polymers close to the glass transition.

6.3.1.2 Structure factor and related quantities

The last paragraph was primarily concerned with the influence of the cooling rate on both the local length scale of a bond vector and the global length scale of the radius of gyration. In order to extend this analysis, one should calculate quantities which provide information about the structural changes on all length scales in the melt, such as the collective static structure factor or the chain's structure factor.

The collective static structure factor for an isotropic system can be defined in the following way[37]:

$$S(q) = \frac{1}{NP} \sum_{n,m=1}^{NP} \left[\langle \exp[i\vec{q} \cdot (\vec{r}_{nc} - \vec{r}_{mc})] \rangle_c \right]_q , \tag{6.13}$$

where $\vec{r}_{nc}$ is the vector to the center of gravity of the n-th monomer in configuration c, $\langle \bullet \rangle_c$ represents the average over all independent configurations defined analogously to eq. (6.12), and the symbol $[\bullet]_q$ stands for the lattice analogue of a spherical average in the continuous reciprocal space.[38]

The result of the simulation for $S(q)$ is plotted in Fig. 6.10 versus q for the intermediate cooling rate $\Gamma_Q = 4 \cdot 10^{-6}$ and for two representative temperatures from the glassy and the liquid phase, i.e., $T = 0.05$ and $T = 1.8$.[38] For these two temperatures the structure factor has the expected shape of a liquid-like system. Since there are no long-range correlations in a liquid, its structure factor exhibits a large first peak which corresponds to the most probable distances between the particles in the liquid and is followed by a sequence of peaks with decreasing intensity, reflecting the loss of spatial correlations.[37] For a polymer liquid one typically finds two important peaks for q-values smaller and equal to that of a bond length, whereby the first one is usually related to interchain interactions and the second, less intense, is attributed to the correlations on the length scale of a bond.[39] Although this simulation models a polymer melt, one can observe only one peak in this q-range, having its maximum around $q = 3.0$, which corresponds to a length of $l \approx 2.0$ in real space, i.e., to the smallest distance that two monomers may have. This observation can be rationalized by the fact that different chains interact via a hard-core potential which is only felt if two monomers try to overlap. Therefore the two experimentally found peaks for polymer liquid are here combined to one peak, the maximum of which is slightly shifted to

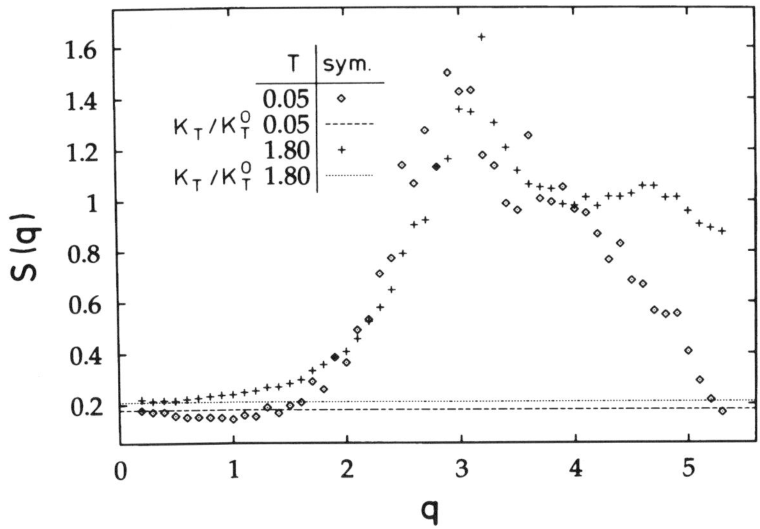

Fig. 6.10 Plot of the collective static structure factor $S(q)$ vs. q for the temperatures and $T = 0.05$ (◇) and $T = 1.8$ (+) and for the intermediate cooling rate $\Gamma_Q = 4 \cdot 10^{-6}$. Additionally, a dashed and a dotted horizontal line are shown which correspond to $\kappa_T/\kappa_T^{\circ}$ calculated by the block-analysis method, respectively (see text). The model is the same as in Fig. 6.7.

lower q-values as the temperature is decreased. This effect is certainly due to stretching of bond vectors caused by the model Hamiltonian. Besides this weak temperature dependence of the maximum the shape of the structure factor stays the same for all temperatures in the low q-range. In particular, no additional peak develops at small q-values (i.e., $q \leq 2.0$) for low temperatures, which would indicate an onset of crystalline ordering. The simulated polymer melt thus conserves the liquid-like structure when it freezes, i.e., it really becomes glassy. In the large q-range ($q \geq 4.0$, which has to be considered with caution due to a systematic deficiency of the spherical averaging procedure on the lattice[38]) the structure factor shows the interesting feature that a small secondary peak quickly loses all its weight as the temperature decreases. This strong temperature dependence must result from subtle interference effects, to which mainly the small bond vectors might have contributed. Since the choice of the model Hamiltonian especially suppresses them, the intensity of the structure factor will reduce in the respective q-range. Another Hamiltonian might therefore affect the shape of the structure factor quite differently for large q-values. This fact is clearly apparent from the structure factor presented in the introduction (Fig 6.2), which resulted from the simulation of the bond-fluctuation method with the Hamiltonian C modeling bisphenol-A-polycarbonate (eq. (6.9); see Section 6.4.2 for more details).[40] In this simulation the collective structure factor was determined at a temperature of approximately twice the Vogel–Fulcher temperature[1] which corresponds to $T \approx 0.3$ in model B,[27] as will be shown below (see Section 6.3.2.2). This temperature thus belongs to a regime where $S(q)$ has already significantly lost in intensity for $q > 4$. Contrary to this behavior the calculated structure factor of bisphenol-A-polycarbonate does not decay to small values in the corresponding q-range (Fig. 6.2). However, the structure factor of bisphenol-A-polycarbonate was calculated after a long relaxation run, which sufficed to equilibrate the polymers on all length scales, so that the history of the cooling process was eliminated, contrary to the simulation with the model Hamiltonian B. Whether the shape of $S(q)$ for large q-values is primarily determined by the choice of the model Hamiltonian or whether the cooling rate also has a crucial influence on it is still unclear, but will be studied by an analysis of the melt's physical aging in the future.

In addition to the behavior of the structure factor at large q-values, also the small q-limit of $S(q)$ is interesting because it is related to the isothermal compressibility κ_T.[37] Moreover, the isothermal compressibility can also be calculated from the density fluctuations in subsystems, using the so-called "block-analysis",[41] by virtue of the fluctuation–dissipation theorem.[37] Therefore there are two independent ways to determine κ_T in the simulation, which should coincide:

$$\lim_{q\to 0} S(q) = \frac{\kappa_T}{\kappa_T^\circ} = \lim_{l\to\infty} \frac{\langle(\rho_l - \langle\rho_l\rangle_l)^2\rangle_l l^3}{\langle\rho_l\rangle_l} . \tag{6.14}$$

In eq. (6.14) the symbol $\langle\bullet\rangle_l$ means the thermal average over a subsystem of linear dimension l and κ_T° is an abbreviation for the isothermal compressibility of the ideal gas, i.e., $\kappa_T^\circ = (T\rho)^{-1}$, whereby ρ is the average density of monomers, i.e., $\rho = \phi/8$. In Fig. 6.10 the result of the block analysis for the two studied temperatures $T = 1.8$ and $T = 0.05$ is included as a dashed and dotted horizontal line, respectively, proving that the small q-limit of $S(q)$ and the calculation of the isothermal compressibility according to the fluctuation–dissipation theorem indeed coincide in our simulation, as required by eq. (6.14).

The full temperature dependence of the isothermal compressibility for the cooling rate $\Gamma_Q = 4 \cdot 10^{-6}$ is shown in Fig. 6.11.[38] Although a first glance at this picture immediately reveals that despite the large statistical effort of our simulation—please note that the averaging is done over about 300 000 monomers—the accuracy of the block-analysis is still too poor to draw quantitative conclusions from it, the statistics suffice for a qualitative interpretation. In the high temperature region ($T \in [0.6, 2.0]$) the ratio κ_T/κ_T° fluctuates around a mean value in the interval between 0.21 and 0.215 (see the dashed line in Fig. 6.11). This high-temperature behavior of the isothermal compressibility can be understood if one uses the equation of

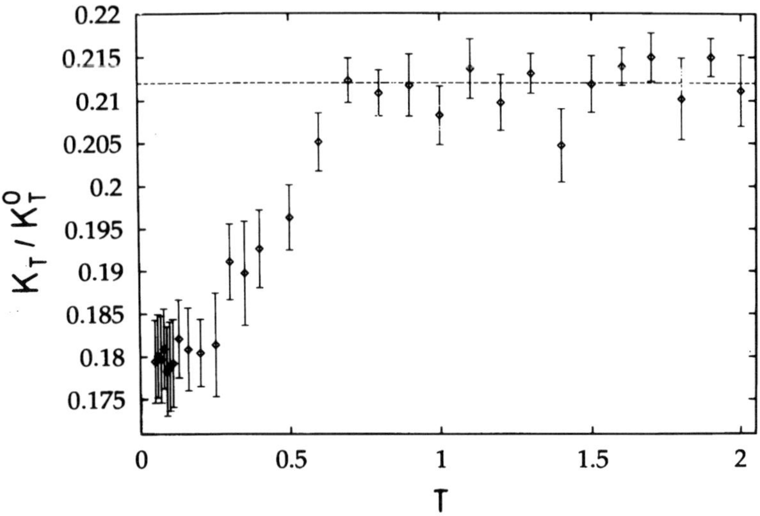

Fig. 6.11 Plot of the temperature dependence of the isothermal compressibility obtained by the block-analysis method (see text). In addition, a dashed horizontal line at $\kappa_{\mathrm{T}}/\kappa_{\mathrm{T}}^{\circ} = 0.212$ is shown which is the high-temperature result from eq. (6.17). The model is the same as in Fig. 6.7.

state for the athermal bond fluctuation model, derived by Deutsch and Dickman,[42] in order to calculate the ratio $\kappa_{\mathrm{T}}/\kappa_{\mathrm{T}}^{\circ}$. The result reads:

$$\left.\frac{\kappa_{\mathrm{T}}}{\kappa_{\mathrm{T}}^{\circ}}\right|_{T=\infty} = \left[\frac{1}{N}\left\{1 + 4\frac{v(N)}{v(1)}\phi_{\mathrm{hs}}\frac{(2-\phi_{\mathrm{hs}})}{(1-\phi_{\mathrm{hs}})^3} + 2\frac{v(N)}{v(1)}\phi_{\mathrm{hs}}^2\frac{(5-2\phi_{\mathrm{hs}})}{(1-\phi_{\mathrm{hs}})^4}\right\}\right]^{-1}. \tag{6.15}$$

In this equation ϕ_{hs} stands for the volume fraction of the hard sphere polymer chains in the continuum of space, and the ratio $v(N)/v(1)$ measures the increase of the excluded volume for a chain of length N in comparison to a monomer. Inserting the appropriate values for $N = 10$ in eq. (6.15) the final result for the isothermal compressibility of the athermal melt becomes: $\kappa_{\mathrm{T}}/\kappa_{\mathrm{T}}^{\circ} = 0.212$, which is included as a dashed horizontal line in Fig. 6.11. This horizontal line describes the high-temperature behavior of the isothermal compressibility very well, emphasizing that the structural properties of the melt are mainly determined by entropic effects in the temperature interval $T \in [0.6, 2.0]$. If one reduces the temperature further the ratio $\kappa_{\mathrm{T}}/\kappa_{\mathrm{T}}^{\circ}$ starts to decrease before it levels off in a narrow temperature range around $T \approx 0.25$ and gets locked at a temperature-independent value for $T \leq 0.2$, a behavior reminiscent of that of $\langle b_{npc}^2 \rangle_{npc}$ (see Fig. 6.7) or of R_{G}^2 (see Fig. 6.8). Close to $T \approx 0.25$ the melt gets trapped in a configuration from which it cannot easily escape, so that *the density fluctuations freeze*. This freezing of the density fluctuation below a certain temperature, which is preceded by a decrease of the isothermal compressibility, has been predicted by the mode coupling approach to the glass transition.[14] The behavior of the ratio κ_T/κ_T° is thus qualitatively consistent with this theory and seems to indicate the existence of a critical glass transition temperature[14] below $T \approx 0.25$. However, one has to be cautious with this kind of interpretation in the framework of the mode coupling theory at the present stage of the simulation, because the melt falls out of equilibrium (around $T \approx 0.25$) if the cooling process becomes too rapid. Therefore an important assumption of the theory, namely that the system is thermally equilibrated over the whole interesting temperature range, is not met. Only long relaxation runs at the respective temperatures can prove whether the freezing of the density fluctuations persists or whether it vanishes as the melt adapts more and more to the external temperature of the heat bath. These relaxation runs have been performed for the three-dimensional model B in order to test the predictions of the mode coupling theory afterwards. The results of this analysis are summarized in Section 6.3.2.3.

Another interesting quantity in this context is the static structure factor of the polymer chains $S_p(q)$. $S_p(q)$ can be defined analogously to the collective static structure factor:

$$S_p(q) = \frac{1}{N} \sum_{n,m=1}^{N} \left[\langle \exp[i\vec{q} \cdot (\vec{r}_{npc} - \vec{r}_{mpc})] \rangle_{pc} \right]_q . \tag{6.16}$$

In this equation the sum runs over all monomers of a chain, and symbols of the averages $\langle \bullet \rangle_{pc}$ and $[\bullet]_q$ have the same meaning as in eqs (6.12) and (6.13), respectively.[38]

Since experiments suggest that the polymers preserve their Gaussian shape[30,34,35] when the melt vitrifies[36] one can try to compare the results of the chain's structure factor with the Debye formula[30,34,35] for some interesting temperatures ranging from the liquid to the glassy region. The Debye formula is given by the following expression[30,34,35]:

$$S_p^{DE}(q) = \frac{2N}{(q^2 R_G^2)^2} \left(\exp(-q^2 R_G^2) - 1 + q^2 R_G^2 \right) . \tag{6.17}$$

The results for R_G^2, discussed in Section 6.3.1.1, will be used in the Debye formula in order to compare it to the simulated values of $S_p(q)$.

The simulated values of $S_p(q)$ as well as the Debye formula exhibit a typical behavior for small and intermediate q-values.[30,34,35] Irrespective of the specific statistics to be applied for the configurational average, the initial decay of the static structure factor is related to the characteristic size of the scattering object, i.e., to the radius of gyration of the polymer in our case[30,34,35]:

$$S_p(q) = N\left(1 - \frac{1}{3}q^2 R_G^2\right) \text{ if } q \ll 2\pi/R_G. \tag{6.18}$$

Therefore the simulated structure factor and the Debye formula must at least coincide in this range of q-values if one uses the simulation results for R_G^2 in eq. (6.17). In addition to the behavior for small q-values there is an intermediate q-range between the q-values corresponding to the length scale of the radius of gyration and that of the mean bond length $\langle b_{npc}^2\rangle_{npc}^{1/2}$ where the static structure factor should become independent of the degree of polymerization, provided that the chains are very long and behave as $S_p(q) \propto q^{-2}$, which is the typical scattering law in this q-regime for Gaussian statistics, i.e.,[30,34,35]:

$$S_p(q) \propto q^{-2} \text{ if } 2\pi/R_G < q \ll 2\pi/\sqrt{\langle b_{npc}^2\rangle_{npc}}. \tag{6.19}$$

Figure 6.12 exemplifies this comparison of the simulated chain's structure factor with the Debye formula for the temperatures $T = 0.05$ and $T = 1.8$ and the slowest studied cooling rate $\Gamma_Q = 4 \cdot 10^{-7}$. If the q-values are very small (i.e., $q \ll 0.1$) the structure factor of the polymer only depends upon the number of scattering monomers and not upon their spatial arrangement, so that $S_p(q)$ is independent of temperature in this range. For larger q-values

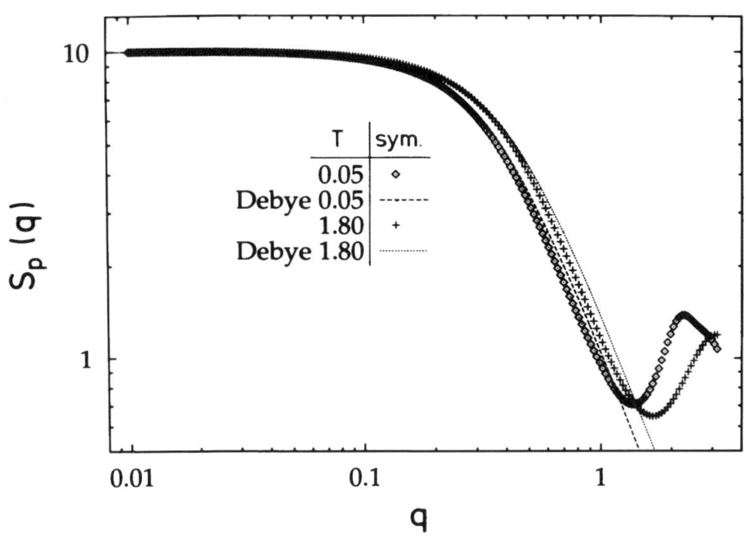

Fig. 6.12 Temperature dependence of the static structure factor of a polymer $S_p(q)$ and comparison of $S_p(q)$ with the Debye formula (eq. [6.17]). The diamonds (◇) and the crosses (+) correspond to the simulation results at $T = 0.05$ and $T = 1.8$, respectively, whereas the Debye formula at the two temperatures $T = 0.05$ and $T = 1.8$ is represented by the dashed and the dotted lines. The model is the same as in Fig. 6.7.

$S_p(q)$ starts to decay as is required by eq. (6.18). Due to the above-mentioned independence of this behavior from the configurational statistics, the validity of eq. (6.18) ceases when the simulated structure factor and the Debye formula begin to deviate from each other. This deviation occurs for $q \approx 0.3$ which is still distinctly smaller than $q = 2\pi/R_G \approx 1.6$[27](see eq. [6.18]). For larger q-values the Debye curve first lies slightly above the simulated structure factors at both temperatures before it crosses them in the interval $q \in [1, 2]$, an interval which probes the length scale of the radius of gyration (i.e., $q = 2\pi/R_G \approx 1.6$). Whereas one does in general expect the Debye formula to become invalid for q-values which are sensitive to changes on the length scale of a bond vector, since a polymer does certainly not exhibit a Gaussian structure on that length scale, the slight deviation of $S_p(q)$ from the Debye formula on the scale of the radius of gyration could not have been predicted in advance. It shows that the lattice chains of the bond fluctuation method, having a length of only $N = 10$, already approximate the character of highly polymerized substances (i.e., chains with $N \to \infty$) extremely well. However, they are still a little bit too small for the scattering law of eq. (6.19) to be observed. The longer the polymers are the larger the radius of gyration becomes and the more the lower limit for the validity of eq. (6.19) is pushed to smaller q-values. Whereas one can only hope to find the scattering law of eq. (6.19) in the range $2\pi/R_G \approx 1.6 < q < 2\pi/\langle b^2_{npc}\rangle^{1/2}_{npc} \approx 2.2$[27] for $N = 10$, for $N = 25$ the lower bound would probably be shifted to $(8\pi^2/5R^2_G(N = 10))^{1/2} \approx 1.0$ and thus be more strongly separated from the region of the microscopic lengths, so that the simulated structure factor and the Debye formula should coincide even better. Indeed, a chain length of $N = 25$ suffices to find the asymptotic scattering law $S_p(q) \propto q^{-2}$, as the analysis of the two-dimensional model A shows.[26] Although a degree of polymerization $N = 10$ is therefore slightly too small to allow for a more quantitative comparison with the Debye formula, the qualitative comparison of $S_p(q)$ with it can explain the temperature dependence of the structure factor. According to eq. (6.17), $S_p^{DE}(q)$ decreases if the radius of gyration increases, so that the shift of the simulation results to lower q-values with falling temperature emphasizes the tendency of the polymers to stretch out not only on the length scale of the bond vector (i.e., for $q \in [2, 2.5]$), but on all others as well. This expansion of the chains during the cooling process is a general result which is independent of the specific choice of the model Hamiltonian and the dimensionality of space.[26,27,38]

6.3.1.3 Chain and cooling rate dependence of the glass transition temperature

From the s-shaped curves of Fig. 6.7 or of Fig. 6.8 one can try to extract the cooling rate dependence of the glass transition temperature T_g if one defines T_g as the intersection point of a linear extrapolation from the liquid and the glassy region.[1,27] A glance at the plots for $\langle b_{npc}^2 \rangle_{npc}$ and R_G^2 shows that only the linear extrapolation from the glassy region is well defined, whereas that from the liquid region is not. To determine $T_g(\Gamma_Q)$ from these quantities would thus introduce systematic errors.[27]

A quantity much better suited for this kind of analysis is the internal temperature T_i of the melt, which is reflected in the actual distribution of the bond vectors on the two available energy levels.[27] As long as the melt is in thermal equilibrium the external heatbath, characterized by the temperature T, controls the population of the two energetic states, whereas this is no longer the case when the melt falls out of equilibrium and finally freezes. Therefore the actually existing energetic distribution of the bond vectors can serve as a "thermometer" to measure the internal temperature of the melt, if one requires T_i to be[27]:

$$T_i := \left[\frac{1}{\epsilon} \ln \frac{g_\epsilon N(0)}{g_0 N(\epsilon)} \right]^{-1} . \tag{6.20}$$

In this equation $N(0)$, $N(\epsilon)$, g_0 and g_ϵ stand for the number of bonds in the ground and the excited state and for the degree of degeneracy of the ground and the excited state, respectively.[27] Figure 6.13 shows a plot of the ratio T/T_i versus temperature. One can see that the linear extrapolations from both the glassy and the liquid region are now well defined, so that T_g may be determined unambiguously. The result of this analysis from T_i, presented in Fig. 6.14, yields a nonlinear relationship between the T_g and $\ln \Gamma_Q$, which can be very well described by the following equation:

$$T_g(\Gamma_Q) = T_K + \frac{A}{\ln(B/\Gamma_Q)} . \tag{6.21}$$

In this equation T_K stands for the freezing temperature at an infinitely slow cooling rate (i.e., $\lim_{\Gamma_Q \to 0} T_g(\Gamma_Q) =: T_K$). The abbrevation T_K for this limit was chosen to remind one of the Kauzmann paradox,[8] which originates from a similar extrapolation procedure. T_K will therefore also be referred to as Kauzmann temperature in the following. In eq. (6.21) T_K is a fit parameter such as A and B. From the fit one obtains: $T_K = 0.17 \pm 0.02$, $A = 0.467 \pm 0.200$ and $B = 5.85 \cdot 10^{-3} \pm 2.5 \cdot 10^{-3}$. The motivation to set up eq. (6.21) stems from the experimentally well established Vogel–Fulcher law for the viscosity $\eta(T)$[1] and the possible definition of the glass transition temperature as that temperature where the simulational observation time t_{sim} becomes comparable to the relaxation time $\tau(T)$ of the considered system.[1] Using therefore the Vogel–Fulcher law with $\eta(T_g) \propto \tau(T_g) \stackrel{!}{=} t_{sim}$ and eq. (6.7) one immediately arrives at eq. (6.21). The numerical result for the Kauzmann temperature coincides within the error bars with the Vogel–Fulcher temperature $T_\circ$ that one obtains by fitting

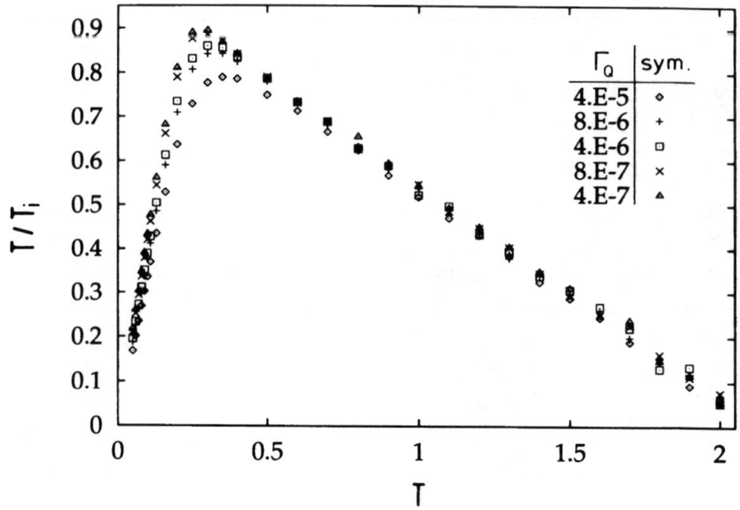

Fig. 6.13 Plot of T/T_{i} vs. T for the five different cooling rates, with the same choices and for the same model as in Fig. 6.7.

the Vogel–Fulcher law to the chain's diffusion coefficient,[27] which will be discussed in more detail in Section 6.3.2.2. Although both extrapolated glass transition temperatures are consistent with each other within the numerical

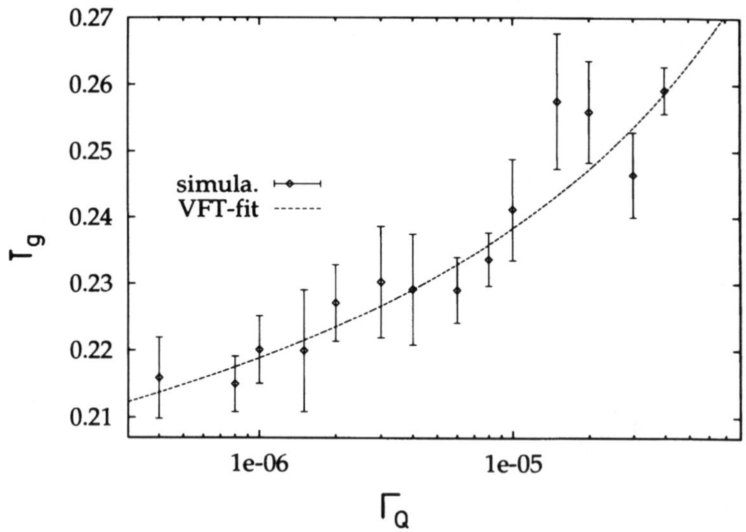

Fig. 6.14 Freezing temperature T_{g} vs. Γ_{Q} obtained from the cooling rate dependence of the internal temperature (see Fig. 6.7). The dashed line represents a nonlinear fit to the data using eq. (6.21).

uncertainties, as it is expected from experiments,[9] their agreement in this simulation could also be accidental, since the value of T_K or T_0 would probably change if additional much slower cooling rates could have been simulated or if the diffusion coefficient could have been reliably determined for much smaller temperatures. A glance at the diffusion data of Fig. 6.19 shows that the diffusion coefficient decreases by about two orders of magnitude in the studied temperature interval. If one compares this to the behavior of an experimental transport coefficient, for instance to the increase of the shear viscosity over its typical value for a fluid in the normal state, the resulting value of the viscosity typically corresponds to a time scale of nanoseconds which is more than ten orders of magnitude smaller than that of the calorimetric glass transition temperature.[14] Since extrapolations will certainly become more and more unreliable the larger the range that they have to cover, the extrapolated results for the Kauzmann and Vogel–Fulcher temperature should not be looked upon as an accurate value for the absolute freezing point of the simulated polymer melt, but rather as an estimation of the temperature regime in which the interesting glass physics of this model might occur.

Although quantitative predictions of the here-attained extrapolations are subject to scepticism due to the limited size of the available temperature or cooling rate interval, the studied cooling rate interval is still large enough to show that the experimentally often observed linear relationship between T_g and $\ln \Gamma_Q$ is only to be found if one confines the simulation to a small range of cooling rates (for instance, to $\Gamma_Q \in [1 \cdot 10^{-6}, 1 \cdot 10^{-5}]$ in this simulation). The larger this range becomes the more the actual relationship should deviate from the linear one, because it predicts that T_g diverges as the cooling procedure approaches a quasi-static and thus thermodynamically well defined process. How large the cooling rate interval has to be in order to find a deviation from the linear behavior depends upon the studied glass-former. Contrary to metallic glasses the glass transition temperature of polymers changes only slightly with the cooling rate.[43] In the case of polyvinylacetate or polymethylmethacrylate it decreases, for instance, by about 8°C[1,3,44] or by about 11°C,[43] respectively, if the cooling rate is reduced by more than three decades. Nevertheless, polymethylmethacrylate already possesses a rather pronounced nonlinear relationship between T_g and $\ln \Gamma_Q$, which may be very well described by eq. (6.21).[43]

In the same way that one can extract the cooling rate dependence of the glass transition temperature it is also possible to determine the influence of the chain length on the freezing behavior of the melt. This influence was studied for the two-dimensional version of model B. The simulations were performed at the same density as in three dimensions (i.e., $\phi = 0.5\bar{3}$) by monitoring the temperature dependence of the above-discussed quantities for several representative chain lengths N from the interval $N \in [3,16]$ while cooling the melt at the fastest rate (i.e., $\Gamma_Q = 4 \cdot 10^{-5}$) used in the three-dimensional case.[45] The results of these simulations show that the chain length qualitatively affects the temperature dependence of the melt in the same way that the cooling rate does. There is a high temperature region where the melt is in thermal equilibrium, followed by a transition region where the influence of the finite cooling speed begins to develop, and a glassy region where the melt is frozen in a state that depends upon the size of the chain length. The properties of this final glassy state are the closer to equilibrium the smaller the chain length is,[45] which stresses how crucially the chain ends dominate the mobility of the polymers and thus their ability to relax.

As for the variation of the cooling rate, one can try to determine the chain length dependence of the glass transition temperature T_g by looking for the intersection point of the linear extrapolations from the liquid and the glassy region in a plot of the ratio T/T_i versus temperature. The result of this analysis from T_i, which is presented in Fig. 6.15, yields a linear relationship between the T_g and the inverse chain length as has been claimed by theory[46] and affirmed in many experiments.[47]

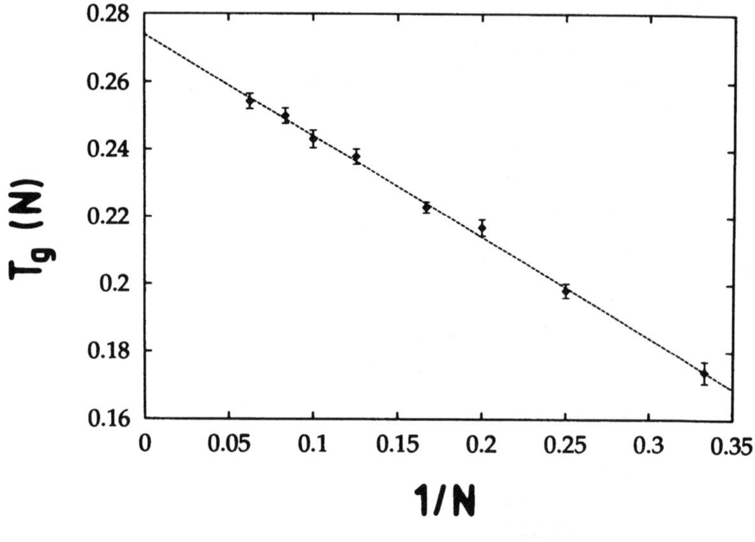

Fig. 6.15 Plot of T_g versus $1/N$ which was determined from the internal temperature of the $2d$ version of model B with $N = 10, \phi = 0.5\bar{3}$, and $\Gamma_Q = 4 \cdot 10^{-5}$.

6.3.2 Dynamic properties of the melt

The discussion of the last subsection and of the experimental investigations on the structural properties of glasses we mentioned in the introduction showed that no sharp distinction between the fluid and the glassy state exists on structural grounds. A distinction between these two states can, however, be made if one compares their typical structural relaxation times, which might for instance be measured by the shear viscosity η. Whereas the viscosity of a liquid in its ordinary state usually varies between $\eta \approx 10^{-2} - 10^{-1}$ poise, corresponding to a time scale of picoseconds,[1,14] its value increases over many orders of magnitude as the liquid becomes more and more undercooled. If the viscosity enters the range of $\eta \geq 10^{13}$ poise the associated structural relaxation time τ becomes macroscopic, i.e., $\tau \approx 10^2$ s or larger. Then the dynamical behavior of the liquid can no longer be distinguished from that of a solid on the time scale of the experiment, so that the temperature, corresponding to $\eta = 10^{13}$ poise, has been defined as the glass transition temperature T_g which has been our main subject in the last subsections. Besides T_g there are two other characteristic temperatures for a fragile glass-former[2,48] of which one lies below and the other above T_g. The one below T_g is the Vogel–Fulcher temperature $T_\circ$ stemming from the empirical Vogel–Fulcher equation, by which one may rather accurately fit the strong increase of the viscosity over more than ten orders of magnitude in the interval $\eta \in [10^2, 10^{13}]$ poise.[1,14,49] The Vogel–Fulcher temperature itself defines the temperature where the viscosity becomes infinitely large, and has thus been looked upon as the transition temperature of the experimentally unreachable underlying thermodynamic liquid–glass phase transition.[1] Contrary to that, the second temperature lies in the experimentally well accessible region of the undercooled liquid corresponding to $\eta \approx 10^2 - 10^4$ poise or to a typical time scale of $\tau \approx 10^{-9}$ s. It is the critical temperature T_c of the mode coupling approach to the glass transition[14] which explains the initial increase of the viscosity up to $\eta \approx 10^3$ poise by the trapping of fluid particles in cages formed by their neighbors[37] as the critical temperature is approached. Which one of these two additional temperatures—if any—is the most relevant for the structural glass transition is still a matter of active scientific debate. However, both temperatures were introduced to characterize or to explain the non-Arrhenius behavior of the structural relaxation time, which is one of the universal features that all fragile glass-formers have in common.[1,14,50,51] Along with this non-Arrhenius-like increase of the relaxation times goes a strong stretching of all structural correlation functions, which is another characteristic trait of glass-forming substances and which can often be adequately described by the empirical Kohlrausch–Williams–Watts formula[1,14,15]:

$$\Phi_K(t) = \Phi_K(0) \exp\left[-(t/\tau_K)^{\beta_K}\right] . \tag{6.22}$$

In the Kohlrausch–Williams–Watts equation the exponent β_K typically ranges between 0.3 and 0.8[1] for structural glasses and is often found to be temperature-independent,[15,52] which means that the structural relaxation time τ_K carries the whole temperature dependence in the Kohlrausch formula. A plot of the correlation function versus the reduced variable t/τ_K

should therefore make the data from different temperatures collapse onto one single master curve. This scaling property is called "time–temperature superposition principle" and believed to be another characteristic feature of the relaxation behavior of structural glass-formers.[1,14]

To what extent the schematic model systems A and B for a polymer melt show this typical relaxation behavior will be addressed in this subsection, by calculating various structural correlation functions that probe the dynamical changes of the melt on different length scales (Section 6.3.2.1). From these correlation functions it is possible to extract relaxation times the temperature dependence of which can be studied and compared to that of transport coefficients, such as the diffusion coefficient. This will be done in Section 6.3.2.2. The final paragraph of this subsection then deals with the calculation of the incoherent intermediate scattering function and its quantitative interpretation in the framework of the idealized mode coupling theory (MCT).[14]

6.3.2.1 Relaxation processes on different length scales

In order to investigate the slowing down of the dynamics on the length scale of the average chain's size one can calculate the auto-correlation function of the polymer's end-to-end distance or of the radius of gyration, which might be defined in the following way:

$$\Phi_{[\mathrm{R_{EE}}]}(t) := \frac{\left\langle (\vec{r}_{Npc}(t) - \vec{r}_{1pc}(t)) \cdot (\vec{r}_{Npc}(0) - \vec{r}_{1pc}(0)) \right\rangle_{pc}}{\left\langle (\vec{r}_{Npc}(0) - \vec{r}_{1pc}(0))^2 \right\rangle_{pc}} , \tag{6.23}$$

$$\Phi_{[\mathrm{R_G}]}(t) := \frac{\sum_{n=1}^{N} \left\langle (\vec{r}_{npc}(t) - \vec{R}_{pc}(t)) \cdot (\vec{r}_{npc}(0) - \vec{R}_{pc}(0)) \right\rangle_{pc}}{\sum_{n=1}^{N} \left\langle (\vec{r}_{npc}(0) - \vec{R}_{pc}(0))^2 \right\rangle_{pc}} . \tag{6.24}$$

Motivated by the experimental results on the glassy dynamics, it is especially interesting to analyze the long time behavior of these correlation functions and to check whether a scaling variable for the time–temperature superposition can be found. A hint in the search for such a scaling variable may be obtained from the Rouse model which provides the simplest description of the dynamics of a polymer melt composed of short chains.[34,53] The Rouse model shows that the long time behavior of the above correlation functions is dominated by the ratio $2\pi^2 Dt/\langle R^2_{\mathrm{EE}}\rangle_{pc}$, whereby D is the diffusion coefficient of a polymer. Since the diffusion coefficient is the long time limit of the mean-square displacement of the chain's center of gravity $\vec{R}_{pc}$, i.e.,

$$D = \lim_{t\to\infty} \left\langle (\vec{R}_{pc}(t) - \vec{R}_{pc}(0))^2 \right\rangle_{pc} /2dt =: \lim_{t\to\infty} g_3(t)/2dt , \tag{6.25}$$

it is tempting to use $2\pi^2 g_3(t)/\langle R^2_{\mathrm{EE}}\rangle_{pc}$ as a generalized time–temperature scaling variable for the two functions eqs (6.23) and (6.24). This kind of time–temperature superposition was tried for the two-dimensional model A in which $\Phi_{[R_{EE}]}$ and $\Phi_{[R_G]}$ were calculated, immediately after the cooling process without any further relaxation, over a time period of $t_{max} = 2.5 \cdot 10^5$ MCS at two different densities $\phi = 0.4$ and $\phi = 0.8$.[54] The simulations were performed at certain representative temperatures extending from the liquid to glassy region, i.e., from $T = 1.0$ to $T = 0.1$. Figure 6.16 shows the result of this analysis for chains with $N = 10$. In fact, the data superimpose nicely on a single master curve for *almost all* times and for both semidilute ($\phi = 0.4$) and glassy system ($\phi = 0.8$), despite the dramatic slowing down of the polymer's mobility with decreasing temperature when going from the semidilute to the glassy system. This means that the ratio $g_3(t)/\langle R_{EE}^2 \rangle_{pc}$ accurately contains all information about the dependence

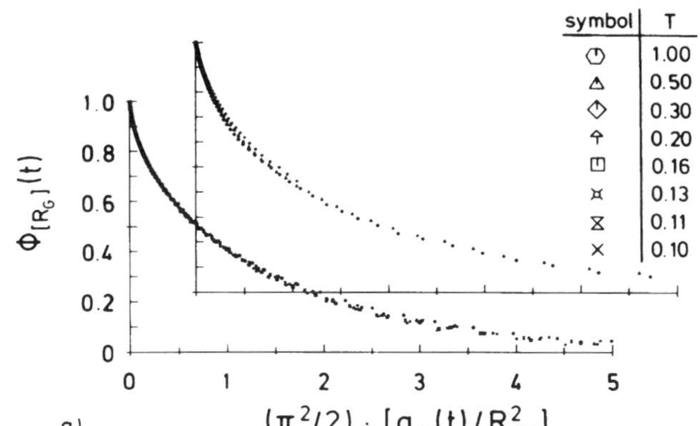

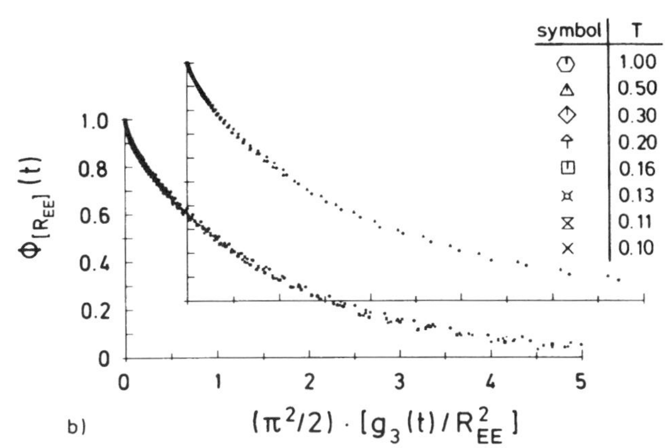

Fig. 6.16 Scaling plot of the auto-correlation functions $\Phi_{[R_{EE}]}(t)$ (a) and $\Phi_{[R_G]}(t)$ (b) versus $g_3(t)/\langle R_{EE}^2 \rangle$ for different temperatures, obtained by a simulation of the 2*d* version of model A. The front (left) part refers to $\phi = 0.4$ and back (right) part to $\phi = 0.8$. The chain length is $N = 10$.

upon temperature, density and release of constraints built up during the cooling process, for those correlation functions to which all Rouse modes[34] contribute. Therefore the mean-square displacement of the chain's center of gravity may be looked upon as the underlying fundamental quantity for the motion of a polymer, so that the theoretical understanding of its time evolution might provide valuable insight in the dynamics of a polymer melt above and close to the glass transition.

Another important length scale in the melt is that of a bond. Information about the dynamics on this local scale may be extracted by calculating, for instance, the bond auto-correlation function[26,54] or a generalized spin auto-correlation function which is defined in the following way[55]: one associates an Ising spin variable s_{npc} with each monomer n of the polymer p in configuration c. This variable is set to $s_{npc} = +1$ if the monomer is unable to move, and to $s_{npc} = -1$ if it is mobile. With this definition and the corollary that $s^2_{npc}(t) \equiv 1$ for all times, a correlation function can be constructed which measures the slowing down of the local mobility of the monomers with decreasing temperature[55]:

$$\Phi(t) := \frac{\langle s_{npc}(t)s_{npc}(0)\rangle_{npc} - \langle s_{npc}(0)\rangle^2_{npc}}{1 - \langle s_{npc}(0)\rangle^2_{npc}}. \tag{6.26}$$

The temperature-dependent time evolution of $\Phi(t)$ was studied for a melt composed of polymers with $N = 10$ and equipped with the Hamiltonian A at a density $\phi = 0.8$ after having equilibrated the melt for about a Rouse time.[55] Contrary to the results described for $\Phi_{[R_{EE}]}$ and $\Phi_{[R_G]}$ the history of the cooling process was eliminated in this simulation before calculating the spin auto-correlation function over about six decades in time for several temperatures from the interval $T \in [0.22, 1.0]$. For this function one can also look for the time–temperature superposition property. Since there is no theory available which would provide an educated guess for a suitable scaling variable, the most natural trial to superimpose the data consists in shifting the curves for different temperatures by an amount $t_{sc}(T)$ until they collapse within a limited time interval at least. The time interval chosen in Ref. 55 was the interval of the final decay of $\Phi(t)$, because the time–temperature superposition property is experimentally observed [14,15,52] and theoretically rationalized[14] for late times. The result of this simple shifting procedure is depicted in Fig. 6.17. It reveals that it is indeed possible to superimpose the last 20–30% of the decay of $\Phi(t)$ onto a single master curve and that the self-similarity of the simulation data increases with falling temperature so that the time zone over which scaling is valid expands. However, such an impressive collapse of the data, as is shown in Fig. 6.16, cannot be obtained by the simple shifting procedure at any temperature, which indicates that the temperature dependence of the whole decay of $\Phi(t)$ is too complicated to be condensed in a single number $t_{sc}(T)$.

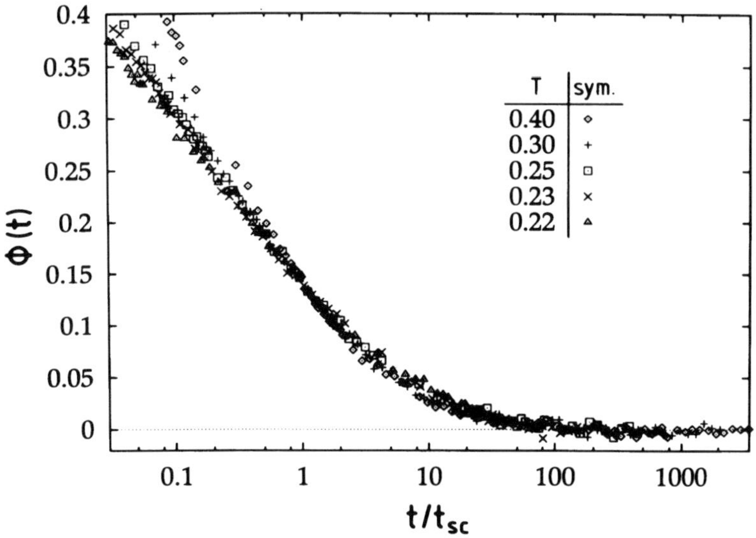

Fig. 6.17 The generalized spin auto-correlation function $\Phi(t)$ for temperatures $T = 0.40, 0.30, 0.25, 0.23$, and 0.22 are plotted versus t/t_{sc} (in logarithmic scale) where $t_{sc} = 130, 400, 1160, 2260$ and 3700 for the respective temperatures. All the curves collapse onto a single curve at late times. This result stems from a simulation of the $2d$ version of model A with $N = 10, P = 200$, and $\phi = 0.8$ after eliminating all cooling rate effects.

6.3.2.2 Relaxation times and diffusion coefficients

A quite general way to extract a relaxation time from a correlation function is to determine the area under the correlation function. In Ref. 55 this method was applied to the above-discussed generalized spin auto-correlation function in order to derive the temperature dependence of the monomeric relaxation time τ. Figure 6.18 compares the result of this analysis to the temperature dependence of the Rouse time τ_R which was assessed by the intersection point of $g_3(t)$ with the mean-square displacement of a monomer measured in the coordinate system of the chain's center of gravity.[55] Although τ and τ_R probe the dynamics of melt on different length scales, their dependence on temperature is remarkably similar. At high temperatures the relaxation times follow an Arrhenius law before they cross over to a Bässler law,[14,56] i.e., to $\tau \propto \tau_R \propto \exp[-A/T^2]$ at low temperatures. For both τ and τ_R the crossover temperature between the two relaxation regimes lies around $T \approx 0.3$ and the value of the constant A, which is a measure of the activation energy, is $A \approx 0.18$. Hence the relaxation time of the chain essentially depends upon temperature in the same way as that of the local monomeric hopping motion, a behavior which is also borne out by experiments on (three-dimensional) polymer glasses.

In Ref. 55 the results of this analysis were interpreted in the following way: although fast quenching of the melt with a finite cooling rate yields a

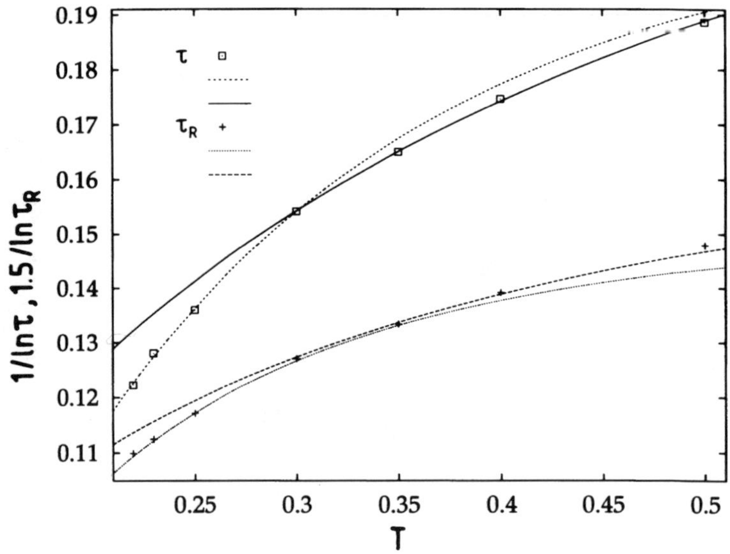

Fig. 6.18 $1/\ln \tau_R$ (□) and $1/\ln \tau$ (+) are plotted against T. The data can be fitted by an Arrhenius law (solid and dashed line) for high temperatures and to a Bässler formula (dashed curve with heavy dashes and dotted line) for low temperatures. The model is the same as in Fig. 6.17.

glass transition at a non-zero temperature (see discussion in the last subsection) the limit of infinitely slow cooling, to which the relaxed data of Ref. 55 actually correspond, seems to shift the thermodynamic glass transition temperature to zero. However, whether this is really the case cannot conclusively be inferred from the simulation data for two reasons: on the one hand, the temperature interval over which the Bässler law fits the simulation data is rather narrow, so that a stronger increase of the relaxation times at even smaller temperatures than predicted by this law may not be excluded. On the other hand, an extension of the available temperature interval for the diffusion coefficient over that used in Ref. 55 opens the possibility for a nonzero thermodynamic glass transition temperature in this two-dimensional model. Figure 6.19 presents these extended results and compares them with the Vogel–Fulcher formula, i.e.,

$$D(T) = D_\infty \exp\left[-\frac{A}{T - T_\circ}\right] . \tag{6.27}$$

Since D_∞ was measured by a simulation at infinite temperature (yielding $D_\infty = 4.7 \cdot 10^{-4}$), eq. (6.27) only possesses two fit parameters, which were determined to be: $A = 0.652 \pm 0.012$ and $T_\circ = 0.097 \pm 0.009$. Hence a nonzero Vogel–Fulcher temperature of $T_\circ \approx 0.1$ results from this fit. However, again one has to realize that the two-dimensional diffusion data only

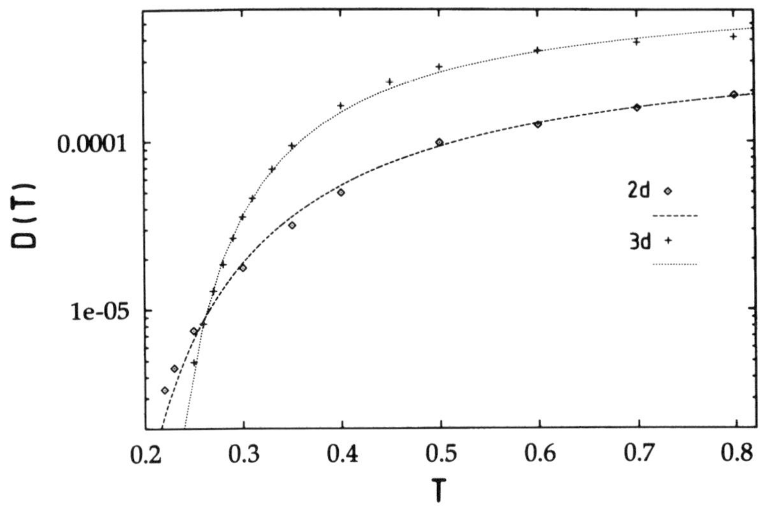

Fig. 6.19 Comparison of the temperature dependence of the two- (◇) and three-dimensional (+) diffusion data, simulated by the $2d$ version of model A with $N = 10, P = 200$, and $\phi = 0.8$, and the $3d$ version of model B with $N = 10, P = 180$, and $\phi = 0.5\bar{3}$. The dashed and the dotted lines represent the best fit by the Vogel–Fulcher formula, yielding $T_\circ \approx 0.097$ ($2d$) and $T_\circ \approx 0.17$ ($3d$), respectively.

decrease by one to two orders of magnitude in the investigated temperature interval, which certainly results in a strong overestimation of the absolute freezing point of this model, as mentioned above. One cannot exclude a vanishing $T_\circ$ then, if it was possible to extend the measurement of the diffusion coefficient to much smaller temperatures. This overestimation problem is not limited to the two-dimensional data, but applies to the three-dimensional results for $D(T)$, which are shown in Fig. 6.19 for the sake of comparison as well. The three-dimensional values for $D(T)$ were obtained from a simulation of model B with chains of length $N = 10$ and at a density of $\phi = 0.5\bar{3}$. Also for these three-dimensional data the Vogel–Fulcher formula works best in comparison with a Bässler or an Arrhenius law, yielding: $D_\infty = 8.61 \cdot 10^{-4} \pm 0.32 \cdot 10^{-4}$, $A = 0.396 \pm 0.041$ and $T_\circ = 0.17 \pm 0.02$. Although the three-dimensional Vogel–Fulcher temperature has to be considered with the same reservations as the two-dimensional one, it is larger than the latter, expressing the general tendency that the three-dimensional diffusion coefficient decreases faster with temperature than in the two-dimensional case. Such an influence of the spatial dimension on the thermodynamic glass transition temperature could have been anticipated from the physical consideration that the configurational constraints are much stronger in three than in two dimensions.

6.3.2.3 *Mode coupling theory*

During the last eight years a new theoretical approach to the structural glass transition, the mode coupling theory (MCT),[14,57] has been developed which claims that the initial increase of the structural relaxation time in the region of the undercooled liquid is caused by a prolongation of the lifetime of the fluid particles in their cages, as the liquid is cooled towards a critical temperature T_c situated well above T_g.[14,57,58] At this critical temperature the dynamics of the undercooled liquid qualitatively changes. For $T \leq T_c$ the original idealized MCT[14,58] predicts that the fluid particles are permanently trapped in their cages, so that density fluctuations (measured for instance by the coherent or incoherent intermediate scattering function in a neutron scattering experiment[33]) suddenly freeze at T_c and thus lead to a spontaneous breaking of ergodicity.[14,58] However, a complete freezing of the density fluctuations has not been observed experimentally up to now, so there is the need to complete the theoretical description by processes that succeed in counterbalancing the localization property of the cages. This can be achieved by the inclusion of so-called hopping processes which restore ergodicity for the whole temperature region around the critical point.[57] The subtle competition of ergodicity breaking and restoring forces close to T_c results in a two-step decay of the coherent or incoherent intermediate scattering function, whereby the first step corresponds to the relaxation of the particles in their cages, and the second step signals the breaking of the cages and with that the onset of the structural relaxation. The time windows where this two-step process develops and where the final structural relaxation takes place are called β- and α-relaxation regime, respectively.[14,57] Although the derivation of the MCT originally started from the theory of simple liquids, its central result, the existence of T_c, is independent of the specific details of the microscopic interaction.[14] It was thus supposed that the applicability of the MCT is not restricted to simple liquids, but that it approximately grasps the essential universal features which dominate the dynamics of all fragile glass formers in the undercooled state. Hence it should also be possible to test the theory for a polymer melt. Such a test of the universal predictions of the MCT was performed for the three-dimensional version of model B at a density of $\phi = 0.5\bar{3}$ and with chains of length $N = 10$.[59] Since this simulation aimed at a quantitative comparison with the idealized version of the MCT, the melt was equilibrated for about $5 \cdot 10^6$ MCS before the incoherent intermediate scattering function $\Phi_q^s(t)$ was calculated and averaged over 28 800 monomers.[59]

For this correlator $\Phi_q^s(t)$ the idealized MCT predicts that it consists of two parts in the β-relaxation regime: the time independent non-ergodicity parameter f_q^{sc}, and a part for which the wave vector and the time dependence factorize, i.e.,

$$\Phi_q^{\mathrm{s}}(t) = f_q^{\mathrm{sc}} + h_q^{\mathrm{s}} c_\varepsilon g(t/t_\varepsilon) \, . \tag{6.28}$$

In eq. (6.28) ε means the reduced distance to the critical temperature, i.e., $\varepsilon := (T_{\mathrm{c}} - T)/T_{\mathrm{c}}$, and h_q^{s} and $c_\varepsilon g(t/t_\varepsilon)$ are called the critical amplitude and the β-correlator, respectively. The scaling function $g(t/t_\varepsilon)$ of the β-correlator can be expanded on the liquid side ($\varepsilon < 0$) as[60]

$$g(t/t_\varepsilon) = \begin{cases} (t_\varepsilon/t)^a - A_1(t/t_\varepsilon)^a + A_2(t/t_\varepsilon)^{3a} - A_3(t/t_\varepsilon)^{5a} & \text{for } t_\circ \ll t \le t_\varepsilon \\ -B(t/t_\varepsilon)^b + (B_1/B)(t_\varepsilon/t)^b & \text{for } t_\varepsilon \le t \le \tau \end{cases} . \tag{6.29}$$

The exponents a and b, ranging in the intervals $a \in]0, 0.5[$ and $b \in]0, 1]$, respectively, are related to each other by the exponent parameter λ which may adopt all values from the interval $\lambda \in [0.5, 1[$.[14] This short and the long time expansion of $g(t/t_\varepsilon)$ match with an accuracy of about 1% on the β-time scale t_ε and thus can be used to construct the full scaling function.[60] With these formulae one can try to fit the simulation data in order to derive the critical temperature by

$$c_\varepsilon = \sqrt{|\varepsilon|} \quad \text{and} \quad t_\varepsilon = t_\circ |\varepsilon|^{-1/2a} = t_\circ \left(\frac{T - T_{\mathrm{c}}}{T_{\mathrm{c}}} \right)^{-1/2a} , \tag{6.30}$$

if one knows the temperature dependence of the microscopic time scale $t_\circ$. This microscopic time scale and the time scale of the α-process were determined from the initial decay of $\Phi_q^{\mathrm{s}}(t)$ by a fit with the Rouse model, and from the final decay by a fit with the Kohlrausch–Williams–Watts formula, respectively, in Ref. 59.

Figure 6.20 shows the results of the simulation for $\Phi_q^{\mathrm{s}}(t)$ which was calculated at the maximum of the static structure factor $q = 2.92$ (measured in units of the reciprocal lattice constant) in the temperature interval $T \in [0.16, 0.21]$. In this temperature interval a two-step process develops which we analyzed by the asymptotic expansions.[29] This analysis splits the studied temperature interval in a high-temperature part ($T \geq 0.19$), where the idealized MCT can account for the relaxation of $\Phi_q^{\mathrm{s}}(t)$ in the β-regime, and a low-temperature part ($T < 0.19$), where it overestimates the freezing tendency of the melt. This behavior is exemplified for the two marginal temperatures $T = 0.19$ and $T = 0.18$ in Figs 6.21 and 6.22, respectively. In the high-temperature region the fit leads to the following results:

$$a = 0.239 \pm 0.030, b = 0.370, \lambda = 0.863 \pm 0.052 > \frac{\pi}{4}; \; f_q^{\mathrm{sc}} = 0.80 \pm 0.03 \, . \tag{6.31}$$

With these parameters it is possible to fit the simulation data over about two to three decades in time for the temperatures $T \geq 0.19$. The results at $T = 0.19$, presented in Fig. 6.21, exemplify how the asymptotic expansions of the idealized MCT are tailored so that they can complete each other in

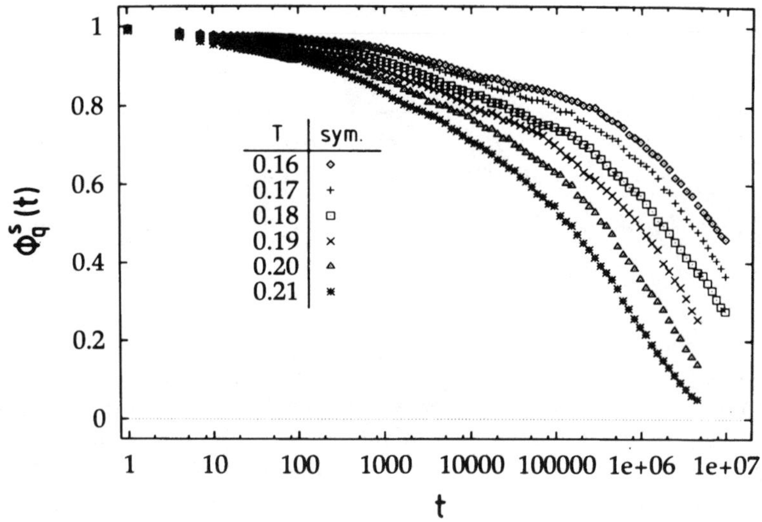

Fig. 6.20 Simulation data of $\Phi_q^s(t)$, calculated at the maximum of the static structure factor $q = 2.92$, for the six different temperatures, i.e., $T = 0.16$ ($\Diamond$), $T = 0.17$ (+), $T = 0.18$ ($\Box$), $T = 0.19$ ($\times$), $T = 0.20$ ($\triangle$), and $T = 0.21$ (*). This result stems from a simulation of the $3d$ version of model B with $N = 10$, $P = 180$ and $\phi = 0.5\bar{3}$ after eliminating most of the cooling rate effects.

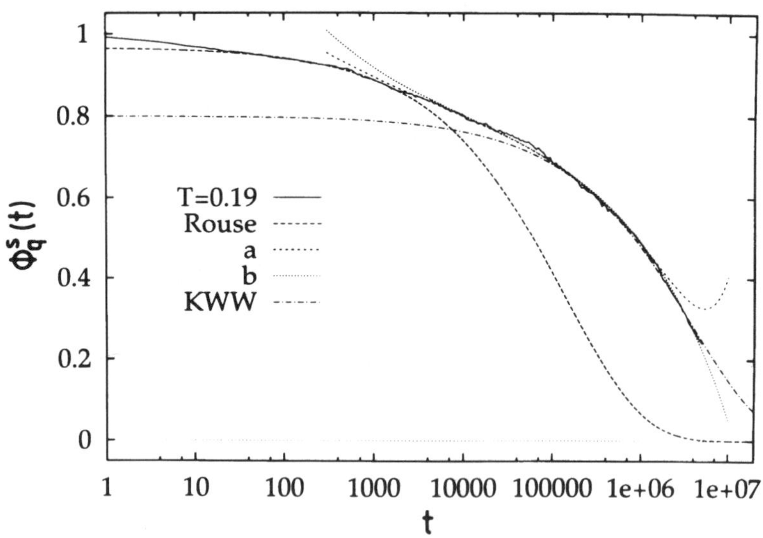

Fig. 6.21 Comparison of the simulation data (solid line) and various fitting formulae at $T = 0.19$. The short time expansion of the Rouse model and the Kohlrausch law are represented by a dashed line with long dashes and by a dashed–dotted line, respectively. The dashed line with the short heavy dashes corresponds to the short time expansion of the scaling function (eq. [6.29]) whereas the dotted line refers to its long time part. The model is the same as in Fig. 6.20.

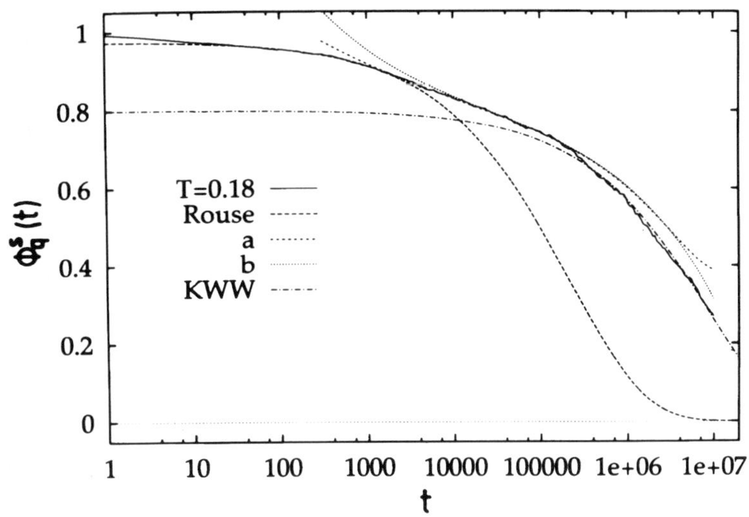

Fig. 6.22 Comparison of the simulation data at $T = 0.18$, with the same fitting formulae and the same choice for the types of the lines as in Fig. 6.21.

the description of the decay of $\Phi_q^s(t)$. First the short time expansion of eq. (6.29) (labeled by the exponent a in Fig. 6.21) starts to fit the data, before the long time expansion (labeled by the exponent b in Fig. 6.21) joins it on the time scale t_ε and then extends the fit to longer times, whereas the short time expansion deviates from the simulation curve. The time window of the β-process, where the MCT applies, is preceded by an interval, where the Rouse theory accounts for the decay of $\Phi_q^s(t)$, and followed by the α-process which can be described by a Kohlrausch law with a temperature-independent exponent $\beta_K = 0.515 \pm 0.020$ (see eq. [6.22]).[59] Therefore the time–temperature superposition principle seems to apply to the long time decay of the correlator $\Phi_q^s(t)$.

If the temperature is reduced to $T = 0.18$ (see Fig. 6.22) only the initial part of the first step of the correlator can adequately be fitted by the short time expansion of eq. (6.29), whereas for longer times $\Phi_q^s(t)$ decays much more strongly than predicted by the idealized MCT. Hence, the idealized theory overestimates the freezing tendency of the polymer melt in this simulation. This result is, however, theoretically expected if one takes the above-mentioned hopping processes into account. In this extended version of the MCT it turns out that the asymptotic expansions of the idealized theory are only applicable to the initial decay of the correlator, whereas more complicated formulae have to be used for longer times.[61] Since one can therefore interpret the decay of $\Phi_q^s(t)$ partly qualitatively and partly quantitatively in the framework of the MCT, the results of the fit for the β-time scale (not shown here) and the critical amplitude (see Fig. 6.23) were used in Ref. 59 to

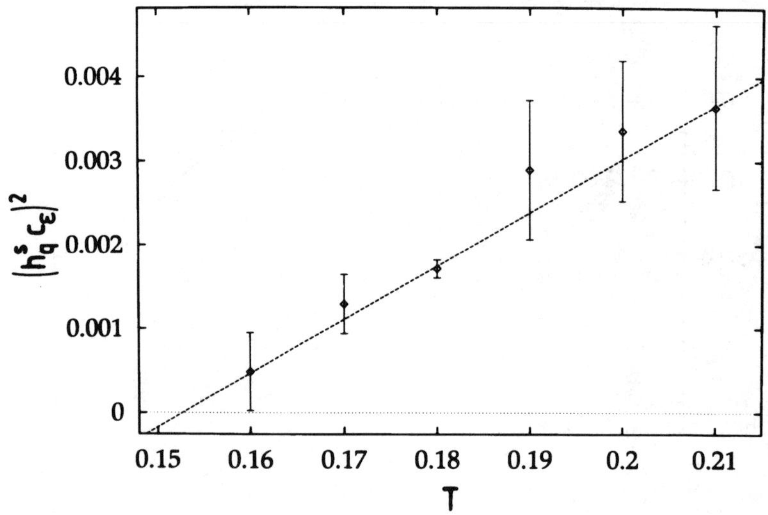

Fig. 6.23 Plot of $(h_q^s c_\varepsilon)^2$ versus T (see eq. [6.30]). The diamonds represent the results of the fit, and the dashed line is a least-square fit, yielding a critical temperature of $T_c = 0.153$.

estimate the critical temperature. From the straight line fits to these data one finds $T_c = 0.146 + 0.006/-0.008$ for t_ε (not shown here) and $T_c = 0.153 + 0.005/-0.009$ for $h_q^s c_\varepsilon$. These values agree with each other within the error bars and can be combined to an average critical temperature of $T_c = 0.150 + 0.005/-0.009$.[59]

6.4 Modeling of specific polymers

6.4.1 How to map naturalistic models to abstract models

As discussed in the introduction there are some rather general features to the glass transition in fragile glasses, and in polymers in particular. We have made use of that finding by devising very simple model systems that nevertheless exhibited the experimental phenomenology and allowed for a very thorough simulational treatment. One of these general features is the possibility to describe the behavior of the melt viscosity as a function of temperature for up to 10 decades in range, using the Vogel–Fulcher law. Let us for a moment view this as sort of a universal feature for polymers with system specific constants. Borrowing an analogy from the treatment of phase transitions we should be able to do a coarse-graining on a miscroscopically detailed model of a polymer that leads to a simplified description of this polymer, losing information on short length and time scales but retaining the constants in the macroscopic laws such as the VF law. The same can be done on any abstract polymer model such as, for instance, the bond fluctuation lattice model. The idea of how to create a lattice model that reproduces the macroscopic behavior of a certain polymer is now to match the two coarse-grained descriptions on the mesoscopic scales. We will have to identify physical properties that capture the structural and dynamic behavior on this mesoscopic scale and map the coarse-grained information about the lattice model onto that of the detailed model by optimizing energy parameters chosen for the lattice model. This way we should end up with a lattice model reproducing the polymers' macroscopic behavior. We modify the old Kuhnian[63] idea of a coarse-graining along the polymer, in that we have to stay below the statistical segment length in our step size to retain information on the species under study. For bisphenol-A-polycarbonate (BPA-PC) we define a coarse-grained chain by connecting points along the backbone one repeat unit apart. This generates coarse-grained bond vectors $\vec{L}$. Their distribution is relatively smooth, as it is produced by six torsional degrees of freedom inside one repeat unit. Figure 6.24[32] shows an example of these distributions for the above-described coarse-graining and several temperatures. At the high temperature of $T = 570$ K the distribution of the coarse-grained bond length shows structure from 7 to 12 Å with a mean value of around 10.5 Å. The mean value slightly increases with decreasing temperature ($T = 420$ K is the glass transition temperature, and $T = 130$ K is deep in the glassy state) whereas the width of the distribution drastically decreases (Fig. 6.24[a]). The corresponding shifts in the mean of the angle Θ between consecutive coarse-grained bonds and in the width of the distribution $P(\Theta)$ are much smaller and not easily resolved on the scale of Fig. 6.24(b). We take the first two moments of the distributions $P(L)$ and $P(\Theta)$ to be a description of the local geometry of the chain. The distributions are obtained by computer-generating random walks on the monomer length scale, using quantum chemically calculated BPA-PC force fields and sampling their statistical properties at the respective temperature.[32] On the bond fluctuation side we coarse-grain over three bonds of the lattice model. Here both distributions can be obtained from a double trimer which can be treated by exact enumeration.[64] The choice of mapping three bonds in the lattice model onto one BPA-PC repeat unit can be motivated by mapping the same number of conformationally relevant degrees of freedom among the two chains. There are six torsions in one BPA-PC repeat unit and we have three bond lengths and three bond angles per bond fluctuation trimer. To match the moments of the coarse-grained bond lengths we have to translate the lattice constant of the simulation to the natural unit of Å. To this end we require the simulation to exhibit the experimental number density

$$\frac{\rho}{M_{\text{monomer}}} = \frac{\phi}{8}\frac{m}{n}\frac{1}{s^3} \tag{6.32}$$

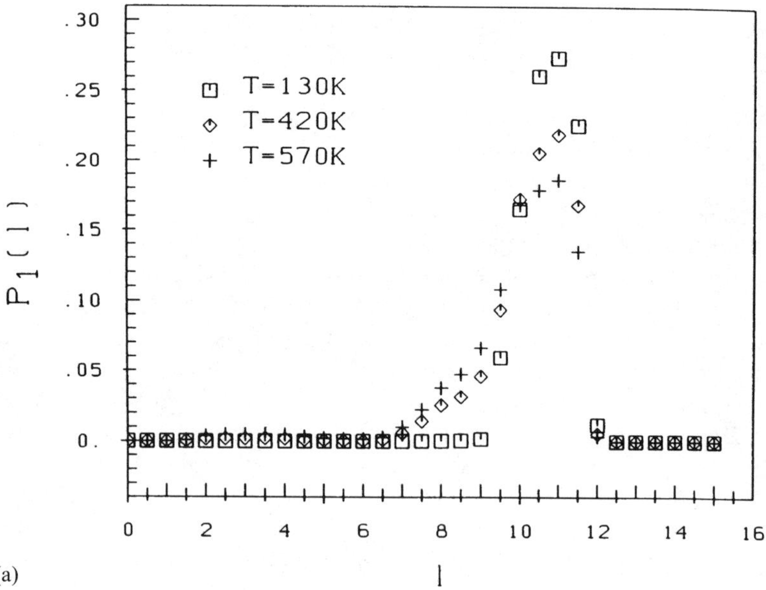

(a)

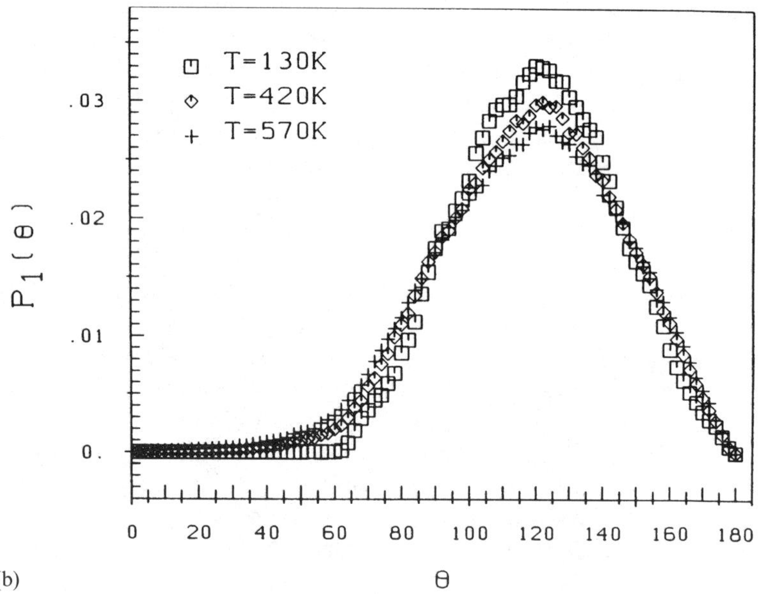

(b)

Fig. 6.24 Probability distributions for the length of coarse-grained bonds (a) and the angle between consecutive coarse-grained bonds (b). Data are for BPA-PC at the indicated temperatures.[32] The subscript 1 refers to a coarse-graining of one repeat unit of the polymer.

(ρ = mass density; $\phi = 0.5$, volume fraction; $n = 1$ $m = 3$). Then s is the conversion factor translating lattice constants into angstrom.

So far we have matched the coarse-grained geometric structure of the chains. To reproduce dynamic properties we must furthermore match time constants on the mesoscopic scale. The rate of conformational changes in a polymer is determined by the rate of barrier crossings in the various torsional potentials along the chain.[32] Applying rate theory we can write these as a product of attempt frequencies and activation factors. This way we could define a mean rate of conformational change as a function of temperature. As we do not yet have a microscopic way of translating the time scale of the simulation (Monte Carlo steps, MCS) to the natural time scale (seconds) we cannot make use of a rate and have to simplify this further by taking the attempt frequencies to be equal. Let E_i be the potential energy at minimum i and $\Delta E_{ij}, j = 1, \ldots n_i$ the heights of the barriers separating minimum i from its n_i neighboring minima j. Then, the probability for a transition from one minimum to another one is

$$\langle P \rangle = \frac{1}{Z} \sum_{i}^{N_{\text{tor}}} \sum_{j=1}^{n_i} \exp(-\beta E_i) \exp(-\beta \Delta E_{ij}) \tag{6.33}$$

with $Z = \sum_i^{N_{\text{tor}}} n_i \exp(-\beta E_i)$. From this we can define a mean barrier as

$$\exp(-\beta \langle \Delta E \rangle) := \langle P \rangle \tag{6.34}$$

The corresponding rate in the lattice model, with its Monte Carlo dynamics, is the acceptance rate for the moves. Here we also drop the arbitrary time constant of the Monte Carlo process and define a mean barrier as

$$\exp(-\beta \langle \Delta E_{\text{BFL}} \rangle) := \langle \min(1, \exp(-\beta\, \Delta E)) \rangle_{\text{trimer}}. \tag{6.35}$$

The average is performed over all possible trimer configurations and jumps of the inner monomer of these trimers, which is again amenable to exact enumeration. Matching the geometric moments and this mean barrier in a nonlinear optimization procedure fixes the constants in Hamiltonian C (eq. [6.9]) as a function of temperature. This nontrivial nonlinear optimization problem is discussed in more detail in Ref. 64.

6.4.2 Modeling bisphenol-A-polycarbonate

The Hamiltonian C is used as input in a dynamic Monte Carlo simulation, where a stepwise cooling procedure is applied and one equilibrates the system to as low a temperature as possible. The system is regarded as equilibrated after the mean center of mass displacement exceeds the value of the mean-squared radius of gyration.[65,66] This time is approximately equal to the Rouse time as the longest relaxation time in the system. The mean-square displacement of the center of mass of the chains is then used to measure the self-diffusion coefficient according to eq. (6.25). As the conformational dynamics of these chains is nicely describable by Rouse theory[67] the self-diffusion coefficient translates into a measurement of the viscosity

$$\eta = \frac{\phi}{8s^3} \frac{\langle R^2 \rangle}{36} \frac{k_B T}{ND} . \tag{6.36}$$

The Rouse theory, furthermore, contains only one rate constant which is connected to the monomeric friction coefficient

$$W = \frac{12}{\pi} \frac{k_B T}{\xi \sigma^2} , \tag{6.37}$$

where ξ is the monomeric friction coefficient and σ is the statistical segment length of the chains. This Rouse rate can be measured by analyzing the mean-square displacement of monomers on short length and time scales

$$\left\langle (\Delta \vec{r}_n(t))^2 \right\rangle = \sigma^2 (W t)^{1/2} . \tag{6.38}$$

The results we want to show were obtained by a somewhat preliminary and simpler version of the mapping,[28] which nevertheless exhibited the properties required in the above description. The structure factor shown in the introduction (Fig. 6.2) was taken at the lowest temperature that still allowed the system to be equilibrated. This corresponds to an undercooled melt of BPA-PC. A comparison to the experimental structure factor of BPA-PC[68,69] does, of course, show some deficiencies in the lattice approach. As the detailed atomistic structure is not accounted for, the position of the amorphous halo is shifted and there is additional intrachain structure in the simulation. The shift exceeds the amount one would expect due to the difference in temperature between the experiment and the simulation, if one took this into account approximately by rescaling affinely with the change in specific volume. An additional peak around 0.6 Å^{-1} due to dipolar interactions between the carbonyl groups is also of course absent in the simulation. The ratio of the intensity at the amorphous halo compared to the $q \to 0$ limit (the compressibility) is, however, correctly reproduced. A closer match between the simulated and real structure could only be obtained by reinserting the atomistic structure into these globally equilibrated simulations. Figure 6.25 shows a comparison of the temperature dependence of the Rouse rate, W, and the chain's self-diffusion coefficient, D, in the form of an inverse activation plot.[40,70,71] Both quantities follow a VF temperature dependence leading to Vogel–Fulcher temperatures around 300 K. This once more shows the strong coupling due to the chain connectivity of the local relaxation (conformation and structure relax by the same type of monomer movements) to the overall conformational relaxation of the chains. To make contact with experiments we have to use eq. (6.36) to calculate the viscosity our simulation would predict for short BPA-PC

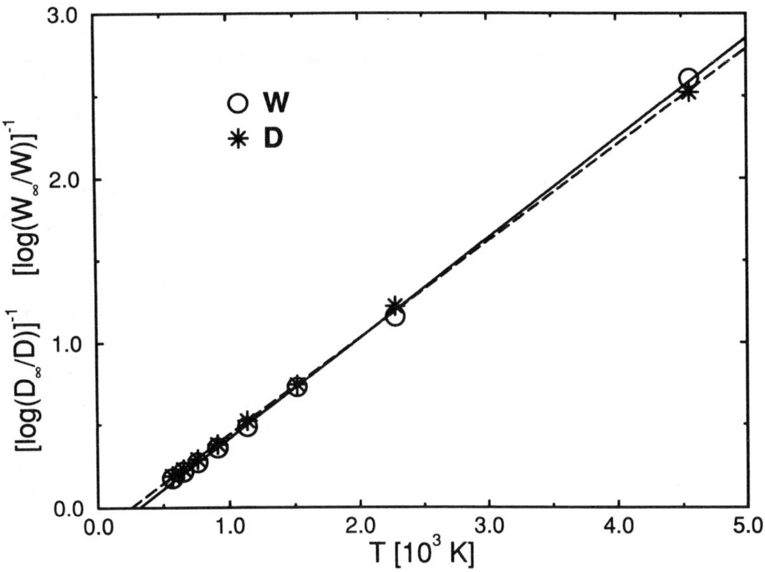

Fig. 6.25 Inverse activation plot for the Rouse rate, W, and the chain self-diffusion coefficient, D. The normalization values for $T = \infty$ for both quantities were measured in independent simulations. VF laws are straight lines extrapolating to a finite value on the abscissa, whereas an Arrhenius law extrapolates to $T = 0$.

chains. Figure 6.26 shows the result for a simulation of chains of length $N = 20$ equilibrated down to 570 K. It is compared to experimental data for BPA-PC oligomers ($N = 20$ corresponds to 7-mers) that were measured at Bayer[72] as a check for the computer simulation results. The Vogel–Fulcher temperatures one obtains from the simulation (321 ± 27 K) and from the experiment (322 K) are virtually identical and the activation energies (simulation: 950 ± 113 K; experiment: 1053 K) agree within the error bars.

Here we have to come back to the change in ensemble imposed by using a lattice model for the simulation of the glass transition. The specific volume in the simulation is matched to the experimental one at 570 K. Keeping it constant at the larger temperatures means that all properties measured there would correspond to experiments done under higher pressure. Irrespective of the fact that the real polymer is of course not stable any longer at the highest temperatures of the simulation, we suppose that the change in pressure required does not exceed a few hundred atmospheres. $T_g(p)$ for BPA-PC changes at about 4.4 K for every 100 atmospheres.[73] This effect then can be expected to be small compared to the uncertainties inherent in the approximations made in the whole mapping procedure. The high degree of quantitative agreement among the Vogel–Fulcher temperatures is certainly fortuitous. A larger effect should be expected for the activation energies, as we cut across the isobars, starting on the high pressure ones. This effect is hard to quantify but may be in part responsible for the larger difference in the measured activation energies. In experiments on polyvinylchloride[74] it was found that the increase in relaxation time at constant volume and that at constant pressure followed very similar Vogel–Fulcher curves. The one at constant pressure showed a stronger deviation from Arrhenius behavior in the intermediate viscosity regime, indicative of a larger activation energy. The deviation seen in the simulation goes in the same direction. On the whole, the rather good agreement between experiment and simulation suggests that the freezing process is mainly dominated by the repulsive part of the atomic interactions[19] (the simulation used only excluded volume interactions). Inclusion of the attractive interactions will allow for a treatment of enthalpic properties and will certainly modify the Vogel–Fulcher parameters, but the latter can be expected to be a minor effect compared to the repulsive part of the interactions. For Fig. 6.26 the vertical scales had to be adjusted by equating the amplitude prefactors in the VF laws found in both simulation and experiment. This translates one Monte Carlo step into $1.5 \cdot 10^{-14}$ s, meaning that the typical time window for the MC simulations would extend from 10^{-14} s to some 10^{-7} s, exceeding that of MD simulations by two orders of magnitude. This means that this

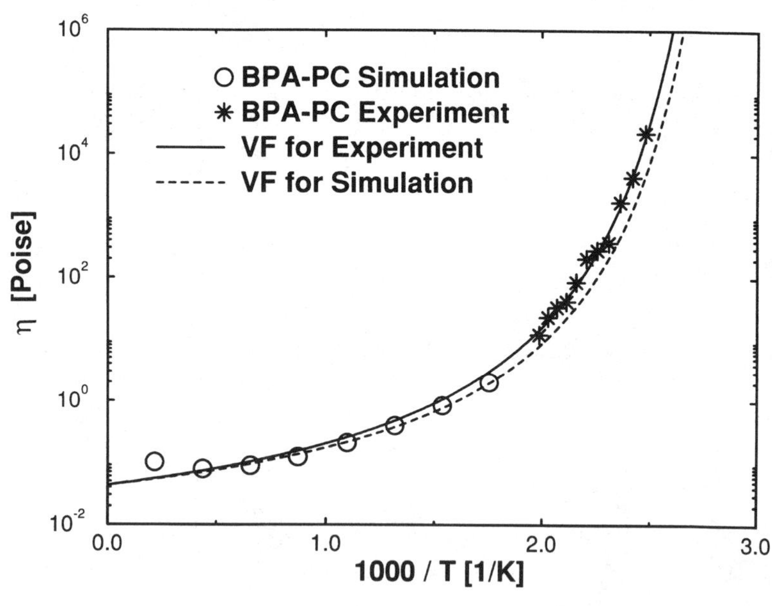

Fig. 6.26 Comparison of the viscosity measured in the simulation of BPA-PC to experimental data for BPA-PC oligomers corresponding to the same degree of polymerization. Also included are the VF laws fitted to the simulational and the experimental data separately. Vertical adjustment is done by identifying the $T \to \infty$ limits, which converts Monte Carlo time into seconds

way of mapping the mesoscopic length and time scales of a real polymer onto a computationally tractable simplified model is a procedure to produce equilibrated macroscopic properties in a temperature range not open to other simulational approaches.

6.5 Summary

We have shown in this chapter how the Monte Carlo simulation technique can be used to study various aspects of the glass transition in polymer melts. The model we employed is the bond fluctuation lattice model, which exhibits a relatively high degree of flexibility, placing it in between continuum models which are computationally more demanding and the standard lattice models where the polymer structure is too much determined by the underlying lattice. Using then the simple Hamiltonians A and B,[26,27,31,38,45,54,55,59,62] which only associate an energy with the bond lengths and leave interaction between the monomers of the melt to be purely repulsive, many experimental features of the glass transition can be reproduced.

These model Hamiltonians generate a competition between the energy of a chain conformation and the density, which makes both the two- and the three-dimensional melt freeze in a liquid-like structure if the relaxation times of the melt become comparable to the time scale of the simulation.[26,27,31,45,55] Therefore the freezing does not take place at a well defined temperature, but it depends upon the choice of the cooling rate and additionally upon the chain length. If one extracts from the simulation data these dependencies of the glass transition temperature T_g, a non-linear relationship results between T_g and the logarithm of the cooling rate[27] whereas the chain length is inversely proportional to T_g.[45] Both functional relationships are observed in experiments.[3,43,44,47]

Other prominent characteristics of the glass transition are the nonexponential, strongly stretched decay of the structural relaxation functions (partly obeying the time–temperature superposition principle) and the Vogel–Fulcher behavior of transport coefficients (as the diffusion coefficient of a chain). Both of these features are present in either the two- or the three-dimensional model.[26,27,54,55,62] During the long relaxation runs to obtain these dynamical results the melt ages considerably, so that a two-step decay of the incoherent intermediate scattering function was discovered in the three-dimensional simulation in a temperature region which belonged to the glassy phase immediately after cooling.[59] This two-step decay could be interpreted within the framework of the mode coupling theory, supplying the first evidence from simulations that this new theoretical approach to the glass transition might also be of relevance for the dynamics of undercooled polymer melts.

As concerns the modeling of a real polymer by the use of a lattice model, we have shown how a mapping on a mesoscopic scale can be defined to match descriptions on different levels of abstraction. Transferring information on mesoscopic length and time scales obtained from a microscopically detailed description into the computationally more easily tractable lattice model, we showed that one can reproduce the Vogel–Fulcher parameters of the polymer under study.

In future it should be possible to extend the analysis of the glass transition of these models in very diverse ways: for instance, by combining the results of the two- and three-dimensional model in order to learn more about the influence of the spatial dimension on the glass transition; or by analyzing in detail the physical aging of the melt; or by calculating suitable dynamical correlation functions in the temperature region where mode coupling effects seem to be important for the three-dimensional model, in order to assess the applicability of the MCT for glass transition in polymer melts. Due to the wealth of available information which is partly hard to obtain or even inaccessible in experiments, the models studied seem to be very promising for further investigation of the static and dynamical aspects of the structural glass transition.

In the modeling of real polymer systems there are several obvious avenues to follow. One is of course the modeling of other polymers, especially structural modifications of BPA-PC, to establish structure–property correlations. Others include the consideration of polydispersity effects and the inclusion of attractive interactions. In the long run the modeling of polymeric substances will only be successful through the synergistic application of different simulation techniques. Thus, in our opinion, the method of choice would be to use the mapping to an abstract model (the procedure is of course not limited to the bond fluctuation model) to generate equilibrated structures on a large scale and to use these structures to reinsert chains described in all chemical detail. They would then have to be re-equilibrated on the short length scales, which is a computationally feasible task with standard MD techniques. A combination of methods with suitable interfaces defined between them should be able to produce information on all interesting length and time scales.

Acknowledgments

The results presented are the outcome of a stimulating and fruitful collaboration with K. Binder, D. W. Heermann, K. Kremer, B. Lobe, N. Pistoor, P. Ray, and H.-P. Wittmann. It is a great pleasure to thank each and all of them. Support from the German Science Foundation under Grant No. SFB262, the German Ministry of Science and Technology under Grant No. 03M4028 and the Bayer AG is gratefully acknowledged. Part of this work could only be done because of a generous grant of computer time at the Höchstleistungsrechenzentrum (HLRZ) at Jülich.

References

1. J. Jäckle, *Rep. Prog. Phys.* **49**, 171 (1986); R. Zallen, *The Physics of Amorphous Solids* (Wiley, New York, 1983).
2. C. A. Angell, in *Relaxation in Complex Systems*, edited by K. L. Ngai and G. B. Wright (National Technical Information Service, US Department of Commerce, Springfield VA, 1985).
3. J. D. Ferry, *Viscoelastic Properties of Polymers* (Wiley, New York, 1980).
4. J. R. Fox and H. C. Andersen, *J. Phys. Chem.* **88**, 4019 (1984).
5. F. E. Luborski (ed.), *Amorphous Metallic Alloys* (Butterworths, London, 1983).
6. A. J. Kovacs, *J. Polym. Sci.* **30**, 131 (1958).
7. K. Sommer, I. Batoulis, W. Jilgeet *et al.*, *Adv. Mater.* **3**, 590 (1991).
8. W. Kauzmann, *Chem. Rev.* **43**, 219 (1948).
9. J. H. Gibbs and E. DiMarzio, *J. Chem. Phys.* **28**, 373 (1958); J. H. Gibbs and E. DiMarzio, *J. Chem. Phys.* **28**, 807 (1958); G. Adam and J. H. Gibbs, *J. Chem. Phys.* **43**, 139 (1965).
10. P. J. Flory, *J. Chem. Phys.* **10**, 51 (1942); M. L. Huggins, *J. Phys. Chem.* **46**, 151 (1942).
11. A. I. Milchev, Comptées rendues de l'Academie bulgare des Sciences **36**, 1415 (1983).
12. H. P. Wittmann, *J. Chem. Phys.* **95**, 8449 (1991).
13. M. H. Cohen and D. Turnbull, *J. Chem. Phys.* **31**, 1164 (1959); M. H. Cohen and G. S. Grest, *Phys. Rev. B* **20**, 1077 (1979); *Phys. Rev. B* **24**, 4091 (1981).
14. W. Götze, in *Liquids, Freezing and the Glass Transition*, edited by J. P. Hansen, D. Levesque and J. Zinn-Justin (North-Holland, Amsterdam, 1990); W. Götze and L. Sjögren, *Reports Prog. Phys.* **55**, 241 (1992).
15. D. Richter, R. Zorn, B. Farago, B. Frick, and L. J. Fetters, *Phys. Rev. Lett.* **68**, 71 (1992); B. Frick, B. Farago and D. Richter, *Phys. Rev. Lett.* **64**, 2921 (1990); D. Richter, B. Frick, and B. Farago, *Phys. Rev. Lett.* **61**, 2465 (1988).
16. R. G. Winkler, P. J. Ludovice, D. Y. Yoon, and H. Morawitz, *J. Chem. Phys.* **95**, 4709 (1991); Do. Y. Yoon, Grant D. Smith, and Tsunetoshi Matsuda, *J. Chem. Phys.* **98**, 10037 (1993).
17. R. J. Roe, D. Rigby, H. Furuya, and H. Takeuchi, *Comput. Polym. Sci.* **2**, 32 (1992); D. Rigby and R. J. Roe, *Macromolecules* **23**, 5312 (1990); D. Rigby and R. J. Roe, *Macromolecules* **22**, 2259 (1989); D. Rigby and R. J. Roe, *J. Chem. Phys.* **89**, 5280 (1988); D. Rigby and R. J. Roe, *J. Chem. Phys.* **87**, 7285 (1987).
18. J. H. R. Clarke and D. Brown, *Mol. Phys.* **58**, 815 (1986); J. H. R. Clarke and D. Brown, *Mol. Simul.* **3**, 27 (1989).
19. C. A. Angell, J. H. R. Clarke, and L. V. Woodcock in *Advances in Chemical Physics*, Vol. 48 (Wiley, New York, 1981).
20. R. D. de la Batie, J.-L. Viovy, and L. Monnerie, *J. Chem. Phys.* **81**, 567 (1984); A. Kolinski, J. Skolnick and R. Yaris, *J. Chem. Phys.* **84**, 1922 (1986).
21. K. Kremer and K. Binder, *Comput. Phys. Rep.* **7**, 259 (1988).
22. K. Binder and A. P. Young, *Rev. Mod. Phys.* **58**, 801 (1986).
23. I. Carmesin and K. Kremer, *Macromolecules* **21**, 2819 (1988).
24. H. P. Wittmann and K. Kremer, *Comput. Phys. Comm.* **61**, 309 (1990); H. P. Wittmann and K. Kremer, *Comput. Phys. Comm.* **71**, 343 (1992).

25. H. P. Deutsch and K. Binder, *J. Chem. Phys.* **94**, 2294 (1991).
26. H.-P. Wittmann, K. Kremer, and K. Binder, *J. Chem. Phys.* **96**, 6291 (1992).
27. J. Baschnagel, K. Binder, and H.-P. Wittmann, *J. Phys. C: Condens. Matter* **5**, 1597 (1993).
28. W. Paul, K. Binder, K. Kremer, and D. W. Heermann, *Macromolecules* **24**, 6332 (1991).
29. K. Binder (ed.), *Monte Carlo Methods in Statistical Physics*, 2nd edition (Springer, Berlin, 1986) and *Applications of the Monte Carlo Method in Statistical Physics* (Springer, Berlin, 1984).
30. P. J. Flory, *Statistical Mechanics of Chain Molecules* (Wiley, New York, 1969).
31. J. Baschnagel, H.-P. Wittmann, W. Paul, and K. Binder in *Trends in Non-Crystalline Solids*, edited by A. Conde, C. F. Conde and M. Millán (World Scientific, Singapore, 1992).
32. J. Baschnagel, K. Binder, W. Paul, *J. Chem. Phys.* **95**, 6014 (1991); J. Baschnagel, K. Qin, W. Paul, and K. Binder, *Macromolecules* **25**, 3117 (1992).
33. S. W. Lovesey, *Thermal Neutron Scattering from Condensed Matter*, Vol. 1,2 (Clarendon Press, Oxford, 1984).
34. M. Doi and S. F. Edwards, *Theory of Polymer Dynamics* (Clarendon Press, Oxford, 1986).
35.] P. G. de Gennes, *Scaling Concepts in Polymer Physics* (Cornell University, Ithaca, NY, 1979).
36. G. D. Wignall, in *Encyclopedia of Polymer Science and Engineering*, Vol. 10 (Wiley, New York, 1987), p. 112.
37. J. P. Hansen and I. R. McDonald, *Theory of Simple Liquids* (Academic Press, New York, 1986); J. P. Boon and S. Yip, *Molecular Hydrodynamics* (McGraw-Hill, New York, 1980).
38. J. Baschnagel and K. Binder, *Physica A* **204**, 47 (1994).
39. B. Frick, D. Richter, and Cl. Ritter, *Europhys. Lett.* **9**, 557 (1989).
40. W. Paul, *AIP Conf. Proc.* **256**, 145 (1992).
41. M. Rovere, D. W. Heermann, and K. Binder, *Europhys. Lett.* **6**, 585 (1988); M. Rovere, D. W. Heermann, and K. Binder, *J. Phys. C* **2**, 7009 (1990); M. Rovere, P. Nielaba, and K. Binder, *Z. Phys. B* **90**, 215 (1993).
42. H.-P. Deutsch and R. Dickman, *J. Chem. Phys.* **93**, 8983 (1990).
43. R. Brüning and K. Samwer, Contribution to the Spring Conference of the German Physical Society (DPG), (Physik-Verlags GMBH, Weinheim, 1992); R. Brüning and K. Samwer, *Phys. Rev. B* **46**, 11318 (1992).
44. A. J. Kovacs, J. M. Hutchinson, and J. J. Aklonis, in *The Structure of Non-Crystalline Materials*, edited by P. H. Gaskell (Taylor and Francis, London, 1977).
45. B. Lobe, J. Baschnagel, and K. Binder, *Macromolecules* **27**, 3658 (1994); B. Lobe, J. Baschnagel, and K. Binder, *J. Non-Cryst. Solids* **172–174**, 384 (1994).
46. T. G. Fox and S. Loshaek, *J. Polym. Sci.* **14**, 315 (1954).
47. A. Rubin and D. Burgin, *Polymer* **16**, 291 (1975); G. Pezzina, F. Zilio-Grandi, and P. Saumartin, *Euro. Polym. J.* **6**, 1053 (1970); J. M. Cowie and P. M. Toporowski, *Euro. Polym. J.* **4**, 621 (1968).
48. T. A. Vilgis, *Phys. Rev. E.* **47**, 2882 (1993).

49. E. W. Fischer, *Physica A* **201**, 183 (1993).
50. G. B. McKenna, in *Comprehensive Polymer Science*, Vol. 2, edited by C. Booth and C. Price (Pergamon, New York, 1989).
51. A. Sharples, in *Polymer Science*, Vol. 1, edited by A. D. Jenkins (North-Holland, Amsterdam, 1972).
52. G. Li, W. M. Du, X. K. Chen, H. Z. Cummins, and N. J. Tao, *Phys. Rev. A* **45**, 3867 (1992).
53. T. P. Lodge, N. A. Rotstein, and S. Prager, in *Advances in Chemical Physics*, edited by I. Prigogine and S. A. Rice (Wiley, New York, 1990).
54. H.-P. Wittmann, K. Kremer, and K. Binder, *Die Makromolekulare Chemie, Theory and Simulations* **1**, 275 (1992).
55. P. Ray, J. Baschnagel, and K. Binder, *J. Phys. Condens. Matter*, **5**, 5731 (1993).
56. H. Bässler, *Phys. Rev. Lett.* **58**, 767 (1987).
57. M. Fuchs, W. Götze, S. Hildebrand, and A. Latz, *J. Phys. C: Condens. Matter* **4**, 7709 (1992).
58. R. Schilling, in *Disorder Effects on Relaxational Processes*, edited by R. Richert and A. Blumen (Springer, Heidelberg, 1994).
59. J. Baschnagel, *Phys. Rev. B* **49**, 135 (1994). J. Baschnagel, *Physica A* **201**, 157 (1994).
60. W. Götze, *J. Phys. C: Condens. Matter* **2**, 8485 (1990).
61. H. Z. Cummins, W. M. Du, M. Fuchs, *et al.*, *Phys. Rev. E* **47**, 4223 (1993).
62. K. Binder, W. Paul, H.-P. Wittmann, J. Baschnagel, K. Kremer, and D. W. Heermann, *Prog. Coll. Poly. Sci.* **91**, 5 (1993).
63. W. Kuhn, *Z. Kolloid,* **76**, 258 (1936); **87**, 3 (1939).
64. W. Paul and N. Pistoor, *Macromolecules* **27**, 1249 (1994).
65. W. Paul, K. Binder, D. W. Heermann, and K. Kremer, *J. Phys. (France) II* **1**, 37 (1991).
66. W. Paul, K. Binder, D. W. Heermann, and K. Kremer, *J. Chem. Phys.* **95**, 7726 (1991).
67. J. Wittmer, W. Paul, and K. Binder, *Macromolecules* **25**, 7211 (1992).
68. L. Cervinka, E. W. Fischer, K. Hahn, *et al.*, *Polymer* **28**, 1287 (1987).
69. C. Lamers, O. Schärpf, W. Schweika, *et al.*, *Physica B* **180-181**, 515 (1992).
70. W. Paul, *J. Non-Cryst. Solids* **172–174**, 682 (1994).
71. A. Schönhals, F. Kremer, A. Hofmann, E. W. Fischer, and E. Schlosser, *Phys. Rev. Lett.* **70**, 3459 (1993).
72. W. Paul, K. Binder, J. Batoulis, B. Pittel, and K. H. Sommer, *Makromol. Chem., Macromol. Symp.* **65**, 1 (1993).
73. *Encyclopedia of Polymer Science and Engineering*, Vol. 7, 2nd edition (Wiley, New York, 1989).
74. J. Koppelmann, in *Proceedings of the Fourth International Congress on Rheology*, Vol. 3, edited by E. H. Lee and A. L. Copley (Wiley, New York, 1965).

7

MONTE CARLO STUDIES OF POLYMER BLENDS AND BLOCK COPOLYMER THERMODYNAMICS

Kurt Binder

7.1 Introduction

For mixtures of simple fluids, it is now possible to simulate phase behavior taking into account full atomistic detail with chemically realistic forces.[1,2] Apart from the region near a critical point, it suffices to simulate small boxes containing atoms or molecules in the order of 10^3 in the respective Monte Carlo[1,2,3] or molecular dynamics[3,4] methods—both of them methods which have been well developed.

These methods work because fluids off critical points are already essentially homogeneous on a length scale of 10 Å, and staying away from the glass transition region[5] the relaxation time needed to reach equilibrium is very small (in the picosecond range). However, for mixtures of long flexible macromolecules the situation is fundamentally different.[6,7] As emphasized in Chapter 1 of this book, already a *single* chain exhibits structure in the length of a chemical bond (≈ 1 Å) to the persistence length (≈ 10 Å) to the coil radius R_{gyr} ($\approx 10^2$ Å). The latter length scale is a lower bound to the correlation length ξ of concentration fluctuations in polymer mixtures. While in small molecule mixtures correlation lengths in the order of 10^3 Å occur in the immediate vicinity of critical points only, in polymer mixtures such large correlation lengths occur in a much larger part of the phase diagram. This is easily understood since in the relation for the correlation length[8] $\xi \approx \hat{\xi}|1 - T/T_c|^{-\nu}$ (T_c being the critical temperature, T the temperature, ν the associated critical exponent) the "critical amplitude" $\hat{\xi}$ is of the order of R_{gyr} rather than the order of the interaction range r_o, as in small molecule mixtures (both for small molecule mixtures and for polymer mixtures, r_o is of the order of a few Å only).[9–12]

The same conclusion emerges when one considers the dynamics of phase separation in experiments where one quenches a polymer blend into the two-phase region underneath the miscibility gap (Fig. 7.1). The initially inhomogeneous state is unstable and fluctuations grow with time t after the quench; their characteristic wavelength $\lambda_m(t)$ (which shows up by a peak in the scattering function $S(q, t)$ at a wavenumber $q = q_m(t) = 2\pi/\lambda_m(t)$) again is typically of the order of[12–14] 10^3 Å.

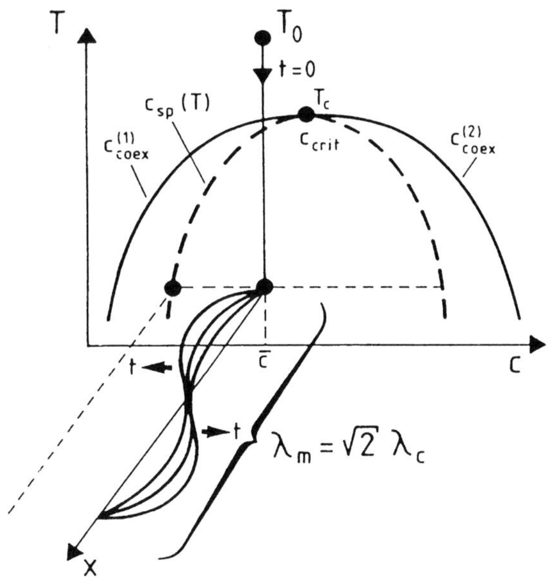

Fig. 7.1 Schematic plot of a quenching experiment at a concentration $\bar{c}$ that leads to a temperature $T < T_c$ in the unstable region (inside the "spinodal curve", shown as broken curve). As the time t after the quench passes, all Fourier components of concentration fluctuations where wavelengths λ exceed a critical wavelength λ_c grow with time (according to the linearized theory of spinodal decomposition, the maximum growth occurs for $\lambda_m = \sqrt{2}\lambda_c$). Growth of these "concentration waves" is indicated for one spatial direction (x). As time t passes, nonlinear effects lead to a coarsening of the A-rich domains (having a concentration according to the left branch of the coexistence curve, $c_{\text{coex}}^{(1)}$) and of the B-rich domains (of concentration $c_{\text{coex}}^{(2)}$), i.e., $\lambda_m(t \to \infty) \to \infty$. (From Binder.[12])

Similar large length scales are found for the thickness of wetting layers near walls that favor one component of the blend[15] and for the length scales characterizing the mesophase ordering in block copolymer melts.[12,16–18] For example, in the "weak segregation regime" of a symmetric diblock copolymer melt ($N_A = N_B = N/2$), i.e., in the disordered phase, the characteristic wavelength λ is in the order of a random coil size, $\lambda \approx R_{\text{gyr}} \propto aN^{1/2}$, where a is the size of an effective segment. But in the "strong segregation regime"[12,16–18] the coils are strongly stretched in the direction perpendicular to the lamellae of the lamellar mesophase (Fig. 7.2),[19] and then the wavelength characterizing the lamellar ordering is much larger, $\lambda \propto R_{\text{gyr}} \propto aN^{2/3}$.

Treating such phenomena on the length scale of 10^3 Å with simulations of models that include all chemical detail is clearly impossible, and thus crudely simplified coarse-grained models must be used.[6,7] The same conclusion emerges when one considers the second factor that controls simulation feasibility (cf. Chapter 1 for a more thorough discussion), namely time scales: long wavelength degrees of freedom relax very slowly! Thus it is an experimental fact that the time scale for the growth of the concentration

(a)

(b)

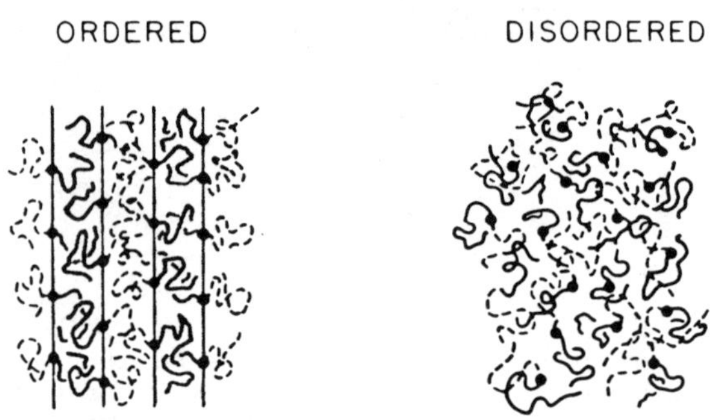

Fig. 7.2 (a) Chemical architecture of a diblock copolymer. A diblock copolymer consists of a polymerized sequence of A-monomers (A-block, shown as full curve) covalently attached at a junction point (dot) to a similar sequence of B-monomers (B-block, shown as dashed curve). (b) The microphase separation transition occurs when a compositionally disordered melt of copolymers (right) transforms to a spatially periodic, compositionally inhomogeneous phase (left) on lowering the temperature (or on increasing the Flory–Huggins χ-parameter,[9] respectively). For nearly symmetric copolymers the ordered phase has the lamellar structure shown, where the junction points order in parallel planes a distance $\lambda/2$ apart. (From Fredrickson and Binder.[19])

fluctuations shown schematically in Fig. 7.1 is in the order of 1 s for many polymer mixtures or even larger.[12–14] Similar slow relaxation occurs for the growth of wetting layers,[15] for the equilibration of well-ordered regions of the block copolymer mesophases when one cools them through the transition temperature,[16] etc.

Fortunately, such very crude models were also used to develop the basic theoretical concepts on these phenomena. A good example is the Flory–Huggins lattice model of polymer blends[20] (Fig. 7.3): Disregarding the local chemical structure of the chains, and any resulting disparity between the size and shape of the subunits, chains are modeled by mutually and self-avoiding walks on a lattice.

In addition, the actual interaction potentials are not explicitly considered, their effect is simply lumped into some "effective" interaction parameters $\epsilon_{AA}, \epsilon_{AB}$ and ϵ_{BB} between the respective nearest neighbor pairs. Also the Leibler[21] theory of mesophase ordering in block copolymers may be thought

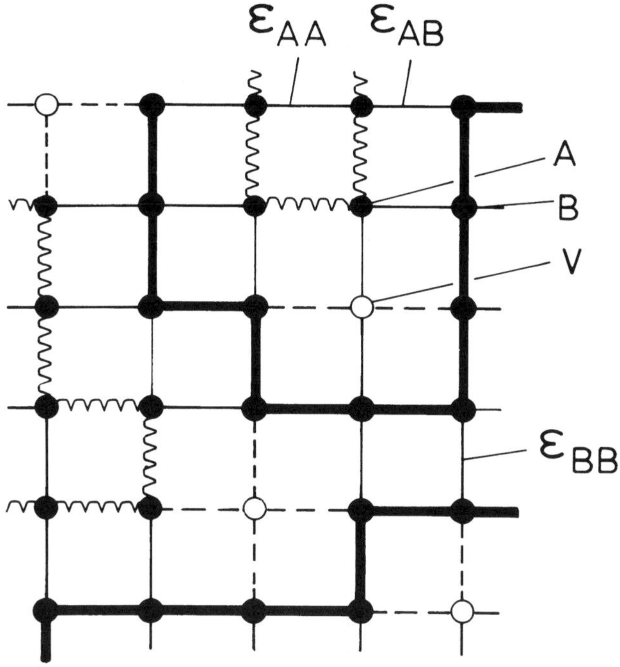

Fig. 7.3 The Flory–Huggins lattice model describes the two types of polymer chains (A,B) as mutually and self-avoiding walks of $N_A(N_B)$ steps on a lattice. Each lattice site is thus either taken by an A-monomer, a B-monomer, or vacant (V). The volume fractions of these monomers are ϕ_A, ϕ_B, and $\phi_v = 1 - \phi_A - \phi_B$. Interaction energies ϵ_{AA}, ϵ_{AB} and ϵ_{BB} are assumed between nearest-neighbor AA, AB and BB pairs, respectively.

of as a continuum approximation to an analogously coarse-grained lattice model of block copolymers.

Why is it useful to simulate models which are as crude as the model of Fig. 7.3? What can we learn from studies such as these? Well, the Flory–Huggins theory[20] of polymer mixtures is a good example to answer that question. This very theory is not only based on such a crude model, but invokes a number of additional uncontrolled approximations. In principle, what one should do is to consider all possible configurations of how to place the two kinds of chains and the vacancies on the lattice (Fig. 7.3). From the total energy E(conf) of each configuration we obtain its statistical weight, the Boltzmann factor $\exp[-E(\text{conf})/k_BT]$ ($k_B =$ Boltzmann's constant). Desired physical properties (coexistence curve, critical temperature T_c, etc.) then follow from the partition function Z and the free energy F,

$$F/k_BT = -\ln Z = -\ln \sum_{\text{conf}} \exp[-E(\text{conf})/k_BT]. \tag{7.1}$$

Of course, it is not possible to carry out this program posed by statistical mechanics exactly, in spite of the crudeness of our model (Fig. 7.3). Thus, Flory and Huggins[20,22,23] proposed an approximation, which will be discussed next.

The excess free energy of the mixture (relative to a pure polymer) is written as the sum of an entropy of mixing term, and an enthalpy term. In the latter, only a reduced energy parameter ϵ enters, which is related to the Flory–Huggins χ parameter

$$\epsilon = \epsilon_{AB} - (\epsilon_{AA} + \epsilon_{BB})/2, \quad \chi = z\epsilon/(k_B T), \tag{7.2}$$

z being the coordination number of the lattice. Then (FH stands for Flory–Huggins; ϕ_A, ϕ_B are the volume fractions of the two types of monomers)

$$F_{FH}/k_B T = \frac{\phi_A \ln \phi_A}{N_A} + \frac{\phi_B \ln \phi_B}{N_B} + \phi_v + \chi \phi_A \phi_B. \tag{7.3}$$

If we had three low molecular weight components ($N_A = N_B = 1$), the first three terms on the right hand side of eq. (7.3) describing the entropy of mixing would be exact, since then each lattice site can be filled independently of its neighbors. For polymers, this entropy expression is approximate only—independent occupation of sites obviously neglects the condition that the chains must not intersect themselves or other chains.

The enthalpy term $\chi \phi_A \phi_B$ is only approximate, too—what would be needed is the energy $\epsilon/k_B T$ multiplied by the probability $P(\phi_A, \phi_B)$ that a site is taken by an A-segment and a nearest neighbor of this site by a B segment. Writing a product expression $P(\phi_A, \phi_B) = \phi_A \phi_B$ obviously neglects correlations in the occupancy of neighboring sites.

A further error is introduced by counting the number of neighboring sites: a segment in the interior of an A-chain can have at most $z - 2$ B-neighbors instead of z B-neighbors—*at least* two neighboring sites are automatically taken by the A-chain itself. Replacing z by $z - 2$ in eq. (7.2), as is sometimes done,[24] has the disadvantage that it is not good for small N_A, N_B.

Now, asking the question, how reliable is this theory with all these approximations? Obviously, this question is not easily answered by comparing the theory to experiments: in most cases, the agreement is not good, but one cannot tell whether the discrepancies are due to inadequacies of the model—after all the model in Fig. 7.3 is a crude caricature of reality—or due to inaccuracies of the approximations. Sometimes such problems are also hidden because too many adjustable parameters are available, so experimental data can be fitted—but one does not know whether the resulting fitted parameters are really meaningful. Obviously, such problems occur for the many extensions of Flory–Huggins theory, where one adds correction terms to it and considers a possible volume-fraction dependence of the χ-parameter.[25]

Since the simulation can be performed on exactly the same model which the analytical theory considers (but avoiding the uncontrolled mathematical approximations of the latter), the simulations provide "benchmarks" against which these theories can be reliably tested. A typical example of this sort of application is shown in Fig. 7.4, where Monte Carlo data[26,27] for the critical temperature of a symmetrical polymer blend ($N_A = N_B = N$) are compared to the Flory–Huggins theory as well as a treatment based on Born–Green–Yvon (BGY) integral equations on a lattice.[28–30] One sees that while both FH and BGY treatments predict a strictly linear variation in agreement with the simulation in this model (the bond fluctuation model at a volume fraction $\phi_v = 0.5$ of vacant lattice sites, interaction range up to a distance of $\sqrt{6}$ lattice spacings at the simple cubic lattice,[26,27] the prefactor in this relation is strongly overestimated by FH, but slightly underestimated by the BGY treatment.[30]

We also note that the same Monte Carlo data[26,27] have helped to sort out an inadequate approximation in the context of the polymer reference interaction site model (PRISM) theory,[31–36] which yielded a relation $T_c \propto \sqrt{N}$ while now $T_c \propto N$ is generally accepted.[26–28,37–41] It has been very difficult to provide convincing experimental evidence[41] on this issue—true symmetrical monodisperse polymer mixtures hardly exist, and the temperature range over which $T_c(N)$ can be studied is limited by the glass transition temperature from below and by chemical instability of the chains from

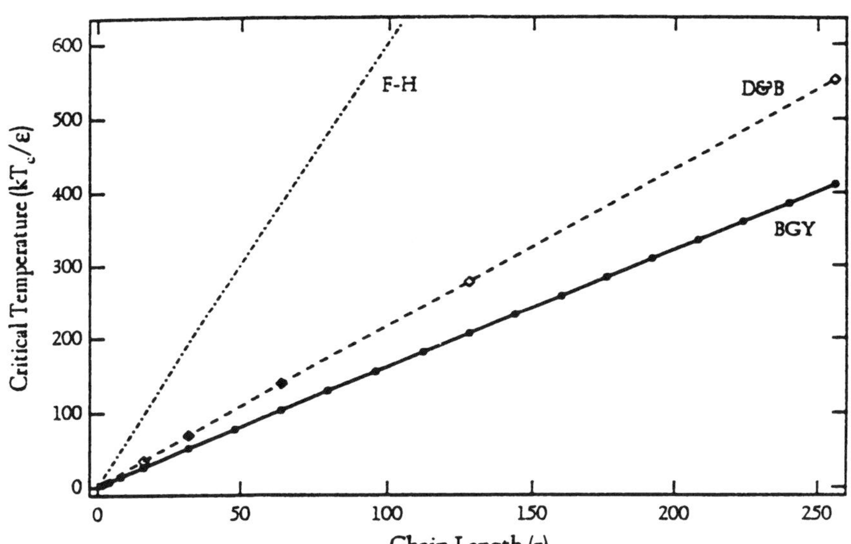

Fig. 7.4 Comparison of the critical temperature versus chain length between the Monte Carlo data of Deutsch and Binder (D&B)[26,27] to the Flory–Huggins (FH) prediction (dot–dashed) and the Born–Green–Yvon (BGY) treatment (full curve). (From Sevian *et al.*[30])

above. Simulations are free of such limitations, of course, and can test such theories much more stringently than the experiment.

In addition, simulations have helped to validate a new version of the PRISM theory,[37–39] which should be a promising starting point (as an off-lattice theory) to allow a much more realistic description of polymer blends rather than the model of Fig. 7.3.

Similar examples that show the fruitful interplay between simulation, analytical theory, and experiments can be given for spinodal decomposition,[6,10–12,42–44] surface enrichment and wetting,[15,45–51] block copolymer ordering,[12,16–19,21,24,52–66] etc. Some of this material will be discussed in later sections.

In this chapter, we focus on simple lattice models of polymers, of the type shown in Fig. 7.3 and the bond fluctuation model,[67,68] and work on simple off-lattice models will also be occasionally mentioned.[2,69,70] But we shall neither address simulations based on continuum generalizations of eq. (7.3), where chains are no longer explicitly considered,[71–74] nor attempt to explain the χ-parameter in terms of simulations of chemically realistic monomers.[75] Both approaches involve many—in our opinion unsolved—questions on methodic aspects that are out of scope here.

7.2 Simulation methodology

7.2.1 Dynamic algorithms and the role of vacancies

Simulations of polymer blends or block copolymers involve two rather distinct aspects: one aspect is the generation of equilibrium configurations of dense polymer melts and the relaxation of the configurations of individual chains;[7,76] this aspect is not essentially different from simulations that deal with one-component polymer solutions and melts, as treated in other chapters of this book. The work described in the present chapter has used dynamic Monte Carlo methods such as combinations of "kink jump" and "crankshaft rotation" algorithms (Fig. 7.5(a))[77–79] or simple hops of effective monomers in randomly chosen lattice directions (in the case of the bond fluctuation model[26,27,40,67,68]) or the "slithering snake" technique.[52–54,55,80,81] All these algorithms need a nonzero concentration of vacancies, of course. In an attempt to work strictly in the limit of $\phi_v = 0$, however, the "bond-breaking method"[82] and the "collective motion algorithm"[24,83–85] have occasionally also been implemented. Apart from the fact that neither of these techniques[86,87] can be mapped to the real dynamics of polymers in melts,[7] even in a coarse-grained sense, and hence cannot be used to study dynamical phenomena such as interdiffusion,[68] spinodal decomposition[42,88] or block copolymer ordering dynamics, these algorithms have certain other drawbacks—the bond-breaking method[82,86] produces some polydispersity; the collective motion method[24,83–85,87] amounts to a very complicated algo-

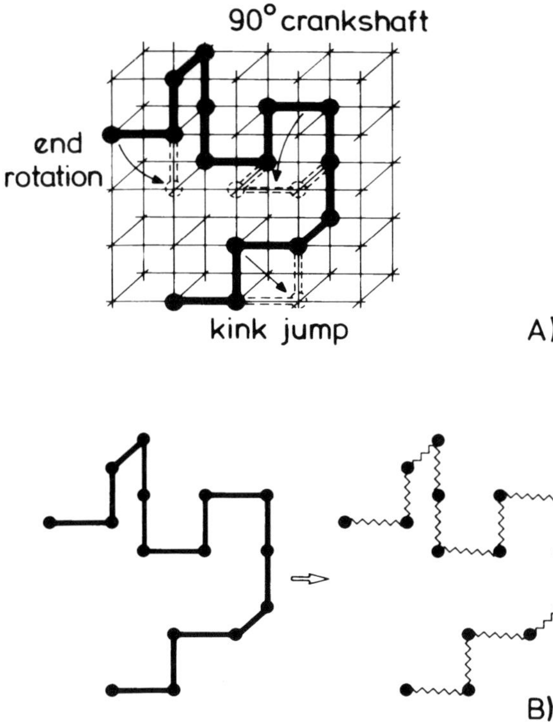

Fig. 7.5 In dynamic Monte Carlo simulations of the original Flory–Huggins model chain configurations are relaxed by end-bond rotations, kink jump motions, and 90° crankshaft moves (A). Only such moves are allowed that do not violate the excluded volume constraint. In a semi-grand-canonical simulation, where the chemical potential difference $\Delta\mu$ between A- and B-monomers is fixed, A-chains are taken out of the system and B-chains are inserted in exactly the same configuration, or vice versa (B). (From Binder.[77])

rithm neither suitable for vectorization nor parallelization and performs relatively slowly and also the problem of its "dynamic correlations" and hence the judgement of statistical errors[89–91] is difficult. In view of these problems, we do not share the view of these authors[24,84,85] that those algorithms offer a great advantage: first of all, physical polymer melts do have a nonzero, albeit small compressibility, and qualitatively this property is only reproduced by a rigid lattice model if it contains a nonzero fraction of vacancies; secondly, the theories which one wishes to test are often easily extended to include a nonzero ϕ_v, and checking the variation of properties with ϕ_v provides an additional test of these theories. Thirdly, all experience tells us[24,68,79,82] that the properties of interest extrapolate smoothly towards their limits reached for $\phi_v \to 0$. Since a comparison of the bond fluctuation model (with $\phi_v = 0.5$ or thereabouts) has in many respects very similar properties[92] to more realistic off-lattice models of dense polymer melts,[93] we consider the presence of vacancies an advantage rather than a disadvantage.

The second aspect is relaxation of the long wavelength degrees of freedom related to phase transitions occurring in the studied systems: for the unmixing of polymer blends, these are long wavelength Fourier components of the volume fractions of species A, B in the system. In conditions where the total numbers of both A-chains and B-chains are kept fixed, the conservation law of the concentration leads to "hydrodynamic slowing down"[92,93] of these Fourier components, i.e., extremely slow relaxation. Near the critical point, the problem becomes even worse due to an anomalous growth of relaxation times ("critical slowing down"),[93,94] and there are severe finite size effects on the transition[95–98] in addition. These problems will be considered in the following subsections.

7.2.2 The semi-grand-canonical technique for polymer blends

For symmetrical polymer blends (as well as weakly asymmetrical ones) the problem of hydrodynamical slowing down of long wavelength concentration fluctuations can be elegantly avoided by carrying out the simulation in the semi-grand-canonical ensemble rather than the canonical ensemble: only the total number of chains $n = n_A + n_B$ is fixed, while the ratio $\phi_A/(1-\phi_V) = n_A N_A/(n_A N_A + n_B N_B)$ fluctuates, in equilibrium with a given chemical potential difference $\Delta\mu = \mu_A - \mu_B$ between the chains (Fig. 7.5(b)) with $\phi_A + \phi_B = 1 - \phi_V = const.$ In the thermodynamic limit, where $n_A, n_B \to \infty$, the different ensembles of statistical mechanics yield completely equivalent results, and for such equilibrium properties it then does not matter whether one computes them in the grand-canonical ensemble (both μ_A and μ_B are given separately, n_A and n_B can fluctuate independently from each other), the semi-grand-canonical ensemble, or the canonical ensemble (where both n_A and n_B would be fixed, while μ_A and μ_B would both be fluctuating). Of course, the grand-canonical ensemble is less useful for dense polymer systems, since it is practically impossible successfully to insert a chain in a randomly chosen configuration in a many-chain system: almost always the excluded volume constraint would be violated, and such moves therefore would be forbidden.

Of course, experiments are done in the canonic ensemble only, and chain "identity switches" as indicated in Fig. 7.5(b) do not occur, but due to this equivalence between the statistical ensembles that distinction does not matter, in the thermodynamic limit. As will be discussed below, finite size effects are rather different in these various ensembles, and since one is usually able to simulate rather small boxes containing less than 10^3chains only, these size effects must be considered carefully. If one is interested in dynamical properties, e.g., the study of initial stages of phase separation (Fig. 7.1), one must use the statistical ensemble that exactly corresponds to the experiment, namely the canonical ensemble. The semi-grand-canonical ensemble for such simulations can only be used to prepare equilibrium initial states,[42] from which the quench (at a temperature T_o, Fig. 7.1) starts.

We first discuss the strictly symmetrical situation, where both $N_A = N_B = N$ and $\epsilon_{AA} = \epsilon_{BB}$, so the two types of chains differ only by their labels. The "order parameter" of the unmixing transition can then be written as

$$m = (n_A - n_B)/(n_A + n_B), \tag{7.4}$$

the volume fractions being

$$\phi_A = (1 - \phi_v)n_A/(n_A + n_B) = (1 - \phi_v)(1 + m)/2, \phi_B = (1 - \phi_v)(1 - m)/2. \tag{7.5}$$

Because of the symmetry of the problem against an interchange of A and B, phase coexistence can occur only for $\Delta\mu = 0$, and the critical point must occur for a critical value of the volume fraction

$$\phi_A^{\text{crit}} = \phi_B^{\text{crit}} = (1 - \phi_v)/2, \;\; m = 0. \tag{7.6}$$

While the partition function in the canonical ensemble is given by eq. (7.1), in the grand-canonical ensemble it is

$$Z_G = \sum_{\text{conf}} \exp[(\mu_A n_A N_A + \mu_B n_B N_B)/k_B T] \exp[-E(\text{conf})/k_B T]. \tag{7.7}$$

For the symmetrical case considered in eqs (7.4)–(7.6), the first exponential factor in eq. (7.7) can be rewritten as

$$\begin{aligned} \exp[(\mu_A n_A N + \mu_B n_B N)/k_B T] = {} & \exp[N(\mu_A + \mu_B)n/(2k_B T)] \\ & \exp[N\Delta\mu m n/(2k_B T)] \end{aligned} \tag{7.8}$$

The number of chains n in a simulation in the semi-grand-canonical ensemble is constant; only the order parameter m (eq. [7.4]) can fluctuate. Therefore the first factor on the right hand side of eq. (7.8) is constant and cancels out from the Monte Carlo averages.

In carrying out the Monte Carlo moves shown in Fig. 7.5 one applies the Metropolis method[1–3,7,76,89,93,99] in a fairly standard way, i.e., the transition probability for the semi-grand-canonical move where one goes from an old configuration (c) to a new configuration (c') via an "identity switch" of a chain, $A \rightleftharpoons B$ (Fig. 7.5(b)), is

$$W(c \to c') = \text{Min}\left\{1, \exp\left[\frac{\Delta\mu(m' - m)Nn}{2k_B T}\right] \exp\left[\frac{(E' - E)}{k_B T}\right]\right\}. \tag{7.9}$$

Here E, m are energy and order parameter in the old configuration, E' and m' refer to the new trial configuration. W is compared to a random number z, uniformly distributed between zero and one: only if W exceeds z is the trial configuration accepted as a new configuration; otherwise it is rejected and the old configuration is counted once more for the averaging. For the standard moves such as shown in Fig. 7.5(a), essentially the same procedure applies, except that $m' = m$ and hence the first exponential factor in eq. (7.9) trivially is unity.

A somewhat delicate question is how often one performs the identity switch (Fig. 7.5(b)) relative to the local moves (Fig. 7.5(a)) needed to equilibrate the chain configurations. For asymmetric compositions ($\phi_B < \phi_A$), which occur in the one phase region in thermal equilibrium choosing $\Delta\mu > 0$, the mean square gyration radii $\langle R_{gA}^2 \rangle$, $\langle R_{gB}^2 \rangle$ of the two types of chain are not identical: one finds[77–79,100,101] that both types of chain contract somewhat in comparison to the noninteracting case ($\epsilon/k_BT = 0$), and the minority chains are distinctly smaller than the majority ones. As a consequence, each chain should relax its configuration by many local motions after each identity switch,[79] before the next identity switch is attempted. In order to avoid any bias and satisfy detailed balance, the choice of beads for the moves in Fig. 7.5(a) and the choice of chains for the identity switch should be random, and the two types of moves should be randomly mixed: note that in the absence of the exponential factor in eq. (7.9) the "*a priori* probability" for the move $c \rightarrow c'$ must equal the *a priori* probability for the inverse move, $c' \rightarrow c$.

A useful transformation is to write the semi-grand-canonical partition function $Z_{SG}(T, \Delta\mu)$ in terms of the density of states $\Gamma(E, m)$, which does not depend on the parameters $(T, \Delta\mu)$ characterizing the considered thermodynamic state:

$$Z_{SG}(T, \Delta\mu) = \int_{-1}^{+1} dm \int dE \exp(-E/k_BT) \exp[N\Delta\mu\, mn/(2k_BT)]\Gamma(E, m). \tag{7.10}$$

The Monte Carlo sampling yields a number $\mathcal{N}$ of configurations that are distributed proportional to $P_{T,\Delta\mu}(E, m)$ given as

$$P_{T,\Delta\mu}(E, m) = \frac{1}{Z_{SG}(T, \Delta\mu)} \exp(-E/k_BT) \exp[N\Delta\mu\, mn/(2k_BT)]\Gamma(E, m). \tag{7.11}$$

In a simulation one may record a "histogram" $H_{T,\Delta\mu}(E, m)$ just counting how often one observes the possible values of E and m. For a large number $\mathcal{N}$ of statistically independent states, this histogram approximates $P_{T,\Delta\mu}(E, m)$. Therefore $H_{T,\Delta\mu}(E, m)$ can be used for a whole range of neighboring values $T', \Delta\mu'$ around $T, \Delta\mu$, by a suitable reweighting[102–104]

$$P_{T',\Delta\mu}(E,m) \approx \frac{Z(T,\Delta\mu)}{Z(T',\Delta\mu')} \exp\left[\frac{\Delta\mu' mNn}{2k_BT'} - \frac{\Delta\mu mNn}{2k_BT}\right] \exp\left(\frac{E}{k_BT} - \frac{E}{k_BT'}\right) \frac{H_{T,\Delta\mu}(E,m)}{\mathcal{N}}. \tag{7.12}$$

This "single histogram extrapolation" is only practical in use for such a range of parameters[104] where the exponentials in eq. (7.12) do not emphasize values of E, m far out on the wings of the histogram H, where the statistical accuracy is bad. But, for polymer mixtures near T_c, the number of chains n in the simulated boxes is often rather small,[108] and then the distributions $P_{T',\Delta\mu'}(E,m)$ are all rather broad and this reweighting works nicely over a reasonably broad range of T and $\Delta\mu$ (Fig. 7.6).

It is often desirable to combine histograms from several simulation runs to get extrapolations over a wider parameter range and to increase the accuracy.[104–108] If s simulations are performed at parameters $\{(T_i, \Delta\mu_i), i = 1, \ldots, s\}$, the generalization of eqs (7.10) and (7.12) for the density of states is $\{\mathcal{N}_i$ is now the number of states recorded in simulation i for histogram $H_{T_i,\Delta\mu_i}(E,m)\}$

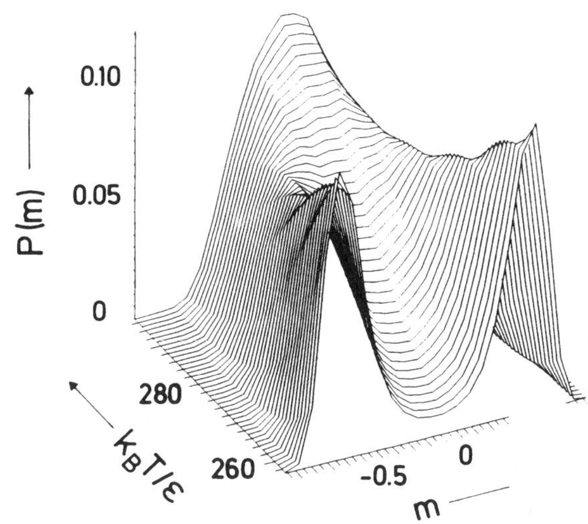

Fig. 7.6 Distribution function $P(m) \equiv \int dEP_{T,\Delta\mu}(E,m)$ of the order parameter m for $\Delta\mu = 0$ plotted over a range of temperatures for $N = 128$, $L = 80$ lattice spacings, $\phi_v = 0.5$, using the bond fluctuation model of symmetric polymer mixtures at a simple cubic lattice and using the extrapolation formula, eq. (7.12), with data from a single temperature run at $k_BT/\epsilon = 266.4$. The number of statistically independent samples was $\mathcal{N} = 16\,800$. (From Deutsch and Binder.[27])

$$\Gamma(E,m) = \sum_{i=1}^{s} w_i(E,m)\mathcal{N}_i^{-1} Z_{SG}(T_i,\Delta\mu_i)\exp[-N\Delta\mu_i mn/(2k_BT_i)] \\ \exp(E/k_BT_i)H_{T_i,\Delta\mu_i}(E,m). \tag{7.13}$$

The weight w_i with which the histogram of the ith simulation enters is determined by minimizing the statistical error of $\Gamma(E,m)$ and is found as[104–106,108]

$$w_i(E,m) = w_i'(E,m)/\sum_{j=1}^{s} w_j'(E,m), \tag{7.14}$$

where

$$w_i'(E,m) = (1+2\tau_i)^{-1}\mathcal{N}_i Z_{SG}^{-1}(T_i,\Delta\mu_i)\exp[N\Delta\mu_i mN/(2k_BT_i)]\exp(-E/k_BT_i). \tag{7.15}$$

Figure 7.7 gives an example where this technique was used to obtain both the absolute value of the order parameter $\langle|m|\rangle$ and its fluctuation "susceptibilities" $S_{\text{coll}}(q=0) = n(\langle m^2\rangle - \langle m\rangle^2)$, $S'_{\text{coll}}(q=0) = n(\langle m^2\rangle - \langle|m|\rangle^2)$ for $\Delta\mu = 0$ in the bond fluctuation model of symmetrical polymer mixtures for a variety of lattice sizes.[40] $S_{\text{coll}}(q)$ describes the scattering of neutrons, x-rays or light under scattering vector q for a state with order parameter $\langle m\rangle$(i.e., volume fractions $\phi_A = (1-\phi_v)(1+\langle m\rangle)/2$, $\phi_B = (1-\phi_v)(1-\langle m\rangle)/2$) in the one phase region. For $\Delta\mu = 0$ we have also $\langle m\rangle = 0$, i.e., these data refer to the critical concentration (Eq. [7.6]). Thus one expects that[8–12]

$$S_{\text{coll}}(q=0) = \hat{C}_+(-t)^{-\gamma}, t = 1 - T/T_c \to 0, \tag{7.16}$$

where $\hat{C}_+$ is a critical amplitude and γ a critical exponent. Of course, for a finite volume it is obvious from the definition of $S_{\text{coll}}(q=0)$ that it can never diverge, where instead it smoothly increases and reaches a saturation value $S_{\text{coll}}(q=0) = n$ (in our normalization) for $T=0$. For the bond fluctuation model shown in Fig. 7.7, where each effective monomer occupies the eight corners of an elementary cell of the sc lattice, the three choices of L shown correspond to $n=108$, 256, and 500, respectively. Therefore it is no surprise that the three curves splay out in the critical region, and are not immediately suitable to read off the critical temperature. Similarly, the quantity $S'_{\text{coll}}(q)$ describes (in the thermodynamic limit) the scattering intensity at a path that follows the coexistence curve of the miscibility gap (Fig. 7.1). For finite L, $S'_{\text{coll}}(q=0)$ has a maximum of finite height, and this maximum is offset from the value of T_c seen in the thermodynamic limit. The same conclusion, that finite size effects mask the location of T_c, is also seen for the smeared-out behavior of the order parameter (Fig. 7.7(a)) and the anomaly of the specific heat[78] (Fig. 7.8). Using finite size scaling theory[95–98] in conjunction with the theoretical values of the critical exponents

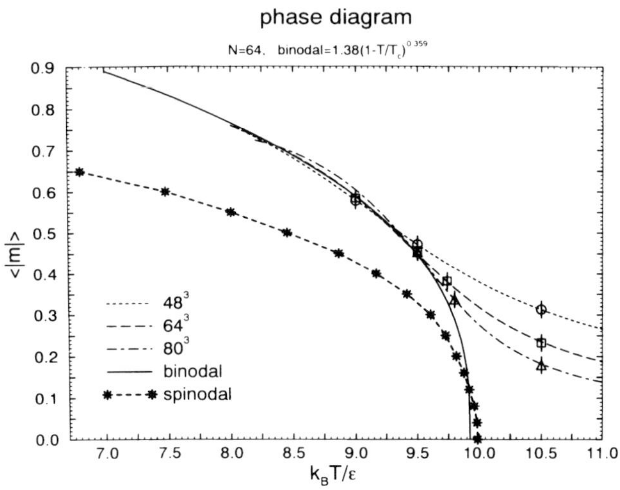

(a)

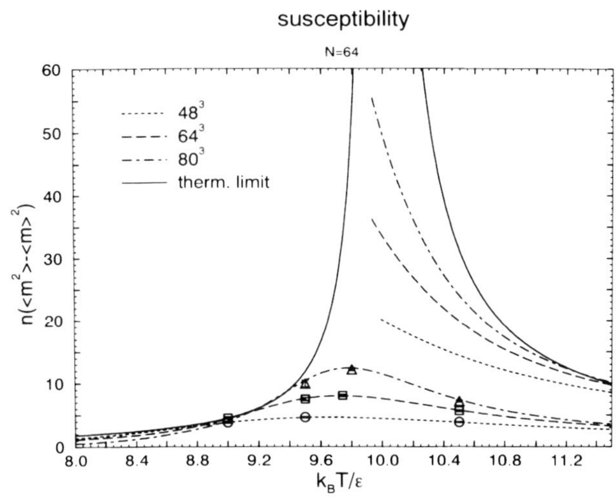

(b)

Fig. 7.7 Order parameter $\langle|m|\rangle$ (a) and order parameter susceptibilities $S_{\text{coll}}(q=0) = n(\langle m^2\rangle - \langle m\rangle^2)$, $S'_{\text{coll}}(q=0) = n(\langle m^2\rangle - \langle|m|\rangle^2)$ (b) plotted vs. temperature for $N = 64$, $\phi_V = 0.5$, and three choices of L as indicated in the figure, using the bond fluctuation model on the simple cubic lattice and a purely repulsive interaction $\epsilon = \epsilon_{AB}$ if two monomers of different kind occur at the shortest possible distance (two lattice spacings), $\epsilon_{AA} = \epsilon_{BB} = 0$. Open symbols (triangles, squares, and circles) show the data directly observed at these temperatures where (always using $\Delta\mu = 0$) the actual simulations were made, including error bars indicating the size of the statistical errors when these exceed the size of the symbols. Broken curves (labeled by the volumes L^3) indicate the multihistogram extrapolation based on the use of all data on $P_T(m)$ at all considered temperatures. In (b) the upper set of curves (only present for $k_BT/\epsilon \geq 10$) refers to $S_{\text{coll}}(q=0)$, while the lower set refers to $S'_{\text{coll}}(q=0)$. Full curves are based on finite size scaling extrapolations towards the thermodynamic limit (see Section 7.2.4). In (a) the broken curve with stars (marked "spinodal") is obtained from a linear extrapolation of $S^{-1}_{\text{coll}}(q=0)$ at constant $\langle m\rangle$ as function of ϵ/k_BT. Note that the order parameter $\langle|m|\rangle$ extrapolated at $\Delta\mu = 0$ to $L \to \infty$ yields the coexistence curve (or "binodal") which near $T_c(k_BT_c/\epsilon = 9.9261)$ is well fitted by $\langle|m|\rangle \cong 1.38(1 - T/T_c)^{0.358}$ in this case. (From Deutsch and Binder.[40])

(for small enough N these should be[9–12] the exponents of the Ising model[8,109]), one can find T_c by extrapolation[78] of the maxima of $S'_{coll}(q=0)$ or the specific heat,[78] (Fig. 7.8). But it is also evident that with a single lattice size[24] only a very rough estimation of the phase transition temperature would be possible. In Section 7.2.4 these finite size effects will be discussed in more detail.

The study of fully symmetric mixtures is relatively simple, since the coexistence curve (and the critical point) occur for $\Delta\mu = 0$, due to the symmetry against interchange of A and B. The problem is more difficult if this symmetry is destroyed, e.g., by making one chain stiffer than the other, choosing an intrachain potential acting differently on the two types of bonds of A-chains and B-chains, or, most simply, using a choice[110–112] $\epsilon_{AA} \neq \epsilon_{BB}$. Choosing $\epsilon_{BB} = -\epsilon_{AB} = -\epsilon, \epsilon_{AA} = \lambda\epsilon_{BB}$ with $\lambda > 0$ one must perform a search in a two-dimensional space $(T, \Delta\mu)$ to locate the critical point $T_c, \Delta\mu_c = \Delta\mu_{coex}(T_c)$, and also locating the phase boundary of the first order transition at $\Delta\mu_{coex}(T < T_c)$ from the A-rich to the B-rich phase is a nontrivial task.

In this case it is useful to record a four-dimensional histogram containing the numbers n_{AA}, n_{BB}, n_{AB} of A–A, B–B and A–B interactions, and the order

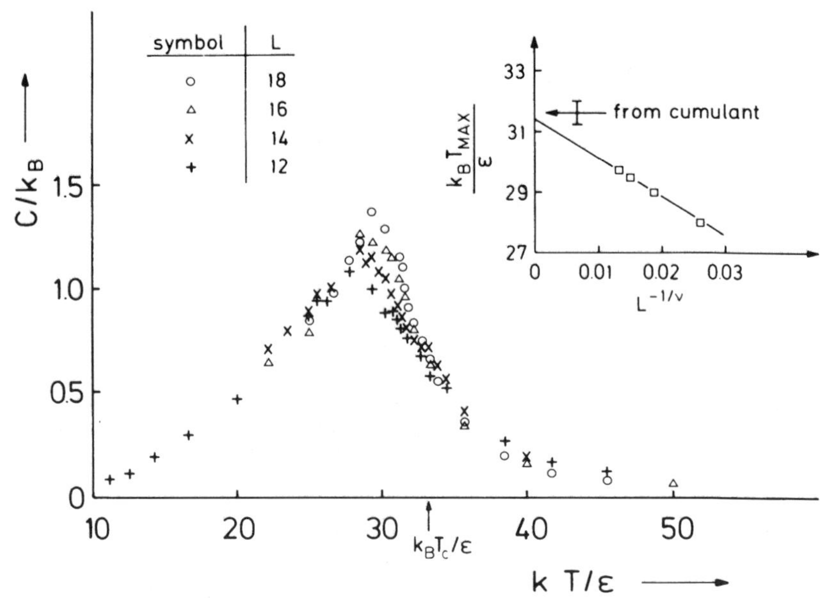

Fig. 7.8 Specific heat per chain C/k_B plotted vs. reduced temperature k_BT/ϵ for the self-avoiding walk model shown in Fig. 7.3 but on the simple cubic lattice, for $N = 32, \phi_V = 0.2, \epsilon_{AB} = 0, \epsilon_{AA} = \epsilon_{BB} = -\epsilon$, and four lattice sizes as indicated. Insert shows that extrapolation of temperatures of the specific heat maxima yields the critical temperature T_c consistent with the cumulant intersection point, if one uses the theoretical correlation length exponent ν of the Ising model,[109] $\nu \approx 0.63$. (From Sariban and Binder.[78])

parameter m. From these histograms, one can obtain all quantities of interest (such as $S'_{\text{coll}}(q=0)$, see Fig. 7.9(a)) in the full $(T, \Delta\mu)$ plane and also carry out extrapolations over a whole regime of values for λ. One sees that $S'_{\text{coll}}(q=0)$ has a maximum along a "ridge" in the $(T, \Delta\mu)$ plane and

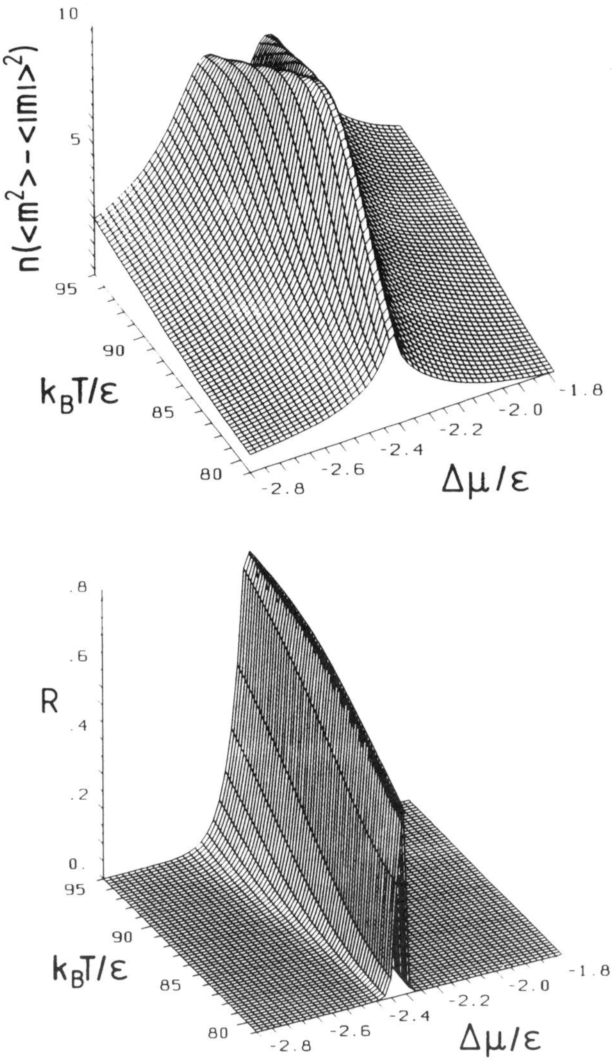

(a)

(b)

Fig. 7.9 The reduced fluctuation of $m, S'_{\text{coll}}(q=0) = n(\langle m^2 \rangle - \langle |m| \rangle^2$ (a) and the ratio R describing phase coexistence (see eq. [7.18]) (b) plotted in the $(T, \Delta\mu)$ plane for an asymmetric polymer mixture (asymmetry parameter $\lambda = 2$) and chain lengths $N_A = N_B = N = 32$ for the bond fluctuation model on the simple cubic lattice (system, size 56^3, $\phi_V = 0.5$). Functions shown are obtained from multihistogram extrapolations using data for $\lambda = 1.0$, 1.1 and 1.5, respectively. (From Deutsch and Binder.[110])

reaches a peak near $(T_c, \Delta\mu_c)$. This ridge already yields a rough estimate for $\Delta\mu_{\text{coex}}(T)$. A more precise estimation is obtained from the probability distribution $P(m)$, which appears to be just as is shown in Fig. 7.6 for a path along the ridge, while for $T < T_c$ either the left peak (representing the B-rich phase) or the right peak (representing the A-rich phase) is quickly suppressed, if one moves at a value $\Delta\mu \neq \Delta\mu_{\text{coex}}(T)$. This observation suggests to define "total probabilities" $P_{\text{A-rich}}, P_{\text{B-rich}}$ for the two phases,

$$P_{\text{A-rich}} = \int_{m^*}^{1} P(m)dm, P_{\text{B-rich}} = \int_{-1}^{m^*} P(m)dm \tag{7.17}$$

m^* being the location of the minimum in between the two peaks of $P(m)$. As an estimate for m^*, one can take the extremum of $S'_{\text{coll}}(q = 0)$. Then we define as ratio R of these two probabilities

$$R \equiv min\{P_{\text{A-rich}}/P_{\text{B-rich}}, P_{\text{B-rich}}/P_{\text{A-rich}}\} \overset{L\to\infty}{\longrightarrow} \begin{cases} 1 & \text{for} \quad \Delta\mu = \Delta\mu_{\text{coex}}(T) \\ 0 & \text{else} \end{cases} \tag{7.18}$$

It is seen (Fig. 7.9(b)) that R is appreciably different from zero only along a rather narrow strip around $\Delta\mu_{\text{coex}}(T)$ in the $(T, \Delta\mu)$ plane. In Section 7.2.4 we shall describe how one uses such data to reliably estimate the coexistence curve.

The most interesting asymmetry, of course, is an asymmetry of chain lengths ($N_A \neq N_B$). In the general case, the semi-grand-canonical technique is not expected to work, however, since the insertion of the longer A-chain replacing the shorter B-chain will almost always be rejected, due to the excluded volume constraint. This is the same difficulty that hampers the measurement of chemical potentials by the Widom test-particle insertion method.[113] Since for short chains this problem can be overcome by the configurational bias Monte Carlo technique,[114–116] it is likely that one can generalize this approach to blends with a small difference in chain lengths. Also the special case where N_A is a simple integer multiple of N_B can be treated[117]: when $N_A = kN_B$, with $k = 2$, 3 or 4, one can replace an A-chain by k B-chains or vice versa. Of course, taking out k B-chains simultaneously and replacing them by an A-chain requires that $2(k - 1)$ chain ends are within the allowed neighborhood for bond vectors, and hence the acceptance rate for such moves rapidly decreases with increasing k. Nevertheless, first results with this technique seem to be rather promising.[117]

7.2.3 *Other ensembles*

As mentioned above, fully grand-canonical ensembles are difficult to use for polymers, due to the problem of inserting chains in a random, unbiased configuration in a dense polymer melt. But of course, the standard canonical ensemble where both n_A and n_B are held fixed can always be used, and has the advantage that it is closest to the experimental reality. In fact, for dynamic studies (interdiffusion,[68,118] early stages of spinodal decomposition[42,80,81,89]) the grand-canonical ensemble would make no sense, while the canonical ensemble can be used.

One can obtain information on the phase behavior in equilibrium from studies such as these in several ways: one approach[117] rests on generalization of techniques for the estimation of chemical potentials in dense polymer systems[114–117,119,120] to chemical potential differences. It turns out that this approach is hampered by slow relaxation effects (which are expected due to both hydrodynamical slowing down[93] and critical slowing down, as discussed above) and hence we are not going into detail here.

Another technique rests on calculating the structure factor $S(q)$. As demonstrated first by Sariban and Binder,[121] though in the framework of a semi-grand-canonical simulation, one can estimate the spinodal curve from a linear extrapolation of $S^{-1}(q \to 0)$ versus ϵ/k_BT: estimating the temperature T where $S^{-1}(q \to 0) = 0$ for various concentrations yields an estimate of the spinodal curve. Figure 7.7(a) did include an estimation of this kind. Although the linearity of this extrapolation is intrinsically a mean-field concept, and thus the critical temperature is slightly overestimated, this technique should yield T_c with relatively small errors for large chain lengths, if one were able to obtain data for $S(q)$ with precision that is good enough.

A rather interesting method,[83] motivated by successful applications to studies of liquid–gas phase separation of Lennard-Jones fluids and other fluids in the canonical ensemble,[122–126] starts from the observation that the temperature variation of the order parameter distribution can be analyzed in subsystems of linear dimension ℓ. Although in the total system (of linear dimension L) the order parameter is conserved, for $\ell \ll L$ neighboring subsystems can freely exchange different types of monomers (and chains). Of course, subsystems must be large enough, so that they still contain many chains.

Figure 7.10 presents, as an example, concentration distributions for a mixture of flexible (A) and stiffer (B) chains of length $N = 10$ on the face-centered cubic lattice. It is seen that the distribution of the volume fraction of the monomers belonging to the stiffer chains resemble nicely a Gaussian around the average volume fraction (chosen as $\bar{\phi}_s = 1/2$ here) at the highest temperature shown, $T = 0.25$, where the stiffness is least pronounced. As T is lowered and the stiffness of the B-chains increases, a double peak structure clearly develops. It is seen that unlike Fig. 7.6 the distributions are very asymmetric, i.e., one peak occurs nearly at $\phi_s = 1$, while the position of the other peak is strongly temperature-dependent. This means that in the regions rich with stiff chains (which are expected to be aligned in a liquid–crystalline type order[127–130]) almost none of the flexible chains are dissolved, unlike the regions rich in flexible chains, which can accommodate

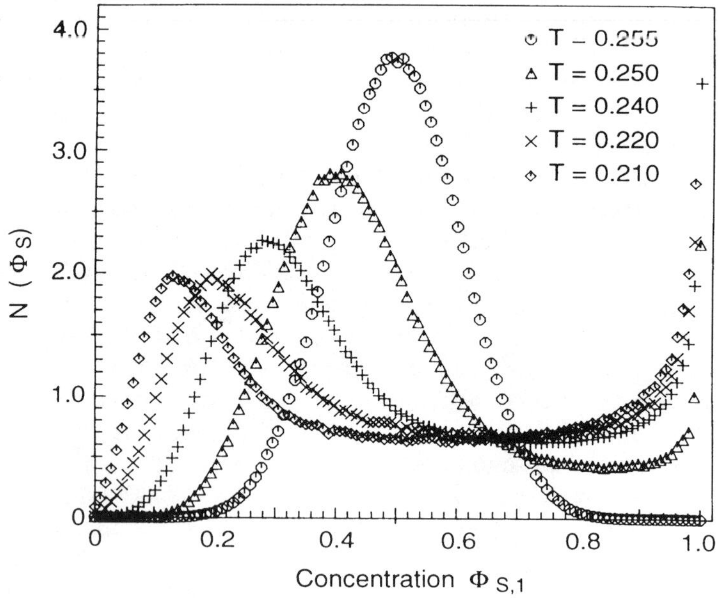

Fig. 7.10 Distribution function $N(\phi_s)$ of the volume fraction of stiff chains at several temperatures, for a subsystem of linear dimension $\ell = 8$ of a total system with linear dimension $L = 30$, using a face-centered cubic lattice with $\phi_V = 0$ and the cooperative motion algorithm. Both chain lengths of the stiff (s) and flexible (f) chains are equal, $N_f = N_s = N = 10$, and their volume fractions are equal and constant, $\bar{\phi}_s = \bar{\phi}_f = 1/2$. For the stiff chains, energies 0, 0.5, 1.0, 1.5 apply for angles $\pi, 2\pi/3, \pi/2$ and $\pi/3$ between successive bonds. Here temperature is measured in energy units. (From Gauger and Pakula.[83])

a certain fraction of stiff chains. One should note that the separation between A- and B-chains in Fig. 7.10 is purely entropy-driven, i.e., no energy between different chains was assumed, irrespective of their character: $\epsilon_{AA} = \epsilon_{BB} = \epsilon_{AB} = 0$. Note that the flat part of the distribution in between the peaks is attributed to interfacial contributions, which are a difficulty of this subsystem approach as discussed by Rovere *et al.*[124]

For small molecule systems, the most popular ensemble to study phase coexistence is the "Gibbs ensemble".[2,131–135] For binary (AB) mixtures below T_c, this amounts to simulating two systems (i.e., two simulation boxes) which can exchange particles (and volume, in the case of an off-lattice model). Thus both systems are in full thermal equilibrium with each other, i.e., they are at the same temperature, pressure and the same values of the chemical potentials μ_A, μ_B. For polymers, the use of this ensemble in its original form meets the same difficulties that grand-canonical simulations do (moves where chains are to be inserted in an already very dense system almost always will be rejected). Of course, for symmetrical mixtures one can resort to a semi-grand-canonical variation of the Gibbs ensemble[69,136]: it is attempted to exchange an A-chain in Box 1 and a B-chain in Box 2. Usually one works only well below T_c, where one box always stays in the A-rich phase and the other box in the B-rich phase (Fig. 7.11). While this type of analysis works well for strongly unmixed phases, the extrapolation into the critical region clearly is somewhat uncertain. But Fig. 7.11 contains at least qualitative evidence that the critical temperature of phase separation is suppressed with decreasing film thickness (for a model where neither component is preferentially attracted towards the confining walls).

7.2.4 *Finite size scaling*

As is well known, sharp phase transitions do occur only in the thermodynamic limit[95–98]: in our case the number n of chains in the considered system should tend to infinity. Computer simulations usually consider very small systems (where n is of the order 10^2 to 10^3), and hence the finite size rounding and shifting of the transition needs to be seriously considered. In fact, Figs 7.6–7.8 provide typical examples: the distribution of the order parameter m (eqs [7.4], [7.5]) of an unmixing transition in a polymer blend changes completely gradually from a single-phase Gaussian above T_c to a

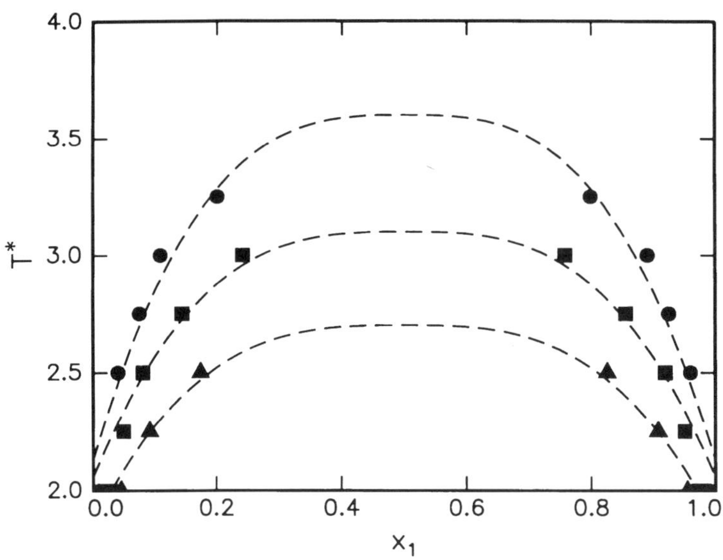

Fig. 7.11 Phase diagrams for a symmetrical off-lattice mixture with $N_A = N_B = N = 20$, where both components are modeled as bead-rod chains, and all nonbonded beads interact with standard Lennard-Jones potentials which are truncated at 2.5σ, for the choice of interaction parameters $\sigma_{AA} = \sigma_{BB} = \sigma_{AB} = \sigma$, $\epsilon_{AA} = \epsilon_{BB} = \epsilon$, $\epsilon_{AB} = 0.9\epsilon$, $T^* = k_BT/\epsilon$, for a monomer density $p^* = \rho\sigma^3 = 0.7$ in each box. Here $x_1 = \phi_A/(\phi_A + \phi_B)$, and data are shown for a bulk system (full dots) and thin films with repulsive walls for thickness 10.5σ (squares) and 5σ (triangles). Lines represent a fit according to $x_1 - x_{1c} \propto (T - T_c)^{1/3}$. (From Kumar *et al.*[69])

double-peak distribution below T_c when the temperature is lowered. While from data from finite systems one can define various pseudo-critical temperatures $T_c(L)$—e.g., the temperature where the specific heat (Fig. 7.8) or $S'_{\text{coll}}(q \to 0)$ (Fig. 7.7(b)) have maxima, or where $P(m)$ starts to develop its central minimum—these temperatures $T_c(L)$ do not agree with each other; they depend on the quantity that is considered and on the linear dimension L.

However, for large enough L where finite size scaling holds, one can show that certain reduced moment ratios should intersect exactly at T_c. This is seen by noting[97] that the linear dimension L scales with ξ, the correlation length of concentration fluctuations[8–12]

$$\xi = \hat{\xi}_\pm |t|^{-\nu}, t \to 0, \tag{7.19}$$

where the two signs $\pm$ of the critical amplitude $\hat{\xi}_\pm$ refer to the sign of $T/T_c - 1 (= -t)$, and ν is the critical exponent ($\nu \approx 0.63$).[109] This statement means that the kth moment $\langle |m|^k \rangle = \int_{-1}^{+1} |m|^k P(m) dm$ of the order parameter distribution $P(m)$ does not depend on L and the temperature distance t from the critical point separately, but rather in the scaled form

$$\langle |m|^k \rangle = L^{-k\beta/\nu} \tilde{M}_k(L/\xi), \tag{7.20}$$

where $\tilde{M}_k(L/\xi)$ is the scaling function and β the critical exponent of the order parameter,

$$\langle |m| \rangle_{L \to \infty} \to \hat{B} t^\beta, t \to 0, \tag{7.21}$$

$\hat{B}$ being the associated critical amplitude. Note that eq. (7.20) reduces for $k = 1$ to eq. (7.21) since $\tilde{M}_1(z \gg 1) \propto z^{\beta/\nu} = L^{\beta/\nu}\xi^{-\beta/\nu}$, using also eq. (7.19). From eq. (7.20) one now concludes that reduced moment ratios (where $k\ell = ij$)

$$U_{ij}^{k\ell} \equiv \langle |m|^k \rangle^\ell / \langle |m|^i \rangle^j = \tilde{U}_{ij}^{k\ell}(L/\xi) \tag{7.22}$$

depend on the ratio L/ξ only. Since $\xi \to \infty$ for $|t| \to 0$, eq. (7.19), for all L the ratios $U_{ij}^{k\ell}$ must yield the same value $\tilde{U}_{ij}^{k\ell}(0)$, and hence plotting $U_{ij}^{k\ell}$ as function of T, all curves for the different choices of L must intersect in one common intersection point, as announced above. The temperature of this intersection point is an estimate for the critical temperature T_c.

Figure 7.12 tests these concepts using data for symmetrical polymer mixtures for $N = 64$ and $N = 256$, in the framework of the bond fluctuation model with interactions between nearest-neighbor effective monomers.[40] For chains not too long ($N \leq 64$) there is indeed a common intersection point, and T_c can be estimated with high accuracy ($k_B T_c/\epsilon = 9.9261$).

At first sight it is surprising that for larger chain length (e.g., $N = 256$, Fig. 7.12(b)) there is not such a well-defined intersection property, but rather, with increasing L the intersection point shifts somewhat to a lower

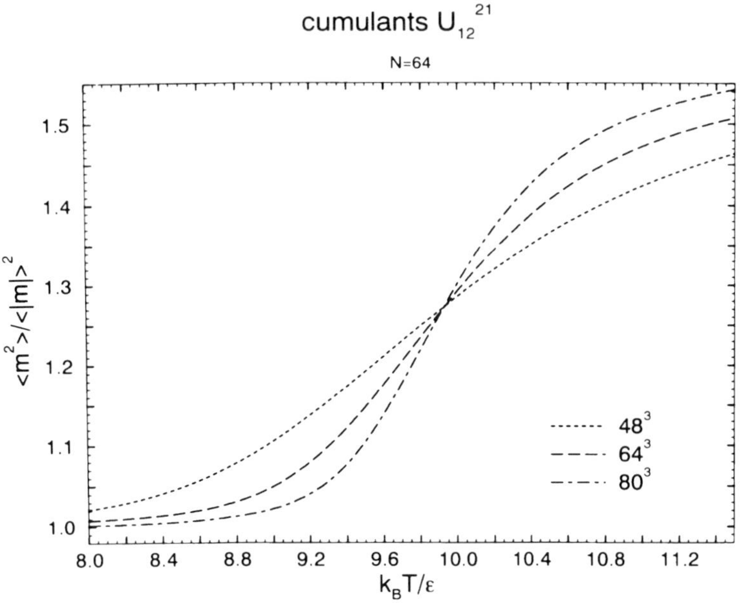

(a)

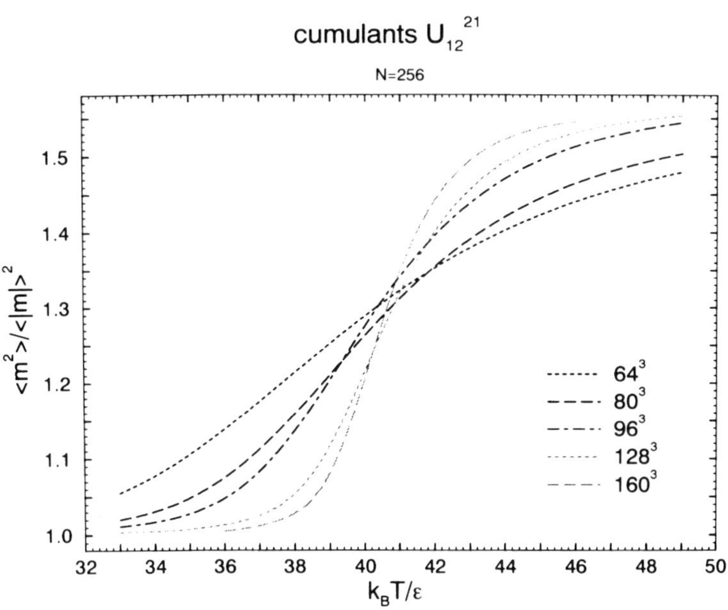

(b)

Fig. 7.12 (a) Ratio of moments $U_{12}^{21} = \langle m^2 \rangle / \langle |m| \rangle^2$ for the symmetric bond fluctuation model of polymer mixtures at $\phi_V = 0.5$ plotted vs. temperature for $N = 64$ (a) and $N = 256$ (b). The model is the same as in Fig. 7.7. Different curves result from histogram extrapolations for different choices of L as indicated in the figure. (From Deutsch and Binder.[40])

value of U_{12}^{21} and to a somewhat lower temperature. It turns out that these deviations from finite size scaling are not due to insufficient equilibration and/or statistics, but reflect a basic physical aspect of the problem: the crossover from Ising-like critical behavior to mean-field critical behavior as $N \to \infty$.[9–12,40,137–140]

This crossover is physically due to the fact that in a dense three-dimensional polymer melt the chain configurations are random-walk-like, i.e., the gyration radius behaves as $R_{\text{gyr}} \approx a\sqrt{N}$, a being the size of an effective monomer, and hence the density of one chain within its own volume ($V_{\text{chain}} \propto R_{\text{gyr}}^3$) is $\rho_{\text{chain}} = N/V_{\text{chain}} \propto a^{-3}N^{-1/2}$. Since the melt density is of order a^{-3}, it follows that there must be monomers of $N^{1/2}$ other chains in the same volume, i.e., each chain interacts with $N^{1/2}$ other "neighbors". This situation is reminiscent of an Ising spin model where each spin has a coordination number $N^{1/2}$, and thus clearly for $N^{1/2} \to \infty$ mean field theory becomes correct.[9]

A more quantitative description of this problem shows that one can describe this crossover by comparing the reduced temperature distance t to the so-called Ginzburg number[40,141,142] Gi, which for the Flory–Huggins theory[20] of a symmetric polymer mixture reduces to[40]

$$Gi \approx 2.3/N. \tag{7.23}$$

For $|t| \gg Gi$ one has mean-field behavior, i.e., the correlation length ξ scales as

$$\xi \propto a\sqrt{N}|t|^{-\nu_{MF}}, \nu_{MF} = 1/2, \tag{7.24}$$

while in the inverse limit ($|t| \ll Gi$) eq. (7.19) holds, with $\hat{\xi}_{\pm} \propto N^{1-\nu} \approx N^{0.37}$. For $|t| \approx Gi$, i.e., $\xi \approx \xi_{\text{cross}} \approx aN$ a smooth crossover between eqs. (7.19) and (7.24) occurs (Fig. 7.13).

Now there is an additional complication: in the mean-field critical region of polymers there is no finite size scaling of the form postulated in eqs (7.20)–(7.22); L does not scale with ξ but rather with a "thermodynamic length" ℓ defined by[40,143,144]

$$\ell \approx [\langle|m|\rangle^{-2} S'_{\text{coll}}(q=0)]^{1/3} \propto aN^{1/3}t^{-2/3}, \tag{7.25}$$

where we have used the critical properties of Flory–Huggins theory[10–12,78]

$$\langle|m|\rangle = \hat{B}_{MF}t^{\beta_{MF}}, \hat{B}_{MF} = \sqrt{3}, \beta_{MF} = 1/2, \tag{7.26}$$

and

$$S'_{\text{coll}}(q=0) = \hat{C}_{-}^{MF}t^{-\nu_{MF}}, \nu_{MF} = 1, \hat{C}_{\pm}^{MF} \propto Na^3. \tag{7.27}$$

As in eq. (7.19), the two signs $\pm$ of the critical amplitude $\hat{C}_{\pm}^{MF}$ refer to the sign of $T/T_c - 1$. Equation (7.25) is made plausible by approximating the

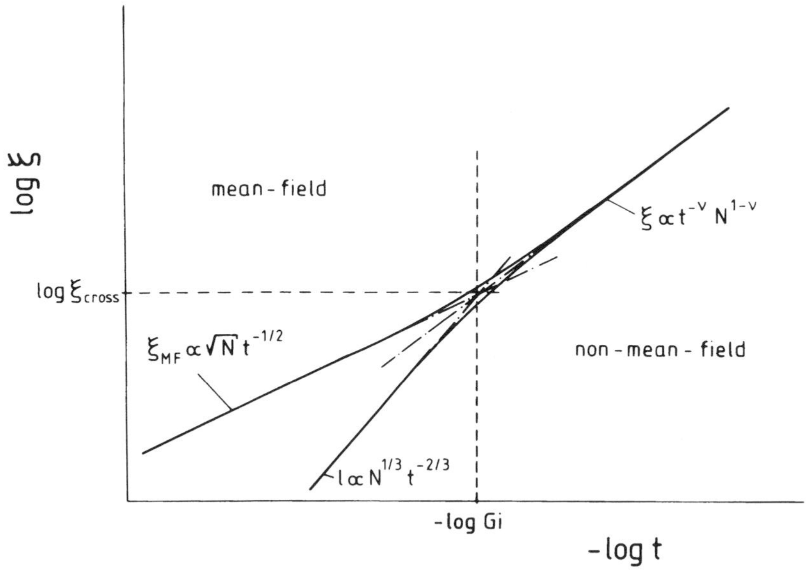

Fig. 7.13 Crossover scaling of characteristic lengths for a polymer mixture. For $t \gg Gi$, i.e., $-\log t$ to the left of $-\log Gi$, one has mean-field behavior, with two characteristic lengths $\xi_{MF} \propto \sqrt{N} t^{-1/2}$ and $l \propto N^{1/3} t^{-2/3}$. These lengths smoothly merge in the crossover regime $t \approx Gi$) and in the non-mean-field critical regime a single characteristic length $\xi \propto N^{1-\nu} t^{-\nu}$ takes over. (From Deutsch and Binder.[40])

two peaks in Fig. 7.6, describing the coexisting A-rich and B-rich phases, as simple Gaussians

$$P(m) \propto \exp\left[-\frac{(m \pm \langle|m|\rangle)^2 L^3}{2S'_{\text{coll}}(q=0)}\right] = \exp\left[-\frac{1}{2}\left(\frac{m}{\langle|m|\rangle} \pm 1\right)^2 (L/\ell)^3\right], \quad (7.28)$$

where we have used the fact that it is the structure factor $S'_{\text{coll}}(q=0)$ that controls the width of the distribution, via a fluctuation relation.

As a consequence of eqs (7.25) and (7.28), in the mean-field critical region of a polymer mixture eq. (7.20) does not hold, and should rather be replaced by

$$\langle|m|^k\rangle = L^{-kv} \tilde{\tilde{M}}_k(L/\ell), v = 3\beta_{MF}/(\gamma_{MF} + 2\beta_{MF}) = 3/4. \quad (7.29)$$

While in the mean-field critical region we hence expect for the maximum value of the scattering function $S'_{\text{coll}}(q=0) = n(\langle m^2\rangle - \langle|m|^2\rangle \propto (L^3/N)$ $(\langle m^2\rangle - \langle|m|\rangle^2)$ a behavior $[S'_{\text{coll}}(q=0)]_{\max} \propto L^{3-2v}/N = L^{3/2}/N$, in the non-mean-field critical regime we should have [from eq. (7.20)] $[S'_{\text{coll}}(q=0)]_{\max} \propto L^{3-2\beta/\nu}/N \approx L^{1.97}/N$, using the Ising model critical exponents[8,109] $(\beta/\nu \approx 0.515)$. While the latter law for short chains is well fulfilled, even long chains such as $N = 256$ and $N = 512$ do not yet show the mean-field exponent 1/2 (Fig. 7.14[a],[40] but rather an "effective exponent" occurs ($v_{\text{eff}} \approx 1.73$) in between the Ising and mean-field values. However, combining data for all chain lengths studied in a crossover scaling plot (Fig. 7.14[b]),[40,145] better evidence for the theoretical exponents can be obtained.

These examples show already that the accurate estimation of T_c and the other critical properties of a symmetrical polymer mixture is a delicate matter, if one wishes to study rather long chains, comparable to experimental molecular weights.[139,140] Note that it would not easily be possible to go to much longer chains (the maximum chain length so far used is[40] $N = 512$) and lattice sizes (the largest lattice used so far is a simple cubic lattice with[40] $160^3 = 4\,096\,000$ sites, containing for $\phi_v = 0.5$ in total 256 000 effective monomers): altogether an effort of about 3000 hours CPU at a CRAY-YMP was spent, with about one half of this time being needed to equilibrate the initial starting configurations of a homopolymer melt for the various lattice sizes.

When the "effective exponent" v (Fig. 7.14[a]) has been obtained as well as T_c (Fig. 7.12), one fits another effective exponent u (which would be equal to $1/\nu$ in the Ising critical regime) by plotting $L^v\langle|m|\rangle$ versus the variable $L^u t$ and choosing u such that an optimal fit on a "master curve" is obtained, since eq. (7.20) can also be written in the form (for $k = 1$) $\langle|m|\rangle L^{\beta/\nu} = \tilde{M}_1(L/\xi) = f_1(L^{1/\nu}t)$, where f_1 is another scaling function. In fact there are other, more direct, methods to estimate u; we refer for a discussion of this approach to the original papers.[27,78,79,108] From the behavior of the master curve $f_1(L^{1/\nu}t)$ for $L^{1/\nu}t \gg 1$, $f_1 = \hat{B}(L^{1/\nu}t)^{\beta}$, one can estimate the critical amplitude $\hat{B}$ of the order parameter, eq. (7.21). In this way finite size scaling theory provides a systematic method to extrapolate data from small lattices towards the thermodynamic limit. This is how the coexistence curve (binodal) described by eq. (7.21) was constructed in Fig. 7.7(a). These techniques carry over to the response functions $S_{\text{coll}}(q = 0)$, $S'_{\text{coll}}(q = 0)$ as well.[27,78,79,108] We are not going into details here, however, because in the general context of critical phenomena such methods are well documented.[89,93,96–98]

It remains to briefly mention how this approach is carried over to asymmetric polymer mixtures, where the critical point can only be located by a search in a two-dimensional space $(T, \Delta\mu)$, see Figs 7.9(a), (b). It was already shown (Fig. 7.9[b]) that the coexistence between A-rich and B-rich phases at $\Delta\mu = \Delta\mu_{\text{coex}}(T)$ can be located rather precisely by the ratio R characterizing the double-peak structure of the distribution function $P(m)$. However, these data do not yet give an indication where the critical point T_c occurs along the ridge in Fig. 7.9(b). This happens because above T_c there is also a value of $\Delta\mu$ where the probabilities $P_{\text{A-rich}}$, $P_{\text{B-rich}}$ defined in eq. (7.17) have equal magnitude, even though $P(m)$ then may have only a single peak.

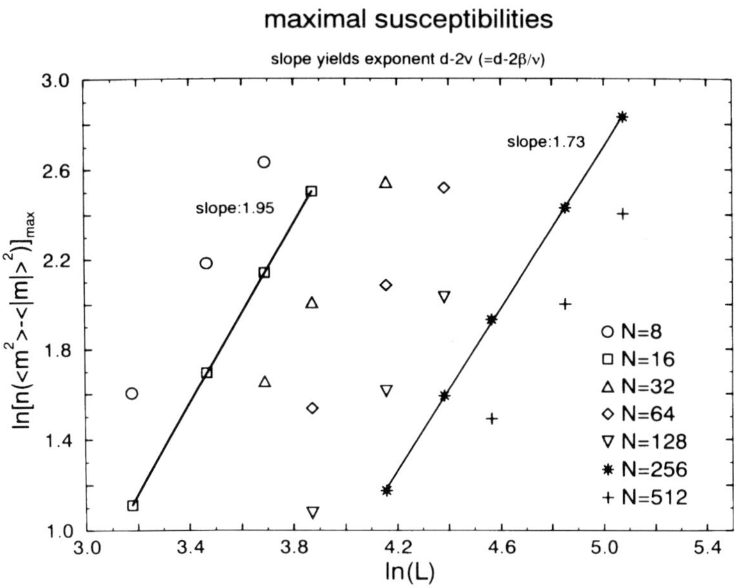

(a)

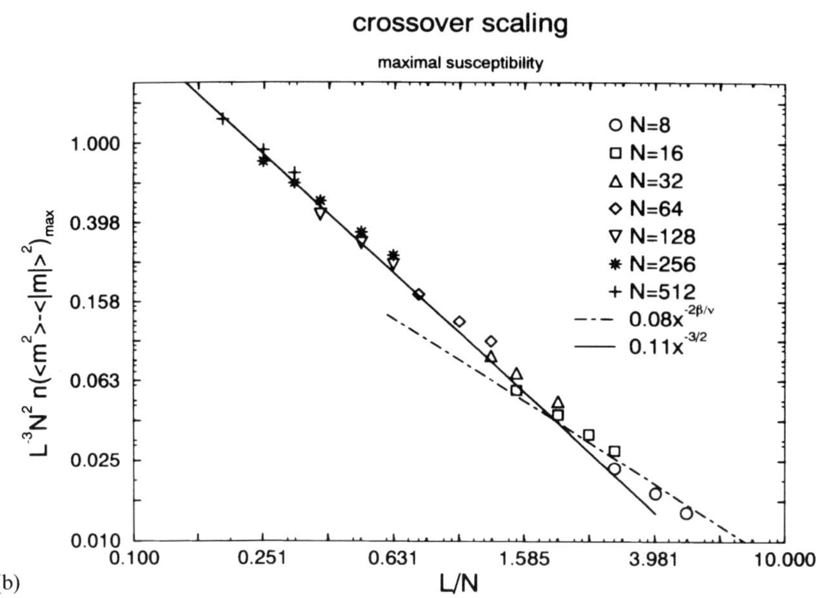

(b)

Fig. 7.14 (a) Log–log plot of the maximum value of the response function $n[\langle m^2\rangle - \langle |m|\rangle^2]$ versus L, for the model of Fig. 7.7(b). Different symbols denote various chain lengths N, as indicated in the figure. Straight lines (with the slopes 1.95 and 1.73, respectively) indicate power-law behavior with "effective" exponents. (b) Same data as in (a) but replotted as a finite size scaling crossover plot, $[S'_{\text{coll}}(q=0)]_{\max}$ normalized by L^3/N^2 is plotted as a function of L/N. Full and broken straight lines indicate fits to power laws which have the theoretical exponents $-2\beta/\nu \approx -1.03$ or $-2v = -3/2$, respectively. (From Deutsch and Binder.[40])

Thus one can view the line of maximum R in Fig. 7.9(b) in an asymmetric mixture as the analog of the temperature axis of a symmetric mixture below and above T_c.

However, the endpoint T_c of the first-order transition line at $\Delta\mu = \Delta\mu_{\text{coex}}(T)$ can now be estimated, looking for an intersection of the moment ratios $U^{ij}_{k\ell}$ for different lattice sizes along a path of maximal R in the $(T, \Delta\mu)$ plane. These ideas[110,111] indeed allow one to estimate T_c with reasonable precision (Fig. 7.15[a]). Using these values of T_c and the order parameter (Δm) in a finite size scaling analysis, one can obtain the coexistence curve and critical concentration very accurately (Fig. 7.15[b]).[110]

The raw data for the volume fractions of the coexisting phases follow in full analogy to eq. (7.17) from

$$m_{\text{A-rich}} = \int_{m^*}^{1} mP(m)dm, m_{\text{B-rich}} = \int_{-1}^{m^*} mP(m)dm, \Delta m = m_{\text{A-rich}} - m_{\text{B-rich}} \tag{7.30}$$

and using the translation from the order parameter m to the volume fractions ϕ_A, ϕ_B (eq. [7.5]) one obtains the phase diagram as shown in Fig. 7.15(b). Again it is clear that the "raw data" have a dramatic size effect, for temperatures $4k_BT/[3+\lambda)\epsilon]$ exceeding 65, and it would be difficult to estimate reliably the critical point if only one lattice size were available.

7.2.5 Technical problems of simulations of block copolymer mesophases

Simulating mesophase formation in block copolymers, one encounters similar problems as for the simulation of macroscopic phase separation in polymer blends. Again the need arises to equilibrate both the local structure of the chain configurations and the long wavelength collective properties.

Minchau, Fried and coworkers[52–55] have combined the "slithering snake" algorithm (Fig. 7.16[a]) with a head–tail interchange move (Fig. 7.16[b]). Their lattice model of the chains is fully analogous to Fig. 7.3, describing chains by self-avoiding walks on the simple cubic lattice, and choosing pairwise interactions $\epsilon_{AA}, \epsilon_{BB}$ and ϵ_{AB} between effective monomers on nearest-neighbor sites. The only distinction now is that a fraction f of the monomers of each chain are A-monomers, and a fraction $1-f$ are B-monomers. This identity of the monomers must be properly taken into account in the moves of Fig. 7.16. For example, in addition to the change of the end link in the standard "slithering snake" algorithm[146,147] one must also shift the A–B junction point along the chain (Fig. 7.16[a]). Both moves in Fig. 7.16 are only accepted with a probability $P = \min\{\exp(-\Delta E/k_BT), 1\}, \Delta E$ being the energy change due to the move.

For block copolymers, the average composition of the system is fixed and equal to f, just as the composition of each individual copolymer chain.

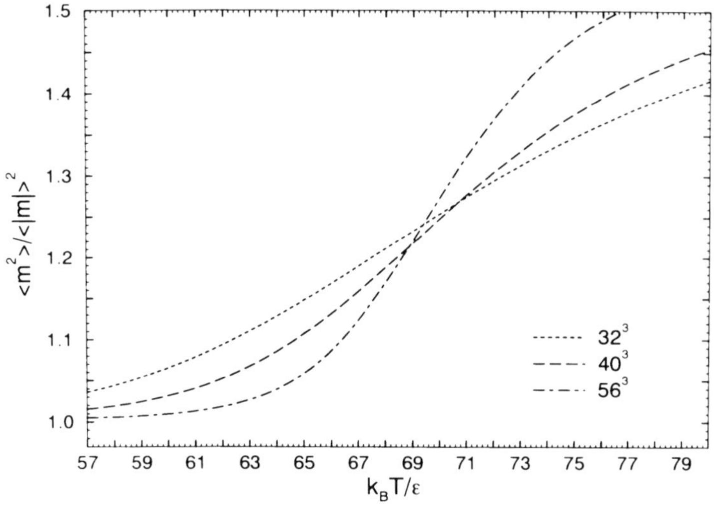

(a)

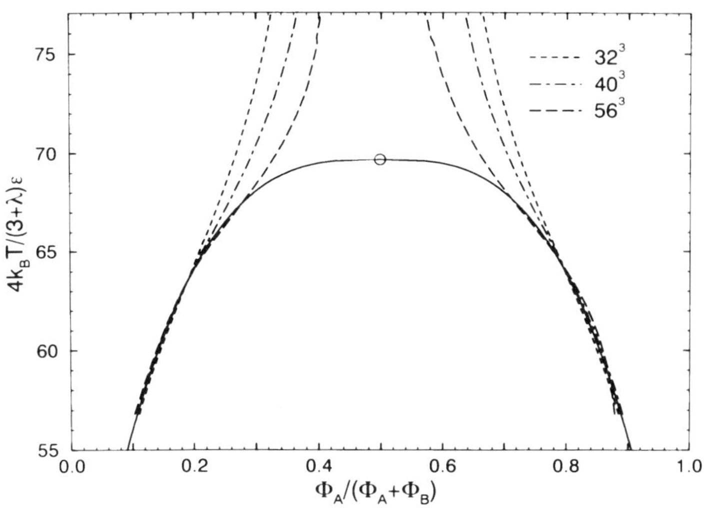

(b)

Fig. 7.15 (a) Plot of $U_{12}^{21} = \langle m^2\rangle/\langle|m|\rangle^2$ versus reduced temperature, along the line of maximal R (eq. [7.18]) in the $(T, \Delta\mu)$ plane, for the same model as shown in Fig. 7.9 ($N_A = N_B = N = 32$, asymmetry parameter $\lambda = \epsilon_{AA}/\epsilon_{BB} = 2$). Three different linear dimensions are shown. From the intersections critical parameters are estimated as $4k_BT_c/[(3+\lambda)\epsilon] = 69.7 \pm 1.0, \Delta\mu_c/\epsilon = -2.327 \pm 0.02, \phi_A^{\rm crit}/(\phi_A + \phi_B = 0.498$. (b) Phase diagram of the asymmetric polymer mixture ($N = 32, \lambda = 2$) in the plane of variables $\{T, \phi_A/(\phi_A + \phi_B)\}$. The dashed lines are the histogram extrapolations for $m_{\rm A\,rich}$ (right) and $m_{\rm B\,rich}$ (left) for the three simulated system sizes. The full line denotes the two branches of the coexistence curve, following from the finite size scaling analysis, and the circle denotes the critical point. (From Deutsch and Binder.[110])

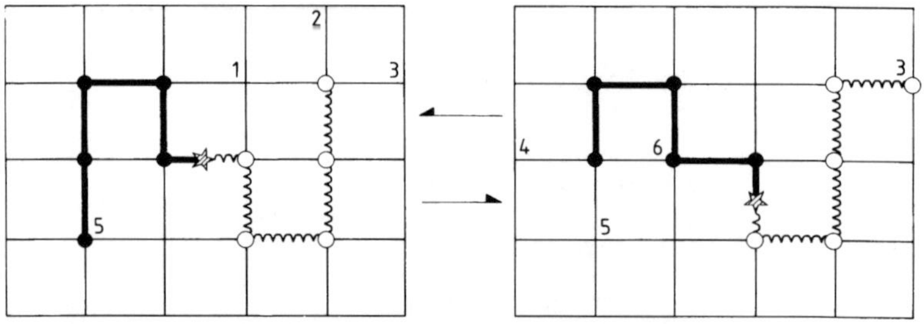

(a)

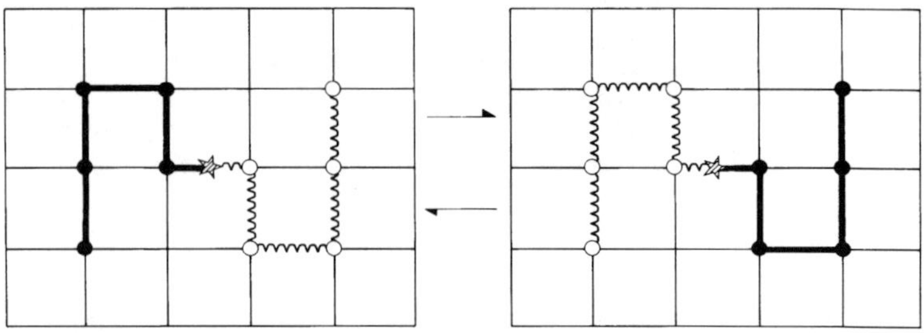

(b)

Fig. 7.16 Examples of a "slithering snake" motion (a) and of an A–B interchange motion (b) for a model of symmetric diblock copolymers ($f = 1/2$) of chain length $N = 10$. The arrows in both figures signify that motions in both directions are considered at random. The lattice sites 1,2,3 are the possible new locations (on a square lattice) for the end monomer of the white block (upper part, left); on the right part of this panel the vacant site labeled 3 has been chosen as the new location for this monomer. The remaining monomers advance along the chain one lattice constant. For the inverse move, the end monomer of the block could move to sites 4 or 5 but not to site 6, due to the excluded volume constraint.(From Fried and Binder.[53])

Hence simulations can be carried out at fixed values of f only; an analog to the semi-grand-canonical technique (where $\Delta\mu$ rather than the composition is fixed) does not exist. As a consequence, it is very important that long-wavelength composition fluctuations relax fast enough. It is this consideration that led Fried *et al.*[52–55] to choose the moves shown in Fig. 7.16: although not microscopically realistic, the "slithering snake" algorithm leads to a particularly fast self-diffusion of the chains: in the absence of backjump correlations, the self-diffusion constant D_N is independent of N, if one measures the time in units of N attempted moves per chain, since in each move the center of gravity moves a distance in the order of the radius divided by N, i.e., $|\Delta\vec{x}_{CG}| \propto a\sqrt{N}/N$, and then $D_N \propto (\Delta\vec{x}_{CG})^2 N \propto a^2$. And the move of Fig. 7.16(b), if it is accepted, should lead to a particularly fast relaxation of Fourier components of the concentration with wavelength comparable to the chain radii and hence comparable to the wavelength of the lamellar ordering.

In practice, the slithering snake move in Fig. 7.16(a) is slowed down if the vacancy concentration ϕ_v is small, and thus Fried *et al.*[52–55] used

$\phi_v \geq 0.2$. With the cooperative motion algorithm, simulations with $\phi_v = 0$ were also performed.[24,84,85] As for the case of polymer blends,[79] one finds that the radii of the chains increase with increasing vacancy content of the system. However, this trend occurs already in the absence of any interactions between A and B monomers,[24,78,148] and thus it is of interest to check for the effect of temperature on the linear dimensions of the radii measured in units of the radii of athermal chains, Fig. 7.17(a). However, within the range of scatter of the data the radii seem to coincide on a master curve for all vacancy concentrations, at least for temperatures $T > T_c$. The phase transition temperatures in this study have been estimated only rather roughly, by locating the specific heat maxima for a single lattice size (Fig. 7.17[b]). Consistent with earlier findings for a different model,[53] there is a decrease of the transition temperature with increasing vacancy content (Fig. 7.17[b]). Since, as will be explained below, a reliable study of the ordered phase for block copolymers is very hard to perform, and finite size effects and finite observation time effects are very difficult to handle for $T < T_c$, it is not clear whether the deviations between data for $\phi_v = 0$ and $\phi_v \neq 0$ for $T < T_c$ in Fig. 7.17(a) are a physical effect, or due to such limitations.

In principle, one could generalize the considerations of Section 7.2.4 to a study of block copolymer mesophase ordering, by suitable identification of the order parameter of the transition. For the lamellar mesophase (Fig. 7.2), this order parameter is written in terms of concentration waves for the local concentration of component A, for instance

$$\phi_A(\vec{x}) = A \exp(iq^*\vec{n} \cdot \vec{x} + \varphi), \; q^* = 2\pi/\lambda^*(T) \tag{7.31}$$

where A is the amplitude and φ the phase of the (complex) order parameter, and the wavevector $\vec{q}^* = q^*\vec{n}$ describes both the wavelength $\lambda^*(T)$ of the lamellar ordering and the orientation of the lamellae ($\vec{n}$ being a unit vector perpendicular to the lamellae). This order parameter obviously is much more complicated than the simple order parameter m of an unmixing transition (eqs [7.4], [7.5].)

From eq. (7.31), the difficulties encountering the study of such ordering phenomena from simulations in small boxes can easily be deduced: for a given choice of box geometry and boundary conditions, the linear dimension L will in general not be commensurate with the wavelength $\lambda^*(T)$—only at special temperatures L can there be an integer multiple of $\lambda^*(T)$, but since this temperature-dependent wavelength is not known beforehand, a serious distortion of the ordering due to the lack of commensurability must be expected.

This fact has been explicitly demonstrated by simulating the block copolymer model of Fig. 7.16 in a $L \times L \times D$ thin film geometry,[149] where at the two repulsive walls of linear dimensions $L \times L$ an energy parameter $\epsilon_{AS} > 0$ was applied, which is repulsive to the A-monomers only. Choosing in the bulk $\epsilon_{AA} = \epsilon_{BB} = 0$, $\epsilon_{AB} > 0$, and $\epsilon_{AS} = \epsilon_{AB}/2$, there is a tendency for the

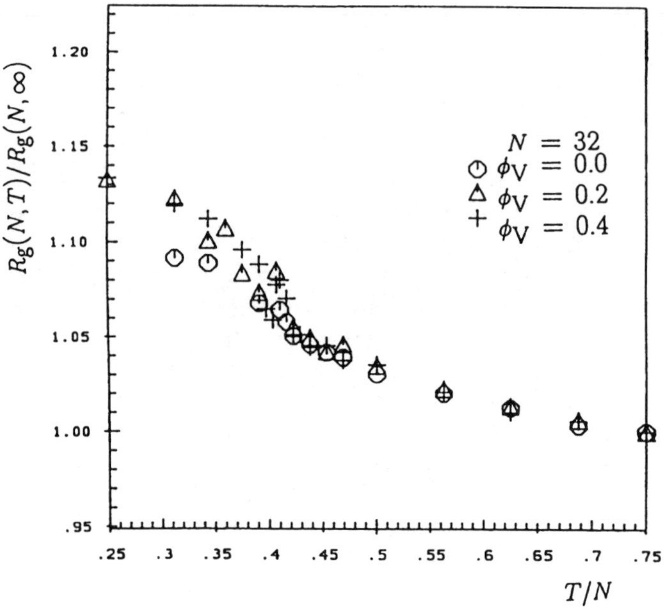

(a)

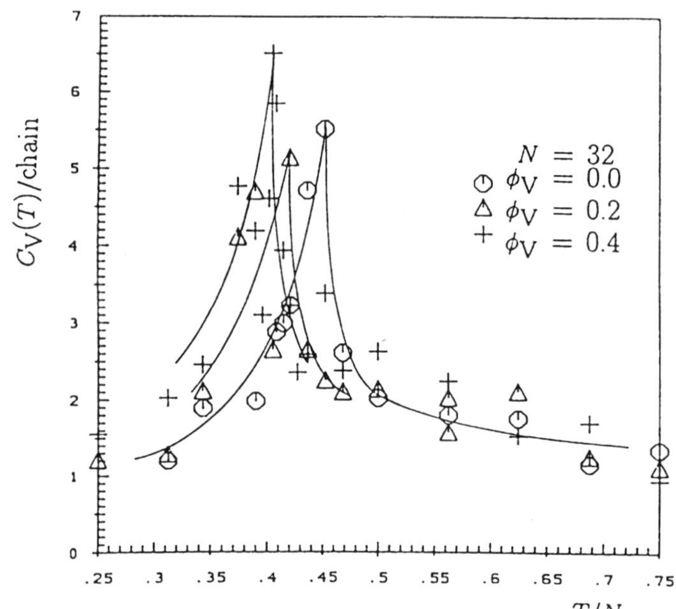

(b)

Fig. 7.17 (a) Radius of gyration for symmetrical block copolymers ($f = 1/2$) of chain length $N = 32$ on the face-centered cubic lattice, for the three different vacancy concentrations $\phi_V = 0, 0.2$ and 0.4, respectively, and normalized by their radii at infinite temperature. Energy parameters and units are chosen as $\epsilon_{AA} = \epsilon_{BB} = 0$, $\epsilon_{AB}/k_B = 1$. All data are for lattice linear dimensions $L = 20$, choosing a bond length $\sqrt{2}$, with 4000 $(1 - \phi_V)$ monomers per system. (b) Specific heat versus temperature for the same model as in (a); curves are guides to the eye only. (From Weyersberg and Vilgis.[24])

block copolymers to form B-rich layers adjacent to the walls, and hence the lamellar ordering occurs preferentially with lamellae oriented parallel to the walls. Assuming thus that the unit vector $\vec{n}$ in eq. (7.31) is perpendicular to the walls, one can check eq. (7.31) directly, by simply recording the profile of the density difference $\rho_A(z) - \rho_B(z)$ as a function of the coordinate z in the direction perpendicular to the walls (Fig. 7.18). Indeed, one can see that for $D = 10$, 18 a wave-like structure of the type assumed in eq. (7.31) develops, with $\vec{n}$ oriented along the z-axis, stabilized by the special boundary conditions described above. Near the ordering temperature it seems that for $D = 10$ a single wavelength $\lambda^*(T)$ just fits, while for $D = 18$ two wavelengths fit.Thus it is no surprise that for $D = 14$ the thickness is incommensurate with this lamellar structure, and thus only a rather weak modulation occurs (simply due to the attraction of B-monomers to the walls which then are removed from an adjacent depletion layer). A closer examination shows[150] that this system with $D = 14$ also orders, but now the vector $\vec{n}$ is tilted relative to the z-axis, and thus $\rho_A(z) - \rho_B(z)$ is no longer sensitive to the ordering. In fact, snapshot pictures of configurations of block copolymer systems in $L \times L \times L$ geometry with periodic boundary conditions sometimes also show a tilting of $\vec{n}$ with respect to the lattice directions.[24,84]

Using such data for the order parameter profiles in thin films for a large range of temperatures T and thickness D, one can draw a contour plot of constant values of the order parameter in the center of the film (Fig. 7.19).[149] One sees that the contours where the order parameter is zero are not simply parallel to the abscissa, but tilted: since the boundary condition at the wall favors a wave trough at the surface, the position of the nth zero then means $(2n - 1)/4$ wavelengths occur up to this position. The first zero, of course, is not included in Fig. 7.19, and thus one wavelength just fits halfway in between the lines of the first and second zero that are included in Fig. 7.19. From this plot one hence can infer which thickness should be used at each temperature to provide an optimum fit of the lamellar structure. For a finite size scaling analysis, however, this would be rather cumbersome, since one would need to work with multiples of the thickness which is optimal at T_c, whereas the precise location of T_c is also not known beforehand.

In this situation, most Monte Carlo work has emphasized the behavior of the collective structure factor[52–55] and of the chain linear dimensions [24,52–55,84,85] in the disordered phase of the bulk copolymer melt. Assigning a spin variable $\sigma(\vec{r})$ that takes the values $\sigma(\vec{r}) = -1, 0, +1$ if lattice site $\vec{r}$ is taken by A, V, B species, respectively, the collective structure factor is defined as the Fourier transform of the spin pair correlation function,

$$S_{\text{coll}}(\vec{q}) = L^{-3} \sum_{i,j} \exp[i\vec{q} \cdot (\vec{r}_i - \vec{r}_j)] \langle \sigma(\vec{r}_i)\sigma(\vec{r}_j) \rangle. \tag{7.32}$$

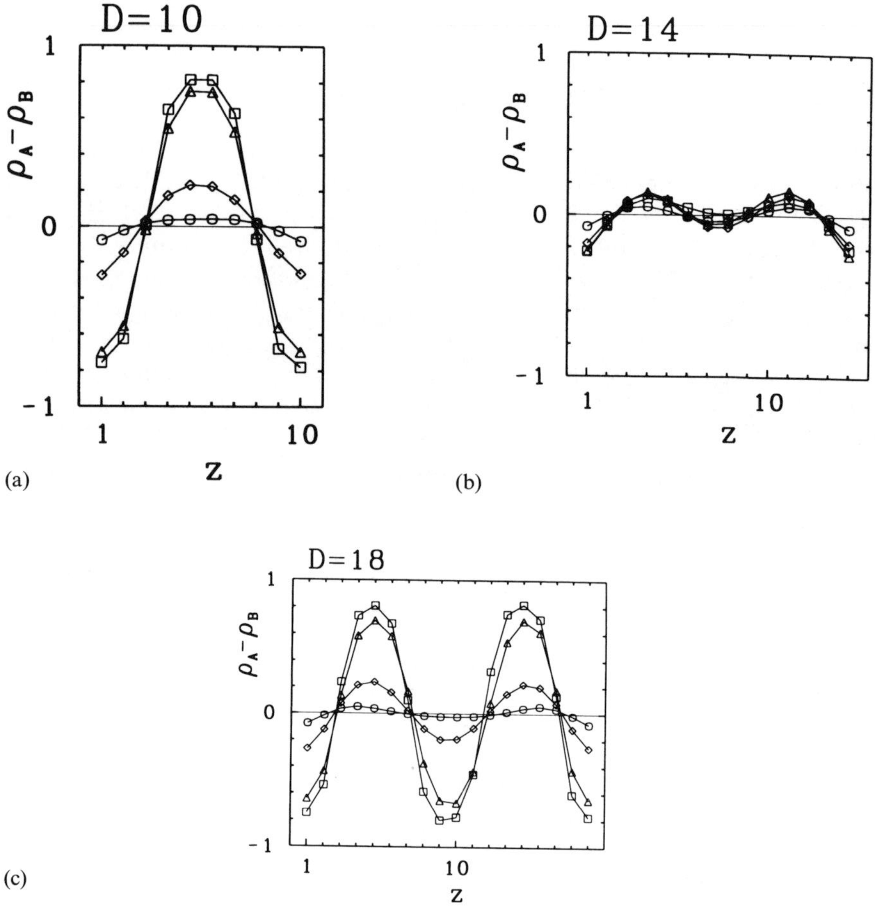

Fig. 7.18 Profiles of the order parameter $\rho_A(z) - \rho_B(z)$ for the symmetric ($f = 1/2$) block copolymer model with $N = 16, \phi_V = 0.2$, and $L \times L \times D$ simple cubic lattices with periodic boundary conditions in the x and y directions, while at the two repulsive walls of size $L \times L$ a repulsive energy of strength $\epsilon_{AS} = \epsilon_{AB}/2$ acts on A-monomers, with units such that $\epsilon_{AB}/k_B = 1$. Linear dimension L in parallel directions is always $L = 16$, while the linear dimension in z-direction is $D = 10$ (a), 14 (b), and 18 (c). Normalized inverse temperatures are indicated by different symbols: $1/T = 0.3$ (circles), 0.4 (diamonds), 0.5 (triangles), and 0.6 (squares). The critical temperature in the bulk is estimated as[53] $1/T_c \approx 0.52 \pm 0.05$. (From Kikuchi and Binder.[149])

This quantity is of particular interest, since it is readily accessible experimentally via small angle scattering techniques[56,57] and also, its behavior is predicted by various theories.[16,21,58,64]

Choosing a simple cubic lattice of linear dimensions L with periodic boundary conditions, obviously only discrete values of $\vec{q}$ are accessible, namely

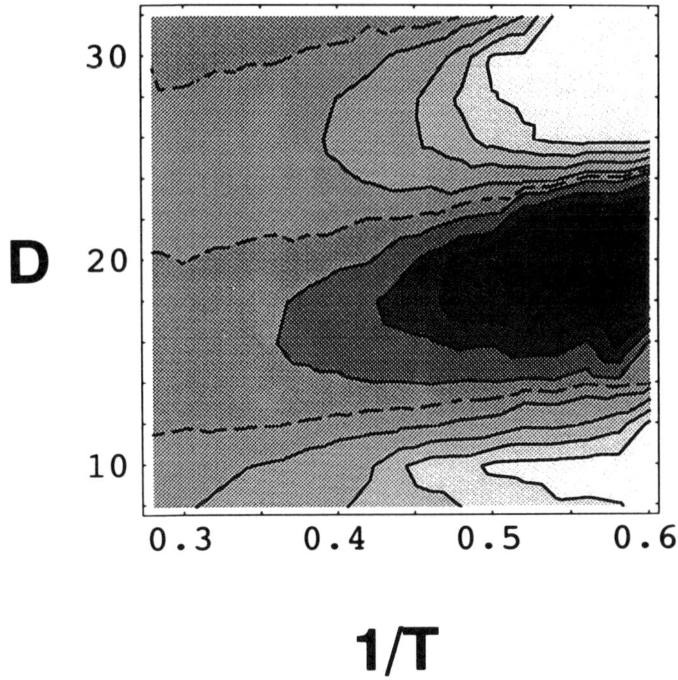

Fig. 7.19 Contour plot for the order parameter $\rho_A(z)$–$\rho_B(z)$ at the center of the films shown in Fig. 7.18 (averaged over the layers $z = D/2$ and $z = D/2 + 1$, choosing all possible even integers for D in the range $8 \leq D \leq 32$), in the plane of variables D and $1/T$. Temperature is measured in units of ϵ_{AB}/k_B. The solid lines denote contours of order parameter values $+0.7$ (enclosing the lightest region), $+0.5, +0.3, +0.1, -0.1, -0.3, -0.5, -0.7$ (enclosing the darkest region). The dashed lines indicate the zeros of the order parameter, $\rho_A(z) - \rho_B(z) = 0$. (From Kikuchi and Binder.[149])

$$\vec{q}_{\vec{\nu}} = \frac{2\pi}{L}(\nu_x, \nu_y, \nu_z); \nu_x, \nu_y, \nu_z = \text{integers}. \tag{7.33}$$

Thus there is also a long wavelength cut-off at $q_{\min} = 2\pi/L$, and also for the absolute value $|\vec{q}_{\vec{\nu}}|$ that is of interest for a spherically averaged structure factor, only values $\sqrt{\nu_x^2 + \nu_y^2 + \nu_z^2} q_{\min}$ are accessible. It makes sense to consider such a spherical average, since in the disordered phase there is no spontaneously broken symmetry and hence no preferred orientation $\vec{n}$, of course. Due to the choice of a lattice, one expects some (unphysical!) lattice anisotropy effects, but they should become appreciable only for uninterestingly large values of $|\vec{q}_{\vec{\nu}}| \gg q^*$, and hence are not further considered here.

Fig. 7.20 shows examples for the type of simulation results that can be obtained.[53,55] No data are available in the small q regime ($qR_g \leq 0.5$) here. Data for large qR_g have not been obtained either, firstly because the calculation of $S_{\text{coll}}(\vec{q})$ then would be too time-consuming, and secondly there

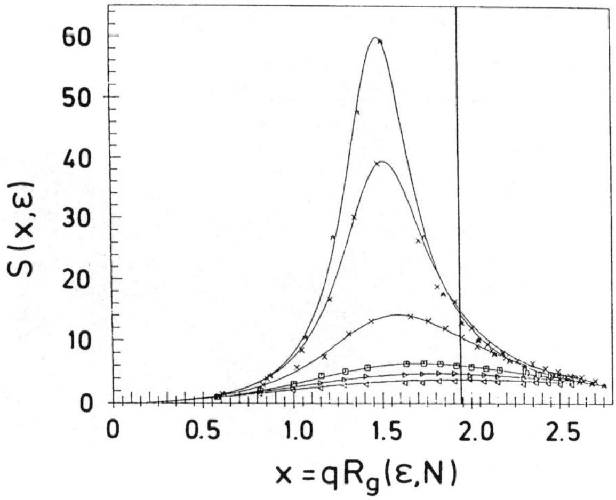

(a)

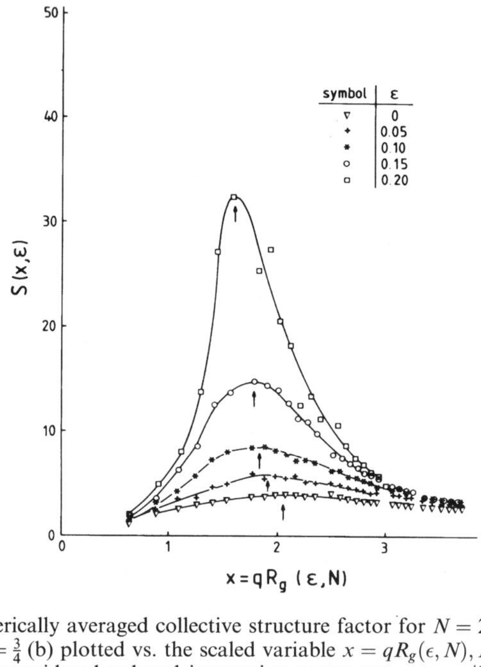

(b)

Fig. 7.20 The spherically averaged collective structure factor for $N = 20, L = 16, f = \frac{1}{2}$ (a) and $N = 40, L = 32, f = \frac{3}{4}$ (b) plotted vs. the scaled variable $x = qR_g(\epsilon, N)$, $R_g(\epsilon, N)$ being the gyration radius at the considered reduced interaction parameter $\epsilon \equiv \epsilon_{AB}/k_BT$, for a simple cubic lattice model of block copolymers of the type shown in Fig. 7.16, with a vacancy content $\phi_V = 0.2$. Curves in case (b) are only guides to the eye, while in case (a) they are least-square fits to a generalized Leibler[21] formula, $NS_{\text{coll}}^{-1}(\vec{q}) = \frac{1}{\alpha}[F(q\tilde{R}_g) - \delta]$, where $F(x)$ is the result for $NS_{\text{coll}}^{-1}(q)$ proposed by Leibler, but with $\alpha, \delta, \tilde{R}_g$ being fit parameters. Different symbols in both parts indicate various values of ϵ as shown in (b), while they are in (a) $\epsilon = 0, 0.05, 0.1, 0.2, 0.3, 0.35$ (from below to above). Note that the Leibler theory (which assumes $R_g = \tilde{R}_g$)[21] predicts the maximum for $f = \frac{1}{2}$ to occur at $x^* \approx 1.95$, independent of ϵ, as indicated by a straight line. For $f = 3/4$, the maximum should occur for $x^* = 2.053$ while the actual maxima are off-set and shift with increasing ϵ to lower values (arrows). (From Fried and Binder.[53,55])

might possibly occur lattice anisotropy artifacts. Even in the restricted wave-number range from $0.5 \lesssim qR_g \lesssim 3.5$ there is some statistical scatter, despite large efforts in computer time. Note that the points shown in Fig. 7.20 include all possible values $|\vec{q}_{\vec{\nu}}|$ in the quoted range, and thus near T_c the precise location of the maximum due to this discreteness of reciprocal space also becomes a problem. As a consequence, the analysis of $S_{\text{coll}}(\vec{q})$ is not suitable for a precise estimation of the microphase separation transition temperature either. As a matter of fact, none of the existing simulations of block copolymer melts[24,52–55,84,85] have yielded a reasonably accurate estimate of T_c so far. Nevertheless these studies have been very useful, as they have indicated qualitative deviations from the standard weak segregation theory[21] and its underlying theoretical concepts (such as the random phase approximation [RPA][9,151]) already in the high-temperature region above T_c. Although this behavior was not originally expected,[16,59,60] it has found experimental verification.[57,66] We shall return to these phenomena in more detail in Section 7.4.

7.2.6 Interfacial structure, surface enrichment, interdiffusion, spinodal decomposition

For simulating bulk properties of materials, one uses periodic boundary conditions in all three spatial directions throughout. However, other boundary conditions are needed for simulating interfacial and surface phenomena.

In models of Ising ferromagnets a common method[152] to stabilize one interface between spin-up and spin-down phases is the use of antiperiodic boundary conditions in one direction of the system, $\sigma(\vec{r}_i + L_\perp \vec{e}_z) = -\sigma(\vec{r}_i)$, where $\vec{e}_z$ is a unit vector in the z-direction (perpendicular to the interface that is stabilized), and $L_\perp$ is the linear dimension of the system perpendicular to the interface. In the x, y-directions parallel to the interface, where one then uses linear dimensions $L_\parallel$ which may be chosen different from $L_\perp$, standard periodic boundary conditions are used ($\sigma(\vec{r}_i + L_\parallel \vec{e}_x) = \sigma(\vec{r}_i)$ if $\vec{e}_x$ is a unit vector in x-direction, etc.). Remembering that the Ising spin model translates into the model of a binary AB mixture (this fact has already been invoked writing the collective structure factor in eq. [7.32] in a pseudospin representation), $\phi_A(\vec{r}_i) = [1 - \sigma(\vec{r}_i)]/2, \phi_B(\vec{r}_i) = [1 + \sigma(\vec{r}_i)]/2$ if $\sigma(\vec{r}_i) = \pm 1$, this antiperiodic boundary condition stabilizes an interface between an A-rich phase and a B-rich phase of a binary (AB) mixture, if one simulates a system at phase coexistence ($\Delta\mu = 0$) for $T < T_c$.

This method can be straightforwardly generalized to lattice models of symmetrical polymer mixtures[153]: an A-chain part of which leaves the box through the boundary at $z = L_\perp$ re-enters the system at $z = 1$ as a B-chain, etc. Comparing such a simulation with a simulation with standard periodic boundary conditions in all directions for the same $L_\parallel \times L_\parallel \times L_\perp$ geometry, one can obtain excess properties due to the interface,[152] such as the interfacial excess (internal) energy U_{int}. From U_{int} one finds the interfacial free energy F_{int} by standard thermodynamic integration techniques[152] {using the relation $U_{\text{int}}(\beta) = \partial F_{\text{int}}(\beta)/\partial\beta$ where $\beta = 1/k_BT$}.

One disadvantage of this technique, of course, is that there is no preferred location of the center of the interface—there is still a translational symmetry that allows the motion of the interface in z-direction arbitrarily. Even if one initializes the system such that at the start of the simulation the interface is in the center of the box (at $z = L_\perp/2$), during the course of the simulation for any finite $L_\parallel$ we expect diffusion of the interface to neighboring positions. In lattice models of metallic alloys, this diffusion is pronounced only for temperatures higher than the interfacial roughening transition[154]: for polymer mixtures such a transition would be a lattice artifact, of course, since in fluid mixtures interfaces are rough due to capillary wave excitations at all temperatures.[155] If one studies interfacial profiles, one can impose a constraint localizing[154] the interface in the center by requiring that the total order parameter of the system is fixed at zero, $m = 0$, so that always an equal number of A-chains and B-chains are in the system.

Another choice for maintaining an A–B interface is to abandon in the z-direction both the periodic and the antiperiodic boundary condition, rather using hard wall boundary conditions there.[68,118] In Ising models, one uses the "fixed spin" boundary condition, where left neighbors of spins with $z = 1$ are always up, and right neighbors of spins with $z = L_\perp$ always down[156]; alternatively, one may choose "boundary fields" at the walls that prefer component B near $z = 1$ and component A near $z = L_\perp$. A variant of this "boundary field technique" has recently been used for block copolymers[149,157] in order to stabilize the lamellar phase, with the A–B interfaces being perpendicular to the interface. In this manner, one can study the interfacial structure between the A-rich and B-rich layers in the lamellar mesophase (Fig. 7.18), as well as the surface enrichment that occurs at the hard walls due to the boundary fields (in Fig. 7.18, species B was favored at both walls) and the density reduction that occurs both near the hard walls and at these A–B interfaces[150] (Fig. 7.21[a]). Such boundary energies preferring one component are also used when one simulates surface enrichment in polymer blends[50,157,158] (Fig. 7.21[b]).

Cifra *et al.*[158] use a wall–monomer interaction $\epsilon_{WA} = \epsilon_{WB}$ but consider different volume fractions $\phi_A \neq \phi_B$ in the bulk of their thin film system, carrying out the simulation in the canonical ensemble, while the data of Fig. 7.21(b)[50] are obtained from semi-grand-canonical techniques. The simulation by Cifra *et al.*[158] is of great interest as a first step towards the modeling of the situation used in experiment, where one normally has two inequivalent interfaces: one is the interface between the thin film of the polymer blend and an adsorbing substrate, and the other is an interface between the film and vacuum or air,[51,159] respectively. Cifra *et al.*[158] achieve this

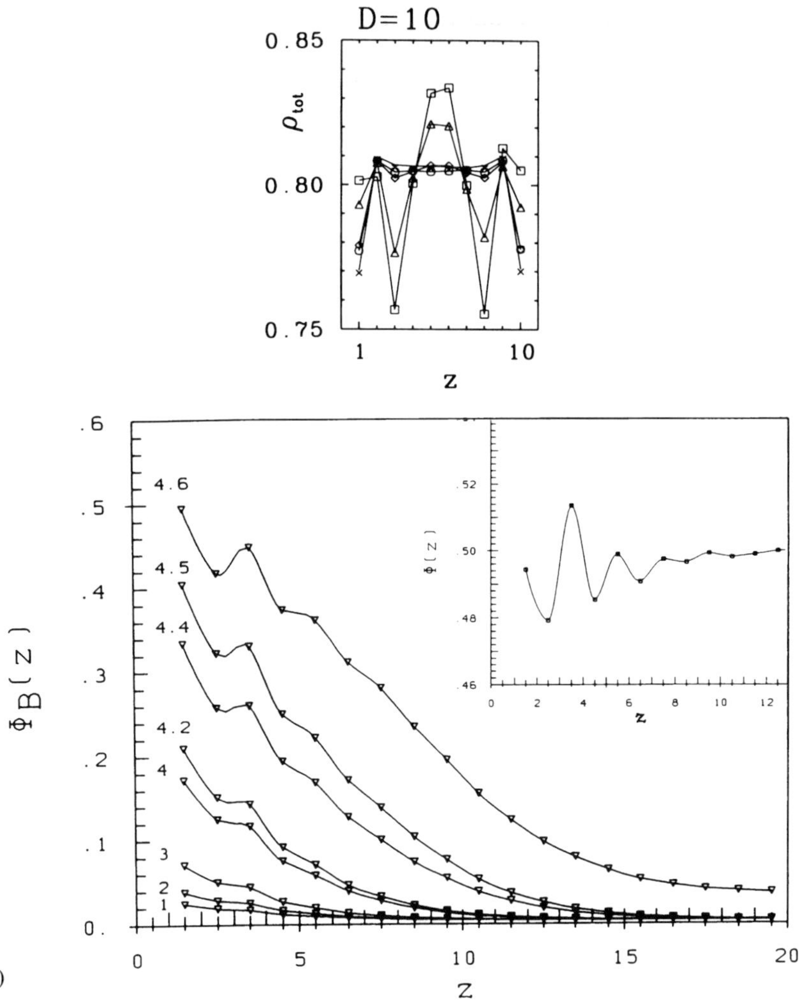

(a)

(b)

Fig. 7.21 (a) Profiles of the total density ρ_{tot} for the symmetric ($f = 1/2$) block copolymer model of Fig. 7.16 with $N = 16, \phi_V = 0.2$ in a $L_{\parallel} \times L_{\parallel} \times D$ geometry with $D = 10$ and periodic boundary conditions in x and y directions, but two hard walls $L_{\parallel} \times L_{\parallel}$ at which a repulsive energy of strength $\epsilon_{AS} = \epsilon_{AB}/2$ acts on A-monomers, for $N = 16, L = 16$, and normalized inverse temperatures $1/T = 0.0$ (crosses), 0.3 (circles), 0.4 (diamonds), 0.5 (triangles) and 0.6 (squares), choosing units such that $\epsilon_{AB}/k_B = 1$. The corresponding order parameter profile has been shown as in Fig. 7.18(a) (From Kikuchi and Binder.[150]) (b) Profile of the volume fraction $\phi_B(z)$ of B-monomers and of the total volume fraction $\phi(z) = \phi_A(z) + \phi_B(z)$ [insert] for the bond-fluctuation model at average volume fraction $\phi = 0.5$, simulating a symmetrical AB mixture with chain length $N = 10$ and temperature $k_B T/\epsilon = 11$ at bulk phase coexistence ($\Delta\mu = \mu_A - \mu_B = 0$). Lattice dimensions chosen were $L_{\parallel} = 100, L_{\perp} = 40$. Interactions in the bulk are $\epsilon_{AA} = \epsilon_{BB} = -\epsilon_{AB} = -\epsilon < 0$, if the distance between monomers is less than $\sqrt{8}$ lattice spacings; interactions at the walls are $\epsilon_{WA} = 0, \epsilon_{WB} = -\mu_1 \epsilon$ if $z = 1$ or $z = L_{\perp}$, respectively. Different curves correspond to various choices of μ_1 as indicated. The critical temperature of the model is[26] $k_B T_c/\epsilon \approx 22.85$ and hence ϕ_B is very small in the bulk of the A-rich mixture at $k_B T/\epsilon = 11$. (From Wang and Binder.[50])

situation by following techniques of Madden[160] where one first confines a film between two walls and then removes one wall.

Interfaces between compatible polymers were also considered,[68,118] in an attempt to study the broadening of interfacial profiles with time due to interdiffusion (Fig. 7.22). While the existing work based on Monte Carlo simulations of the bond fluctuation models proves the feasibility of such simulations, which are desirable since they mimic the experimental proce-

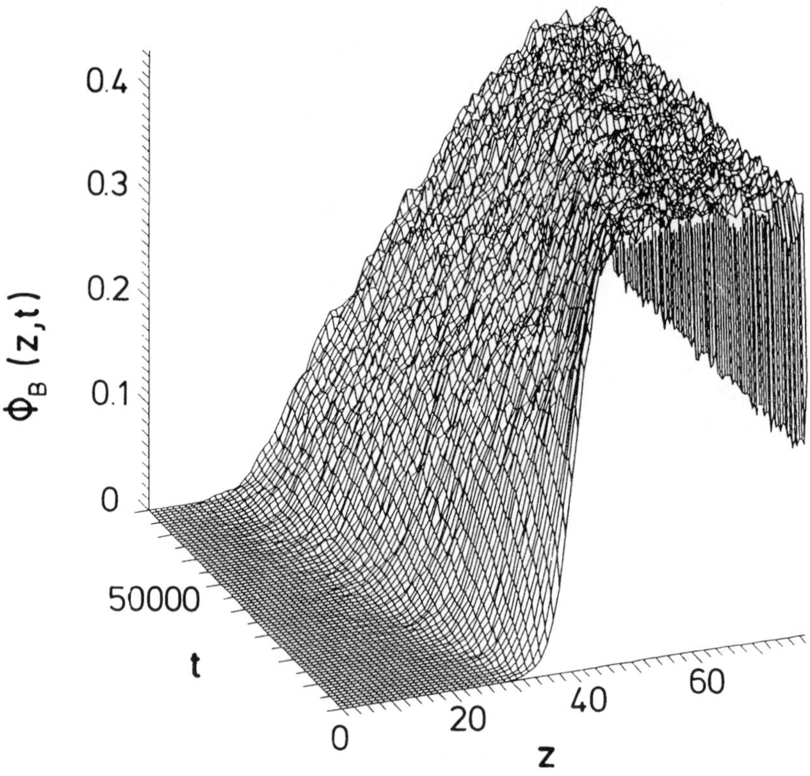

Fig. 7.22 Simulation of the time evolution of the concentration profile $\phi_B(z, t)$ of polymers across an A–B interface, using the bond fluctuation model of polymers on the simple cubic lattice, chain lengths $N_A = N_B = 20$, a fraction $\phi_V = 0.58$ of vacant sites, jump rates $\Gamma_A = \Gamma_B = 1$ for both types of monomers, and an attractive energy ϵ_{AB} between unlike monomers which are less than three lattice spacings apart from each other, at a temperature $k_BT/\epsilon_{AB} = -18/5$. In the initial configuration of a $20 \times 20 \times 80$ lattice with two hard walls at $z = 0$ and at $z = 81$ and periodic boundary conditions in x, y-directions all chains with center of gravity positions $z \geq 40$ are considered as B-chains, while all chains with center of gravity positions at $z < 40$ are treated as A-chains. Time t is measured in attempted Monte Carlo steps per monomer (MCS). To gain statistics, 48 systems are run in parallel at a vector processor, and averaged together. From the broadening of the profile with increasing time one can infer the interdiffusion constant, following corresponding experimental work.[161] (From Deutsch and Binder.[68])

dure[161]—and there are still many questions open—one must stress the important caveat that the Monte Carlo dynamics completely lacks the hydrodynamic forces. Velocity fluctuations are coupled to the order parameter fluctuations in a fluid binary mixture and contribute to the interdiffusion, as is well known from the study of critical dynamics in fluid binary mixtures,[94,162] and this coupling needs to be taken into account for a calculation of the interdiffusion constant.[163] The Monte Carlo dynamics treats the polymers as if one were to consider interdiffusion in a rigid solid structure. Here we thus actually encounter a point where the simple lattice models, the successes of which have been given so much emphasis in this chapter, do miss some important physics! In principle, a molecular-dynamics (MD) simulation of bead-spring models[164] (see also Chapters 3–5 of the present book) would be a much more suitable approach to simulate interdiffusion, but we are not aware of any MD counterparts of Fig. 7.22. Due to the slowness of the interdiffusion process and the need to sample over many chains to get sufficient statistics, however, such studies would be computationally very demanding.

Similar problems concern the study of phase separation kinetics in polymers.[10–14] Due to the slowness of this process, Monte Carlo simulations,[42,80,81,88] could only study the early stages of spinodal decomposition, even for very short chains. During this initial growth of concentration fluctuations, the hydrodynamic effects mentioned above are less important,[165] although there is broad evidence that the hydrodynamic couplings need to be taken into account during the late stages.[12–14] The quantity that one wishes to study is the dependence of the collective structure factor $S_{\text{coll}}(\vec{q}, t)$ (eq. [7.32]) as a function of time t after the quench to the considered state inside the miscibility gap. In addition to problems with the discrete character of reciprocal space (eq. [7.33])—which are rather serious here since one is really interested in a long wavelength limit (small q)—there is the problem that $S_{\text{coll}}(\vec{q}, t)$ exhibits the so-called "lack of self-averaging"[166]: the relative statistical error of $S_{\text{coll}}(\vec{q}, t)$ obtained from n observations decreases only as $\sqrt{2/n}$, irrespective of the size of the simulated system (it does not decrease with increasing volume L^d). Conversely, the statistical error of quantities such as the order parameter or internal energy decreases inversely proportional to the square root of the volume, if one stays outside of the regions where phase transitions occur. Thus reasonably accurate data for $S_{\text{coll}}(\vec{q}, t)$ inevitably need a large number of samples n.

7.3 Results for polymer blends

In the previous section, the methodic aspects of computer simulations of phase transitions in polymer blends were emphasized. In the present section, we briefly review some of the most important results that have been obtained so far.

7.3.1 Test of the Flory–Huggins theory and of the Schweizer–Curro theory

Already in the introduction it was pointed out that the Flory–Huggins approximation[20] for the excess free energy of mixing (eq. [7.3]) involves many doubtful approximations, the accuracy of which is hard to ascertain *a priori*. In view of the fact that a fit to experimental data usually involves adjusting both the concentration dependence and the temperature dependence of the χ-parameter, the feasibility of such a fitting procedure clearly is not a sensitive test of the theory.

One of the basic consequences of eq. (7.3) for a symmetric mixture is the condition for the critical temperature T_c {remember $\epsilon = \epsilon_{AB} - (\epsilon_{AA} + \epsilon_{BB})/2$}

$$\frac{1}{2}N(1-\phi_V)\chi_c = \frac{z\epsilon(1-\phi_V)N}{2k_BT_c} = 1. \tag{7.34}$$

Thus one finds that T_c is proportional to the chainlength N and the volume fraction $1-\phi_V$ of sites that are not vacant, and only one combination ϵ of the three energy parameters ϵ_{AA}, ϵ_{AB} and ϵ_{BB} enters.

The first test of these predictions, for the model of Fig. 7.3, revealed surprising discrepancies (Fig. 7.23).[78,79,167] It is seen that the Flory–Huggins theory strongly overestimates the critical temperature. Obviously, the theory gets worse the larger the vacancy concentration. For low vacancy content ($\phi_V \to 0$) there is still a discrepancy by a factor of two,[82] and although this discrepancy is reduced when one considers instead the approximation proposed by Guggenheim,[168] the latter clearly is also not exact.

For low polymer concentrations ($1-\phi_V \lesssim 0.4$) it is also no longer true that T_c depends on ϵ_{AA}, ϵ_{AB} and ϵ_{BB} only via the single parameter ϵ, since then the three curves in the upper part of Fig. 7.23 should superimpose. In contrast, the strong increase of T_c^{FH}/T_c for the purely repulsive case ($\epsilon_{AB} \neq 0$, $\epsilon_{AA} = \epsilon_{BB} = 0$, crosses in the upper part of Fig. 7.23) is qualitative evidence for the prediction[169–174] that polymer blends in a common good solvent become compatible in the semidilute concentration regime. This happens because the short-range repulsive interactions are effective only on length scales larger than the screening length ξ_s for the excluded volume interactions, which scales with the volume fraction ϕ of monomers as[9]

$$\xi_s \propto \phi^{-\nu_e/(3\nu_e-1)} \approx \phi^{-0.77}, \nu_e \approx 0.59, \tag{7.35}$$

where ν_e is the exponent describing coil dimensions in good solvents ($\langle R^2_{\text{gyr}}\rangle \propto N^{2\nu_e}$). This means that over length scales of the order of ξ_s successive monomers of a chain can be grouped into "blobs", such that the N_B monomers inside a blob $\{\xi_s \propto N_B^{\nu_e}$, i.e., $N_B(\phi) \propto \phi^{-1/(3\nu_e-1)}\}$ essentially have no interactions with monomers of other chains. As a consequence, a chain of length N can be rescaled into a chain of $N_{\text{eff}} = N/N_{\text{blob}}(\phi) \propto N\phi^{1/(3\nu_e-1)}$ blobs, with a renormalized interaction $\chi_{\text{eff}}(\phi)$ between blobs instead of the

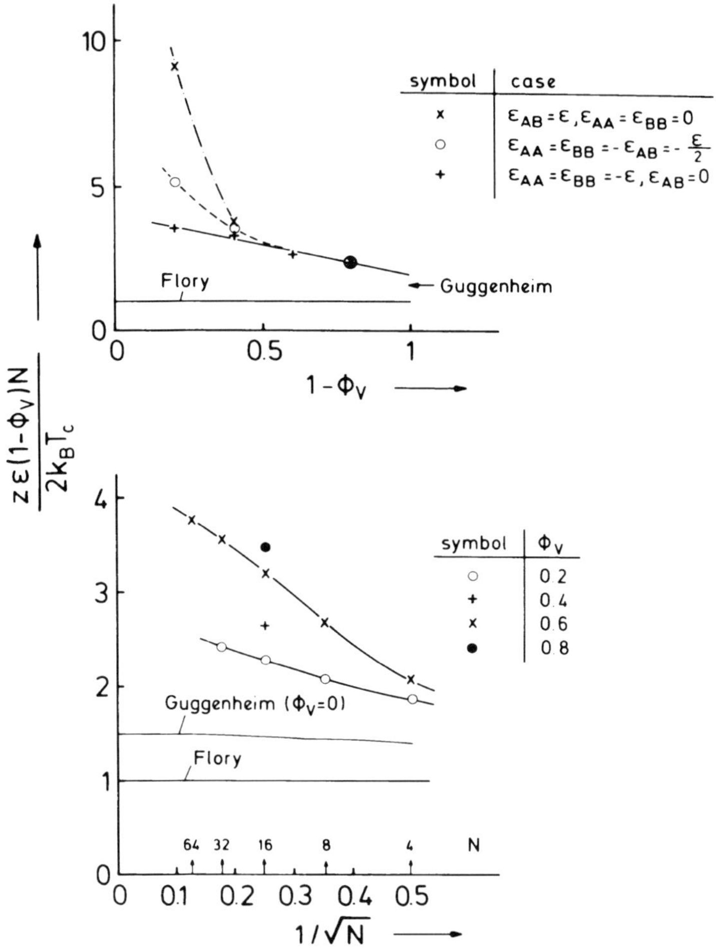

Fig. 7.23 Ratio between the Flory–Huggins critical temperature, $T_c^{FH} = N(1 - \phi_V)z\epsilon/(2k_B)$ and the actual critical temperature T_c for the self-avoiding walk model of polymer mixtures on the simple cubic lattice (Fig. 7.3) plotted versus concentration $1 - \phi_V$ of sites taken by monomers (upper part) and versus the inverse square root of the chain length (lower part). Upper part refers to $N = 16$ (for $1 - \phi_V = 0.8$, all three choices of the energy parameters coincide; this is marked by a solid dot). Curves are only drawn to guide the eye. Both the Flory approximation, eq. (7.34), which implies $T_c^{FH}/T_c \equiv 1$, i.e., a horizontal straight line, and the Guggenheim approximation[168] (which is available for $\phi_V = 0$ only) are indicated. (From Sariban and Binder.[79])

original interaction χ (eq. [7.2]) between monomers. Renormalization group theory predicts[169–174]

$$\chi_{\text{eff}}(\phi) \propto \phi^{x/(3\nu_e - 1)}, x \approx 0.22 \tag{7.36}$$

which implies that the number of unfavorable contacts vanishes in the semi-dilute limit. This reduction of the effective χ-parameter due to reduction of the binary contacts is tested in Fig. 7.24,[175] where the critical temperature in units of this effective interaction parameter is plotted versus $N/N_{\text{blob}}(\phi)$. It is seen that the data are reasonably compatible with this blob picture.

The fact that T_c^{FH}/T_c increases with increasing N (Fig. 7.23, lower part) has been interpreted by Schweizer and Curro[34,37] as a piece of evidence for their PRISM (polymer reference interaction site model) prediction[31] that $T_c \propto N^{1/2}$ due to correlation effects: then $N/T_c \propto N^{1/2} \to \infty$ as $N \to \infty$. Indeed, on the basis of short chains as studied by Sariban and Binder,[78,79] a conclusion such as that cannot be ruled out. However, Deutsch and Binder[26,27] studying the bond fluctuation model[67,68] provided rather clear evidence (Fig. 7.4) that the original PRISM result $T_c \propto N^{1/2}$ is incorrect, and one rather has $T_c \propto N$ as in Flory–Huggins theory[20] (although the prefactor in the relation $T_c \propto N$ is much lower than predicted[20]). This conclusion that $T_c \propto N$ holds is corroborated by a more recent version of the PRISM theory.[38,39]

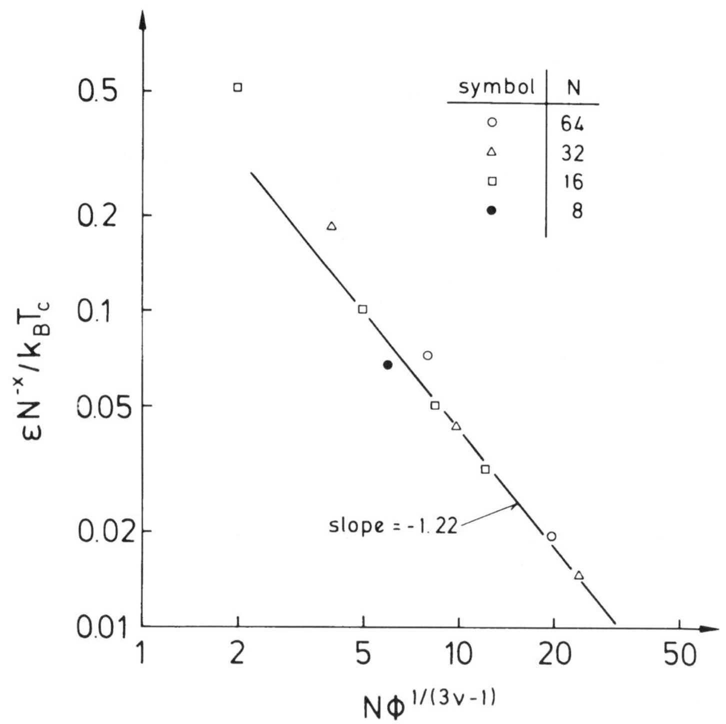

Fig. 7.24 Normalized critical temperature $\epsilon N^{-x}/k_B T_c$, using $x \cong 0.22$[169] for the model of Fig. 7.3 plotted vs. $N\phi^{1/(3\nu_e - 1)}$ (this means a rescaling of N with the number $N_{\text{blob}}(\phi)$ of effective segments per blob, cf. Ref. 9). Straight line shows the asymptotic exponent of the scaling function for large N, $1 + x \approx 1.22$. This plot includes data in the regime $0.15 < 1 - \phi_V < 0.4$, and $8 \leq N \leq 64$. (From Sariban and Binder.[175])

The increase of N/T_c (Fig. 7.23) with increasing N can be attributed to a correction due to the free ends of the chain,

$$k_B T_c(N)/\epsilon = AN + B, \tag{7.37}$$

where both constants A, B are expected to be model-dependent. While a positive constant B (as found for the bond fluctuation model at $\phi_V = 0.5$ when $\epsilon_{AB} = -\epsilon_{AA} = -\epsilon_{BB} = \epsilon \neq 0$ if neighboring monomers are at distances 2 or $\sqrt{5}$ or $\sqrt{6}$ lattice spacings, respectively, Fig. (7.4)) is physically plausible because a monomer at a chain end may have more neighbors from other chains than a monomer in the chain interior, a negative constant B may occur as well (Fig. 7.25).[40] Fig. 7.25 also shows that very large N may be needed until $T_c(N)/N$ becomes independent of N.

For all models studied (Figs 7.4, 7.23, 7.25) the Flory–Huggins theory[20] overestimates the critical temperature $T_c(N)$ by a rather large factor. It turns out that this overestimation is simply due to an overestimation of the number of nearest-neighbor contacts between monomers of different chains (Fig. 7.26).[79,176] We distinguish here between the number of contacts $n_c(AB)$ per

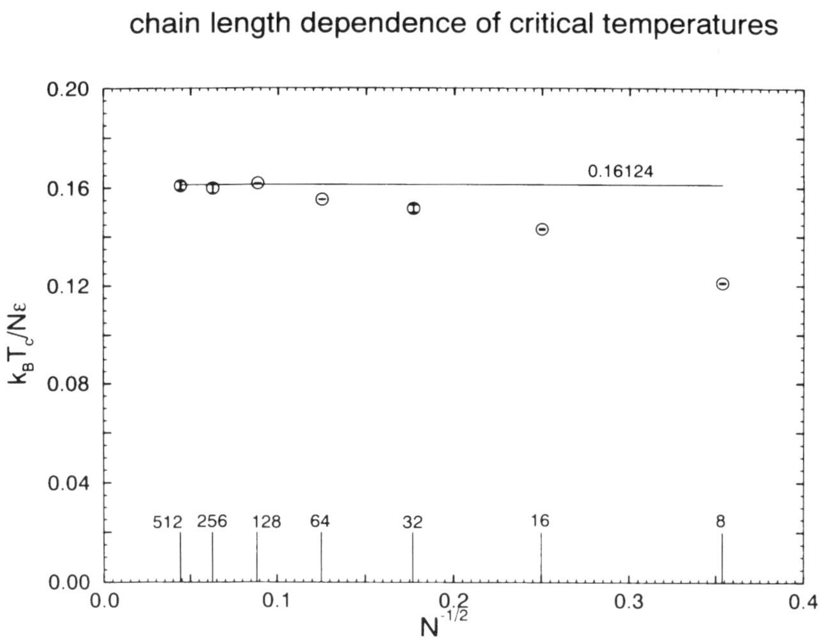

Fig. 7.25 Variation of the observed normalized critical temperature $k_B T_c(\epsilon N)$ with $N^{-1/2}$, for the bond fluctuation model on the simple cubic lattice where $\epsilon = \epsilon_{AB}$, $\epsilon_{AA} = \epsilon_{BB} = 0$, $\phi_V = 1/2$, and the interaction is nonzero only if two monomers are at the closest possible distance (two lattice spacings). Vertical bars along the abscissa indicate the chain lengths for which calculations were made. Note that in this case the effective coordination number[40] is $z_{\text{eff}} \approx 5$, and hence eq. (7.34) would predict $k_B T_c(\epsilon)/N \approx 1.25N$, while the actual data are well fitted by $k_B T_c(\epsilon)/N \approx 0.161 - 0.227/N$. (From Deutsch and Binder.[40])

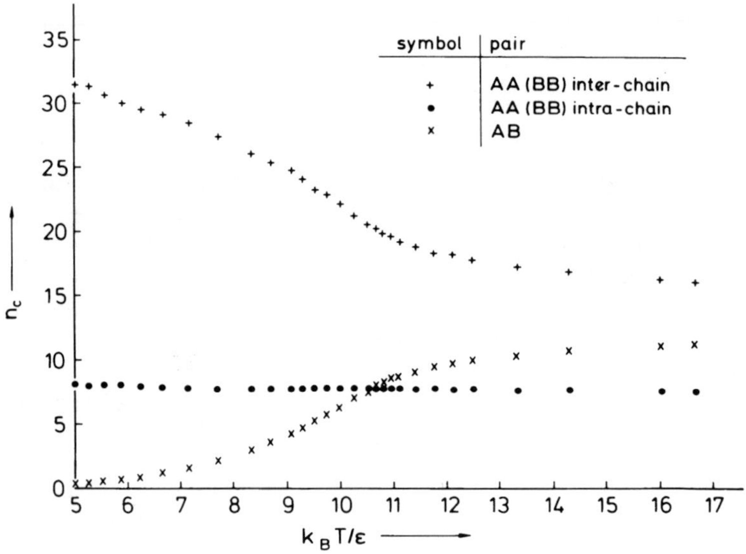

Fig. 7.26 Number of contacts n_c between different types of nearest-neighbor pairs plotted vs. temperature, for the model shown in Fig. 7.3 and the simple cubic lattice, using parameters $N = 32, \phi_V = 0.6, \epsilon_{AB} = 0, \epsilon_{AA} = \epsilon_{BB} = -\epsilon/2$. The data refer to $\Delta\mu = 0$, i.e., the system is at critical composition (at $T < T_c = 10.9\epsilon/k_B$ it then corresponds to a system exactly at the binodal or in a two-phase mixed state, respectively). (From Sariban and Binder.[176])

chain involving different monomers and between contacts of the same type $n_c(AA) + N_c(BB)$ involving different chains, as well as the corresponding intrachain contacts. It is clear that the latter can only drive a collapse of chains, but not a phase separation between different chains. Therefore it is only the total number of interchain contacts that should be compared with the corresponding number of Flory–Huggins theory,[20] which is $n_c^{tot}(FH) = zN(1 - \phi_V)$, or the Guggenheim theory,[168] $n_c^{tot}(G) = [(z-2)N + 2](1 - \phi_V)$. For the example shown in Fig. 7.26, one would conclude $n^{tot}(FH) \approx 77$, $n_c^{tot}(G) \approx 54$, while the actual count is $n_c^{tot} = n_c(AB) + n_c(AA) + n_c(BB) \approx 27$ [interchain contacts] (at high temperatures) to $n_c^{tot} \approx 32$ (at low temperatures), i.e., n_c^{tot} is lower than estimated by a factor of two to three. To some extent this reduction in the actual contact number is due to the intrachain contacts (there are seven to eight such contacts for the example shown in Fig. 7.26), and in part due to the effect that vacancies are more effective in lowering n_c^{tot} due to small-scale blob effects than via the factor $(1 - \phi_V)$. If one would correct the Flory–Huggins theory by introducing a reduced effective coordination number consistent with the actual number n_c^{tot} of interchain–monomer contacts, it would overestimate the critical temperature only slightly, similar to the mean-field approximation of the standard Ising model. Thus we refute the (unfounded) suggestion of Guevara-Rodriguez *et al.*[177] that the Monte Carlo critical temperatures should be raised by a factor of two when comparing to the Flory and Guggenheim or cluster variation[178] theories.

7.3.2 Critical phenomena and the Ising–mean field crossover

In Section 7.2.4 it was shown that via a finite size scaling analysis a meaningful extrapolation of simulation data to the thermodynamic limit is possible, and in this way one can extract estimates for both critical exponents (β, γ) and amplitudes $\{\hat{C}_+(N), \hat{C}_-(N)$ and $\hat{B}(N)\}$ of the collective scattering function above T_c (eq. [7.16]) and below T_c $\{S_{\text{coll}}(q=0) = \hat{C}_-(N)t^{-\gamma}$, $t = 1 - T/T_c \to 0\}$ or the order parameter $\{\langle |m| \rangle = (\phi_A^{\text{coex}} - \phi_A^{\text{crit}})/\phi_A^{\text{crit}} = \hat{B}(N)t^{\beta}\}$, respectively. While for short enough chains $(N \leq 32)$, data both for the simple self-avoiding walk model of Fig. 7.3 and for the bond fluctuation model are nicely consistent with the expected critical exponents for the three-dimensional Ising models[8,109] $(\beta \approx 0.325, \gamma \approx 1.241)$, for $N \geq 64$ one rather finds "effective exponents" due to the gradual crossover from Ising critical behavior to mean-field behavior expected for $N \to \infty$[9–12,40] (see Fig. 7.13). Thus already the coexistence curve in Fig. 7.7(a) involves an effective exponent $\beta_{\text{eff}} \approx 0.358$, and similarly the scattering function in Fig. 7.7(b) involves another effective exponent $\gamma_{\text{eff}} \approx 1.247$. While these numbers are still relatively close to the Ising values, larger deviations are found[27,40] for the larger N. But it is also obvious when one plots these effective exponents extracted from the size-behavior at T_c from data around a mean lattice linear dimension L_m against the appropriate scaling variable L_m/N (see Fig. 7.14), that these deviations are just due to a gradual crossover from one universality class to another (Fig. 7.27).[40]

Here also the behavior of the amplitude ratio $\hat{C}_+^{\text{eff}}/C_-^{\text{eff}}$ is included, which has the universal value 2 in the mean-field limit and 4.9 in the Ising limit.[179] Unfortunately, there is considerable scatter in these data, and thus Fig. 7.27 should not be considered as a reliable estimation of a crossover scaling function: more powerful computers than currently available would be needed to accurately obtain such crossover scaling functions.

This conclusion is reiterated by a study[40] of the crossover scaling of the order parameter (Fig. 7.28[a]) and collective scattering function (Fig. 7.28[b]) plotted versus the reduced temperature distance from T_c. But these data are qualitatively similar to corresponding experimental data,[140] and confirm the conclusion that polymer mixtures with short chain lengths have no mean-field critical behavior, and one must use very long chains to see pure mean-field like behavior because the crossover region is very wide. In this respect these data reject early statements[9,10,137–139] suggesting that for polymer mixtures non-mean-field effects are rather unimportant. While the simulations[40] cannot match the accuracy of the corresponding experiment,[140] they have the advantage that data both for

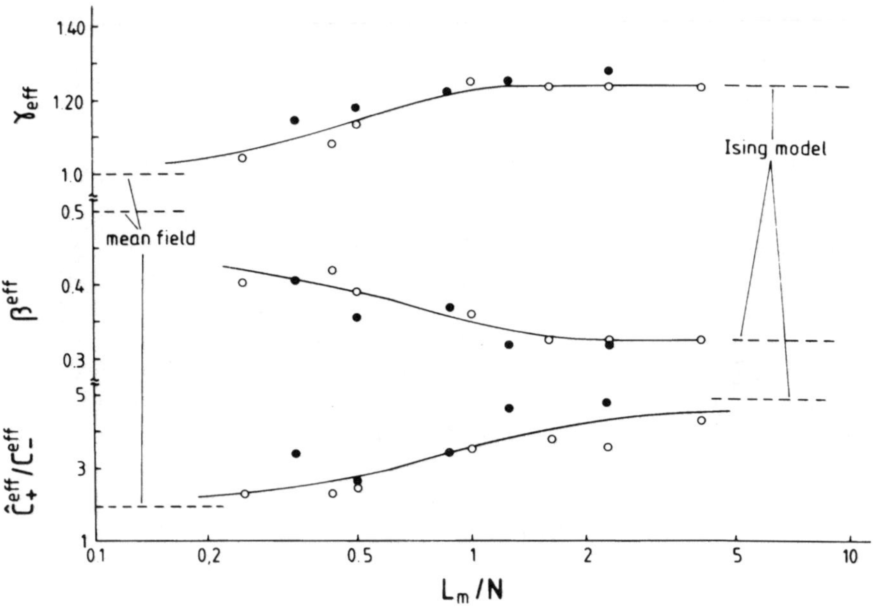

Fig. 7.27 Semi-log plot of the effective exponent $\gamma_{\rm eff}$ of the collective scattering function $S_{\rm coll}(q \to 0)$ (upper part), of the order parameter exponent $\beta_{\rm eff}$ (middle part) and the ratio of critical amplitudes $\hat{C}_+^{\rm eff}/\hat{C}_-^{\rm eff}$(lower part) versus the crossover scaling variable L_m/N, which is proportional to the ratio of characteristic lengths $L_m/\xi_{\rm cross}$. L_m is the mean size of the lattices used in the finite size scaling analysis. Asymptotic limits are shown as dashed straight lines. Open circles refer to the bond fluctuation model which has $\epsilon_{AA} = \epsilon_{BB} = 0, \epsilon_{AB} \neq 0$ for effective monomers two lattice spacings apart (Fig. 7.25), full dots to the version of the bond fluctuation model with[27] $\epsilon_{AB} = -\epsilon_{AA} = -\epsilon_{BB} = \epsilon$ and $z_{\rm eff} \approx 14$ (critical temperatures of this model are included in Fig. 7.4). Curves through the points are guides to the eye only. (From Deutsch and Binder.[40])

$T > T_c$ and $T < T_c$ are available, and the simulated polymers are strictly monodisperse and ideally symmetric. In contrast, the experiment[140] possibly suffers from effects due to asymmetry of the two polymers, polydispersity, etc; the fact that several chemically different polymer blends must be combined in the analysis in order to have a wide enough variation of the (effective) chainlength N (with T_c being in a physically accessible range) complicates the analysis further. This example hence nicely illustrates the fact that although both simulation and experiment have their own limitations, taking their complementary evidence together, significant progress can be obtained.

7.3.3 Asymmetric mixtures

As discussed in Sections 7.2.2 and 7.2.3, the studies of asymmetric mixtures are still quite scarce and so far restricted to few types of asymmetry and

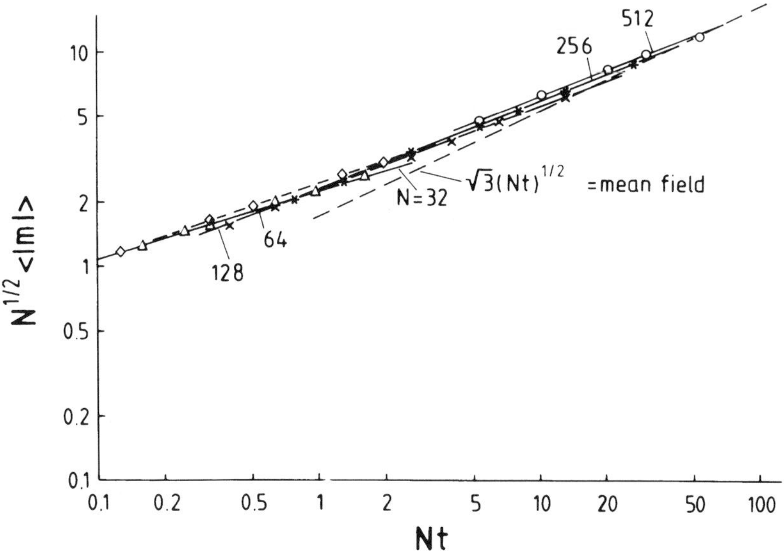

(a)

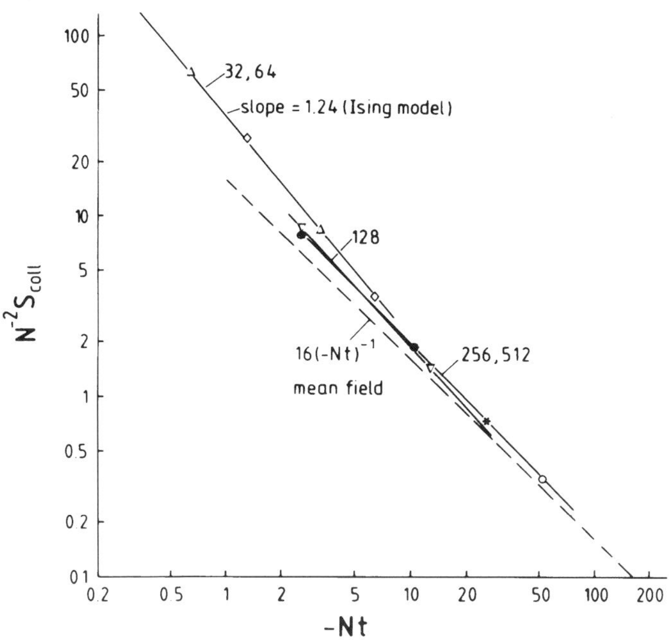

(b)

Fig. 7.28 (a) Log–log plot of $N^{1/2}\langle|m|\rangle$ vs. Nt, which is proportional to the variable t/Gi controlling the Ginzburg criterion (Fig. 7.14), using the fits resulting from finite size scaling. For each N the power law $\langle|m|\rangle = \hat{B}_{\text{eff}}(N)t^{\beta_{\text{eff}}}$ yields a straight line, which is always shown around $t \approx 0.01$ only. (b) Log–log plot of $N^{-2}S_{\text{coll}}(q=0)$ vs. $N(-t)$, for $T > T_c$, using the fits resulting from finite size scaling, $S_{\text{coll}} = \hat{C}_{+}^{\text{eff}}(N)(-t)^{-\gamma_{\text{eff}}}$, for $-t$ around $t \approx -0.01$. Numbers at the straight lines show the chain lengths N used. The mean-field limits are indicated as broken curves. Note that the coexistence curves are almost parallel to the T-axis (From Deutsch and Binder.[40])

rather short chain lengths only. Here we give some further details for the model already discussed in Figs 7.9 and 7.15, where still $N_A = N_B = N$, but $\epsilon_{AA} = \lambda\epsilon_{BB}$ with $\lambda \neq 1$. This model was studied again with the motivation to check whether Flory–Huggins theory[20] accounts for asymmetry effects correctly.[110] In fact, eq. (7.34) can be rewritten for $\epsilon_{AB} = -\epsilon_{BB} = \epsilon$ as $k_B T_c^{FH}/\epsilon = zN(1 - \phi_V)(3 + \lambda)/4$. Therefore the temperature scale in Figs 7.15(b) and 7.29 is normalized such that in the Flory–Huggins approximation the critical temperatures would occur for the same abscissa value. Indeed one finds that for $N = 32$

$$4k_B T_c/[(3 + \lambda)\epsilon] = 69.7 \pm 1 \tag{7.38}$$

for $1 \leq \lambda \leq 5$; thus, although Flory–Huggins theory overestimates the constant in eq. (7.38) by about a factor 3 (since $z \approx 14$ for the model of Figs 7.15 and 7.29), it describes the dependence of T_c on the asymmetry parameter λ correctly.[110,111] We also emphasize that for $N = 32$ the critical behavior of the model is nicely consistent with Ising exponents, irrespective of λ. Also

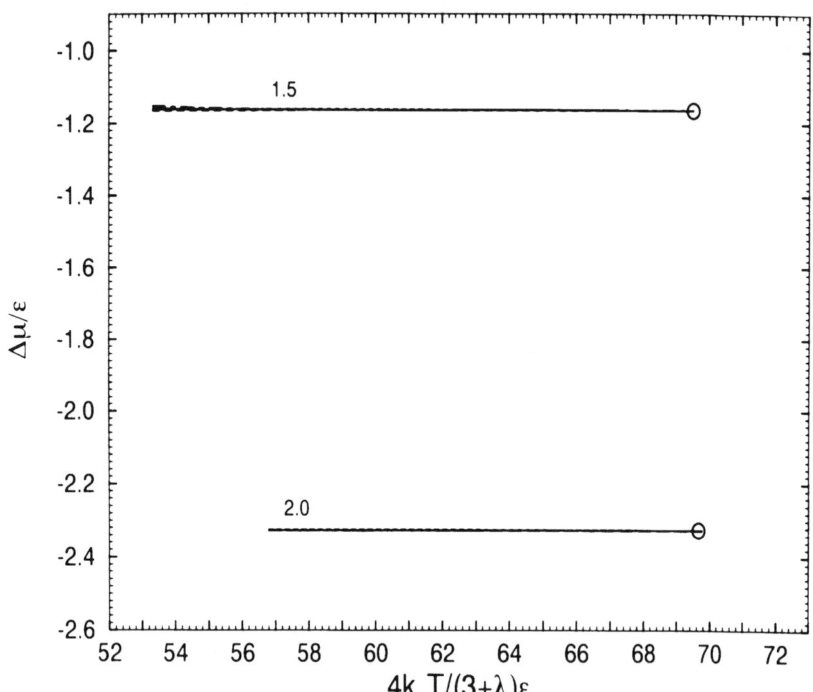

Fig. 7.29 Phase diagrams in the $(T, \Delta\mu)$ plane for the bond fluctuation model on the simple cubic lattice with asymmetric energy parameters $(\epsilon_{AA} = \lambda\epsilon_{BB}, \epsilon_{BB} = -\epsilon_{AB} = -\epsilon)$ for $N = 32, \phi_V = 0.5$ and two choices of λ. First-order transitions describe the coexistence of A-rich and B-rich phases, as shown in the T, ϕ-plane in Fig. 7.15(b). Critical points are denoted as circles. (From Deutsch and Binder.[110])

the model with $N_A \neq N_B$ that has recently been studied[117] is compatible with Ising exponents.

7.3.4 *Chain conformations in blends*

A common tool for the study of concentration inhomogeneities in polymer blends is the small angle scattering of light, neutrons or x-rays.[9–14] The standard tool for interpreting the scattering intensity of such experiments is a formula proposed by de Gennes[9] using the so-called random phase approximation (RPA).[9,151] In its simplest version for an incompressible mixture, which is modeled by a Flory–Huggins lattice with $\phi_V = 0$, this RPA result is

$$S_{\text{coll}}^{-1}(\vec{q}) = [\phi_A S_A(\vec{q})]^{-1} + [\phi_B S_B(\vec{q})]^{-1} - 2\chi, \tag{7.39}$$

where $\phi_A + \phi_B = 1$ and $S_A(\vec{q})$, $S_B(\vec{q})$ are the single-chain structure factors as one finds them in pure A (B) melts. Describing them by the Debye function[9,20]

$$S_A(\vec{q}) = N_A f_D(x_A), \quad x_A = \langle R^2_{\text{gyr,A}} \rangle q^2, \quad f_D(x) = \frac{2}{x}[\exp(-x) - 1 + x], \tag{7.40}$$

and analogously for $S_B(\vec{q})$, one usually assumes that the structure factors $S_A(\vec{q})$, $S_B(\vec{q})$ (and the mean square gyration radii $\langle R^2_{\text{gyr,A}}, \rangle$, $\langle R^2_{\text{gyr,B}}, \rangle$) do not depend on the thermodynamic state of the mixture (as described by χ, ϕ_A or temperature T and chemical potential difference $\Delta\mu$, respectively), but are simply the corresponding quantities of the pure melt. It should be noted that eqs (7.39) and (7.40) are also a starting point to construct an effective free energy functional for inhomogeneous states in polymer blends,[10,180,181] to consider interfacial problems,[46–49,182–184] spinodal decomposition,[10–12,180] etc.

Thus it is of interest to check the accuracy of eq. (7.39) by testing the basic assumption that the mean square gyration radii are independent of the state of the mixture (Fig. 7.30).[100] It is seen that for $T > T_c$ this assumption still is a reasonable approximation, since deviations from this approximation are in the order of several percent only. However, considering a mixture of asymmetric composition ($\phi_A \gg \phi_B$) as done in Fig. 7.30, one notes that the chain contraction of the minority component is distinctly stronger than for the majority component (Fig. 7.30[a]). This effect becomes particularly dramatic for dilute mixtures, where one may consider states with $T \ll T_c$ without entering the miscibility gap. Such isolated chains in an energetically very unfavorable matrix tend to contract strongly, as is obvious from Fig. 7.30(b): the chains tend to gradually collapse as T decreases and thus maximize favorable (i.e., attractive) interactions ϵ_{BB} (or minimize unfavorable, i.e., repulsive ϵ_{AB}-interactions, respectively). Fig. 7.31 demonstrates

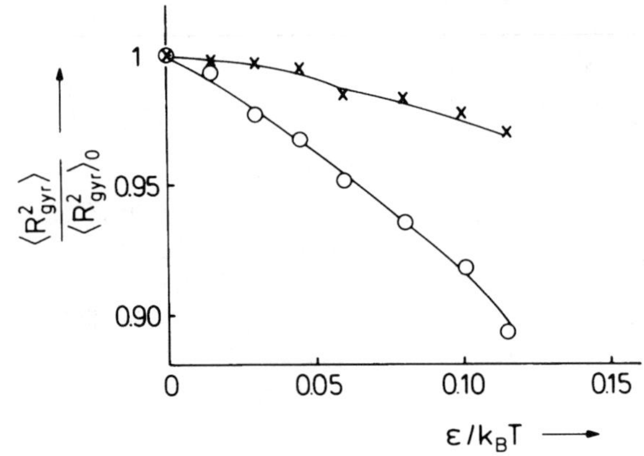

(a)

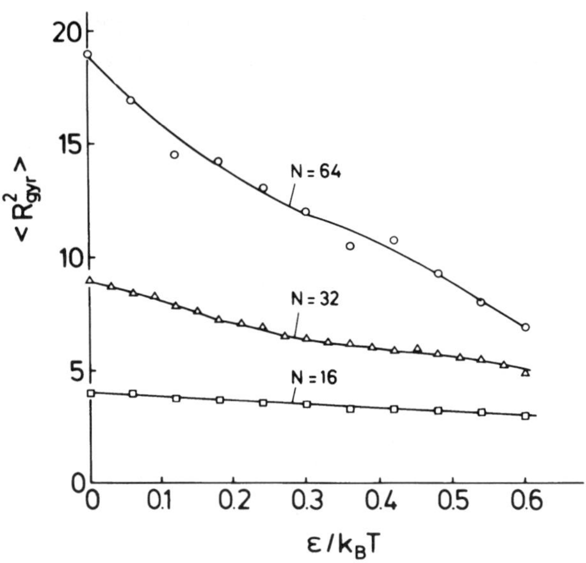

(b)

Fig. 7.30 (a) Normalized mean-square gyration radius at constant $\phi_A/(1-\phi_V) = 0.9$ plotted vs. inverse temperature for the model of Fig. 7.3 on the simple cubic lattice, choosing $N = 32, \phi_V = 0.6, \epsilon_{AB} = 0, \epsilon_{AA} = \epsilon_{BB} = -\epsilon$. The crosses refer to the majority component (A), the circles to the minority component (B). The largest value of ϵ/k_BT shown corresponds to a state at the coexistence curve. (From Sariban and Binder.[79]) (b) Mean-square gyration radius of an isolated B chain in an A-rich matrix, for the same model as in (a) but three chain lengths. Note that the inverse critical temperatures for these mixtures are $\epsilon/k_BT \approx 0.166(N=16), 0.092(N=32)$, and $0.049(N=64)$, respectively.[79] (From Sariban and Binder.[100])

that for $N\epsilon/k_BT \gtrsim 5$, i.e., $T_c/T \gtrsim 3$, isolated chains are already in a collapsed configuration ($\langle R^2\rangle_{\epsilon/k_BT} \propto \langle R^2_{\text{gyr}}\rangle_{\epsilon/k_BT} \propto N^{2/3}$ while $\langle R^2\rangle_0 \propto \langle R^2_{\text{gyr}}\rangle_0 \propto N$).

The behavior in Figs 7.30(b) and 7.31 signifies a serious breakdown of the simple RPA, of course—remember that the simple RPA would imply hor-

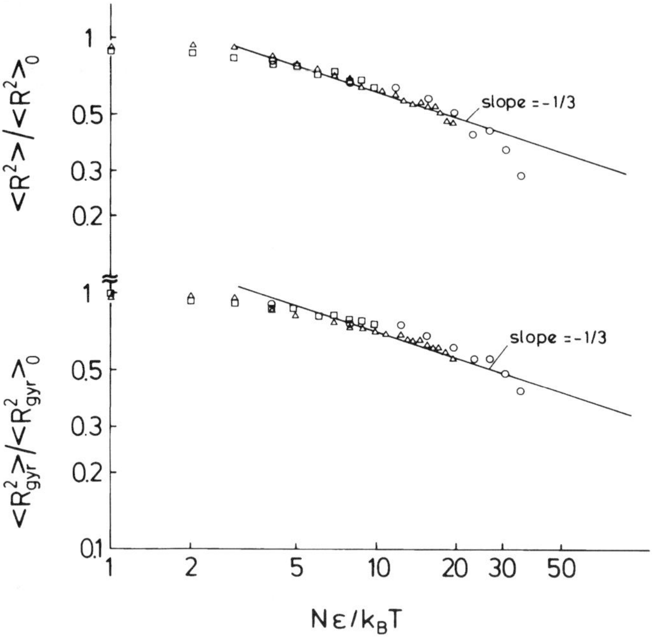

Fig. 7.31 Data such as shown in Fig. 7.30(b) replotted in scaled form on a log–log plot vs. $N\epsilon/k_BT$. Three values of N are included, $N = 16$ (squares), $N = 32$ (triangles) and $N = 64$ (circles). Upper part refers to the mean-square end-to-end distance and lower part to the mean-square gyration radius of the minority chain, in both cases normalized by the value of the noninteracting limit, $\epsilon/k_BT \to 0$. Straight lines with slope $-1/3$ indicate the collapsed behavior. (From Sariban and Binder,[100] in replotted form.)

izontal straight lines in all these plots. While qualitatively such deviations can be interpreted in terms of the screening of interactions resulting from a perturbation theory based on the Edwards[185] Hamiltonian,[24,61,186,187] a quantitative description of data such as shown in Figs 7.30 and 7.31 is still a challenge to theory!

7.3.5 Interdiffusion and phase separation kinetics

As already discussed in Section 7.2.6, studies of interdiffusion have focused on the nonequilibrium behavior where in a compatible (AB) system a layer of pure A-polymers is brought on a layer of pure B-polymers, and one studies the broadening of the resulting interfacial profile between A- and B-monomers with time (Fig. 7.22).[68,118] These studies have attempted to address theoretical ideas about the relation between interdiffusion (D_{int}) and self-diffusion constants (D_s) in blends, and to check for the so-called "thermodynamic factor". In fact, if one extends simple mean-field concepts to the dynamics of interdiffusion in polymer mixtures,[10,11] one finds that the effect of the interactions (as described by the Flory–Huggins parameter χ) for a symmetric blend ($N_A = N_B = N$) can be cast in the simple form[10,11,68]

$$D_{\text{int}}(\chi)/D_{\text{int}}(\chi = 0) = 1 - 2\phi_A\phi_B N\chi. \tag{7.41}$$

Here the right-hand side of the equation is simply the "thermodynamic factor" which hence should be linear in the variable $N\chi$. The Monte Carlo test of eq. (7.41) revealed[68] a rather pronounced curvature of this ratio $D_{\text{int}}(\chi)/D_{\text{int}}(\chi = 0)$ as function of χ, however; also the short chains accessible (N = 10, 20, and 40, respectively) did not conform to the expected scaling in terms of the variable $N\chi$. This problem certainly deserves further study, however, since the chain lengths used are very short, and there are also possible systematic errors: in principle, one should deduce $D_{\text{int}}(\chi)$ from interdiffusion of layers which differ in their concentrations $\phi_A^{(1)}, \phi_A^{(2)}$ only a little (rather than having $\phi_A^{(1)} = 0, \phi_A^{(2)}/(\phi_A^{(2)} + \phi_B^{(2)}) = 1$, as done in the available studies.[68,118] Already in Section 7.2.6 it was pointed out that one cannot compare such studies to experiment directly, since the hydrodynamic interactions left out in the Monte Carlo work do have important physical consequences.

Similar restrictive comments need to be made about the existing simulations of spinodal decomposition in polymer blends. First studies addressed the two-dimensional case.[80,81] In this case, there is no chain interpenetration, and since polymer mixtures in $d = 2$ dimensions are not predicted to become mean-field-like for $N \to \infty$, one expects the same behavior as in small molecule mixtures, and this is what has been found.[80,81] Using the slithering snake algorithm,[146,147] the dynamics of the model is neither realistic at short times (where the chain dynamics should be described by Rouse-model type motions[188,189]) nor at late times (where hydrodynamic effects play a role, as discussed above,[163] at least in $d = 3$ dimensions).

Here we only briefly discuss studies[42] of early stages of spinodal decomposition for the model of Fig. 7.3 on the simple cubic lattice, where chain motions as shown in the upper part of Fig. 7.5 are used: this model is expected to yield a Rouse-like dynamics. Since the available chain lengths are very short ($N \leq 32$) and the model that has been studied has a relatively high vacancy content ($\phi_V = 0.6$), the chains are definitely not entangled. Even for this limit, where the chain dynamics is much faster than it would be for dense melts of highly entangled chains[189] only rather deep quenches ($T_c/T \gtrsim 2$) are accessible, because otherwise the kinetics would be too slow because of critical slowing down[94]; and in addition the critical wavelength $\lambda_c = 2\pi/q_c$ of spinodal decomposition[10–13] would be rather large and necessitate the simulation of very large systems. From these remarks it is obvious that even when one uses crudely simplified coarse-grained models, the simulation of spinodal decomposition of polymer blends for shallow quench depths ($T_c/T \approx 1.1$) and long chains ($10^2 \leq N \leq 10^3$), as they are used in experiment,[13,14] would be prohibitively difficult. Only for relatively short chains and deep quenches have such studies thus far been at all feasible.[42,121]

Figure 7.32 shows the time evolution of the equal-time collective structure $S(q,t)$ and compares it to predictions of the linearized theory of spinodal decomposition.[10,11,180,190] While the latter predicts roughly correctly the position q_m, where $S(q,t)$ in the early stages develops a peak, one cannot identify a critical wavenumber q_c which would have the predicted independence of time, $S(q_c,t) = \text{const}$: this would imply that the curves shown in Fig. 7.32(a) should have a common time-independent intersection point. And while the linearized theory[10,11,180,190] predicts an exponential increase of the intensity with time, the observed increase (Fig. 7.32[b]) is linear in time or even slower. In this respect, the behavior found in this simulation[42] is more similar to small-molecule mixtures[44,191,192] rather than mixtures of polymers with high molecular weight, where one does find the predicted[10,11] initial regime of exponential growth.[13,14] This discrepancy is not unexpected, since the simulation deals with very short chains and relatively high vacancy content. Figure 7.33 then compares the normalized critical growth rate $q^{-2}dS(q,t/dt|_{t=0}$ with the corresponding theoretical prediction.[10,42] Although the general trend is similar, the simulation results do not yield any clear evidence for the critical wavenumber q_c where the growth rate τ_q^{-1} of the structure factor $S(q,t)$ vanishes, according to the linearized theory.[10,190–192] Figure 7.33 shows that for deep quenches the "Cahn plot",[190–192] where $q^{-2}dS(q,t)/dt|_{t=0}$ is plotted versus q^2, is nonlinear already on the level of the linearized theory, unlike the case of shallow quenches ($T_c/T - 1 \ll 1$).

Thus the curvature of the plot in Fig. 7.33(a) is no evidence against the usefulness of the linearized theory. We also emphasize that both simulation and theory predict that for deep quenches the growth of concentration fluctuations starts on wavelengths less than a coil size ($q^2\langle R^2_{\text{gyr}}\rangle > 1$). At later times, however, coarsening sets in (Fig. 7.34) and the characteristic wavenumber $q_1(t)$ becomes steadily smaller as time t after the quench increases. At late times, however, the Monte Carlo simulation of quenching experiments should not be trusted, since it does not account for the effect of hydrodynamic interactions, as mentioned above. Finally we mention that in deep quenches there is also a pronounced chain contraction that occurs during the very early stages of the quench.[42]

7.3.6 Surfaces of polymer blends and wetting transitions

In Section 7.2.6 we have already discussed that choosing a $L_\parallel \times L_\parallel \times L_\perp$ thin film with two $L_\parallel \times L_\parallel$ hard walls a distance $L_\perp$ apart it is simply possible to study surface enrichment of polymer blends, by choosing a wall potential that prefers one component (Fig. 7.21[b]). If we choose conditions such that for a symmetric mixture ($N_A = N_B = N$) the system is at a

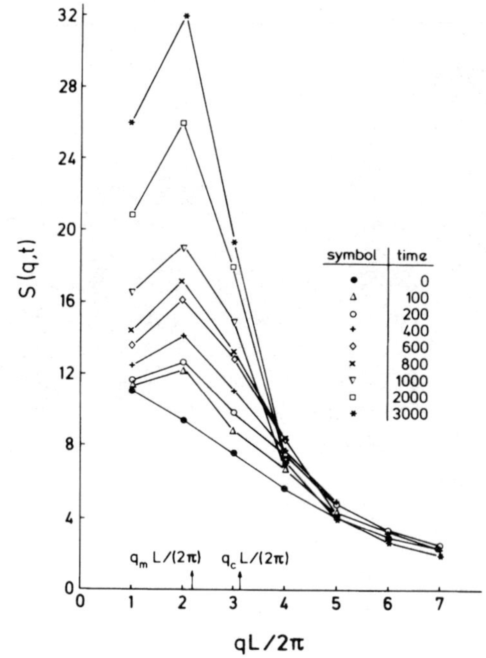

(a)

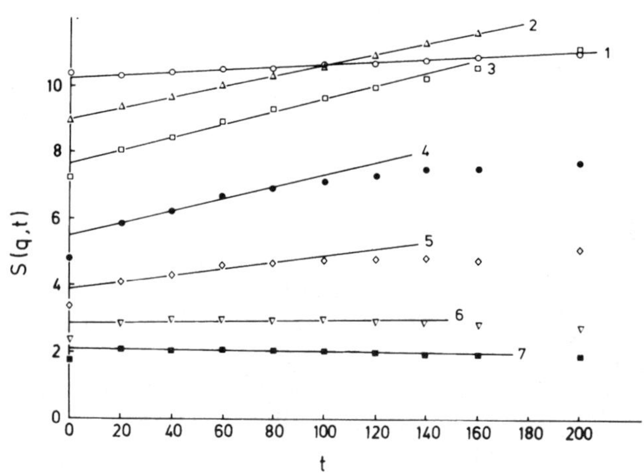

(b)

Fig. 7.32 (a) Collective structure factor $S(q,t)$ for a symmetric polymer mixture ($N_A = N_B = N = 32, \phi_V = 0.6$, the model of Figs 7.3 and 7.5 on the simple cubic lattice with $\epsilon = \epsilon_{AB}, \epsilon_{AA} = \epsilon_{BB} = 0$ is used) plotted vs. wavenumber q at various times t after a quench from $\epsilon/k_BT = 0$ to $\epsilon/k_BT = 0.20$, at critical concentration $\phi_A = \phi_B = (1-\phi_V)/2$. Choosing $\vec{q}$ in a lattice direction, $S(q,t)$ is defined only for the discrete values $q_\nu = (2\pi\nu)/L, \nu = 1, 2, 3, \ldots$ Thus only the points at the integer values of $qL/(2\pi)$ are significant, and only for clarity have these points been connected by straight lines. Note that $\epsilon/k_BT_c = 0.105$ and $L = 40$ was used. Time is measured in attempted Monte Carlo steps per monomer. Arrows show the predictions of the linearized theory of spinodal decomposition[10] for the wavenumber of maximum growth (q_m) and the critical wavenumber (q_c). (b) Same as (a) but $S(q,t)$ plotted vs. time for the seven smallest values q_ν (data are labeled by ν). Slopes of the straight lines indicate estimation of the initial growth rate $dS(q,t)/dt|_{t=0}$. (From Sariban and Binder.[42])

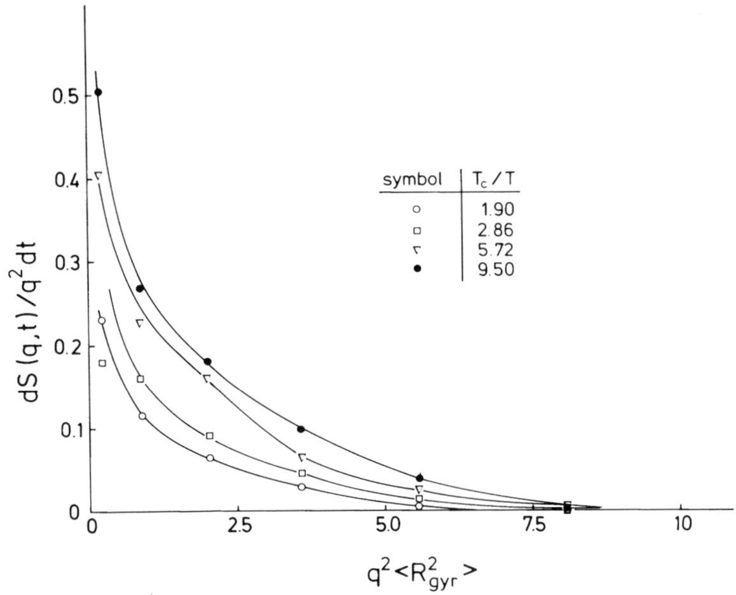

(a)

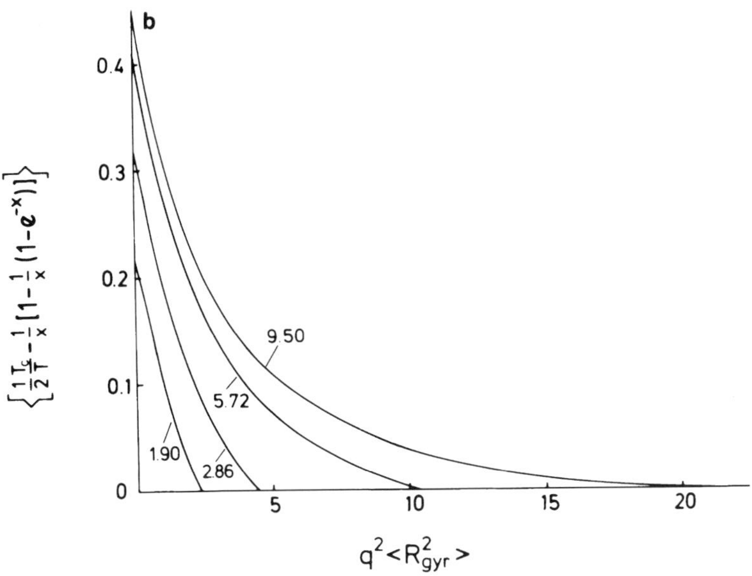

(b)

Fig. 7.33 (a) Initial growth rate for quenches that lead through the critical point and four different ratios of T_c/T (as indicated in the figure) plotted vs. $q^2\langle R^2_{\text{gyr}}\rangle$. Data for $dS(q,t)/dt$ are estimated, as shown in Fig. 7.32 (b). (b) Plot of the normalized relaxation rate τ_q^{-1}/q^2 of the linearized theory of spinodal decomposition for quenches that lead through the critical point $(\phi_{\text{crit}}, T_c)$ to a temperature T versus $x = q^2\langle R^2_{\text{gyr}}\rangle$. The theory of Ref. 10 yields $\tau_q^{-1}/q^2 = \tau_o\{\frac{1}{2}T_c/T - \frac{1}{x}[1 - \frac{1}{x}(1 - \exp(-x))]\}$; the time scale factor τ_o in the figure has been absorbed in the ordinate scale. Various curves are plotted for the same values of T_c/T as part (a). (From Sariban and Binder.[42])

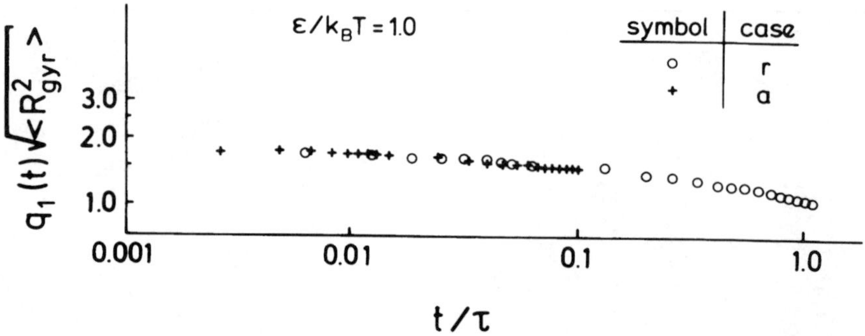

Fig. 7.34 Log–log plot of the first moment $q_1(t)$ of the structure factor $S(q, t)$ versus time. Here $q_1(t)$ is normalized with the gyration radius $\langle R^2_{\rm gyr}\rangle^{1/2}$, and the time t is normalized with a "diffusion time" $\tau \equiv \langle R^2_{\rm gyr}\rangle / D_{\rm eff}(t), D_{\rm eff}(t)$ being an effective self-diffusion constant of the chains. The data refer to $N = 32, \phi_V = 0.6, \phi_A = \phi_B = (1 - \phi_V)/2$ and $k_B T/\epsilon = 1$ (i.e., the deepest of the quenches shown in Fig. 7.33), both for the case of repulsive interactions $(r)[\epsilon_{AB} = \epsilon, \epsilon_{AA} = \epsilon_{BB} = 0]$ and for attractive (a) interactions $(\epsilon_{AB} = 0, \epsilon_{AA} = \epsilon_{BB} = -\epsilon)$. (From Sariban and Binder.[42])

temperature T below the critical temperature T_c at phase coexistence $(\Delta\mu = 0)$ in the A-rich phase, one may observe a wetting transition.[46–50,182,193,194] This wetting transition (which was recently also observed experimentally[159] shows up in the Monte Carlo simulation (Fig. 7.21[b]) as a divergence of the excess density ϕ_s of the component (B) that is enriched at the wall, $\phi_s = \sum_{z=1}^{L_\perp/2}[\phi_B(z) - \phi_\infty]$, where the bulk density $\phi_B(z \to \infty) = \phi_\infty$ is taken in practice as the density in the center of the film, as long as the system is still "nonwet" (= incompletely wet,[194] respectively). Figure 7.35(a) shows that this wetting transition can be first order (then ϕ_s discontinuously jumps from a finite value to infinity) or second order (then ϕ_s should have a logarithmic divergence when one approaches the transitions from the incompletely wet side[46,194]). For a first-order wetting transition, the local concentration near the wall $(\phi_1 = \phi_B(z = 1))$ jumps discontinuously to a value which is in excess of the other branch of the coexistence curve, while for a second-order wetting transition ϕ_1 varies perfectly smooth (Fig. 7.35[b]). The resulting phase diagram (Fig. 7.36) disagrees in one important aspect with mean-field theory,[46] as it yields a second-order wetting transition at rather low temperatures and a first-order wetting at higher temperatures, while mean-field theory predicts a second-order wetting transition near the bulk critical point only. The reason for this discrepancy is unclear. The change of the chain linear dimensions near the wall has also been studied for this model,[157] but it turns out that there is little difference between A and B chains near the wall, despite the surface enrichment: only in the bulk are B-chains distinctly smaller, since they are rather dilute and the chain contraction effect discussed in Section 7.3.4 occurs.

(a)

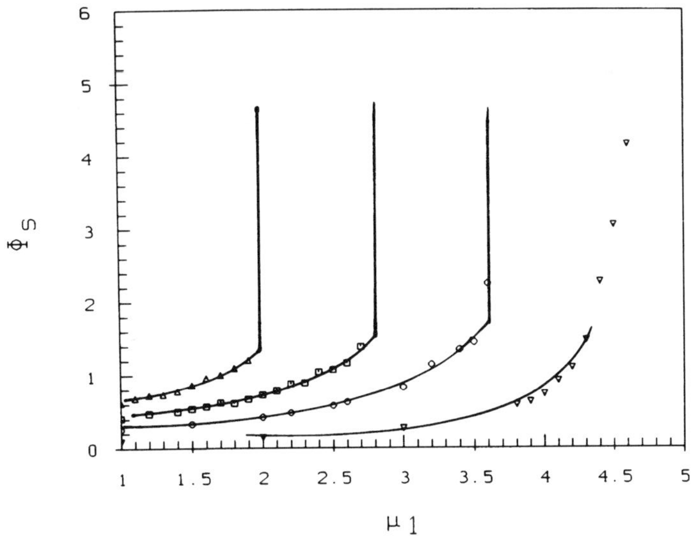

(b)

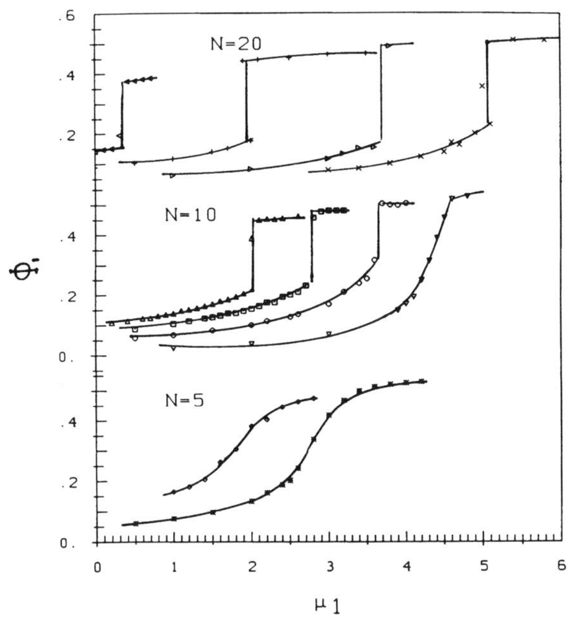

Fig. 7.35 (a) Surface excess density ϕ_s of a symmetric polymer mixture of chain length $N_A = N_B = N = 10$, using the bond fluctuation model with $\phi_V = 0.5$ on the simple cubic lattice and $\Delta\mu = 0$, linear dimensions $L_\parallel = 100, L_\perp = 40$. Interactions in the bulk are $\epsilon_{AA} = \epsilon_{BB} = -\epsilon_{AB} = -\epsilon < 0$, if the distance between monomers is less than $\sqrt{8}$ lattice spacings; interactions at the walls are $\epsilon_{WA} = 0, \epsilon_{WB} = -\mu_1\epsilon$ if $z = 1$ or $z = L_\perp$, respectively. Data are shown as function of μ_1 for $k_BT/\epsilon = 11$ (inverted triangles), 15 (circles), 17 (squares) and 18.5 (triangles on their baseline), respectively. (b) Surface concentration ϕ_1 of the enriched species vs. surface chemical potential μ_1, at bulk concentration ϕ_∞ for the model of part (a) but three different chain lengths. Values of ϕ_∞ (and corresponding temperatures k_BT/ϵ) are, from right to left: for $N = 20, 0.01(25), 0.022(29), 0.048(35), 0.095(39)$; for $N = 10$, 0.004(11), 0.018(15), 0.033(17), 0.047(18.5); and for $N = 5, 0.014(8), 0.038(10)$. (From Wang and Binder.[50])

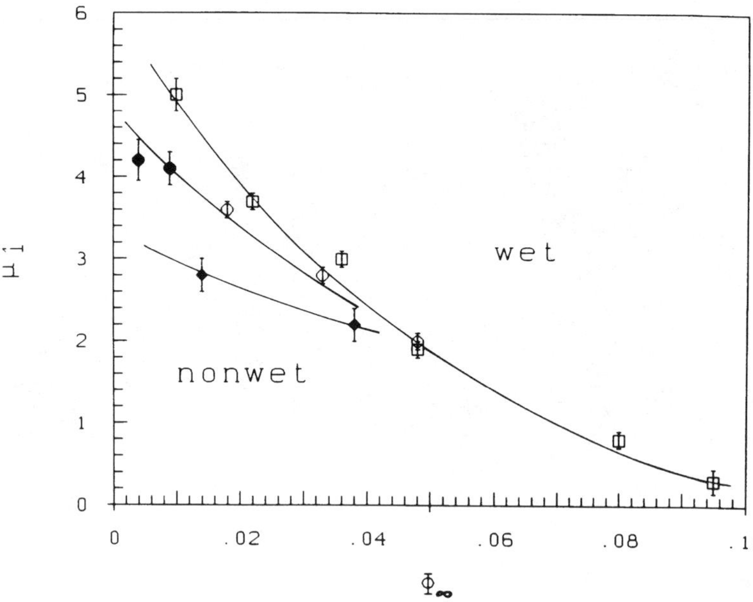

Fig. 7.36 Phase diagram of the wetting transition, in the plane of variables μ_1 and ϕ_∞, for the model of Figs 7.21 and 7.35. Different symbols denote different chain lengths: $N = 5$ (diamonds), $N = 10$ (circles), and $N = 20$ (squares). The solid symbols indicate that the transition is second-order. Note that the bulk critical concentration is $\phi_\infty = 1/4$ (since $\phi_V = 1/2$), and thus all data are far from bulk criticality. Curves are guides to the eye only. (From Wang and Binder.[50])

Cifra *et al.*[158] studied surface enrichment effects in blends for $N = 30$ (using the model of Fig. 7.3 on the simple cubic lattice) at various compositions in the one-phase region of the blend. They made the choice $\epsilon_{WA} = \epsilon_{WB} = \epsilon_{AA} = \epsilon_{BB} = -0.5$ and varied ϵ_{AB}, choosing $\epsilon_{AB} < 0$ as well to ensure miscibility in the bulk. Although $\epsilon_{WA} = \epsilon_{WB}$ in their study, they find that the majority component is enriched both at the hard wall and at the interface polymer–vacuum which they also study. This enrichment is all the more pronounced the more asymmetric their blend composition is, but occurs only if $\epsilon = \epsilon_{AB} - (\epsilon_{AA} + \epsilon_{BB})/2$ is nonzero. Their model contains a nontrivial total density profile as well, which adjusts itself as a result of the energy parameters mentioned above. For the chosen parameters, Cifra *et al.*[158] find vacancy concentrations in the order of 5–15% in the bulk of their films and argue that the resulting nonzero finite compressibility of their model has the typical value for melts of linear polymers. Cifra *et al.*[158] argue that their results are also in contradiction with the simple mean-field theory, and they also study the change of coil dimensions near the wall (note that for their choice of attractive *A–B* interactions there is an increase of radii[195] with increasing interaction strength in the bulk, in contrast to the model with attractive *AA*, *BB* interactions discussed in Section 7.3.4).

Clearly, all these simulations of surface effects on polymer blends[50,157,158] are very stimulating, but at this point they may be considered to be first steps only. The effect of long range polymer wall interactions[48] should also be considered.

7.4 Results for block copolymers

While the simulation of phase transitions in polymer blends dates back[78,167] to 1987 and the necessary tools (finite size scaling, etc.) are well developed, studies of mesophase formation in block copolymers have only just begun,[24,52–55,84,85] and only rather short chain lengths have been accessible ($N \leq 60$). We consider this work as rather preliminary, and hence this section can be relatively short.

7.4.1 Test of the Leibler theory

First studies[52–54] addressed the transition of symmetrical block copolymers ($f = \frac{N_A}{N_A+N_B} = \frac{1}{2}$) from the disordered melt to the lamellar mesophase (Fig. 7.2), choosing the model of Fig. 7.16 (see Section 7.2.5 for a discussion) and focusing attention on the collective structure factor (Fig. 7.20[a]). According to the Leibler theory[21] it has the explicit form

$$NS_{\text{coll}}^{-1}(q) = F(x) - 2\chi N, \quad F(x) = \frac{x^4}{4}\left(\frac{x^2}{4} + e^{-x^2/2} - \frac{1}{4}e^{-x^2} - \frac{3}{4}\right)^{-1}, \tag{7.42}$$

where $x = q\langle R_g^2\rangle^{1/2}$ here, and $\langle R_g^2\rangle$ the mean square gyration radius of the total polymer. From eq. (7.42) it is easy to show that the minimum of $F(x)$ occurs at $x^* \cong 1.95$, and therefore $S_{\text{coll}}(q)$ has a maximum at a value independent of χ and N when one chooses x as the abscissa scale, as done in Fig. 7.20(a). It is clear that there are rather pronounced deviations from this prediction.

A further consequence results when we note that χ should simply be proportional to $\epsilon \equiv \epsilon_{AB}/k_BT$ for this model: one then predicts that $S_{\text{coll}}^{-1}(q^*)$ should simply decrease linearly with the product ϵN, and $S_{\text{coll}}^{-1}(q^*) = 0$ then determines the microphase separation transition temperature. In this section we have redefined the meaning of ϵ, in order to have a notation consistent with the original literature.[53,54] Figure 7.37 demonstrates that there is pronounced curvature on a plot of $NS_{\text{coll}}^{-1}(q^*)$ versus ϵN, rather than a linear decrease. This behavior is reminiscent of both experimental findings[56,66] and theoretical predictions due to Fredrickson and Helfand,[58] who have taken fluctuation corrections to the Leibler theory

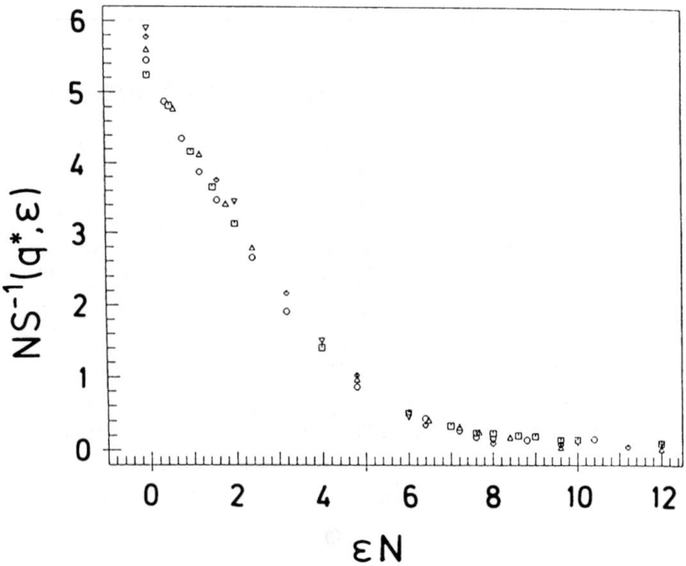

(a)

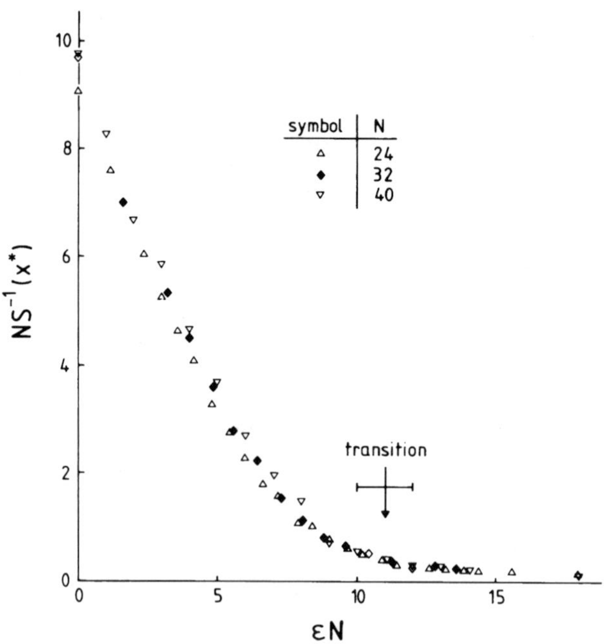

(b)

Fig. 7.37 The inverse collective structure factor $NS_{\text{coll}}^{-1}(q^*)$ is plotted vs. the product ϵN (note $\epsilon = \epsilon_{AB}/k_BT$ here), for $f = \frac{1}{2}$ (a) and $f = \frac{3}{4}$ (b). Several choices for N are included (in case [a], these are $N = 16$, circles; 20, squares; 24, triangle on base; $N = 32$, diamonds; and $N = 40$, inverted triangles). In case (a), the transition is estimated to occur for $\epsilon N \approx 8.3 \pm 0.7$, while in case (b) it occurs for $\epsilon N = 11 \pm 1$ (arrow). (From Fried and Binder.[53,55])

into account. They predict that for $f = 1/2$ the transition should be a fluctuation-induced first-order transition, while according to the Leibler theory it is second-order for $f = 1/2$ and first-order for $f \neq 1/2$. However, due to finite-size rounding of the transition the weak singularity in $S_{\text{coll}}^{-1}(q^*)$ that can be seen experimentally[66] is not visible in the simulations at all (Fig. 7.37). Neither can one deduce accurate transition temperatures from the specific heat maxima[24] (Fig. 7.17[b]) without finite size scaling analysis, and hence a significant test of the Fredrickson–Helfand theory[58] is not yet possible (this would also require much longer chains than are so far available). We also note that this theory implies that a simple scaling in terms of the variable ϵN is not valid. While the data of Fig. 7.37 are suggestive of such a scaling, the small but systematic deviations prevent one from drawing conclusions that are too definite. Note also that the range of chain lengths N available in Fig. 7.37 is simply too small for a significant test of this point.

As will be discussed in the next section, the shift of q^* with increasing ϵN (Fig. 7.20) can be interpreted as resulting from an onset of chain stretching. While experimentally[57] it was suggested that the transition from Gaussian coils to stretched chains that occurs at a temperature T^* above T_c is rather sharp, the more complete simulation data on this point[54] show that this stretching of the chains sets in very gradually.

7.4.2 Chain conformations and the breakdown of the random phase approximation (RPA)

It is one of the huge advantages of simulations that the interplay between coil conformations and thermodynamic properties can be studied in detail, while a similarly complete experimental study would be very cumbersome to perform. It turns out that already in the disordered phase of block copolymers, distinct changes of the polymer configuration with increasing interaction strength occur: they were first detected by simulations[52] and later[57] confirmed by experiment; although one can understand the existence of these phenomena by theoretical arguments qualitatively,[24] we are not aware of a quantitatively reliable theory yet that could explain these phenomena.

Figure 7.38 summarizes some of the main results[53–55] (see also Fig. 7.17[a]). One sees that while the gyration radius increases only by about 7% from the noninteracting case to the transition, an increase of 15% occurs for the distance R_{AB} between the center of gravities of the A-block and the B-block. This behavior is due to a gradual stretching of the chains; the block copolymer conformation deforms from a Gaussian coil towards a dumbbell-like conformation. If one were formally to associate a positive charge to A and a negative charge to B, the chain conformation change would imply the formation of a rather large dipole moment. While due to this gradual chain stretching the total gyration radius increases, the gyration

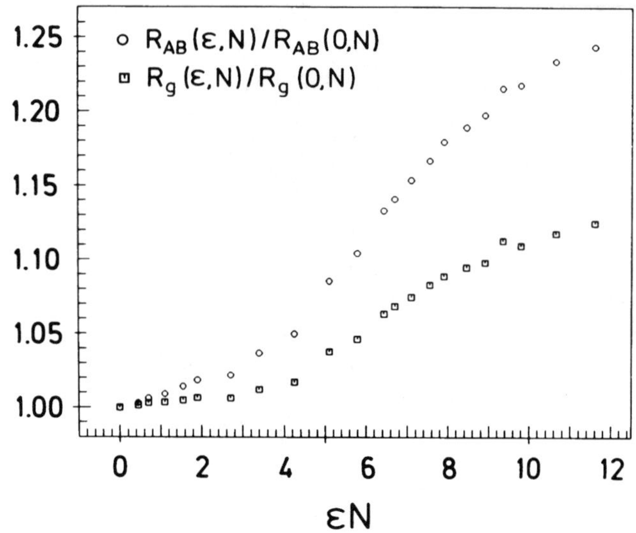

(a)

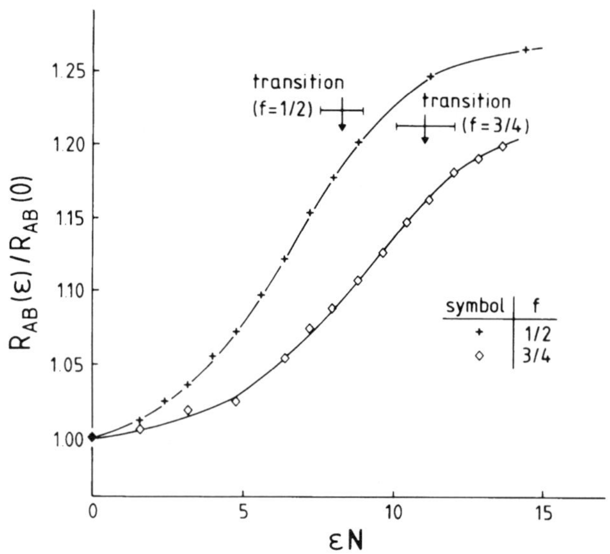

(b)

Fig. 7.38 (a) Radius of gyration $R_g \equiv \langle R_{\text{gyr}}^2 \rangle^{1/2}$ and distance $R_{AB} \equiv \langle (\vec{R}_{CM}^A - \vec{R}_{CM}^B)^2 \rangle^{1/2}$ between the center of mass $\vec{R}_{CM}^A$ of the A-block and the center of mass $\vec{R}_{CM}^B$ of the B-block are shown as a function of ϵN where $\epsilon = \epsilon_{AB}/k_B T$ and the model of Fig. 7.16 with $\phi_V = 0.2$ on the simple cubic lattice is used. Data for different ϵ and N are averaged together at the same values of the product of ϵN to gain statistics. (From Fried and Binder.[53]) (b) Distance R_{AB} normalized by its value of $\epsilon = 0$ plotted vs. ϵN, both for $f = 1/2$ and $f = 3/4$. Arrows show the location of the transition temperatures. (c) Radius of gyration R_g of the total chain, of the A-block (R_g^A) and of the B-block (R_g^B) plotted vs. ϵN for $f = 3/4$. Three different chain lengths are indicated by different symbols. Arrow shows the transition to the cylindrical mesophase. (d) Plot of the normalized principle radii $(R_i/R_g)^2_{\epsilon N}/(R_i/R_g)^2_{g}$ vs. ϵN, for $f = 3/4$ and three choices of N as indicated. (Parts [b]–[d] from Fried and Binder.[55])

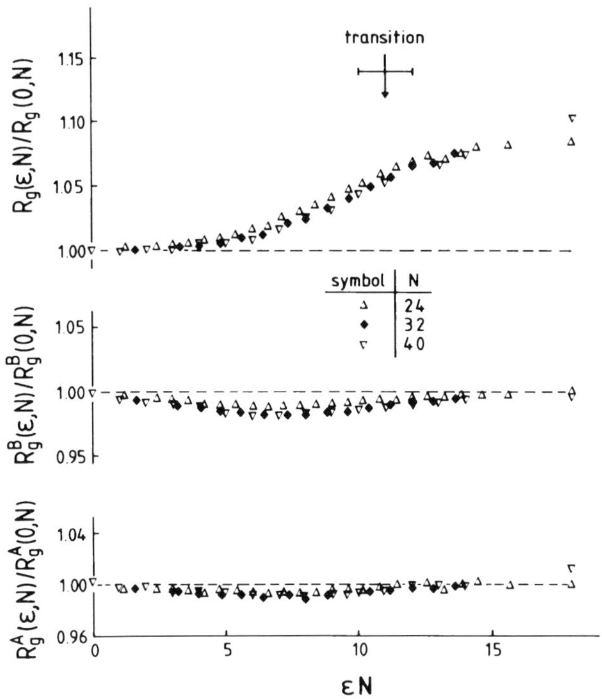
transition
symbol
N
24
32
40
$R_g(\varepsilon,N)/R_g(0,N)$
$R_g^B(\varepsilon,N)/R_g^B(0,N)$
$R_g^A(\varepsilon,N)/R_g^A(0,N)$
εN

(c)

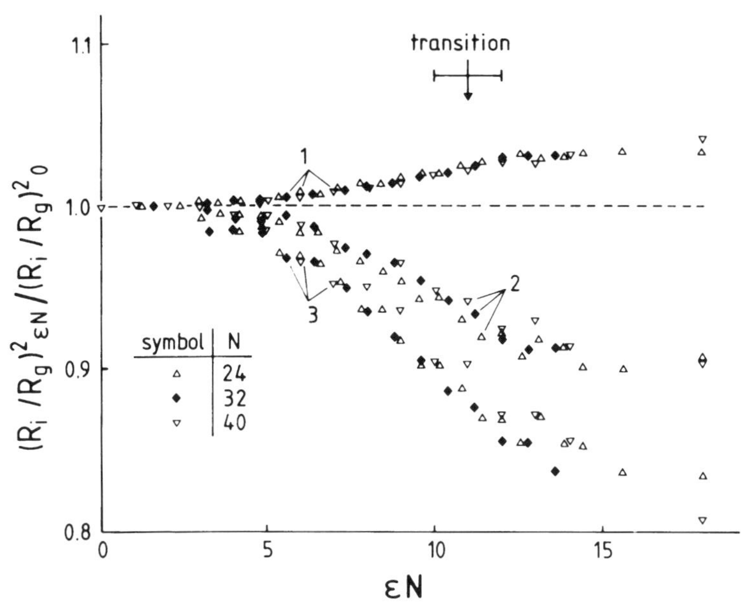
transition
1
2
3
symbol
N
24
32
40
$(R_i/R_g)^2_{\varepsilon N}/(R_i/R_g)^2_0$
εN

(d)

radii of the individual blocks contract initially (Fig. 7.38[c]). Only in the transition region does the stretching tendency begin to take over, also for the individual blocks. This chain stretching also means that the shape of the polymer configuration gets more and more elongated, as a study of the principal radii of the chains reveals[54,55] (Fig. 7.38[d]). One radius increases while the other two decrease. While for $f = 1/2$ one finds the shrinking of the other two to be identical, for $f = 3/4$ (where an asymmetric dumbbell forms!) the smallest radius has the most pronounced reduction.

We emphasize at this point that all these pretransitional chain stretching effects found in the simulations are at variance with the simple RPA treatment as embodied in Leibler's theory[21]: this simple form of the RPA would mean only that all the ratios shown in Fig. 7.38 should be strictly unity, irrespective of ϵ and N. Thus the simulations reveal rather drastic deviations from RPA, a hitherto somewhat unexpected result, since for dense melts the RPA has been taken as basically exact by many researchers. In this context, it is also important to recall that these deviations from the RPA should not be attributed to the vacancy content of the model (recall that Fig. 7.17(a) has studied this behavior for variable vacancy content, and behavior independent of ϕ_V for $T > T_c$ was found[24]).

It is now possible to account qualitatively for the dumbbell formation seen in Fig. 7.38, by a theory[24] that computes effective renormalized interactions from the Edwards Hamiltonian. However, a quantitative description of these data presented in Figs 7.17(a) and 7.38 is still lacking. While experimental results[57,66] have also been interpreted in terms of chain stretching, it is important to recall that experiments up to now could measure $S_{\text{coll}}(q)$ only: an independent measurement of both q^* and R_g for the same material has thus far not been feasible. This fact again illustrates one advantage of simulation, namely that it makes a much more detailed insight possible in that all quantities of interest can be "measured" simultaneously from one model system.

7.4.3 Asymmetric block copolymers; ring polymers

For linear diblock copolymers with asymmetric composition ($f \neq 1/2$) the lamellar mesophase (Fig. 7.2) competes with the (hexagonal) cylindrical structure and with cubic structures.[16–18,21] In the hexagonal structure, the blocks of the minority (B)-component form long B-rich cylindrical rods which are oriented parallel to each other in the A-rich matrix, and form a triangular lattice in the plane perpendicular to the orientation vector of the cylinders. The occurrence of this ordering has been inferred[55] from Monte Carlo simulations of lattice models with $f = N_A/(N_A + N_B) = 3/4$ (also at compositions near this value[196]), from an inspection of "snapshot pictures" of the chain configurations. For still smaller B-content one expects the formation of micelles,[17] i.e., the block copolymers gather together in objects like star polymers with many arms, the B-rich spherical cores of these "stars" then arrange in a cubic lattice in the ordered phase.[16–18] So far we are not aware of any Monte Carlo work on these phases, nor have any studies been feasible addressing other, more complicated, ordered phases of block copolymers.[16,62]

For asymmetric composition already in the disordered phase the peak position q^* of the collective structure factor depends on f (cf. also Figs 7.20[a], [b]) and this problem has already been addressed by Leibler's[21] RPA theory. Figure 7.39 shows that in the noninteracting limit the theory predicts this position q^* correctly, but with increasing strength of the interaction ϵN the peak position $q^*(T)$ decreases, and in the transition region $x^* = q^* \langle R^2_{\text{gyr}} \rangle^{1/2}$ has only about 80% of its value in the noninteracting limit. Unfortunately, neither an accurate location of the transition temperature nor a study of the order of the transition has been possible so far. The value of ϵ at the transition increases[55,196] with increasing deviation of f from 1/2, which is qualitatively consistent with theoretical predictions.[16,21,58,63] Unfortunately, the Monte Carlo data are too crude to allow any quantitative test of the theory. Also Fig. 7.39 reveals small but systematic deviations from the scaling with the variable ϵN, but again the data are too crude to

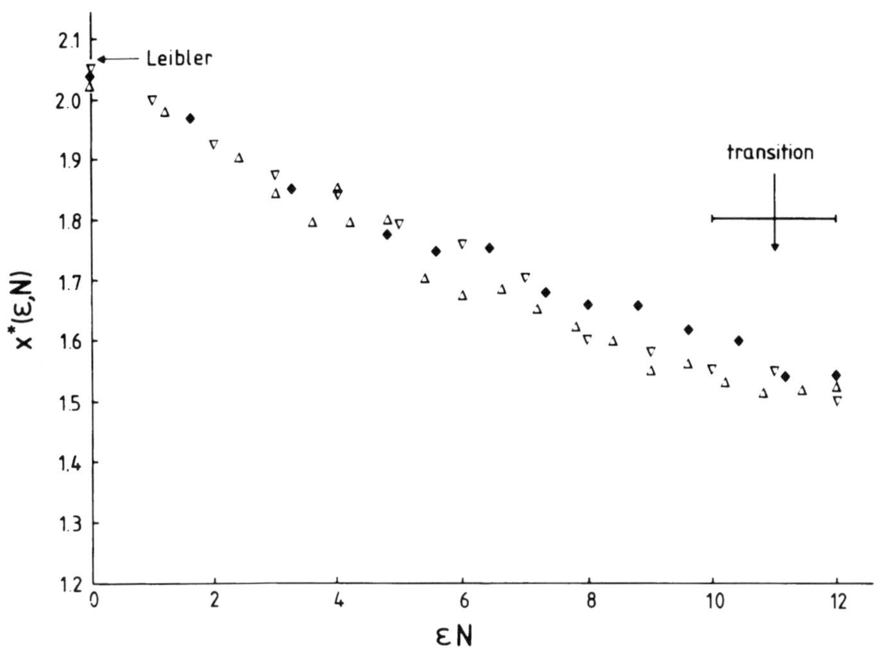

Fig. 7.39 Peak position $x^*(\epsilon, N) = q^* R_g(\epsilon, N)$ where $S(q, \epsilon)$ has its maximum plotted vs. ϵN for $f = 3/4$ and $N = 24$ (triangles on base), $N = 32$ (diamonds) and $N = 40$ (inverted triangles). Here $\epsilon \equiv \epsilon_{AB}/k_B T > 0$, $R_g = \langle R^2_{\text{gyr}} \rangle^{1/2}$. (From Fried and Binder.[55])

allow a statement whether this is due to the fluctuation corrections of Fredrickson and Helfand.[58]

A very interestingt study concerns microphase separation in topologically constrained ring polymers[85]: here the two types of chains (A, B) are covalently bonded at both ends to form closed loops (avoiding catenation, so no rings are permanently locked together). Here, already in the noninteracting case, the coils do not have Gaussian statistics; due to the topological constraint the scaling exponent ν in the relation $R_g \propto N^{\nu}$ is not $\nu = \frac{1}{2}$ but $\nu \approx 0.40$–0.46 in a ring polymer melt.[85,197,198] Consequently, one might expect the deviations from RPA to be much more dramatic than for linear polymers. But it turns out that the ratio of critical temperatures $T_c^{\text{linear}}/T_c^{\text{cyclic}} \approx 1.7 \pm 0.1$ (which could only be estimated very roughly, since the same problems as discussed in Sections 7.2.5 and 7.4.1 apply) is roughly in agreement with the RPA prediction.[199] Weyersberg and Vilgis[85] also find that the structure is lamellar with a wavelength smaller by a factor of 0.51–0.55 than for corresponding linear polymers.

7.4.4 Block copolymers in reduced geometry: thin films, interfaces, etc.

Already in Sections 7.2.5 and 7.2.6 we mentioned that one way to stabilize the direction of the orientation vector describing the lamellar order is the choice of a thin-film geometry. Choosing a boundary field that favors the B-component on both walls, the order fits to the film geometry only if there is a special commensurability condition satisfied between film thickness D and wavelength $\lambda(T)$ of the lamellar order, namely[149] $D = n\lambda(T)$, where n is an integer. Figures 7.18 and 7.19 have given evidence for this effect, and surface enrichment of the B-component at the walls was also seen in the disordered phase, as well as the enrichment of vacancies both at the walls and at the A–B-interfaces (Fig. 7.21[a]).[150]

It is interesting to analyze in more detail the phenomena that occur when the above commensurability condition is not satisfied. Figure 7.40 presents some snapshot pictures where only the A-monomers are shown as black dots, neither B-monomers nor bonds connecting the monomers being shown. It is known from a study of the order parameter profiles (such as shown in Fig. 7.18) that the intrinsic wavelength of lamellar order that the system wants to form is about $\lambda \approx 9$ lattice spacings. Choosing $D = 30$, with a slight distortion ($\lambda = 10$ instead of $\lambda = 9$) three wavelengths fit into the film, if one orients the lamellae parallel to the wall (Fig. 7.40[a]). On the other hand, for $D = 14$ choosing either one wavelength ($\lambda = 14$) or two wavelengths ($\lambda = 7$) would be too strongly distorting the natural order that the system wants to form: then a better alternative is to choose a perpendicular rather than parallel orientation of the lamellae, since with $L = 24$ and $n = 3$ only a slight distortion results ($\lambda = 8$). This snapshot picture shows that the profiles shown in Fig. 7.18(b) do not mean the system stays disordered at low temperatures; they only imply that the biasing effect of the wall is not strong enough to enforce orientation of the lamellae parallel to the film.

For $D = 24$ now, the system could form lamellae with $\lambda = 8$ and $n = 3$ with orientations either parallel or perpendicular to the walls, having chosen $L = 24$ as well. Figure 7.40(c) shows that in the simulation a domain state resulted instead, where two lamellae run parallel to the walls in the upper part of the film, and three perpendicular to the walls in the lower part. Thus with these choices of boundary conditions and linear dimensions one could also study domain interfaces in the ordered structure of block copolymers. So far this direction of research has not been exploited.

A very interesting problem is the adsorption of block copolymers at the interface between incompatible homopolymers,[200–207] which lower the interfacial tension and therefore act as compatibilizing agents in such blends. This phenomenon has been studied theoretically (e.g., Refs 200–203, 207), experimentally (e.g., Ref. 206), and by Monte Carlo simulation.[204] In the last work the A, B homopolymers are not included explicitly in the simulation, however, and their existence shows up only indirectly via suitable energy parameters which differ in the A-phase (for $z > L/2$) from those in the B-phase (for $z < L/2$). The A–B interface hence is sharp on the scale of the lattice spacing and treated as strictly localized. Wang *et al.*[204] treat L^3 lattices with lattice sizes up to $L = 50$ and up to 400 chains of composition $N_A = N_B = 10$ or variable f with $N = 20$ up to $f = 3/4$, and discuss the description of the block copolymer adsorption at the A–B interface in terms of Langmuir-type isotherms.

7.5 Discussion

In this chapter simulations of the thermodynamics of binary blends of linear polymers and of diblock copolymers were reviewed, with an emphasis on work devoted to testing elementary theories such as the Flory–Huggins theory of blends, the Leibler theory of block copolymer mesophase formation, and the random phase approximation. We emphasize that despite the length of this chapter, a complete coverage of the subject was not attempted—thus, for instance, we have neither treated simulations of block copolymer micellization[205] nor microphase separation in interpenetrating polymer networks.[208] The strategy of this article was to explain the kind of questions that one can answer with Monte Carlo simulations of polymer blends and block copolymer melts as well as the specific techniques that one uses in this context (semi-grand-canonical ensemble, histograms, finite size scaling, etc.) by describing typical "case studies" in some detail. For simplicity, these examples were for the most part taken from work in the group of the author. But recently there has been growing activity on these subjects in many other groups, for example particularly relevant work on

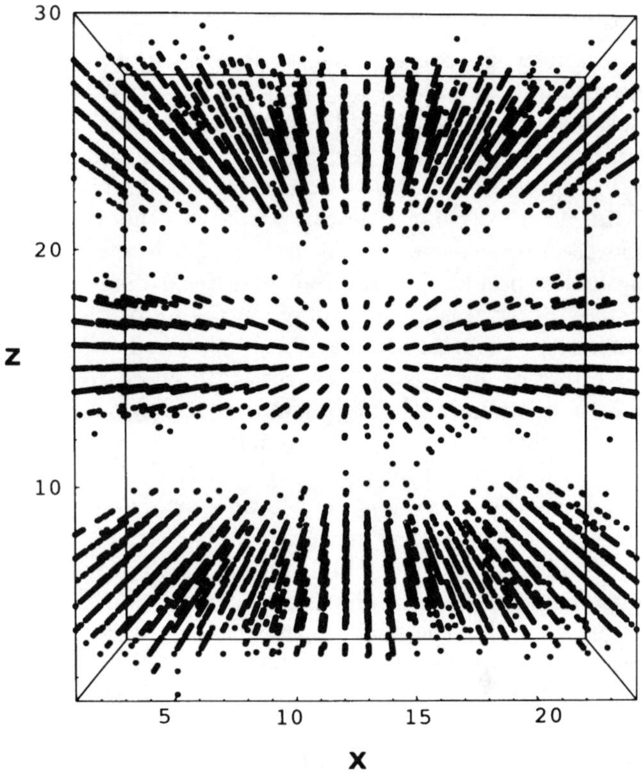

(a)

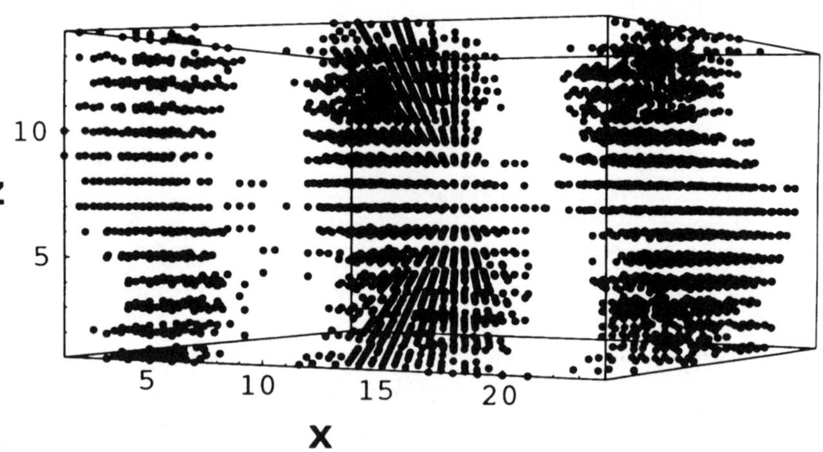

(b)

Fig. 7.40 Snapshot picture of typical configurations of the symmetrical block copolymer model (chain length $N = 16, f = 1/2, \phi_v = 0.2$) of Fig. 7.16 in a thin film geometry ($L \times L \times D$, with $L = 24$ and a repulsive interaction $\epsilon_{AS} = \epsilon_{AB}/2$ at the two hard $L \times L$ walls). All snapshots refer to $\epsilon_{AB}/k_BT = 0.6$, and show only A-monomers as block dots, while neither B-monomers nor vacancies are shown, and also the bonds connecting the monomers are not displayed. Part (a) refers to $D = 30$, part (b) to $D = 14$, part (c) to $D = 24$. (From Kikuchi and Binder.[150].

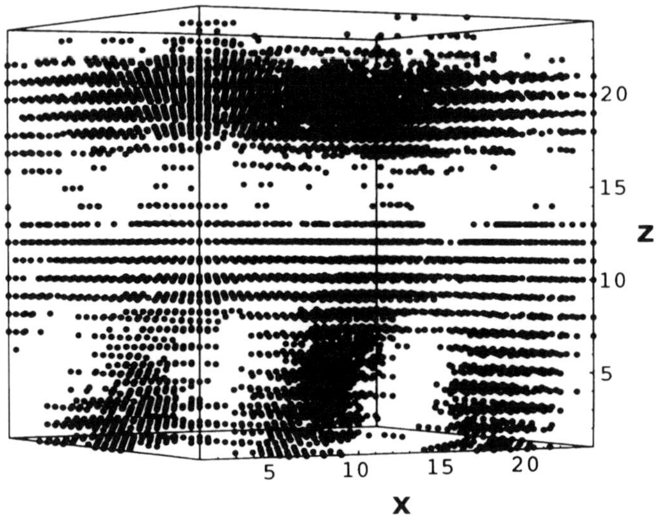

(c)

block copolymer simulations can be found in a very recent paper by Larson.[209]

It has been shown that the study of simplified lattice models has given valuable insight on various questions debated in the literature. For example, the simulations have shown that the Schweizer–Curro prediction that the critical temperature T_c of symmetrical blends scales with $\sqrt{N}$ is incorrect, and thus they have helped to find an improved version of this valuable integral equation theory which does not suffer from this defect. While already Flory–Huggins theory correctly predicts $T_c \propto N$, it severely overestimates the constant of proportionality in this relation, and thus there is still need for better theories. The Monte Carlo work provides benchmarks against which these theories can be tested stringently, since all model parameters in the simulations are precisely fixed, and one can perform a comparison between simulation and theory that is not obscured by unknown adjustable parameters. Similarly, the simulations have given valuable hints to chain conformational changes induced by the interactions, such as pretransitional stretching of chains in block copolymer melts.

For some problems, however, the shortness of the chains is still a limitation, as well as the lattice structure (which prevents the study of critical dynamics, later stages of spinodal decomposition, etc., where the fluid nature of real polymer melts and blends is essential). While future work could clearly address questions such as mixtures of block copolymers and homopolymers, asymmetry effects (different stiffness of chains, etc.) and interfacial structure, one needs to develop complementary molecular dynamics methods for the study of dynamical phenomena in mixtures and block copolymer melts.

Acknowledgments

The material presented in this review has developed in the course of a fruitful collaboration with A. Sariban, H.-P Deutsch, H. Fried, B. Minchau, M. Kikuchi, J.-S. Wang, and M. Müller. It is a great pleasure to thank them. Furthermore, the author has to thank particularly A. Weyersberg, A. Gauger, S. K. Kumar, and J. E. G. Lipson for allowing him to include material from their recent research (Refs 24, 30, 69, 83; Figs 7.4, 7.10, 7.11, 7.17). He also is very much indebted to numerous colleagues for stimulating interactions, in particular K. S. Schweizer, F. S. Bates, J. Klein, E. J. Kramer, G. H. Fredrickson, G. Meier, D. Schwahn, D. W. Heermann, K. Kremer, and W. Paul. The work from the author's group that is reviewed here has been supported by the Deutsche Forschungsgemeinschaft (DFG-Bi 314/3), by the Max-Planck-Institut für Polymerforschung Mainz, the Materialwissenschaftliches Forschungszentrum (MWFZ) at the University of Mainz, and—last but not least—by generous grants of computer time at the CRAY-YMP supercomputer at the Höchstleistungsrechenzentrum (HLRZ) Jülich and at the Siemens–Fujitsu VP 100 at the Regionales Hochschulrechenzentrum (RHRK) Kaiserslautern, respectively. Finally, thanks are due to B. Dünweg for a careful reading of the manuscript.

References

1. D. Levesque and J. J. Weis, in *The Monte Carlo Method in Condensed Matter Physics*, edited by K. Binder (Springer, Berlin, 1992), p. 121.
2. A. Z. Panagiotopoulos, *Mol. Simul.* **9**, 1 (1992).
3. M. P. Allen and D. J. Tildesley, *Computer Simulation of Liquids* (Clarendon Press, Oxford, 1987).
4. G. Ciccotti and W. G. Hoover (eds) *Molecular Dynamics Simulations of Statistical Mechanical Systems* (North-Holland, Amsterdam, 1986).
5. J. Jäckle, *Rep. Progr. Phys.* **49**, 171 (1986).
6. K. Binder, *Makromol. Chem., Macromol. Symp.* **50**, 1 (1991).
7. K. Binder, in *Computational Modeling of Polymers*, edited by J. Bicerano (M. Dekker, New York, 1992), p. 221.
8. H. E. Stanley, *An Introduction to Phase Transitions and Critical Phenomena* (Oxford University Press, Oxford, 1971).
9. P. G. de Gennes, *Scaling Concepts in Polymer Physics* (Cornell University Press, Ithaca, NY, 1979).
10. K. Binder, *J. Chem. Phys.* **79**, 6387 (1983).
11. K. Binder, *Colloid Polym. Sci.* **265**, 273 (1987).
12. K. Binder, *Adv. Polymer Sci.* **112**, 181 (1994).
13. T. Hashimoto, *Phase Transitions* **12**, 47 (1988).
14. T. Hashimoto, in *Materials Science and Technology, Vol. 12: Structure and Properties of Polymers*, edited by E. L. Thomas (VCH, Weinheim, 1993), p. 251.

15. U. Steiner, E. Eiser, J. Klein, A. Budkowski, and L. J. Fetters, *Science* **258**, 1126 (1992).
16. F. S. Bates and G. H. Fredrickson, *Annu. Rev. Phys. Chem.* **41**, 525 (1990).
17. R. A. Brown, A. J. Masters, C. Price, and X. F. Yuan, in *Comprehensive Polymer Science*, edited by G. Allen (Pergamon Press, Oxford, 1989), p. 155.
18. E. Helfand and Z. R. Wasserman, in *Developments in Block Copolymers I*, edited by I. Goodman (Applied Science Publishing, New York, 1982).
19. G. H. Fredrickson and K. Binder, *J. Chem. Phys.* **91**, 7265 (1989).
20. P. J. Flory, *Principles of Polymer Chemistry* (Cornell University Press, Ithaca, NY, 1953).
21. L. Leibler, *Macromolecules* **13**, 1602 (1980).
22. M. J. Huggins, *J. Chem. Phys.* **9**, 440 (1941).
23. P. J. Flory, *J. Chem. Phys.* **9**, 660 (1941).
24. A. Weyersberg and T. A. Vilgis, *Phys. Rev. E.* **48**, 377 (1993).
25. R. Koningsveld, A. Kleintjens, and E. Nies, *Croat. Chim. Acta* **60**, 53 (1987).
26. H.-P. Deutsch and K. Binder, *Europhys. Lett.* **17**, 697 (1992).
27. H.-P. Deutsch and K. Binder, *Macromolecules* **25**, 6214 (1992).
28. J. E. G. Lipson, *Macromolecules* **24**, 1334 (1991).
29. J. E. G. Lipson, *J. Chem. Phys.* **96**, 1418 (1992).
30. H. M. Sevian, P. K. Brazhnik, and J. E. G. Lipson. *J. Chem. Phys.* **99**, 4112 (1993).
31. K. S. Schweizer and J. G. Curro, *Phys. Rev. Lett.* **60**, 809 (1988).
32. K. S. Schweizer and J. G. Curro, *J. Chem. Phys.* **88**, 7242 (1988).
33. K. S. Schweizer and J. G. Curro, *J. Chem. Phys.* **91**, 5059 (1990).
34. K. S. Schweizer and J. G. Curro, *Chem. Phys.* **149**, 105 (1990).
35. J. G. Curro and K. S. Schweizer, *Macromolecules* **23**, 1402 (1990).
36. J. G. Curro and K. S. Schweizer, *Macromolecules* **24**, 6736 (1991).
37. K. S. Schweizer and J. G. Curro, *J. Chem. Phys.* **94**, 3986 (1991).
38. A. Yethiraj and K. S. Schweizer, *J. Chem. Phys.* **97**, 5927 (1992).
39. A. Yethiraj and K. S. Schweizer, *J. Chem. Phys.* **98**, 9080 (1993); K. S. Schweizer and A. Yethiraj, *J. Chem. Phys.* **98**, 9053 (1993).
40. H.-P. Deutsch and K. Binder, *J. Phys. (France) II* **3**, 1049 (1993).
41. M. D. Gehlsen, J. H. Rosedale, F. S. Bates, C. D. Wignall, and K. Almdal, *Phys. Rev. Lett.* **68**, 2452 (1992).
42. A. Sariban and K. Binder, *Macromolecules* **24**, 578 (1991).
43. F. S. Bates and P. Wiltzius, *J. Chem. Phys.* **91**, 3258 (1989).
44. K. Binder, in *Materials Science and Technology, Vol. 5: Phase Transformations of Materials*, edited by P. Haasen (VCH, Weinheim, 1991) p. 405.
45. M. Tirrell and E. E. Parsonage, in *Materials Science and Technology, Vol. 12: Structure and Properties of Polymers*, edited by E. L. Thomas (VCH, Weinheim, 1993), p. 654.
46. I. Schmidt and K. Binder, *J. Phys. (France)* **46**, 1631 (1985).
47. I. Carmesin and J. Noolandi, *Macromolecules* **22**, 1689 (1989).
48. R. A. L. Jones, *Phys. Rev. E* **47**, 1437 (1993).
49. Z. Y. Chen, J. Noolandi, and D. Izzo, *Phys. Rev. Lett.* **66**, 727 (1991).
50. J.-S. Wang and K. Binder, *J. Chem. Phys.* **94**, 8537 (1991).
51. R. A. L. Jones, L. J. Norton, E. J. Kramer *et al.*, *Europhys. Lett.* **12**, 41 (1990).

52. B. Minchau, B. Dünweg, and K. Binder, *Polym. Commun.* **31**, 348 (1990).
53. H. Fried and K. Binder, *J. Chem. Phys.* **94**, 8349 (1991).
54. H. Fried and K. Binder, *Europhys. Lett.* **16**, 237 (1991).
55. K. Binder and H. Fried, *Macromolecules* **26**, 6878 (1993).
56. F. S. Bates, J. H. Rosedale, G. H. Fredrickson, and C. J. Glinka, *Phys. Rev. Lett.* **61**, 2229 (1988).
57. K. Almdal, J. H. Rosedale, F. S. Bates, G. D. Wignall, and G. H. Fredrickson, *Phys. Rev. Lett.* **65**, 1112 (1990).
58. G. H. Fredrickson and E. Helfand, *J. Chem. Phys.* **87**, 697 (1987).
59. T. Ohta and K. Kawasaki, *Macromolecules* **19**, 2621 (1986).
60. K. Kawasaki, T. Ohta, and M. Kohrogu, *Macromolecules* **21**, 2972 (1988).
61. R. Borsali and T. A. Vilgis, *J. Chem. Phys.* **92**, 3610 (1990).
62. M. Olvera de la Cruz, A. M. Mayes, and B. W. Shift, *Macromolecules* **25**, 944 (1992).
63. J. Melenkovitz and M. Muthukumar, *Macromolecules* **24**, 4199 (1991).
64. J. L. Barrat and G. H. Fredrickson, *J. Chem. Phys.* **95**, 1281 (1992).
65. H. Tang and K. F. Freed. *J. Chem. Phys.* **96**, 8621 (1992).
66. B. Stühn, R. Mutter, and T. Albrecht, *Europhys. Lett.* **18**, 427 (1992).
67. I. Carmesin and K. Kremer, *Macromolecules* **21**, 2819 (1988).
68. H.-P. Deutsch and K. Binder, *J. Chem. Phys.* **94**, 2294 (1991).
69. S. K. Kumar, H. Tang and I. Szleifer, *Mol. Phys.* **81**, 867 (1994).
70. S. K. Kumar, in *Computer Simulation of Polymers*, edited by E. A. Colbourn (Longman, Harlow, 1993).
71. A. Chakrabarti, R. Toral, J. D. Gunton, and M. Muthukumar, *Phys. Rev. Lett.* **69**, 2072 (1989).
72. A. Chakrabarti, R. Toral, J. D. Gunton, and M. Muthukumar, *J. Chem. Phys.* **92**, 6899 (1990).
73. A. Chakrabarti, R. Toral, and J. D. Gunton, *Phys. Rev.* **A 44**, 6503 (1991).
74. Y. Oono and M. Baliana, *Phys. Rev. Lett.* **61**, 1109 (1988).
75. C. F. Fan, B. D. Olafson, M. Blanco, and S. L. Hsu, *Macromolecules* **25**, 3667 (1992).
76. K. Kremer and K. Binder, *Comput. Phys. Reports* **7**, 259 (1988).
77. K. Binder, *Colloid Polym. Sci.* **266**, 871 (1988).
78. A. Sariban and K. Binder, *J. Chem. Phys.* **86**, 5859 (1987).
79. A. Sariban and K. Binder, *Macromolecules* **21**, 711 (1988).
80. A. Baumgärtner and D. W. Heermann, *Polymer* **27**, 1777 (1986).
81. B. M. Forrest and D. W. Heermann, *J. Phys. (France) II* **1**, 909 (1991).
82. O. F. Olaj, M. Wimmer, and G. Zifferer, *Makromol. Chem., Rapid Commun.* **11**, 451 (1990).
83. A. Gauger and T. Pakula, *J. Chem. Phys.* **98**, 3548 (1993).
84. A. Gauger, A. Weyersberg, and T. Pakula, *Makromol. Chem., Theory Simul.* **2**, 531 (1993).
85. A. Weyersberg and T. A. Vilgis, *Phys. Rev. E* **49**, 3097 (1994).
86. O. F. Olaj and W. Lantschbauer, *Makromol. Chem., Rapid Commun.* **3**, 847 (1982).
87. J. Reiter, T. Edling, and T. Pakula, *J. Chem. Phys.* **93**, 837 (1990).
88. A. Sariban and K. Binder, *Polym. Commun.* **30**, 205 (1989).

89. K. Binder and D. W. Heermann, *Monte Carlo Simulation in Statistical Physics*, 2nd Edition (Springer, Berlin, 1992).
90. H. Müller-Krumbhaar and K. Binder, *J. Statist. Phys.* **8**, 1 (1973).
91. N. Madras and A. Sokal, *J. Statist. Phys.* **50**, 109 (1988).
92. D. Forster, *Hydrodynamic Fluctuations, Broken Symmetry, and Correlation Functions* (Benjamin, London, 1983).
93. K. Binder, in *Phase Transitions and Critical Phenomena*, Vol. 5b, edited by C. Domb and M. S. Green (Academic Press, London, 1976) p. 1.
94. P. C. Hohenberg and B. I. Halperin, *Rev. Mod. Phys.* **49**, 435 (1977).
95. M. E. Fisher and M. N. Barber, *Phys. Rev. Lett.* **28**, 1516 (1972).
96. K. Binder, *Z. Physik B* **43**, 119 (1981).
97. V. Privman (ed.), *Finite Size Scaling and Numerical Simulation of Statistical Systems* (World Scientific, Singapore, 1990).
98. K. Binder, in *Computational Methods in Field Theory*, edited by H. Gausterer and C. B. Lang (Springer, Berlin, 1992) p. 59.
99. N. Metropolis, A. W. Rosenbluth, M. N. Rosenbluth, A. H. Teller, and E. Teller, *J. Chem. Phys.* **21**, 1087 (1953).
100. A. Sariban and K. Binder, *Makromol. Chem.* **189**, 2357 (1988).
101. A. Sariban, *Macromolecules* **24**, 578 (1991).
102. Z. W. Salsburg, D. J. Jacobsen, W. Fickett, and W. W. Wood, *J. Chem. Phys.* **30**, 64 (1959).
103. A. M. Ferrenberg and R. H. Swendsen, *Phys. Rev. Lett.* **61**, 2635 (1988).
104. R. H. Swendsen, J.-S. Wang and A. M. Ferrenberg, in *The Monte Carlo Method in Condensed Matter Physics*, edited by K. Binder (Springer, Berlin, 1992) p. 75.
105. C. H. Bennett, *J. Comput. Phys.* **22**, 245 (1976).
106. A. M. Ferrenberg and R. H. Swendsen, *Phys. Rev. Lett.* **63**, 1195 (1988).
107. J. P. Valleau and D. N. Card, *J. Chem. Phys.* **57**, 5457 (1972).
108. H.-P. Deutsch, *J. Stat. Phys.* **67**, 1039 (1992).
109. J. C. Le Guillou and J. Zinn-Justin, *Phys. Rev.* **B 21**, 3976 (1980).
110. H.-P. Deutsch and K. Binder, *Makromol. Chem., Macromol. Symp.* **65**, 59 (1993).
111. H.-P. Deutsch, *J. Chem. Phys.* **99**, 4825 (1993).
112. K. Binder, K. Vollmayr, H.-P. Deutsch, J. D. Reger, and M. Scheucher, *Int. J. Mod. Phys.* **C3**, 1025 (1992).
113. B. Widom, *J. Chem. Phys.* **39**, 2808 (1963).
114. I. Siepmann, *Mol. Phys.* **70**, 1145 (1990).
115. D. Frenkel and B. Smit, *Mol. Phys.* **75**, 983 (1992).
116. J. J. de Pablo, M. Laso, and U. W. Suter, *J. Chem. Phys.* **96**, 6157 (1992).
117. M. Müller, Dissertation, Johannes-Gutenberg-Universität Mainz, Germany, in preparation.
118. W. Jilge, I. Carmesin, K. Kremer, and K. Binder, *Macromolecules* **23**, 5001 (1990).
119. M. Müller and W. Paul, *J. Chem. Phys.* **100**, 719 (1994).
120. S. Kumar, I. Szleifer, and A. Panagiotopoulos, *Phys. Rev. Lett.* **66**, 2935 (1991).
121. A. Sariban and K. Binder, *Colloid Polym. Sci.* **267**, 469 (1989).

122. M. Rovere, D. W. Heermann, and K. Binder, *Europhys. Lett.* **6**, 585 (1988).
123. M. Rovere, D. W. Heermann, and K. Binder, *J. Phys: Condensed Matter* **2**, 7009 (1990).
124. M. Rovere, P. Nielaba, and K. Binder, *Z. Physik B: Condensed Matter* **90**, 215 (1993).
125. D. Marx, P. Nielaba, and K. Binder, *Phys. Rev. Lett.* **67**, 3127 (1991).
126. D. Marx, P. Nielaba, and K. Binder, *Phys. Rev. B* **47**, 7788 (1993).
127. P. J. Flory, *Proc. R. Soc. Lond. A* **236**, 60 (1956).
128. D. Y. Yoon and A. Baumgärtner, *Macromolecules* **17**, 2864 (1984).
129. A. Baumgärtner, *J. Chem. Phys.* **84**, 1905 (1984).
130. A. L. Rodriguez, H.-P. Wittmann, and K. Binder, *Macromolecules* **23**, 4327 (1990).
131. A. Z. Panagiotopoulos, *Mol. Phys.* **61**, 813 (1987).
132. A. Z. Panagiotopoulos, N. Quirke, M. Stapleton, and D. Tildesley, *Mol. Phys.* **63**, 527 (1988).
133. B. Smit, Ph. de Smedt, and D. Frenkel, *Mol. Phys.* **68**, 931 (1989).
134. B. Smit and D. Frenkel, *Mol. Phys.* **68**, 951 (1989).
135. B. Smit and D. Frenkel, *J. Chem. Phys.* **94**, 5663 (1991).
136. G. C. A. M. Mooij, D. Frenkel, I. Szleifer, and A. Z. Panagiotopoulos, preprint
137. P. G. de Gennes, *J. Phys. Lett. (France)* **38**, L 44 (1977).
138. J. F. Joanny, *J. Phys. A* **11**, L 117 (1978).
139. S. F. Bates, I. H. Rosedale, P. Stepanek *et al.*, *Phys. Rev. Lett.* **65**, 1839 (1990).
140. G. Meier, D. Schwahn, K. Mortensen, and S. Janssen, *Europhys. Lett.* **22**, 577 (1993).
141. V. L. Ginzburg, *Soviet. Phys. Solid State* **2**, 1824 (1960).
142. M. A. Anisimov, S. B. Kiselev, J. V. Sengers, and S. Tang, *Physica A* **188**, 487 (1992).
143. K. Binder, M. Nauenberg, V. Privman, and A. P. Young, *Phys. Rev. B* **31**, 1498 (1985).
144. K. Binder, *Z. Physik B* **61**, 13 (1985).
145. K. Binder and H.-P. Deutsch, *Europhys. Lett.* **18**, 667 (1992).
146. A. K. Kron, *Polym. Sci. USSR* **7**, 1361 (1965).
147. F. T. Wall and F. Mandel, *J. Chem. Phys.* **63**, 4592 (1975).
148. W. Paul, K. Binder, D. W. Heermann, and K. Kremer, *J. Physique (France) II* **1**, 37 (1991).
149. M. Kikuchi and K. Binder, *Europhys. Lett.* **21**, 427 (1993).
150. M. Kikuchi and K. Binder, *J. Chem. Phys.* **101**, 3367 (1994).
151. G. Jannink and P. G. de Gennes, *J. Chem. Phys.* **48**, 2260 (1968).
152. K. K. Mon, S. Wansleben, D. P. Landau, and K. Binder, *Phys. Rev. B* 39, 7089 (1989).
153. J. Reiter, G. Zifferer, and O. F. Olaj, *Macromolecules* **23**, 224 (1990).
154. F. Schmid and K. Binder, *Phys. Rev. B* **46**, 13553, 13565 (1992).
155. J. Rowlinson and B. Widom, *Molecular Theory of Capillarity* (Clarendon Press, Oxford, 1982).
156. H. J. Leamy, G. H. Gilmer, K. A. Jackson, and P. Bennema, *Phys. Rev. Lett.* **30**, 601 (1973).

157. J.-S. Wang and K. Binder, *Makromol. Chem., Theory and Simulations* **1**, 49 (1992).
158. P. Cifra, F. E. Karasz, and W. J. Mac Knight, *Macromolecules* **25**, 4895 (1992).
159. U. Steiner, E. Eiser, J. Klein, A. Budkowski, and L. J. Fetters, *Science* **248**, 1126 (1992).
160. W. G. Madden, *J. Chem. Phys.* **87**, 1405 (1987).
161. A. W. Jordan, R. C. Ball, A. M. Donald *et al.*, *Macromolecules* **21**, 235 (1988).
162. G. H. Fredrickson and F. S. Bates, *J. Chem. Phys.* **85**, 633 (1986).
163. M. Doi and A. Onuki, *J. Phys. (France) II* **2**, 1631 (1992).
164. K. Kremer and G. S. Grest, *J. Chem. Phys.* **92**, 5057 (1990).
165. K. Kawasaki and T. Ohta, *Progr. Theor. Phys.* **59**, 62 (1978).
166. A. Milchev, K. Binder, and D. W. Heermann, *Z. Phys. B* **63**, 521 (1986).
167. A. Sariban, K. Binder, and D. W. Heermann, *Colloid Surf. Sci.* **265**, 273 (1987); *Phys. Rev. B* **35**, 6873 (1987).
168. E. A. Guggenheim, *Proc. R. Soc. Lond. A* **183**, 201, 231 (1945).
169. J. F. Joanny, L. Leibler, and R. Ball, *J. Chem. Phys.* **81**, 4640 (1984).
170. L. Schäfer and Ch. Kappeler, *J. Phys. (Paris)* **46**, 1853 (1985).
171. D. Broseta, L. Leibler, and J. F. Joanny, *Macromolecules* **20**, 1935 (1987).
172. A. Onuki and T. Hashimoto, *Macromolecules* **22**, 879 (1989).
173. L. Schäfer and Ch. Kappeler, *Colloid Polym. Sci.* **268**, 995 (1990).
174. L. Schäfer, U. Lehr, and Ch. Kappeler, *J. Phys. (France) I* **1**, 211 (1991).
175. A. Sariban and K. Binder, *Colloid Polym. Sci.* **272**, 1474 (1994).
176. A. Sariban and K. Binder, *Colloid Polym. Sci.* **266**, 389 (1988).
177. F. de J. Guevara-Rodríguez, F. Aguilera-Granja and R. Kikuchi, *J. Phys. (France) II*, **4**, 589 (1994).
178. R. Kikuchi, *Phys. Rev.* **81**, 988 (1951).
179. V. Privman, A. Aharony, and P. C. Hohenberg, in *Phase Transitions and Critical Phenomena*, Vol. 14, edited by C. Domb and J. L. Lebowitz (Academic Press, New York, 1991).
180. P. G. de Gennes, *J. Chem. Phys.* **72**, 4756 (1980).
181. P. Pincus, *J. Chem. Phys.* **75**, 1996 (1981).
182. I. Nakanishi and P. Pincus, *J. Chem. Phys.* **79**, 997 (1983).
183. J. F. Joanny and L. Leibler, *J. Phys. (Paris)* **39**, 951 (1978).
184. K. Binder and H. L. Frisch, *Macromolecules* **17**, 2928 (1984).
185. S. F. Edwards, *Proc. Phys. Soc. Lond.* **88**, 265 (1965).
186. T. A. Vilgis and R. Borsali, *Macromolecules* **23**, 3172 (1990).
187. M. G. Brereton and T. A. Vilgis, *J. Phys. (Paris)* **50**, 245 (1989).
188. P. E. Rouse, *J. Chem. Phys.* **21**, 127 (1953).
189. M. Doi and S. F. Edwards, *Theory of Polymer Dynamics* (Clarendon Press, Oxford, 1986).
190. J. W. Cahn, *Acta Metall.* **9**, 795 (1961).
191. J. D. Gunton, M. San Miguel, and P. S. Sahni, in *Phase Transitions and Critical Phenomena*, Vol. 8, edited by C. Domb and J. L. Lebowitz (Academic Press, London, 1983) p. 267.
192. S. Komura and H. Furukawa (eds) *Dynamics of Ordering Process in Condensed Matter* (Plenum, New York, 1988).

193. J. W. Cahn, *J. Chem. Phys.* **66**, 366 (1977).
194. S. Dietrich, in *Phase Transitions and Critical Phenomena*, Vol. 12, edited by C. Domb and J. L. Lebowitz (Academic Press, London, 1988) p. 1.
195. P. Cifra, F. E. Karasz, and W. J. Mac Knight, *J. Polym. Sci., Polym. Phys.* **26**, 2379 (1988).
196. A. Weyersberg, Dissertation (University of Mainz, Germany, 1993, unpublished).
197. M. E. Cates and J. M. Deutsch, *J. Phys. (Paris)* **47**, 2121 (1986).
198. S. Geyler and T. Pakula, *Makromol. Chem., Rapid Commun.* **9**, 617 (1988).
199. J. F. Marko, *Macromolecules* **26**, 1442 (1993).
200. J. Noolandi and K. M. Hong, *Macromolecules* **15**, 482 (1981); **17**, 1531 (1984).
201. L. Leibler, *Makromol. Chem., Macromol. Symp.* **16**, 1 (1988).
202. J. Noolandi and T. A. Kavassalis, in *Molecular Conformation and Dynamics of Macromolecules in Condensed Systems*, edited by M. Nagasawa (Elsevier, Amsterdam, 1988) p. 285.
203. K. R. Shull and E. J. Kramer, *Macromolecules* **23**, 4769 (1990); K. R. Shull, E. J. Kramer, G. Hadzioannou, and W. Tang, *Macromolecules* **23**, 4780 (1990).
204. Y. Wang and W. L. Mattice, *J. Chem. Phys.* **98**, 9881 (1993); Y. Wang, Y. Li, and W. L. Mattice, *J. Chem. Phys.* **99** 4068 (1993).
205. Y. Wang, W. L. Mattice, and D. H. Napper, *Langmuir* **9**, 66 (1993).
206. S. H. Anastasiadis, I. Gancarz, and J. T. Koberstein, *Macromolecules* **21**, 1449, 2980 (1988).
207. A. N. Semenov, *Macromolecules* **25**, 4967 (1992).
208. M. Schulz and K. Binder, *J. Chem. Phys.* **98**, 655 (1993).
209. R. G. Larson, *Mol. Simul.* **13**, 321 (1994); *Macromolecules* **27**, 4198 (1994).

8

SIMULATION STUDIES OF POLYMER MELTS AT INTERFACES

D. Y. Yoon, M. Vacatello, and G. D. Smith

8.1 Introduction

The structures and properties of polymer melts at various interfaces have been the subject of numerous studies in recent years due to the critical role interfaces play in such important polymer applications as thin films, coatings, lubricants, adhesives and composite matrix materials. Experimentally, development and application of new techniques, such as the surface force apparatus,[1–4] grazing incidence X-ray scattering,[5] and neutron reflectivity,[6] have generated many new and unexpected results. These experiments allow one to investigate the structures and dynamic properties of polymer thin films and interfaces at small dimensions down to the 1 nm range. At such small dimensions, the conventional continuum concepts of polymer structure–property relaxationships in bulk states are not applicable. For example, recent surface force apparatus experiments indicate that solid surfaces cause even the low viscosity Newtonian liquids of low molecular weight polymers to become solidified or rubbery at the interfaces.[2–4]

Such novel experimental findings have prompted very active theoretical studies of polymer interfaces that encompass both analytical models[7–9] and simulation methods in order to obtain a fundamental understanding of both the equilibrium and dynamic properties of polymer melts confined by various interfaces. Because of the inherent difficulties encountered in analytical approaches, simulations have been the more popular theoretical approach.

To date, simulations studies have employed lattice chains,[10–13] bead chains,[14–20] and atomistic polymethylene chains of both united atoms,[21–31] and explicit atoms.[31,32] However, the implications of the various approximations adopted in representing polymer chains on simulations results and the connections between the various models have not been thoroughly investigated.

In this review, we have chosen to focus mainly on the atomistic chain simulations, and will describe in detail equilibrium and dynamic properties of polymer melts at interfaces predicted by these simulations. The ability of the simulations to provide insight into the equilibrium and dynamic behavior of the interfaces as seen experimentally will be discussed. A brief description of systems of bead chains will also be presented, as these models also take into account the off-lattice continuum nature of the interface that is critical to many properties but is difficult to handle in a lattice model. For equilibrium properties, it has been demonstrated that simulation results for lattice chain melts at interfaces are well reproduced by the Scheutjens–Fleer lattice theory.[13] The connection between lattice and off-lattice models will therefore be explored briefly by comparing results of atomistic simulations with predictions of the Scheutjens–Fleer theory.[25] For dynamic properties, there exists no analytical theoretical model, and hence, simulation results will become important in formulating key steps and testing new theoretical approaches to describe polymer dynamics at interfaces.

8.2 Systems of atomistic chains

8.2.1 General considerations

Computer simulation studies of melts of chain molecules at interfaces, in which the chain molecules are modeled at the atomic level (that is, with realistic chain structure, geometry and interatomic interactions), have been limited to relatively short hydrocarbon chains. Section 8.2 reviews the general features of these models and the predicted equilibrium and dynamic properties of the chain molecules in the interfacial regions. The main aim of the studies discussed here is to elucidate the packing of the monomers at interfaces, the associated orientational correlations of chain bond vectors, and anisotropy of the molecular motions, on a nanometer length scale (Fig. 8.1). The following sections describe the models and systems that have been employed for this purpose. Section 8.2.2 provides a discussion of the various models and methods that have been employed for the atomistic chain simulations. Monte Carlo studies of liquid *n*-tridecane ($C_{13}H_{28}$) in broad slits with solid surfaces modeled as impenetrable walls,[21] and in a broad and a narrow slit in equilibrium,[22] with the surfaces modeled as ordered atomistic layers and including the effects of diluents,[23] are discussed in Sections 8.2.3 and 8.2.5, respectively. Molecular dynamics simulations of liquid tridecane, *n*-octacosane ($C_{28}H_{28}$) and *n*-hexacontane ($C_{60}H_{122}$) chains between neutral[24] and attractive[25] unstructured solid surfaces are presented in Section 8.2.4. Section 8.2.4 also includes molecular dynamics simulations of liquid *n*-hexadecane ($C_{16}H_{34}$) between strongly attractive atomistic surfaces under various external loads[29] and of systems of *n*-octane and 2-methylheptane between weakly attractive surfaces for various surface separations at constant chemical potential.[27,28] Systems of alkane liquids with two free surfaces,[26] or with a solid surface and a free surface[29] are discussed in Section 8.2.6. Recently, explicit atom simulations of liquid *n*-tridecane have been performed for atomistic surfaces at constant volume[31] and constant pressure,[32] and for systems with free surfaces.[32] These results are presented in Section 8.2.7. Finally, a comparison is presented between the predictions of equilibrium properties of polymer melt

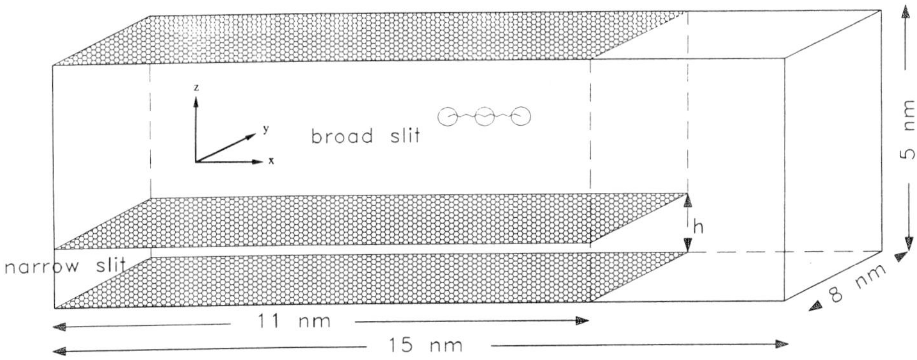

Fig. 8.1 Schematic view of the basic unit cell adopted in the Monte Carlo simulations of Refs 22 and 23. The shaded areas represent layers of hexagonally packed units.

interfaces from the Scheutjens–Fleer lattice theory and atomistic simulations in Section 8.2.8.[25]

8.2.2 Models and methods

8.2.2.1 Monte Carlo simulations

The system simulated in Ref. 21 consisted of a box with periodic boundary conditions in the x and y directions. The two faces of the box perpendicular to the z axis were impenetrable, in the sense that they cannot be crossed by the C–C bonds. The n-tridecane ($C_{13}H_{28}$) molecules were modeled as sequences of 13 point units (united atom or UA approximation) with fixed bond lengths (0.153 nm) and fixed bond angles (112°). Units belonging to different chains or belonging to the same chain but separated by more than three skeletal bonds interacted through a 12-6 Lennard-Jones potential with minimum energy -0.14 kcal mol^{-1} at a distance of 0.4 nm. The interactions were truncated at 0.5 nm, and the potential energy was shifted to be zero at this distance. As in Ref. 33, the nonbonded interactions of units separated by three skeletal bonds in the same chain were separately computed by using a torsional potential of the form

$$V(\phi) = (V_m/2)[x(1 - \cos\phi) + (1 - x)(1 - \cos 3\phi)] \tag{8.1}$$

with $V_m = 4.0$ kcal mol^{-1} and $x = 0.163$. These values yield 0.5 kcal mol^{-1} and 3.6 kcal mol^{-1} for the energy difference and the barrier height, respectively, between the trans and gauche state.

The system contained 2885 n-tridecane chains in a box with edges 15.0 nm along both the x and y directions in the film plane, while the separation of the plates, that is the box edge along the z axis, was 5.0 nm. The initial model was generated by introducing at random the chains in the basic cell one at a time, in such a way that two nonbonded units could not come within 0.32 nm of each other. The system was then equilibrated at room temperature ($T = 300$ K) by standard Monte Carlo techniques utilizing reptation moves. That is, a tridecane molecule was selected at random at each step, one of its two terminal units was removed and a new unit was randomly added at the opposite chain end. The new configuration was accepted when $r \leq \exp(-\Delta E/kT)$, with r being a random number in the [0, 1] range and ΔE the total energy change. When r was greater than $\exp(-\Delta E/kT)$, the original configuration was restored and counted and a new reptation attempt was started. The results were averaged over 10 equilibrium models obtained in sequence and separated by 10^6 accepted reptation steps.

A good indication that the force field adopted in the calculations provides an adequate description of the structural properties of liquid tridecane and that the Monte Carlo method used to equilibrate the system is statistically unbiased comes from examination of the properties of molecules with center of mass at distances from the walls greater than 1.5 nm. These properties were found to be in remarkable agreement with those calculated for ideal unperturbed rotational isomeric state (RIS) chains,[34] showing that the chains in contact with the walls can be considered in equilibrium with the unperturbed bulk liquid and that the perturbation due to the presence of the walls does not extend more than 1.5 nm into the liquid.

Monte Carlo simulations of liquid tridecane in narrow and broad slits in equilibrium were performed using a three-dimensionally periodic box with edges 15.0 nm, 8.0 nm and 5.0 nm in the x, y and z directions, respectively.[22] The solid surfaces were modeled as planar arrays of hexagonally packed units perpendicular to the z axis. These arrays are infinitely extended parallel to the y axis, while they are only extended from $x = 0$ to $x = 11.0$ nm in the x direction. The basic box is illustrated in Fig. 8.1. A tridecane chain in the all trans conformation is also shown in the correct proportions, with the circles around some of the units having a radius of 0.2 nm. Two separate calculations were performed, with $h = 1.2$ nm in the first case, and $h = 1.0$ nm in the second case. Due to the periodicity along the z axis, the region of the box between $x = 0$ and $x = 11.0$ nm consists of two slits of thickness 1.2 nm and 3.8 nm in the first calculation, and of thickness 1.0 nm and 4.0 nm in the second calculation.

As shown in Ref. 21, tridecane molecules at a distance from the nearest surface greater than approximately 1.5 nm are essentially unperturbed. Hence, more than 25% of the volume of the basic box is filled with unperturbed bulk liquid in the calculations of Ref. 22. The simulation system is set up such that molecules close to the solid surfaces, included those in the narrow slits, are in equilibrium with this reservoir of bulk liquid. In particular, the density in the narrow slits is not established *a priori*, but is determined by the balance of all the interactions at play. Since the r.m.s. end-to-end distance and r.m.s. radius of gyration of unperturbed tridecane chains at

300 K are 1.13 nm and 0.394 nm,[21] respectively, the assumption is made that molecules located between $x = 2.0$ nm and $x = 9.0$ nm in Fig. 8.1 are representative of molecules between surfaces infinitely extended in both the x and the y directions.

The simulations in Ref. 22 were performed using the UA approximation (see above) with the methods and the force field already described. The same 6-12 potential was used for the mutual interactions of the methyl and methylene groups and for their interactions with the atomic units comprising the solid surfaces. A total of 1297 tridecane molecules was introduced in the basic cell one at a time in such a way that two nonbonded units (including those of the solid surfaces) could not come within 0.32 nm from each other. The system was then equilibrated at 300 K using the reptation technique. The results shown in the next sections represent the average over 10 different equilibrium models obtained in sequence and separated by more than 800 000 accepted reptations. It was verified that the tridecane molecules were able to diffuse during the simulation from the narrow slits to the wider slits and vice versa.

Similar techniques were used in Ref. 23 to simulate systems of tridecane chains in the presence of small amounts of a diluent component. As in Ref. 22, the tridecane chains were represented as sequences of point units with fixed bond lengths and angles, while the diluent molecules were represented by isolated point units. The same 6-12 potential was utilized for the nonbonded interactions of all the units in the system, including the diluent and the units constituting the solid surfaces. In all, 1190 tridecane chains and 819 diluent molecules (5 wt% of the total units) were introduced in the basic cell such that two nonbonded units could not approach within a distance less than 0.32 nm. The system was then thoroughly equilibrated at 300 K using reptations for the tridecane molecules and local displacements for the diluent. When a given diluent molecule was selected for an attempted move, trial coordinates were randomly generated inside a sphere of 0.1 nm radius centered on the old position. The results shown in the Section 8.2.5 represent, for each calculation, the average over 12 different equilibrium models obtained in sequence and separated by approximately one million accepted Monte Carlo steps.

8.2.2.2 Molecular dynamics simulations

Molecular dynamics simulation techniques maintain certain advantages over other simulation techniques. The primary advantage of the molecular dynamics techniques is that during a simulation the time evolution of a system follows a reversible trajectory through phase space. As a result, dynamic properties of the system can be determined directly. Expression of the potential energy of a system of molecules in terms of simple intramolecular and intermolecular potential functions allows for the calculation of the force acting on each force center, or "atom", of the system. Through simple numerical integration, the position and velocity of each atom can be determined as a function of time. The integration time step is limited by the highest-frequency motions in the system, as the forces are usually assumed to be constant during an integration step. Constraining high-frequency motions such as bond stretches allows for use of larger integration time steps. Simulations of this type are at constant energy (e.g., Ref. 26).

It is usually desired to perform molecular dynamics simulations at constant temperature. In the simulations considered here, constant temperature is accomplished in one of two ways. In the first, stochastic forces and associated frictional forces are introduced which act individually on each atom of the system. This approach is hereafter referred to as stochastic dynamics (SD) simulation. The mean-square magnitude of the stochastic forces, which are purely random and Gaussian, is proportional to the temperature of the system, as described in Ref. 24. A canonical ensemble is simulated if the friction coefficient γ for the frictional forces is chosen such that $1/\gamma$ is much smaller than the total simulation time. The resulting damping forces do influence dynamic quantities, however, and this is the primary drawback of the SD method. The SD method was used in Refs 24, 25 and 31. The second method for constant temperature simulations involves either direct scaling of atomic velocities, as in Refs 29 and 30, or the inclusion of an additional "temperature" degree of freedom to the system by the Nosé method,[35] as applied in Refs 31 and 32. Such simulations are hereafter referred to as molecular dynamics (MD) simulations. Of the MD methods considered, only the Nosé method yields a true canonical ensemble.

Both SD and MD simulations of alkanes melts confined by solid surfaces (solid/liquid interfaces) and MD simulations of liquid alkanes at free surfaces (liquid/vapor interfaces) have been performed. The alkane molecules were represented by realistic atomistic force fields with constrained bond lengths. In all cases except Refs 29 and 30, the bond angle flexibility was maintained and in all cases the torsional flexibility was maintained. In most simulations the methyl and methylene groups were represented by single, spherically symmetric Lennard-Jones (LJ) force centers, i.e., the united atom (UA) approximation. Results from simulations which explicitly include the pendant hydrogen atoms as individual force centers, which we refer to as the explicit atom (EA) representation, will also be discussed.

The MD and SD systems considered are divided into three classes: constant density, constant chemical potential and constant pressure. In constant density systems periodic boundary conditions are applied in the x and y direction, and the chains are confined between two parallel solid surfaces or plates perpendicular to the z axis. The density of these systems is determined by the dimensions of the periodic system and the number of molecules contained in the system and is fixed during the simulations. The constant density systems simulated in Ref. 24 consisted of a series of n-alkanes

(C_nH_{2n+2}, $n = 13$, 28 and 60) confined between neutral, flat solid surfaces. In the z direction, structureless, purely repulsive smooth surfaces were placed with a separation of 3.5 nm. The range of the potential was limited to σ, the van der Waals diameter of the UA force centers. After equilibration, SD simulations of 3.74 ns, 11.22 ns, and 3.74 ns, respectively, for systems of 77 tridecane chains at 300 K, 36 octacosane chains at 400 K, and 68 hexacontane chains at 400 K, were performed with an integration time step of 3.74 fs. The x and y dimensions were such that the densities at the center of the films matches experimental melt densities for tridecane at 300 K and for octacosane and hexacontane at 400 K.[36]

Constant density simulations were also performed in Ref. 25. Here, the effect of including attractive interactions between the surface and the monomers was studied by considering systems where every monomer experiences the attractive potential (attractive surface simulations) and systems in which only the chain ends experience the attractive potential (sticky end simulations). The attractive surface simulations were performed on an ensemble of 49 tridecane chains with a surface separation of 3.0 nm. The sticky end simulations were performed on an ensemble of 40 octacosane chains with a surface separation of 4.0 nm. The temperatures, densities, integration scheme and potential functions were as described above for Ref. 24, with a SD simulation time of 7.5 ns for both systems. The additional attractive surface–monomer potential was of the form

$$V_{\text{att}}(z) = -\frac{\epsilon_{\text{att}}}{z}\left\{1 - 10\left[\frac{z}{z_{\text{c}}}\right]^3 + 15\left[\frac{z}{z_{\text{c}}}\right]^4 - 6\left[\frac{z}{z_{\text{c}}}\right]^5\right\} \tag{8.2}$$

where z is the separation of the monomer from a surface, and z_c is the cutoff distance for the potential (0.8 nm). The polynomial form of the potential yields a smooth decay of the potential to zero at the cutoff. The factor ϵ_{att} was chosen to be 50 kcal nm/mol. The function corresponds to the interaction of a particle charged with 0.25 e.u. and a flat metal surface. The total potential between the surface and a monomer has a minimum at about 0.25 nm with a magnitude of -1.2 kcal/mol. This attraction is much greater than the monomer–monomer attraction (0.14 kcal/mol maximum), but somewhat less than the -1.7 kcal/mol per monomer which has been estimated as the potential minima for a $C_{24}H_{50}$ molecule on a graphite surface.[37]

The final constant density system considered involves a comparison of SD simulation results for an UA and an EA tridecane system.[31] The systems studied were identical to the system of Ref. 24, with the exception of the structure of the surfaces. Here the surface topography, instead of being structureless, reflected a (111) face of an fcc solid with the nearest-neighbor distance between particles being 0.4 nm. The surface particles interacted with the tridecane atoms as united atoms in the UA simulations and as carbon atoms in the EA simulations, and hence constitute a weakly attractive surface.

The simulations described above were performed at constant density, i.e., a volume was imposed on the system irrespective of the resulting pressure or chemical potential. MD simulations performed at constant chemical potential, where the confined liquid is in equilibrium with a vapor or bulk liquid phase, have also been performed. Simulations with free surfaces, i.e., with vapor/polymer interfaces, allow for the study of the equilibrium liquid–vapor interface structure and the calculation of the surface tension, a thermodynamic property fundamental to the understanding of the behavior of a material at interfaces. An MD study of the equilibrium liquid–vapor interface structure and surface tension of thin films of *n*-decane and *n*-eicosane ($C_{20}H_{42}$) has been performed in Ref. 26. The system studied consisted of a box with periodic boundary conditions in all directions. The liquid polymer, however, while fully occupying the x and y dimensions, occupied only a fraction of the system in the z direction, resulting in two liquid–vapor interfaces. The liquid phase ranged from about 4.0 to 7.0 nm in thickness. Simulations were performed at 400 K for both decane and eicosane, with additional decane simulations at 300 K. A similar system of tridecane molecules, using a well calibrated EA force field,[38] has been studied at 400 K and 300 K in Ref. 32.

Systems with a solid and free interface, which can be thought of as thin films of alkane melts lying on a solid substrate, have been simulated in Ref. 30. Here the system was periodic in the x and y directions only. The systems consisted of *n*-hexadecane at 350 K for films of 1.0, 2.0 and 4.0 nm in nominal thicknesses. The surface was modeled after a gold (001) plane. The well-depth parameter for the monomer–surface atom interaction was 0.429 kcal/mol, indicating a relatively strongly attractive surface.

The final constant chemical potential configuration considered represents a film of alkane melts sandwiched between two solid plates in Refs 27 and 28. The system was periodic in the x and y directions, but only a portion of the surfaces in the y direction was occupied by the solid substrate. Constant chemical potential was maintained using the reservoir method, where the liquid bubbles which form at the edge of the substrates are in equilibrium with a vapor phase which interacts across the periodic boundaries. The surfaces were modeled as static surfactant crystalline monolayers which interacted with the alkanes as united atom CH_2 and CH_3 groups (weakly attractive surface). Both *n*-octane and 2-methylheptane systems were studied as a function of surface separation.

Constant pressure, or external load, systems have been studied in Refs 29 and 32. In Ref. 29, *n*-hexadecane melt films in a semidroplet configuration were considered, where the surfaces and liquid were periodic in the x direction but limited in extent in the y direction, with the entire substrate being occupied in the x direction but only partially occupied in the y direction.

The system was studied as a function of external load on the surfaces. As molecules which detached from the droplet were removed from the simulation, the system was not in equilibrium with the vapor phase. The solid substrates were modeled after crystalline solids with both weak attraction ($\epsilon_3 = 1$) and strong attraction ($\epsilon_3 = 3$) between the surface atoms and the polymer segments, which were treated as UA monomers; here ϵ_3 is the well-depth of the surface–monomer potential and is measured in units of the well-depth of the monomer–monomer potential. Bond angles were constrained in the simulation. In Ref. 32, constant pressure simulations of liquid tridecane were performed for a system periodic in the x and y directions. The surface structure was that of the (111) face of an fcc crystalline solid. Here, an explicit atom representation of the alkane chains was used. Results are presented for surface atom–polymer atom interactions equal to those of the carbon–carbon and carbon–hydrogen interactions for carbon and hydrogen atoms, respectively (a weakly attractive surface) for films nominally 4 nm thick at 450 K.

8.2.3 Liquid n-*tridecane near impenetrable walls by Monte Carlo simulations*

The first calculation on atomistic systems of chain molecules near solid surfaces was performed in Ref. 21 modeling the surfaces as two impenetrable walls placed at a distance much greater than the molecular dimensions. The overall chain density of the model has been chosen such that the local density far from the walls was equal to that of the bulk liquid at 300 K. Since all the properties of molecules far from the walls were found to match those of unperturbed chains, as described by the rotational isomeric state (RIS) model at this temperature,[34] the chains in contact with the impenetrable walls can be considered to be in equilibrium with the unperturbed bulk liquid.

8.2.3.1 Segmental density distribution

Figure 8.2 plots the reduced segmental density ρ, defined as the ratio of the density in sublayers of thickness 0.05 nm parallel to the wall surfaces divided by the density of liquid tridecane, as a function of the distance (z) of the sublayer from the nearest wall surface. The presence of a wall perturbs the segmental density distribution over a distance smaller than 1.5 nm, giving rise to a large maximum at $z = 0.13$ nm followed by a series of progressively less intense minima and maxima. The overall aspect of the curve is that of a damped cosine function with periodicity very close to 0.4 nm and shifted 0.13 nm along the z axis. Hence, the two impenetrable surfaces placed at $z = 0$ and at $z = 5.0$ nm have the same effects of two densely packed and ordered layers of methylene units at $z - 0.27$ nm and at

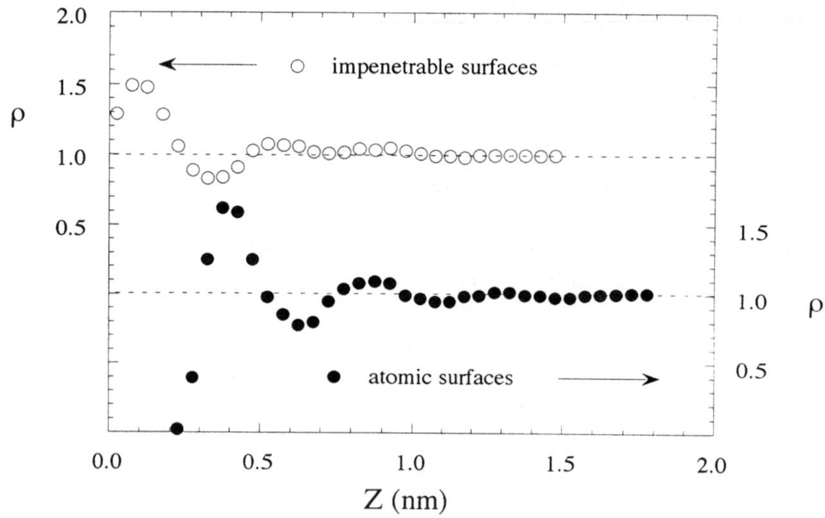

Fig. 8.2 The reduced segmental density, ρ, as a function of the distance from the nearest wall for impenetrable surfaces (open circles, redrawn from Ref. 21) and atomistic surfaces (filled circles, redrawn from Ref. 22).

$z = 5.27$ nm. In other words, the system simulated in this calculation is equivalent to a sample of tridecane melts between two parallel surfaces of the latter kind separated by 5.54 nm.

The segmental distribution along the z axis near the surfaces has also been examined for the different chain units along the chain. All the distribution curves show a maximum, approximately located at $z = 0.1(k - 1)$ nm, k being the sequence number of the chain unit considered ($1 \leq k \leq 7$ for tridecane). This indicates that methylene units in the middle of the chains ($k = 7$) tend to avoid the walls. In practice, 48% of the units in the first sublayer of thickness 0.05 nm adjacent to the walls are methyl groups or the first methylene units adjoining them. The tendency of the chain ends to accumulate at the walls is simply explained by considering that internal chain segments in the proximity of the walls experience greater orientational and conformational constraints than segments at the chain ends (see Section 8.2.3.3).

8.2.3.2 Molecular distribution, size and shape

The reduced density of molecular centers of mass (cm), that is the density of centers of mass divided by the corresponding bulk value, shows a large maximum at z between 0.1 and 0.3 nm (Fig. 8.3). This distinct maximum shows that most of the chain segments adjacent to the walls belong to the practically flat two-dimensional chains. The distribution of units of chains

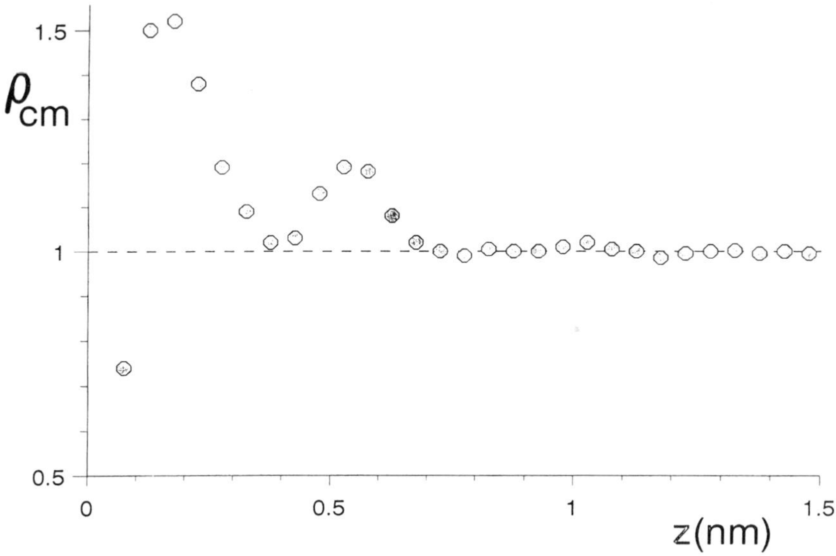

Fig. 8.3 The reduced density of centers of mass, $\rho_{\rm cm}$, versus the distance from the nearest wall. (Redrawn from Ref. 21.)

that belong to the second peak of molecular mass center profile, located between 0.4 and 0.6 nm from the walls, is nearly unperturbed. However, terminal portions of these chains extend up to the first segmental peak layer and contribute to the high density of units in the latter. In particular, approximately 1/3 of the methyl chain ends in the first layer belong to chains with center of mass in the second maximum. The enhanced density of methyl ends in the immediate proximity of the walls is entirely explained by the contribution of the terminal portions of these chains.

8.2.3.3 Order parameters and conformation

Figure 8.4 plots two order parameters, s_c and s_n, for two-bond chain segments with center-of-mass in the various sublayers of thickness 0.05 nm. The chain axis order parameter s_c is defined to be $(1/2)(3 < \cos^2\theta > -1)$, θ being the angle between the z axis and the vectors connecting units of the same chain separated by two bonds; the segment plane order parameter s_n is the corresponding order parameter with respect to the z axis of vectors perpendicular to the direction of the two-bond segments and contained in that two-bond segment plane.

Both s_c and s_n are negative in the first sublayer in contact with the walls for obvious geometrical reasons. However, in the region of z corresponding to the first segment density maximum of Fig. 8.2, s_c is negative and s_n is positive. Hence, the high-density packing in this region is realized with

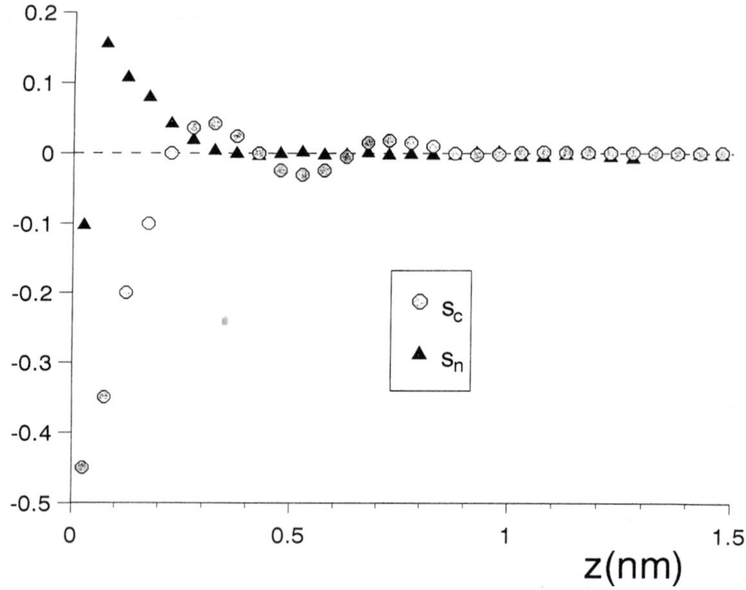

Fig. 8.4 The order parameter along the z axis of the vectors connecting units $j-1$ and $j+1$ in the same chain (s_c) and of the vectors perpendicular to the latters and contained in the plane of $j-1$, j and $j+1$ units (s_n) as a function of the distance from the nearest wall. (Redrawn from Ref. 21.)

two-bond segment chain axis aligned parallel to the surface, but with their planes preferentially oriented perpendicular to it. Moreover, this orientational ordering is different for internal and for terminal two-bond chain segments. That is, the average value of s_c for segments at $z < 0.3$ nm is -0.12 in the case of terminal segments, while it is -0.23 for segments in the middle of the chains. The fact that internal chain segments in the proximity of the walls experience greater orientational constraints than the external chain segments seems to be the major reason for the observed tendency of the former to avoid the walls and the higher concentration of methyl chain ends in the first sublayers near the walls.

The perturbation of the conformational characteristics due to the constraints of the surfaces is most noticeable in the first sublayer (0.05 nm thickness), where the fraction of trans conformations is as high as 0.80 for obvious geometrical reasons, as compared with 0.63 in the bulk liquid. However, in the other regions the chains are apparently able to adapt to the strongly anisotropic environment near the walls without any substantial changes of their local-scale conformations. Figure 8.5 plots the intermolecular segmental orientational correlation parameter $s_{ij}(r_{ij}) = (1/2)(3 < \cos^2 \theta_{ij}(r_{ij}) > -1)$, where $\theta_{ij}(r_{ij})$ is the angle between any two-bond segments i and j belonging to different chains and separated by a distance r_{ij}. The circles have been calculated considering only two-bond segments

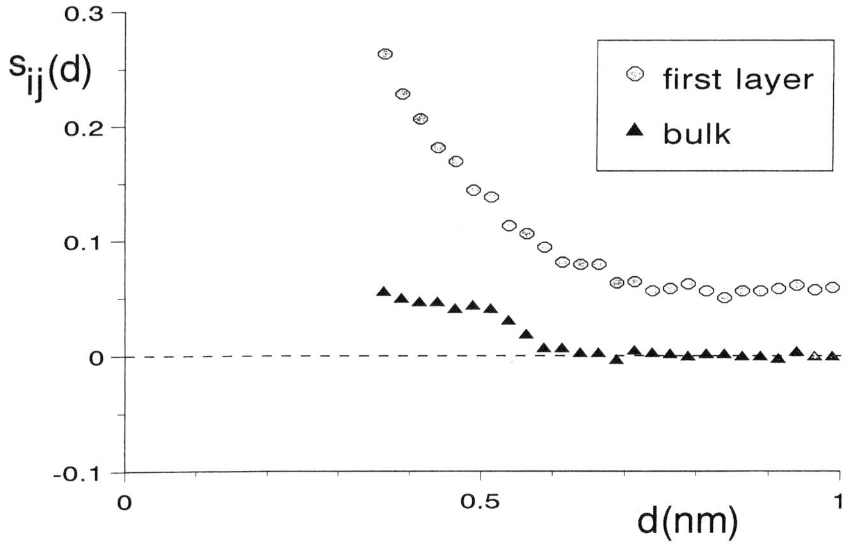

Fig. 8.5 The intermolecular orientational correlation parameter of two bond segments, $s_{ij}(d)$, in the first layer of 0.2 nm thickness adjacent to the walls and in a similar layer far from the walls. (Redrawn from Ref. 21.)

with center-of-mass within a layer of thickness 0.2 nm adjacent to the walls, while the triangles refer to a similar layer far from the walls (bulk). Figure 8.5 confirms that orientational correlations in bulk liquid *n*-alkanes are small and limited to first-neighboring segments at most.[33] The behavior in the first layer near a solid surface is quite different, since the orientational correlation between neighboring segments is significantly higher than the limiting value (0.06) imposed by the common (partial) alignment parallel to the surface. This implies that the nearly two-dimensional chains contained in the first layer tend to exhibit significant orientational correlations among the neighboring chain segments. The lateral dimension of such correlated regions, however, does not appear to be large.

8.2.4 N-Alkane systems near neutral and attractive surfaces by SD and MD simulations

United atom (UA) atomistic computer models of liquid tridecane, octacosane ($C_{28}H_{58}$) and hexacontane ($C_{60}H_{122}$) confined by flat, neutral (i.e., purely repulsive) solid surfaces have been studied in Ref. 24. Similar systems of liquid tridecane and octacosane between ordered atomistic surfaces, and between flat surfaces that are attractive with respect to all the chain segments or with respect to the chain ends only, have been studied in Ref. 25. In these cases, the thickness of the films is such that the density at their centers matches the experimental density of the corresponding melt at the simulation temperatures (300 K for tridecane, 400 K for octacosane and hexacontane). Hexadecane films between weakly and strongly attractive solid surfaces for systems in which the film thickness is allowed to adjust according to the applied load have been simulated in Ref. 29. In Refs 27 and 28, *n*-octane and 2-methylheptane films were simulated at constant chemical potential as a function of surface separation for surfactant monolayer (weakly attractive) surfaces.

8.2.4.1 Segmental distribution, order parameters and conformations

Figure 8.6 shows the normalized segmental density, the order parameters s_c and s_n and the fraction of trans bonds, P_{trans}, as a function of the separation z from the flat surface for a tridecane melt at 300 K between neutral surfaces (dotted lines) and between attractive surfaces (solid lines), for the segment–surface potentials shown in Fig. 8.7.[25] The behavior of these prop-

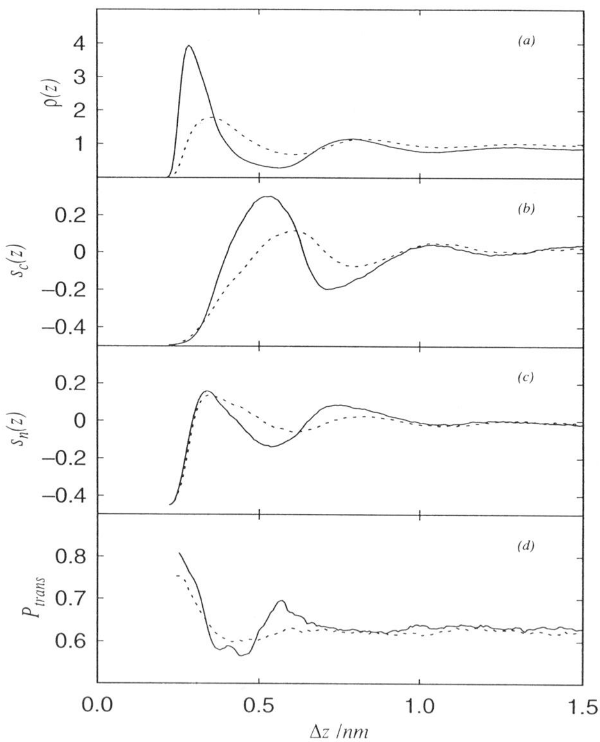

Fig. 8.6 The normalized segmental density, the order parameters s_c and s_n and the fraction of trans bonds, P_{trans}, as a function of z for a tridecane melt at 300 K between neutral surfaces (dotted lines) and between attractive surfaces (solid lines). (Redrawn from Ref. 25.)

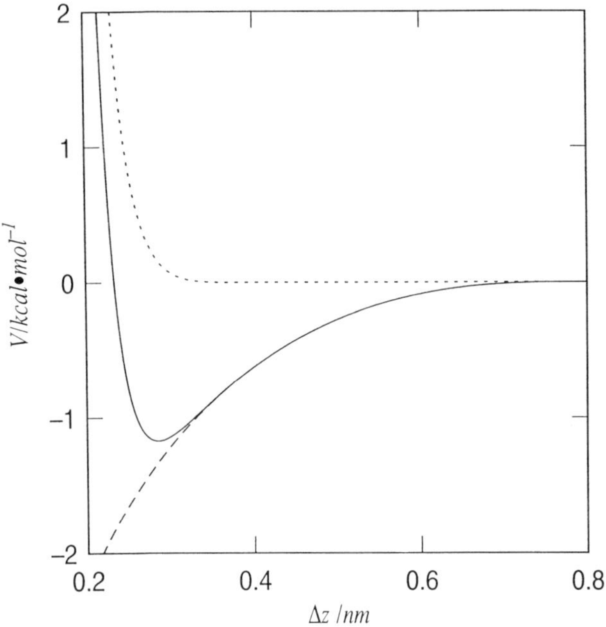

Fig. 8.7 The potential energy of an attractive surface. The dotted line represents the Lennard-Jones potential for the flat surfaces, the dashed line represents the attractive part (eq. 8.2), and the solid line is for the resulting potential.

erties in the film with the neutral surfaces is found to be in very good agreement with the results obtained near impenetrable walls by Monte Carlo simulations in Ref. 21. In particular, starting from a surface, the segmental density (Fig. 8.6) shows a series of progressively less intense maxima and minima with a periodicity close to 0.4 nm, as observed in Fig. 8.2. However, the first maximum is obviously located near $z = 0.4$ nm, as compared to $z = 0.13$ nm for the impenetrable walls of Ref. 21, which simply disallowed any segment placement with its center beyond the surface planes. A similar shift is seen in the plots for s_c and s_n, all the other features practically coinciding with those reported in Ref. 21. The same is true for the fraction of trans conformations, showing significant deviations with respect to the unperturbed distribution only for segments located very close to the surfaces. None of the properties shown in Fig. 8.6 are substantially changed when the neutral flat surfaces are substituted by neutral ordered atomistic surfaces (in Ref. 21 and Section 8.2.5.1 below).

As expected, the segmental density in the first or adsorbed layer for attractive surfaces is significantly higher than for neutral surfaces, as shown in Fig. 8.6. Although strongly attractive when compared with neutral walls, the potential plotted in Fig. 8.7 are somewhat less attractive than a graphite surface.[37] The first maximum of the segment density is observed at $z = 0.3$ nm, very close to the minimum of the attractive potential between surface and chain units (0.29 nm). At this density maximum, the order parameter s_c indicates an almost perfect two-dimensional alignment of the chain segments parallel to the surface, while the fraction of trans bonds is greater than the unperturbed bulk value. Hence, the first layer of chains in contact with attractive surfaces is much more ordered than near neutral surfaces (see Section 8.2.4.2). Furthermore, the first maximum in the density is followed by a deep depletion region and by a second less intense maximum. Both the depletion region and the second maximum are more pronounced than those near neutral walls. Analogous results have been obtained in Ref. 29, in which the degree of layering was found to be much greater for strongly attractive surfaces and to increase with increasing external load. The orientation of two-bond segments caused by the surfaces is also found to persist farther into the melts with increased adsorption strength. For the weakly attractive surfaces, the preferential orientation is found to decay very rapidly, essentially vanishing for distances greater than 0.5 nm from the nearest surface. In contrast, greater oscillations in chain axis orientation, corresponding to preferential alignment parallel to the surfaces in high density regions separated by preferentially perpendicular alignment to the surfaces in a region of low density, are observed near strongly attractive surfaces.

Systems in which the chain-ends are preferentially attracted by the surfaces yield a much higher fraction of chain-ends in the first segmental layer. This results in a partial destruction of the tendency to form high segmental density layers near the surfaces, since the chain-ends take up more volume than the other units.[25]

8.2.4.2 Molecular distribution, size and shape

Like the segmental properties, the molecular properties calculated for liquid tridecane near neutral surfaces are in excellent agreement with those obtained near impenetrable walls by Monte Carlo simulations in Refs 21 and 22. In particular, the distribution of molecular centers of mass shows a maximum at $z = 0.4$ nm, which becomes much more intense near attractive walls, where the corresponding density of chain centers of mass is nearly six times greater than in the bulk. Furthermore, the maximum occurs at $z = 0.3$ nm, coincident with the minimum of the interaction potential between surfaces and chain units. Hence, most of the monomer segments adjacent to the attractive surfaces belong to flat, two-dimensional chains. This is shown in more detail in Fig. 8.8 by investigating the content of trains, tails and loops, which are defined as follows, focusing on the skeletal bonds as the basic units for the atomistic chains:

- The thickness of the interface is chosen to be 0.7 nm.
- Trains are successive bonds in the interface.

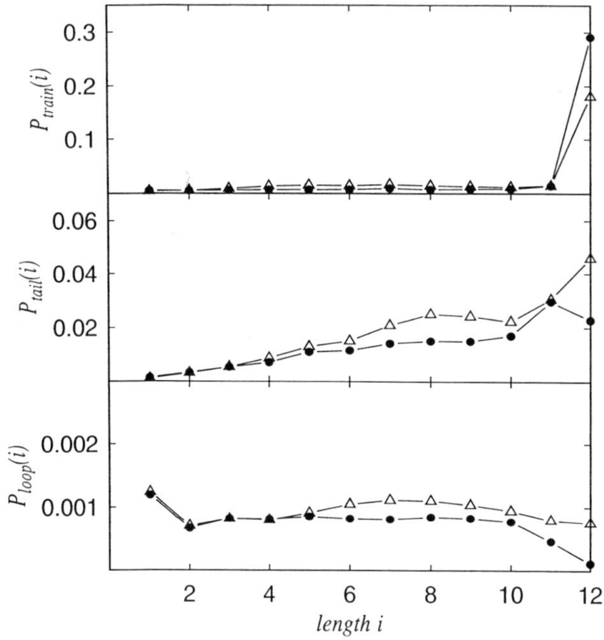

Fig. 8.8 The probability distribution of trains (a), tails (b) and loops (c) defined by eq. 8.3 found in tridecane melts at 300 K in the system of neutral surfaces (triangles) and of attractive surfaces (circles).

- Tails are sequences of bonds which contain chain-ends located outside the interface and the other end of the sequence connected to a monomer within the interface.
- Loops are sequences of bonds which are out of the interface, with both ends of the sequences connected to monomers in the interface.
- If a train connects to a tail or loop, the division point is defined by the orientation of the last bond belonging to the interface. If the angle between the z-axis and the axis defined by the last two bonds is smaller than 45°, the last bond of this train is switched to tail or loop.
- The frequency of the occurrence of train, tail and loop sequences were counted and averaged over the simulation period, and the weighted probability was calculated, defined by

$$P(i) = f(i)i/N \tag{8.3}$$

where N is the total number of bonds in the simulation box and $f(i)$ is the number of the sequences of i bonds in length. Thus, $P(i)$ corresponds to the probability of a bond to be found in a certain sequence of length i.

Figure 8.8 reveals that the fraction of tridecane molecules totally contained in a layer of thickness 0.7 nm near a surface (i.e., trains of 12 bonds) is strongly enhanced for the attractive surfaces. This enhancement in long trains is accompanied by a reduction of both shorter trains and longer tails. Similar results have been obtained in Ref. 29, where the tendency of entire *n*-alkane molecules, or large sequences of methylene units, to be adsorbed on the solid surfaces increases dramatically as the strength of the surface–chain attraction is increased. This tendency was also found to increase with increasing external load.

Very different behavior was found in Ref. 25 near surfaces with preferential attraction for the chain-ends of *n*-$C_{28}H_{58}$ melts. As shown in Fig. 8.9, longer trains were found to decrease and shorter trains increase in this case, with a simultaneous increase of long tails and loops. This behavior is well explained by the abundance of chain-ends in the first segmental layer close to the surfaces, which somewhat excludes the other chain units from this region. These characteristics due to sticky chain-ends for *n*-$C_{28}H_{58}$ melt systems are reproduced quite well by Scheutjens–Fleer self-consistent-field lattice theory (see Section 8.2.8).[25]

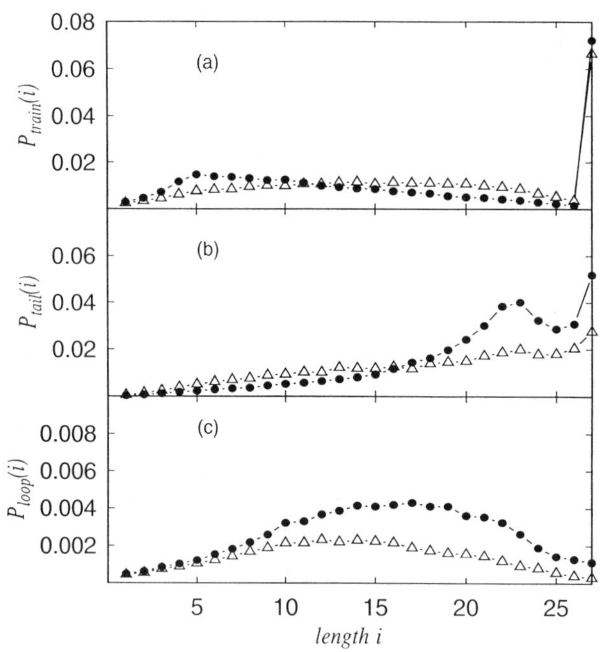

Fig. 8.9 The probability distribution of trains (a), tails (b) and loops (c) defined by eq. 8.3 found in the systems of octacosane melts at 400 K with neutral surfaces (triangles) and sticky ends (circles).

8.2.4.3 Effects of chain branching

Recently, simulations of *n*-octane and 2-methylheptane have been performed at constant chemical potential as a function of surface separation.[27,28] Comparison of the results for *n*-octane and 2-methylheptane yields insight into the effects of chain branching on the segmental/molecular layers of *n*-alkanes and on solvation forces. Solvation forces, denoted as forces between the solid surface and molecules comprising the central region of the confined film, as a function of surface separation are shown in Fig. 8.10. The *n*-octane solvation forces show oscillations with a period of about 0.4 nm, the same period as is seen in the layering and modulation of the segment density profile. This oscillatory behavior qualitatively reproduces what is seen in surface force apparatus experiments.[1] The slight amount of chain branching present in 2-methylheptane can be seen to reduce significantly the oscillatory nature of the solvation forces. Branching also significantly reduces the oscillations in the density profiles for separations which correspond to an integral number of layers at small separation distances (two or three layers). In these cases, it appears that the bulkiness of the chain-end in 2-methylheptane significantly interferes with the ability of the alkane molecules to fit into the dense layers.[28]

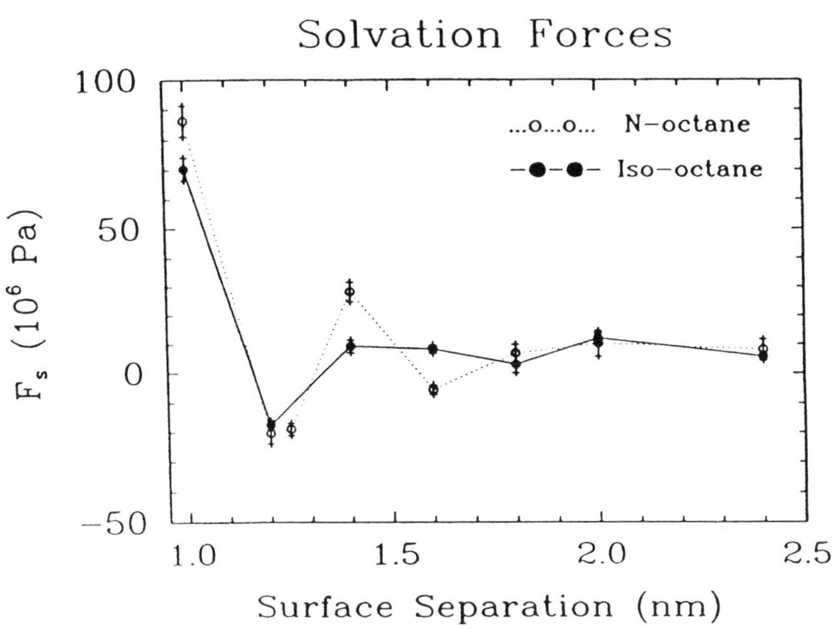

Fig. 8.10 Solvation forces as a function of separation between solid surfaces for *n*-octane (dotted line) and 2-methylheptane (solid line) at 297 K. Surfaces separations are the distances between the centers of the outermost layers of atoms forming the solid surfaces minus one C–C bond length (0.153 nm). (Redrawn from Ref. 28.)

8.2.4.4 Dynamic properties

Because of the anisotropic character introduced into the system by the presence of the solid surfaces, the molecular mobility of chains near the surfaces is expected to be anisotropic. Moreover, this anisotropy will be a function of the distance of the molecular center of mass from the nearest surface. This effect is illustrated in Figs 8.11 and 8.12. Figure 8.11 plots the apparent diffusivity D_{ap}

$$D_{\text{ap}}(z_{\text{cm}}) = \frac{1}{6t_c}\left\langle [\mathbf{r}_{\text{cm}}(t+t_c) - \mathbf{r}_{\text{cm}}(t)]^2 \right\rangle \tag{8.4}$$

for tridecane molecules near neutral and attractive surfaces; here z_{cm} is the separation of the chain center-of-mass from the surface, $\mathbf{r}_{\text{cm}}$ is the position vector of the center of mass and $t_c = 750\,\text{fs}$. The apparent diffusivity D_{ap}, reflecting chain mobility, is seen to be nearly independent of z_{cm} near neutral surfaces, while it shows a substantial decrease near attractive surfaces.[25] Figure 8.12 shows the normalized parallel component (upper curve) and the perpendicular component of D_{ap}, with respect to the surface, for n-$C_{28}H_{58}$ (octacosane) chains near neutral surfaces.[24] In the central region,

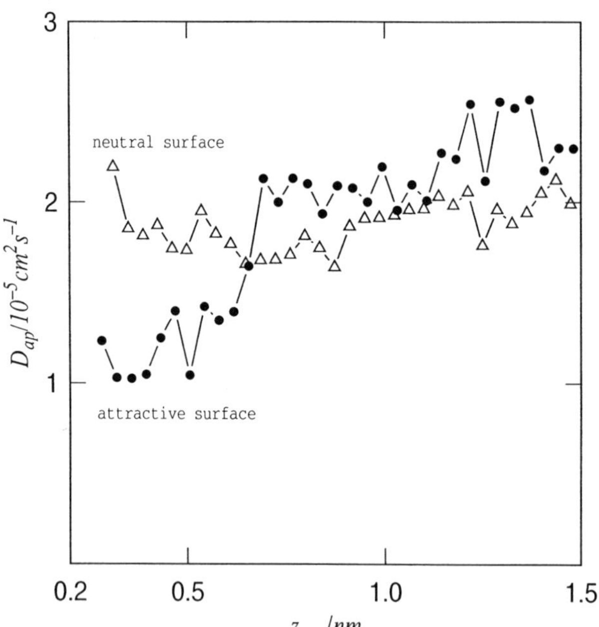

Fig. 8.11 The apparent diffusion constant as a function of the distance between the surface and the center of mass of a molecule in a system of tridecane melts at 300 K confined between neutral surfaces (triangles) and attractive surfaces (circles). (Redrawn from Ref. 25.)

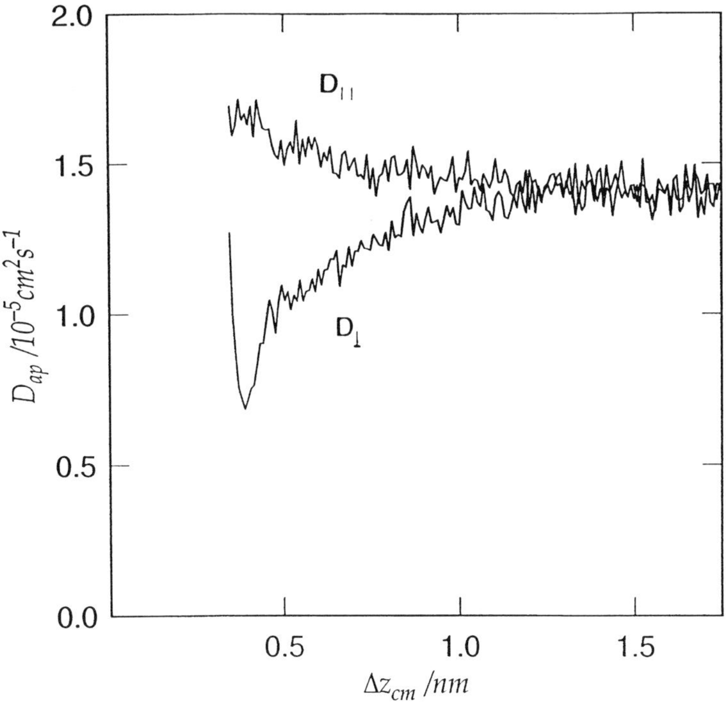

Fig. 8.12 The normalized x–y (parallel) components and the z (perpendicular) component of D_{ap} for octacosane chains near neutral surfaces. (Redrawn from Ref. 24.)

the apparent diffusivity shows isotropic, bulk behavior, as was also seen for the static properties. As one approaches a surface, the diffusivity in the direction perpendicular to the surface is decreased, while that parallel to the surface is enhanced. As a result, the overall diffusivity is practically independent of z_{cm}. The decrease of chain mobility, as shown by D_{ap}, near attractive surfaces is due to a decreased diffusivity both perpendicular and parallel to the surfaces, the latter effect being probably due to the enhanced density and hence reduced free volume found near the surfaces.

8.2.5 Liquid tridecane in a narrow and a broad slit in equilibrium

The molecular arrangements and conformation of liquid tridecane at 300 K in broad slits of thickness 4.0 nm and 3.8 nm, in equilibrium with very narrow slits of thickness 1.2 nm and 1.0 nm, respectively, have been studied in Ref. 22, adopting the simulation set-up shown in Fig. 8.1. Similar systems containing 5% of a diluent having the size and the interactions of a methylene group have also been simulated in Ref. 23. The systems are such that the liquid between the solid surfaces can be considered in equilibrium with a tank of bulk liquid tridecane at room temperature and pressure (see Fig. 8.1). Also, in order to check the effects of the structure of the surfaces in such narrow slits, the surfaces are comprised of two-dimensional arrays of units having the same size and interaction parameters of a methylene group, packed as in the (111) face of the fcc lattice.

8.2.5.1 Segmental distribution

In view of the results of Ref. 21, molecules located in the center of the broad slits and molecules located between $x = 11.0$ nm and $x = 15.0$ nm (see Fig. 8.1) and far from the borders of the solid surfaces are expected to be representative of the unperturbed bulk liquid. In fact, these regions are characterized by a density equal to that of liquid tridecane at room temperature (750 kg/m^3), and by molecular properties practically identical to those of unperturbed chains at the same temperature. This shows that the molecular arrangements near one of the surfaces in the broad slits are not influenced by the presence of the other surface, such that they are representative of the situation of liquid tridecane near a single, atomistic surface. Furthermore, the calculations in Ref. 22 and those in Ref. 25 lead to the same conclusion that the range of influence of the surfaces on the structural organization of the liquid does not depend on the atomistic topographic details of surface structures.

The normalized density (ρ) of units in sublayers of thickness 0.05 nm in the broad slits is plotted as a function of the distance (z) from the nearest surface in Fig. 8.2. Only units with x between 2.0 nm and 9.0 nm are considered. Figure 8.2 shows the series of progressively less intense maxima and minima with a periodicity slightly greater than 0.4 nm, the first maximum being centered at $z = 0.4$ nm. The intensities of the various maxima and minima are generally comparable to those observed for impenetrable walls, with the exception of the first maximum, which is slightly more intense near structured surfaces than near impenetrable walls. Hence, the atomistic structure of the surfaces has a fairly minor influence on the behavior of the segmental density, limited to the first layer of units at most.

Figure 8.13 plots the normalized density of units in the slit of thickness 1.2 nm, calculated as before, as a function of z. The dotted curve shows the behavior expected in the absence of cooperative effects due to the simultaneous presence of two surfaces. That is, it shows the $z < 0.6$ nm zone of Fig. 8.2 in the first half-slit, and its mirror image in the second half-slit. The tendency of the tridecane molecules to form densely packed segmental layers near the solid surfaces is slightly increased in a slit of thickness 1.2 nm, while it is almost totally destroyed when the slit thickness is reduced to 1.0 nm (not shown). In particular, the two density maxima corresponding to the first layers near the surfaces are more intense in Fig. 8.13 than that in Fig. 8.2, while the intermediate minimum is deeper. On the contrary, the maxima are

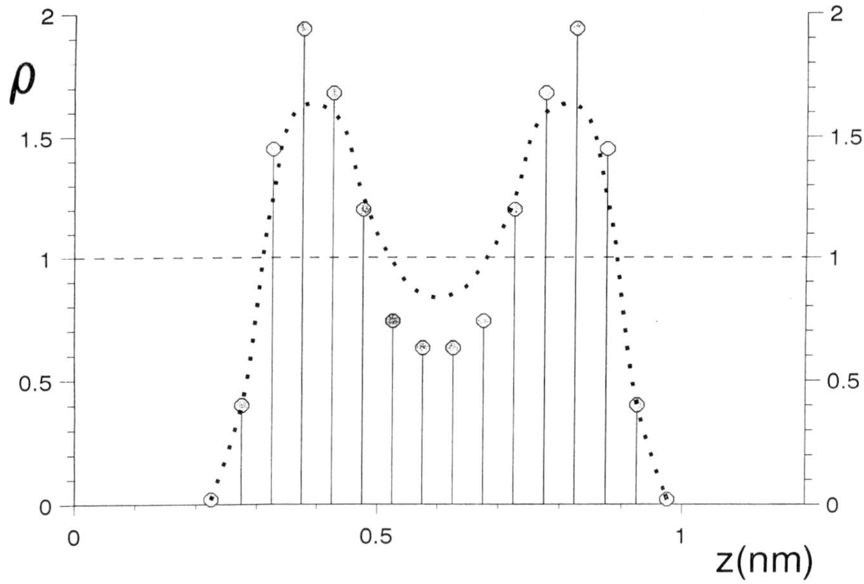

Fig. 8.13 The normalized density of units as a function of z in a narrow 1.2 nm slit. The dotted curves represent the behavior expected in the absence of cooperative effects due to the simultaneous presence of the two plates. (Redrawn from Ref. 22.)

less intense for the 1.0 nm slit than in Fig. 8.2, and the intermediate minimum is practically absent. There is an associated difference in the total density of units in the two slits. In fact, the total number of methyl and methylene groups in the narrow slit of 1.2 nm thickness is twice the corresponding number in a layer of thickness 0.6 nm near one of the surfaces in the broad slits. For the slit of 1.0 nm thickness, the overall density is approximately 10% less than the value expected in the absence of cooperative effects of the two surfaces. A simple explanation for this behavior is that two layers of methylene units can be accommodated quite well in a slit of thickness 1.2 nm, while there is no room for them in a slit of thickness 1.0 nm.

As found in Ref. 21, the density of chain-ends near the surfaces is in all cases higher than the density of any other chain unit. For instance, the methyl groups are 1.7 times more abundant than any of the internal chain units (i.e., unit 5, 6 or 7) in the region $z < 0.4$ nm of the broad slits. This effect is much less pronounced in the narrow slits, where the corresponding ratio is approximately 1.3.

8.2.5.2 Liquid tridecane in the presence of a diluent

Simulation studies show that the chain-ends tend to be more abundant in the first segmental layer in contact with the solid surfaces, both in narrow and in broad slits. This effect originates from the higher restrictions imposed by the presence of the surface on the orientational and conformational freedom of internal chain segments as compared with those of terminal segments. Therefore, the addition of a small amount of a short-chain component (diluent) can be expected to influence the properties of a melt of chain molecules at various interfaces.

The calculations in Ref. 23 show that the distribution of total units (tridecane plus diluent) obtained in the case of the pure liquid is not modified by the inclusion of small amounts of a diluent. In particular, the tendency of the tridecane molecules to form densely packed segmental layers is somewhat increased in a slit of thickness 1.2 nm (in Fig. 8.1 with $h = 1.2$ nm) in comparison with the situation near a single solid surface, as was found for the system without a diluent. The fraction of chain-ends and the fraction of diluent molecules in layers of thickness 0.05 nm parallel to the xy plane are shown in Fig. 8.14, plotted along the z direction as an average for the units between $x = 2$–9 nm in Fig. 8.1. Both fractions show a pronounced increase near the surfaces, the average fraction of diluents in the first 0.4 nm being as high as 0.10, which is more than twice the bulk value of 0.045. The tendency to concentrate near the surfaces is greater for the diluent than for the chain-ends. For instance, the average fraction of chain-ends in the same 0.4 nm region is 0.20, that is about 1.4 times larger than the corresponding bulk

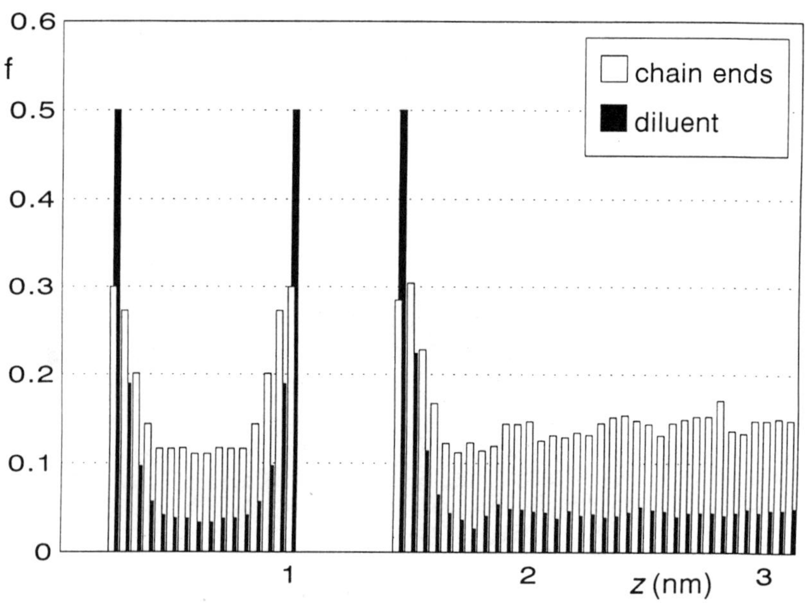

Fig. 8.14 The fraction of chain-ends and diluents in sublayers of 0.05 nm thickness for a system with 1.2 nm slit, plotted along the z axis, at $x = 2$-9 nm region, in the system shown in Fig. 8.1 with $h = 1.2$ nm. (Redrawn from Ref. 23.)

value (0.146). In other words, the diluent molecules replace part of the chain-ends in the first segmental layer in contact with the surfaces, thus decreasing both the number of chain molecules in these regions and the probability that the tridecane chains form two-dimensional train configurations.

The preferential adsorption of chain-ends and diluents at solid surfaces differs in an important aspect. The preferential adsorption of chain-ends takes place on a local scale, in that this effect is absent over the length scale of the radius of gyration. For example, the proportion of chain-ends and internal chain units in the first 0.6 nm near a surface is not much higher than $2/13 = 0.154$. This proportion is obviously 0.154 in narrow slits and in the bulk liquid of tridecane. In contrast, the adsorption of diluent molecules corresponds to a net migration of diluents from the bulk region to the narrow slits and to the interfacial region of the broad slits. In fact, the fraction of diluent molecules far from the solid surfaces in the two simulated systems is as small as 0.045, indicating that the equilibration of a liquid mixture of this kind with solid surfaces leads to a segregation of diluents, even without any preferential interactions with surfaces.

8.2.6 Systems with free surfaces

Simulations of systems with free surfaces, i.e., vapor/melt interfaces, allow for the study of the equilibrium liquid–vapor interface structure and the calculation of the surface tension, a thermodynamic quantity fundamental to the understanding of the behavior of a material at interfaces. The liquid–vapor interfaces of films of *n*-decane and *n*-eicosane have been recently simulated in Ref. 26 using the UA approximation. The simulations predict that although the total monomer density profile decreases monotonically as one proceeds from the film center toward the vapor phase, the chain center of mass and central segment density profiles are strongly peaked at the point where the total monomer density begins to decrease. These features are illustrated in Fig. 8.15. The outer edge of the interface is dominated by chain-ends, as has also been observed near solid surfaces (see above). With decreasing temperature and increasing chain length, the width of the interfacial region decreases, while the amount of local segregation increases. It is also found that the chains are slightly flattened parallel to the interface in the outer regions, while they are slightly elongated and aligned tendentiously perpendicular to the interface in the region corresponding to the peak of the center of mass profile. These effects can be understood in terms of the large energy penalty associated with bringing a chain into the region of decreasing monomer density.[26]

Surface tensions have been calculated from the virial tensor as

$$\gamma = \frac{1}{2A}(2V_{zz} - V_{xx} - V_{yy}) \tag{8.5}$$

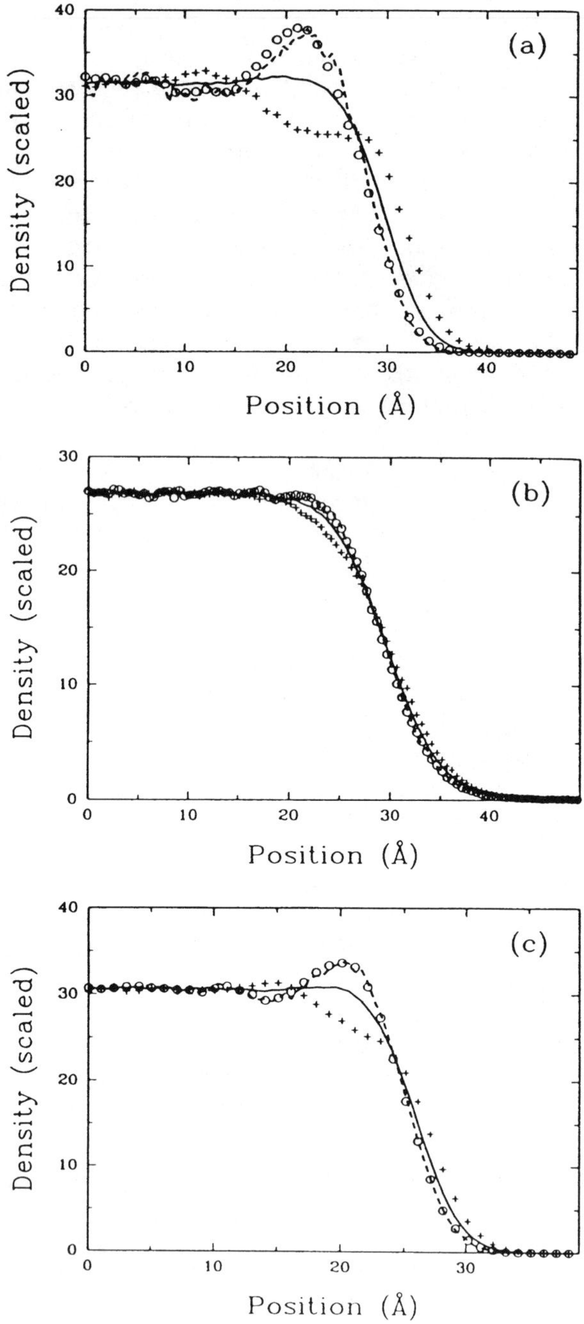

Fig. 8.15 Scaled density profiles of the end segments (cross), middle segments (open circle), chain centers of mass (- - -), and total hydrocarbon (—) of (a) eicosane at 400 K, (b) decane at 400 K and (c) decane at 300 K. (Redrawn from Ref. 26.)

where γ is the surface tension and A is the total surface area of the two interfaces. The components of the atomic virial tensor are $V_{\alpha\alpha} = (1/2)\sum\sum f^{\alpha_{ij}} r^{\alpha_{ij}}$, where the double sum is over all the atoms in the system, and $f^{\alpha_{ij}}$ and $r^{\alpha_{ij}}$ are the components of the force and distance vectors between atoms i and j, respectively. The molecular virial $V_{\alpha\alpha} = (1/2)\sum\sum f^{\alpha_{ab}} r^{\alpha_{ab}}$, where the force is now calculated between molecules a and b and the distance vector is between the centers of mass of the molecules, can also be used. As forces beyond the cut-off distance (0.976 nm in Ref. 26) are not included, the surface tension or the virial tensor must be corrected for this truncation. The calculated surface tensions with truncation corrections for *n*-decane and eicosane were overestimated by 20 to 50% compared to experimental values. The effect of including hydrogen atoms explicitly on the calculated surface tensions is discussed in Section 8.2.7.

A system with a solid and a free interface can be thought of as a thin film of material lying on a solid substrate. *N*-hexadecane films of thickness 1.0, 2.0 and 4.0 nm, with the solid surface modeled after a gold (001) plane, have been simulated in Ref. 30. The resulting segment density profiles and the diffusion constants, both parallel and perpendicular to the interfaces, are shown in Fig. 8.16. The segment density profiles show strong layering effects, with high density layers separated by lower density depletion regions. The liquid/vapor interface shows a monotonic decrease in segmental density, as already shown for systems with two free surfaces[26] (see above). For the thick film (4.0 nm), monomer segregation or layering is not apparent beyond 1.8 nm from the solid surface, even for the highly attractive surface used in these simulations. A significant increase in the fraction of torsions in the trans state was found in the adsorbed layer. As discussed above for *n*-alkanes confined between two solid surfaces, this effect has been found to be dependent upon the nature of the surface/polymer interactions and is enhanced for attractive surfaces.

For the thin film (1.0 nm), diffusion perpendicular to the surfaces is essentially zero, as seen in Fig. 8.16. However, the molecules remain highly mobile parallel to the surfaces. For the thick film, the diffusion constant profiles indicate a drop in molecular mobility both parallel and perpendicular to the surfaces as one approaches the solid surface, indicating a drop in overall mobility. This is in contrast to systems confined between neutral surfaces, where the decrease in the perpendicular mobility is much less dramatic and the parallel mobility is actually enhanced at the surfaces (see Fig. 8.12). It is qualitatively consistent with results for systems confined between attractive surfaces, although in Ref. 25 the decrease in mobility at the surface was significantly less than seen here. This discrepancy may be due to the difference in surface topography, indicating that for strongly attractive surfaces, atomistic surface topography may have a much greater effect on the mobility than flat, structureless surfaces. These differences emphasize the importance of the nature of the interactions between polymer and solid

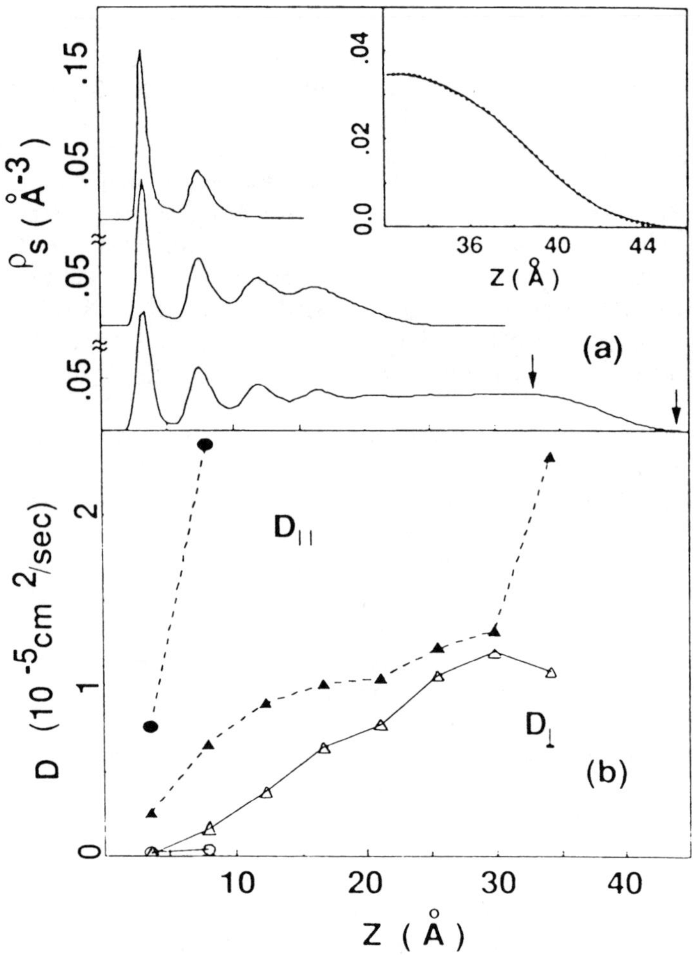

Fig. 8.16 (a) Segment density profiles for three thicknesses of adsorbed *n*-hexadecane films at 350 K, plotted vs. distance for the solid surface, with the original at the center of atoms in the top layer of the solid substrate. An error-function fit for the tail region (marked by arrows) of the thickest film is shown in the inset. (b) Diffusion constants in the directions parallel and perpendicular to the surface plotted vs. distance. Circles and triangles correspond to the thinnest (about 0.1 nm) and thickest (about 4.0 nm) films, respectively. Solid symbols denote parallel diffusion and empty ones perpendicular diffusion. (Redrawn from Ref. 30.)

surfaces on the chain dynamics at the interfaces. However, even for an atomistic, strongly attractive surface, the chains maintain a rather high degree of mobility, indicating that a true solidification is not occurring. In contrast to the solid/polymer interface, an increased mobility parallel to the interface can be seen in the polymer/vapor interfacial region. If one takes the perpendicular mobility of the diffusivity in the 4.0 nm film at 2.5 nm from the solid surface as representative of the bulk diffusivity, a total diffusion constant of about 3×10^{-5} cm^2/s is obtained. This value is about three times larger than the experimental value for hexadecane at 350 K.[36] The discrepancy is indicative of the limitations of using the UA approximation in obtaining quantitative dynamical results.[31]

8.2.7 *Explicit atom simulations of* n-*alkanes at interfaces*

The UA approximation, where methyl and methylene groups are represented as single, spherically symmetric Lennard-Jones centers, has proved to be adequate for gleaning qualitative information about the interface between alkanes and solid surfaces. However, the UA force field must be calibrated, primarily by adjusting the van der Waals radius of the united atoms, for alkanes of different chain lengths in order to reproduce properties such as X-ray scattering patterns, P-V-T behavior and self-diffusion rates in the melts.[31,38,39] Even when accurately calibrated with respect to these properties, the UA force field may not reproduce well chain conformation and local chain dynamics.[38] In contrast, a single explicit atom (EA) force field, taking account of all the hydrogens atoms explicitly, can accurately reproduce these properties for various molecular weights, the price being obviously paid in terms of simulation CPU time.[31,38] Both UA and EA calculations have recently been performed to simulate a system of tridecane chains between planar arrays of carbon atoms packed as in the (111) face of a fcc solid (weakly attractive surface).

Figure 8.17 compares the monomer density, the two-bond order parameters s_c and s_n and the trans bond fraction profiles obtained for the UA system at 300 K and EA systems at 300 K and 450 K. The tendency toward layering of the monomers is much stronger in the EA system, where the high density layers and low density depletion regions are much better defined. Indeed, a third layer, not seen in the UA simulations, appears in Fig. 8.17. The center of mass density profile (Fig. 8.18) shows that the tendency toward a greater degree of segregation exists in the EA system also on a molecular level. Additionally, the orientational order parallel to the surfaces in the region of high segment density and perpendicular to the surfaces in the depletion regions is significantly more pronounced in the EA system. However, the influence of the surfaces on the local chain conformations is limited to segments (from the trans bond fraction profile) very close to the solid surfaces in both the UA and EA chains. On the other hand, the z-component of the characteristic ratio in Fig. 8.18 ($C_n = \langle R^2 \rangle / (n-1)l^2$, where n the number of monomers, $\langle R^2 \rangle$ the mean square end-to-end distance, and l is the bond length) reveals that the two-dimensional character of the chains, seen in the first layer for both the UA and EA simulations, extends to the second molecular layer only in the EA simulation.

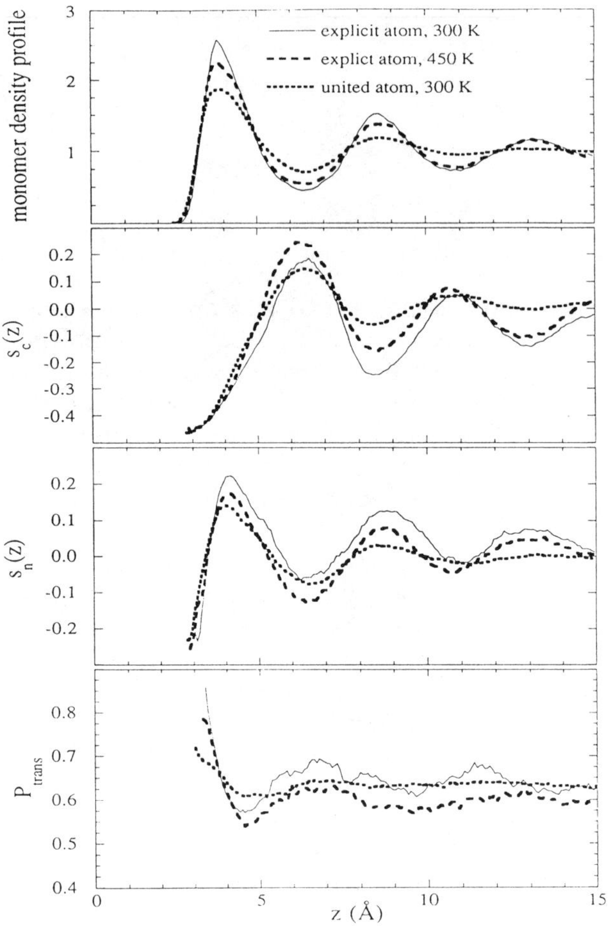

Fig. 8.17 Monomer density profile (normalized to the bulk value), order parameters, and trans bond fraction profile for confined *n*-tridecane melts using an united atom force field at 300 K and an explicit atom force field at 300 and 450 K. (Redrawn from Ref. 31.)

The cause of the difference in both monomer and molecular segregation and orientation observed between the UA and the EA system in Ref. 31 is at least in part due to calibration effects. In fact, the force fields employed in Ref. 31 were not calibrated to reproduce the P-V-T behavior of tridecane. Subsequent simulations of bulk liquid tridecane with well calibrated force fields show that the simulated structure factors depend significantly on the force field parameters for the UA system, while the results for the EA system exhibit only minor dependence on parameters.[38] In this regard, it should be noted that details of structural and dynamical properties of chain molecules at interfaces depend not only on the interactions between the suface atoms

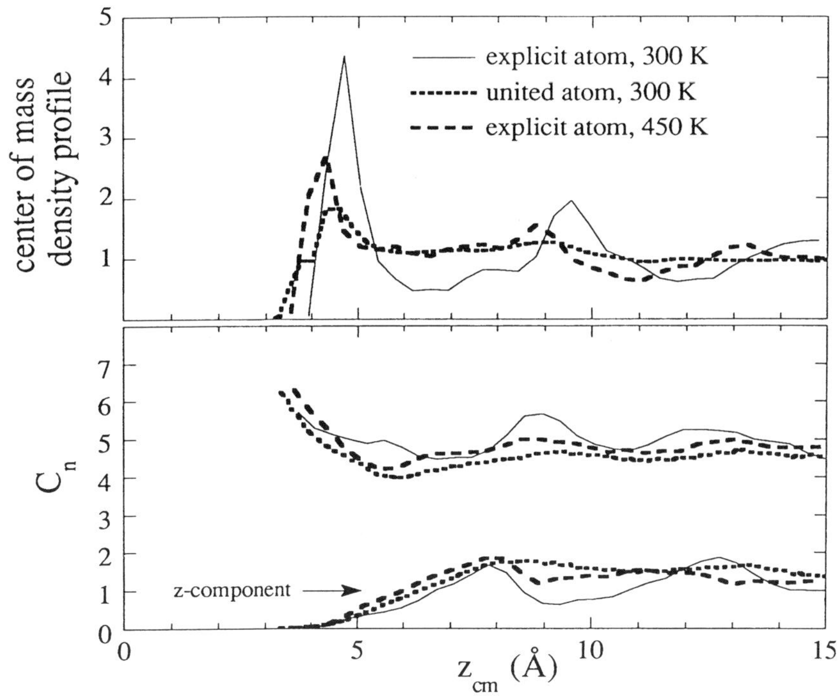

Fig. 8.18 Center of mass density profile (normalized to the bulk value) and characteristic ratio C_n for confined *n*-tridecane melts at using an united atom force field 300 K and an explicit atom force field at 300 and 450 K. (Redrawn from Ref. 31.)

and chain segments, but also on the accurate description of the polymer–polymer interaction.

Constant pressure simulations of tridecane confined between atomistic, weakly attractive parallel plates using a calibrated EA force field have been performed for films of nominal 4 nm thickness as 450 K.[32] The most interesting difference between these and earlier UA simulations is the extension of ordering effects near the surfaces farther into the bulk, similar to the effects seen in the previous constant volume EA simulations (see Fig. 8.17). The monomer density profile and order parameter for the constant pressure EA simulations at 450 K are very close to those for the constant volume EA simulations. The apparent diffusivity perpendicular to the surfaces shows the characteristic decrease as one approaches the surfaces, while the overall mobility is essentially unaffected. This is consistent with previous results obtained for surfaces which are not strongly attractive. The bulk value of the diffusion constant of $3.3 \times 10^{-5}\,\mathrm{cm}^2/\mathrm{s}$, obtained from the diffusivity at the center of the 4 nm film, agrees reasonably well with the experimental value of $5.0 \times 10^{-5}\,\mathrm{cm}^2/\mathrm{s}$. This is in contrast to some UA simulations, where the diffusion constant for molecules in the center of the film is significantly greater than the experimental value (see above).

Surface tensions for alkanes where hydrogen atoms are considered explicitly have been recently determined for tridecane at 300 K and 400 K.[32] Values of 30.5 dyne/cm and 23.4 dyne/cm, including truncation corrections, were obtained, respectively, at 300 K and 400 K, with uncertainties of about 3 dyne/cm. These values were obtained using a truncation distance of 0.9 nm. Values obtained using a truncation distance of 0.6 nm did not differ significantly from these values. Experimental values are 25.4 dyne/cm and 16.7 dyne/cm at 300 K and 400 K, respectively.[40] The differences between simulation and experimental values are comparable with those reported in Ref. 26 for united atom alkanes. Apparently, the systematic overestimation of the surface tension obtained by simulations using simple two-body nonbonded interactions, which has also been observed in simple Lennard-Jones fluids,[41] is not alleviated by explicit treatment of the hydrogen atoms. Therefore, it seems necessary to include the three-body forces to accurately predict the experimental values, as was done successfully for Ar and Xe systems.[42]

8.2.8 Comparison of atomistic simulations with Scheutjens–Fleer lattice theory

Recently, the results of SD simulations for a UA model of octacosane (n-$C_{28}H_{58}$) chains with neutral and sticky chain-ends to surfaces have been compared with the predictions of Scheutjens–Fleer lattice theory.[25] For this purpose the octacosane melts between two flat surfaces separated by 3.5 nm spacing were compared with a lattice chain system. The system consisted of a 90% filled cubic lattice consisting of seven layers (layers 1 and 7 being the surface layers), each layer corresponding to the dimensions of about 0.45 nm observed between monomer density peaks. The chains were eight segments in length since a lattice segment of about 0.45 nm corresponds to around 3.5 CH_2 monomer units. The attractive energy for the end segments to the surfaces was set at 1 kcal/mol, and calculations were performed at 400 K. The reader is referred to Ref. 13 and the references contained therein for details of the SCF calculations. The resulting train probabilities are compared with simulation results in Fig. 8.19. The decrease in the number of long trains and the increase in the number of short trains for the sticky end system is reflected in the SCF calculations. The dramatic increase in the probability of a train consisting of an entire chain (length n) with respect to a train of length $n - 1$ is also predicted by the SCF calculations, as is the fact that the probability of a train of length n is not significantly changed by the presence of sticky ends. For the SCF lattice calculations with the neutral surface, the increase in the probability of a train of length n relative to a train of length $n - 1$ is simply the result of the

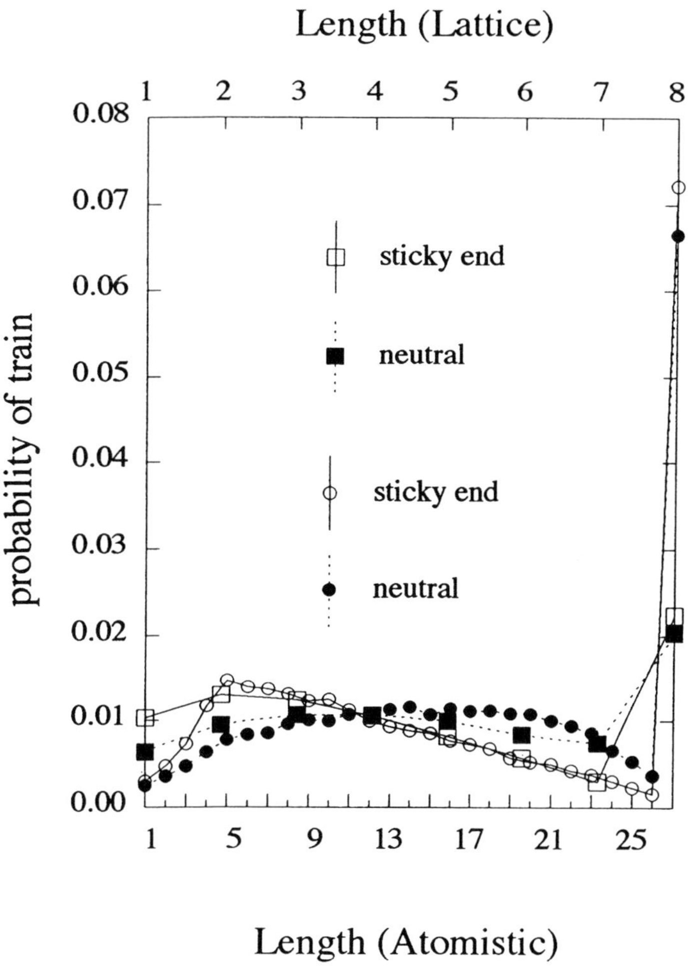

Fig. 8.19 The probability distribution of trains for octacosane melts at 400 K with neutral surfaces and sticky ends, compared to SCF lattice theory predictions. The SCF probabilities have been scaled by (8/27) for comparison. (Redrawn from Ref. 25.)

fact that there are more lattice sites available on the surface for the *n*th unit (4) than off the surface (1). For the case of sticky ends, the probability of the *n*th unit being on the surface is dramatically increased by the attractive energy, but as the probability of long train sequences is reduced by the high concentration of chain-ends on the surface, the net result is that the probability of a train of length *n* is not significantly affected by the attractive potential.

Although the SCF lattice theory reproduces chain conformations quite well, the restricted nature of the lattice precludes prediction of features

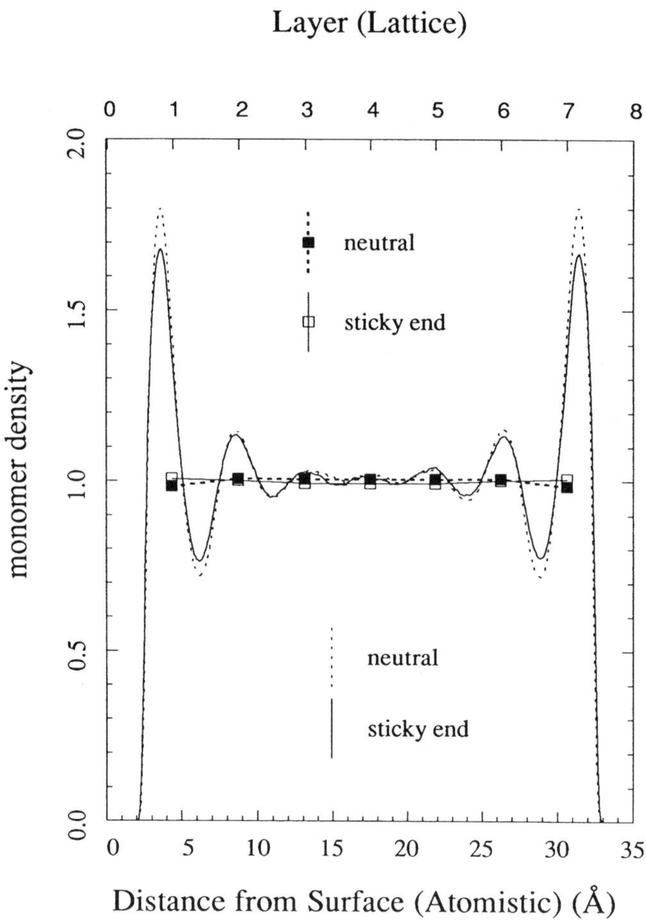

Fig. 8.20 Monomer density profile for octacosane melts at 400 K with neutral surfaces and sticky ends, compared to SCF lattice theory predictions (squares). (Redrawn from Ref. 25.)

which occur primarily due to the continuum nature of the atomistic model. This can be seen in Fig. 8.20, where the monomer density profiles for octacosane from simulation and SCF lattice theory are compared.

8.3 Systems of bead chains

8.3.1 General considerations

As discussed in Section 8.2, the atomistic systems have been limited to rather low molecular weight hydrocarbon systems. In attempting to gain general understanding for high molecular weight polymer systems, one popular development has been modeling the polymer chains as long necklaces of beads, each bead representing an isodiametric unit of a real polymer chain. The general features of the equilibrium arrangements near impenetrable surfaces have been studied by Monte Carlo calculations on liquids of freely jointed chains with realistic nonbonded interactions among the beads.[14,15] More recently, analogous methods have been applied to a system mimicking the dimensions, shape and interactions of the chains in a polyethylene melt.[16] The Monte Carlo method has also been used to simulate systems of short chains of hard spheres between impenetrable walls at various densities,[17] and mixtures of these chains with monomers in broad and narrow slits.[18] The dynamic properties of dense systems of bead-spring model chains with up to 30 monomers between structureless solid surfaces have been studied by molecular dynamics techniques.[19] The same model chains have been studied in narrow slits with structured surfaces under shear.[20]

8.3.2 Models and methods

8.3.2.1 Monte Carlo calculations

In the simulations of Refs 14–16, the polymer chains are represented as necklaces of beads connected by links of length σ. Beads belonging to different chains or belonging to the same chain but separated by two or more links interact through a 12-6 Lennard-Jones potential with minimum energy $E(\sigma) = -0.42$ kcal/mol, truncated at the distance σ. The value selected for $E(\sigma)$ is such to mimic the interactions among isodiametric units in polyethylene when σ is taken to be 0.46 nm. The model chains are freely jointed in the calculations of Refs 14 and 15, while the chain flexibility has been reduced in Ref. 16 by adding a bending potential of the form $E(\theta) = (K_\theta/2)(\theta-\pi)^2$, with θ the angle between consecutive links and $K_\theta = 0.82$ kcal/mol. The value of $\langle \cos\theta \rangle$ predicted on the sole basis of this bending potential is approximately 0.46 at 400 K, and is not modified by the inclusion of Lennard-Jones interactions between beads separated by two bonds. Hence, an unperturbed model chain of N beads subject to this bending potential is expected to have the same shape and dimensions of an unperturbed polyethylene chain comprising approximately 3.5 N methylene groups.

In each calculations, the initial model has been generated at random such that the centers of two nonbonded units could not come within 0.65σ from each other. The systems have been then equilibrated at 400 K using reptation and crankshaft movements. The latter mechanism, which cannot be used in atomistic calculations, consists in the rotation of a randomly selected sequence of one or more beads around the vector connecting the two adjoining beads in the same chain; when the selected sequence terminates with one of the chain-ends, it is rotated around a random vector. After each attempt, the resulting trial model is accepted or rejected according to the outcome of a standard Boltzmann test (see above). The size and the composition of the systems that have been modeled are listed in Table 8.1. In all cases, the periodicity is 50 in the x and y directions, while the systems are constrained along the z axis by two impenetrable walls at the distances given in the third column of Table 8.1, mimicking two ordered and well packed parallel layers of beads with centers at a distance approximately greater by 2σ. Therefore, the simulated systems represent reasonably long chain lengths and a very high bead density nearly $1.0\sigma^{-3}$.

8.3.2.2 Molecular dynamics calculations

The bead-spring model of polymer chains utilized for molecular dynamics calculations in Ref. 19 consists of beads interacting through a truncated and shifted 12-6 Lennard-Jones potential

$$U_{\mathrm{bb}}(r) = 4\epsilon\left(\left(\frac{\sigma}{r}\right)^{12} - \left(\frac{\sigma}{r}\right)^{6} + \frac{1}{4}\right) \quad \text{for } r \leq 2^{1/6}\sigma \tag{8.6}$$

and $U_{\mathrm{bb}}(r) = 0$ for $r > 2^{1/6}\sigma$. Consecutive beads in the same chain are connected by finite extensibility nonelastic springs with spring potential $U_{\mathrm{s}} = k_{\mathrm{s}} \ln[1 - (r/R_0)^2]$, where r is the actual spring extension, $k_{\mathrm{s}} = 33.75\epsilon$ and $R_0 = 1.54\sigma$. The basic cell is periodic in the x and y directions with edges 8.0σ, while the two faces of the cell perpendicular to the z axis are two structureless walls interacting with the beads through a 10-4 potential of the form to represent weakly attractive, strongly attractive, and neutral interactions. The density of the simulated systems is approximately $0.65\sigma^{-3}$, and the temperature is in all cases equal to ϵ/k, with k the Boltzmann constant. The initial bead coordinates have been obtained by lattice Monte Carlo calculations, followed by short constant temperature MD simulations

Table 8.1 *Size and composition of the systems of bead chain polymers simulated in Refs 14–16.*

System	Beads/chain	Wall separation	No. of chains	Bending potential (K_θ kcal/mol)
A	200	50	625	0
B	100	50	1 250	0
C	100	10	250	0
D	100	5	125	0
E	100	50	1 250	0.85
F	50	50	2 500	0

with scaling of the bead velocities. Chains in the range of 6–30 beads have been simulated in the slits of 6σ–20σ in width.

8.3.3 Results

8.3.3.1 Bulk chains

Since the wall separation of 50σ, used in most of the calculations in Refs 14–16, is several times larger than the unperturbed radius of gyration of the chains, chains with center of mass located far from the walls are expected to assume their unperturbed geometry and conformation. This is seen in Table 8.2, showing the r.m.s. end-to-end distance ($\langle R^2\rangle^{1/2}$), the r.m.s. radius of gyration ($\langle R_G^2\rangle^{1/2}$) and the characteristic ratio ($C_n = \langle R^2\rangle/(N-1)\sigma^2$) of chains with center of mass in the central region of systems with wall separation equal to 50σ. The last column shows the characteristic ratio (C_n') obtained when σ is taken to be 0.46 nm and each bead is considered to consist of 3.5 CH_2 groups, with the C–C bond length 0.153 nm. A value of the order of 1.5 for C_n is approximately what one would expect for nearly unperturbed chains in the first three cases. Remarkably, the value of C_n' found far from the walls for the chains with bending potential (system E) is practically equal to the experimental result for unperturbed polyethylene chains at 400 K.[34] Also, the ratio $\langle R^2\rangle/\langle R_G^2\rangle$ is in all cases very close to 6, showing that the distribution of units around the center of mass follows a Gaussian statistics. In particular, this indicates that system E constitutes a good description of the molecular arrangements of a melt of polyethylene chains confined between two walls.

8.3.3.2 Segmental density and order parameters

In agreement with the results obtained for short chain hydrocarbons near solid surfaces, the calculations in Refs 14–16 show that the composition of the first layer of beads in contact with the walls is substantially different from the composition of the bulk liquid. For instance, the density of beads

Table 8.2 *Average dimension of the chains in the central region of systems with wall separation equal to* 50σ.

Beads/chain	$\langle R^2\rangle^{1/2}$ (σ)	$\langle R_G^2\rangle^{1/2}$ (σ)	$\langle R^2\rangle/\langle R_G^2\rangle$	C_n	C_n'
200	18.14	7.54	5.79	1.60	4.13
100	12.40	5.09	5.94	1.54	3.98
50	8.64	3.52	6.01	1.49	3.85
100 (system E)	16.05	6.50	5.93	2.58	6.68

in a layer of thickness σ in contact with one of the walls is between 1.1 and 1.2 times the bulk value. This high density is associated with a large increase of the fraction of chain-ends, which is as high as 1.5 times the bulk value in this layer. When examined in greater detail, the bead density profiles show an intense maxima in proximity to the walls, followed by progressively less intense minima and maxima with a periodicity equal to σ. This behavior, though similar to that observed for small molecule fluids,[43,44] is damped considerably for the polymer systems, presumably due to the unfavorable entropic effect of constraining chain configurations at the interfaces.

The calculations in Refs 14–16 show that the perturbation of the bead density does not extend into the liquid more than approximately 2σ from the walls. A similar result is obtained for the order parameter with respect to the z axis of the links between consecutive chain beads. The order parameter is obviously negative for links with centers very close to the walls. It becomes slightly positive at a distance along the z axis of 1.5σ and assumes its isotropic value of 0 at the distance of *ca.* 2σ. The same behavior has been observed in all the calculations performed, irrespective of chain length, wall separation and chain flexibility. These findings are very similar qualitatively to the results obtained for liquid *n*-alkane systems (see above). The simulation results are also in good agreement with the predictions of the mean field lattice models of Helfand[7] and Scheutjens–Fleer.[8] Therefore, one can conclude that the perturbation due to the wall on the packing and orientation of the chain segments is limited to the first or the first two layers of beads immediately adjacent to the walls. The packing of the beads and the disorder of the chain links in all the other layers are identical to those in the unperturbed bulk liquid.

8.3.3.3 Chain statistics near the surfaces

As a consequence of the previous results, one expects that the chain statistics is perturbed only for chains having beads in contact with the walls, all the other chains being practically unperturbed. Figure 8.21 plots the distribution along the z axis (D_{cm}) of the centers of mass of the chains having beads in the first layer of thickness in contact with the walls for the polyethylene-like bead chains in system E of Table 8.1. Each chain has been given a weight equal to the number of beads contained in the first layer. About 90% of all the beads in contact with the walls are seen to belong to chains with center of mass at a distance from the walls smaller than the radius of gyration. As far as the nature of this perturbation is concerned, the r.m.s. end-to-end distance and the r.m.s. radius of gyration of chains in the proximity of the walls are not substantially different from their unperturbed bulk values. However, these chains tend to be flattened in the x–y plane. This is seen in Fig. 8.22, plotting the reduced z-component of the radius of gyration ($\langle R^2_{G,z}\rangle$) of the polyethylene-like chains in system E (with

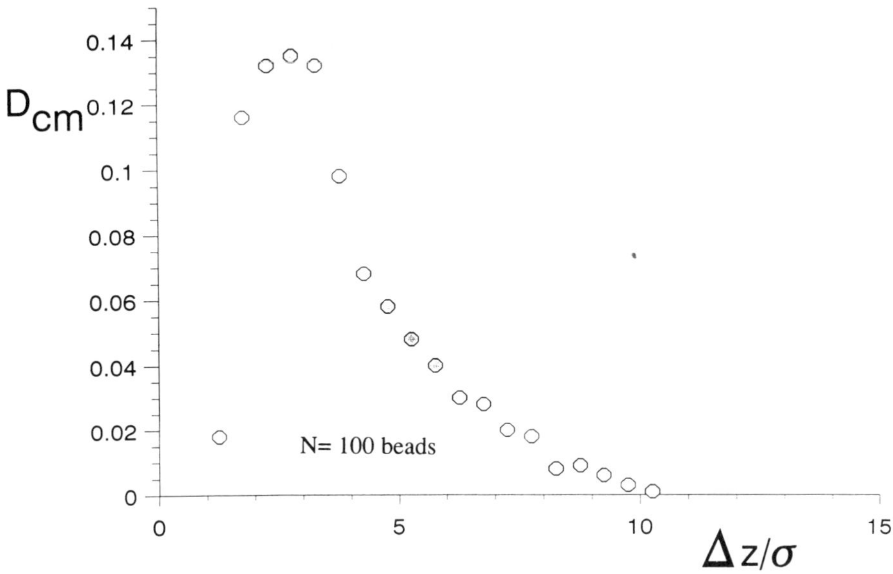

Fig. 8.21 The distribution, D_{cm}, of centers of mass of chains having beads in the first layer of thickness near the walls in system E listed in Table 8.1. Each chain has been given a weight equal to the number of beads in the first layer.

$N = 100$), as a function of the distance of their center of mass from the nearest wall. For the same system, Fig. 8.22 also plots the reduced density of centers of mass (ρ_{cm}). Both quantities are normalized to their unperturbed bulk values. For z between 2σ and 5σ, where ρ_{cm} shows a maximum, $\langle R^2_{G,z} \rangle$ is still less than one half of its unperturbed value. Hence, most of the chain segments located within the radius of gyration from the surface belong to the polymer chains which are substantially flattened parallel to the surface. Analogous results have been obtained for long freely jointed chains,[14,15] for short chains of hard spheres,[18] and for the bead-spring chains.[19]

The gradual increase of $\langle R^2_{G,z} \rangle$ in Fig. 8.22 corresponds to a gradual evolution of the molecular shape from nearly two-dimensional structures near the walls to three-dimensional Gaussian random-coils at a distance from the walls close to the unperturbed radius of gyration. The molecular shape of chains with center of mass close to the walls, when examined in details, can be described as an apparently random combination of train and loop sequences. The average length of train sequences (that is, a sequence of beads located in the first layer of thickness σ near a wall) is 4.1 beads for chains of 100 beads in the absence of a bending potential, and increases to 4.6 beads for the polyethylene-like bead chains simulated in system E, which corresponds to approximately 16 methylene units. Hence, one should take

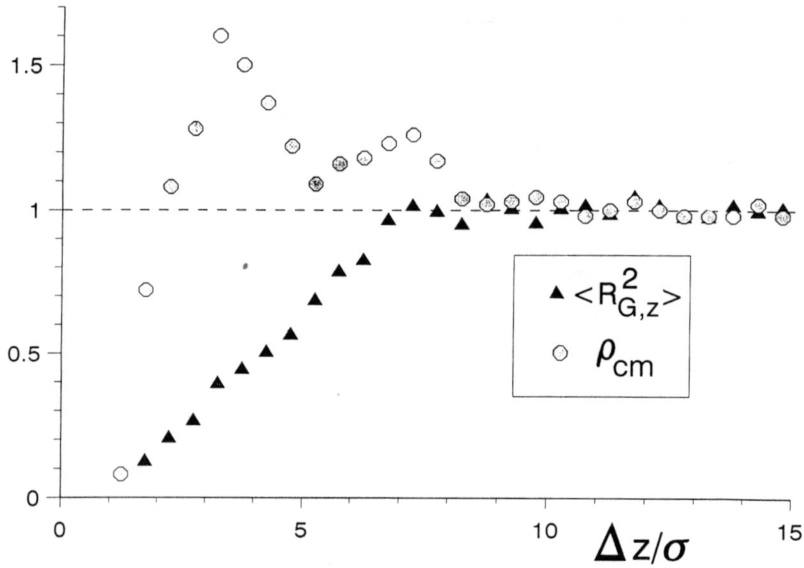

Fig. 8.22 The reduced z-component of the mean-square radius of gyration ($\langle R^2_{G,z} \rangle$) and the reduced density of centers of mass (ρ_{cm}) as a function of the distance from the nearest wall

care in relating the results of atomistic simulations and experiments on short chains to high molecular weight polymer systems, since the entire length of most n-alkane systems investigated does not exceed the average train sequence length predicted for the polymer.

8.3.3.4 Dynamic properties

The influence of the solid surface on chain dynamics has been investigated in Ref. 19 by monitoring both the chain center of mass diffusion and the longest relaxation time of chains as a function of the distance of the chain center of mass from a solid surface. The longest relaxation time of the chain is given by the end-to-end vector autocorrelation function and is directly related to the viscosity. As with static properties, it was found that the dynamic behavior of chains in the central region of the thick slit (width $= 20\sigma$) was unperturbed by the presence of solid surfaces.

Near the surfaces, the diffusivity showed the typical anisotropic behavior as the center of mass of the chains approached a solid surface. For neutral walls, the overall apparent diffusivity showed essentially no dependence on the chain center of mass distance from a wall. In these cases, mobility perpendicular to the wall decreased significantly as the wall was approached, but this effect was offset by an increase in mobility parallel to the walls. In the case of strongly attractive walls, the mobility parallel to the walls showed no enhancement near the walls, and the mobility perpendicular showed a greater decrease than for neutral walls, resulting in a decrease in the total mobility near the attractive surfaces. This effect is attributed to the increase in segment density near the surfaces for cases with strongly attractive surface–bead interactions.

Because significant displacement of the center of mass of the chains occurs over the time scale of the longest relaxation time of the chains, the longest relaxation time is not truly a local property. For neutral surfaces, no dependence of the longest relaxation time on the initial center of mass position of the chains was seen. For strongly attractive surfaces, a significant increase in the longest relaxation time was seen for shorter chains (10 beads) in the vicinity of the solid surfaces. This effect decreases significantly for longer chains (20 beads), again indicating that the longest relaxation time is not a local property. The significant increase in the longest relaxation time for the shorter chains near the strongly attractive surface does indicate, however, that such effects may occur in higher molecular weight systems given a high enough degree of molecular layering. It is worth noting that the degree of segmental and molecular segregation appears to be greater in the atomistic systems.

8.4 Conclusions

In this chapter the main focus has been on interfaces of alkane melts; but, as has already been noted above, related questions show up in the context of various other models in the literature as well.[45–48] Experiments have shown that polymer melt interfaces possess many interesting equilibrium and dynamic properties that are not easily understood. Atomistic simulations have been able to provide significant insight into much of this observed behavior. Atomistic simulations indicate the presence of several layers of relatively high segmental density separated by depletion regions of low density. The oscillations in solvation forces observed experimentally as a function of film thickness have been shown to be a directed result of this segmental layering. The layer nearest the surface is a region of relatively high segmental density that consists of segments of molecules lying nearly parallel to the surface (trains). For *n*-alkane molecules, these trains often consist of entire molecules. For high molecular weight systems, the results of bead-chain simulations indicate the average length of train sequences to be approximately 16 methylene units for polyethylene. Therefore, one should take caution in relating the results of short chain systems to the long polymer chains, since the effects of the segmental layering and train sequences on physical properties may vary strongly with the overall chain length.

While every off-lattice simulation predicts the general same behavior, the persistence of the layering from the surfaces and the degree of orientation within the layers depends strongly upon the details of the model. It is therefore apparent that the quantitative predictions require an accurate representation of the polymer and surface structure and the interactions. That quantitative predictions of surface properties are possible is supported by the reasonably good agreement between predicted and experimental surface tensions for *n*-alkane melts. From an analytical approach, the distribution of trains, tails and loops predicted by the atomistic simulations is reproduced quite well by the Scheutjens–Fleer lattice theory. The modulation of local density, however, is not predicted by the lattice theory.

Despite the success in explaining (at least at a qualitative level) the equilibrium properties of polymer melt interfaces, simulations have not contributed significantly to explaining many of the dynamic properties of these systems, except perhaps to indicate what is not occurring at these interfaces. Although significant increase in orientation of segments in the surface layers is seen, there is no indication of epitaxial crystallization, even for strongly attractive surfaces. Additionally, while a decrease in molecular mobility is seen near strongly attractive surfaces, the decrease does not indicate solidification (or vitrification) of the polymers in contact with surfaces. Therefore, the dramatic changes in dynamic properties seen experimentally for thin polymer films remain an open area for further investigation by simulation techniques.

Acknowledgments

The authors are truly indebted to our collaborators who made key contributions to the progress covered in this article, particularly to S. K. Kumar, R. G. Winkler, T. Matsuda, F. Auriemma, P. Ludovice and R. L. Jaffe. Financial assistance from IBM Italy, IBM Germany, UNITIKA Ltd, Eloret Institute, and the NATO Fellowship is acknowledged.

References

1. H. K. Christenson, D. W. R. Gruen, R. G. Horn, and J. N. Israelachvili, *J. Chem. Phys.* **87**, 1834 (1987).
2. J. van Alsten and S. Granick, *Phys. Rev. Lett.* **61**, 2570 (1988).
3. M. L. Gee, P. M. McGuiggan, J. N. Israelachvili, and A. Homola, *J. Chem. Phys.* **93**, 1895 (1990).
4. H.-W. Hu, G. A. Carson, and S. Granick, *Phys. Rev. Lett.* **68**, 2346 (1991).
5. B. J. Factor, T. P. Russell, and M. F. Toney, *Macromolecules* **26**, 2847 (1993).
6. T. P. Russell, *Mater. Sci. Rep.* **5**, 171 (1990). M. Stamm, in *Physics of Polymer Surfaces and Interfaces*, edited by I. C. Sanchez (Butterworth-Heinemann, Boston, 1992), p. 163.
7. E. Helfand, *Macromolecules* **9**, 307 (1976).
8. J. M. H. Scheutjens and G. J. Fleer, *J. Phys. Chem.* **83**, 193 (1979).
9. D. N. Theodorou, *Macromolecules* **21**, 1400 (1988).
10. W. G. Madden, *J. Chem. Phys.* **87**, 1405 (1987).

11. G. ten Brinke, D. Aussere, and G. Hadziiounnou, *J. Chem. Phys.* **89**, 4374 (1988).
12. K. F. Mansfield and D. N. Theodorou, *Macromolecules* **22**, 3143 (1989).
13. G. D. Smith, D. Y. Yoon, and R. L. Jaffe, *Macromolecules* **25**, 7011 (1992).
14. S. K. Kumar, M. Vacatello, and D. Y. Yoon, *J. Chem. Phys.* **89**, 5209 (1988).
15. S. K. Kumar, M. Vacatello, and D. Y. Yoon, *Macromolecules* **23**, 2189 (1990).
16. M. Vacatello and D. Y. Yoon, *Makromol. Chem., Macromol. Symp.* **48/49**, 349 (1991).
17. R. Dickman and C. K. Hall, *J. Chem. Phys.* **89**, 3168 (1988).
18. A. Yethiraj and C. K. Hall, *J. Chem. Phys.* **91**, 4827 (1989).
19. I. Bitsanis and G. Hadziioannou, *J. Chem. Phys.* **92**, 3827 (1990).
20. P. A. Thompson, G. S. Grest, and M. O. Robbins, *Phys. Rev. Lett.* **68**, 3448 (1992).
21. M. Vacatello, D. Y. Yoon, and B. C. Laskowski, *J. Chem. Phys.* **93**, 779 (1990).
22. M. Vacatello and F. Auriemma, *Makromol. Chem., Theory Simul.* **2**, 77 (1993).
23. M. Vacatello, *Makromol. Chem., Theory Simul.* (in press).
24. R. G. Winkler, T. Matsuda, and D. Y. Yoon, *J. Chem. Phys.* **98**, 729 (1993).
25. T. Matsuda, G. D. Smith, R. G. Winkler, and D. Y. Yoon, *Macromolecules* **28**, 165 (1995).
26. J. G. Harris, *J. Phys. Chem.* **96**, 5077 (1992).
27. Y. Wang, K. Hill, and J. G. Harris, *J. Phys. Chem.* **97**, 9013 (1993).
28. Y. Wang, K. Hill, and J. G. Harris, *Langmuir* **9**, 1983 (1993).
29. M. W. Ribarsky and U. Landman, *J. Chem. Phys.* **97**, 1937 (1992).
30. T. K. Xia, J. Ouyang, M. W. Ribarsky, and U. Landman, *Phys. Rev. Lett.* **69**, 1967 (1992).
31. D. Y. Yoon, G. D. Smith, and T. Matsuda, *J. Chem. Phys.* **98**, 10037 (1993).
32. G. D. Smith and D. Y. Yoon (in preparation).
33. M. Vacatello, G. Avitabile, P. Corradini, and A. Tuzi, *J. Chem. Phys.* **73**, 548 (1980).
34. P. J. Flory, *Statistical Mechanics of Chain Molecules* (Interscience, New York, 1969).
35. S. Nosé, *J. Chem. Phys.* **81**, 512 (1984).
36. H. Ertl, H. Dullien, and F. A. L. Dullien, *AICHE J.* **19**, 1215 (1973).
37. R. Hentschke, B. L. Schurmann, and J. P. Rabe, *J. Chem. Phys.* **96**, 6213 (1992).
38. G. D. Smith and D. Y. Yoon, *J. Chem. Phys.* **100**, 649 (1994).
39. S. Toxvaerd, *J. Chem. Phys.* **93**, 4290 (1990).
40. J. J. Jasper, *J. Phys. Chem. Ref. Data* **1**, 841 (1972).
41. J. K. Lee, J. A. Barker, and G. M. Pound, *J. Chem. Phys.* **60**, 1976 (1974).
42. J. A. Barker, *Mol. Phys.* **80**, 815 (1993).
43. J. R. Henderson and F. van Swol, *Mol. Phys.* **51**, 991 (1984).
44. I. K. Snook and D. Henderson, *J. Chem. Phys.* **68**, 2134 (1978).
45. A. Yethiraj and C. K. Hall, *Macromolecules* **23**, 1865 (1990).
46. K. F. Mansfield and D. N. Theodorou, *Macromolecules* **24**, 6283 (1991).
47. J.-S. Wang and K. Binder, *J. Phys.* I *(France)* **1**, 1583 (1991).
48. I. A. Bitsanis and G. ten Brinke, *J. Chem. Phys.* **99**, 3100 (1993).

9

COMPUTER SIMULATIONS OF TETHERED CHAINS

Gary S. Grest and Michael Murat

9.1 Introduction

The study of tethered polymer chains[1] is an area which has received increasing attention in recent years. These are systems in which one or both ends of the chain are constrained in their motion because they are attached to a d dimensional surface. This surface could be a point or small central core ($d = 0$) as in the case of a many-arm star polymer, a line ($d = 1$) as in the case of a comb polymer, or a flat surface ($d = 2$) as in the case of a polymer brush.[2] Polymers attached to themselves to form a polymer network or a tethered membrane are also examples of tethered chain systems. An interesting example of a tethered membrane is the spectrin/actin membrane skeleton of the red blood cell skeleton.[3,4] A schematic illustration of these four examples of tethered chain is shown in Fig. 9.1. Additional interest in tethered chains is due to their technological applications in colloidal stabilization and lubrication.[5,6]

The distinctive feature of tethered chain systems is the presence of a new length scale which is not present in other systems. This new length scale is the distance s between tethering or grafting points. When the tethering density is high, nearby chains crowd each other, forcing the polymer to stretch out away from the grafting site in order to gain interaction energy. The stretching, however, is accompanied by a penalty in terms of the elastic energy, due to a decrease in configurational entropy. The more highly curved the surface, the more accessible volume the chain has as a function of the distance from the tethering site, and the stretching becomes less significant. For chains tethered to a line or a flat surface, this stretching is strong enough to change the scaling of the end-to-end distance R and the radius of gyration R_G with the chain length N, from that of a free chain in a comparable solvent. While polymers attached to themselves to form a polymer network or membrane are not locally stretched in the same manner, the tethering can give rise to interesting structures which globally increase in size with the total number of monomers N in ways which are quite different than predicted by simple scaling theories.

This interesting property that tethered chains could be strongly stretched was first noted by Alexander[7] in his study of chains end-grafted onto a flat plane. De Gennes[8] and Cantor[9] stressed the importance of tethered chains in

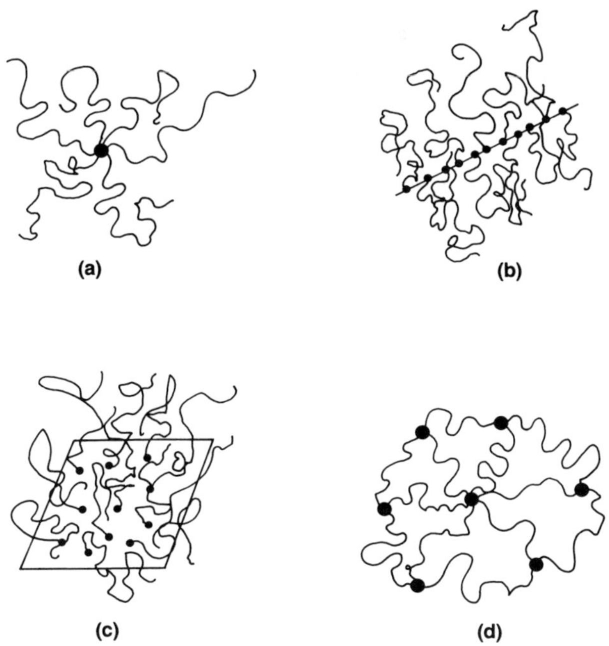

Fig. 9.1 Examples of topologies of polymers tethered to (a) a point, (b) a line, (c) a plane, and (d) themselves.

understanding the self-assembly of diblock copolymers in a selective solvent. Daoud and Cotton[10] and Birshtein *et al.*[11] later generalized the Alexander approach to star polymers. For chains tethered to a d dimensional object, these theories which are based on a blob model suggest that for moderate grafting densities ρ_a, the monomer density $\rho(r)$ decreases from the grafting surface as

$$\rho(r) \sim \rho_a^{(3\nu-1)/2\nu} r^{(1-3\nu)(2-d)/2\nu}, \tag{9.1}$$

where ν is the correlation length exponent for a dilute chain, $R \sim N^{\nu}$. For a good solvent $\nu = 0.588$ in three dimensions[12] and 1/2 in a Θ solvent. For a star polymer ($d = 0$) in a good solvent, the density decays as $r^{-1.30}$, while for a polymer brush (d = 2), the density is constant up to a height h. As will be seen below that eq. (9.1) works quite well for a star polymer but not for a brush. This scaling theory also predicts that the end-to-end distance R increases more rapidly with N than for a free chain for $d > 0$,

$$R \sim N^{2\nu/(2-d+d\nu)} \rho_a^{(1-\nu)/(2-d+d\nu)}. \tag{9.2}$$

For a star polymer, for $N \gg 1$, there is sufficient space for the arms and R scales as N^{ν}, exactly as for a free chain. However for chains grafted to a line or a flat surface, R increases more rapidly. For a polymer brush $R \sim N$, independent of solvent quality. While these simple predictions are not always valid, they give a clear indication of how tethering polymers can change their properties. Since these scaling theories were worked out, there has been an extensive amount of work using numerical and analytic self-consistent field calculations, renormalization group techniques, computer simulations and experiments investigating various aspects of tethered polymer chains. Here we will review this work.

The concept of tethered chains can be used to bring together a number of diverse fields. As an example consider the simple case of polymers tethered to a point. While many-arm star polymers can be made by chemical reactions, star-like aggregates also occur for diblock copolymers in a selective solvent, polymers with one ionic end group in a low dielectric media and for end-functionalized polymers grafted to a small colloidal sphere. When the length of each arm is long, the size of all four of these systems has the same dependence on number of arms f and chain length N. Similarly the scaling properties of a polymer brush made from diblock copolymers in a selective solvent or grafted end-functionalized polymers are the same. This means that one can elucidate many of the common global features of these systems without a detailed analysis of their local structure. By considering tethered systems as a group and not isolating polymers attached to a point from, say, those attached to a plane, one can arrive at a better understanding of each.

In this paper we will review how simulations on relatively simple coarse-grained models have been helpful in developing a more complete understanding of tethered chain systems. In the next section, we discuss the various methods, both Monte Carlo (MC) and molecular dynamics (MD) which have been applied to these systems. Because at least one end of each chain is constrained, some methods commonly used for free chains have to be modified. In addition, the extra constraints and high monomer density near the tethering point often lead to very slow relaxation processes. In the following four sections, we will review the properties of polymers attached to d-dimensional surfaces and to themselves starting with star polymers in Section 9.3. In Section 9.4, we consider polymers tethered to a line, both rigid and flexible. Then, in Section 9.5, we discuss polymer brushes for which the most extensive amount of work has been done. In Section 9.6, we discuss tethered membranes in which linear polymers are connected to form a D dimensional surface. The case $D = 2$ has generated a lot of interest in recent years. Finally in Section 9.7, we present a brief summary of the important results and discuss the outlook for the future of simulations in this area. In all cases, we have tried to compare the simulation results to both theory and experiment for a variety of solvent qualities.

9.2 Models and methods

In this section we will review the basic models and methods that have been used to simulate tethered chain systems. Before beginning a simulation, one obviously must decide whether to use Monte Carlo or molecular dynamics. If one decides on MC, then one has a further choice, namely whether to do the simulation on a lattice or in the continuum. It is also important to decide how detailed a microscopic model to use. For the long time, large-distance phenomena which are of interest here, a realistic, microscopic model for the polymer would require too much computer time. Instead, we use simple coarse-grained models, which incorporate the essential physics such as the noncrossability of the chains. In this section, we will briefly present an overview of these techniques and discuss some of the advantages and disadvantages of each, particularly as applied to tethered systems. Because of the additional constraints due to the tethering, some methods, particularly some standard lattice MC methods do not work as well as they do for linear chains. In addition to these two simulation methods, we will describe how one numerically solves the self-consistent field (SCF) equations, since this method has been very important in understanding the properties of polymer brushes and block copolymers.

In the MC method the subsequent configurations of the polymers are generated stochastically. The collective motion of a chain is modeled by the acceptance of attempted new configurations. Under certain circumstances one can use such an approach not only for static properties but also for the investigation of the dynamics of the system.[13–21] In order to use MC methods for the simulation of the dynamical properties of polymers, one needs a method which is based on local stochastic moves. For a nonreversal random walk (RW), that is a random walk with excluded volume for nearest and next-nearest neighbors along the chain only, it can be shown[22] that this method reproduces the Rouse model.[23] In such a case the local configurations and the attempted moves are the same as in the case in which one includes the excluded volume interaction between all monomers. In order to fulfill this requirement, it is necessary to structure the algorithm according to the following simple rules.[15] First, one should choose a monomer at random, then choose one of the possible moves at random and finally check whether the attempted move is allowed or not. If one includes only excluded volume interaction, then all moves that fulfill the excluded volume constraints are accepted. If additional interactions are included, then the allowed moves are accepted according to a Boltzmann weight.[15] This approach can be applied to both lattice and off-lattice polymers. The natural unit for a time-step then is one attempted move per monomer, which measures time up to an unknown but constant prefactor. For a discussion see the chapter on entanglement effects in polymers by Kremer and Grest, Chapter 4 of this volume.

In addition to MC methods, molecular dynamics has also been a very powerful method for studying the properties of liquids.[24–27] In this method, one simply solves Newton's equations of motion for each monomer. The total momentum and energy of the system are conserved quantities and the simulations are in what is known as a microcanonical ensemble. However to sample the entire phase space there must be an exchange of momentum and energy of each polymer chain with its surroundings. For large, dense systems with many chains, this is not a problem as is the case for dense polymer melts.[28] However, for dilute chains one has to be more careful. In this case, one has two ways to proceed. The first is to take the solvent molecules (monomers or very small chains) explicitly into account. This is how one studies hydrodynamic effects.[29,30] However because one typically needs as many as 20 times or more solvent particles per monomer, this makes the algorithm very inefficient and it can only be used for small systems.[31] An alternative method is to treat the solvent as a continuum. In the simplest approximation, the solvent acts as a heat bath for the polymer and produces a viscous drag when the polymer moves. While this neglects any hydrodynamic contributions, adding only these two terms to Newton's equations turns out to produce a very efficient algorithm for studying the structure of tethered polymers and membranes. Because of the random noise term, the chains satisfy free-draining Rouse dynamics[23] at long times, similar to MC. The method is particularly effective at high density, where the MC methods run into difficulty.

9.2.1 Lattice models

9.2.1.1 Standard lattice algorithms

MC simulations for lattice polymers in which a flexible polymer chain is modeled by a self-avoiding random walk (SAW) on a periodic lattice have been widely used to study the properties of linear and ring polymers, both for dilute chains as well as for entangled chains at moderate density.[13–21] In this case, the walk is a succession of N steps subject to the condition that no lattice site is visited more than once. Typically one uses the diamond lattice or the simple cubic lattice. Figure 9.2 illustrates the typical moves one includes for the simple cubic lattice. If one only includes the end-bond and two-bond kink jump[32] alone, this method is very inefficient since the kink-jump move does not create any new bond vectors connecting the neighboring bonds, but only exchanges the two neighboring bond vectors.[15] In this case new bonds can only diffuse into the chain from the ends, causing an artificial slowing down of the dynamics. A ring polymer, for instance, could never equilibrate if one includes only moves which exchange bonds, such as the 2-bond move in Fig. 9.1 and the relaxation would be very slow for stars or brushes because one end is fixed. Thus it is essential to include

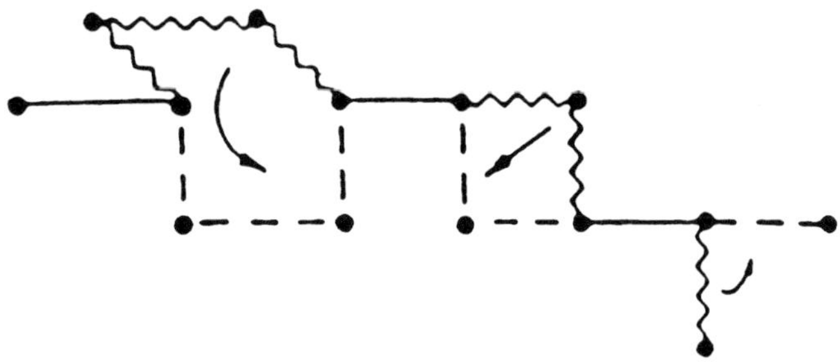

Fig. 9.2 Illustration of some of the possible moves on the simple cubic lattice for the standard kink jump algorithm.

other moves, such at the 90° crankshaft motion.[22,33,34] This choice of moves reproduces the Rouse dynamics for the nonreversal RW.[15] The lattice model as shown in Fig. 9.2 also allows for some blocked configurations,[35] which in principle the system can never reach or get out from. Though they are not important for low densities, care must be taken that they are not in the initial configurations from which one starts the simulation. For chain lengths of practical interest, Sariban and Binder[36] have shown that the statistical weight of these forbidden configurations are negligibly small and can be safely ignored. However as the chain length increases, these blocked configurations become more and not less important. In some cases, it is helpful to introduce additional moves[37] which move the bond a finite distance along the chain, without rotating it by moving two bond vectors in opposite directions along the chain. These moves are purely exchanges which speed up the local diffusion of bond-vectors along the chain. While they do not create new bond directions as the 3-bond crankshaft move shown in Fig. 9.2, they help to equilibrate the system more quickly, particularly for dense systems.

The advantages of working on a lattice are clear. Since one is dealing with a discretized system, one can use integer arithmetic instead of floating point. This means that determining the distance between two monomers can be done very quickly leading to a very fast update procedure. For large enough memory, as is standard for modern computers, one can store the entire lattice and check the excluded volume constraint simply by checking the occupation of the lattice sites. However, in order to move the chains a significant amount, one needs a reasonable acceptance rate of the attempted moves typically larger than 0.1. Consequently, since one usually attempts to move several monomers simultaneously, one is confined to relatively low densities. This simple algorithm becomes difficult to equilibrate in regions where the density is high, as for example in a melt[16,37] of linear chains, near the center of a star polymer or for a polymer brush at high density.[38,39]

Another advantage of MC methods is the that one can use nonlocal moves which are not dynamically correct to equilibrate the system rapidly. One simple example of such a move is the slithering snake or generalized reptation motion, first discussed by Kron[40] and Wall and Mandel.[41] In this move, a monomer from one end of the chain is removed and added to the other end of the chain in a random orientation. However in many cases, such moves are not very useful for tethered chains, since the nonlocal move typically violates the constraints. One exception to this was developed by Dickman and Hong[42] in their simulations of polymer brushes. In their lattice SAW (self-avoiding walk) model for end-grafted brushes, they used a combination of local bond-flips, which permit a kink to diffuse along the chain, and the nonlocal slithering snake motion. However, since the grafting point should remain fixed, the whole chain is then translated to bring the new end of the chain back to the correct grafting site. Both the bond-flip and the slithering snake moves are rejected if they violate the nonoverlap or wall constraints. Unfortunately much of the advantage of the nonlocal move is lost, since a global search has to be made to check for possible overlaps when the chain is translated. Another nonlocal move is to transport beads from kinks or chain ends along the chain contour to another position along the chain.[43–45] This method, known as the cooperative motion algorithm, is much more efficient than algorithms which only use local motions and can be used in situations such as ring polymers or end-grafted brushes, where the original slithering snake algorithm is not applicable. Since the beads are moved cooperatively along a closed path, the method is the only method that works at very high densities, where other lattice methods fail. It has been applied to study end-grafted brushes in a melt and at high density.[46,47] Another nonlocal method, developed by Murat and Witten[48] to study cross-linked polymers in which the motion of the ends is restricted, has been applied by Baljon-Haakman and Witten[49] to study associating polymers. Finally, we should mention global pivoting schemes,[15,50–54] in which a chain monomer is chosen at random and the entire chain segment beyond that site is rotated rigidly so that the rotated section again fits on the lattice. For polymers attached to an impenetrable surface, the original pivot algorithm can be modified[35,52,53] so that the section of the chain not attached is consistently pivoted with respect to the section which is attached. Also one can apply the pivot moves only to interior sections of the chain in cases when either one or both ends are tethered.[35]

In addition to dynamic MC methods, in which one uses time averages along stochastic trajectories in phase space, static MC methods are often used to generate polymer chains by constructing walks either step by step or by larger units. Each successively constructed chain is independent of the previous one. This method has been widely used to study the properties of dilute chains under a variety of solvent conditions. While a review of these methods is beyond the scope of this chapter, we would like to point out a few examples where they have been used to study tethered chain systems. For a complete review of these methods, the reader should see Ref. 15. Mazur and McCrackin,[55,56] Kolinski and Sikorski,[57] Lipson *et al.*[58–60] and Barrett and Tremain[61] have investigated branched polymers, such as stars and combs, using the inverse restricted Rosenbluth–Rosenbluth[62] MC algorithm. However while this method works well near the Θ point, it is known to be dangerous for even moderate N in a good solvent.[63,64] For a good solvent, Batoulis and Kremer[64] used a newer, modified version of the dimerization method of Suzuki[65] and Alexandrowicz[66] to study good solvent stars with $f \leq 6$ arms on a fcc lattice with overall polymerization $fN \sim 400$. Using this improved method, they were able to reduce the uncertainty in exponents such as $\gamma(f)$ which describes the dependence of the partition function $Z(N)$ on $N, Z(N) \sim N^{\gamma-1}$ for large N. The dynamical methods cannot determine γ. At the Θ temperature, T_θ, Batoulis and Kremer[67] used the inverse restricted sampling method for $f \leq 12$ and showed that the T_θ for a star polymer is the same as for a linear chain. Ohno and Binder[68] have recently developed a completely stochastic (unbiased) sampling method and studied star polymers with N up to 125 and $f = 20$ in two dimensions. In a related work, Lescanec and Muthukumar[69] used an off-lattice kinetic SAW algorithm to study comb-burst and starburst molecules.[70–73]

9.2.1.2 Bond fluctuation algorithm

In order to circumvent some of the difficulties of the standard lattice MC methods, particularly at high density, one might try to construct a chain of small spheres and allow the bond angle and maybe even the bond length to vary. Off-lattice simulations in which the bond angle but not the length is allowed to vary, the pearl necklace model,[13,14,74] as well as models in which both the bond angle and length are allowed to fluctuate, the bead-spring model,[19] have become very popular recently, particular for studying dense melts and tethered chain systems. However before discussing these models in detail, it is important to mention a second class of lattice models first used by Carmesin and Kremer.[75] This method is known as the bond fluctuation algorithm and combines ideas from standard lattice MC methods with the notion that one should allow the bond angle and the bond length to vary to produce a very efficient algorithm. Because the method only involves local moves, it is useful for investigating dynamic as well as static properties. Figure 9.3 illustrates the method as it was used to study a two-dimensional polymer melt.[75] Each monomer consists of 2^d lattice sites. In addition to the excluded volume interaction, the bond length l is restricted to a maximum extension to avoid bond crossing. On the square lattice, one has the constraint that $1 < l \leq \sqrt{13}$. For $d = 3$, the situation is slightly more complicated. In this case a set of 108 different bonds are allowed.[76] Since each

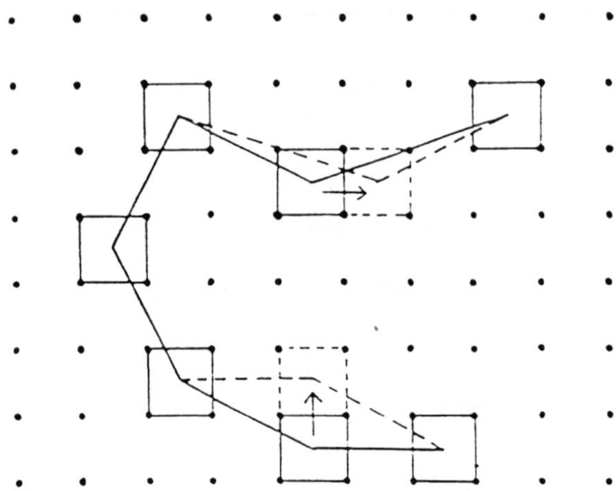

Fig. 9.3 Illustration of the bond fluctuation method for a $2d$ linear polymer. Possible moves of the two monomers are indicated.

monomer occupies 2^d sites, but every jump only requires 2^{d-1} empty sites, the method works at relatively high densities. It also suffers less from the nonergodicity problems due to blocked configurations than the standard methods. The method has been widely used to study two-[75] and three-dimensional[77] polymer melts, at densities as large as 0.8 and 0.5, respectively. This is effectively a much higher density than for the same number of occupied sites using the standard lattice models. The method also works well for branched and tethered polymers and has been applied to study end-grafted polymer brushes by Lai and Binder,[78–80] Dickman and Hong,[42] and Dickman and Anderson[81] and star polymers by Su *et al.*[82,83] The model has also been used to study polymer networks[84,85] and glasses.[86,87] A variant of this model in which each monomer consists of only one site on a sc lattice and the bond lengths are allowed to fluctuate among the values $1, 2^{1/2}$ and $3^{1/2}$ has been used by Shaffer[88,89] to study polymer brushes and melts. Excluded volume interactions are enforced by tracking the locations of the midpoints of all bonds on a secondary lattice.

One advantage of the bond fluctuation method is that it works well on a variety of platforms, including vector computers,[76] massively parallel computers,[77] and super scalar processors.[78,79] For a dense polymer melt in three dimensions, the vectorized code gives *ca.* 1.7×10^6 attempted moves per second for one processor on the Cray YMP for a volume fractions of $\phi = 0.41$. On massively parallel computers,[77] the program has only been run to date in such a way that each processor ran one independent system. While the method should be parallelizable across multiple processors, this has not been done yet.

9.2.1.3 *Starting the simulation*

One final issue in regard to lattice simulations is how does one start the simulation so that the system is not trapped in configurations which it cannot get out of. This has been handled in a number of different ways. The simplest procedure is to start the system in an ordered state. For linear chains in a melt, this can be done by stretching out the chains or folding them in a simple pattern. However this method is not very efficient since the system is far from equilibrium. Often this procedure is not practical for tethered chains, due to topological constraints. Cases where it has been used are for star polymers, where one can begin with all arms fully extended or for end-grafted polymer brushes, with all chains aligned perpendicular to the grafting plane. An alternative way is to grow the chains with the correct topological constraints taking the self-avoidance into account whenever possible.[77,78] If one is careful to avoid highly knotted structures, which might become artificially favored during the growth process, then the simulation should be ergodic. One can then use the standard simulation procedures to produce configurations which are self-avoiding. At high density, the cooperative motion algorithm,[43,44] is a very efficient way to go from a configuration where some of the bonds overlap to a self-avoiding configuration. Chakrabarti and Toral[38] simultaneously grew and equilibrated chains[90] in their work on polymer brushes. Chains already in the system are equilibrated using elementary moves that involve bead jumps and crankshaft motion. In addition, after every MC step, an attempt is made to insert an additional chain (represented as a SAW) in which one end is attached randomly to a grafting site on the surface. This trail insertion fails if the new chain overlaps with chains already present. By this method, they can reach moderate surface coverages, in a reasonable amount of CPU time. The method can easily be generalized to other tethered chain systems in which only one end of the chain is tethered. For higher densities, it is possible to grow the additional chains, monomer by monomer, instead of trying to insert an entire chain. This will be particularly important for long, branched chains.

9.2.2 *Off-lattice models*

Although the bond fluctuation method can be applied to a wide variety of problems, as noted above, the use of a lattice inevitably has some limitations. As discussed above, ergodicity problems[15,17,35] are expected to become severe, particularly at high density or in the presence of random obstacles.[91] For very high density, the lattice simulations often show precursors of the glass transition which result from artificially long-lived high density states. Many tethered chain systems are very inhomogeneous with high density regions near the tethering points and lower-density ones away from the tethering region. An example of this is a multi-arm star or starburst molecule. While the lattice simulation methods may be very efficient at handling the low-density regions, they can often be very inefficient at relaxing the fluctuations in the higher-density regions. In a lattice model, it is also very difficult to determine certain quantities such as the osmotic pressure.[42] In addition, the lattice structure may give rise to artifacts at high density as observed by Lai and Binder[79] in their simulations of polymer brushes. For these reasons, it is often desirable to carry out the simulation off-lattice, using either MC or MD.

The easiest way to include excluded volume effects is to put spheres centered at each connection point on the chain. The spheres can either be hard[92, 93, 94] or soft. For soft spheres, a Lennard-Jones interaction is often used, where the interaction between monomers is

$$U^0(r) = \begin{cases} 4\epsilon\left[\left(\frac{\sigma}{r}\right)^{12} - \left(\frac{\sigma}{r}\right)^{6} - \left(\frac{\sigma}{r_c}\right)^{12} + \left(\frac{\sigma}{r_c}\right)^{6}\right] & r \le r_c; \\ 0 & r > r_c, \end{cases} \tag{9.3}$$

Here ϵ characterizes the strength of the interaction, σ is the unit of length and r_c is the interaction cut-off. The potential is applied between all possible pairs of beads. In many cases,[13,28,95–109] the interaction is truncated at $r_c = 2^{1/6}\sigma$, such that the potential is purely repulsive. This is an efficient model for studying good solvent conditions. Since the potential is close to a hard sphere, we refer to this case as athermal for convenience. To introduce the effect of solvent quality, the range of the interaction is often extended. The cut-off was set to infinity by Baumgärtner[110] in his study of the collapse transition and Sheng *et al.*[111] in their studies of the vapor–liquid phase diagram for linear chains and by Freire *et al.*[112] in their study of star polymers. However for many chain systems, it is computationally more efficient to truncate the potential at a finite value, such as $r_c = 2.5\sigma$.[113–120] Thus by changing T, it is possible to vary the relative importance of the monomer–monomer attraction and change the effective quality of the solvent without explicitly introducing a solvent. To determine T_θ, one can measure the mean-squared radius of gyration $\langle R_G^2 \rangle$ and/or the end-to-end distance $\langle R^2 \rangle$,

$$\begin{aligned} \langle R_G^2 \rangle &= \frac{1}{N}\left\langle \sum_{i=1}^{N} (\mathbf{r}_i - \mathbf{r}_{cm})^2 \right\rangle, \\ \langle R^2 \rangle &= \langle (\mathbf{r}_1 - \mathbf{r}_N)^2 \rangle, \end{aligned} \tag{9.4}$$

for a variety of N and T. Here $\mathbf{r}_i$ is the position of ith monomer and $\mathbf{r}_{cm}$ is the center-of-mass of the chain. T_θ is the temperature near which both of these quantities scale as $NF(N^\phi \tau_r)$, where $\tau_r = (T - T_\theta)/T_\theta$ is the reduced temperature and $\phi = 1/2$ in $d = 3$. At T_θ, both $\langle R_G^2 \rangle$ and $\langle R^2 \rangle \sim N$. The exact value of T_θ depends sensitively on the details of both the bonded and non-bonded interactions, particularly the range of interaction.[110,111,114,115,118] In a good solvent, simulations with the purely repulsive potential and those with $r_c > 2^{1/6}\sigma$ are equivalent, though the former are significantly faster.

The bond length can be either be of fixed length (pearl necklace model) or allowed to vary (bead-spring model). In the former, rigid links of length l connect the monomers. The local motion of a chain is achieved through random rotation of bonds around the axis connecting nearest-neighbor beads. If the beads are hard spheres, one accepts the move if it does not violate the excluded volume condition. For a soft sphere, one has to compute the energy change δU associated with the move and accept the move with probability 1 if $\delta U < 0$ and probability $\exp(-\delta U/k_B T)$ if $\delta U > 0$ where k_B is the Boltzmann constant. This procedure can be proved to satisfy detailed balance and ensures that one generates configurations with the correct Boltzmann weight. In order to obtain a reasonable acceptance rate, it is often necessary to restrict the maximum size of the attempted bond rotation to very small values for high density. In addition, if the diameter of the hard sphere or the unit of length in the Lennard-Jones interaction σ is less than about $0.9l$, then one has to explicitly check that the rotation does not cut bonds. This additional check, which must be done every time-step is very time-consuming and diminishes the efficiency of this method considerably. At high density, it is often useful to supplement the basic crankshaft move by a cooperative relaxation[121] of several bonds on a chain. In their work on polymer brushes in a poor solvent, Weinhold and Kumar[119] removed sections of the chain and regrew them in a manner that samples low-energy conformations preferentially. In this extra move, a random bond is chosen where the chain is cut and each bead is regrown in sequence by sampling a number of random vectors. The entire move is then accepted or rejected using a criterion[121] which is based on the Rosenbluth and Rosenbluth[62] importance sampling idea.

In the spirit of coarse-grained models, there is no reason to keep the bond length fixed. In fact, there are a number of advantages, as noted in the discussion on the bond fluctuation model, to allowing the bond length to vary. In a MC simulation, this enables one to use a very simple set of moves, namely choose a monomer at random and attempt to displace it a random amount in a random direction. The move will be accepted or rejected based on the Boltzmann weight as in the pearl necklace model. In MD simulations, it is significantly more efficient to allow the bond length to vary than to have to apply constraints to keep the bond length fixed.[25] The actual form of the interaction is not that important, as long as the maximum extent of the bond is small enough that bonds crossing is inhibited. If the nonconnected monomers interact with the Lennard-Jones interaction, eq. (9.3), then as long as the maximum extent of the bond is less than about 20%, the probability of bond crossing is less than 10^{-6} and no further checks for bond crossing is necessary. The simplest interaction between connected monomers is a harmonic potential of the form,

$$U^{ch}(r) = \frac{k}{2}(r - l_o)^2 \tag{9.5}$$

where l_o is the mean bond length and k is the spring constant. If $l_o = \sigma$, then k should be greater than about $500k_BT/\sigma^2$ so as to inhibit bond crossing. k is usually chosen to depend linearly on T so that the fluctuations in the bond length are independent of T. To make the model more flexible locally, k can be decreased but then the maximum extent of the bond must be limited by introducing a finite extensibility into $U^{ch}(r)$. In a MC simulation, one can simply introduce an upper cut-off, l_{max}, and a lower cut-off, l_{min}, such that for $r < l_{min}$ and $r > l_{max}$, $U^{ch}(r) = \infty$.[91,122–128] Since one has the extra parameter, l_{max}, to limit the maximum extent of the bond, the spring constant k can be quite weak or even vanish, as done in several simulations on tethered membranes.[122–128] The harmonic potential can also be modified to include a maximum cut-off. One widely used model which does this is the finite extendible nonlinear elastic (FENE) model first introduced by Bird *et al.*[129] In this model, the connected monomers interact with the Lennard-Jones interaction plus a second interaction of the form[129]

$$U^{ch}(r) = \begin{cases} -0.5kR_0^2 \ln\left[1 - \left(\frac{r}{R_0}\right)^2\right] & r \le R_0; \\ \infty & r > R_0. \end{cases} \tag{9.6}$$

This second, attractive interaction has its minimum at $r = 0$. Thus the anharmonic spring tries to pull the connected monomers close together while the Lennard-Jones repulsion at short distances, pushes them apart. In several early simulations using this model,[130] the spring constant k was too small and the upper cut-off R_0 too large to ensure that the chains did not cross. A careful study[28] of the parameter space revealed that a reasonable set of parameters which allowed for a rather flexible spring that does not stretch too much are $k = 25 - 30k_BT/\sigma^2$ and $R_0 = 1.5\sigma$. This gives an average bond length of 0.97σ. Another possible tethering potential has been used by Abraham and coworkers[100,101,107,108,113] in their studies of tethered membranes, in which the potential vanished for a range of lengths and diverged at the upper and lower limits as a power law.

The MD simulations which we will discuss here are based on the bead-spring model[131] in which each monomer is weakly coupled to a heat bath. For a discussion of simulations on more realistic models for polymers, see the article by Clarke in Chapter 5 in this volume. Denoting the total potential of monomer i by U_i, the equation of motion for monomer i of mass m is given by

$$m\frac{d^2\mathbf{r}_i}{dt^2} = -\nabla U_i - m\Gamma\frac{d\mathbf{r}_i}{dt} + \mathbf{W}_i(t). \tag{9.7}$$

Here Γ is the bead friction which acts to couple the monomers to the heat bath and $\mathbf{W}_i(t)$ describes the random force acting on each bead. The strength of the random force is coupled to the bead friction by the fluctuation–dissipation theorem,

$$\langle \mathbf{W}_i(t) \cdot \mathbf{W}_j(t') \rangle = \delta_{ij}\delta(t-t')6k_BmT\Gamma. \tag{9.8}$$

For a single, free polymer, the Einstein relation leads to an overall diffusion constant $D_o = k_BT/\Gamma N$. For this method to be efficient, the value of Γ cannot be too small, since then the coupling to the heat bath will be too weak and the system will not sample phase space very well. However, if the coupling is too strong, then the viscous damping term and the random force term dominate over the inertia term in eq. (9.7). The motion will then be dominated by Langevin dynamics even for very early times and there would be very little cooperative motion of the monomers which is important, particularly in dense systems. In this limit the algorithm would also be very ineffective. For a given bead friction, the motion of a monomer for $t < \Gamma^{-1}$ is undamped, subject only to the chain constraint, while for $t \gg \Gamma^{-1}$ the motion is Rouse-like.[28]

Provided that Γ is not so large that the inertia term becomes irrelevant, the equations of motion can be integrated with any standard algorithm. Third- and fifth-order predictor–corrector[132] and velocity–Verlet[133] algorithms have been tested.[19] For Lennard-Jones interaction, eq. (9.3), between all monomers and the FENE model between connected monomers, the equations of motion for each monomer can be stably intergrated with a time step Δt between 0.006–0.012τ, where $\tau = \sigma(m/\epsilon)^{1/2}$ is the unit of time in Lennard-Jones units. Γ in the range $0.5 \leq \Gamma\tau \leq 2.0$ gives good results. For the purely repulsive potential, $r_c = 2^{1/6}\sigma$, and $T \sim \epsilon/k_B$, the velocity–Verlet algorithm was found[19] to be stable for a time step Δt about a factor of 2 larger than for the predictor–corrector algorithm. For longer-range interactions, it is necessary to reduce Δt. Also, the higher T, the smaller Δt. Note that since this is a coarse-grained model, τ cannot be mapped directly to a microscopic time.[28] In the original implementations of the code, Gaussian distributed noise in eq. (9.8) was used.[28,98,99] However Dünweg and Paul[134] have recently pointed out that the Gaussian noise term can be replaced by equally distributed random numbers, which have the same mean and second moment as required by eq. (9.8). Since simulations using both types of noise give the same results, we now use equally distributed random numbers, saving 5–15% in CPU time, depending on the range of interaction. An additional advantage of including the coupling to the heat bath is that one can use a larger time step Δt than if $\Gamma = 0$. This more than compensates for the extra computer time needed to generate the pseudo-random numbers.

Another advantage of off-lattice simulations is the ability to carry out the simulations at constant pressure in addition to constant volume as has widely been done in the study of bulk fluids and solids.[25] For chains tethered to a surface, constant surface pressure Π_a simulations can be very efficient particularly for studying a wide range of grafting densities. For a polymer attached to a two-dimensional flat surface, constant pressure MD simulations can be carried out following the method suggested by Andersen[135] and Parrinello and Rahman[136] for three-dimensional systems. In this method, the size of the simulation cell is not kept constant during the simulation, but rather allowed to vary to keep the pressure constant. Therefore, in addition to the equations of motion for the monomers, extra equations are introduced to describe the time dependence of the lengths of the simulation cell. For polymers attached to a $d = 2$ surface, L_x and L_y, parallel to the grafting surface are allowed to vary.[116] A mass M_w is assigned to the walls perpendicular to the x and y directions. The applied surface pressure Π_a is balanced by the internal osmotic pressure $\mathcal{P}_\alpha$ in the $\alpha = x$ and y directions,

$$\mathcal{P}_\alpha A = \sum_i m\dot{r}_{i\alpha}^2 + \sum_i \sum_{j>i} r_{ij\alpha} f_{ij\alpha}. \tag{9.9}$$

Here m is the monomer mass, $r_{i\alpha}$ is the α coordinate of particle i, $r_{ij\alpha} = r_{i\alpha} - r_{j\alpha}$, and $f_{ij\alpha}$ is the force between monomers i and j. The "potential energy" associated with the surface area is $\Pi_a A$ and the extra kinetic energy is

$$\mathcal{K}_A = \frac{1}{2} M_w \sum_{\alpha=1}^{2} \dot{L}_\alpha^2. \tag{9.10}$$

The shape, as well as the size of the simulation cell may vary during the simulation. In order to keep the simulation cell a square, one can also apply the average osmotic pressure $\bar{\mathcal{P}}_a = (\mathcal{P}_1 + \mathcal{P}_2)/2$ in both directions. One advantage of the constant pressure simulation method is that the grafting density ρ_a can be changed dynamically by simply varying Π_a.

Because we are interested in systems with rather short-range interactions with a well-defined cut-off, it is inefficient to examine all of the pairs of particles ij at each time-step to determine which ones have a nonzero force between them. An alternative is to divide the simulation cell into smaller cells of size $r_c + r_s$, where r_s is a small skin of order 0.3–0.5σ. Now one simply has to check pairs within neighboring cells to determine which pairs ij have a nonzero force. The extra skin r_s allows one to create a list of neighbors,[137] which only has to be updated every 10–20 time steps. Combining the link cell with a Verlet neighbor table is the most efficient method for doing any off-lattice simulation with short-range interactions. To obtain peak vector performance, one has to do some additional organization in order to take advantage of the hardware architecture. For inhomogeneous systems, like star polymer, brushes and tethered membrane, we used the vectorized link cell method discussed by Grest *et al.*[138] For homogeneous systems, the layer link cell method[138,139] is more efficient. The difference is that the layer link method requires more work to set up the Verlet neighbor table so that the DO loop which sums all the forces on particle i can be vectorized. However for inhomogeneous systems, this extra work exceeds the gain in vectorizing this one loop. MD codes can also be parallelized using link cell methods.[140] On the new superscalar workstations, the most efficient algorithm is to follow the procedure discussed in Ref. 138 for the link cell method, but simplify the DO loop where the forces between pairs ij are determined. In this case, one can determine the distance between all pairs which are in the list and use a simple GOTO statement to bypass the rest of the computation if $r > r_c$. In the vectorized code, the same calculation has to be done in a more roundabout way in order for the loop to vectorize.[138]

For off-lattice simulations, how one starts the simulation is usually not as much a concern as on a lattice. One simple technique is to use a simple MC procedure to grow RW chains with the appropriate bond length and some restriction to avoid backfolding.[22,28] This can be done very simply by requiring that $|\vec{r}_{i-1} - \vec{r}_{i+1}| > r_o$ where r_o is of order 0.3–1.0σ. To remove the overlap, one can either use any of the MC procedures described above or carry out a standard MD simulation for a few thousand steps with a softer potential which does not diverge at short distance,[28] before turning on the actual potential.

From the previous discussion the question arises as to which method is best to use for studying tethered chain systems. The answer depends a lot on the particular system under consideration and whether it is better to work in the continuum or not and whether it is acceptable to have stochastic dynamics on all time scales. On a lattice, the best method is clearly the bond fluctuation method. For off-lattice, MD using a bead-spring model is very efficient particularly on vector supercomputers. On super scalar workstations, there are some indications that MC for the bead-spring model is also efficient.[91] The bond fluctuation algorithm as well as the MD are highly vectorized, while the MC bead-spring simulations have not been vectorized as of yet. Comparisons between the various methods are difficult, particularly between lattice and off-lattice simulations since the models and methods are so different. In some cases, such as polymer brushes and tethered membranes, all three of these methods are equally applicable and it is more of an individual choice which method to use. However in some applications, such as a constant pressure simulation,[135,136] or for polymers under shear, one should use a continuum model. Similarly, if one is interested in the behavior of gels or polymer networks under swelling or elongation or the forces between polymer brushes, then a continuum simulation using MD is probably the most appropriate way to proceed. Continuum simulations, either MC or MD, are also essential for simulations on the more detailed, microscopic polymer models.[141,142] For a dense melt of linear chains, a detailed comparison is possible and has been discussed by Kremer and Grest[143] in Chapter 4 of this volume.

9.2.3 Numerical solution of SCF equations

Although not a simulation method, the numerical solution of the self-consistent mean-field (SCF) equations[144–148] is a very powerful technique applicable to a variety of tethered chain systems. In the SCF models, the excluded volume interaction is incorporated not by excluding segments from occupied positions but by using a potential field created by the configurations of all the other segments. The potential field is in turn a function of the segment distribution which must be determined self-consistently. The chain statistics are then described as those of a RW under an average potential field. While the method neglects fluctuations of the potential, it includes fluctuations in the single chain. Unlike the MC and MD methods described above, equilibrium is guaranteed and it is very fast, even on small computers. The method is suitable for complex systems and requires fewer assumptions than analytic self-consistent methods.

While the method is not limited to a lattice, the derivation of the equations is more transparent on a lattice. As in standard lattice MC, the polymer chain is made up of repeat units of length a, the lattice spacing. The goal is to compute the polymer chain distribution function $\psi(s, \mathbf{r}, \mathbf{r}')$, which describes the probability that a polymer chain of segment length s starting at $\mathbf{r}'$ is at position $\mathbf{r}$. Often, the system has some sort of symmetry that reduces the problem to a one-dimensional one. The distribution function can then be written in terms of z and z', $\psi(s, z, z')$, and is determined by the self-consistent potential generated from all possible polymer configurations which are consistent with some set of boundary conditions. For example, for a polymer end-grafted to a surface, the boundary condition imposed is that one end is attached to the surface which is otherwise impenetrable. The set of polymer configurations is made finite by imposing a spatial lattice and discretizing the path of a chain from $\mathbf{r}(s)$ to a sequence of lattice positions labeled by $\{s_i\}$. Since it is assumed that the interaction energies (polymer–polymer, polymer–solvent and polymer–surface) are evaluated in a mean-field approximation, they depend only on the average density $\rho(z)$ of the polymer and the solvent volume fraction in each layer. One can then write an equation for $\psi(s, z, z')$ inductively as[149]

$$\psi(s, z) = e^{-V(z)/k_B T}\left[\frac{1}{6}\psi(s-1, z-1) + \frac{1}{6}\psi(s-1, z+1) + \frac{2}{3}\psi(s-1, z)\right]. \tag{9.11}$$

That is, to arrive at z in s steps, a chain must have been at one of the adjoining layers on the previous step. The factors 1/6 and 2/3 are for a simple cubic lattice. If one imposes appropriate boundary conditions on $\psi(1, z)$, then one can iteratively solve eq. (9.11) for a given $V(z)$. For chains end-grafted to a planar wall, $V(0) = \infty$ and $\psi(1, z) = \delta_{z,1}$.

The monomer density $\rho(z)$ for a given $\psi(s, z, z')$ is the normalized weight for a chain to travel from its origin $z' = 0$ through the point z in s steps to some other endpoint z', which it must reach in $N - s$ steps. For a one component brush in a solvent, the origin is the grafting surface and as such

$$\rho(z) = \frac{\int dz' \int ds \psi(s, z, 0)\psi(N - s, z, z')}{\int dz' \psi(N, 0, z').} \tag{9.12}$$

The self-consistency comes in because ψ depends on $\rho(z)$ through $V(z)$. This set of equations can then be solved by a relaxation technique. For diblock copolymers and blends, additional distribution functions are necessary, but the general form of the solution is similar. Scheutjens and Fleer[148] were the first to exploit the numerical solution of the SCF equations on a lattice for polymers in solution using eq. (9.11). It is possible now to generate an extremely large number of conformations and study N as large as 100 000. Their method has been widely applied to study adsorption of polymers,[148,150] block copolymers,[151] and polyelectrolytes[152] as well as polymers end-grafted on a flat or curved surface.[153–160] It has been subsequently modified to study homopolymer melts near a surface,[161] surface segregation of polydispersed chains,[162] diblock copolymers at an interface between immiscible homopolymers[163] and between two parallel walls,[164] lamellar blends of a diblock copolymers and homopolymers,[165] and end-grafted brushes in equilibrium with a blend of adsorbing and nonadsorbing polymers.[166] In the standard lattice model approach, the density of the monomers is assumed to be dependent only on the distance from the surface. However these equations can be generalized to include a second dimension, parallel to the surface as shown by Huang and Balazs[167] in their study of polymer brushes in a poor solvent.

The SCF equations were originally written[144–147] in terms of a diffusion equation for $\psi(s, \mathbf{r}, \mathbf{r}')$. If one expands $\exp(-V(\mathbf{r})/k_BT) = 1/(1 + V(\mathbf{r})/k_BT)$, which is exact in the large N limit, then it is easy to see that eq. (9.11) is simply the discrete representation of the modified diffusion equation of Edwards,[144,145]

$$\frac{\partial \psi}{\partial s} = \frac{a^2}{6}\nabla^2 \psi + \frac{V(z)}{k_BT}\psi. \tag{9.13}$$

The polymer chain is then treated as a continuous curve and space is discretized into steps of length a, the segment size.[145] After an initial guess for the distribution function, often a RW, the local segment density is calculated and inserted into the partial differential equation, leading to the calculation of a new distribution function. This procedure is iterated as in the lattice version, until the difference between the new and previous iteration is less than some predetermined amount. This method has been used to study diblock copolymers in spherical and lamellar phases,[146] and homopolymer and diblock copolymer mixtures.[147] End-grafted polymer brushes attached to a flat surface in a solvent have been studied by Muthukumar and Ho[168] and Whitmore and Noolandi[169] and to a curved surface by Dan and Tirrell.[170]

A numerical method, based upon the statistical thermodynamic theory of Ben-Shaul *et al.*[171] for amphiphilic aggregates, has been generalized to polymeric systems by Carignano and Szleifer.[172–174] The theory is a mean-field approximation in which one looks at a central chain in the mean field of the other chains and the solvent. Unlike the SCF methods, where the chains are taken as RWs on a lattice, more microscopic details of the chains are included by treating the interaction energies within a chain more precisely by generating a large number of representative conformations of a single chain. This has been done using either a rotational isomeric state model or one of the potentials described above. The constraint of volume filling, for an incompressible system of solvent and polymer segments, leads to a mean-field equation for the concentration profile of the polymer segments. Although derived using a thermodynamic theory, the resulting equations are very similar to the SCF equations and can be applied to both lattice and off lattice polymers.

9.3 Polymers tethered to a point

Star polymers are one special class of branched polymers, in which one end of each linear chain is tethered to a small central core to form a single molecule.[175–183] Many arm stars with a narrow molecular weight (M_W) distribution can now be made by chemical reactions with up to 56 arms for polyisoprene[180] and up to 270 arms for polybutadiene.[181] Polymers with one ionic end group naturally associate into star-like agglomerates in a low dielectric media.[184,185] Asymmetric diblock copolymers in a selective good solvent for one of the blocks can also form star-like micelles in dilute solution. These have been studied extensively both theoretically[147,186–191] and experimentally.[192–196] The scaling picture[10,11] for stars has been used to understand how end-grafted chains stabilize colloidal particles.[197] While most stars which have been studied have equal length arms of chemically identical chains, it is also possible to make asymmetric stars in which the length of each arm is not the same[179] or where the arms are made of more than one polymer type.[198] Halperin[199] has considered theoretically the case where the arms of the star are made of two chemically distinct polymers. In the dilute limit, light and neutron scattering and viscometry have been used to give several measures of the size of the star, including the radius of gyration, the hydrodynamic radius and the thermodynamic radius. The dynamics of the entire star as well as of any individual arm have been studied by neutron spin-echo scattering[200,201] and dielectric relaxation.[202] The crossover from a dilute to a semidilute solution of stars up to 18 arms has been studied by small angle neutron scattering,[203–205] light scattering[206] and small angle X-ray scattering.[196]

Theoretically, many of the same techniques which work well for linear chains can also be applied to star polymers, including some exact results,[207] scaling,[10,11,208] and renormalization group.[209–211] Daoud and Cotton[10] and Birshtein and Zhulina[11] generalized the de Gennes scaling[212] model for linear polymers to star polymers. In this approach, the star consists of three regions, an inner melt-like extended core region, an intermediate region resembling a concentrated solution and an outer semidilute region. In this outer region, a blob model is used to describe the overall structure of a star. Since the volume accessible to a given chain increases with the distance r from the center of the star, the monomer volume fraction $\rho(r)$ is expected to be a decreasing function of r. Each arm can be seen as a succession of growing spherical blobs as shown in Fig. 9.4. Within one blob each arm behaves as an isolated chain. At a given distance r from the center, a sphere of radius r is cut by f arms. The star looks like a semidilute solution with a screening length $\xi(r)$, where $\xi(r)$ is a function of r and the number of arms f. Each blob contains monomers of a single chain. Since f blobs cover the sphere of radius r, the blob radius is

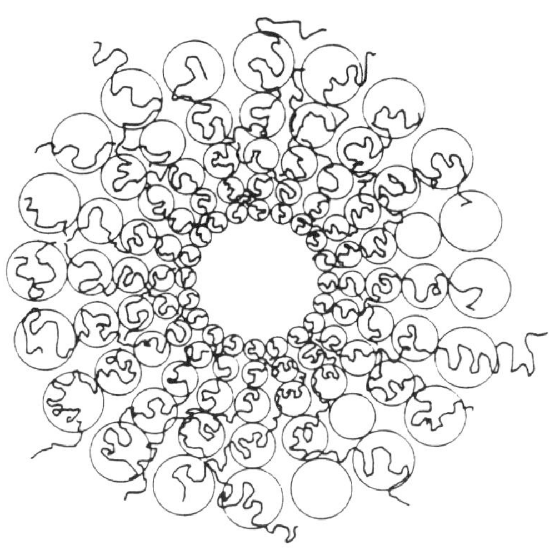

Fig. 9.4 A representation of scaling model[10,11] of a many-armed star, in which each arm is made of a succession of blobs of size ξ increasing from the center of the star to the outside. (From Ref. 197.)

$$\xi(r) \sim rf^{-1/2}. \tag{9.14}$$

As in any semidilute solution, each blob contains $n(r) \sim \xi(r)^{1/\nu}$ monomers. Here ν is the correlation length exponent, which is 0.588 for a good solvent in three dimensions.[12] From this picture, one can easily verify that $\rho(r)$, the number density of monomers falls off as

$$\rho(r) \sim n(r)/\xi(r)^3 \sim f^{(3\nu-1)/2\nu} r^{(1-3\nu)/\nu} (= f^{0.65} r^{-1.30}), \tag{9.15}$$

which is the same as eq. (9.1) for $d = 0$. Qualitatively, this scaling behavior is nicely illustrated in Fig. 9.5(a), in which we present a projection of a typical configuration of a star polymer with $N = 50$ monomers per arm for $f = 10$, 30 and 50. This simulation was carried out using the MD method for monomers interacting with a Lennard-Jones interaction, eq. (9.3), truncated at $r_c = 2.5\sigma$. The good solvent results are for $T = 4.0\epsilon/k_B$. This figure nicely shows the density falloff and corresponding increase in $\xi(r)$ with increasing r, as expected. A similar projection for the purely repulsive Lennard-Jones interaction is shown in Ref. 96. For the mean-squared average center-to-end distance $\langle R^2 \rangle$ and radius of gyration $\langle R_G^2 \rangle$,

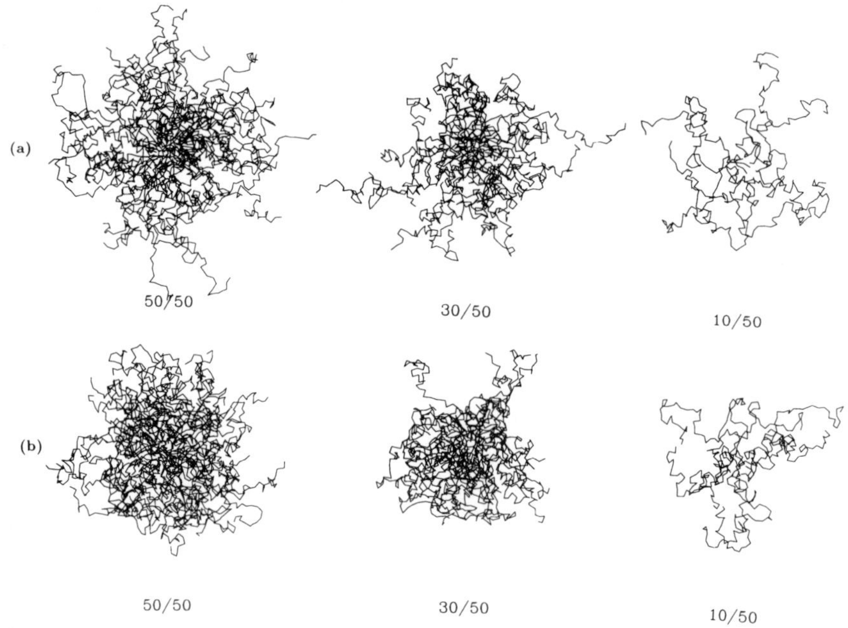

Fig. 9.5 Projection of a typical configuration of a star of $f = 10$, 30 and 50 arms with $N = 50$ monomers per arm for (a) a good solvent ($T = 4.0\epsilon/k_B$) and (b) a Θ solvent ($T = T_\theta = 3.0\epsilon/k_B$). The pictures give an impression of the increasingly homogeneous density in the $f = 30$ and 50 arms cases, while the 10 arms star is obviously governed much more by single-chain properties. These are from MD simulations with a Lennard-Jones interaction, eq. (9.3), cut-off at $r_c = 2.5\sigma$ between nonbonded monomers.

$$\langle R^2 \rangle \sim \langle R_G^2 \rangle \sim N^{2\nu} f^{1-\nu}. \tag{9.16}$$

In a good solvent, $\langle R^2 \rangle \sim N^{1.18} f^{0.41}$. As the solvent quality is decreased and one approaches T_θ, the range over which this result is valid decreases.[213] At T_θ (Fig. 9.5[b]), each arm can be pictured as a succession of ideal spherical blobs and $n(r) \sim \xi(r)^2$ and eq. (9.15) is replaced by

$$\rho(r) \sim n(r)/\xi(r)^3 \sim f^{1/2} r^{-1}. \tag{9.17}$$

This last result is also valid in the intermediate region, though the crossover from the outer to the intermediate region has not been observed.

9.3.1 Star polymers in a good solvent

Simulations, both MC and MD, have been used to test these scaling predictions and to determine other properties of a star polymer, including the static structure factor in the dilute limit. At present, it is not possible to simulate a melt or even a semi-dilute solution of many-arm star polymers due to the long relaxation times. For few-arm stars ($f \lesssim 12$) MC methods are clearly most efficient, while for large number of arms, MD methods work very well. For small f, the density of monomers of the star is low almost everywhere and static MC methods in which one generates the chains by constructing walks can be used.[55,58,61,64,67,214] Using this method, Batoulis and Kremer[67] were able to make very accurate estimates of the exponent γ as well as $\rho(r)$ and $\langle R_G^2 \rangle$ for $f \leq 6$ in a good solvent. Dynamic MC also works well in this limit, particularly if one invokes nonlocal moves, such the pivot algorithm.[52,215] However as f increases, the interior becomes very dense and many of these methods fail or become inefficient. In this case, one can use either MD methods or a local stochastic MC method such as the bond fluctuation method on a lattice[82,83] or a simple off-lattice MC in which one attempts to move one monomer at a time. It is also possible to use nonlocal moves in the dilute, outer regions of the star and local moves near the interior, though this has not been done to the best of our knowledge.

In a good solvent, the results from various groups agree very well[52,55,57,58,61,64,82,83,96,97,112,216,217,218,219], though most of these are for small f. For small f, the tethered end of each arm can easily be attached to a single point. However, for large f, either the size of the central region must be increased or the maximum length of the bond between the central site and the first monomer must be increased.[96] Toral and Chakrabarti[94] have studied the crossover from star to brush as the radius of curvature of this sphere increases (see Section 9.5). In the dynamic methods, both MC and MD, the initial state of a star can be easily generated by growing chains from a point or a small sphere. Usually these simulations are initialized by simply letting the chains overlap in the initial configuration and equilibrating them, as discussed in Section 9.2, until the excluded volume conditions are all satisfied. As an example, results for $\langle R_G^2\rangle/\langle R_{G1}^2\rangle$ and $\langle R^2\rangle/\langle R_{G1}^2\rangle$, where $\langle R_{G1}^2\rangle$ is the radius of gyration for a single polymer chain of N monomers is shown in Fig. 9.6. The off-lattice MC simulations of Freire *et al.* [112] and the MD simulations of Grest *et al.*[96,117] are presented. Experimental results for polystyrene and polyisoprene[180] are also shown. Note that by normalizing $\langle R_G^2\rangle$ by $\langle R_{G1}^2\rangle$ for a single arm, the results for both the simulations and experiment fall on the same curve. By dividing out of the single chain radius of gyration, one can compare results from different simulations as well as different experiments. Even though both simulations were carried out for coarse-grained models, in which no local bending and torsion terms were included, the simulations describe the experimental results very well. As expected from the scaling theories, the global properties of the star are universal. The solid line in Fig. 9.6 indicates the predicted asymptotic power, 0.41. Note that the limiting scaling form is reached very early. Comparison of experimental results with renormalization group calculations are reviewed in Ref. 220. The density profile $\rho(r)$ scaled by $f^{0.65}$ is shown in Fig. 9.7 for four values of f ranging from 4 to 50, with $N = 100$.[117] The measured slope is approximately 1.30 ± 0.03 for $f \geq 10$, in excellent agreement with the expected value 1.30. The rapid decay for large r is due to the finite chain length. As clearly seen, $\rho(r)$ scales with the number of arms f as predicted by scaling theory. This is in agreement with earlier simulations of Grest *et al.* [96] for $N = 50$ for a range of f and Batoulis and

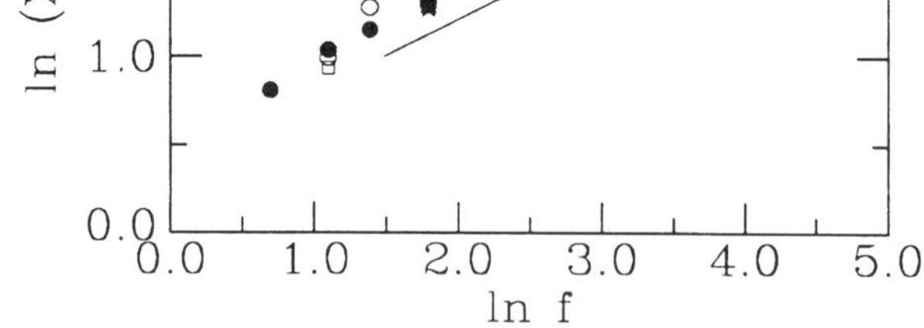

Fig. 9.6 Radius of gyration $\langle R_G^2\rangle/\langle R_{G1}^2\rangle$ (lower curve) and average squared end-to-end distance $\langle R^2\rangle/\langle R_{G1}^2\rangle$ (upper curve) versus f for a good solvent. $\langle R_{G1}^2\rangle$ is the mean-squared radius of gyration for a single polymer chain. The solid circles are data for polystyrene and polyisoprene from Ref. 180. The crosses are from the off-lattice MC simulations of Freire *et al.* [112] for $N = 49$ or 55. The other symbols are from MD simulations[96,117] for monomers interacting with a purely repulsive Lennard-Jones interaction, eq. (9.3), at $T = 1.2\epsilon/k_B$ for $N = 100(\circ)$ and at $T = 4.0\epsilon/k_B$ for $r_c = 2.5\sigma$ for $N = 50(\triangle)$ and 100 ($\square$). The solid line has slope of 0.41. (From Ref. 117.)

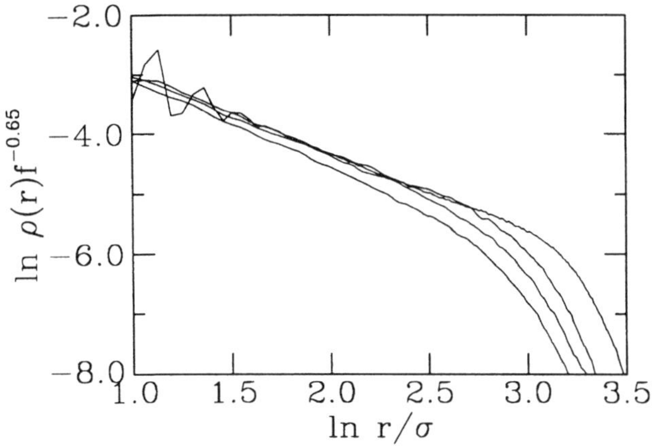

Fig. 9.7 Log–log plot of the density $\rho(r)$ scaled by $f^{(3\nu-1)/2\nu} = f^{0.65}$ versus r/σ for the $f = 50$, 20, 10 and 4-arm stars with $N = 100$ monomers per arm in a very good solvent. The larger f, the further the curves extend in r. These results are for $T = 1.2\epsilon/k_B$ for a purely repulsive Lennard-Jones interaction, eq. (9.3), truncated at $r_c = 2^{1/6}\sigma$. (From Ref. 117.)

Kremer[64] for $3 \leq f \leq 6$. Using a SCF model, Dan and Tirrell[170] and Wijmans and Zhulina[159] found that $\rho(r)$ scaled as predicted by eq. (9.15). As seen in Fig. 9.7, the scaling is valid even at very short distances from the center indicating that, at least for these stars, the simulations clearly exhibit the scaling predicted for a swollen star and there is no need to consider the core region. This core region is important for micellar stars and for chains grafted onto a small colloidal particle.

Because the interior of the star can become quite dense as f increases, one would expect that the free end of the chain is excluded from the core region by simple steric effects. The expected width of the distribution $P(R)$ of center-to-end distances R can be estimated from the scaling model,[10,11] by considering the distance R of a chain confined in a narrow cone of opening $\theta = 2f^{-1/2}$, since a cone subtends a solid angle which is $1/f$ of the sphere. The end-to-end distance of such a chain fluctuates as though it were confined in a straight tube of diameter $R\theta$. The free energy of this chain, relative to an unconfined chain, is of order $k_BTR/(R\theta)$ or k_BT per blob of size $R\theta$. Decreasing R to zero costs a further energy of this order. Thus the free energy associated with a small fluctuation ΔR of R away from its average value is of order $k_BTR/(R\theta)[\Delta R/R]^2$. Thus thermal fluctuations in ΔR, of energy $\sim k_BT$, are expected to be of order $\langle R\rangle\theta^{1/2} = 2^{1/2}\langle R\rangle f^{-1/4}$. For $f \gg 1$, $P(R)$ should approach a Gaussian shape of width $\Delta R \ll \langle R\rangle$. For small r, Ohno and Binder[208] found that in the scaling limit $P(R) \sim (R/\langle R\rangle)^{\theta(f)}$, where $\theta(f)$ is related to the exponent $\gamma(f)$ (see Section 9.2). For large f, $\theta(f) \sim f^{1/2}$ in three dimensions,[208] and $P(R)$ will appear to have an exclusion zone, although strictly speaking it does not. Li and Witten[221] used a variational approach to minimize the free energy with respect to the free end distribution and the stretching profiles of the polymer chains. Their results suggest a very large exclusion zone for the free ends and a different functional form for $P(R)$ for large f. The free ends would be restricted to the last 6% of the layer height and $P(R)$ would not be Gaussian. However as seen in Fig. 9.8(a), $P(R)$ for a 20- and a 50-arm star for N = 100 in a very good solvent are Gaussian as predicted by the scaling theory.[117] For smaller f, as shown in Fig. 9.8(a) for $f = 10$, there is some deviation from the Gaussian, particularly for very small f, as the

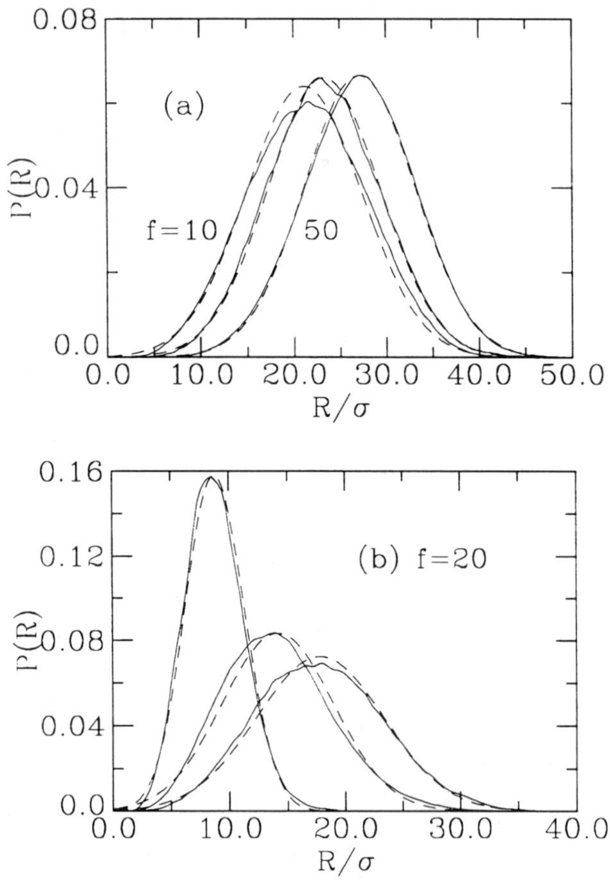

(a)

(b)

Fig. 9.8 Distribution function $P(R)$ for the center-to-end distances versus R for (a) a star with $f = 10$, 20 and 50 arms in good solvent and (b) a $f = 20$-arm star for $T = 2.0$, 3.0 and $4.0\epsilon/k_B$ (from left to right).[117] The results for (a) are for a purely repulsive Lennard-Jones potential at $T = 1.2\epsilon/k_B$, while the results shown in (b) are for the potential, eq. (9.3), truncated at $r_c = 2.5\sigma$. The solid line is the raw data, while the dashed line is a Gaussian with the same width and standard deviation as the data.

Gaussian overestimates the number of free ends near the origin. As the quality of the solvent decreases, $P(R)$ becomes somewhat nonsymmetric, with the center shifted slightly to the center to that of the Gaussian.[117] This result is in agreement with the numerical SCF calculation of Wijmans and Zhulina[159] for polymers attached to the surface of a highly curved sphere.

While the center of the distribution $P(R)$ is clearly dependent on the number of arms, the absolute width depends only weakly on f.[117] The width depends mostly on the solvent quality and N. Thus the relative width decreases roughly as $1/\langle R\rangle \sim f^{-0.2}$. This behavior is consistent consistent with the simple theoretical arguments[97] discussed above and with the variational calculation of Li and Witten.[221] They predict that the relative width of the end distribution does not decrease to zero but reaches a finite asymptote of about six percent. Since the observed widths are much larger than this, stars with several times larger number of arms than those that have been used so far are necessary to approach the predicted relative width. Is is also important to note that for a highly curved cylinder where they were able to also solve the SCF equations exactly, the size of the exclusion zone decreased significantly compared to the variational approach. In addition the variational approach appears to work less well as the dimension of the surface being tethered to decreases presumably because the strong stretching ansatz which they use is not as applicable. Thus this lack of agreement is not too surprising and if the theory is correct, it applies only when the number of arms is very large. Finally, because many arms are attached to a central core, steric effects which are not included in these calculations must play an important role. Thus while the scaling theory is clearly an oversimplification, it describes the distribution of free end very well for f in the range often studied experimentally.

Experimentally, small angle neutron scattering[200–204,222–224] is a useful way to determine not only the size of the star but also something about its internal structure. The measured scattering intensity, $I(q)$, as a function of the scattering wave vector $q = 4\pi/\lambda \sin(\theta/2)$, where λ is the incident neutron wavelength and θ is the scattering angle, can be represented by the product $S(q)P(q)$. Here $S(q)$ represents the interparticle structure factor determined by the interaction potential between the star molecules, while $P(q)$ represents the form factor of the individual stars determined by the structure of the star. In a dilute solution, the interactions between stars can be neglected, $S(q) = 1$ and the intensity $I(q) \sim P(q)$. Usually $P(q)$ is determined by selectively labeling a small fraction of the stars so that $S(q)$ can then be deduced from $I(q)/P(q)$. Benoit[225] derived $P(q)$ for a Gaussian star in 1953. The influence of excluded volume interactions were incorporated by Alessandrini and Carignano.[226] $P(q)$ can also be determined from the simulations,

$$P(q) = \frac{1}{Nf} | \sum_i \exp(i\mathbf{q} \cdot \mathbf{r}_i)|^2, \tag{9.18}$$

where the sum is over all Nf monomers. Note that in most papers on polymers the intra- and interparticle correlations are not separated and are simply referred to as the structure factor $S(q)$ which is proportional to $I(q)$. For a star, there are at least three characteristic lengths which must be considered: the radius R of the star, the correlation length $\xi_{\max}$ of the largest blob and the monomer size σ. For $qR < 1, P(q)$ has the normal Guinier behavior: $P(q) = Nf\,[1 - q^2\langle R_G^2\rangle/3 + \mathcal{O}(q^4\langle R_G^2\rangle^2)]$. For $\xi_{\max}^{-1} < q < \sigma^{-1}$, the scattering can be understood by covering the polymer with spheres of radius q^{-1}. In the absence of strong correlations between the positions of the spheres, these scatter incoherently. The resulting scattering intensity $NfP(q)$ is the number of spheres times the structure factor of a single sphere. In the star, the majority of these imaginary spheres are contained within blobs much larger than the spheres. The correlations between spheres are thus nearly the same as in a simple excluded volume of polymer. This gives a fractal scattering law $P(q) \sim q^{-1/\nu}$, independent of f and N. In between these two limits, $P(q)$ must fall from about Nf to about $(\xi_{\max}/\sigma)^{1/\nu}$. While the simplest way to connect these two regimes, is by a power-law decay q^{-d}, the actual scattering is more complex.[96] In this regime, q^{-1} is larger than the largest blob, and the polymer chain structure is invisible to the scattering. The dominant scattering is from the relatively sharp (see Fig. 9.7) outer boundary of the star, which gives rise to a faster decrease in the scattering envelope, q^{-d-1}. For large f, oscillations superimposed on this Porod envelope should also be seen. Results for $P(q)$ are shown in Fig. 9.9 for star polymers with $N = 50$ and $f \leq 50$. For high q, one finds the expected scaling of $P(q)$, though for large f, the observed $\nu = 0.65 \pm 0.05$, which is larger than expected. This is only a finite size effect due to the fact that N is only 50.[117] It is also possible to measure other partial structure factors by labeling only the inner $N/2$ monomers of each arm or the outer $N/2$ monomers. These partial structure factors plus the single-arm $P(q)$ were investigated in Ref. 96 and compared to small angle neutron scattering data by Richter *et al.*[200]

9.3.2 Star polymers in a Θ and poor solvent

At the Θ point, the self-repulsion of the monomers, due to the excluded volume, is just compensated for by the interactions with the solvent. While this tricritical point is well understood for linear chains, less is known about the properties of stars at T_θ. Candau *et al.*[227] assumed that at T_θ all the arms can interpenetrate each other completely. This means that all virial coefficients vanish and the size of the star is given by the Zimm–Stockmayer[228] equation,

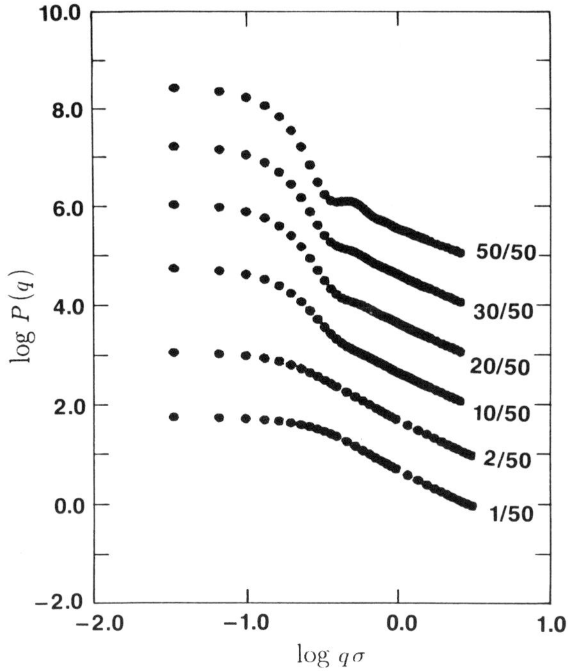

Fig. 9.9 Form factor $P(q)$ versus $q\sigma$ for four-star polymers in a good solvent with $f=10$–50 for $N=50$ simulated by MD for a purely repulsive Lennard-Jones interaction between nonbonded monomers from Ref. 96. Also shown are the results for two linear polymers with 50 and 100 monomers. The data have been offset for clarity.

$$g(f) = \lim_{N\to\infty} \frac{\langle R_G^2 \rangle}{f\langle R_{G1}^2 \rangle} = (3f-2)/f^2 \tag{9.19}$$

While this is obviously an oversimplification, it turns out to work quite well. As seen from eq. (9.17), the Daoud and Cotton scaling argument gives $g(f) \sim f^{-1/2}$. However since at T_θ, the second virial coefficient between a pair of arms vanishes, a third arm still gives a repulsive interaction.[11] Batoulis and Kremer[67] have argued that because of this effect, two, but not more than two, arms of the star can share a blob of diameter $\xi(r)$ at T_θ and one needs more arms than in a good solvent to observe the Daoud–Cotton scaling. They estimated that the RW results, eq. (9.19), should cross over to the scaling result for $5 \leq f \leq 10$.

To study the importance of three body interactions on stars, Batoulis and Kremer[67] carried out high precision MC simulations of linear and star polymers on the fcc lattice for $1 \leq f \leq 12$ and $N \leq 900$ using the inverse restricted sampling method.[62] These simulations extended earlier simulations on smaller stars by Mazur and McCrackin.[55] They found that T_θ

was independent of f for $N \to \infty$. Similar results were found experimentally for polyisoprene stars in dioxane[180] where the temperature at which the second virial coefficient A_2 vanishes for small f and N is lower than T_θ for a linear chain but increased as N increased. In the limit of large M_W, T_θ for the star is the same as for a linear chain.

The g-factor obtained from simulations and experiment are compared to the RW result in Table 9.1. Lattice MC results,[67] MD results on the bead-spring model,[117] and experimental data for polystyrene[177] in cyclohexane and polyisoprene[180] in dioxane are presented. Comparison of experimental and earlier MC results to renormalization group calculations are presented in Ref. 220. Since the Θ-point is only known approximately for the off-lattice bead-spring model, the MD results are probably not as accurate as the lattice MC results of Ref. 67. Results from the off-lattice MC simulations of Freire *et al.*[112] for $f = 6$ and 12 also agree very well with the two simulations listed, while their results for $f = 18$ are somewhat too low. Zifferer[215] found $g = 0.640$, 0.385, and 0.285 for $f = 4$, 8 and 12, respectively, using a pivot algorithm on the tetrahedral lattice. For small f, the RW result gives a reasonable approximation for g, though for larger f the RW result underestimates the size of the star. Note that all of these results are in contrast with those of Bruns and Carl[214] who interpret their MC results in terms of an f-dependent Θ point and, at least for $4 \leq f \leq 6$, g is within their numerical accuracy essentially the same as the RW result. For $f \gtrsim 8$ the polyisoprene data deviate significantly from both the simulations and the

Table 9.1 *g values for $T = T_\theta$.*

f	RW^a	MC^b	MD^c	PS^d	PI^e
3	0.778	0.79	0.73	0.76	
4	0.625	0.68	0.69	0.63	0.69
5	0.520	0.55	0.55		
6	0.444	0.48	0.45	0.45	0.47
8	0.344	0.39	0.40		0.44
10	0.280		0.34		0.42
12	0.236	0.28		0.27	0.37
18	0.160			0.22	0.29
20	0.145		0.20		

[a] Random walk, eq. (9.19), Ref. 228.
[b] Monte Carlo results of Batoulis and Kremer.[67]
[c] Molecular dynamics results of Grest[117] for $T = T_\theta = 3.0\epsilon/k_B$ for $N = 100$.
[d] Polystyrene (PS) results for $f = 4$ and 6 from Ref. 176 and for $f = 3$, 12 and 18 from Ref. 177.
[e] Polyisoprene (PI) results from Ref. 180.

Where available g was determined from extrapolations for large molecular weight when possible, otherwise the data for the longest arms were used.[67]

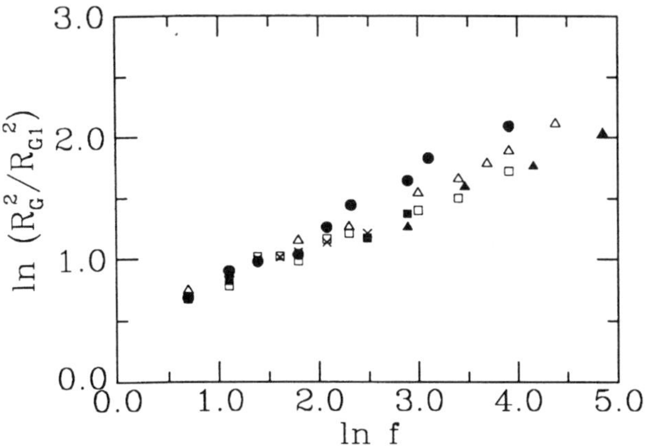

Fig. 9.10 Radius of gyration $\langle R_G^2 \rangle / \langle R_{G1}^2 \rangle$ versus f for a Θ solvent. The solid circles are for polyisoprene in dioxane,[180] the solid squares are for polystyrene in cyclohexane[177] and the solid triangles are for polybutadienes in dioxane.[178,182,183] The crosses are from the MC simulations of Batoulis and Kremer[67] extrapolated to large N. The open triangles ($N = 50$) and squares ($N = 100$) are from MD simulations[117] in which nonbonded monomers interact with a Lennard-Jones interaction, eq. (9.3), cut-off at $r_c = 2.5\sigma$ for $T = T_\theta = 3.0\epsilon/k_B$. (From Ref. 117.)

polystyrene data. The reason for this difference is unclear. In Fig. 9.10, $\langle R_G^2 \rangle / \langle R_{G1}^2 \rangle$ is plotted versus f for the MC simulations of Batoulis and Kremer[67] and the MD simulations in which the monomers interact with a Lennard-Jones interaction, eq. (9.3), truncated at 2.5σ for $N = 50$ and 100.[117] Experimental data for polystyrene,[177] polyisoprene[180] and polybutadiene[178,182,183] are also shown. For small $f \lesssim 8$, the results for all three experimental systems agree reasonably well with both the MC and MD results, though for larger f the results from the three experiments and the simulations scatter significantly. The polyisoprene data can be fitted reasonably well with the scaling form over the entire range of f, while the polystyrene and polybutadiene data as well as the MD results suggest that scaling can only be reached for much larger values of f. This is in contrast to good solvent results where the data for different experimental systems agree very well. The disagreement for this supposedly universal value between the three data sets is surprising and unexplained. Some of it may be due to finite chain length effects as seen in the simulation results for $N = 50$ and 100 and from the uncertainty in determing the Θ temperature for both simulation and experiment.

The monomer density $\rho(r)$ for a Θ solvent was found by Grest[117] to agree nicely with the scaling prediction, $f^{1/2}r^{-1}$. As in a good solvent the free ends are excluded from the center of the star. The distibution of free end for stars with $f = 20$ for a good, Θ and poor solvent is shown in Fig. 9.8(b). Note that the distributions are approximately symmetric, with a slight shift towards the center of the star. A Gaussian form is a pretty good fit, particularly for large f. However for small f, it overestimates the number of free ends near the origin.[117] As the radius of curvature of the central core increases, the distributions are no longer described by a Gaussian as discussed in Section 9.5.6.

9.3.3 Relaxation of star polymers

The scaling picture of a star can also be used to predict its dynamic relaxation processes.[97] There are at least three qualitative distinct relaxation processes for a star, which occur on different time scales and only weakly couple to each other. While all three of these mechanisms also occur for linear polymers, in a star they are easily separable.

First the star relaxes via an overall shape fluctuation or elastic modes. This time scale is that of cooperative diffusion over the star size R. A second process is rotation diffusion of the object. For linear polymers, these first two have the same relaxation time, up to constants of order unity. However for large f, the shape fluctuations relax progressively faster. Hence these two processes are expected to have the same dependence on N but a different dependence on f. The rotational diffusion is slower since it is not enhanced by the pressure within the star, as are the elastic modes. The third process, and by far the slowest, is the disentangling of two or more interwined arms. Such a configuration can easily survive the shape or rotational relaxations. The larger f, the better one can distinguish these processes. If excluded volume interactions are ignored, as done by Zimm and Kilb[229] in the very first study of the dynamics of branched polymers, then these three different relaxation mechanisms could not have been distinguished.

The fastest process is the shape fluctuations, which can be measured directly by studying the fluctuations of the inertia tensor $\mathbf{M}$ which is given by

$$M_{\alpha\beta} = \frac{1}{Nf}\sum_{i=1}^{fN}(\mathbf{r}_i - \mathbf{r}_{cm})_\alpha(\mathbf{r}_i - \mathbf{r}_{cm})_\beta, \tag{9.20}$$

where $\mathbf{r}_{cm}$ is the center of mass of the whole star. Defined this way, $\langle R_G^2\rangle = \langle M_{xx} + M_{yy} + M_{zz}\rangle$. The typical shape fluctuations are then given by the time autocorrelation function of the elements of $\mathbf{M}, R_G^2, R_{Ga}^2$, or R^2.[82,96,97,219] For example, the autocorrelation function of $C_R(t)$ is defined by

$$C_R(t) = (\langle R(t)R(0)\rangle - \langle R\rangle^2)/(\langle R^2\rangle - \langle R\rangle^2). \tag{9.21}$$

Results for $C_R(t)$ for stars in a good solvent are presented in Fig. 9.11 for a series of four stars with $N = 50$.[96] The relaxation times, τ_{el} for the elastic modes, which are determined by the slope at long times on this semilog plot are essentially the same. Su *et al.*[82] found similar results for 3- and 4-arm

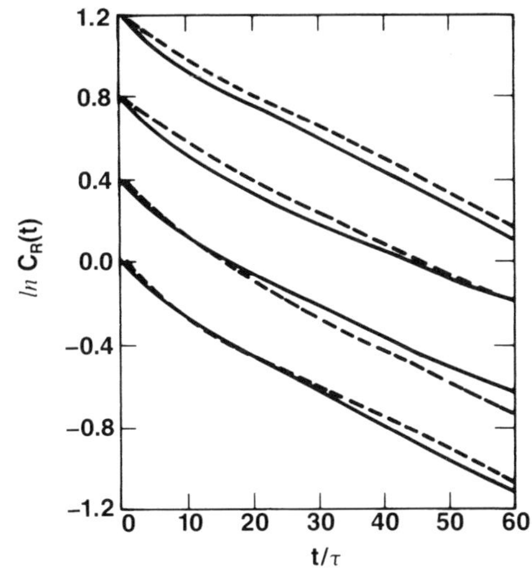

Fig. 9.11 Autocorrelation function $C_R(t)$ for the end-to-center distance (solid lines) and single-arm radius of gyration (dashed lines) for four stars in a good solvent. The curves have been displaced by 0.4 for clarity. Note that the relaxation times are essentially the same for all four values of f. (From Ref. 96.)

stars using the bond-fluctuation method. This curious result that τ_{el} is nearly independent of f is in contrast with the independent strand model of Zimm and Kilb[229] which predicts that $\tau \sim f$ but can be explained using the scaling picture.[96] Consider the fluctuations in R, the center-to-end distance for a single arm. The fluctuations in the total length arise from independent fluctuations of order ξ within each blob. Thus R fluctuates by an amount of order $\xi(R/\xi)^2$. For a star, the largest length available is given by the size of the largest blob, which has a diameter $\xi_{max} \sim Rf^{-1/2}$. The bulk of the monomers are to be found in this largest blob.[10,11] Thus the relative amplitude of the fluctuations in $\langle R^2 \rangle$ falls off as $\xi_{max}/R \sim f^{-1/2}$. The longest relaxation time τ_B of such a local fluctuation is given by the Rouse time of an isolated chain of $n_{B\max}$ monomers, $\tau_B \sim \xi^2_{max} n_{B\max}$, where $n_{B\max} \sim \xi^{1/\nu}_{max}$. In the scaling picture, $n_{B\max} \sim Nf^{-1/2}$ and thus $\tau_B \sim (Nf^{-1/2})^{(1+2\nu)}$. This describes the initial stage of the local relaxation. In order to produce an overall shape fluctuation, a density fluctuation must diffuse a distance of the order of the diameter of the star, which is R. The diffusion constant for a semidilute solution is given by ξ^2 divided by the local relaxation time τ_B. Thus τ_{el} is of order,

$$\tau_{el} \sim \tau_B (R/\xi_{max})^2 \sim N^{1+2\nu} f^{1-(1+2\nu)/2}. \tag{9.22}$$

In a good solvent, $\tau el \sim N^{2.18} f^{-0.09}$, while at the T_θ, $\tau_{el} \sim N^2 f^0$. Thus the dependence of τ_{el} on f is expected to be very weak, in agreement with the results shown in Fig. 9.11 for a good solvent for $N = 50$. Similar results for chains of length $N = 100$ were observed by Grest[117] for stars in both a good and a Θ solvent. Even for undiluted stars, Boese *et al.*[202] found using dielectric relaxation that the relaxation times of the center-to-end distance for *cis*-polyisoprene stars depended imperceptibly on f for $3 \leq f \leq 18$.

The next relaxation time is for the entire object to rotate or to move a distance comparable to its own diameter, since the same time is needed for the star to move its own distance or to make a complete rotation. For an assembly of Nf objects subjected to independent random forces as in the Rouse[23] model, the diffusion constant is given by $D \sim (Nf)^{-1}$. Within the diffusion time τ_D, the system moves a distance of about its own diameter,

$$\tau_D \sim R^2/D \sim NfN^{2\nu}f^{1-\nu} \sim N^{1+2\nu}f^{2-\nu}. \tag{9.23}$$

This time is smaller than for a linear chain of Nf monomers[212] but much larger than τ_{el}. Since D is very small, it is difficult to measure in a simulation and has not been done. The rotational diffusion time, which should be comparable to τ_D, can be analyzed by studying[97] the time autocorrelation function of the center-to-end vector $\mathbf{R}$ or the autocorrelation of the squares of the second-order spherical harmonics of the angles at which the principal axes of the ellipsoid are oriented with respect to a fixed-coordinate system.[82] However even for small f,[82] there was no linear region in the semilog plots of $C(t)$ for the times that can presently be simulated, making it impossible at present to test eq. (9.23).

Even when a star moves its own distance, the topological arrangement of two arms may be largely intact. A simple way to think about this is to imagine two interwined arms which can follow the shape fluctuations of the star without changing its topological character. Grest *et al.*[97] argued that this time should scale as

$$\tau_e \sim \exp(cf^{1/2}), \tag{9.24}$$

where c is a constant. To measure this time, one needs to observe the relaxation of some quantity which survives both the shape fluctuations and rotation of the entire object, but does not survive the disentangling fluctuations. One such sensitive quantity is the scalar product of two arbitrary center-to-end vectors: $\mathbf{R}_i \cdot \mathbf{R}_j$, where $\mathbf{R}_i$ is the vector from the vertex to the free end of the ith arm. This dot product is largest for nearby arms, which are the most likely to interwine. The product for two arms which were initially nearby relaxes to zero as these disentangle. Quantitatively, one can define the entanglement correlation function $C_e(t)$ as

$$C_e(t) = \frac{1}{f(f-1)} \sum_{i=1}^{f} \sum_{j \neq i}^{f} \langle [\mathbf{R}_i(0) \cdot \mathbf{R}_j(0)][\mathbf{R}_i(t) \cdot \mathbf{R}_j(t)] \rangle. \tag{9.25}$$

This angular correlation can only decay after two arms which are strongly entangled have disentangled. Overall rotation of the star does not affect $C_e(t)$. Simulation results[97] for $C_e(t)$ clearly demonstrated a strong dependence of τ_e on f, but the decay was too slow to determine τ_e accurately.

The three relaxation times discussed above are for the Rouse[23] model, which applies to almost all the simulations on stars. However, experimentally hydrodynamic effects are important.[200,201] At present it is only possible to include solvent molecules explicitly in a simulation for very small stars ($Nf \sim 50$). Smit *et al.*[31] have studied a 3-arm star with $N = 6$ in the presence of a solvent. While the introduction of hydrodynamic effects changes two of the relaxation times, there remain three distinct times for a star. In the Zimm[230] model, one can show[97] that $\tau_{\text{el}} \sim N^{3\nu} f^{(2-3\nu)/2}$ and $\tau_D \sim N^{3\nu} f^{3(1-\nu)/2}$. As in the Rouse case, τ_{el} and τ_D have the same N-dependence but very different f dependences. The prediction for τ_e remains unchanged. In a good solvent, $\tau_{\text{el}} \sim N^{1.77} f^{0.125}$ and $\tau_d \sim N^{1.77} f^{0.62}$. As studies of hydrodynamic effects on linear chains has only been feasible in the past few years,[29,30] it is not surprising that little has been done on simulating many arm stars with a solvent explicitly present. At present at least another order of magnitude in computer speed will be necessary for a serious study of the hydrodynamics of stars. Experimentally there are a number of interesting results[200,201] for which simulations can be very useful in helping to interpret when such simulations are more feasible.

9.4 Polymers tethered to a line

Polymers attached to a linear backbone form another class of tethered chains, which are intermediate between the stars and brushes. Long-chain comb polymers are branched polymers in which branches of length N are attached to a flexible polymer chain.[231] The branches can either be equally spaced or random. When the branches are long and closely spaced, excluded volume interactions among the tethered side chains can significantly stiffen the central contour.[232] Though such "bottlebrush" polymers have been synthesized,[233] so far the backbone has been substantially shorter than the side branches. In this case, the structure will not be very different than for a star polymer in which the branches (arms) are attached to a central point.[234] Diblock copolymers in a selective solvent can also form cylindrical micelles that have similar structures.[192,235]

In this section, we present three examples of polymers tethered to a line. First, we test the scaling predictions for polymers tethered to a straight backbone which is inflexible. In this case, the scaling theory[236] predicts that the height of a bottlebrush should scale as $N^{0.74} \rho_l^{0.26}$ in a good solvent, where ρ_l is the number of branches per unit length. This is in between the N^ν for a star and N for a planar brush (Section 9.5). We then discuss how the length and density of the branches increase the persistence length of the backbone, and how $\langle R_G^2 \rangle$ for long branches attached to a short, flexible backbone, compares to $\langle R_G^2 \rangle$ for a star with an equal number of arms. For large N and small f, attaching the branches to a central point or a flexible line is equivalent, though for large f, $\langle R_G^2 \rangle$ for the star increases slower with f than for chains attached to a line.

9.4.1 Polymers tethered to an inflexible line

The scaling theory discussed above for stars can also be applied to chains grafted to a line.[236] Each branch or arm is again treated as a series of spherical blobs. Since the volume accessible at a radial distance r from the grafting line increases, the monomer density decreases and the size of the blobs increases with r. At a distance r from the grafting line, a cylinder of length L has $\rho_l L$ blobs of radius $\xi(r)$ covering a cross-sectional area of Lr. Therefore the blob radius,

$$\xi(r) \propto (r/\rho_l)^{1/2}. \tag{9.26}$$

Each blob contains $\xi^{1/\nu}$ monomers and the monomer density falls off as

$$\rho(r) \propto \xi(r)^{1/\nu}/\xi(r)^3 \propto (r/\rho_l)^{(1-3\nu)/2\nu}. \tag{9.27}$$

For a good solvent, $\rho(r) \sim (r/\rho_l)^{-0.65}$, while in a Θ solvent, $\rho(r) \sim (r/\rho_l)^{-1/2}$. This is significantly slower than for a star, reflecting the fact that the volume available to the chains increases with r faster for a star than for a cylinder. Figure 9.12 shows a typical configuration for polymers grafted to a line for three different values of ρ_l for 50 chains of length $N = 50$ in a good solvent.[104] The polymers are projected on the plane perpendicular to the grafting line. The figure is taken from a MD simulation with a purely repulsive Lennard-Jones interaction between nonbonded monomers. In comparison to stars, the chains are much more stretched. The average height h of the bottlebrush can be derived by noting that $\rho_l N \sim \int_0^h \rho(r) r dr$,

$$h \propto N^{2\nu/(1+\nu)} \rho_l^{(1-\nu)/(1+\nu)}. \tag{9.28}$$

In a good solvent $h \propto N^{0.74} \rho_l^{0.26}$.

Li and Witten[221] have used a variational approach to determine the properties of polymers in solution grafted to convex surfaces when only binary interactions are present. For a strongly curved cylinder, they were also able to solve the equations exactly. For polymers grafted to a line, both methods gave the result that the free chain ends were strongly excluded from a region near the grafting line. In their variational approach, they found that the exclusion zone was out to 79% of the brush height h, while in the exact solution the exclusion zone was significantly less, 42%. The variational approach predicted that $\rho(r)$ would decay as eq. (9.27) in the exclusion

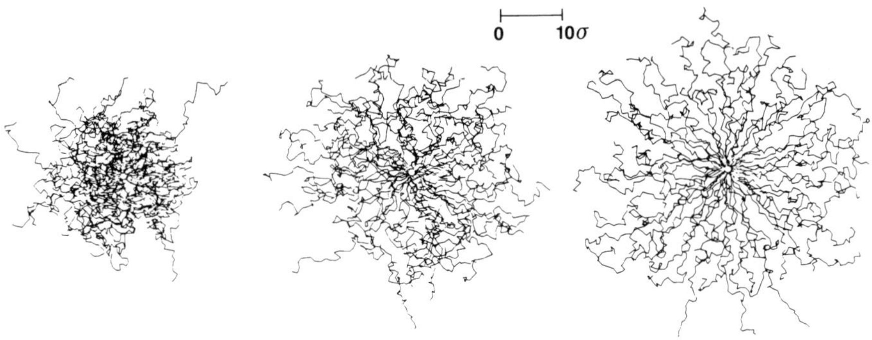

Fig. 9.12 Typical configurations of systems of 50 chains of length $N = 50$ grafted on a line at linear density (a) $\rho_l\sigma = 0.38$, (b) 3.14, (c) 6.28 in a good solvent. The chains are projected on the plane perpendicular to the grafting lines. Note that these are not star polymers. (Results are from Ref. 104.)

zone and more rapidly thereafter. The exact solution for $\rho(r)$ is in terms of Gauss' Hypergeometric function and has the power law form, eq. (9.27), for $r \ll h$. For general r, $\rho(r)$ decays somewhat less rapidly, particularly in the mid-region of the brush.

Both MD simulations[104] and numerical SCF analysis[170] found that in a good solvent the monomer density profiles $\rho(r)$ scaled with $\rho_l^{0.65}$ and were insensitive to N in the power law regime as expected. However both also found indications that $\rho(r)$ decayed more slowly than eq. (9.27), with the exponent of the power law being closer to 0.5 than to 0.65. This difference may be an indication that Li and Witten[221] are correct in their prediction of a somewhat slower decay for $\rho(r)$ in the central portion of the brush, since this would likely appear as a lower power law, particularly considering that the largest value of N considered was 300.[170] To clarify this point, it would be necessary to study much longer chains in order to distinguish a power law from the exact solution of the SCF equations. In addition, Dan and Tirrell[170] also found that the brush height as a function of N reached the asymptotic limit predicted, while Murat and Grest[104] observed that h scaled with ρ_l as predicted.

In the simple blob picture, the free ends are assumed to be within the last blob. Analytic solutions of the SCF equations, both for polymers in solution[221] and in a melt[237] strongly suggest that there is an extensive region from which the free ends are excluded for a strongly curved cylinder. However as will be discussed in Section 9.5, this exclusion zone vanishes as the radius of curvature of the surface goes to infinity. As seen in Section 9.3, there clearly exists an exclusion zone in a star, though it is significantly narrower than predicted by Li and Witten.[221] The existence of an exclusion zone for the free ends has been studied by both simulations and numerical SCF for the present case. In Fig. 9.13, we plot the monomer end density $\rho_e(r)$ for four values of ρ_l from our MD simulations with a purely repulsive Lennard-Jones interaction.[104] For low ρ_l, the chains are not highly stretched for this chain length ($N = 50$) and as a result there is no exclusion zone. As ρ_l increases an exclusion zone appears. However at such high ρ_l, steric packing constraints also play a role, as the monomer density is very large for several monomer diameters away from the grafting line, forcing the free ends away from this region. To really validate the theory, it would be necessary to study much longer chains at coverages comparable to $\rho_l\sigma = 0.38$. Dan and Tirrell[170] and Wijmans and Zhulina[159] also looked for a dead zone using a numerical SCF theory. Though the smallest radius of curvature for the cylinder presented by Dan and Tirrell was $R_c = 5\sigma$ at a surface coverage $\rho_a\sigma^2 = 0.3$, which corresponds to $\rho_l\sigma = 9.4$, their results for $\rho_e(r)$ are very similar in shape to curves c and d in Fig. 9.13. Their studies included chains of length $50 \leq N \leq 300$ for $\rho_a \geq 0.10$. To quantify the size of this zone, Dan and Tirrell defined the width x_Q of the exclusion zone as the distance from the surface where the density of the free ends reaches 5%. They showed that for small ratios of $R_c/h\,(\leq 0.2), x_Q$ grew linearly with h, consistent with the analytical predictions. However because of the large ρ_a

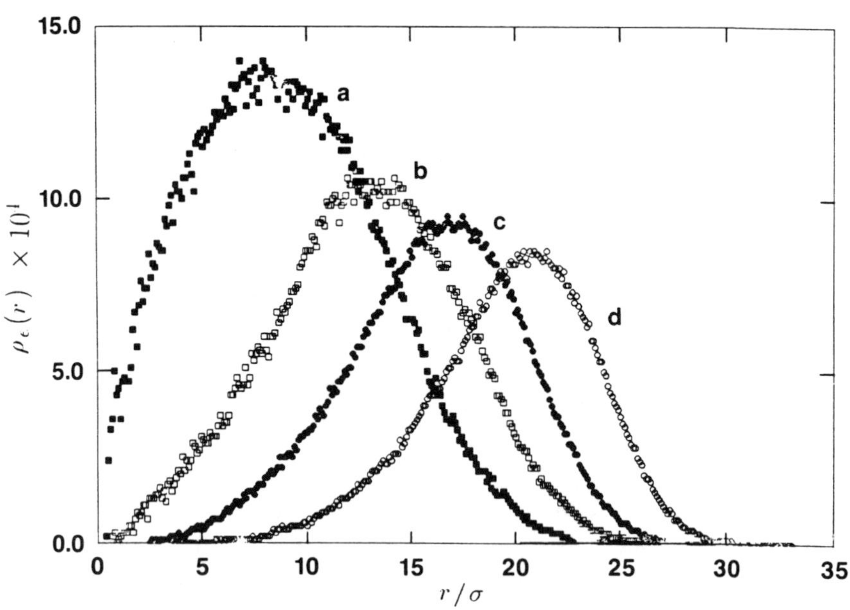

Fig. 9.13 Distribution of the free ends for chains attached to a line as a function of the distance r/σ from the grafting line for (a) $\rho_l\sigma = 0.38$, (b) 1.51, (c) 3.14 and 6.28. Results are from the MD simulations of Ref. 104 for a purely repulsive Lennard-Jones interaction between nonbonded monomers.

used, their results cannot be used to prove that the analytic theories are correct. The recent work of Wijmans and Zhulina[159] for higher curvature (small R_c) and longer chains ($N = 1000$) shows a dead zone even for $\rho_l\sigma = 0.6$, but not for 0.03. These results strongly support the analytic SCF calculations, in that they are first to identify numerically an exclusion zone for such low coverages and long chain lengths that only binary contacts and not steric effect are important.

9.4.2 Polymers tethered to a flexible line

The addition of branches to a flexible linear chain increases its persistence length. However because the local density is also increased, the overall size of a comb polymer is actually reduced relative to a SAW chain having the same number of total monomers. Experimentally there are only limited data on long combs[231] and there have been no scattering experiments using labeled monomers to measure the persistence length of a comb as a function of the number and length of the branches. The data which are available are for the size of a comb which can be compared to the size of the chain without branches. Theoretically, there have been several analytic studies using a Gaussian model[238] which showed significant discrepancies with MC results for combs[56,60] in contrast to the results of similar comparisons for stars with only a few arms[58] and H-combs.[59,239] Fredrickson[240] found, using a Flory theory, that there are three scaling regimes which describe the size of the polymer. For $\rho_l \ll N^{-9/10}$, the branches are so far apart that the backbone excluded volume dominates the branch excluded volume and the flexible coil is unperturbed. For intermediate ρ_l, $N^{-9/10} \ll \rho_l \ll N^{-3/5}$, the excluded volume of the branches dominates that of the backbone and $R \sim \rho_l^{2/5} N^{9/25} N_b^{3/5}$, where N_b is the length of the backbone. For $\rho_l \gg N^{-3/5}$, the system is more like a semiflexible rod. In this limit, the persistence length is very large and the fact that the backbone is flexible is only important at very long length scales as for cylindrical micelles. The overall size has the same scaling with N as for an inflexible line, $R \sim \rho_l^{17/20} N^{3/4} N_b^{3/5}$. These different regimes have not been tested by simulations. The only MC simulations on dilute comb polymers have either used quite short side branches, $N \leq 9$,[58] or a short backbone chain, $N_b = 20$.[241] Lipson[60] compared her MC results for the end-to-end distance of the comb as a function of the weight fraction of material in the backbone w_{bb} to experimental data and found reasonable agreement, even though the branches were arranged regularly in the simulations while experimentally they were randomly attached. However $\langle R^2 \rangle_{\text{comb}} / \langle R^2 \rangle_{\text{SAW}}$ does not depend only on w_{bb} but also on the length of the branch. Short, closely spaced branches stretch the comb backbone more than long widely spaced branches, though the effect may not too large. In the high density regime, for a given $w_{bb}(\rho_l N \text{constant})$, $R \sim N^{-1/10}$ within Flory theory. Clearly more simulations with longer branches and a range of branching densities as well as experiments on labeled chains are needed before a complete understanding of the properties of comb polymers can be achieved.

In the high branching regime, there has been limited progress in synthesizing molecules with axisymmetric branching around a central backbone contour.[233,234,242] Though such polymers have been synthesized, the backbone has not been substantially longer than the side chains. Tsukahara[234] has determined the size of 14 such polymers and found that they were not very different from stars of the same M_W. However as the length of the backbone increases, the bottlebrush polymer, which has no reason to remain spherically symmetric, will become elongated in order to alleviate overcrowding. This crossover from star-like to a more elongated, axially symmetric polymer can easily be studied by computer simulations. Since the density in the interior of the bottlebrush is high, MD is an ideal method for simulating such systems. Here we present some new MD results in a good solvent for monomers interacting with a Lennard-Jones interaction truncated at $r_c = 2.5\sigma$. Figure 9.14 shows a projection of a typical configuration of a 40-arm star and bottlebrush with $N = 50$ monomers per arm. Note that the star is symmetric while the bottlebrush is more elongated. For this case, the ratio of $\langle R_G^2 \rangle$ for the bottlebrush to that of the star is 1.3. In Fig. 9.15, results for $\langle R_G^2 \rangle$ versus f for the two types of polymers are presented. Note that as expected, for small $f \lesssim 20$ for $N = 50$, $\langle R_G^2 \rangle$ for the two cases is the same, in agreement with Tsukahara.[234] However for larger f, $\langle R_G^2 \rangle$ for the bottlebrush increases significantly faster than for a star. In this case the crossover is at $f \approx N/2$, though additional simulations on larger systems are necessary before one can determine how general this result is.

9.5 Polymeric brushes

The structure formed when one end of a chain is tethered to a $d = 2$ surface, is referred to as a "grafted layer" or more commonly as a "brush". Brushes can be made by attaching a functional group to one end of a chain that can then bind to the surface. The binding energy can either be quite high (several hundred k_BT), in which case the end is chemically attached,[243] or of order ten k_BT, in which case it is physi-adsorbed.[244,245] Diblock copolymers, in which one of the blocks (usually the shorter one) adsorbs strongly to the surface, while the other does not, also form brushes,[154,246–253] as well as a diblock copolymer at the air–liquid interface of a selective solvent.[254–257] A qualitatively similar system is formed by a symmetric diblock copolymer in a lamellar phase in the strong segregation limit, where the junctions between the two segments of the copolymer lie on a plane.[258] Although the junction points are not fixed to the surface, and are free to move, this movement is

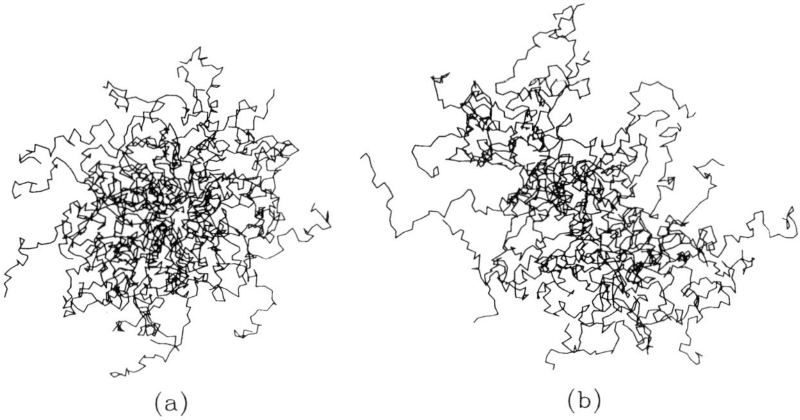

Fig. 9.14 Projection of a typical configuration of two branched polymers both with $f = 40$ arms and $N = 50$ monomers per arm for a good solvent. The results are at $T = 4.0\epsilon/k_B$ for monomers interactions with a Lennard-Jones interaction truncated at $r_c = 2.5\sigma$. Polymer (a) has the topology of a star, with all 40 arms attached to a small central core, while (b) is a bottlebrush made of 40 arms which are tethered to a flexible backbone.

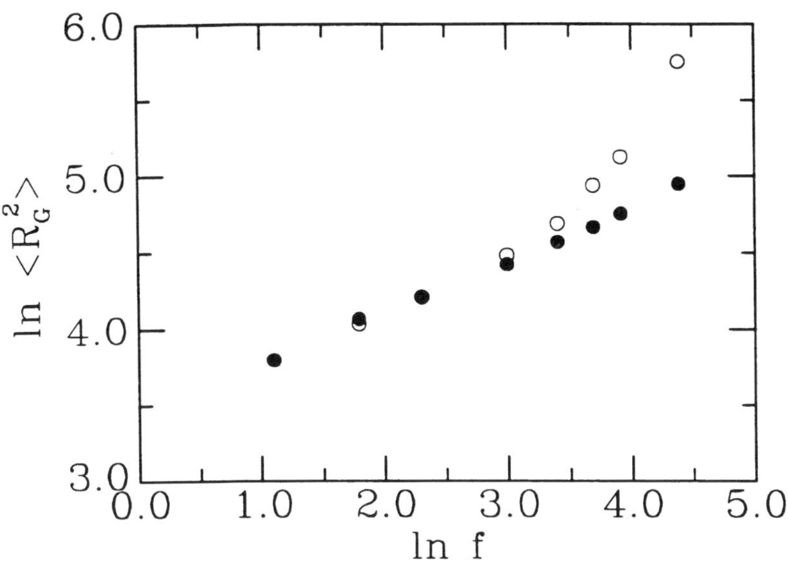

Fig. 9.15 Radius of gyration $\langle R_G^2 \rangle$ for a polymer made of f arms each of length $N = 50$ attached to a small central sphere (•) or to a flexible line (○). The simulations were carried out under good solvent conditions at $T = 4.0\epsilon/k_B$.

strongly damped by the repulsive interaction with other chains. This system, which closely resembles a pair of brushes brought into contact, has been reviewed by Binder.[21,259]

When the grafting density ρ_a increases above a critical density, the chains overlap and stretch away from the grafting surface to avoid overcrowding. The stretching effect of tethering is much stronger for a brush than for polymers grafted on a point or on a line. The reason for this stretching and the amount of stretching depend upon the environment of the brush.[2] One can differentiate between wet and dry brushes, depending upon the presence or absence of a solvent. In a wet brush, the stretching is due to the attraction between the polymer segments and the solvent. The amount of stretching is determined both by the quality of the solvent and by the binding energy of the end group to the grafting surface. In a dry brush, the driving force for stretching is the incompressibility of the polymer chains, which prevents overfilling of the space occupied by the brush.

A large number of experimental, theoretical and numerical studies on brushes have been published during the last several years, including two excellent review articles.[1,2] In this section, we shall discuss simulations on brushes and related systems, and compare their results with experiments and theoretical predictions. Although there has been some recent interest in brushes with chains of different chemical composition,[260–265] we shall limit our discussion to brushes whose chains are of identical chemical composition.

9.5.1 Brushes in good solvents

As first noted by Alexander[7] and de Gennes,[8] grafting polymer chains in a good solvent densely onto a surface leads to a deformation of the chains from their equilibrium configurations. This deformation is a result of the competition between the entropic elastic energy of the chain and the monomer–monomer interaction. Alexander[7] assumed that the chains are uniformly stretched and that the density of the monomers is constant up to a height h from the grafting surface. The free energy can then be written using a Flory approximation as the sum of elastic stretching energy of Gaussian chains and binary monomer–monomer interactions.[8] Minimizing this free energy with respect to h, one finds that h scales as

$$h \sim N\rho_a^{1/3}. \tag{9.29}$$

One can alternatively view the brush as a stack of blobs, as was done for stars and polymers grafted on a line except that brushes are even simpler since the blob size is independent of the distance from the surface. This approach gives $h \sim N\rho_a^{(1-\nu)/2\nu}$. Using $\nu = 0.588$ for a good solvent, one obtains $h \sim N\rho_a^{0.35}$. However, it has been common practice in the literature to use the Flory value $\nu = 3/5$ and $h \sim N\rho_a^{1/3}$, identical to eq.(9.29). This result is rather striking, as the thickness of the grafted layer increases *linearly* with N, significantly different than the $N^{0.59}$ for self-avoiding walks and star polymers and the $N^{0.74}$ for polymers grafted to a line.

Numerical SCF calculations[153–160,168] have also been applied to the study of end-grafted chains. These showed that the density of the monomers is not uniform. Skvortsov *et al.* [155] found that their numerical results can be fitted with a parabolic decay of the monomer density from a maximum near the grafting surface to zero at the brush height. Later Milner *et al.*[266,267] and Zhulina *et al.*[268] independently proposed an analytical theory based upon an observation by Semenov.[269] When the stretching is strong, a chain of length N whose free end is at a specified point, fluctuates very little around the "most favorable" configuration than minimizes the classical action. The relative amplitude of this fluctuation diminishes with increasing N. As $N \to \infty$, this configuration dominates. For binary interactions, analytical solutions for the properties of the brush can be obtained. The brush height was found to have the same scaling with N and ρ_a as predicted by Alexander,[7] although the numerical coefficient is slightly larger. This agreement in the overall size scaling confirms the SCF assumption that the stretching diminishes the fluctuations around the most favorable configuration. The monomer density $\rho(z)$ in this theory varies with the distance z from the grafting distance as

$$\rho(z) = C_1\rho_a^{2/3} - C_2(z/N)^2, \tag{9.30}$$

where both C_i depend only on the excluded-volume parameter. This parabolic density profile is valid in the limit of very long N and moderate ρ_a, where binary interactions dominate. In order to demonstrate the validity of the analytical theory, we show in Fig. 9.16 the density profile (eq. [9.30]) for $\rho_a\sigma^2 = 0.1$ and several values of N, together with the corresponding numerical SCF calculations.[157] The agreement, obtained without any adjustable parameters, is excellent. The size of the "tail" region at the outer edge of the brush, where there is a small difference between the theory and the numerical calculations, decreases with increasing N.

Another important difference between the scaling approach and the SCF theory is the location of the free ends of the chains. Unlike the implicit assumption of the scaling analysis that all the free ends are at the outer boundary of the brush, the density of the free ends is nonzero everywhere within the brush. For moderate ρ_a,[266]

$$\rho_e(z) = C_3\rho_a^{-2/3}N^{-2}z\left[1 - (z/h)^2\right]^{1/2}, \tag{9.31}$$

where h is the value of z at which the monomer density vanishes. For high ρ_a, Milner *et al.*[266] showed that $\rho_e(z)$ diverges at $z = h$ as

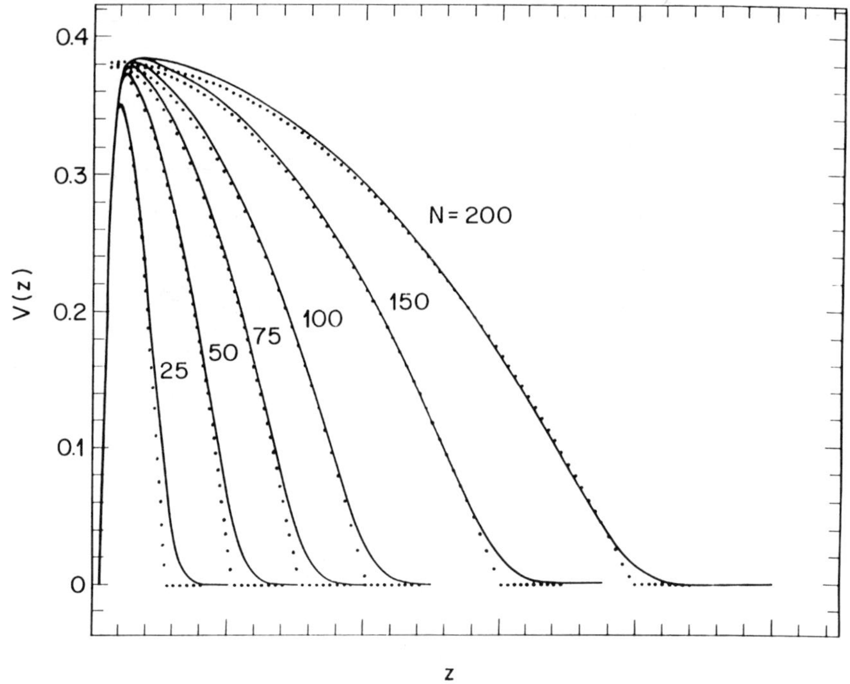

Fig. 9.16 Monomer density profile for $\rho_a = 0.1$ and various values of N within the SCF approximation. The solid lines are from numerical calculations based on the Scheutjens–Fleer theory. The dotted curve is from eq. (9.30). Note that the ordinate is the potential field $V(z)$, which is proportional to the density profile. (From Ref. 157.)

$$\rho_e(z) = C_4 \rho_a^{-2/3} N^{-2} z \left[1 - (z/h)^2\right]^{-1/2}. \tag{9.32}$$

Thus, at the high density limit, most of the ends are concentrated near the outer edge of the brush, as assumed by Alexander[7] and de Gennes.[8]

Numerous experimental studies and computer simulations have been carried out during the last several years to check the results of these two theoretical approaches. Some of the experimental work was aimed at studying the scaling of h by measuring the force between two brushes.[244,246–249,251] The inner structure of the brushes has been probed by small angle neutron scattering[243,270–272] and neutron reflectivity.[252,253,255,256,273] All these studies, as well as the simulations, gave results that were consistent with the analytical SCF predictions.

Figure 9.17 shows typical configurations of an off-lattice brush consisting of 50 chains with $N = 50$, with three values of ρ_a. The value of h increases significantly with increasing ρ_a. The scaling $h \sim N\rho_a^{1/3}$ has been confirmed in several simulations.[38,78,98] These showed that provided ρ_a is above the critical overlap density $\rho_a^* \sim N^{-6/5}$, the brush size, as measured by the first

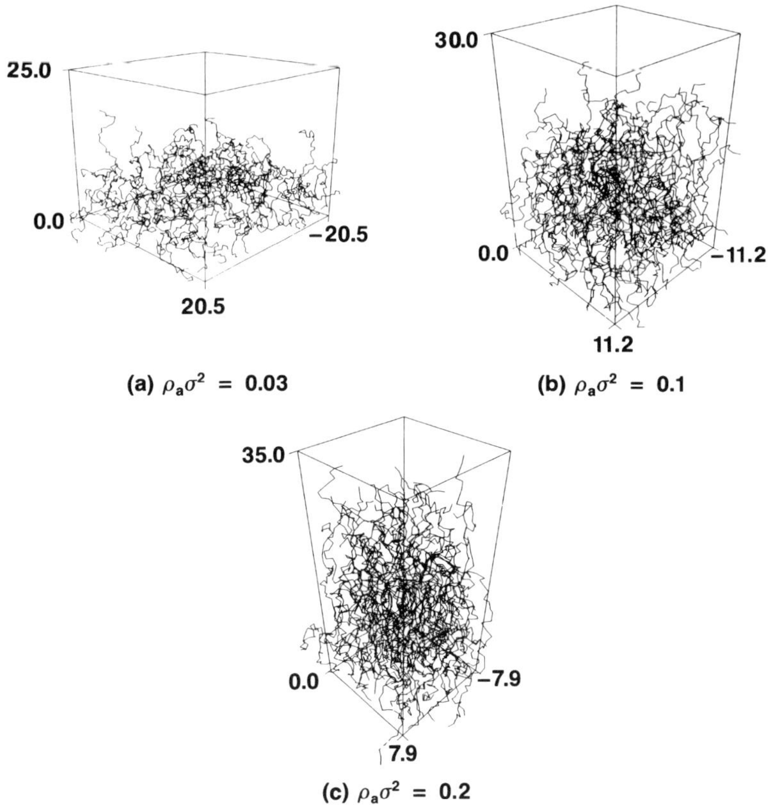

Fig. 9.17 Typical configurations of brushes with $N = 50$ and different values of ρ_a, taken from Ref. 98. Each of the 50 chains in the brushes is grafted onto a randomly chosen point on the lower surface. Periodic boundary conditions are used in the horizontal plane.

moment of the density distribution $\langle z \rangle$ or the z-contribution to the mean-squared radius of gyration, $\langle R^2_{Gz} \rangle$, approaches the predicted scaling form, as seen in Fig. 9.18, curve (a). When $\rho_a < \rho^*_a$, h is almost independent of ρ_a, as expected in the mushroom regime. Since $\rho^*_a \sim N^{-6/5}$, the scaling variable evaluated at the lower crossover, $N\rho_a^{*1/3} \sim N^{3/5}$, is not universal but increases with N in agreement with the results shown in Fig. 9.18, curve (a). For large ρ_a, Grest[116] has shown that there is another N-dependent threshold, ρ_{a1}, above which the brush size scales as $N\rho^x_a$ with x approximately 1/2. This behavior, seen in curve (a) of Fig. 9.18, is in agreement with Raphaël, Pincus and Fredrickson[213] who showed that at high surface densities where three body interactions dominate over two body terms, there is a second scaling regime with $x = 1/2$. Experimentally this regime is difficult to study, as such high grafting densities are not easily accessible. The limited

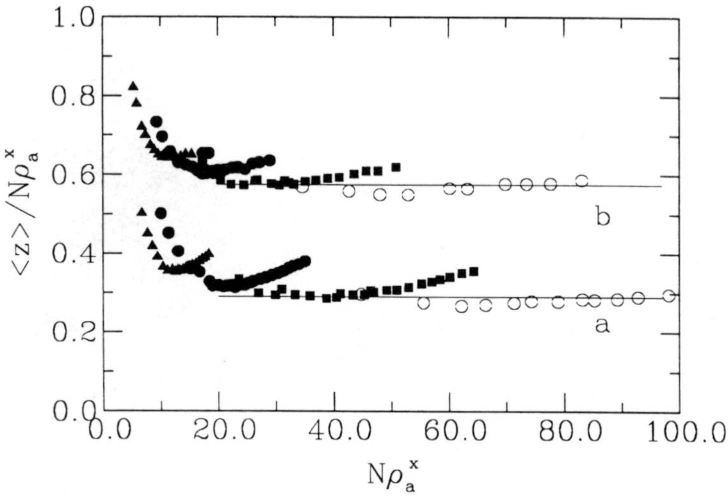

Fig. 9.18 Average monomer height $\langle z \rangle / N\rho_a^x$ versus $N\rho_a^x$ for brushes in (a) a good solvent with $x = 1/3$ and (b) a Θ solvent with $x = 1/2$. The results for Θ are shifted by 0.2 for clarity. (From Ref. 116.)

width of the scaling regime, $\rho_a^* < \rho_a < \rho_{a1}$ within which $\langle z \rangle \sim N\rho_a^{1/3}$, also makes it difficult to test the dependence of h on ρ_a experimentally.[256] If the chains are not long enough, $\rho_a^* \sim \rho_{a1}$ and there is no regime in which the data scale as predicted theoretically, as seen for $N = 25$. For intermediate length chains, the scaling regime is very limited, as for $N = 50$ in Fig. 9.18, curve (a). For $N = 50$ and 100, h scales almost linearly[116] with ρ_a for a wide range of ρ_a from the mushroom regime to $\rho_a \sigma^2 \approx 0.30$ and 0.11, respectively. This may explain the data of Factor *et al.*[256] for PS in a good solvent for M_w up to 175 000, who found that h varied approximately linearly with ρ_a over the entire range studied. Only for long chains is ρ_a^* significantly less than ρ_{a1} and the scaling prediction valid over an extended range.

Force measurements between two brushes yield experimental information on the brush size, since the brushes first interact when the distance between their respective grafting surfaces is $2h$. Taunton *et al.*[244] applied this method to brushes consisting of polystyrene chains attached to a mica surface via a zwitterionic head group. The chains were deposited onto the surface from a concentrated solution of the polymer. The measurement of the contact distance gave the seemingly surprising result that the brush height scaled as $N^{0.6}$ rather than as the expected N. Similar behavior was also observed in brushes consisting of diblock copolymers, with only one of the blocks adsorbing on the surface.[247,248,250] However, unlike the assumption of the theory or the simulations, in which the grafting density is constant, these experiments were conducted under conditions of constant binding energy of order $10k_BT$ between the functional end-group and the grafting surface. The as well as the simulations, gave results that were consistent with the analytical SCF predictions.

Figure 9.17 shows typical configurations of an off-lattice brush consisting of 50 chains with $N = 50$, with three values of ρ_a. The value of h increases significantly with increasing ρ_a. The scaling $h \sim N\rho_a^{1/3}$ has been confirmed in several simulations.[38,78,98] These showed that provided ρ_a is above the critical overlap density $\rho_a^* \sim N^{-6/5}$, the brush size, as measured by the first moment of the density distribution $\langle z \rangle$ or the z-contribution to the mean-squared radius of gyration, $\langle R_{Gz}^2 \rangle$, approaches the predicted scaling form, as seen in Fig. 9.18, curve (a). When $\rho_a < \rho_a^*$, h is almost independent of ρ_a, as expected in the mushroom regime. Since $\rho_a^* \sim N^{-6/5}$, the scaling variable evaluated at the lower crossover, $N\rho_a^{*1/3} \sim N^{3/5}$, is not universal but increases with N in agreement with the results shown in Fig. 9.18, curve (a). For large ρ_a, Grest[116] has shown that there is another N-dependent threshold, ρ_{a1}, above which the brush size scales as $N\rho_a^x$ with x approximately 1/2. This behavior, seen in curve (a) of Fig. 9.18, is in agreement with Raphaël, Pincus and Fredrickson[213] who showed that at high surface densities where three body interactions dominate over two body terms, there is a second scaling regime with $x = 1/2$. Experimentally this regime is difficult to study, as such high grafting densities are not easily accessible. The limited width of the scaling regime, $\rho_a^* < \rho_a < \rho_{a1}$ within which $\langle z \rangle \sim N\rho_a^{1/3}$, also makes it difficult to test the dependence of h on ρ_a experimentally.[256] If the chains are not long enough, $\rho_a^* \sim \rho_{a1}$ and there is no regime in which the data scale as predicted theoretically, as seen for $N = 25$. For intermediate length chains, the scaling regime is very limited, as for $N = 50$ in Fig. 9.18, curve (a). For $N = 50$ and 100, h scales almost linearly[116] with ρ_a for a wide range of ρ_a from the mushroom regime to $\rho_a\sigma^2 \approx 0.30$ and 0.11, respectively. This may explain the data of Factor *et al.*[256] for PS in a good solvent for M_w up to 175 000, who found that h varied approximately linearly with ρ_a over the entire range studied. Only for long chains is ρ_a^* significantly less than ρ_{a1} and the scaling prediction valid over an extended range.

Force measurements between two brushes yield experimental information on the brush size, since the brushes first interact when the distance between their respective grafting surfaces is $2h$. Taunton *et al.*[244] applied this method to brushes consisting of polystyrene chains attached to a mica surface via a zwitterionic head group. The chains were deposited onto the surface from a concentrated solution of the polymer. The measurement of the contact distance gave the seemingly surprising result that the brush height scaled as $N^{0.6}$ rather than as the expected N. Similar behavior was also observed in brushes consisting of diblock copolymers, with only one of the blocks adsorbing on the surface.[247,248,250] However, unlike the assumption of the theory or the simulations, in which the grafting density is constant, these experiments were conducted under conditions of constant binding energy of order $10k_BT$ between the functional end-group and the grafting surface. The

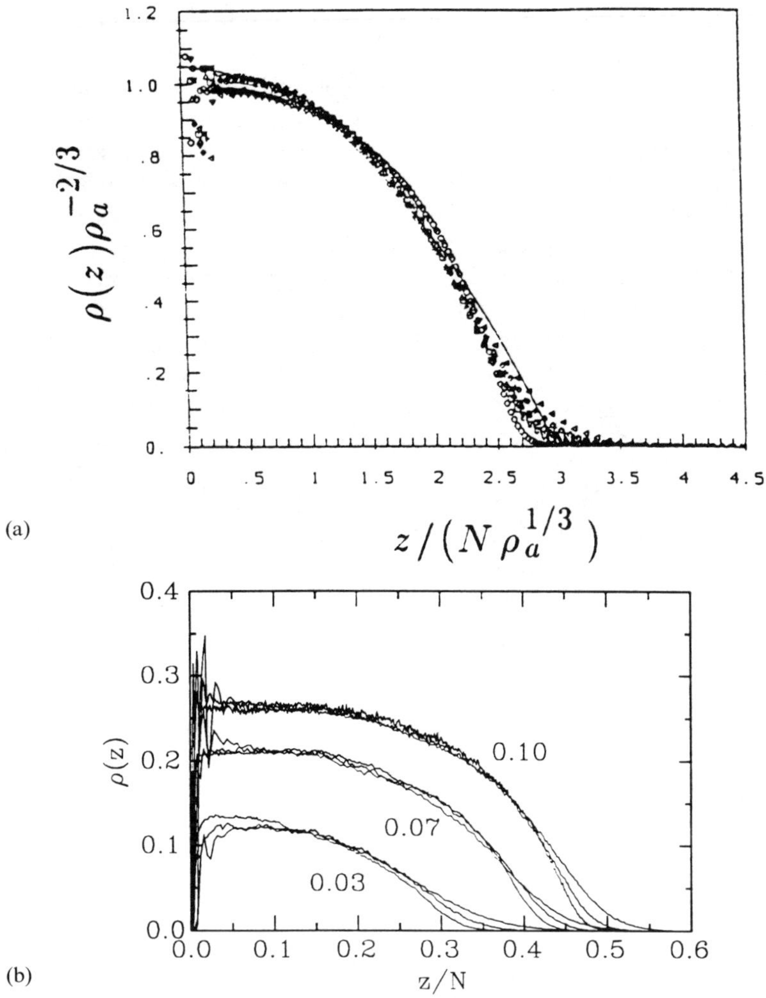

Fig. 9.19. (a) Scaling plots of monomer density for a brush in a good solvent. From MC simulations of Lai and Binder.[78] The different symbols refer to simulations with $30 \leq N \leq 60$ and $0.05 \leq \rho_a\sigma^2 \leq 0.2$. See Ref. 274 for further details. The solid curve is the SCF prediction. (b) Monomer number density $\rho(z)$ for brushes in a good solvent for chains of length $N = 50$, 100 and 200 for $\rho_a\sigma^2 = 0.03$, 0.07 and 0.10. (Taken from MD simulations of Ref. 115.)

density of chains on the surface is then determined by a balance between the stretching energy and the energy of the attachment.[244] To estimate ρ_a within the scaling picture,[7,8] note that the chains consist of (N/n_B) blobs, of size $\xi \sim \rho_a^{-1/2}$. Each k_BT of binding energy can support about one blob.[244] For a fixed number of blobs, the larger N, the larger each blob. Since the size of a blob is related to the number of monomers in it, $\xi \sim n_B^{\nu}$, $h \sim N^{1-2\nu/3} = N^{3/5}$,

Carignano and Szleifer[172–174] proposed a new statistical thermodynamic theory of polymeric brushes, based upon the probability distribution function of chain conformations, which depends on the local osmotic pressure. The interactions between the monomers within a chain are taken into account "exactly" by generating a large number of representative conformations. Mean-field approximation is used to evaluate the interactions between the chains. The resulting equations are then solved numerically. The theory was applied both to off-lattice and lattice chains. Calculations for the former were carried out for the same systems that were used in the MD simulations of Murat and Grest,[98] to allow for detailed comparison. Excellent qualitative and quantitative agreement was observed for all the systems, including those with very high grafting densities. This observation supports the adequacy of the mean-field approximation to grafted polymeric systems in good solvents, in which chain fluctuations at large length scales are suppressed by the strong stretching.

In contrast to the surface force experiments, neutron reflectivity and small angle neutron scattering give information on the inner structure of the brushes. These have provided additional support for the parabolic density profile predicted by the SCF theory. Field *et al.*[252] determined the density profile of polystyrene-poly(ethylene oxide) (PS-PEO) block copolymers adsorbed from deuterated toluene (which is a good solvent for PS) onto quartz. The PEO block is much shorter than the PS blocks, and adsorbs strongly to quartz. Neutron reflectivity data for brushes with chains of M_W between 80 000 and 502 000 were fitted to density profiles of the form $A - Bz^n$. A and B are determined by the volume fraction of the polymer at the interface, and the layer thickness of the adsorbed polymer. Except for the highest molecular weight chain, the best fits were obtained for $1.9 \le n \le 2.6$, with an uncertainty of about 0.5. This agrees well with the prediction of a parabolic profile. Brushes with $M_W = 502\,000$ gave $n = 4.6 \pm 1.5$. This large exponent was attributed to the low segment density associated with this high molecular weight brush, whose scattering length density profile becomes indistinguishable from that of the solvent away from the grafting surface. As a result, the fit becomes less sensitive to n. The fitted profiles, however, correspond to reflectivity curves that have an oscillatory behavior. The oscillations can be damped by the addition of an exponential tail to the parabolic profile. Equally satisfactory fits were obtained by using a trial form such as an error function. This gives a density profile identical to the parabolic one, except that it vanishes gradually at the outer edge of the brush, without requiring an additional correction. Neutron reflectivity methods were also applied by Cosgrove *et al.*[280] to brushes with small N. Their results were in agreement with the parabolic profile, as well as with profiles obtained using Scheutjens–Fleer SCF lattice model.[148,154] Additional experimental evidence for the parabolic density profile came from the small angle neutron scattering experiments on PS chains grafted onto porous silica.[243,272] By analyzing the monomer–monomer and the monomer–solid surface structure factors, they showed that both are consistent with a parabolic density profile with an exponential tail. Furthermore, they observed that at large scattering vectors (small distances) scattering is dominated by concentration fluctuations. The blob size that was identified from the behavior in this regime was found to increase linearly with the interanchor distance, as expected. This showed that the small-scale behavior is dominated by density fluctuations (as in the scaling picture), while the large-scale behavior is free from these fluctuations (as in a mean-field description).

In addition to structural information, simulations also allow one to calculate the surface osmotic pressure $\Pi_a = \partial f_c/\partial \rho_a^{-1}$ within a brush, where f_c is the free energy per chain. In off-lattice simulations, one can calculate this quantity through the evaluation of the virial[25] from the forces between monomers. Although Π_a can be determined in a constant ρ_a simulation, Grest[116] carried out a constant Π_a simulation and determined the dependence of Π_a on ρ_a. Figure 9.20(a) shows Π_a for brushes in a good solvent. Π_a is scaled by $N\rho_a^{5/3}$, which is the SCF prediction[2] for this quantity. The prediction of the scaling theory[7,8] for Π_a in a good solvent is $\Pi_a \sim N\rho_a^{11/6}$. The data in Figure 9.20(a) are best fitted by $\Pi_a \sim N\rho_a^y$, with $y = 2.5 \pm 0.2$. Simulation result for y is larger by about 1 than the predictions of both theories. A possible reason[116] for the discrepancy may be the finite volume of the monomers which is not included in either theory. On the other hand, the value of y is very close to the exponent 2.4 found in the statistical thermodynamic theory of Carignano and Szleifer,[172] which takes into account the finite volume of the monomers. Another possible explanation is that the systems used in the simulations are not yet in the scaling limit, since the average monomer density ϕ within the brush was quite high, $\phi\sigma^3 \gtrsim 0.1$ in almost all cases. In order to check this possibility, simulations with much lower ρ_a and higher N (of order 1000) are needed. Such simulations are not feasible presently. These results are also consistent with the results of Kent *et al.*,[255] who studied PDMS-PS diblock copolymers in a selective solvent (ethyl-benzoate), in which the PDMS block spreads at the liquid–air interface and the PS block is submerged in the solvent. The grafting density in these experiments can then be changed by applying a surface pressure. Thus, constant Π_a simulations represent this experimental situation better than constant ρ_a. Kent *et al.*[257] found that Π_a increases much more rapidly with ρ_a than predicted by the SCF and scaling theories. One recent simulation by Shaffer[88] seems to be in variance with the above results. Shaffer's lattice MC simulations for the free energy seem to agree well with the predictions of scaling and SCF theory for low and moderate densities and short ($N \leq 60$) chainlengths. The reasons for the observed differences are not understood.

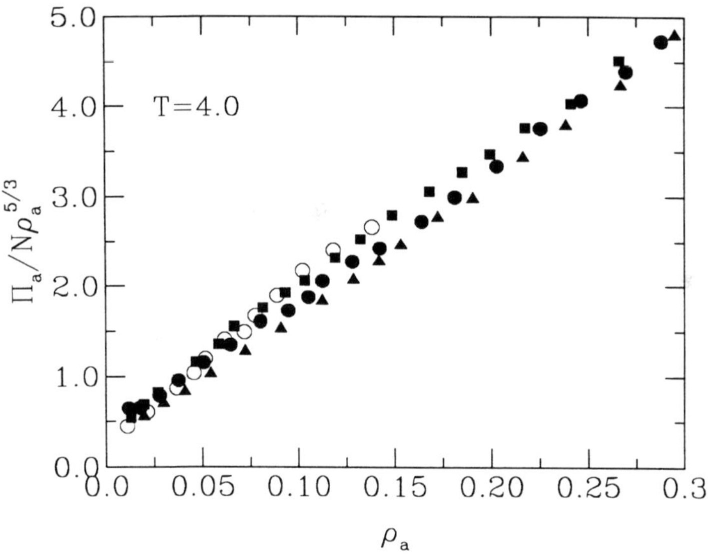

(a)

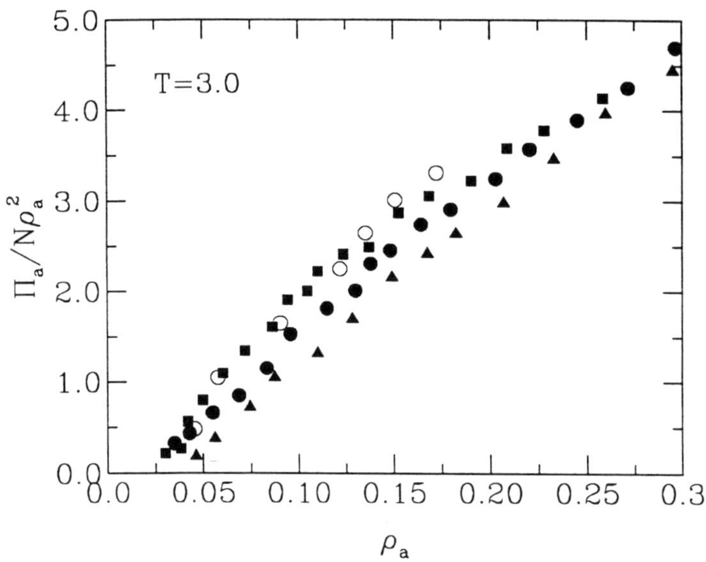

(b)

Fig. 9.20 (a) Scaling plot of $\Pi/N\rho_a^{5/3}$ versus ρ_a in a good solvent at $T = 4.0$ for chain lengths $N = 25$(▲), 50(•), 100(■) and 200 (○). (b) Scaling plot of $\Pi/N\rho_a^2$ versus ρ_a in a Θ solvent ($T = 3.0$). Symbols are the same as in (a). (From Ref. 116.)

Almost all the simulations mentioned in this section have dealt with brushes in which the grafting head is firmly attached to the surface, at a point which cannot move within the grafting surface. In some experimental situations, such as diblock copolymers at the air–liquid interface of a selective solvent,[254–257] the grafting end is free to move, as long as it stays close to the surface. Simulations that allow the grafting site to move within the grafting surface[78,116] ("annealed" brushes), have shown that the structure of these annealed brushes is identical to those of brushes in which the ends are firmly attached. However two small differences have been observed. The first one is that at fixed Π_a, the additional force between the grafting site and the first monomer gives a slightly different equilibrium value for ρ_a.[116] However, this difference decreases as N^{-1}, as expected for an end-effect. An additional difference is observed in the dynamics of the brushes, as discussed in subsection H on time-dependent phenomena in brushes.

9.5.2 Brushes in Θ and poor solvents

As the quality of the solvent decreases, the structure of the brushes changes considerably. Halperin[281] carried out a Flory-type mean-field analysis of the collapse of grafted layers as the solvent quality is reduced. For nonoverlapping chains (mushroom regime), the transition was found to be similar to that of free chains, in which the scaling of the chain size changes from $N^{3/5}$ in a good solvent to $N^{1/3}$ in a poor one. For higher densities (overlapping chains), he found that this transition is replaced by a "weak collapse", where the brush dimension diminishes, but the chains still remain stretched, $h \sim N$. Shim and Cates[276] and Zhulina *et al.*[282] generalized the analyses of Milner *et al.*[266] and Zhulina *et al.*,[268] valid for good solvent conditions, to solvents of lower quality. For a Θ solvent, they found an elliptical density profile, with a vertical asymptote at $z = h$. Their results are consistent with the numerical SCF studies of Wijmans *et al.*[158] who studied the effect of finite chain length for $T = T_\theta$. For poorer solvents, the density profile is expected to be quite flat up to the outer edge of the brush, where it drops discontinuously to zero. Ross and Pincus[283] calculated the structure factor of brushes within the random phase approximation. They found that the brush collapses continuously as the solvent quality is reduced, with the density fluctuations in the plane parallel to the grafting surface damped by the positive osmotic pressure in the brush. In a twice-grafted layer (chains with both ends attached to two parallel plates), which cannot undergo a uniform collapse, the chains form "bundles" with monomer-rich and monomer-poor regions.[283,284] Yeung *et al.*,[285] Huang and Balazs,[167] Williams[286] and Tang and Szleifer[287] on the other hand, predict such lateral instabilities in singly grafted layers for finite chain lengths, with the spatial dimensions of the fluctuations depending upon chain length, solvent quality, and grafting density.

Brushes under variable solvent conditions have been simulated by Lai and Binder[79] using a bond fluctuation model and by Grest and Murat[115] using MD. The Θ temperature for the bond fluctuation model was calculated from an independent study of a single chain to be $T_\theta = 1.9\epsilon$. The MD simulations used a FENE potential, eq. (9.6), between bonded monomers and a LJ potential truncated at $r_c = 2.5\sigma$ to represent the attractive interactions of the monomers, as described in detail in Section 9.2. The Θ temperature was found to be $T_\theta = 3.0 \pm 0.1$.[115]

Dramatic changes are found to occur as the temperature (quality) of the solvent is decreased. Figure 9.21 shows snapshots of three chains out of 40 in a brush at two temperatures. While the chains are strongly stretched at the higher T, they shrink considerably at the lower one, which is below T_θ. Curve (b) in Fig. 9.18 shows the normalized brush height scale as $N\rho_a^x$ at T_θ, with the predicted[281,282] exponent $x = 1/2$. As in a good solvent, this scaling is observed only in a limited range of ρ_a. Low grafting densities are characterized by the "mushroom" behavior in which the height is insensitive to the grafting density. One should note, however, that the critical coverage ρ_a^* at which the chains first overlap is itself dependent upon the solvent quality. ρ_a^* scales as N^{-1} for a Θ solvent, compared to $N^{-6/5}$ for a good solvent. For $\rho_a > \rho_{a1}$, height increases somewhat more rapidly than predicted, due to the high monomer densities reached in this regime.[116] For $T < T_\theta$, the brush height increases roughly linearly[115] with ρ_a, as predicted.[282] However, in this regime, $\rho_a^* \sim N^{-2/3}$, so that very long chains are needed to observe substantial stretching in simulations. Furthermore, as will be discussed later in this subsection, brushes with $\rho_a < \rho_a^*$ undergo a phase separation under constant ρ_a conditions.[79,115] Note that chains which overlap under good solvent conditions may be in the mushroom regime as the T is lowered. Due to

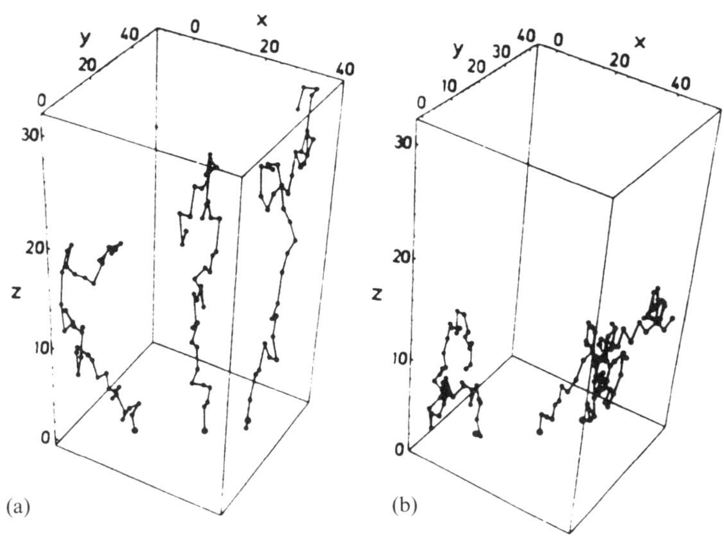

Fig. 9.21 Typical configurations for three of the 40 chains for a polymer brush for two temperatures, (a) $T = 3.3$ and (b) $T = 1.4$. Here $N = 40$ and $\rho_a = 0.1$. The first is in a good solvent above $T_\theta = 1.9$, while the second is in a poor solvent. The results are from the MC simulations of Lai and Binder.[79]

these limitations, it is difficult to observe the predicted height scaling in simulations, unless one uses much larger N.

The qualitative change in the brush structure with the solvent quality is also manifested in the monomer density profile, as demonstrated in Fig. 9.22 for brushes with $N = 200$ at constant grafting densities of $\rho_a \sigma^2 = 0.03$ and 0.1, in the vicinity of T_θ. The maximum extent of the density profile decreases with decreasing T, with a corresponding increase in the monomer density within the brush. Similar results are found in the constant Π_a simulations of Grest,[116] except that ρ_a changes smoothly with T, such that the height remains approximately constant. A quantitative comparison between the simulation results and the analytical SCF predictions for the density profile reveals a much worse agreement at T_θ than at high temperatures (good solvent). Analytical SCF theory predicts[276,282] that the monomer density decays to zero as an ellipse,

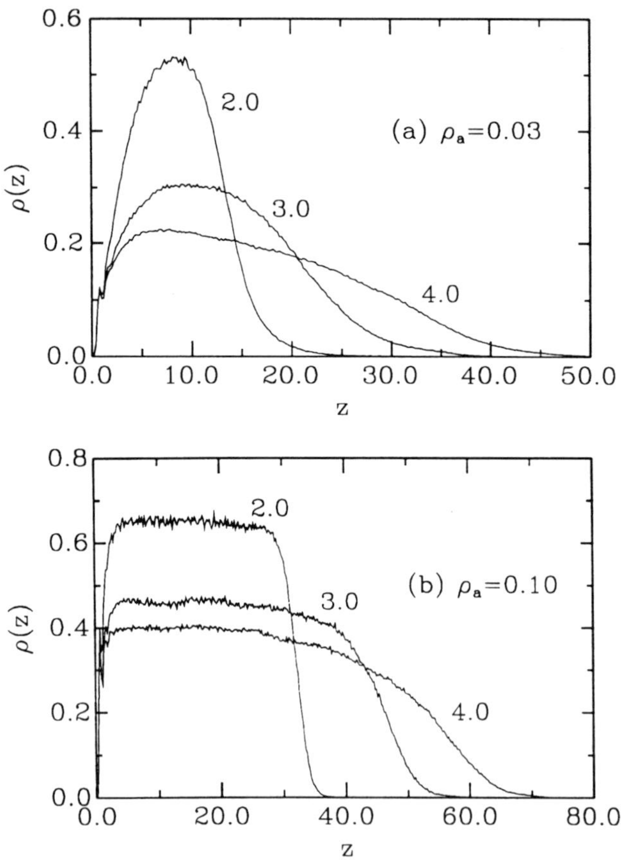

Fig. 9.22 Monomer number density $\rho(z)$ versus z for a chain of length $N = 200$ at $T = 2.0\epsilon/k_B$, $T = T_\theta = 3.0\epsilon/k_B$ and $T = 4.0\epsilon/k_B$ for (a) $\rho_a \sigma^2 = 0.03$ and (b) $\rho_a \sigma^2 = 0.10$. (From Ref. 115.)

$$\rho(z) = C_6 \rho_a^{1/2} \left[1 - C_7 (z/N\rho_a^{1/2})^2\right]^{1/2}. \tag{9.33}$$

Both MC[79] and MD[115] simulations find that the density cannot be fitted by this functional form. The agreement does not seem to get better with increasing chain length. While the scaling suggested by eq. (9.33), namely, that $\rho(z)/\rho_a^{1/2}$ versus $z/N\rho_a^{1/2}$ should exhibit a data collapse, seems to be confirmed in the MC simulations,[79] the plots are for a very limited range of N (between 30 and 50) and of ρ_a($\rho_a\sigma^2$ between 0.075 and 0.125). Even so, the scaling plot shows some systematic deviations. These were magnified in the MD simulations[115] that were run using a larger range of chain lengths (between 50 and 200). Similar deviations are observed in the scaling plots of the structure factor.[115]

The lack of agreement between the simulations and the analytical SCF predictions is also reflected in the density of the free ends. Zhulina *et al.*[282] found that $\rho_e(z)$ grows linearly with z from the grafting surface up to the brush height, at which point it drops sharply to zero. Although a rough linear increase is observed in the simulations[79,115] at short distances from the grafting surface, the drop to zero is found to be very smooth, probably due to the finite length of the chains. The scaling behavior suggested by the SCF prediction seems also not to hold, especially when data from a wide range of N and ρ_a are taken into account.[115]

The reason for the disagreement is presumably the relatively large monomer densities reached within the brush. In principle, for very long chains and very low ρ_a, one may reach a situation in which the monomer density is small even near T_θ. This regime has not been reached in any of the simulations. On the basis of the mapping[115] of the coarse-grain model treated in the simulations to the experimental systems,[244] it is difficult experimentally to satisfy both the conditions of strong chain overlap and dominance of binary interactions, implicit in the analytical SCF study.[282] This is consistent with the numerical SCF calculations of Whitmore and Noolandi.[169]

Below T_θ, simulations[79,115,119] show that the monomer density becomes almost uniform up to the brush height, which grows roughly linearly with $N\rho_a$. The monomer density in this regime becomes quite high, close to the melt density, making the simulations difficult to perform. In MC simulations, the acceptance rate of the MC moves becomes very low, while in MD simulations, chain relaxation slows down considerably. However, for moderate N, equilibration can be reached, but needs to be checked carefully[115] by verifying that identical results are obtained by different paths.

In their experiments on grafted poly(dimethylsiloxane) chains, Auroy and Auvray[271] changed the solvent quality using two different methods: by modifying the chemical composition of a binary mixture, and by changing T around T_θ of a single solvent. In the former method, the brush height is found to change dramatically (by a factor of about three) at a critical composition. This effect is shown to be due to the "preferential solvation" of the good solvent by the polymer chains.[288,289] This phenomenon has not been studied yet by computer simulations, since such a study would require explicit representation of a large number of solvent molecules, demanding considerable computer resources. The method of changing T induced a much more gradual variation of the brush height as a function of T. Perahia *et al.*[273] found, using neutron reflectivity, that h is insensitive to the solvent quality. Their experiments were carried out on PS chains grafted to silicon with a binding energy of $\sim 8k_BT$, in equilibrium with a dilute solution of chains. In this case, the equilibrium grafting density is also dependent upon the solvent quality and is obtained when the chemical potential of the brush equals that of the solution. Equilibrium is determined by a balance between the stretching energy and binding energy.[273] A similar situation is found at constant Π_a simulations.[116] While the experimental situation is different than that of the constant Π_a simulations, it is interesting that both give a brush height which is nearly independent of solvent quality. The agreement suggests that experimentally, the surface osmotic pressure is probably approximately constant as the solvent quality is varied. However, we are not aware of any argument which would suggest that equal chemical potential for the brush and solvent would lead to a constant or nearly constant surface osmotic pressure.

An interesting phenomenon observed in the simulations is the separation of the brush into monomer-rich and monomer-poor regions[79,115] below T_θ for fixed ρ_a, as shown in Figs 9.23(a) and (b). This phenomenon is attributed to the instability of the grafted layer to tangential fluctuations in a sufficiently poor solvent,[285–287] and arises since the overlap concentration ρ_a^* increases as solvent quality decreases. For fixed ρ_a, the brush may be stretched in a good solvent, but be below ρ_a^* in the poor solvent regime. In this case, the brush will try to phase separate. However, as the grafting site is firmly anchored, the best that the system can do is phase separate locally. A similar result was found in a two-dimensional numerical mean-field model[167] in which the monomer density is *a priori* assumed to depend upon a coordinate parallel to the grafting surface as well as on the coordinate z perpendicular to it. As expected, at a given ρ_a, this phase separation disappears as N is increased.[79,115,287] Furthermore, the chain length at which the transition from the "dimpled" layer to a "uniform" state decreases as the chains overlap more. For instance, for $\rho_a\sigma^2 = 0.03$, the layer is inhomogeneous even for $N = 200$. When $\rho_a\sigma^2$ is increased to 0.1, brushes with chains as short as 50 monomers do not exhibit a strong phase separation, as shown in Fig. 9.23(c). Results of Ref. 285 confirm this observation. This lateral structure in a poor solvent has also been observed recently using atomic force microscopy.[290,291] Israels *et al.*[292] have shown that this phase separation can be used to produce a pH-controlled gating by fabricating a surface with regions where grafted chains are absent. A similar type of microscropic phase separation can occur if the free ends of the chain have

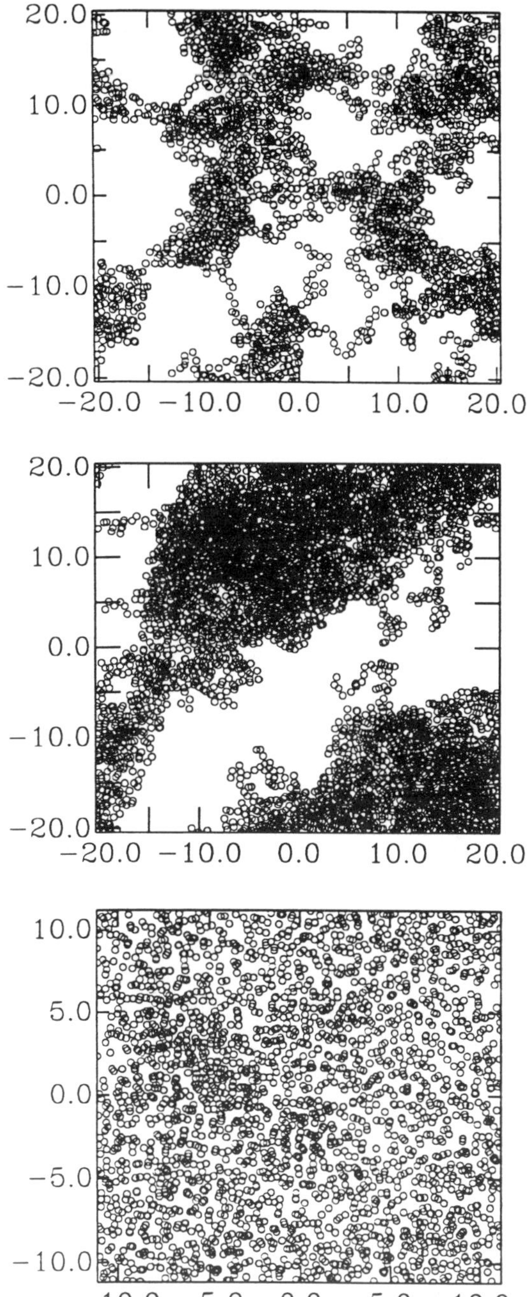

(a)
(b)
(c)

Fig. 9.23 Projection of all monomers onto the $z = 0$ plane to show the phase separation at $T < T_\theta$ for various N and ρ_a. (a, top) $N = 50$ and $\rho_a = 0.03$; (b, middle) $N = 100$ and $\rho_a = 0.03$; (c, bottom) $N = 50$ and $\rho_a = 0.1$. (From Ref. 115.)

an associating group attached.[293,294] One should note that the phase separation should be observed only for brushes in which the chains are chemically adsorbed to the surface, with no free chains in the solution. If the brush is in equilibrium with polymers in solution, the free chains will phase separate to fill the exposed regions. One also does not expect to observe phase separation under constant surface pressure conditions. For fixed Π_a, ρ_a increases continuously as T decreases and the system does not phase separate. This was observed in constant Π_a simulations.[116] While such a uniform and continuous collapse is consistent with a number of theoretical calculations,[276,281–283] it should be noted that these studies did not allow for the possibility that the brush may become unstable to tangential density fluctuations in a poor solvent. Finally, Tang *et al.*[295] have studied the phase behavior of tethered polymers with lateral mobility in poor solvents and mapped out the phase diagram.

As in the good solvent case, the dependence of the osmotic pressure on the surface coverage under Θ solvent conditions, as found in simulations,[116] disagrees with both scaling results[7] and analytical SCF theory.[2] Both these theories predict the osmotic pressure to increase as $\Pi_a \sim \rho_a^2$. The simulation results, shown in Fig. 9.20(b), are best fitted by $\Pi_a \sim N\rho_a^y$, with $y = 3.0 \pm 0.2$. As in good solvents, y is larger by about 1 than the scaling and SCF predictions. It agrees however with the value 3.0 found by Carignano and Szleifer[172] in their statistical thermodynamic model. Reasons for this deviation are identical to those suggested for the good solvent case.

9.5.3 Attractive grafting surfaces

In most of the brush simulations, there is no adsorption of monomers, since the wall–monomer interaction is taken to be short-ranged and purely repulsive. This repulsive interaction leads to a depletion layer few monomers wide near the wall, where the monomer density is lower. As T is lowered, the monomers reduce their free energy by being closer to each other than to the wall, increasing the size of this exclusion zone.[115] In some experiments,[273,296] on the other hand, a slight excess layer of the monomers near the surface is observed. Effects of an attractive wall on the brush structure have been discussed by Alexander.[7] When the purely repulsive wall–monomer potential is replaced with a contact attraction or an attractive well,[38,154] or with an attractive Lennard-Jones potential[115,116] a similar surface excess is found in the simulations as well, provided the magnitude of the attractive interaction is of the order of the monomer–monomer interaction parameter. Qualitatively similar results were obtained in numerical SCF calculations.[154]

The competition between the monomer–wall attraction and the binding energy of the end group may lead to interesting phenomena.[7,297] When the magnitude of the former is much smaller than that of the anchoring energy, but is of order the monomer–monomer interaction, and ρ_a is low, there is enough room at the surface for all the monomers to be next to the wall. This leads to a flat structure, the so-called "pancake". However, as ρ_a increases, the monomers will be pushed away from the grafting surface, and a brush-like layer will be formed. This transition is predicted[7,297] to be a first-order phase one. Chakrabarti[298] studied this pancake-to-brush transition in a MC simulation. He used an interaction parameter $\chi_s = 0.5k_BT$ to represent the difference between the energy of adsorption of a monomer and that of a solvent molecule. Chains with $N = 50$ and 100 at $0.0002 \leq \rho_a\sigma^2 \leq 0.04$ were studied. Figure 9.24 shows the mean height of the layers (the first moment of the density profile) as a function of ρ_a for the two chain lengths. At low ρ_a, the height is independent of N and corresponds to a monolayer. Above a critical ρ_a, a rapid increase of the height is observed. The increase is faster for the longer chains. The same qualitative transition to the brush behavior is observed in the density of the monomers and of the free ends as well. This result is consistent with the experiments of Ou-Yang and Gao.[299] However, the limited range of N does not allow for the determination of the order of the transition.

Other effects of an interacting surface on the brush structure have also been studied using a SCF approach[300,301] and MC simulations.[38] Both treatments find that in the presence of an attractive surface (within the brush regime), the brush height decreases as $h/h_0 = (1 - \lambda)^{1/3}$, where h_0 is the brush height for a neutral surface and λ is a measure of the strength of

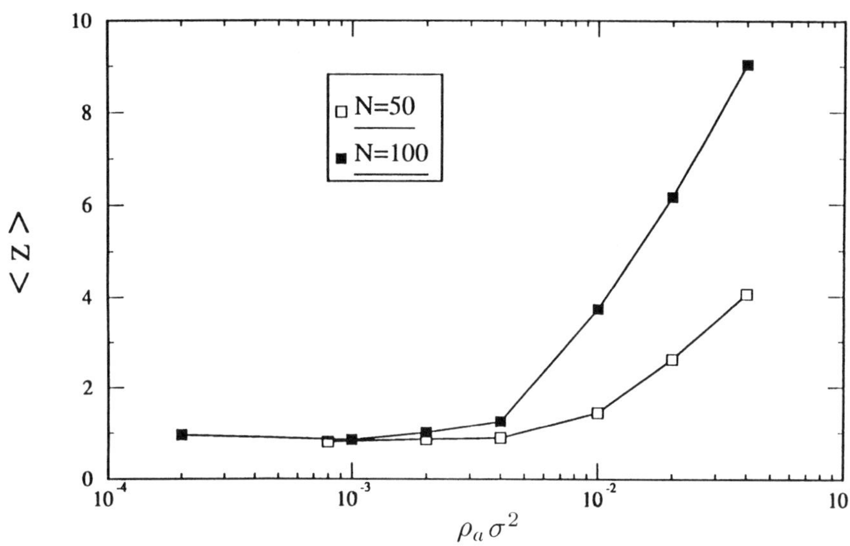

Fig. 9.24 The thickness of the grafted layer with an attractive surface as a function of the grafting density ρ_a for two values of the chain length, $N = 50$ and 100. (From Ref. 298.)

the surface interaction. The SCF predictions[301] for the amount of the surface monomer excess due to the attractive interactions also compares well with the numerical results. This agreement between the simulations and the SCF approach is yet another demonstration of the validity of neglecting chain fluctuations, in these strongly stretched chains. When the monomer–wall interaction is repulsive, with a wide potential barrier, the end density within the repulsive region is found to vanish (dead zone), although the monomer density is still finite,[38] consistent with the predictions of Ref. 301.

9.5.4 *Polydispersity effects*

Until now, we considered brushes in which N is equal for all chains. Some degree of polydispersity is to be expected in most experimental systems. Extensions of the theory show that polydispersity in a brush may have significant implications. For instance, forces between brushes are found to extend farther even with small polydispersity.[302–305] Klushin and Skvortsov[306] showed that the height fluctuations that are of order $\langle \delta h^2 \rangle \sim N^2$ in monodisperse brushes are strongly depressed even for a relatively narrow chain length distribution. Milner *et al.*[303] found that for a uniform distribution of chain lengths, the density profile may change completely and lose its parabolic character. This was verified in the MC simulations of Chakrabarti and Toral,[38] who simulated brushes with a uniform distribution of chain lengths between 17 and 80, and showed quantitative agreement with the predicted density profile of Ref. 303.

The case of a bimodal chain length distribution has received a more quantitative treatment, both in the context of SCF theory[303,307,308] and simulations.[38,309] SCF theory predicts that in a bidisperse brush, the free ends of the shorter chains (of length N_s) are confined to a height h_s from the grafting surface. The free ends of the longer chain (of length N_l), on the other hand, reside between h_s and a larger height h_l. This complete segregation is valid only for very long chains. It is further assumed that both chains are under strong stretching conditions. For an overall grafting density ρ_a,

$$h_s = h_0\left(1 - p^{2/3}\right)^{1/2} \quad \text{and} \quad h_l = h_0\left(1 + np^{1/3}\right), \tag{9.34}$$

where p is the fraction of the longer chains, $h_0 \sim N_s \rho_a^{1/3}$ is the height of a monodisperse brush of the shorter chains at ρ_a and $n = N_l/N_s - 1$ is the relative difference in the chain lengths. h_l is the total brush height. This segregation leads to a bimodal monomer density distribution, which is identical to the distribution of a monodisperse brush with chains of length N_s and ρ_a up to the distance h_s from the grafting surface. Beyond that distance, the density profile has a more complicated form, with a discontinuous derivative at $z = h_s$. Similar functional behavior is exhibited by the distribution of the free ends.

Tests of these predictions were carried out using MC simulations. Chakrabarti and Toral[38] studied the case of a brush with $N_l = 98$ and $N_s = 49$, $p = 05$. They found good agreement with the distribution functions predicted by the SCF analysis,[303,307] both for the monomers and for the free ends. The discontinuity in the slopes of both distributions was smoothed by the finite length of the chains; nevertheless an incipient kink was apparent. Lai and Zhulina[309] undertook a comprehensive comparison with the SCF predictions, using a bond-fluctuation model. They simulated brushes with $20 \leq N_s \leq 40$ and $40 \leq N_l \leq 80$ at various ρ_a, p and n. The scaling of the heights h_s and h_l, as well as the scaling of the first moment of the distribution of the free ends of both types of chains, were found to be consistent with the SCF theory.[303,307] Similar agreement is found for the monomer density distribution, with the "kink" becoming more prominent as the chain lengths get larger. When the different contributions to the total density from both chains are plotted separately, it is found that the profile of the monomers from the short chain is independent of the length of the longer chain, as predicted by the theory. Some deviation from the predicted density profile is seen near the outer edge of the brush, probably due to insufficient stretching of these segments. The effect of finite N has been investigated by Dan and Tirrell[308] using a numerical SCF theory.

Unlike the assumption of Refs 303 and 307 that the free ends of the two species of chains are completely segregated, interpenetration of the free ends of each chain into the domain of the other is clearly observed in the simulations[38,309] and numerical SCF.[308] Figure 9.25 shows the end distribution of short chains with $N_s = 20$ and of long chains with $N_l = 40$ and 60 taken from the simulations of Ref. 309. Also shown is the end distribution of monodisperse chains of length N_s at this grafting density. The interpenetration seems to be asymmetric, with more of the short chains penetrating into the outer layer than vice versa. The penetration of the longer chains into the short chain region decreases with increasing N_l, indicating that this is a finite N effect. The free end distribution of the short chains is also found to be independent of N_l, similar to the monomer density. Other properties, like the average position of the ith monomer, and the orientation vector of the ith bond also agree reasonably well with the SCF analysis, especially for the larger N_l. The discrepancies found originate from the interpenetration of the chain ends, a finite size phenomenon not accounted for in the analytic SCF theory.

9.5.5 Interaction between brushes

Two brushes repel each other as they are brought into contact. This repulsion, which is a result of the osmotic interaction between the polymer segments, is the basis for colloidal stabilization.[5,197,310] It has also been utilized to probe the brush structure.[244–249,251] Most of these studies have been on

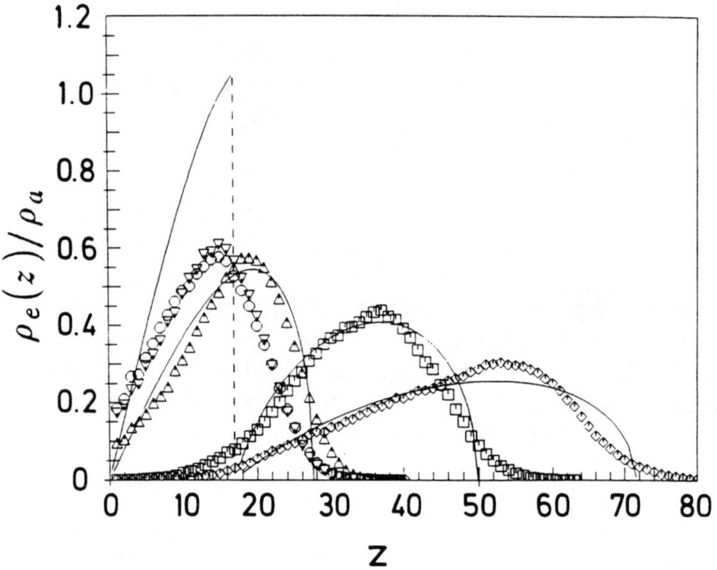

Fig. 9.25 Free end distribution for brushes with $N_s = 20$ and $\rho_a = 0.1$. The brush contains an equal number of long and short chains. (○)$N_l = 40$, distribution for short chains; (□)$N_l = 40$, distribution for long chains; (△) monodisperse brush; (◇)$N_l = 60$, distribution for short chains; (▽)$N_l = 60$, distribution for long chains. (From Ref. 309.)

symmetric brushes, in which both surfaces have the same surface coverage and molecular weight. Recently Watanabe and Tirrell[311] have studied asymmetric brushes, in which the two surfaces have equal ρ_a but different N or different ρ_a and different N but about the same weight per unit area of polymer. They also studied the case where there is chemical asymmetry of the grafted polymers.

Simulations in which two brushes are pushed towards each other give detailed information about these interactions, including the force–distance profile, changes in the monomer density as a result of compression and the amount of interpenetration of the two brushes. This last quantity is usually ignored in theoretical treatments, as it is expected to vanish for infinitely long chains.[312] In experimentally relevant systems,[115] however, the amount of interpenetration will be finite due to the finite length of the polymer chains. Interpenetration affects the interaction between grafted surfaces, as well as the rheological behavior of a collection of such surfaces,[245] and can be calculated directly in a computer simulation.

Simulations of the interaction between two parallel brushes in a good solvent have been carried out using MD[99] and MC[42,74,81,93,313] methods. The MD study of Murat and Grest[99] concentrated on the interpenetration of two symmetric brushes as well as on the force between them as a function of their separation. The force profile was compared to experimental results obtained for terminally attached polystyrene chains. van Lent *et al.*[156] and Wijmans *et al.*[160] applied numerical SCF theory to study the effect of free polymer on the structure and interaction between brushes. In the former, the interest was in determining the effect of the free polymer chains on colloid stabilization. Chakrabarti *et al.*[313] studied interpenetration in both symmetric and asymmetric brushes, comparing the results in detail to numerical SCF study of Shim and Cates.[314] Dan and Tirrell[308] found that the interpenetration of bimodal brushes and equivalent length monodispersed brushes was very similar. MC studies of the interaction between two brushes were carried out by Dickman and Hong[42] and Dickman and Anderson[81] using a novel technique to measure the force. Carignano and Szleifer[173] determined the force between plates using their statistical thermodynamic model for both linear chains and 3-arm stars grafted to the surface at one end.

As two brushes are brought into contact, two processes are found to occur concurrently: interpenetration and compression. When the distance D between the brushes equals twice the maximum extent h_{ext} of each, the parabolic profiles of both brushes begin to overlap and the density increases everywhere between the two grafting surfaces. At small separations the density becomes almost uniform in the gap between the grafting surfaces. However, for low to moderate compressions, the density profile of each brush remains roughly parabolic. This behavior is observed in all the simulations mentioned here, as well as in the numerical SCF calculations of Muthukumar and Ho.[168] The amount of interpenetration at a given separation decreases as the N increases. One can quantify the amount of interpenetration in several ways. The quantity $I(D)$, where

$$I(D) = \int_{D/2}^{D} \rho_1(z)dz \Big/ \int_0^D \rho_1(z)dz. \tag{9.35}$$

was suggested in Ref. 99. Here $\rho_1(z)$ is the contribution of one of the brushes to the overall density and z is the distance measured from the grafting surface of that brush. Results of the SCF theory of Milner *et al.*[266] together with a scaling argument by Witten, Leibler and Pincus[315] can be used to show[99] that $I(D)$ has the following scaling form with N and ρ_a,

$$I(D)N^{-2/3}h_{\text{par}}^{4/3} \sim x^{-4/3}(1 - x^3), \tag{9.36}$$

where $x = D/2h_{\text{par}}$. Here h_{par} is the brush height derived from the parabolic density profile predicted by the SCF theory. In addition to giving the interpenetration as a function of the separation for a given brush, this expression also shows that the $I(D)$ at a given separation decreases as $N^{-2/3}\rho_a^{-4/9}$. Since the decay with N is rather slow, the amount of interpenetration is expected to be significant for experimental systems. Figure 9.26 shows this functional behavior, together with the MD data from different systems with $N = 50$

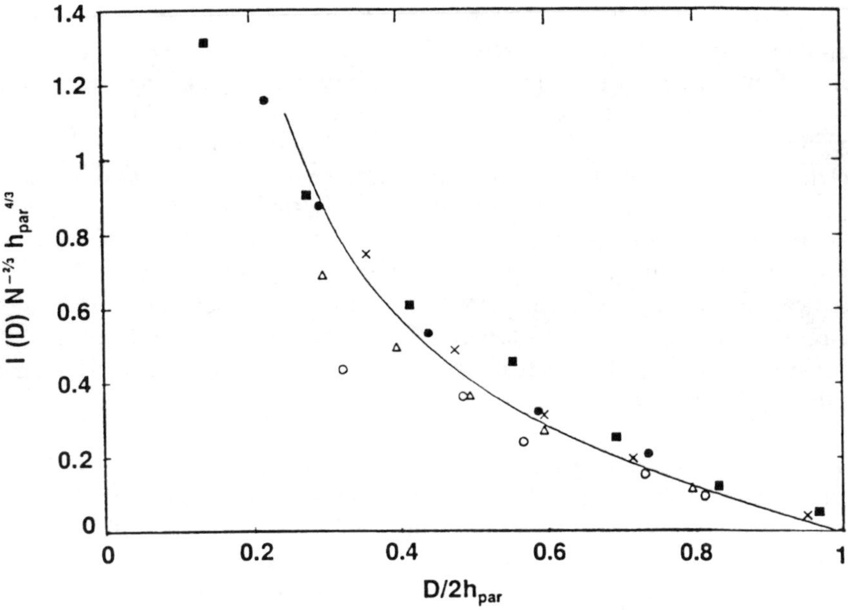

Fig. 9.26 Interpenetration $I(D)$ scaled as in eq. (9.36) for several cases. $N = 100$, $\rho_a = 0.03$ (•); $N = 50, \rho_a = 0.03$ (■); $N = 50, \rho_a = 0.05$ (×); $N = 50, \rho_a = 0.1$ (△); $N = 50$, $\rho_a = 0.2$ (○). (From Ref. 99.)

and 100 and $0.03 \leq \rho_a\sigma^2 \leq 0.20$.. The data are found to collapse reasonably well for small compressions. Even though $N = 50$ is barely in the scaling limit (see Fig. 9.18[a]), the results for $N = 50$ behave similarly to those for $N = 100$, which has a wider range of scaling behavior. The data for $\rho_a\sigma^2 = 0.2\,(> \rho_{a1})$ deviate from this curve at lower values of compression, than those for $\rho_a\sigma^2 = 0.1$. This can be qualitatively understood by noting that when the overall monomer density approaches unity, the density profile is almost uniform and further decrease of D causes a uniform compression rather than an increase in the interpenetration. As such high densities are achieved at smaller compressions the higher ρ_a, $I(D)$ curves for those cases saturate first.

Following Shim and Cates,[314] Chakrabarti *et al.*[313] used a different measure for the amount of interpenetration,

$$\beta(D) = \frac{D}{\theta_1\theta_2}\sum_{z=0}^{D}\rho_1(z)\rho_2(z). \tag{9.37}$$

Here θ_i ($i = 1, 2$ for the two brushes) is the sum over $\rho_i(z)$ over all layers z. With this definition, total interpenetration corresponds to $\beta(D) = 1$. Results from both lattice and off-lattice MC simulations were compared with the numerical SCF calculation[314] for identical N and ρ_a. Simulations for asymmetric brushes (cases for which the two brushes have different length chains and/or surface coverages) were also included. The agreement is found to be better for high compressions. It should be noted however, that these simulations dealt with quite short chains ($N = 50$ for off-lattice and $N \leq 40$ for lattice), and therefore the amount of interpenetration is significantly higher than in the other simulations mentioned for comparable compressions. For instance, the density of the monomers of one of the brushes at the grafting surface of the other is finite for $D \simeq h_{\text{ext}}$, while it decays to zero long before reaching the other surface for $N = 100$ (Ref. 104) and for $N = 80$ (Ref. 42) at the same compression.

The force between grafted polymers in a good solvent is repulsive at all separations. There are also no hysteresis effects, indicating fast equilibration and lack of entanglement. This was observed both in experiments[244–249,251] and simulations.[42,81,99] In MD simulations, forces between the brushes can be calculated in a straightforward manner. As the forces between the monomers are known at each time step, one can calculate the osmotic pressure from the virial.[25] The force per unit area $f(D)$ between the two plates is then found by subtracting the pressure at $D = 2h_{\text{ext}}$. The force between the brushes is found to relax very rapidly, in contrast to other equilibrium properties, which take much longer to relax. Some experiments measure the interaction energy $E(D)$,[248]

$$E(D) = \int_{2h_{\text{ext}}}^{D} f(D')dD'. \tag{9.38}$$

rather than the force per unit area. Both scaling arguments[312,316] and SCF theory[266] predict that this energy should scale as $E(D) \sim h\rho_a^x \tilde{E}(D/2h)$, with $x = 3/2$ and $4/3$ respectively. The difference originates from the dependence of the osmotic pressure on the monomer density in the two theories. These predictions are obtained assuming that there is no interpenetration, so that bringing the two brushes into contact results only in their compression. The repulsive force is then calculated from the increase in the free energy due to the compression. The two theories also predict different scaling functions $\tilde{E}(x)$. Figure 9.27 shows the MD simulation results for $E(D)$ scaled as suggested by the scaling theory. The energies for $N = 50$, $\rho_a\sigma^2 = 0.01$ and the two different values of N at $\rho_a\sigma^2 = 0.03$ collapse to a single curve. The other cases converge to this curve only for $D/2h_{\text{ext}} \to 1$, although all the curves have the same basic form. The inset in Fig. 9.27 shows the results of Taunton *et al.*[244] from experiments with terminally attached polystyrene chains of various M_W. The solid line is a smooth curve passing through the simulation data points for $\rho_a\sigma^2 = 0.03$. Since the energy scale in simulations is arbitrary, this line is vertically shifted to lie on the experimental points for comparison. The overall agreement between the simulation and

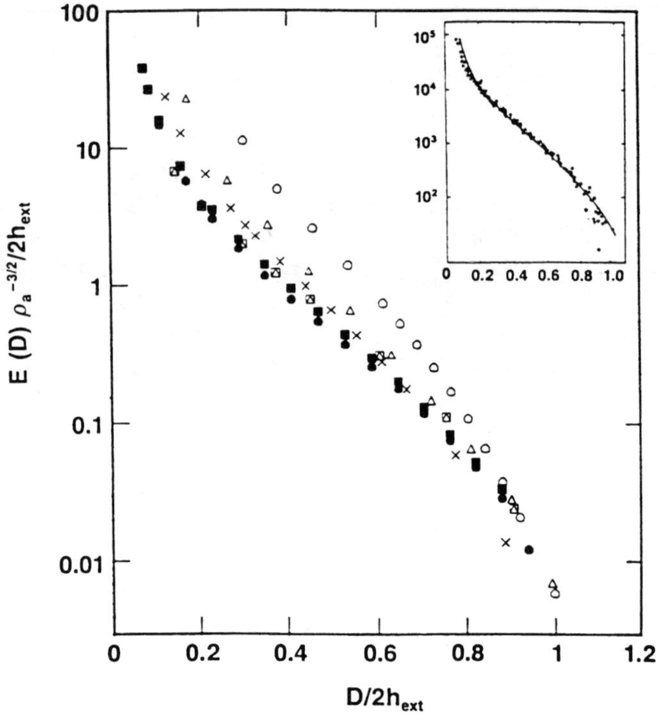

Fig. 9.27 Energy per unit area of interaction between plates. The symbols are identical to those in Fig. 9.26. Additional results for $N = 50$ and $\rho_a = 0.01$ are shown as triangle inside square. The inset shows results of Taunton *et al.* [244] (data points) and the MD simulation results[99] (full line) for $N = 100$, $\rho_a = 0.03$. The simulation results are vertically shifted by an arbitrary amount. (From Ref. 99.)

the experimental results is excellent. (This fit between the MD simulations and experiments at intermediate values of surface coverage has been utilized[115] to map the simulation to real physical systems.) We emphasize that the fit is achieved using no adjustable parameters, except the vertical shift due to the arbitrary energy scale of our simulations. The length scale used in the scaling of D is taken from simulations of a single brush and not from those of interacting brushes.

Although the data collapse shown in Fig. 9.27 is with the functional form predicted by the scaling analysis, a comparable collapse of the data is obtained using the prediction of the SCF theory. This is also the case with experiments.[244,317] Therefore it is difficult to confirm either one of the two theories using force measurements only. Only by probing the inner structure of the brushes can one decide which of the two theories describes the situation better. This has been done by both experiments and simulations, as explained earlier in this chapter.

We expect the agreement between the simulations and the theoretical predictions to break down at high compressions when ϕ, the average density of the monomers between the surfaces, approaches unity. At this limit, the osmotic pressure diverges logarithmically with $1 - \phi$, rather than increase as a power of ϕ, as assumed by both theories. In Fig. 9.27 we indeed see this deviation for $\rho_a\sigma^2 = 0.10$ and 0.20 at $D/2h_{\text{ext}} \approx 0.6$ and 0.8 respectively, since this dense regime is already reached at these compressions. Such a trend, however, is not seen in the experiments,[244–246,249] which are at relatively low ρ_a or in simulations with a lower ρ_a. Because experiments were done for a system of only a few k_BT binding energy, ρ_a decreased with increasing N, as noted in subsection 9.5.1, and consequently high values of ρ_a were not achieved.

Calculation of the inter-brush force in lattice simulations is not as straightforward as in off-lattice simulations where the forces are easy to measure. On a lattice, Dickman and Hong[42] devised a statistical mechanical method for calculating the force in terms of the variation of the configurational probabilities of chains when the separation of the grafting surfaces is changed from D to $D - 1$. The method was applied to two-dimensional,[42] and more recently, to three-dimensional[81] bond fluctuation simulations of brushes. The resulting force profiles were found to be consistent with the SCF predictions,[266] except at large separations (small compressions), due to the tail in the density profile of finite length chains. No quantitative comparisons with experiment were carried out.

With decreasing separation of the brushes, the distribution of the free ends of the brushes also undergoes qualitative changes. The distribution gets wider, with significant penetration into the range of the opposite brush. Interestingly enough, the free ends migrate preferentially towards the depletion zone near the grafting surfaces, where the density of the monomers is low. The distribution function of the free ends of *each* brush increases near *both* grafting surfaces; that is, the free end of a chain preferentially visits both grafting surfaces. This is observed in both MD[99] and MC[42,81,313] simulation.

9.5.6 *Brushes on curved surfaces*

In many applications, including colloid stabilization,[5,197,310] polymers are end-grafted onto large particles which are intermediate between the planar brush and a star. Curved brushes, with either spherical or cylindrical symmetry, can also be formed by asymmetric diblock copolymers in the strong segregation limit.[258] When the grafting surface curves towards the polymers (concave curvature), the volume available to the polymer chains is lowered compared to a flat surface. In this case, the chains are stretched throughout the brush and the brush is qualitatively unaltered from the flat case.[237,269] MC simulations on a tetrahedral lattice for chains inside a spherical cavity have been carreid out by Limpouchová and Procházka.[318] Here we will be concerned only with convex curvature for which the available volume grows with the distance from the grafting surface, and the chains become less stretched. The structure of the brush is then altered considerably. The limiting cases of vanishing radius of curvature for spherical and cylindrical grafting surfaces are a star polymer and chains tethered on a rigid line, respectively. These were already discussed in previous sections. Here we discuss the intermediate case corresponding to the crossover from these highly curved surfaces to a flat one.

Simulations of curved brushes have been carried out by Murat and Grest[104] for cylindrical surfaces using MD and by Toral and Chakrabarti[94] for spherical surfaces using off-lattice MC (pearl necklace). Both studies treated cases in which the radius of curvature R_c is of the order h. The monomer density profile for three cases is shown in Fig. 9.28. The density is given as a function of the distance from the grafting surface. For the case of the largest radius of curvature ($R_c = 20\sigma$), the density profile is practically indistinguishable from the parabolic profile obtained for a flat brush. As the radius decreases, the density profile changes its shape, going over to one with a steeper increase near the surface. The maximum density that is attained somewhere close to the surface decreases as well. An identical behavior is seen in the numerical SCF study of Dan and Tirrell[170] and Wijmans and Zhulina[159] except that the layering near the surface observed in Fig. 9.28, does not appear in the SCF study. For spherical brushes, a similar transition from a concave monomer density profile (for $R_c = 5\sigma$) to a qualitatively parabolic one (for $R_c = 11.18\sigma$) is observed for $N = 50$ chains, as shown in Fig. 9.29, where the MC results of Toral and Chakrabarti[94] and the statistical thermodynamic results of Carignano and Szleifer[174] are shown. Note the excellent agreement between the two.

SCF treatment for a flat brush had shown that the free ends of the brushes are not excluded from any region in the brush; there is a finite probability of finding them anywhere within the brush. For stars, on the other hand, the free ends of the chains are pushed outwards, due to the fast decay of the monomer density at the outer regions. This leads to a "dead zone" for the free ends, whose size is a finite fraction of the star's radius.[97] The question naturally arises whether such a "dead zone" exists for brushes with finite radii of curvature. Ball *et al.*[237] predicted the existence of a dead zone in cylindrical brushes in a melt while Wijmans and Zhulina[159] predicted similiar behavior for chains immersed in a low molecular weight solvent. In a melt, a SCF analysis[237] found that in the limit of very high-molecular-weight chains grafted at a small ρ_a, there is a dead zone whose thickness grows from zero for flat brushes to a fraction $2/\pi$ of the brush height for strongly curved cylinders. For intermediate curvatures, the size of the dead zone is not negligible; when the cylinder radius equals the brush height, the thickness of this zone is about one-third of the layer thickness. Li

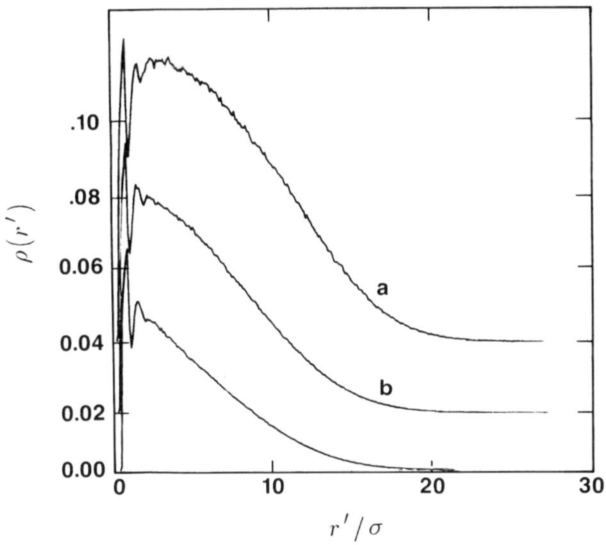

Fig. 9.28 Monomer density profile $\rho(r')$ of brushes with $N = 50$ and $\rho_a = 0.03$, grafted onto a cylinder of radii (a) $R_c = 20$, (b) $R_c = 5$ and (c) $R_c = 2$. Each curve is shifted vertically with respect to the previous one by 0.02 for clarity. r' is the distance from the surface of the grafting cylinder. (From Ref. 104.)

and Witten[221] applied a variational approach to the problem of polymers grafted on convex surfaces and extended the treatment of Ball *et al.*[237] to strongly curved brushes in a solvent. Their work confirmed the existence of a finite dead zone for brushes in a good solvent, although the addition of the solvent was found to decrease the size of the dead zone.

The free end density as a function of the distance from the grafting surface shows no indication of a dead zone for various cylindrical[104] and spherical[94] systems in which the ratio h/R_c is between 0.4 and 3.5. Indications of a dead zone appear in the simulations[94,104] only when the radius of the grafting surface is very small compared to the brush height. Thus there seems to be a discrepancy between the simulations and the analytical SCF analysis for intermediate values of the radius of curvature. Numerical SCF calculations of Dan and Tirrell[170] and Wijmans and Zhulina[159] with comparable values of N and ρ_a also showed no immediately apparent exclusion zone, except for the very high N, high curvature limit. (See Section 9.4 for a discussion of the dead zone at this extreme limit.) As noted by Toral and Chakrabarti,[94] it may be difficult to observe an exclusion zone in simulations with finite radii of curvature, as the width of this zone may be comparable to the length scale over which finite chain length effects in the density are important. In order to observe such a zone, it may be necessary to use chains of length much larger than the ones used ($N \leq 100$) in the present simulations.

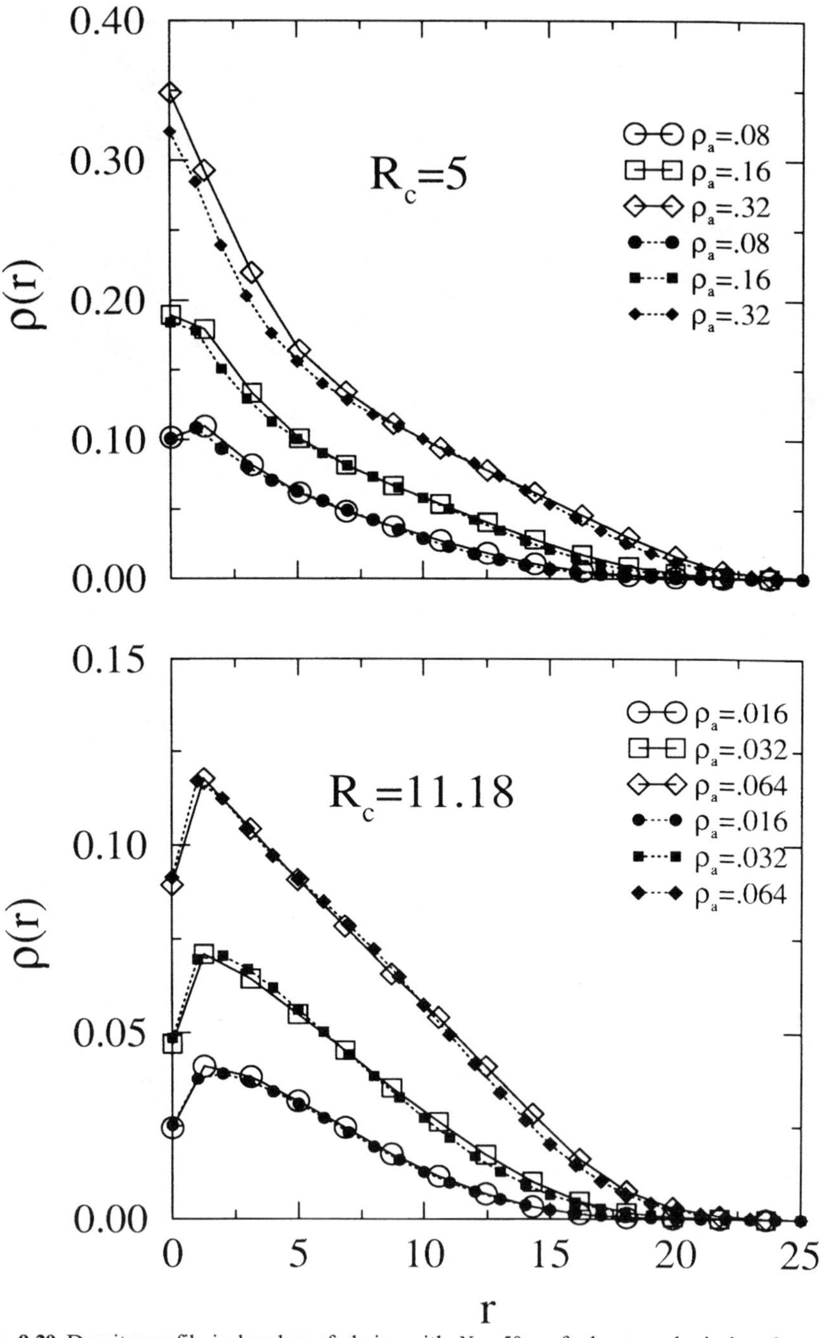

Fig. 9.29 Density profile in brushes of chains with $N = 50$ grafted onto spherical surfaces of radius $R_c = 5$ and 11.18 at various grafting densities ρ_a. The closed symbols are from the MC simulations of Toral and Chakrabarti[94] and the open symbols from the statistical thermodynamic theory of Carignano and Szleifer.[174] (From Ref. 174.)

9.5.7 Brushes without a solvent

All the preceding discussion in this section was related to brushes in which the monomer–monomer interactions are mediated by a low molecular weight solvent. This type of a grafted layer which contains a high concentration of the solvent, is referred to as a "wet brush". When grafted chains are in contact with a melt of shorter, chemically identical chains of length $P < N$, the free chains screen the excluded volume interactions and reduce the overlap coverage to $\rho_a^* \simeq N^{-6/5}P^{-2/5}$.[8,213,319,320] For $\rho_a \simeq P^{-2}$, the P chains are almost completely expelled from the brush and $h \sim N\rho_a$. This case is commonly referred to as a "dry brush". In the intermediate regime, $h \sim NP^{-1/3}\rho_a^{1/3}$. The driving force for the stretching of the chains is then the incompressibility of the chains. When there are no free chains within the brush, the density is nearly uniform, independent of the distance from the grafting plane. Compared to the wet brushes, there have been fewer theoretical[8,46,47,160,166,213,269,319–321] and experimental[322–324] studies of dry brushes. There have been no brush simulations in which the solvent molecules are explicitly taken into account, even for $P = 1$. The only simulations of dry brushes use a repulsive wall at height $h \sim N\rho_a$ to achieve melt densities in the brush monomers.

Simulations of dry brushes are also rather difficult to equilibrate, due to the high monomer densities within the brush, which give rise to very long equilibration times. Although MD has been applied to a melt of chains in the bulk,[28] there have been no MD simulations of brushes in the melt regime, with or without solvent chains. In standard MC algorithms, the acceptance rate of the attempted moves becomes extremely low in this regime. However, the method of cooperative rearrangements,[43,44] in which kinks and chain-ends are moved along the chain, is relatively efficient in this regime. Pakula and Zhulina[46] have used this method to study a dry brush in contact with a repulsive wall. They also studied the case where the other wall can adsorb the free end of the chain.[47]

In a system of dry layers formed by chains of length N at a sufficiently high grafting density, the chains will stretch considerably with respect to their unperturbed dimension. Much larger N are needed in order to stretch a dry brush than a wet brush, due to the overall large monomer density. The stretching is not homogeneous, with the segments near the grafted wall more stretched than the ones in the outer regions. Following Semenov's SCF analysis for block copolymer melts,[269] Pakula and Zhulina[46] used a mean-field free energy functional in terms of the local stretching and of the distribution of the free ends. Unlike the case of good solvents, their functional contains only the elastic term, since the monomer–monomer interactions are completely screened out. Minimizing this functional, with the constraints that the density is uniform everywhere in the brush and the chain length is conserved, they obtained

$$\rho_e(z) = \frac{z}{h}\left[h^2 - z^2\right]^{-1/2} \tag{9.39}$$

for the free end distribution and

$$E(z, z') = \frac{\pi}{2N}\left[z'^2 - z^2\right] \tag{9.40}$$

for the stretching dz/dn at point z of a chain whose free end is at z'. Using these quantities, they were able to calculate various characteristics of the melt brushes, such as the mean and the mean-squared height of the free ends, the orientational order parameter, and the spatial distribution of the centers of mass of the chains. These were then compared with the results of simulations.

Simulations[46] of brushes with $40 \le N \le 160$ and $0.25 \le \rho_a \le 0.5$ showed satisfactorily good agreement with the predictions. For these systems, the ratio of the brush height to unperturbed height of the chains is between 1.6 to 6.3, so that the brushes are moderately stretched. For comparison, nongrafted chains between two neutral walls with a separation of h were also simulated. The stretching of the chains is accompanied by their orientation. The order parameter, defined as $\eta = (3\langle \cos^2\theta\rangle - 1)/2$, where θ is the angle between the end-to-end vector of a chain and the normal to the grafting surface, is shown in Fig. 9.30 for both grafted and ungrafted chains. Chains whose centers of mass are close to the grafting wall align parallel to it. The same effect is seen for nongrafted chains as well. Far away from the walls, the nongrafted chains are randomly oriented, while grafted chains with the free ends far away from the wall become more oriented. Local orientation of

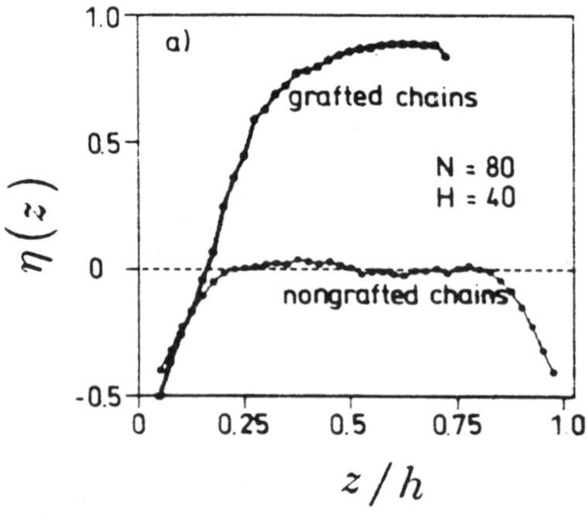

Fig. 9.30 Order parameter of the end-to-end vector of grafted and nongrafted chains in the melt, as a function of the position of the center of mass of the chain. The chains are of length $N = 80$. Nongrafted chains are confined between two neutral walls at a distance of 40 lattice units from each other. (From Ref. 46.)

the chain segments is also found to be consistent with the predictions based upon the mean-field analysis. Distributions of the free ends and of the centers of mass show increasingly better agreement with the analytical results as N increases.

Another system at melt density that was recently simulated is that of a brush in which the free ends are weakly adsorbing. The functional end-group interacts with a wall at a distance $H > h$ from the grafting wall, with h being the height of the brush without functional end-groups. Adsorption of the free ends onto the wall (with an adsorption energy ϵ of order $k_B T$ or larger) may lead to a "bridging" effect where a fraction p of the the chains are connected to both walls. This bridging effect has also been studied by Johner and Joanny[325] for brushes in a good solvent, using an extension of the SCF method[266] developed for brushes without the interacting free ends. They found that p remains small except for unrealistically large adsorption energies, and that there exists a "dead zone" near the adsorbing wall from which free ends are excluded. The existence of the dead zone leads to a very slow bridging kinetics.[325] Zhulina and Pakula[47] treated a similar problem under melt conditions, using both SCF analysis and MC simulations with the cooperative rearrangement algorithm.[43,44] They found that bridging is governed not by ϵ alone but rather by ϵ/μ_0, where $\mu_0 \simeq N\rho_a^2$ is the chemical potential of the brush with no interacting free ends. Therefore, increasing N at constant ϵ decreases the fraction of chains attached to both walls, in agreement with the results of Johner and Joanny[325] for brushes in a good solvent. Noticeable bridging is expected to take place only for adsorption energies $\epsilon \sim N$. In this case, the fraction p of the bridging chains is found to increase sharply with ϵ/μ_0, with a possible phase transition at zero ϵ/μ_0. A dead zone for the unattached free ends is observed in the melt case as well. The thickness of the free zone increases with increasing p (or equivalently, with increasing ϵ/μ_0). The simulations reproduced the mean-field results quite well, with the agreement becoming better with increasing N.

9.5.8 Time-dependent phenomena

Most of the simulation work on brushes has concentrated on structural properties. Much less work has been done in studying time-dependent phenomena. Following Halperin, Tirrell and Lodge,[1] it is convenient to divide the time-dependent phenomena into *kinetic*, which includes the assembly and dissolution of the chains in the brush, and *dynamic*, which describes the shape fluctuations of the chains, and their response to external perturbations.

As already mentioned, the binding energy of the grafting end group is very often several $k_B T$ and the chains are in dynamic equilibrium with free chains in the solution. In equilibrium, ρ_a is determined from the balance between the elastic energy and the binding energy. Lai [80] was the first to attempt such a simulation, though only for short chains, $N \leq 20$. Even for these short chains, the equilibration time turned out to be of order 10^7 to 10^8 MC steps/monomer. More recently, Zajac and Chakrabarti[326] have studied the kinetics of brush formation for chains up to 200 monomers. Both of these simulations showed that when one starts from a situation in which the grafting surface is bare, and all the chains are in the solution, the initial rate of adsorption is fast and the kinetics is governed by diffusion. Later on in the adsorption process, the chains have to overcome the barrier set by the chains already adsorbed. The equilibrium coverage attained at very long times increases with the binding energy and decreases with N. A quantitative comparison with a theoretical[327,328] treatment of this process was not possible in Lai's work since the theory assumes a reservoir of chains in the solution, while his simulation was performed with a fixed number of chains. In contrast, Zajac and Chakrabarti did their simulations in a situation where the brush was in contact with a bulk reservoir. They found that even though the density profiles scale, they were not parabolic due to the presence of free chains which penetrated into the brush layer to some extent. For the chain lengths and binding energies used, they found that their chains were not strongly stretched. In both simulations, when the initial configuration is such that all the chains in the system are artificially grafted to the surface, the same final grafting density and monomer density profile are obtained, demonstrating that equilibrium is established in both cases. A related problem is to understand the kinetics of expulsion of a single chain detached from the surface of a brush.[329,330] This gives some important insight into the full kinetics problem, since expulsion of a chain is one aspect of the interchange of chains in solution with those on the surface. Wittmer *et al.*[330] find that a chain cut off from the wall is expelled at a constant center of mass velocity. This velocity decreases as the inverse of the chain length and increases with grafting density ρ_a. In the early stages of the expulsion, the tension of the monomers close to the wall relaxes and the chain retracts. The rinsing of the brush by a pure solvent has also been studied by simulation.[326,330]

The simulations of Lai[80] and Zajac and Chakrabarti[326] also provide valuable insight into the phenomenon of the replacement of the long chains by the shorter ones, as observed in some recent experiments,[331–333] and predicted theoretically.[334,335] When a solution of short chains are added to a brush of long chains, the shorter chains replace the longer ones, reaching equilibrium coverages that depend upon the length of the two chains and the relative binding energies. The driving force for this replacement is the larger elastic energy of the longer chains. When the binding energy of the end-group to the surface is the same for both the long and short chains, the higher elastic energy of the longer chains leads to their detachment from the surface. The longer chains are also found to be highly stretched in the inner layer of the brush (which contains mostly the short chains) and less stretched at the outer layer, resembling the structure observed in polydispersed brushes with fixed grafting density.[309] In the limit of large end adsorption energy, all but the shortest species are excluded from the brush, irrespective of the solution composition and the width of the molecular weight distribution.[335] The simulations of Zajac and Chakrabarti[326] are the first to study this exchange in a simulation.

The *dynamics* of brushes is easier to deal with in the simulations than the *kinetics*. In fact, even while calculating the equilibrium properties of the brush, one routinely calculates the equilibration times to make sure that the simulations are carried out on time scales much longer than the equilibration times. A convenient way to calculate the relaxation time is to evaluate the autocorrelation function $C_x(t)$, defined in eq. (9.21). Here x is any structural quantity, such as R_G, R or their z components, R_{Gz} and R_z. The MD simulations of brushes in a good solvent[98] showed that the relaxation times derived from the decay of these autocorrelation functions increases with N as $\tau_x \sim N^a \rho_a^b$, with a and b in the range 2.4–3.1 and 0.8–1.1 respectively. Bond fluctuation MC simulations[78] gave a similar result, with $a = 3.0 \pm 0.1$ and $b = 0.83 \pm 0.08$. A scaling argument,[98] based on the idea that the monomers within the blobs obey Rouse dynamics[23] and a chain relaxes when the fluctuations within the blobs diffuse through the whole chain, gave $\tau_x \sim N^2 \rho_a^{-1/6}$. This expression, which is similar to the one obtained for the relaxation of a chain restricted to move in a tube,[336] is in contradiction with the simulation results. This argument had also been suggested for the relaxation time of the individual arms of stars[97] (see Section 9.3), where it was found to be consistent with the simulation results. The difference between the two cases is that in a star the largest (outermost) blobs dominate the relaxation process, whereas in the brush, all the blobs contribute equally to the fluctuations. Klushin and Skvortsov[337] proposed instead that τ_x is determined by the fluctuations of the end-to-end distance of the chains. These fluctuations have a mean-squared average magnitude $\langle \delta h^2 \rangle \sim h^2 \sim N^2 \rho_a^{2/3}$. Excluding hydrodynamic interactions (as is the case in simulations), the diffusion constant D of a chain of N monomers is proportional to N^{-1}.[279] This gives $\tau_x \sim \langle \delta h^2 \rangle / D \sim N^3 \rho_a^{2/3}$ for the relaxation times, in reasonable agreement with the simulations, although the exponent b is slightly smaller than the values observed in the simulations. Lai and Binder[78] suggested that the MC results should be corrected by the dependence of the monomer jump rate on ρ_a. This correction changes the value of b from 0.83 to about 0.6, which is slightly smaller than the prediction of Klushin and Skvortsov. It should be noted that the N^3 dependence is not related to the reptation or the entanglement of the chains; it follows from the fact that a chain-end has to traverse a distance of order N before the chain completely relaxes.

Johner and Joanny[325] and Wittmer *et al.*[330] used the Zimm model in their work based upon a SCF theory which includes the entropy of chain-end distribution, and found a relaxation time of order $\tau_x \sim h^3 \sim N^3\rho_a$. Although this is consistent with the simulation results, a direct comparison cannot be made, as this theory takes into account the hydrodynamic interactions within the correlation blobs, which is not included in the simulations. The collective relaxation dynamics of a swollen brush in a good solvent has recently been studied by Farago *et al.* [338] by neutron spin-echo spectroscopy. They observe a multi-decay time relaxation which can be described by treating the brush as a semidilute polymer solution with a varying concentration profile. Another method of investigating the dynamics is by nmr. Blum and coworkers have used this technique to study the effect of solvent quality on the dynamics of chain monomers.[339]

At the Θ point, the chain relaxation is found to be slower. Lai and Binder[79] found $\tau_x \sim N^3\rho_a^{1.6}$. Applying the Klushin and Skvortsov argument[337] with $h \sim N\rho_a^{1/2}$ in a Θ solvent, one finds $\tau_x \sim N^3\rho_a$. Therefore, the dependence of the relaxation times on ρ_a is stronger in the simulations than the scaling arguments of Klushin and Skvortsov. Even with the correction[79] for the dependence of the jump rate on ρ_a, the exponent of ρ_a is still expected to be at least 1.2. Interestingly enough, the Zimm model[325,330] gives an exponent of 3/2 for the grafting density, very close to the simulation results.

Unlike the autocorrelation functions that are related to the overall relaxation of the whole chain, the mean-squared displacements

$$g_i^z(t) = \langle (z_i(t) - z_i(0))^2 \rangle, \tag{9.41}$$

and

$$g_i^\perp(t) = \langle (x_i(t) - x_i(0))^2 \rangle + \langle (y_i(t) - y_i(0))^2 \rangle \tag{9.42}$$

give information on the local dynamics of the monomer *i*. Lai and Binder studied the behavior of the displacements for both good[78] and Θ-solvents.[79] In a good solvent and for intermediate times, $g_i^z(t) \sim t^{1/2}$, typical of Rouse dynamics, except for monomers very close to the grafting site. At times of order τ_x, this displacement saturates. For the monomers at the free end of the chains, the saturation value of g^z is of order h^2. Monomers close to the grafting site have almost no time regime which satisfies Rouse-like dynamics, due to the restricting effect of grafting. Motion of the chains in the *xy*-directions exhibits a behavior similar to the *z* motion. However, for chains that have grafting groups that are mobile in the plane, $g_i^\perp(t)$ crosses over to free diffusion ($g^\perp(t) \sim t$) at the same time that $g_i^z(t)$ saturates. One can then in principle calculate a diffusion constant for the lateral motion by extrapolating $g_i^\perp(t)/4t$ to infinite times. Reliable values of the diffusion constant cannot be derived from existing data, due to the very limited

$g^{\perp}(t) \sim t$ regime observed in the present simulations. Simulations with much longer chains are needed to evaluate the diffusion constant and its dependence on N and ρ_a accurately. The MC study of Milik *et al.* [39] on short chain brushes with mobile adsorbed heads exhibit similar dynamic behavior, with a transition from a Rouse regime to free diffusion, through a limited $t^{1/4}$ regime which the authors attribute to entanglement effects. This explanation does not seem plausible, however, as both in this study and in Ref. 78, the slowing down is most pronounced for the end monomers, whereas chain entanglement is expected to most affect the chain center. In a Θ solvent,[79] the behavior of $g_i^z(t)$ is qualitatively similar to the good solvent case. The transverse motion, $g_i^{\perp}(t)$, on the other hand, is found to behave as $t^{1/4}$, indicating possible entanglement effects at the high monomer densities observed at this solvent condition.

9.6 Polymers tethered to themselves

In the previous three sections, we reviewed the properties of chains tethered to a d-dimensional surface. Another interesting case is when the polymers are tethered to themselves. Probably the most familiar example of chains tethered together is that of crosslinked rubber,[340,341] in which a melt of long chains is randomly crosslinked, either by chemical reaction or radiation. Above a critical concentration of crosslinkers, the system is a solid and has a well defined shear modulus. Random crosslinked rubbers have a broad distribution for the number of monomers between crosslinks and length of dangling ends.[342] These dangling ends are similar to the arms of a star polymer in that they are tethered at one end. However they are elastically inactive and do not contribute to the modulus. While random crosslinked rubbers are easy to make, most theoretical work is on ideal networks, in which the chemical distance between crosslinks is a delta function and there are no dangling ends. Both types of networks are reviewed by Kremer and Grest[143] in Chapter 4 of this book and will not be discussed further here.

While polymer networks which are produced by crosslinking chains in a melt or semi-dilute solution are clearly three-dimensional, it is also possible to imagine other circumstances, where chains tethered to each other form a fractal. In 1984, Cates[343] introduced the term "polymeric fractal" to describe systems which are made of flexible polymer chains at short length scales but have an arbitrary self-similar connectivity at large distances. A linear chain is the simplest example of a polymeric fractal. Other examples include swollen gelation/percolation clusters at the percolation threshold p_c, Sierpiński gaskets and branched polymers. Such fractals are interesting because they have no inherent rigidity and are locally very flexible. In equilibrium, they take on shapes which often have a very different fractal dimension d_f than when they were first constructed. Tethered, self-avoiding membranes in which monomers are connected to form a regular D dimensional surface embedded in d dimensions[122,344] can also be classified as polymeric fractals though they were not first perceived this way. The reason is that the global properties of tethered membranes are the same if one connects linear chains together to form a D dimensional array[107–109] (see Fig. 9.1[d]) or if each of these chains is of length one. While the local flexibility of the membrane depends on the length of the chains, the overall global properties do not.

In this section, we review some of the recent progress in simulating polymeric fractals, with particular emphasis on tethered membranes made of linear polymer segments connected together to form a two-dimensional surface.[107–109] After a brief review of the theory, we present results from a number of groups which show that two-dimensional tethered membranes remain flat and do not crumple. We then consider the effect of changing the solvent quality by adding attractive interactions between nonbonded monomers. While there is clear evidence for a collapsed phase[113] at low T, the nature of the crossover from flat to compact state remains unclear.

9.6.1 Flory theory

As a first approximation, it is convenient to neglect self-avoidance and use a Gaussian approximation[212,345] to estimate the fractal dimension d_f which relates the number of monomers N of the fractal to R_G, $R_G^{d_f} \sim N$. Within this approximation, it is straightforward to show[343,344] that d_f depends only on the object's connectivity through its spectral dimension $\tilde{d}$,

$$d_{fo} = \frac{2\tilde{d}}{(2 - \tilde{d})}. \tag{9.43}$$

Here the subscript o indicates that the excluded volume interactions are absent. For a linear polymer $\tilde{d} = 1$, eq. (9.43) reproduces the result for a Gaussian chain, namely $d_{fo} = 1/\nu_o = 2$.[212] Levinson[346] found that it also works well for triangular Sierpiński gaskets ($\tilde{d} = 2 \ln 3/\ln 5 \approx 1.365$)[347] embedded in dimension $3 \leq d \leq 8$. For percolation clusters at p_c, $\tilde{d} \simeq 4/3$,[348–350] and $d_{fo} = 4$, which has been confirmed by Grest and Murat[102] using MD. For tethered membranes, $\tilde{d} = D = 2$ and $d_{fo} = \infty$. A more detailed analysis by Kantor *et al.* [122,344] found that $R_{Go} \sim \ln N$ which they confirmed by MC simulations.

Self-avoidance can then be introduced at the level of Flory theory[345] as for linear polymers by balancing the elastic (entropic) free energy of the *phantom* object without self-avoidance with the mean-field estimate of the excluded volume interaction,[343]

$$\frac{F}{k_B T} = \frac{1}{2}\left(\frac{R_G}{R_{Go}}\right)^2 + \frac{1}{2}v\left(\frac{N}{R_G^d}\right)^2 R_G^d. \tag{9.44}$$

Here v is the excluded volume parameter. By minimizing eq. (9.44) and using eq. (9.43) to relate R_{Go} to N, the Flory estimate for d_f is

$$d_f = \frac{(d+2)\tilde{d}}{(\tilde{d}+2)}. \tag{9.45}$$

Thus even in the case when self-avoidance is important, d_f depends only on the spectral dimension $\tilde{d}$ and the dimension of space d.

Despite the fact that the Flory argument is a simple mean-field theory, it works very well for a number of cases, often producing estimates for d_f which differ from the exact result by only a few percent or less for $d < d_{\text{uc}}$. Here $d_{\text{uc}} = 4\tilde{d}/(2 - \tilde{d})$ is the upper critical dimension above which self-avoidance is irrelevant.[343,351–354] For a linear polymer eq. (9.44) works extremely well for $d \leq d_{\text{uc}} = 4$. In three dimensions, the theory predicts $\nu = 1/d_f = 3/5$, which is very close to the best renormalization groups estimates[12] of 0.588. Percolation clusters at the percolation threshold are another example where eq. (9.45) works well. Since $\tilde{d} \simeq 4/3$ in all dimensions,[348–350] eq. (9.45) predicts that $d_f \simeq 2(d+2)/5$, independent of the dimension in which it was generated. In $d = 3, d_f \simeq 2$. Daoud *et al.*[355] also found that for gelation clusters generated just below p_c $(d_f \simeq 2.5) d_f$ was reduced to 2, after the excluded volume interactions from the nearby clusters are screened (removed). Their arguments were identical to those presented here, though they did not express their result in terms of $\tilde{d}$. Since gelation and percolation clusters are believed to be in the same universality class,[356] this is a good test case for eq. (9.45). Adam *et al.*,[357] using light scattering, found that the undiluted gelation clusters have $d_f = 2.5 \pm 0.09$, while Bouchaud *et al.* [358] found using small angle neutron scattering, that the diluted clusters have $d_f = 1.98 \pm 0.03$. Recently, we[102] carried out a detailed MD simulation study of percolation clusters generated near p_c with both two- and three-dimensional connectivity embedded in three dimensions and found that d_f was 2.0 within our numerical resolution. Any deviations from Flory theory were very small.

While the Flory approximation works well for some systems, this is not always the case. Two such examples are Sierpiński gaskets[359] in $d = 2$ and two-dimensional tethered membranes in $d = 3$. For tethered membranes, the Flory-level theory predicts that $d_f = 5/2$ for $\tilde{d} = D = 2$ and $d = 3$. This state is commonly referred to as a crumpled membrane since the relaxed membrane should be isotropic. This expectation that the membrane would crumple was supported by renormalization group calculations[122,351–353] and early MC simulations[122] on small systems. However a number of more detailed MC[123–128] and MD[100–103,107–109] simulations found that the membranes remained flat, $R_G^2 \sim N$, and did not crumple. Instead the largest two eigenvalues of the moment of inertia tensor **M** scaled as N, while the smallest eigenvalue scaled as N^{ζ}, with $\zeta < 1$. Various estimates for ζ are discussed below.

It is interesting to consider in more detail why some systems, in particular those for which $\tilde{d}$ is close to 1, crumple and others do not. The lack of a crumpling transition in $d = 3$ for two-dimensional tethered membranes has recently been explained[101] in terms of an implicit bending rigidity which is induced by the self-avoidance requirement even when such a term is not present in the microscopic Hamiltonian. If bending rigidity is relevant, then the Flory theory is not expected to work. To support this argument, Abraham and Nelson[101] simulated tethered membranes with only second-neighbor-excluded volume interactions. They found that this is sufficient to produce a flat phase even when further neighbor interactions are turned off. Additional support of this idea is given by Abraham[107] and Petsche and Grest[109] who found that a network of flexible linear polymers connected to form a two-dimensional tethered membrane remains flat even for long polymer chains. Removing sites[102] or bonds[360] randomly from the membrane also did not cause it to crumple. Kroll and Gompper's[361] finding that a membrane of flexible tethered strings[362] in which self-avoidance is enforced by ensuring that the elementary triangles are impenetrable is asymptotically flat also supports this idea. In all of these cases, the membrane remains flat even though the local bending rigidity is quite low, suggesting that the interactions inducing the long-range bending rigidity, whatever they are, must be relevant under renormalization. Had this not been the case, then above a critical length for the polymer chains, the induced bending rigidity would fall below the critical value necessary to keep the membranes flat[363–365] and the membrane would crumple.

To understand why excluded volume interactions do not generate a comparable bending rigidity for gelation/percolation clusters near p_c or for linear chains, it is informative to consider the relevancy of higher interaction terms in the Hamiltonian. Since renormalization group approaches explicitly take into account the two-body repulsive interactions, it seems reasonable to assume that higher order terms are responsible for generating the implicit bending rigidity that keeps a tethered membrane flat. At least within Flory theory[352,354] it is straightforward to determine for a given $\tilde{d}$ and d whether the n-body interaction v_n is relevant or not. The line where v_n becomes relevant is

$$\tilde{d} = \frac{2(n-1)d}{2n+d}. \tag{9.46}$$

Two-body terms are irrelevant for $d > d_{\text{uc}}$ as mentioned above. Three-body terms are irrelevant for $\tilde{d} > 4d/(6+d)$. For polymeric fractals embedded in $d = 3$, three-body terms are irrelevant for $\tilde{d} < 4/3$. This explains why there is no induced bending rigidity for linear chains and gelation/percolation clusters near p_c. This may also explain why these systems satisfy eq. (9.45) so well. However, for tethered membranes Flory theory suggests that four-body interactions are relevant below six-dimensions and must be taken into account.[354] In fact, three-, four- and five-body interactions are relevant and six-body terms are marginal in $d = 3$, according to Flory theory. If three-body terms are sufficient to generate an effective bending rigidity, then Flory theory would predict that above $d = 6$ membranes should crumple, while if four-body terms are necessary to generate the bending rigidity, then the crumpling would occur above $d = 4$. While there is no reason to take the Flory theory very seriously as to the exact dimension at which a tethered membrane will crumple, it does suggest that crumpled membranes may exist, albeit in dimensions higher than three. Grest[103] carried out MD simulations for two-dimensional tethered membranes embedded in $3 \leq d \leq 8$ dimensions which showed that for $d \geq 5$, the membranes do in fact crumple. Recent MD simulations[366] in $d = 4$ with a hard-core bead diameter σ as small as 10% of the bond length failed to find a crumpled state. For $d = 5$, they also find that the membrane is crumpled. These results are in agreement with recent analytic results by Goulian[367] who used a Gaussian variational approximation and Le Doussal[368] who did an expansion in large embedding space dimension d. They found that the flat phase is stable for $d = 3$ and tethered membranes crumple only for $d > 4$ for $D = 2$ in agreement with Grest.[103] For $d > 4$, their results for d_f are much closer to the simulation results than the Flory theory, eq. (9.44). Guitter and Palmeri[369] used a variational approach for large d and found that the membranes with $D = 2$ in $d = 3$ lie at the boundary of the always-flat phase. A phase diagram for polymeric fractals for $(\tilde{d}, d)$ summarizing where the crumpled and flats occur and where self-avoidance is relevant is presented in Ref. 103. This diagram follows an earlier one by Levinson.[346]

9.6.2 *High-temperature flat phase*

The simplest model of a tethered membrane is composed of purely repulsive spheres which are connected together to form a planar triangulated network. In MC simulations, the spheres are taken as hard spheres, while in MD simulations, they interact with a purely repulsive Lennard-Jones interaction, eq. (9.3). A variety of tethering potentials have been used, as discussed in Section 9.2, usually for a hexagonal sheet of size L containing $N = (3L^2 + 1)/4$ monomers. These systems are often referred to as "open" since the perimeter is free. To minimize finite size effects, some simulations[126] have been done on "closed" systems in which the monomers are connected to form a spherical shell. Abraham[370] used periodic boundary conditions and a computational cell which was allowed to vary in size using a constant-pressure MD technique. More recently, simulations have been carried out for membranes in which linear chains of n monomers are connected together to form a hexagonal lattice[107,108] or triangular lattice.[109] In the former case, the simulations are for open membranes in which the chains are connected at three-point vertices. In the latter case, the membrane is closed and all of the vertices are sixfold except for 12 which are five-fold, in order to form a closed vesicle. An illustration of an open membrane with $n = 8$ is shown in Fig. 9.31. The largest simulations have been with $N = 29\,420$ for open membranes and 25 002 for closed membranes, both for $n = 8$. In this review we consider the case where the hard sphere diameter or Lennard-Jones length σ is comparable to the bond length between tethered monomers. There have been several studies[100,124,128] which have considered the case in which σ is smaller than the bond length. In all these cases, the membrane was asymptotically flat, for nonzero σ, provided the repulsive interaction acted between all monomers independent of their chemical distance.[128]

In Fig. 9.32, a typical configuration is shown for a tethered membrane with $N = 4219$ particles ($L = 75, n = 0$) from the MD simulations of Ref. 101. Note that in two directions the membrane is very large while quite thin in the third direction, consistent with the fact that the two largest eigenvalues of **M** scale as N and $\zeta < 1$. In Fig. 9.33, similar results are shown for closed membranes with N approximately 4000 for three values of n.[109] Note that for small n, the membrane retains some memory of its initial icosahedral topology. This system has a relatively large local bending rigidity due to the large connectivity, similar to the open membrane in Fig. 9.32. For larger N, the shape will become more spherical. For $n = 8$ and 16, however, the local bending rigidity is significantly reduced and there is little evidence of the icosahedral topology even for systems as small as a few thousand

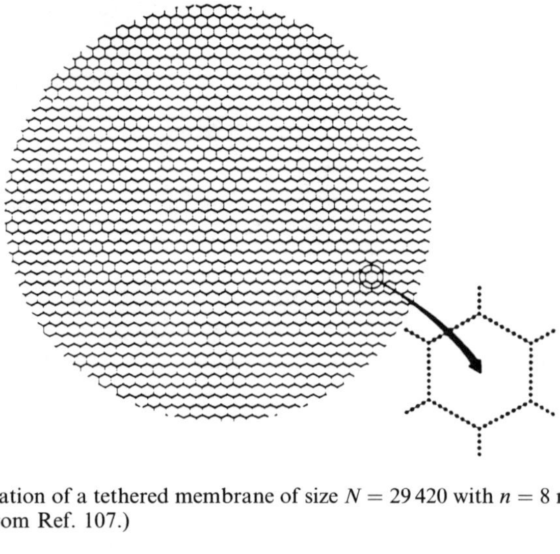

Fig. 9.31 Illustration of a tethered membrane of size $N = 29\,420$ with $n = 8$ monomers between each vertex. (From Ref. 107.)

Fig. 9.32 Typical configuration of an open self-avoiding tethered membrane of 4219 monomers ($n = 0$) from Ref. 101. The tethering bonds are drawn between bonded neighbors, whose hard-core size is not shown.

monomers. A configuration for a larger membrane, $N = 16\,002$ is presented in Ref. 109. To demonstrate that these membrane are flat, consider how $\langle R_G^2 \rangle$ scales with L. In Fig. 9.34, results from Abraham[107] are presented for three values of $n = 0$, 1, and 9. Note that, independent of n, $\langle R_G^2 \rangle^{1/2} \sim L$. Similar results were found by Petsche and Grest for closed membranes.[109]

One of the original motivations for studying large membranes made of linear chains was to investigate whether reducing the local bending rigidity would cause the membrane to crumple. Abraham[107] showed very conclusively in his study of open membranes that the order parameter,[363] defined as the ratio of $\langle R_G^2 \rangle$ to the square of the membrane size, vanishes only as the mass fraction of the tethered membrane vanishes. Petsche and Grest[109] found similar results for closed membranes. In Fig. 9.35, we plot $\langle R_G^2 \rangle / L_o^2$ versus the mass fraction, for both the open[107] and closed[109] membranes. Here L_o is a measure of the membrane size, which for an open membrane is its length in the perfectly flat state and for a closed vesicle is the length along an edge in the initial icosahedral configuration. Since for a crumpled membrane, the order parameter vanishes, the results strongly support the conclusion mentioned above that tethered membranes remain flat, even when the holes are quite large and the mass fraction is quite low.

It is not necessary to use a finite-range hard-core repulsion to ensure self-avoidance, as is done in the standard bead-spring models for tethered

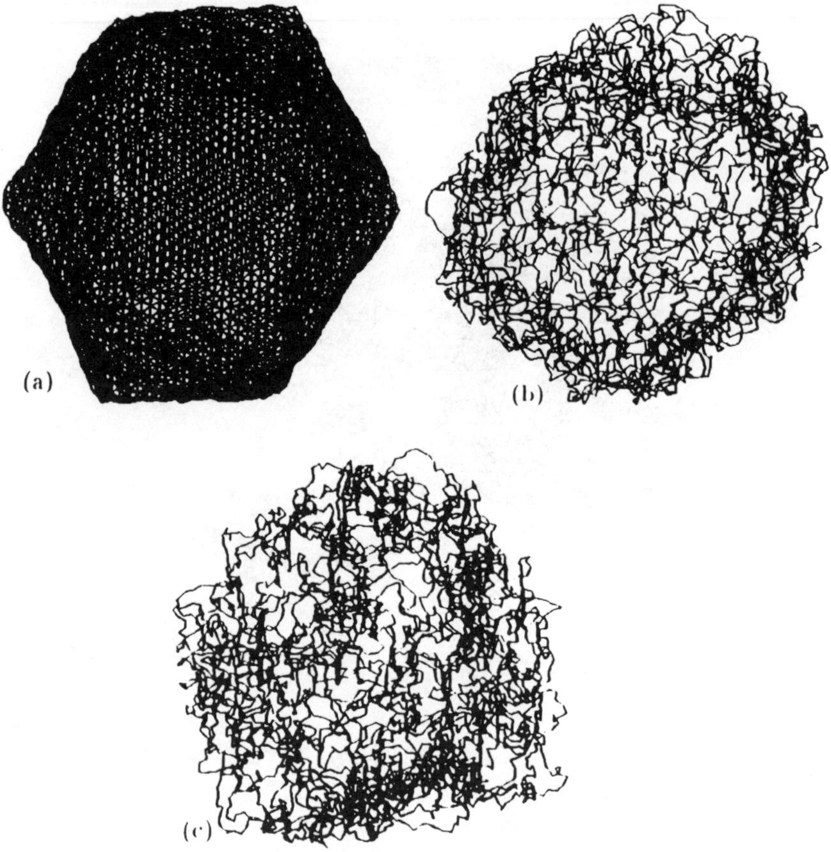

Fig. 9.33 Typical configuration for a closed self-avoiding membrane for (a) $N = 4002$ ($n = 0$), (b) $N = 4\,002$ ($n = 8$) and (c) $N = 4412$ ($n = 16$).

membranes. Self-avoidance can also be guaranteed by requiring that the elementary surface triangles do not intersect.[361,362] This results in a very flexible surface which can fold in on itself without any cost in energy. The resulting surface is therefore much rougher than those constructed with hard spheres and small n, more like what one would expect for bead-spring models in the limit of large n. Kroll and Gompper[361] found that such a surface is asymptotically flat in disagreement with earlier statements by Baumgärtner[362] that this system crumpled. This result is also consistent with the data shown in Fig. 9.35, which shows that the membranes remain flat for all finite n.

While it is now commonly accepted that self-avoiding membranes without attractive interactions remain flat and do not crumple, the flat phase still exhibits some interesting properties. For a membrane containing N monomers, the transverse displacements have an amplitude $\langle h^2 \rangle \sim N^{\zeta}$ where

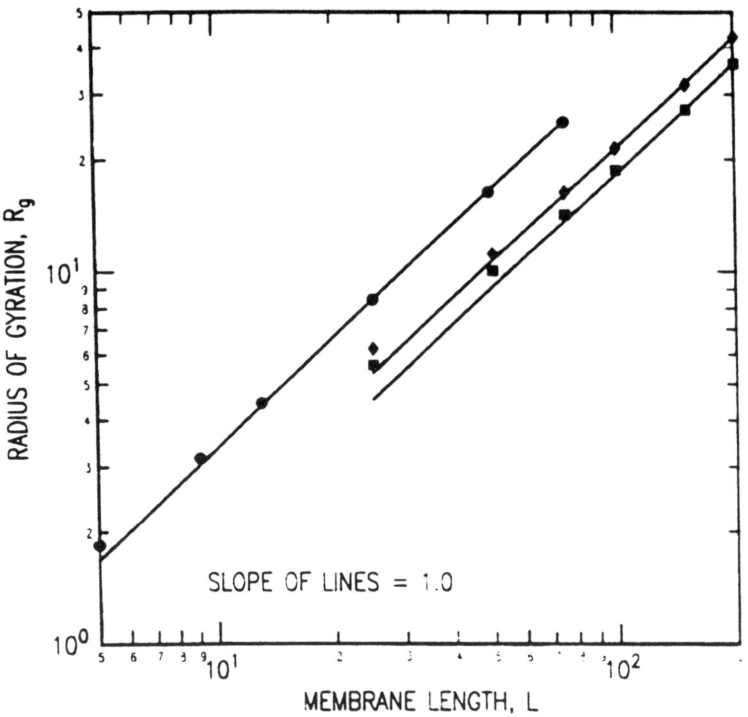

Fig. 9.34 The mean radius of gyration as a function of membrane size L for membrane with n monomers between each vertex. Here $n = 0$ (•), $n = 1$ (◆) and $n = 8$ (■). Note that the compact membrane ($n = 0$) has the same slope as the perforated membranes. The slope of each line is 1.0.

$1/2 \leq \zeta < 1$.[371] The smallest eigenvalue of **M** is a convenient way to measure $\langle h^2 \rangle$ for an open membrane. The scale-dependent bending rigidity κ in the flat phase scales as $N^{\eta/2}$, with $\eta = 2 - 2\zeta$, while the scale-dependent shear modulus scales as $N^{-\eta_\mu/2}$, with $\eta_\mu = 4\zeta - 2$.[372,373] Nelson and Peliti[372] carried out a one-loop, self-consistent calculation and found that $\zeta = 1/2$, which would mean that the membrane would have a finite shear modulus on large scales. Recently, Le Doussal and Radzihovsky[374] carried out a self-consistent screening approximation which improved on the Nelson–Peliti theory by allowing nontrivial renormalization of the elastic moduli. They found $\zeta = 0.59$ and that the Poisson ratio is $-2/(D+2)$. Early simulations on open, triangular networks ($n = 0$), gave ζ in the range of 0.63–0.65,[101,125,375] slightly larger than predicted by Le Doussal and Radzihovsky. However some earlier estimates of ζ were closer to 1/2. Lipowsky and Girardet[371] carried out MC simulations for a continuum model of a solid-like elastic sheet and found that their data were consistent with $\zeta = 1/2$. They did not give any error bars, but did claim that their

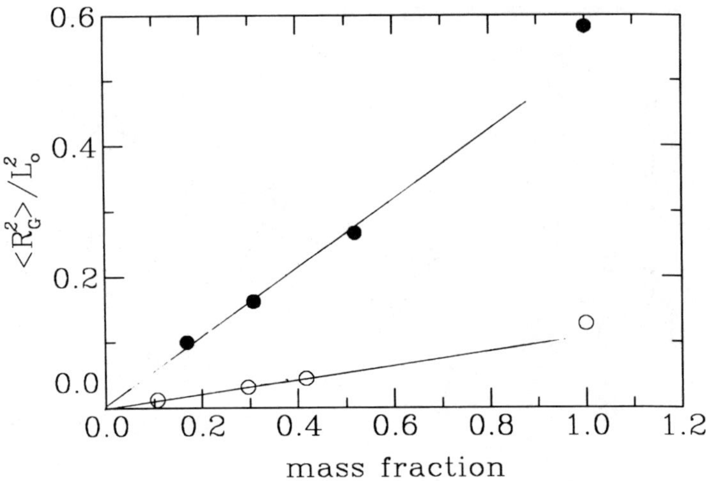

Fig. 9.35 Order parameter as a function of mass fraction for various n. The open circles are from Abraham[107] and the closed circles are from Petsche and Grest.[109] Here L_o is a measure of the membrane size, which for the open membrane is its length when it is perfectly flat and for a closed vesicle it is the length along an edge in the initial icosahedral configuration.

results were inconsistent with ζ in the range 0.63–0.65. However their data could not rule out ζ slightly larger than 1/2. They associated the difference with earlier simulations with a crossover effect since for membranes with $n = 0$, the actual values of $\langle h^2 \rangle$ are relatively small and it is difficult to determine ζ accurately. Abraham[370] has pointed out that the larger values for ζ may also be due to finite size effects since all these bead-spring type simulations were performed on open membranes with a free perimeter. Abraham replaced the free-perimeter with periodic boundary conditions and found $\zeta = 0.53 \pm 0.03$, consistent with Lipowsky and Girardet[371] but still lower than the best theoretical estimate of 0.59.[374]

Another way to reduce finite size effects is to study closed vesicles. The first simulations on closed vesicles $(n = 0)$ were by Komura and Baumgärtner.[126] For a closed vesicle in the flat phase, all three eigenvalues of $\mathbf{M}$ should scale linearly with N and a new measure of the height fluctuations and ζ are needed. At first thought, one might try to measure $\langle h^2 \rangle$ from the fluctuations in the mean-square radius of gyration,[126]

$$\chi_{R_G} = [\langle(\delta R_G^2)^2\rangle]^{1/2} \equiv [\langle R_G^4 \rangle - \langle R_G^2 \rangle^2]^{1/2}. \tag{9.47}$$

However, unlike an open membrane, a spherical shell or tethered vesicle cannot bend without being stretched.[376] That is, there is a linear coupling between the out-of-plane undulation modes and the in-plane phonon modes which causes a strong suppression of out-of-plane fluctuations at long length scales. This leads to new scaling behavior[376] for χ_{R_G} which is different from $\langle h^2 \rangle$. Zhang *et al.*[376] found that $\chi_{R_G} \sim N^{\zeta_1}$, with $\zeta_1 = (2 - \eta/2)/(2 + \eta)$. They also showed that the fluctuations in the volume scaled as $\chi_V \sim N^{3/(2+\eta)}$. In Fig. 9.36, results[109] for χ_{R_G} versus N are shown for $n = 0$ and 8 for closed vesicles in which the monomers interact with a purely repulsive Lennard-Jones potential. Least-square fits to the slope give $\zeta_1 = 0.55 \pm 0.02$ for $n = 0$ and $\zeta_1 = 0.56 \pm 0.02$ for $n = 8$. This gives $\zeta = 0.57 \pm 0.02$ for $n = 0$ and $\zeta = 0.59 \pm 0.02$ for $n = 8$. Zhang *et al.*[376] simulated the same model and found that $2\zeta_1 = 1.31$ for $n = 0$, which gives $\zeta = 0.59 \pm 0.02$. They also measured χ_V and found $\eta = 0.79 \pm 0.03$ and $\zeta = 0.60 \pm 0.02$. Within the statistical uncertainty, both these simulations give ζ in very good agreement with the theoretical predictions of $\zeta = 0.59$ by Le Doussal and Radzihovsky.[374]

A second, important, unresolved issue concerns the interpretation of recent scattering experiments on polymerized membranes. Laser scattering measurements[377] of the static structure function $S(q)$ for graphite oxide crystalline membranes suggests that they are crumpled and not flat as indicated by computer simulations. For a flat membrane, one expects that for large enough systems, $S(q) \sim q^{-2}$ while for a crumpled membrane $S(q) \sim q^{-2.5}$. While Wen *et al.*[377] do observe a power law decay with a slope of 2.54, one should not interpret this as convincing evidence for a crumpled phase, since these experiments are on randomly oriented membranes. Recently, experiments on this system have been repeated by Naranjo *et al.*[378] using freeze-fracture electron microscopy and static light scattering. They find that the membranes are flat with some curling at the edges. Abraham and Goulian[108] have shown that for a wide range

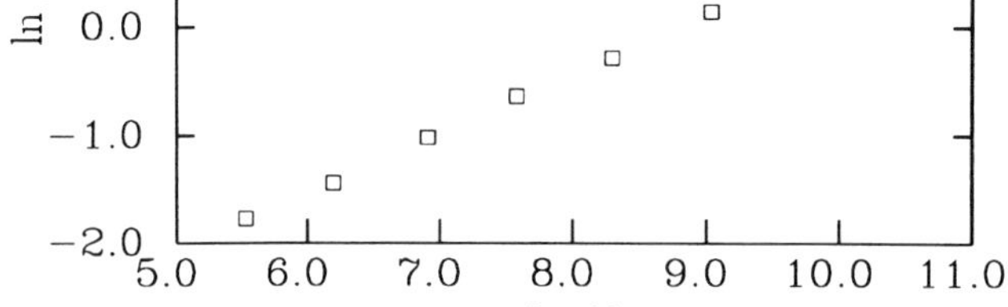

Fig. 9.36 Fluctuations in the mean-squared radius of gyration χ_{R_G}, eq. (9.47), versus N for a closed vesicle with $n = 0$ and 8. (From Ref. 109.)

of q, the isotropically averaged $S(q) \sim q^{-2.35}$ for open, tethered membranes containing as many as 29 420 monomers. They found no evidence for a q^{-2} regime even though these membranes are definitely flat. The expected q^{-2} regime which must be present for large enough samples is apparently suppressed and only the rough regime[379] in which $S(q) \sim q^{-3+\zeta}$ is observed. This interpretation is consistent with the computer simulations of Abraham and Goulian[108] on open membranes where $\zeta \approx 0.64$. It also agrees with recent X-ray and light-scattering experiments[380,381] on isolated red blood cell (RBC) skeletons in high salt in which a q^{-2} regime at low q was followed by a $q^{-2.35}$ regime for larger q, consistent with the behavior expected for a flat membrane. While it is fairly certain that the interactions in the red blood cell skeletons are purely repulsive, it is possible that attractive interactions[382] are important for the graphite oxide crystalline membranes and they are in fact crumpled. Additional experiments are needed to clarify whether attractive interactions play a role in this case. Recently, Stupp *et al.*[383] have synthesized two-dimensional polymers which may be very useful in experimentally testing the results discussed here.

While the graphite oxide crystalline membranes[377,378] are open membranes with a large local bending rigidity, induced by the close packing of monomers, the RBC skeleton is a closed vesicle, roughly spherical, made up of flexible worm-like chains in a triangulated network. The RBC skeleton is attached to the cytoplasmic side of the liquid lipid cell membrane[3,4] and can be isolated by detergent treatment. The spectrin tetramers which make up the chains have a contour length of $\approx$ 2000 Å and a persistence length of 100–200 Å in high ionic strength buffer. Mammalian RBC skeletons contain $\sim$ 70 000 triangular meshes. Obviously, this system is too complex to simulate directly on the computer. However, the essential properties of the membrane on intermediate length scales can be obtained from a study of a coarse-grained bead-spring model which has the same topology. In high salt, the actin oligomers have a mean spacing of 400–500 Å, or about 1/5–1/4 of their fully extended length. Using this as a means of mapping to a bead-spring model, Petsche and Grest[109] estimated that each linear chain should contain about 50–60 monomers. However even this is too large to simulate for more than a few hundred triangles since the largest systems one can study in a few hundred hours of Cray time is about 25 000–30 000 monomers. The largest systems studied to date have been for $n \leq 16$. Larger systems will have to wait for the next generation of supercomputers. However, in spite of the fact that the largest samples are about 50 times smaller than the experimental systems and contain only excluded volume interactions, the cross-sections of the simulated membranes compare very favorably with the shape of RBC skeletons.[381] Results for $S(q)$ are also in agreement with recent X-ray and light scattering studies of Schmidt *et al.*.[381]

9.6.3 *Effect of attractive interactions*

In the previous section, we saw that in a good solvent with no attractive interactions, the membrane is flat due to an entropic bending rigidity. However, what happens when the solvent quality is lowered is not completely clear. Will there be a Θ point which separates the high T flat phase and a low T phase or will there be a range of intermediate T where the attractive interactions balance the entropic bending rigidity to produce a phase in which the membrane is crumpled? What is the nature of the low temperature phase? Abraham and Kardar[113] have conjectured that the high T flat phase is separated from a low T collapsed phase by either a sequence of folding transitions or by a crumpled phase. There is presently evidence for both scenarios. Here we will briefly review the three simulations which have addressed these points. Additional simulations in this area would be helpful.

The first question to address is the nature of the low-temperature phase. For a very poor solvent, one would expect that the monomers would try to collapse into a compact state with $d_f = 3$. However as noted by Kantor *et al.*[122] finding such a state is nontrivial due to the self-avoidance constraint. Abraham and Nelson[101] turned the attractive part of the Lennard-Jones interaction on ($r_c = 2.5\sigma$) and found that at low $T = 1.4\epsilon/k_B$, the membrane collapsed, $\langle R_G^2 \rangle \sim N^{1/3}$. Thus at least for membranes up to size $N = 4219$, the monomers were able to find a low-temperature collapsed configuration without violating the excluded volume constraint. Whether this is possible or the concerns raised by Kantor *et al.*[122] in regard to the problem of finding a collapsed state are valid for much larger systems is unclear. When T was raised to $3.5\epsilon/k_B$, the flat phase was recovered.

At intermediate T, Abraham and Kardar[113] found that the membrane underwent a series of folding transitions. All of their simulations were for open membranes with $n = 0$. At $T_1(L)$, which depended strongly on the size of the membrane, the membrane folded once. A crease neatly divided the membrane in half. The folded structure had nearly the same width as the unfolded one, so that the in-plane density was twice as large. On further cooling, a second fold occurred at $T_2(L)$ and the membrane divided itself into roughly four equal parts folded together. For the system sizes studied ($L \leq 75$), no additional transitions were observed. Finite size effects are expected to play an important role particularly for range of L which can presently be handled on present day computers. Abraham and Kardar found that for the largest system studied, $T_1(75) \simeq 3.2\epsilon/k_B$ and $T_2(75) \simeq 2.8\epsilon/k_B$, while for $L = 49$, $T_1(49) \simeq 2.9\epsilon/k_B$ and $T_2(49) \simeq 2.2\epsilon/k_B$. For small L, there is a competition between the free-energy gain from folding ($\sim bL^2/2$) and the energy loss due to creasing ($\sim \epsilon_c L$). For small $b < 0$, a finite membrane remains flat until b is sufficiently negative to compensate for the crease energy. The singly folded membrane then has a higher energy barrier to create a second crease, thereby lowering T_2 relative to T_1 for small L. However for large L, Abraham and Kardar[113] predict that there will be a sequence of folding transitions for finite L which converge with each other as $L \to \infty$.

The unfolding transition of singly folded membranes is closely related to the unbinding transition of two distinct surfaces.[384] Using this analogue, Abraham and Kardar[113] developed a scaling picture for the change in free energy for a single fold. They found that for sufficiently large membranes, the energy loss of creasing is actually irrelevant compared to the entropy loss of edge fluctuations. Thus the folding temperature and the temperature for unbinding two membranes should be very close. From simulations of the unbinding of a bimembrane, they estimated that the single fold transition would occur at $T_1 \simeq 5.7\epsilon/k_B$ for $L \to \infty$, compared to $3.2\epsilon/k_B$ for $L = 75$. At these high temperatures, the creases would presumably be more rounded to minimize the cost of bending. The sharp creases found for $L \leq 75$ are due to the relatively low T of the transition for finite L.

Liu and Plischke[382] reported MC results for a hard sphere model in which a longer range attractive potential was added. Though the details of the interaction and the simulation method were different than that used by Abraham and Kardar,[113] they expected to find a sequence of folding transitions separating the flat and collapsed phases. However Liu and Plischke found that the equilibrium states were not characterized by a series of folds. Instead they observed that for a range of intermediate T, the membrane appeared to crumple. That is, $\langle R_G^2 \rangle \sim N^{2/d_f}$, with $d_f \simeq 2.5$. For the range of L studied, $7 \leq L \leq 33$, the static structure factor $S(q)$ scaled nicely with qL^{2/d_f} for three intermediate temperatures. For high T, the membrane was flat while for low T, it collapsed. However since the largest N was 817, it is not possible to determine whether, for larger N, this intermediate range would collapse to a single temperature, like the Θ point for linear chains, or whether there actually exists a range of T where the attractive interactions counterbalance the entropic bending rigidity to produce a stable crumpled phase. More recently, Grest and Petsche[120] studied closed membranes in which the attractive interactions were turned on by extending r_c to 2.5σ. In this case the low temperature phase is collapsed and not folded. They find no evidence for an extended range of T where the membrane is crumpled. For $n = 4$, there is a first order transition from the high temperature flat phase to the intermediate temperature crumpled phase with considerable hysteresis. The flat phase could be supercooled to about $3.0\epsilon/k_B$, while the low temperature phase was stable up to $3.25\epsilon/k_B$. The simulations for $n = 8$ showed an entirely different behavior. In this case, there appeared to be a continuous or very weak first order transition from the high temperature flat phase to the low temperature collapsed phase at $T = 2.89 \pm 0.03\epsilon/k_B$. Assuming the transition is continuous, a scaling analysis suggests that at the transition there is an intermediate state which has a fractal dimension $d_f \simeq 2.4$, somewhat smaller but close to the value predicted by the Flory theory for a crumple membrane.

9.7 Conclusions

Tethered chains comprise a very active field of research in which experi ments, theory and simulations continue to interact strongly with on another. In this chapter, we have tried to emphasize this collaborative effort even though our primary interest was in computer simulations of thes systems. We hope we have achieved this goal.

The constraint of tethering induces new structures that involve new lengtl scales which are not present for free chains in solution. The crowding o other chains near the tethering points often forces the polymer to stretch ou away from the grafting site in order to gain interaction energy. This stretching is accompanied by a penalty in elastic energy due to a decrease in configurational entropy. The interplay of these two contributions leads to interesting structures which depend strongly on the dimension d of the tethering surface. Tethering to a flat surface induces the greatest amount of stretching since the volume accessible to the chains does not increase as the distance from the grafting site increases, as occurs for chains grafted to a point or a line. As shown here, simulations have been a very valuable tool in clarifying the properties of these systems and bridging the gap between experiment and theory.

We believe that we have shown in this review that simulations have been helpful in improving our understanding of many of the basic properties of tethered chain systems. However a lot of interesting work remains to be done. All of the work presented here has been for chain length N typically in the range of 50–100 monomers, though a few simulations were carried out for N as large as 200. For good and Θ solvents this seems to be a reasonable length, in that it is large enough to reach the predicted scaling regime, as shown in Fig. 9.18 for the height of a brush. However in a poor solvent, much longer chains are needed (probably $N \gtrsim 1000$) to access the scaling regime in a brush. This is because as the solvent quality decreases, the excluded volume interactions are screened and the overall monomer density increases substantially. For a given N, the overlap density is significantly larger in a poor solvent compared to a good solvent. For N of order 100, the chains are simply not very stretched unless one goes to very large grafting densities. This is one area of interest for the future. However a more interesting problem involves the explicit introduction of solvent molecules, either small molecules of size one monomer or polymer chains of length P. Then one can study hydrodynamic effects as well as the exclusion of the solvent as the chainlength P increases. For a star polymer or brush in contact with linear chains of length $P < N$, we estimate that systems of approximately 20 000 to 30 000 monomers are needed which should take of order a hundred hours on a Cray-YMP per phase point (N, P, ρ_a). Considering the recent advances in computer speeds this is not unreasonable. Studies of hydrodynamic effects would require much larger systems and longer times, due to large finite size effects,[29,30] but should be possible within the next few years. Interaction of star and other branched polymers in solution will require substantially more time due to their slow diffusion and slow relaxation and are probably not feasible in the next few years.

Acknowledgment

One of the authors (M. M.) wishes to thank Exxon Research for their hospitality during the final stages of writing this review.

References

1. A. Halperin, M. Tirrell, and T. P. Lodge, *Adv. Polym. Sci.* **100**, 31 (1991).
2. S. T. Milner, *Science* **251**, 905 (1991).
3. B. Alberts, D. Bray, J. Lewis *et al.*, *The Molecular Biology of the Cell* (Garland, New York, 1983).
4. A. Elgsaeter, B. Stokke, A. Mikkelsen, and D. Branton, *Science* **234**, 1217 (1986).
5. D. H. Napper, *Polymeric Stabilization of Colloidal Dispersions* (Academic Press, London, 1983).
6. G. J. Fleer, M. A. Cohen Stuart, J. M. H. M. Scheutjens, T. Cosgrove, and B. Vincent, *Polymers at Interfaces* (Chapman and Hall, London, 1993).
7. S. Alexander, *J. Phys.* (*Paris*) **38**, 983 (1977).
8. P.-G. de Gennes, *Macromolecules* **13**, 1069 (1980).
9. R. Cantor, *Macromolecules* **14**, 1186 (1981).
10. M. Daoud and J. P. Cotton, *J. Phys.* (*Paris*) **43**, 531 (1982).
11. T. M. Birshtein and E. B. Zhulina, *Polymer* **25**, 1453 (1984); T. M. Birshtein, E. B. Zhulina, and O. V. Borisov, *Polymer* **27**, 1078 (1986).
12. J. C. LeGuillou and J. Zinn-Justin, *Phys. Rev. B* **21**, 3976 (1980).
13. A. Baumgärtner, *Ann. Rev. Phys. Chem.* **35**, 419 (1984).
14. A. Baumgärtner, in *Applications of the Monte Carlo Method in Statistical Physics*, edited by K. Binder (Springer, Berlin, 1984).
15. K. Kremer and K. Binder, *Comp. Phys. Rept.* **7**, 259 (1988).
16. J. Skolnick and A. Kolinski, in *Advances in Chemical Physics*, Vol. 78, edited by I. Prigogine and S. A. Rice (Wiley, New York, 1990), p. 223.
17. K. Binder, in *Molecular Level Calculations of the Structure and Properties of Non-Crystalline Polymers*, edited by J. Biscerano (Marcel Dekker, New York, 1992), p. 221.
18. G. S. Grest and K. Kremer, in *Elastomeric Polymer Networks*, edited by J. E. Mark and B. Erman (Prentice Hall, Englewood Cliffs, 1992), p. 164.
19. K. Kremer and G. S. Grest, *J. Chem. Soc. Faraday Trans.* **88**, 1707 (1992).
20. K. Kremer, in *Computer Simulation in Chemical Physics*, edited by M. P. Allen and D. J. Tildesley (Kluwer Academic, Amsterdam, 1993), p. 397.

21. K. Binder, *Adv. Polym. Sci.* **112**, 181 (1994).
22. K. Kremer, *Macromolecules* **16**, 1632 (1983).
23. P. E. Rouse, *J. Chem. Phys.* **21**, 1272 (1953).
24. F. F. Abraham, *Adv. Phys.* **35**, 1 (1986).
25. M. P. Allen and D. J. Tildesley, *Computer Simulation of Liquids* (Clarendon Press, Oxford, 1987).
26. R. W. Hockney and J. W. Eastwood, *Computer Simulation using Particles* (Adam Hilger, Bristol, 1988).
27. G. Coccotti, D. Frenkel, and I. R. McDonald, *Simulations of Liquids and Solids* (North Holland, Amsterdam, 1987).
28. K. Kremer and G. S. Grest, *J. Chem. Phys.* **92**, 5057 (1990); **94**, 4103 (1991) erratum.
29. C. Pierleoni and J.-P. Ryckaert, *Phys. Rev. Lett.* **66**, 2992 (1991); *J. Chem. Phys.* **96**, 8539 (1992).
30. B. Dünweg and K. Kremer, *Phys. Rev. Lett.* **66**, 2996 (1991); *J. Chem. Phys.* **99**, 6983 (1993).
31. B. Smit, A. Van der Put, C. J. Peters, J. de Swaan Arons, and J. P. L. Michels, *J. Chem. Phys.* **88**, 3372 (1988).
32. P. H. Verdier and W. H. Stockmayer, *J. Chem. Phys.* **36**, 227 (1962); P. H. Verdier, *J. Chem. Phys.* **45**, 2122 (1966).
33. H. J. Hilhorst and J. M. Deutch, *J. Chem. Phys.* **63**, 5153 (1975); H. Boots and J. M. Deutch, *J. Chem. Phys.* **67**, 4608 (1977).
34. M. T. Gurler, C. C. Crabb, D. M. Dahlin, and J. Kovac, *Macromolecules* **16**, 398 (1983); C. Stokeley, C. C. Crabb, and J. Kovac, *Macromolecules* **19**, 860 (1986).
35. N. Madras and A. D. Sokal, *J. Stat. Phys.* **50**, 109 (1988); N. Madras, A. Orlitsky, and L. A. Shepp, *J. Stat. Phys.* **58**, 159 (1990).
36. A. Sariban and K. Binder, *Macromolecules* **21**, 711 (1988).
37. J. Skolnick, R. Yaris, and A. Kolinski, *J. Chem. Phys.* **88**, 1407 (1988); J. Skolnick and R. Yaris, *J. Chem. Phys.* **88**, 1418 (1988).
38. A. Chakrabarti and R. Toral, *Macromolecules* **23**, 2016 (1990); A. Chakrabarti, P. Nelson, and R. Toral, *Phys. Rev. A* **46**, 4930 (1992).
39. M. Milik, A. Kolinski, and J. Skolnick, *J. Chem. Phys.* **93**, 4440 (1990).
40. A. K. Kron, *Polym. Sci. USSR* **7**, 1361 (1985).
41. F. T. Wall and F. Mandel, *J. Chem. Phys.* **63**, 4592 (1975); F. Mandel, *J. Chem. Phys.* **70**, 3984 (1979).
42. R. Dickman and D. C. Hong, *J. Chem. Phys.* **95**, 4650 (1991).
43. R. Pakula, *Macromolecules* **20**, 679 (1987); T. Pakula and S. Geyler, *Macromolecules* **20**, 2909 (1987).
44. S. Geyler, T. Pakula, and J. Reiter, *J. Chem. Phys.* **92**, 2676 (1990); J. Reiter, T. Edling, and T. Pakula, *J. Chem. Phys.* **93**, 837 (1990); A. Gauger, A. Weyersberg and T. Pakula, *Makromol. Chem., Theory Sim.* **2**, 531 (1993).
45. O. Jagodzinski, E. Eisenriegler, and K. Kremer, *J. Phys. I (France)* **2**, 2243 (1992).
46. T. Pakula and E. B. Zhulina, *J. Chem. Phys.* **95**, 4691 (1991).
47. E. B. Zhulina and T. Pakula, *Macromolecules* **25**, 754 (1992).
48. M. Murat and T. A. Witten, *Macromolecules* **23**, 520 (1990).

49. A. R. C. Baljon-Haakman and T. A. Witten, *Macromolecules* **25**, 2969 (1992).
50. M. Lal, *Mol. Phys.* **17**, 57 (1969).
51. B. MacDonald, N. Jan, D. L. Hunter, and M. O. Steinitz, *J. Phys. A* **18**, 2627 (1985).
52. G. Zifferer, *Makromol. Chem.* **191**, 2717 (1990); *Makromol. Chem., Theory Simul.* **1**, 55 (1992).
53. T. C. Clancy and S. E. Webber, *Macromolecules* **26**, 628 (1993).
54. A. D. Sokal, see Chapter 2 of this volume.
55. J. Mazur and F. McCrackin, *Macromolecules* **10**, 326 (1977).
56. F. L. McCrackin and J. Mazur, *Macromolecules* **14** , 1214 (1981).
57. A. Kolinski and A. Sikorski, *J. Polym. Sci., Polym. Chem. Ed.* **20**, 3147 (1982); *J. Polym. Sci., Polym. Chem. Ed.* **22**, 97 (1984).
58. J. E. G. Lipson, S. G. Whittington, M. K. Wilkinson, J. L. Martin, and D. S. Gaunt, *J. Phys. A* **18**, L469 (1985); M. K. Wilkinson, D. S. Gaunt, J. E. G. Lipson, and S. G. Whittington, *J. Phys. A* **19**, 789 (1986); S. G. Whittington, J. E. G. Lipson, M. K. Wilkinson, and D. S. Gaunt, *Macromolecules* **19**, 1241 (1986).
59. J. E. G. Lipson, D. S. Gaunt, M. K. Wilkinson, and S. G. Whittington, *Macromolecules* **20**, 186 (1987).
60. J. E. G. Lipson, *Macromolecules* **24**, 1327 (1991); *Macromolecules* **26**, 203 (1993).
61. A. J. Barrett and D. L. Tremain, *Macromolecules* **20**, 1687 (1987).
62. A. W. Rosenbluth and M. N. Rosenbluth, *J. Chem. Phys.* **23**, 356 (1955).
63. J. Batoulis and K. Kremer, *J. Phys. A* **21**, 127 (1988).
64. J. Batoulis and K. Kremer, *Macromolecules* **22**, 4277 (1989).
65. K. Suzuki, *Bull. Chem. Soc. Jpn.* **41**, 538 (1968).
66. Z. Alexandrowicz, *J. Chem. Phys.* **51**, 561 (1969).
67. J. Batoulis and K. Kremer, *Europhys. Lett.* **7**, 683 (1988).
68. K. Ohno and K. Binder, *J. Stat. Phys.* **64**, 781 (1991).
69. R. L. Lescanec and M. Muthukumar, *Macromolecules* **23**, 2280 (1990); *Macromolecules* **24**, 4892 (1991).
70. P. G. de Gennes and H. Hervet, *J. Phys. Lett. (France)* **44**, L351 (1983).
71. D. A. Tomalia, H. Baker, J. Dewald *et al.*, *Macromolecules* **19**, 2466 (1986); D. A. Tomalia, V. Berry, M. Hall, and D. M. Hedstrand, *Macromolecules* **20**, 1167 (1987).
72. D. A. Tomalia, A. M. Naylor, and W. A. Goddard III, *Angew. Chem., Int. Ed. Engl.* **29**, 138 (1990).
73. M. L. Mansfield and L. I. Klushin, *J. Phys. Chem.* **96**, 3994 (1992); *Macromolecules* **26**, 4262 (1993); M. L. Mansfield, *Polymer* **35**, 1827 (1994).
74. R. Toral, A. Chakrabarti, and R. Dickman, *Phys. Rev.* **E50**, 343 (1994).
75. I. Carmesin and K. Kremer, *Macromolecules* **21**, 2819 (1989); *J. Phys. (Paris)* **51**, 915 (1990).
76. H.-P. Wittmann and K. Kremer, *Comp. Phys. Comm.* **61**, 309 (1990); *Comp. Phys. Comm.* **71**, 343 (1992) erratum.
77. W. Paul, K. Binder, D. W. Heermann, and K. Kremer, *J. Phys. II (France)* **1**, 37 (1990); *J. Chem. Phys.* **95**, 7726 (1991).
78. P.-Y. Lai and K. Binder, *J. Chem. Phys.* **95**, 9288 (1991).

79. P.-Y. Lai and K. Binder, *J. Chem. Phys.* **97**, 586 (1992).
80. P.-Y. Lai, *J. Chem. Phys.* **98**, 669 (1993).
81. R. Dickman and P. E. Anderson, *J. Chem. Phys.* **99**, 3112 (1993).
82. S.-J. Su, M. S. Denny, and J. Kovac, *Macromolecules* **24**, 917 (1991).
83. S.-J. Su and J. Kovac, *J. Phys. Chem.* **96**, 3931 (1992).
84. M. Schulz and J. U. Sommer, *J. Phys. Chem.* **96**, 7102 (1992); J. U. Sommer, M. Schulz and H. L. Trautenberg, *J. Chem. Phys.* **98**, 7515 (1993).
85. M. Schulz and K. Binder, *J. Chem. Phys.* **98**, 655 (1993).
86. H.-P. Wittmann, K. Kremer, and K. Binder, *J. Chem. Phys.* **96**, 6291 (1992).
87. J. Baschnagel, K. Binder, and H.-P. Wittmann, *J. Phys. Cond. Matter* **5**, 1597 (1993).
88. J. S. Shaffer, *Phys. Rev. E* **50**, R683 (1994).
89. J. S. Shaffer, *J. Chem. Phys.* **101**, 4205 (1994).
90. A. Kolinski, J. Skolnick, and R. Yaris, *J. Chem. Phys.* **84**, 1922 (1987); *J. Chem. Phys.* **86**, 1567 (1987).
91. A. Milchev, W. Paul, and K. Binder, *J. Chem. Phys.* **99**, 4786 (1993); I. Gerroff, A. Milchev, W. Paul, and K. Binder, *J. Chem. Phys.* **98**, 6526 (1993).
92. A. Baumgärtner and K. Binder, *J. Chem. Phys.* **71**, 2541 (1979); A. Baumgärtner, *Z. Physik* **42**, 265 (1981).
93. L. J. Gallego, C. Rey, and M. J. Grimson, *Mol. Phys.* **74**, 383 (1991).
94. R. Toral and A. Chakrabarti, *Phys. Rev. E* **47**, 4240 (1993).
95. I. Webman, J. L. Lebowitz, and M. H. Kalos, *Phys. Rev. B* **21**, 5540 (1980).
96. G. S. Grest, K. Kremer, and T. A. Witten, *Macromolecules* **20**, 1376 (1987).
97. G. S. Grest, K. Kremer, S. T. Milner, and T. A. Witten, *Macromolecules* **22**, 1904 (1989).
98. M. Murat and G. S. Grest, *Macromolecules* **22**, 4054 (1989).
99. M. Murat and G. S. Grest, *Phys. Rev. Lett.* **63**, 1074 (1989); in *Computer Simulation of Polymers*, edited by R. J. Roe (Prentice Hall, Englewood Cliffs, 1991), p. 141.
100. F. F. Abraham, W. E. Rudge, and M. Plischke, *Phys. Rev. Lett.* **62**, 1757 (1989).
101. F. F. Abraham and D. R. Nelson, *J. Phys. (France)* **51**, 2653 (1990); *Science* **249**, 393 (1990).
102. G. S. Grest and M. Murat, *J. Phys. (France)* **51**, 1415 (1990).
103. G. S. Grest, *J. Phys. I (France)* **1**, 1695 (1991).
104. M. Murat and G. S. Grest, *Macromolecules* **24**, 704 (1991).
105. E. R. Duering, K. Kremer, and G. S. Grest, *Phys. Rev. Lett.* **67**, 3531 (1991).
106. G. S. Grest, K. Kremer, and E. R. Duering, *Europhys. Lett.* **19**, 195 (1992); *Physica A* **194**, 330 (1993).
107. F. F. Abraham, in *Microscopic Simulations of Complex Hydrodynamic Phenomena*, edited by M. Mareschal and B. L. Holian (Plenum, New York, 1992), p. 361.
108. F. F. Abraham and M. Goulian, *Europhys. Lett.* **19**, 293 (1992).
109. I. B. Petsche and G. S. Grest, *J. Phys. I (France)* **3**, 1741 (1993).
110. A. Baumgärtner, *J. Chem. Phys.* **72**, 871 (1980); *J. Chem. Phys.* **73**, 2489 (1980).

111. Y.-J. Sheng, A. Z. Panagiotopoulos, S. K. Kumar, and I. Szleifer, *Macromolecules* **27**, 400 (1994).
112. J. J. Freire, J. Pla, A. Rey, and R. Prats, *Macromolecules* **19**, 452 (1986); J. J. Freire, A. Rey, and J. G. de la Torre, *Macromolecules* **19**, 457 (1986); A. Rey, J. J. Freire, and J. G. de la Torre, *Macromolecules* **20**, 342 (1987).
113. F. F. Abraham and M. Kardar, *Science* **252**, 419 (1991).
114. S. K. Kumar, *J. Chem. Phys.* **96**, 1490 (1992).
115. G. S. Grest and M. Murat, *Macromolecules* **26**, 3108 (1993).
116. G. S. Grest, *Macromolecules* **27**, 418 (1994).
117. G. S. Grest, *Macromolecules* **27**, 3493 (1994).
118. V. I. Harismiadis and I. Szleifer, *Mol. Phys.* **81**, 851 (1994).
119. J. D. Wienhold and S. K. Kumar, *J. Chem. Phys.* **101**, 4312 (1994).
120. G. S. Grest and I. B. Petsche, *Phys. Rev.* **E50**, R1737 (1994).
121. D. Frenkel, G. C. A. M. Mooij, and B. Smit, *J. Phys. Cond. Matter* **3**, 3053 (1991); J. I. Siepmann and D. Frenkel, *Mol. Phys.* **75**, 59 (1992); J. J. dePablo, M. Laso and U. W. Suter, *J. Chem. Phys.* **96**, 6157 (1992).
122. Y. Kantor, M. Kardar, and D. R. Nelson, *Phys. Rev. Lett.* **57**, 791 (1986); *Phys. Rev. A* **35**, 3056 (1987).
123. M. Plischke and D. Boal, *Phys. Rev. A* **38**, 4943 (1988).
124. D. Boal, E. Levinson, D. Liu, and M. Plischke, *Phys. Rev. A* **40**, 3292 (1990).
125. J.-S. Ho and A. Baumgärtner, *Phys. Rev. Lett.* **63**, 1324 (1989); A. Baumgärtner and J.-S. Ho, *Phys. Rev. A* **41**, 5747 (1990).
126. S. Komura and A. Baumgärtner, *Phys. Rev. A* **44**, 3511 (1991).
127. G. Gompper and D. M. Kroll, *J. Phys. I (France)* **1**, 1441 (1991); *Europhys. Lett.* **15**, 783 (1991).
128. Y. Kantor and K. Kremer, *Phys. Rev. E* **48**, 2490 (1993).
129. R. B. Bird, R. C. Armstrong, and O. Hassager, *Dynamics of Polymeric Liquids*, Vol. 1 (Wiley, New York, 1977).
130. M. Bishop, M. H. Kalos, and H. L. Frisch, *J. Chem. Phys.* **72**, 3328 (1980); D. Ceperly, M. H. Kalos, and J. L. Lebowitz, *Macromolecules* **14**, 1472 (1981).
131. G. S. Grest and K. Kremer, *Phys. Rev. A* **33**, 3628 (1986).
132. C. W. Gear, *Numerical Initial Value Problems in Ordinary Differential Equations* (Prentice-Hall, Englewood Cliffs, NJ, 1971).
133. W. C. Swope, H. C. Andersen, P. H. Berens, and K. R. Wilson, *J. Chem. Phys.* **76**, 637 (1982).
134. B. Dünweg and W. Paul, *Int. J. Mod. Phys.* **2**, 817 (1991).
135. H. C. Andersen, *J. Chem. Phys.* **72**, 2384 (1980).
136. M. Parrinello and A. Rahman, *Phys. Rev. Lett.* **45**, 1196 (1980).
137. L. Verlet, *Phys. Rev.* **159**, 98 (1967).
138. G. S. Grest, B. Dünweg, and K. Kremer, *Comp. Phys. Comm.* **55**, 269 (1989).
139. D. Rapaport, *Comp. Phys. Rep.* **9**, 1 (1988).
140. S. Plimpton, *J. Comp. Phys.* **117**, 1 (1995) and references therein.
141. J. H. R. Clarke, see Chapter 5 of this volume.
142. For a review see *Computer Simulation of Polymers,* edited by R. J. Roe (Prentice-Hall, New York, 1990).
143. K. Kremer and G. S. Grest, see Chapter 4 of this volume.
144. S. F. Edwards, *Proc. Phys. Soc. London* **85**, 613 (1965).

145. A. K. Dolan and S. F. Edwards, *Proc. R. Soc. Lond. A* **337**, 509 (1974); *Proc. R. Soc. Lond.* **343**, 427 (1975).
146. E. Helfand and Z. R. Wasserman, *Macromolecules* **9**, 879 (1976); *Macromolecules* **11**, 960 (1978); *Polym. Eng. Sci.* **17**, 582 (1977).
147. K. M. Hong and J. Noolandi, *Macromolecules* **14**, 727 (1981); *Macromolecules* **16**, 1083 (1983); J. Noolandi and K. M. Hong, *Macromolecules* **15**, 482 (1982); *Macromolecules* **16**, 1443 (1983).
148. J. M. H. M. Scheutjens and G. J. Fleer, *J. Phys. Chem.* **83**, 1619 (1979); *J. Phys. Chem.* **84**, 178 (1980); *Macromolecules* **18**, 1882 (1985).
149. E. A. DiMarzio and R. J. Rubin, *J. Chem. Phys.* **55**, 4318 (1971).
150. J. M. H. M. Scheutjens, G. J. Fleer, and M. A. Cohen Stuart, *Colloid Surf.* **21**, 285 (1986); G. J. Fleer, J. M. H. M. Scheutjens, and M. A. Cohen Stuart, *Colloid Surf.* **31**, 1 (1988).
151. B. van Lent and J. M. H. M. Scheutjens, *Macromolecules* **22**, 1931 (1989).
152. M. A. Cohen Stuart, G. J. Fleer, J. Lyklema, W. Norde, and J. M. H. M. Scheutjens, *Adv. Colloid Interface Sci.* **34**, 477 (1991).
153. S. Hirz, M.Sc. thesis, University of Minnesota, 1987.
154. T. Cosgrove, T. Heath, B. van Lent, F. Leermakers, and J. Scheutjens, *Macromolecules* **20**, 1692 (1988).
155. A. M. Skvortsov, I. V. Pavlushkov, and A. A. Gorbunov, *Polym. Sci. USSR* **30**, 487 (1988).
156. B. van Lent, R. Israels, J. M. H. M. Scheutjens, and G. J. Fleer, *J. Colloid Interface Sci.* **137**, 380 (1990).
157. S. T. Milner, *J. Chem. Soc. Faraday Trans.* **86**, 1349 (1990).
158. C. M. Wijmans, J. M. H. M. Scheutjens, and E. B. Zhulina, *Macromolecules* **25**, 2657 (1992).
159. C. M. Wijmans and E. B. Zhulina, *Macromolecules* **26**, 7214 (1993).
160. C. M. Wijmans, E. B. Zhulina, and G. J. Fleer, *Macromolecules* **27**, 3238 (1994).
161. D. Theodorou, *Macromolecules* **21**, 1391 (1988); *Macromolecules* **21**, 1400 (1988); *Macromolecules* **21**, 1411 (1988).
162. A. Hariharan, S. K. Kumar, and T. P. Russell, *Macromolecules* **23**, 3584 (1990).
163. K. R. Shull and E. J. Kramer, *Macromolecules* **23**, 4769 (1990); K. R. Shull, E. J. Kramer, G. Hadziioannou, and W. Tang, *Macromolecules* **23**, 4780 (1990).
164. K. R. Shull, *Macromolecules* **25**, 2122 (1992).
165. K. R. Shull and K. I. Winey, *Macromolecules* **25**, 2637 (1992).
166. K. R. Shull, *J. Chem. Phys.* **94**, 5723 (1991).
167. K. Huang and A. C. Balazs, *Macromolecules* **26**, 4736 (1993).
168. M. Muthukumar and J.-S. Ho, *Macromolecules* **22**, 965 (1989).
169. M. D. Whitmore and J. Noolandi, *Macromolecules* **23**, 3321 (1990).
170. N. Dan and M. Tirrell, *Macromolecules* **25**, 2890 (1992).
171. A. Ben-Shaul, I. Szleifer, and W. M. Gelbart, *J. Chem. Phys.* **83**, 3597 (1985).
172. M. A. Carignano and I. Szleifer, *J. Chem. Phys.* **98**, 5006 (1993); *J. Chem. Phys.* **100**, 3210 (1994).
173. M. A. Carignano and I. Szleifer, *Macromolecules* **27**, 702 (1994).
174. M. A. Carignano and I. Szleifer (to be published, 1995).

175. W. Burchard, *Adv. Polym. Sci.* **48**, 1 (1983).
176. J. Roovers and S. Bywater, *Macromolecules* **5**, 385 (1972); *Macromolecules* **7**, 443 (1974).
177. J. Roovers, N. Hadjichristidis, and L. J. Fetters, *Macromolecules* **16**, 214 (1983); K. Huber, W. Burchard, and L. J. Fetters, *Macromolecules* **17**, 541 (1984).
178. P. M. Toporowski and J. Roovers, *J. Polym. Sci., Polym. Chem. Ed.* **24**, 3009 (1986).
179. N. Khasat, R. W. Pennisi, N. Hadjichristidis, and L. J. Fetters, *Macromolecules* **21**, 1100 (1988).
180. B. J. Bauer, L. J. Fetters, W. W. Graessley, N. Hadjichristidis, and G. F. Quack, *Macromolecules* **22**, 2337 (1989).
181. J. Roovers, P. Toporowski, and J. Martin, *Macromolecules* **22**, 1897 (1989).
182. L.-L. Zhou, N. Hadjichristidis, P. M. Toporowski, and J. Roovers, *Rubber Chem. Tech.* **65**, 303 (1992).
183. J. Roovers, L.-L. Zhou, P. M. Toporowski, M. van der Zwan, H. Iatrou, and N. Hadjichristidis, *Macromolecules* **26**, 4324 (1993).
184. N. S. Davidson, L. J. Fetters, W. G. Funk, W. W. Graessley, and N. Hadjichristidis, *Macromolecules* **21**, 112 (1988).
185. G. Broze, R. Jerome, P. L. Teyssie, and C. Marco, *Macromolecules* **16**, 1771 (1983).
186. L. Leibler, H. Orland, and J. C. Wheeler, *J. Chem. Phys.* **79**, 3550 (1983).
187. L. Leibler and P. A. Pincus, *Macromolecules* **17**, 2922 (1984).
188. E. B. Zhulina and T. M. Birshtein, *Polym. Sci. USSR* **27**, 570 (1985).
189. A. Halperin, *Macromolecules* **20**, 2943 (1987).
190. M. R. Munch and A. P. Gast, *Macromolecules* **21**, 1360 (1988).
191 R. Nagarajan and K. Ganesh, *J. Chem. Phys.* **90**, 5843 (1989).
192. Z. Tuzar and P. Kratochvil, *Adv. Colloid Interface Sci.* **6**, 201 (1976); in *Surface and Colloid Science*, Vol. 15, edited by E. Matijević (Plenum, New York, 1993), p.1.
193. C. Price, *Pure Appl. Chem.* **55**, 1563 (1983).
194. H. Watanabe and T. Kotaka, *Macromolecules* **17**, 342 (1984).
195. L. J. M. Vagberg, K. A. Cogan, and A. P. Gast, *Macromolecules* **24**, 1670 (1991); K. A. Cogan, M. Capel, and A. P. Gast, *Macromolecules* **24**, 6512 (1991).
196. G. A. McConnell, A. P. Gast, J. S. Huang, and S. D. Smith, *Phys. Rev. Lett.* **71**, 2102 (1993).
197. T. A. Witten and P. Pincus, *Macromolecules* **19**, 2509 (1986).
198. H. Iatrou and N. Hadjichristidis, *Macromolecules* **26**, 2479 (1993); *Macromolecules* **25**, 4649 (1992).
199. A. Halperin, *J. Phys. (France)* **49**, 131 (1988).
200. D. Richter, B. Stuhn, B. Ewen, and D. Nerger, *Phys. Rev. Lett.* **58**, 2462 (1987); D. Richter, B. Farago, L. J. Fetters, J. S. Huang, and B. Ewen, *Macromolecules* **23**, 1845 (1990).
201. D. Richter, B. Farago, J. S. Huang, L. J. Fetters, and B. Ewens, *Macromolecules* **22**, 468 (1989).
202. D. Boese, F. Kremer, and L. J. Fetters, *Macromolecules* **23**, 1826 (1990).

203. K. Huber, S. Bantle, W. Burchard, and L. J. Fetters, *Macromolecules* **19**, 1404 (1988).

204. W. Dozier, J. Huang, and L. J. Fetters, *Macromolecules* **24**, 2810 (1991).

205. L. Willner, O. Jucknischke, D. Richter, B. Farago, L. J. Fetters, and J. S. Huang, *Europhys. Lett.* **19**, 297 (1992).

206. M. Adam, L. J. Fetters, W. W. Graessley, and T. A. Witten, *Macromolecules* **24**, 2434 (1991).

207. K. De'Bell and T. Lookman, *Rev. Mod. Phys.* **65**, 87 (1993).

208. K. Ohno and K. Binder, *J. Chem. Phys.* **95**, 5444 (1991); *J. Chem. Phys.* **95**, 5459 (1991).

209. A. Miyake and K. F. Freed, *Macromolecules* **16**, 1228 (1983); *Macromolecules* **17**, 678 (1984); J. F. Douglas and K. F. Freed, *Macromolecules* **17**, 1854 (1984).

210. C. H. Vlahos and M. K. Kosmas, *Polymer* **25**, 1607 (1984).

211. B. Duplantier, *Phys. Rev. Lett.* **57**, 941 (1986); *Europhys. Lett.* **8**, 677 (1988); B. Duplantier and H. Saleur, *Phys. Rev. Lett.* **57**, 3179 (1986); *Phys. Rev. Lett.* **59**, 539 (1987).

212. P. G. de Gennes, *Scaling Concepts in Polymer Physics* (Cornell University Press, Ithaca, NY, 1979).

213. E. Raphaël, P. Pincus and G. H. Fredrickson, *Macromolecules* **26**, 1996 (1993); M. Aubuy, G. H. Fredrickson, P. Pincus, and E. Raphaël, *Macromolecules* **28** (1995).

214. W. Bruns and W. Carl, *Macromolecules* **24**, 209 (1991).

215. G. Zifferer, *Makromol-Chem., Theory Simul.* **2**, 653 (1993); *Macromol-Theory Simul.* **3**, 163 (1994).

216 W. Mattice, *Macromolecules* **13**, 506 (1980)

217 B. H. Zimm, *Macromolecules* **17**, 795 (1984).

218 M. Bishop and J. H. R. Clarke, *J. Chem. Phys.* **90**, 6647 (1989).

219 A. Sikorski, *Makromol. Chem., Theory Simul.* **2**, 309 (1993).

220. J. F. Douglas, J. Roovers, and K. F. Freed, *Macromolecules* **23**, 4168 (1990).

221. H. Li and T. A. Witten, *Macromolecules* **27**, 449 (1994).

222. K. Huber, W. Burchard, S. Bantle, and L. J. Fetters, *Polymer* **28**, 1990 (1987); *Polymer* **28**, 1997 (1987).

223. A. T. Boothroyd, G. L. Squires, L. J. Fetters *et al.*, *Macromolecules* **22**, 3130 (1989).

224. L. Willner, O. Jucknischke, D. Richter *et al.*, *Macromolecules* **27**, 3821 (1994).

225 H. Benoit, *J. Polym. Sci.* **11**, 507 (1953).

226 J. L. Alessandrini and M. A. Carignano, *Macromolecules* **25**, 1157 (1992).

227. F. Candau, P. Rempp, and H. Benoit, *Macromolecules* **5**, 627 (1972).

228. B. H. Zimm and W. H. Stockmayer, *J. Chem. Phys.* **17**, 1301 (1949).

229. B. H. Zimm and R. W. Kilb, *J. Polym. Sci.* **37**, 19 (1959).

230. B. H. Zimm, *J. Chem. Phys.* **24**, 269 (1954).

231. J. Roovers, *Polymer* **16**, 827 (1975); *Polymer* **20**, 843 (1979); J. Roovers and P. Toporowski, *J. Polym. Sci., Polym. Phys. Ed.* **18**, 1907 (1980).

232. Z.-G. Wang and S. A. Safran, *J. Chem. Phys.* **94**, 679 (1991).

233. G. F. Meijs and E. Rizzardo, *Rev. Macro. Chem. Phys.* **C30**, 305 (1990).

234. Y. Tsukahara, K. Mizuno, A. Segawa, and Y. Yamashita, *Macromolecules* **22**, 1546 (1989).
235. K. I. Winey, E. L. Thomas, and L. J. Fetters, *J. Chem. Phys.* **95**, 9367 (1991); *Macromolecules* **25**, 2645 (1992).
236. T. M. Birshtein, O. V. Borisov, E. B. Zhulina, A. R. Khokhlov, and T. A. Yurasova, *Polym. Sci. USSR* **29**, 1293 (1987).
237. R. C. Ball, J. F. Marko, S. T. Milner and T. A. Witten, *Macromolecules* **24**, 693 (1991).
238. M. Kurata and M. Fukatsu, *J. Chem. Phys.* **41**, 2934 (1964); E. F. Casassa and G. C. Berry, *J. Polym. Sci., Polym. Phys. Ed.* **4**, 881 (1966); G. C. Berry *J. Polym. Sci., Polym. Phys. Ed.* **6**, 1551 (1968); W. C. Forsman, *Macromolecules* **1**, 343 (1968); H. Galina, *Macromolecules* **16**, 1479 (1983).
239. M. Bishop and C. J. Saltiel, *J. Chem. Phys.* **98**, 1611 (1993); *J. Chem. Phys.* **99**, 9170 (1993).
240. G. H. Fredrickson, *Macromolecules* **26**, 2825 (1993).
241. A. Gauger and T. Pakula, *Macromolecules* **28**, 190 (1995).
242. N. Jalal and R. Duplessix, *J. Phys. (France)* **49**, 1775 (1988).
243. P. Auroy, L. Auvray, and L. Léger, *J. Phys. Cond. Matter* **2**, SA317 (1990); *Phys. Rev. Lett.* **66**, 719 (1991); *Macromolecules* **24**, 5158 (1991); *Macromolecules* **24**, 2523 (1991); P. Auroy, Y. Mir, and L. Auvray, *Phys. Rev. Lett.* **69**, 93 (1992).
244. H. J. Taunton, C. Toprakcioglu, L. J. Fetters, and J. Klein, *Nature* **332**, 712 (1988); *Macromolecules* **23**, 571 (1990).
245. J. Klein, D. Perahia, and S. Warburg, *Nature* **352**, 143 (1991); J. Klein, E. Kumacheva, D. Mahalu *et al.*, *Nature* **370**, 634 (1994).
246. G. Hadziioannou, S. Patel, S. Granick, and M. Tirrell, *J. Am. Chem. Soc.* **108**, 2869 (1986); M. Tirrell, S. Patel, G. Hadziioannou, *Proc. Natl. Acad. Sci.* **84**, 4725 (1987).
247. M. A. Ansarifar and P. F. Luckham, *Polymer* **29**, 329 (1988).
248. H. J. Taunton, C. Toprakcioglu, and J. Klein, *Macromolecules* **21**, 3333 (1988).
249. S. S. Patel and M. Tirrell, *Ann. Rev. Phys. Chem.* **40**, 597 (1989).
250. A. P. Gast and M. R. Munch, *Polymer Comm.* **30**, 324 (1989).
251. E. Parsonage, M. Tirrell, H. Watanabe, and R. G. Nuzzo, *Macromolecules* **24**, 1987 (1991).
252. J. B. Field, C. Toprakcioglu, R. C. Ball *et al.*, *Macromolecules* **25**, 434 (1992).
253. J. B. Field, C. Toprakcioglu, L. Dai *et al.*, *J. Phys. II (France)* **2**, 2221 (1992).
254. S. Granick and J. Herz, *Macromolecules* **18**, 460 (1985).
255. M. S. Kent, L.-T. Lee, B. Farnoux, and F. Rondelex, *Macromolecules* **25**, 6240 (1992).
256. B. J. Factor, L.-T. Lee, M. S. Kent, and F. Rondelez, *Phys. Rev. E* **48**, R2354 (1993).
257. M. S. Kent, L.-T. Lee, B. J. Factor, F. Rondelez, and G. Smith, *J. Phys. II (France) Colloq.* **3**, C8–49 (1993); *J. Chem. Phys.* **102**, (1995).
258. F. S. Bates, *Science* **251**, 898 (1991).
259. K. Binder, see Chapter 7 of this volume.
260. A. Halperin, *Europhys. Lett.* **4**, 439 (1987).

261. J. F. Marko and T. A. Witten, *Phys. Rev. Lett.* **66**, 1541 (1991); *Macromolecules* **25**, 296 (1992).
262. J. F. Marko and A. Chakrabarti, *Phys. Rev. E* **48**, 2739 (1993).
263. H. Dong, *J. Phys. II (France)* **3**, 999 (1993); H. Dong, J. F. Marko, and T. A. Witten, *Macromolecules* **27**, 6428 (1994).
264. P.-Y. Lai, *J. Chem. Phys.* **100**, 3351 (1994).
265 G. Brown, A. Chakrabarti, and J. F. Marko, *Europhys. Lett.* **25**, 239 (1994).
266. S. T. Milner, T. A. Witten, and M. E. Cates, *Macromolecules* **21**, 2610 (1988); *Europhys. Lett.* **5**, 413 (1988).
267. S. T. Milner, Z.-G. Wang, and T. A. Witten, *Macromolecules* **22**, 489 (1989).
268. E. B. Zhulina, O. V. Borisov, and V. A. Pryamitsyn, *J. Colloid Interface Sci.* **137**, 495 (1990).
269. A. N. Semenov, *Sov. Phys. JETP* **61**, 733 (1985).
270. T. Cosgrove, *J. Chem. Soc. Faraday Trans.* **86**, 1323 (1990).
271. P. Auroy and L. Auvray, *Macromolecules* **25**, 4134 (1992).
272. P. Aurot and L. Auvray, *J. Phys. II (France)* **3**, 227 (1993).
273. D. Perahia, D. Weisler, S. K. Satija *et al.*, *Phys. Rev. Lett.* **72**, 100 (1994).
274. P.-Y. Lai and E. B. Zhulina, *J. Phys. II (France)* **2**, 547 (1992).
275. One of the earliest computer simulations of brushes was carried out by Cosgrove *et al.*[154] These authors generated self-avoiding walks that were restricted by the grafting surface. These simulations had the drawback that they were essentially single-chain simulations. The effect of finite grafting density was introduced by placing periodic boundary conditions in the grafting plane. Thus each chain was actually simulated in an environment of identical chains.
276. D. F. K. Shim and M. E. Cates, *J. Phys. (Paris)* **50**, 3535 (1989).
277. P.-Y. Lai and A. Halperin, *Macromolecules* **24**, 4981 (1991).
278. M. Laradji, H. Guo, and M. J. Zuckerman, *Phys. Rev. E* **49**, 3199 (1994).
279. M. Doi and S. F. Edwards, *The Theory of Polymer Dynamics* (Clarendon Press, Oxford, 1986).
280. T. Cosgrove, T. G. Heath, J. S. Phipps, R. M. Richardson, *Macromolecules* **24**, 94 (1991).
281. A. Halperin, *J. Phys. (Paris)* **49**, 547 (1988).
282. E. B. Zhulina, O. V. Borisov, V. A. Pryamitsyn, and T. M. Birshtein, *Macromolecules* **24**, 140 (1991).
283. R. S. Ross and P. Pincus, *Europhys. Lett.* **19**, 79 (1992).
284. A. Halperin and E. B. Zhulina, *Europhys. Lett.* **15**, 417 (1991); *Macromolecules* **24**, 5393 (1991).
285. C. Yeung, A. C. Balazs, and D. Jasnow, *Macromolecules* **26**, 1914 (1993).
286. D. R. M. Williams, *J. Phys. II (France)* **3**, 1313 (1993).
287. H. Tang and I. Szleifer, *Europhys. Lett.* **28**, 19 (1994).
288. P.-Y. Lai and A. Halperin, *Macromolecules* **25**, 6693 (1992).
289. J. H. van Zanten, *Macromolecules* **27**, 5052 (1994); *Macromolcules*, **27**, 6797 (1994).
290. S. J. O'Shea, M. E. Welland, and T. Rayment, *Langmuir* **9**, 1826 (1993).
291. W. Zhao, G. Krausch, M. H. Rafailovich, and J. Sokolov, *Macromolecules* **27**, 2933 (1994).

292. R. Israels, D. Gersappe, M. Fasolka, V. A. Roberts, and A. C. Balazs, *Macromolecules* **27**, 6679 (1994).
293. W. Li and A. C Balazs, *Mol. Simul.* **13**, 257 (1994).
294. D. Gersappe, M. Fasolka, S. Jacobson, and A. C. Balazs, *J. Chem. Phys.* **100**, 9170 (1994).
295. H. Tang, M. A. Carignano, and I. Szleifer, *J. Chem. Phys.* **102**, 3404 (1995).
296. T. Cosgrove, T. G. Heath, K. Ryan, and B. van Lent, *Polym. Commun.* **28**, 64 (1987); T. Cosgrove, T. G. Heath, and K. Ryan, *Langmuir* **10**, 3500 (1994).
297. C. Ligoure, *J. Phys. II (France)* **3**, 1607 (1993).
298. A. Chakrabarti, *J. Chem. Phys.* **100**, 631 (1994).
299. H. D. Ou-Yang and Z. Gao, *J. Phys. II (France)* **1**, 1375 (1991).
300. A. Johner and C. M. Marques, *Phys. Rev. Lett.* **69**, 1827 (1992).
301. J. F. Marko, A. Johner, and C. M. Marques, *J. Chem. Phys.* **99**, 8142 (1993).
302. S. T. Milner, *Europhys. Lett.* **7**, 695 (1988).
303. S. T. Milner, T. A. Witten, and M. E. Cates, *Macromolecules* **22**, 853 (1989).
304. S. Dhoot, H. Watanabe, and M. Tirrell, *Colloid Surf.* **A86**, 47 (1994).
305. C. F. Laub and J. T. Koberstein, *Macromolecules* **27**, 5016 (1994).
306. L. I. Klushin and A. M. Skvortsov, *Macromolecules* **25**, 3443 (1992).
307. T. M. Birshtein, Y. V. Liatskaya, and E. B. Zhulina, *Polymer* **31**, 2185 (1990).
308. N. Dan and M. Tirrell, *Macromolecules* **26**, 6467 (1993).
309. P.-K Lai and E. B. Zhulina, *Macromolecules* **25**, 5201 (1992).
310. M. A. Cohen Stuart, T. Cosgrove, and B. Vincent, *Adv. Colloid Interface Sci.* **24**, 143 (1986).
311. H. Watanabe and M. Tirrell, *Macromolecules* **26**, 6455 (1993).
312. P.-G. de Gennes, *C. R. Acad. Sci. Paris* **300**, 839 (1985).
313. A. Chakrabarti, P. Nelson, and R. Toral, *J. Chem. Phys.* **100**, 748 (1994).
314 D. F. Shim and M. E. Cates, *J. Phys. (Paris)* **51**, 701 (1990).
315. T. A. Witten, L. Leibler, and P. Pincus, *Macromolecules* **23**, 824 (1990).
316. S. Patel, M. Tirrell, and G. Hadziioannou, *Colloid Surf.* **31**, 157 (1988).
317. Authors of Ref. 244 found equally satisfactory agreement between their experiments and the predictions of either the SCF or the scaling theories if both ρ_a and the scale of the force are taken as adjustable parameters. However, when known material parameters and measured grafting densities are used,[302] the parabolic profile, with a small polydispersity (M_W/M_N) is found to account for the force profile with no adjustable parameters, whereas the step function profile (scaling analysis) had too short a range for the onset of the force.
318. Z. Limpouchov and K. Procházka, *Collect. Czech. Chem. Comm.* **58**, 2291 (1993); *Collect. Czech. Chem. Comm.* **59**, 803 (1994).
319. L. Leibler, *Makromol. Chem., Macromol. Symp.* **16**, 1 (1988).
320. E. B. Zhulina, O. V. Borisov, and L. Brombacher, *Macromolecules* **24**, 4679 (1991).
321. M. Aubouy and E. Raphaël, *J. Phys. II (France)* **3**, 443 (1993).
322. R. A. L. Jones, L. J. Norton, K. R. Shull *et al.*, *Macromolecules* **25**, 2359 (1992).

323. A. Budkowski, U. Steiner, J. Klein, and L. J. Fetters, *Europhys. Lett.* **20**, 499 (1992); A. Budkowski, J. Klein, U. Steiner, and L. J. Fetters, *Macromolecules* **26**, 2470 (1993).
324. X. Zhao, W. Zhao, X. Zheng *et al.*, *Phys. Rev. Lett.* **69**, 776 (1992).
325. A. Johner and J. F. Joanny, *Europhys. Lett.* **15**, 265 (1991); *J. Chem. Phys.* **96**, 6257 (1992); *J. Chem. Phys.* **98**, 1647 (1993).
326. R. Zajac and A. Chakrabarti, *Phys. Rev. E* **49**, 3069 (1994).
327. C. Ligoure and L. Leibler, *J. Phys. (France)* **51**, 1313 (1990).
328. A. Johner and J. F. Joanny, *Macromolecules* **23**, 5299 (1990).
329. A. Halperin and S. Alexander, *Macromolecules* **22**, 2403 (1989); A. Halperin, *Europhys. Lett.* **8**, 351 (1989).
330. J. Wittmer, K. Binder, A. Johner, and J. F. Joanny, *J. Chem. Phys.* **101**, 4379 (1994).
331. J. Klein, K. Kamiyama, H. Yoshizawa *et al.*, *Macromolecules* **25**, 2062 (1992).
332. E. Kumacheva, J. Klein, P. A. Pincus, and L. J. Fetters, *Macromolecules* **26**, 6477 (1993).
333. H. D. Ou-Yang and Z. Gao, in *Colloid–Polymer Interactions: Particulate, Amphiphilic and Biological Surfaces*, edited by P. L. Dubin and P. Tong (American Chemical Society, Washington, DC, 1993), p. 70.
334. S. T. Milner, *Macromolecules* **25**, 5487 (1992).
335. N. Dan, *Macromolecules* **27**, 2310 (1994).
336. K. Kremer and K. Binder, *J. Chem. Phys.* **81**, 6381 (1984).
337. L. I. Klushin and A. M. Skvortzov, *Macromolecules* **24**, 1549 (1991).
338. B. Farago, M. Monkenbusch, D. Richter *et al.*, *Phys Rev. Lett.* **71**, 1015 (1993).
339. F. D. Blum, B. R. Sinha, and F. C. Schwab, *Macromolecules* **23**, 3592 (1990); B. R. Sinha, F. D. Blum, and F. C. Schwab, *Macromolecules* **26**, 7056 (1993).
340. L. R. G. Treloar, *The Physics of Rubber Elasticity* (Clarendon Press, Oxford, 1975).
341. P. J. Flory, *Proc. R. Soc. Lond. A* **351**, 351 (1976); S. F. Edwards and T. A. Vilgis, *Rep. Prog. Phys.* **51**, 243 (1988).
342. G. S. Grest and K. Kremer, *Macromolecules* **23**, 4994 (1990).
343. M. E. Cates, *Phys. Rev. Lett.* **53**, 926 (1984); *J. Phys. (France)* **46**, 1059 (1985).
344. For a review see the two articles by D. R. Nelson and Y. Kantor in *Statistical Mechanics of Membranes and Interfaces*, edited by D. R. Nelson, T. Piran, and S. Weinberg (World Scientific, Singapore, 1989).
345. P. Flory, *Statistical Mechanics of Chain Molecules* (Interscience, New York, 1969).
346. E. Levinson, *Phys. Rev. A* **43**, 5233 (1991).
347. Y. Gefen, A. Aharony, B. B. Mandelbrot, and S. Kirkpatrick, *Phys. Rev. Lett.* **47**, 1771 (1981).
348. S. Alexander and R. Orbach, *J. Phys. Lett. (France)* **43**, L625 (1982).
349. R. Rammal and G. Toulouse, *J. Phys. Lett. (France)* **44**, L13 (1983).
350. C. J. Lobb and D. J. Frank, *Phys. Rev. B* **30**, 4090 (1984); J. G. Zabolitzky, *Phys. Rev. B.* **30**, 4077 (1984); H. J. Herrmann, B. Derrida, and J. Vannimenus, *Phys. Rev. B* **30**, 4080 (1984); D. C. Hong, S. Havlin, H. Herrmann, and H. E. Stanley, *Phys. Rev. B* **30**, 4083 (1984).

351. M. Kardar and D. R. Nelson, *Phys. Rev. Lett.* **58**, 1298 (1987).
352. B. Duplantier, *Phys. Rev. Lett.* **58**, 2733 (1987).
353. J. A. Aronowitz and T. C. Lubensky, *Europhys. Lett.* **4**, 395 (1987).
354. T. Hwa, *Phys. Rev. A* **41**, 1751 (1989).
355. M. Daoud, F. Family, and G. Jannink, *J. Phys. Lett. (France)* **45**, 199 (1984).
356. D. Stauffer, *J. Chem. Soc. Faraday Trans. 2* **72**, 1354 (1976).
357. M. Adam, M. Delsanti, J. P. Munch, and D. Durand, *J. Phys. (Paris)* **48**, 1809 (1987).
358. E. Bouchaud, M. Delsanti, M. Adam, M. Daoud, and D. Durand, *J. Phys. (Paris)* **47**, 1273 (1986).
359. E. Duering and Y. Kantor, *Phys. Rev. B* **40**, 9443 (1989).
360. M. Plischke and B. Forcade, *Phys. Rev. A* **43**, 2056 (1991).
361. D. M. Kroll and G. Gompper, *J. Phys. I (France)* **3**, 1131 (1993).
362. A. Baumgärtner, *J. Phys. I (France)* **1**, 1549 (1991); A. Baumgärtner and W. Renz, *Europhys. Lett.* **17**, 381 (1992).
363. Y. Kantor and D. R. Nelson, *Phys. Rev. A* **38**, 4020 (1987).
364. R. L. Renken and J. B. Kogut, *Nucl. Phys. B* **342**, 753 (1990).
365. R. G. Harnish and J. F. Wheater, *Nucl. Phys. B* **350**, 861 (1991).
366. S. J. Barsky and M. Plischke, *Phys. Rev.* **E50**, 3911 (1994).
367. M. Goulian, *J. Phys. II (France)* **1**, 1327 (1991).
368. P. Le Doussal, *J. Phys. A* **25**, L469 (1992).
369. E. Guitter and J. Palmeri, *Phys. Rev. A.* **45**, 734 (1992).
370. F. F. Abraham, *Phys. Rev. Lett.* **67**, 1669 (1991).
371. R. Lipowsky and M. Girardet, *Phys. Rev. Lett.* **65**, 2893 (1990).
372. D. R. Nelson and L. Peliti, *J. Phys. (France)* **48**, 1085 (1987).
373. R. Lipowsky, *Europhys. Lett.* **7**, 255 (1988).
374. P. Le Doussal and L. Radzihovsky, *Phys. Rev. Lett.* **69**, 1209 (1992).
375. S. Leibler and A. C. Maggs, *Phys. Rev. Lett.* **63**, 406 (1989).
376. Z. Zhang, H. T. Davis, and D. M. Kroll, *Phys. Rev. E* **48**, R651 (1993).
377. X. Wen, C. W. Garland, T. Hwa *et al.*, *Nature* **355**, 426 (1992).
378. M. S. Spector, E. Naranjo, S. Chiruvolu and J. A. Zasadzinksi, *Phys. Rev. Lett.* **73**, 2867 (1994).
379. M. Goulian, N. Lei, J. Miller, and S. K. Sinha, *Phys. Rev. A* **46**, R6170 (1992).
380. K. Svoboda, C. F. Schmidt, N. Lei *et al.*, *Mat. Res. Soc. Symp.* **248**, 63 (1992).
381. C. F. Schmidt, K. Svoboda, N. Lei *et al.*, *Science* **259**, 952 (1993).
382. D. Liu and M. Plischke, *Phys. Rev. A* **45**, 7139 (1992).
383. S. I. Stupp, S. Son, H. C. Lin, and L. S. Li, *Science* **259**, 59 (1993).
384. R. Lipowsky, *Phys. Rev. Lett.* **62**, 705 (1989); S. Grotehans and R. Lipowsky, *Phys. Rev. A* **41**, 4574 (1990).

INDEX

23rd Feb 05